AF615192

MONOGRAPH OF THE SPATHIDIIDA (CILIOPHORA, HAPTORIA)

MONOGRAPHIAE BIOLOGICAE

VOLUME 81

Series Editor

H. J. Dumont

Aims and Scope

The Monographiae Biologicae provide a forum for top-level, rounded-off monographs dealing with the biogeography of continents or major parts of continents, and the ecology of well individualized ecosystems such as islands, island groups, mountains or mountain chains. Aquatic ecosystems may include marine environments such as coastal ecosystems (mangroves, coral reefs) but also pelagic, abyssal and benthic ecosystems, and freshwater environments such as major river basins, lakes, and groups of lakes. In-depth, state-of-the-art taxonomic treatments of major groups of animals (including protists), plants and fungi are also elegible for publication, as well as studies on the comparative ecology of major biomes. Volumes in the series may include single-author monographs, but also multi-author, edited volumes.

The titles published in this series are listed at the end of this volume.

Monograph of the Spathidiida (Ciliophora, Haptoria)

Vol I: Protospathidiidae, Arcuospathidiidae, Apertospathulidae

by

WILHELM FOISSNER
University of Salzburg, Austria

and

KUIDONG XU
University of Salzburg, Austria

A C.I.P. Catalogue record for this book is available from the Library of Congress.

ISBN-13 978-1-4020-4210-2 (HB)
ISBN-13 978-1-4020-4735-0 (ebook)

Published by Springer,
P.O. Box 17, 3300 AA Dordrecht, The Netherlands.

www.springer.com

Printed on acid-free paper

Contents

Preface

This book continues our series of ciliate monographs, several of which have been published or are in preparation: class Colpodea (FOISSNER 1993); suborder Hypotrichia (BERGER 1999, 2005; volume 3 in preparation); and order Oligotrichida (AGATHA, in preparation). Furthermore, there are several thematic monographs available, viz., on the ciliates used as bioindicators in water quality assessment (FOISSNER et al. 1991, 1992, 1994, 1995); on the ecology and taxonomy of limnetic plankton ciliates (FOISSNER et al. 1999); on soil ciliates from Namibia (FOISSNER et al. 2002); and on the generic names of the ciliates (AESCHT 2001). Altogether, we described and revised over 1000 ciliate species in these monographs, that is, circa one tenth of the species described, but only one thirtieth of the diversity proposed (FOISSNER 1999b, c, 2004a). Certainly, this is far from being complete, but our detailed revisions will be long-lived, and we are optimistic to do some further monographs, even in a time where molecular biology overwhelms most other biological disciplines.

The family Spathidiidae belongs to the subclass Haptoria, a group of rapacious, "lower" holotrichs. The family comprises about 200 described species, most belonging to the time-honoured genus *Spathidium*. Several colleagues doubted the validity of so many *Spathidium* species (BUITKAMP 1977b, FINLAY et al. 1996, WENZEL 1955). However, our monograph shows not only the validity of most described species, but adds 50 new species discovered in over 500 samples from terrestrial biotopes worldwide. Thus, the spathidiids are as diverse as proposed by KAHL (1930a, b) 75 years ago! Now, they consist of over 250 species distributed in four families and 20 (!) genera, several of which will be described in this monograph or were established rather recently. About half of the species have been described or redescribed with modern methods, and thus each needs an average of eight printed pages in the revision. Accordingly, the over 200 species will be not squeezed into a single, large volume with 1000 pages, but they are split into two parts which form a harmonic unit, but can be used also independently. Further, the split facilitates publication, which was considerably delayed because we had to perform basic investigations on ontogenesis, conjugation and resting cysts as well as to describe nearly 100 populations half of which represented new species.

Our monograph is also a first attempt to standardize ciliate species descriptions by fixing and/or quantifying as many features as possible, for instance, the shapes of body and extrusomes, distances, and other "abouts". Indeed, most ciliate descriptions are inprecise using, for instance, different names for the same shape. Botanists recognized the resulting problems very early and thus established descriptive baselines many years ago (STEARN 1992).

We earnestly hope that our revision will be of use for a long time, not only to taxonomists, but also to ecologists and molecular biologists who not yet recognized the bioindicative capacity of protozoa (FOISSNER 1987a, 1997b) and the many interesting species this group contains.

Salzburg, August 2005

Wilhelm Foissner
Kuidong Xu

Acknowledgements

We are indebted to many persons for concrete assistance during the preparation of this monograph. Specifically, we should at least mention our deep gratitude to Dr. Eva HERZOG, Dr. Brigitte MOSER, Miss Birgit PEUKERT, Mr. Andreas ZANKL, and Dr. Wolf-Dietrich KRAUTGARTNER for their technical assistance, ranging from help with the scanning electron microscope to typing of the manuscript; to Dr. Remigius GEISER for nomenclatural advice; to Mr. Fritz SEYRL (Bundesstaatliche Studienbibliothek Linz) and the library staffs at the University of Salzburg for their patient assistance in locating rare or obscure publications; to FOISSNER´s former assistants, Prof. Dr. Helmut BERGER and Dr. Sabine AGATHA, for stimulating discussions about alpha-taxonomy and systematics; to Judith TERPOS from Springer (formerly Kluwer) publishers for her editorial advice and patience; and to the Series Editor, Prof. Dr. Henri J. DUMONT (The State University of Ghent, Belgium) for accepting this voluminous monograph.

The thousands of figures comprising the plates are either originals or reproductions from widely scattered sources in the vast protozoological literature of the past 150 years. Specific acknowledgements are generally made in the explanations to the figures: there are named the authors of the papers in which the illustrations originally appeared. Likewise, those colleagues who provided samples, often from remote and/or special habitats, are acknowledged in the individual species descriptions.

Last but not least, we wish to acknowledge the indispensable financial aid from the Austrian Science Foundation (FWF) which, after the recommendation of peer-reviewers, granted the project with a research assistant (Dr. Kuidong XU) and a part-time technician.

A General Section

As explained in the preface, the monograph has been split into two volumes. This applies also to the general section. In volume I, the general morphology, the life cycle, and the principal investigation methods are described, while volume II will deal mainly with ecology and geographic distribution, phylogeny, evolution, and classification. Further, it will contain a user-friendly key to all species.

1 Morphology and Principal Terms

In this section, the principal morphology and terminology will be explained, using simple but well-defined figures. Indeed, this is a first step to the urgently needed general methodology of ciliate descriptions. Partly, figures and terms were taken from STEARN (1992), who reviewed the terms used in descriptive botany. There are still some uncertainties, also in botanical terminology, which should be clarified in a more comprehensive treatment of the matter.

For general ciliate terminology, we refer to the excellent compilations of CORLISS (1979) and MARGULIS et al. (1993), while the general ciliate morphology is exhaustively treated in CORLISS (1979) and PUYTORAC (1994).

1·1 Size and shape, morphometry (Fig. 1–3, 5)

The spathidiids range from about 60 × 10 µm to 400 × 50 µm in vivo. The volume of one of the largest species, that is, *Epispathidium securiforme* is 35 times larger than that of one of the smallest species, that is, *Edaphospathula minor*. This is a small range when compared to those found in colpodids (200 000, FOISSNER 1993) and oxytrichid stichotrichs (140, BERGER 1999). We have used seven categories of size (as reflected in body length; arbitrarily if unrealistically set up with non-overlapping ranges) as follows: very small, 10–30 µm; small, 30–60 µm; moderately small, 60–90 µm; medium sized, 90–150 µm; large, 150–300 µm; moderately large, 300–500 µm; very large, > 500 µm.

According to the name, spathidiids should be spatula-shaped. Unfortunately, this hardly applies (Fig. 1–3, 5)! Most are very narrowly to broadly bursiform with a more or less slanted anterior end and a rounded posterior. Others are cylindroidal, vermiform, clavate, axe-shaped etc. Thus, a huge variety of shapes exists, and the ratio of body length to body width ranges from about 1:1 to 30:1. Most species are slightly to distinctly flattened laterally, some are even leaf-like. Many of the terrestrial species are small and/or slender, as is typical for soil organisms in general (FOISSNER 1987a). However, those living in mosses and leaf litter may be large, for instance, the common *Epispathidium amphoriforme*. The shape is stabilized by bundles of cortical microtubules (WILLIAMS et al. 1981), but usually the cortex remains flexible and the shape may thus strongly deform in over- or under-nourished cells. Theronts and trophonts occur, but true polymorphism is lacking. Likewise, metaboly and pronounced contractility do not occur because myonemes are absent.

Invariably, the oral apparatus occupies the anterior body end, forming a more or less

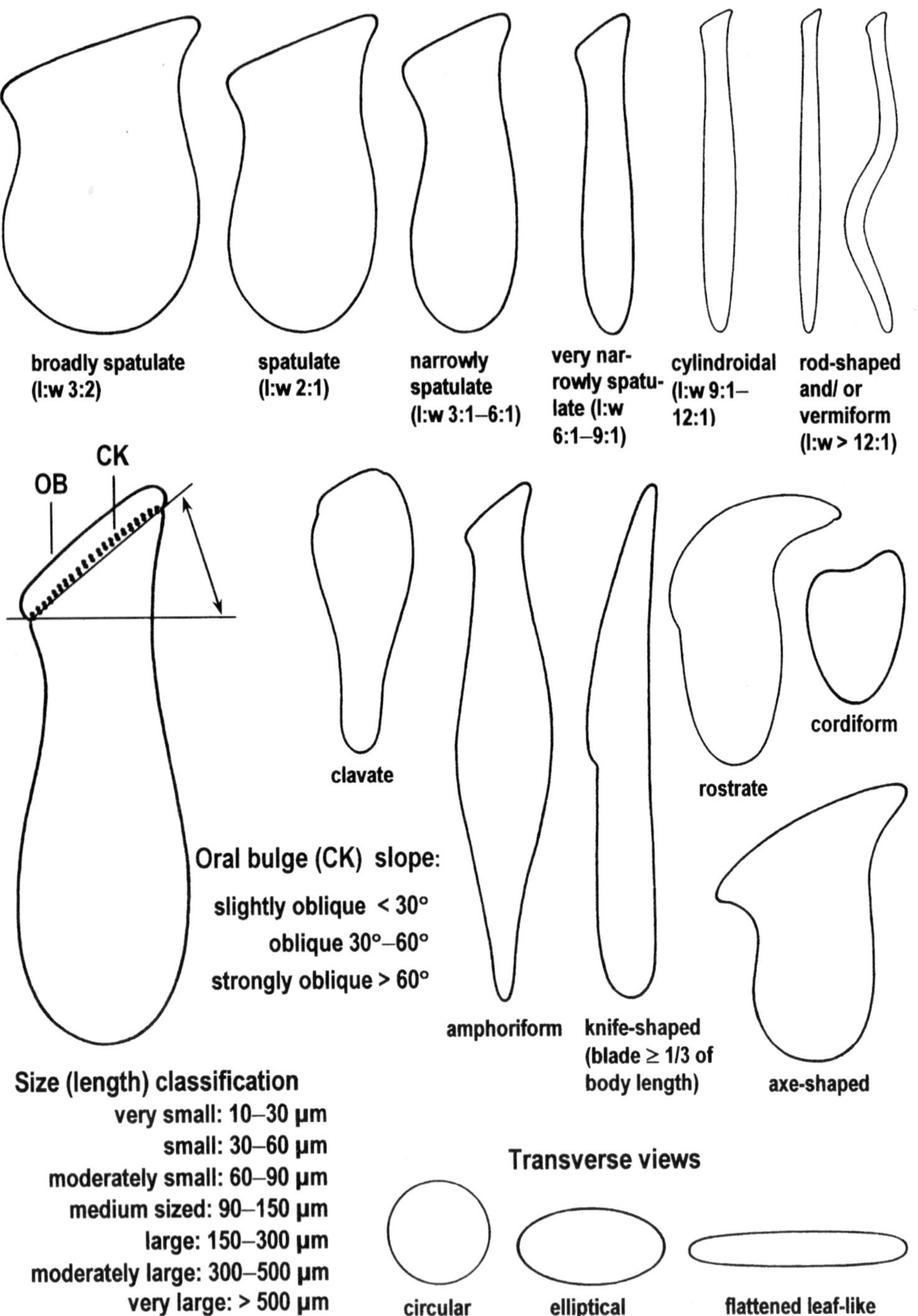

Fig. 1 Classification (terminology) of body shape in lateral and transverse view, size (length), and slope of oral bulge in spathidiid ciliates. CK – circumoral kinety, l:w – ratio of body length to width, OB – oral bulge.

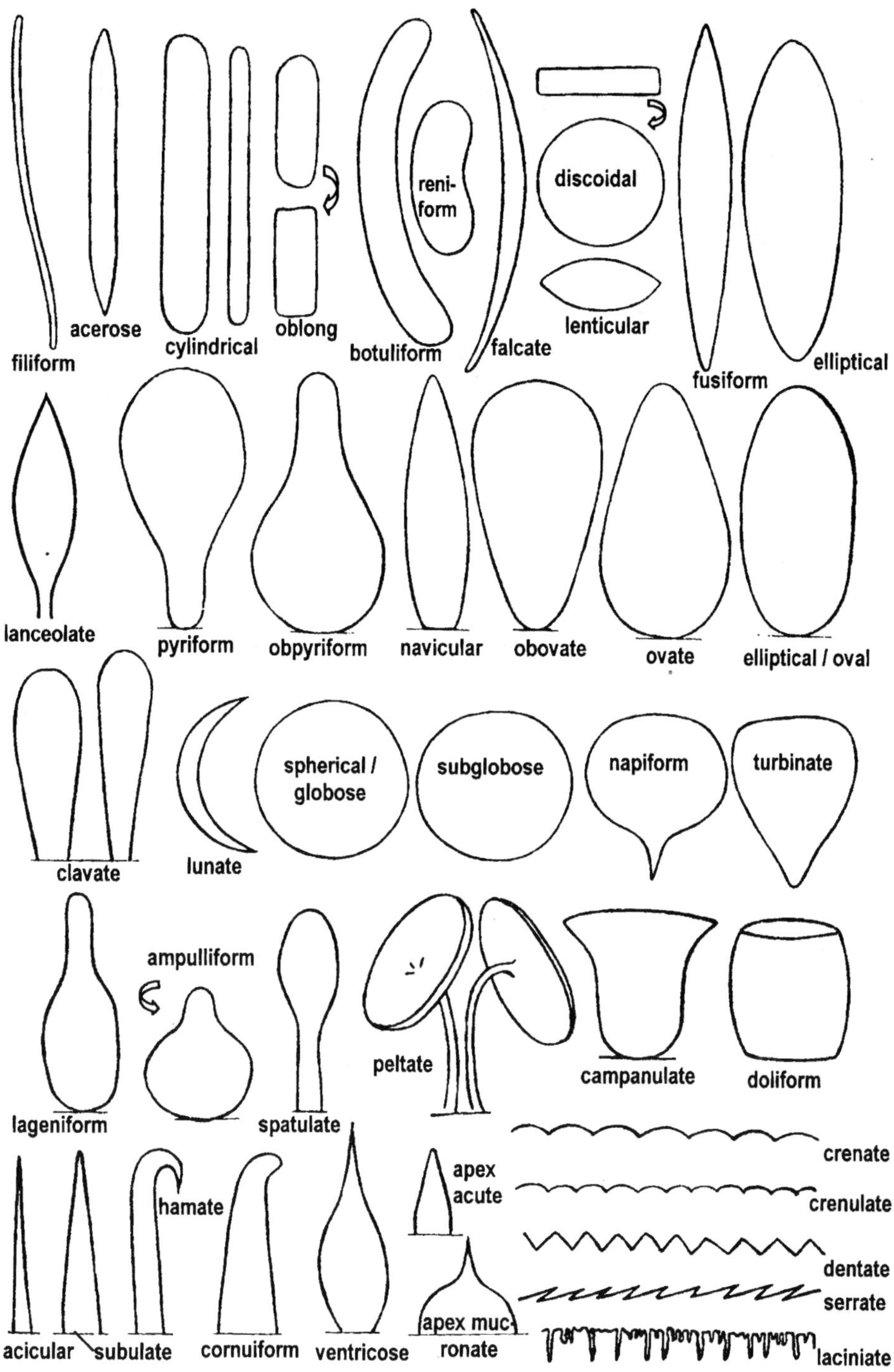

Fig. 2 Terminology of shapes/outlines. From STEARN (1992), modified.

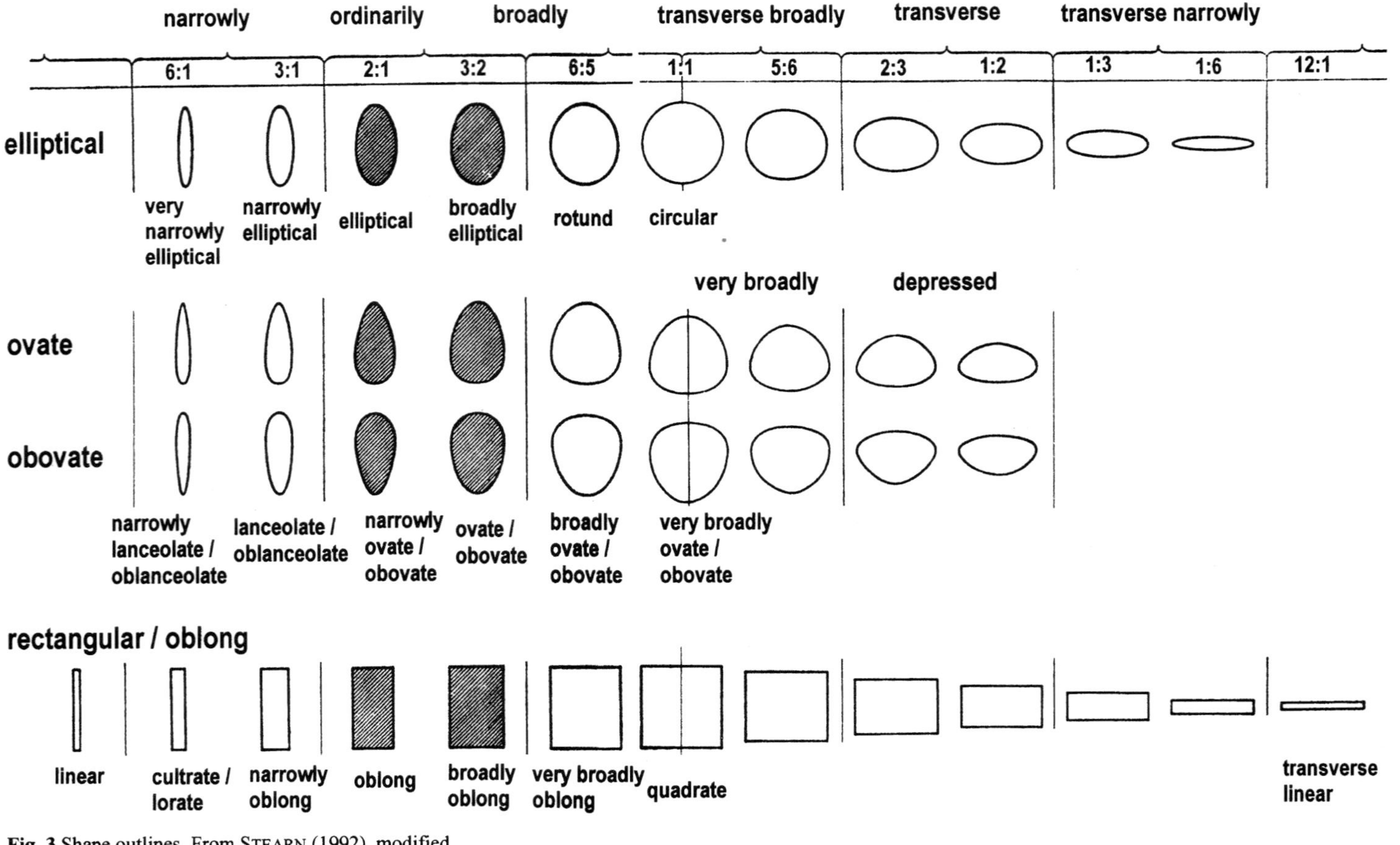

Fig. 3 Shape outlines. From STEARN (1992), modified.

distinct bulge rarely extending to posterior body end (genus *Rhinothrix*). The bulge, respectively, the anterior body end is more or less slanted. Thus, bulge length is a composite of body width and bulge slope. We have used three categories of bulge slope as follows: slightly oblique, < 30°; (ordinarily) oblique, 30–60°; strongly oblique, > 60° (Fig. 1).

Detailed morphometrics are available from about 150 populations belonging to circa 100 species, mainly due to the investigations of FOISSNER (1984), FOISSNER et al. (2002), and the present monograph. These data show that the mean coefficients of variation are very similar to those found in colpodids, stichotrichine hypotrichs, and other kinetofragminophoran ciliates (BERGER 1999, FOISSNER 1984, 1993, FOISSNER et al. 2002). Accordingly and in the contrast to the widespread assumption (e.g., WENZEL 1955), spathidiids are as variable or stable as other ciliate groups. The impression of a special variability of the spathidiids is caused by observations of a few species in pure cultures, where variability is indeed high – not only in spathidiids but also in most other ciliates. Malformed specimens can survive and sometimes even reproduce in pure cultures due to the optimal food supply, while they usually die in nature. The number of ciliary rows usually varies only slightly, that is, between 5–10% and is thus one of the most important features for species recognition. Most other characteristics vary between 10% and 20%, while body length:width ratio and the number of macronucleus nodules in multinucleate species have high coefficients of variation between 20 and 40%. The morphometric data give important information about the stability of features and their significance for species recognition. Thus, all descriptions should be accompanied by morphometrics of at least the main features.

1·2 Nuclear apparatus (Fig. 4)

Spathidiids have a single macronucleus and micronucleus or many macronucleus nodules and several micronuclei. The shape of the macronucleus ranges from globular to a long, tortuous strand, which may be band-like flattened, a curious feature described only recently (FOISSNER et al. 2002). The micronuclei may be globular, ellipsoidal, lenticular, narrowly ovate, or bluntly fusiform. Altogether, we have distinguished 16 macronuclear and 10 micronuclear configurations (Fig. 4). This high diversity makes the nuclear apparatus to one of the most important features for species recognition. The patterns are as stable or as variable as those of other ciliates and are sometimes obscured by post-divisional, post-conjugational, or ontogenetic reorganization processes (MOORE 1924a, b, XU & FOISSNER 2004). **When in doubt, look at very early dividers which invariably show the "real" nucleus pattern.** So far, chromatin extrusion has not been described.

The macronucleus contains globular, oblong, or irregular dense masses about 1–5 µm in size. Usually, they are recognizable in vivo and impregnate deeply with protargol. There is some indication that these inclusions represent nucleoli: (i) the central mass of the macronucleus of *Colpoda steinii* and *Dileptus* sp. deeply impregnates with protargol and represents a compound nucleolus, according to cytochemical and electronmicroscopical investigations (FOISSNER 1993, RAIKOV 1982); (ii) chromatin bodies are usually smaller than 1 µm and numerous, while nucleoli are often larger than 1 µm and comparatively rare (RAIKOV 1982). These two features apply to the macronuclear structures impregnating with various protargol methods. Thus, we designate these

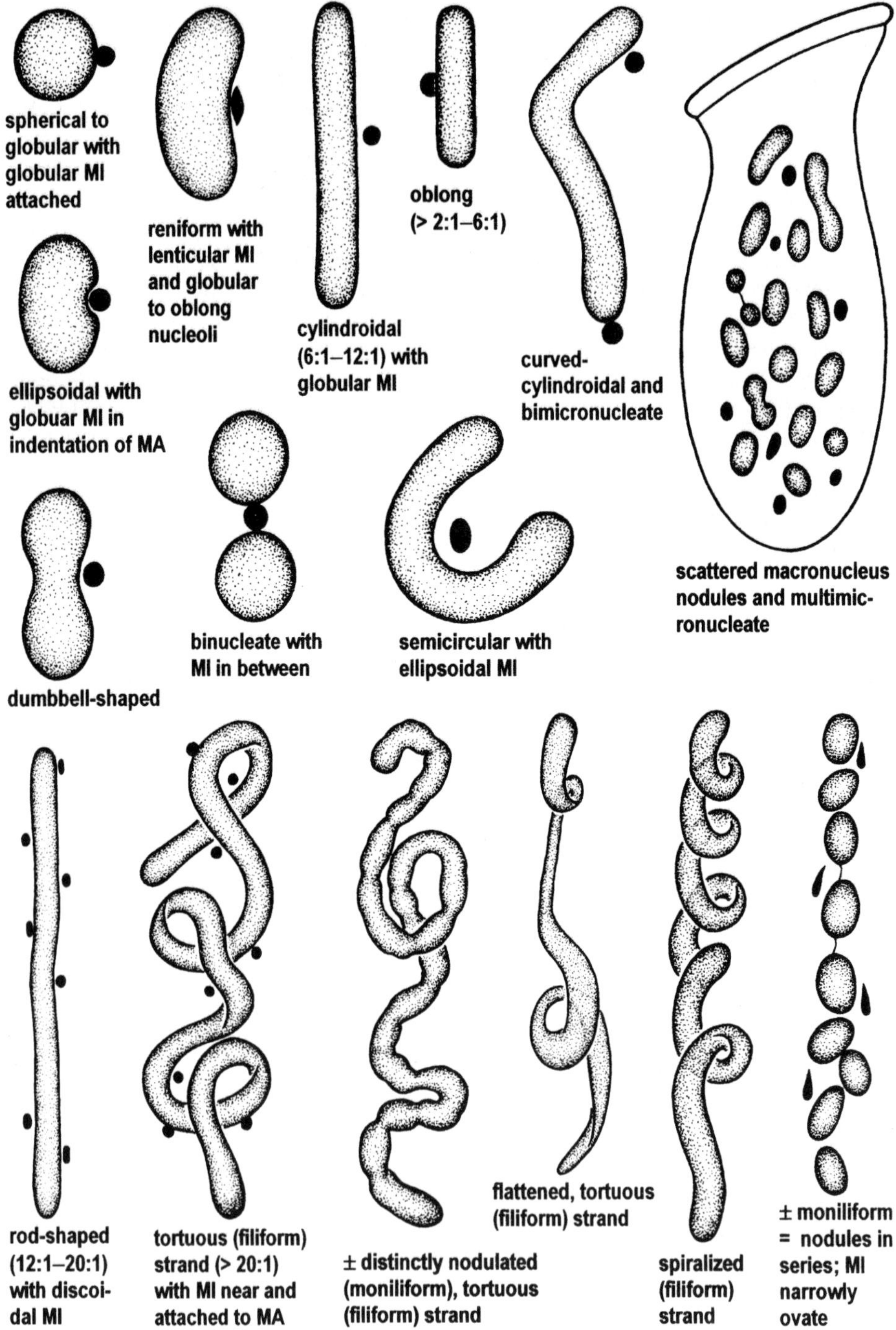

Fig. 4 Shapes of macronucleus and micronucleus in spathidiid ciliates. Nucleoli shown only in the reniform type. MA – macronucleus, MI – micronucleus.

structures as "nucleoli", being aware that this needs confirmation by cytochemical investigations. See BOHATIER et al. (1978) for a brief electronmicroscopical description of the nuclear apparatus of *Epispathidium amphoriforme*.

1·3 Contractile vacuole and cytopyge (Fig. 5)

Most spathidiids have a single contractile vacuole in posterior body end; few have a second contractile vacuole with distinct excretory pores underneath the dorsal brush, that is, above mid-body, for instance *Arcuospathidium bulli*; and some have several contractile vacuoles, each with their own excretory pores, in a dorsal and/or ventral row (Fig. 5). The bivacuolate pattern evolved independently in *Spathidium* and *Arcuospathidium*, while the multivacuolate pattern is diagnostic of the genus *Supraspathidium* (FOISSNER 2003c). Usually, the vacuole is a simple blister surrounded by smaller collecting vesicles during the diastole. Some species of doubtful classification have a more or less conspicuous collecting canal which supplies the main vacuole, for instance, *Spathidium latissimum* and *Supraspathidium teres* (Vol. II).

The fluid collected by the contractile vacuole is expelled via one or several ordinary excretory pores. Usually, the pores are scattered in the posterior pole area, sometimes they are slightly subterminal on the ventral or dorsal side, and in a few tailed species they are far subterminal, for instance, in *Spathidium apospathidiforme* and *Apospathidium atypicum* (Vol. II).

The cytopyge is known in only few species, where it is in the posterior area, that is, near the excretory pores of the contractile vacuole. In some species, for instance, *Rhinotrix porculus*, the fecal mass traverses the contractile vacuole when it is expelled.

1·4 Extrusomes (Fig. 5–7)

Shape, size, and arrangement of the extrusomes are highly diverse and are thus a main diagnostic feature of the individual species (Fig. 5–7). This was already emphasized by KAHL (1930a, b). Accordingly, the extrusomes must be carefully studied in vivo because they often become distorted in silver preparations or do not impregnate at all. Indeed, the extrusome features are so important that species cannot be recognized without this information.

Most or even all spathidiids are predators. Thus, they have toxicysts, except of some curious species which lack toxicysts at all. About one third of the species has two shape and/or size types of toxicysts, and some even have three kinds, for instance, *Rhinothrix porculus*. Further, (likely) all spathidiids have at least one kind of mucocysts, appearing as "cortical granules"(Fig. 7).

The toxicysts are studded in the oral bulge and scattered in the cytoplasm; additionally, they may be attached to the somatic cortex. For determining the shape, size, and arrangement of the toxicysts only fully developed (mature) toxicysts may be used, that is, those which are anchored to the oral bulge or somatic cortex (Fig. 5, 6); cytoplasmic toxicysts are frequently not fully developed. This is evident, inter alia, from their impregnation capacity: anchored toxicysts usually do not impregnate with protargol, while various cytoplasmic developmental stages often impregnate rather deeply.

Shape and arrangement of the mucocysts are much less diverse, at least in the light

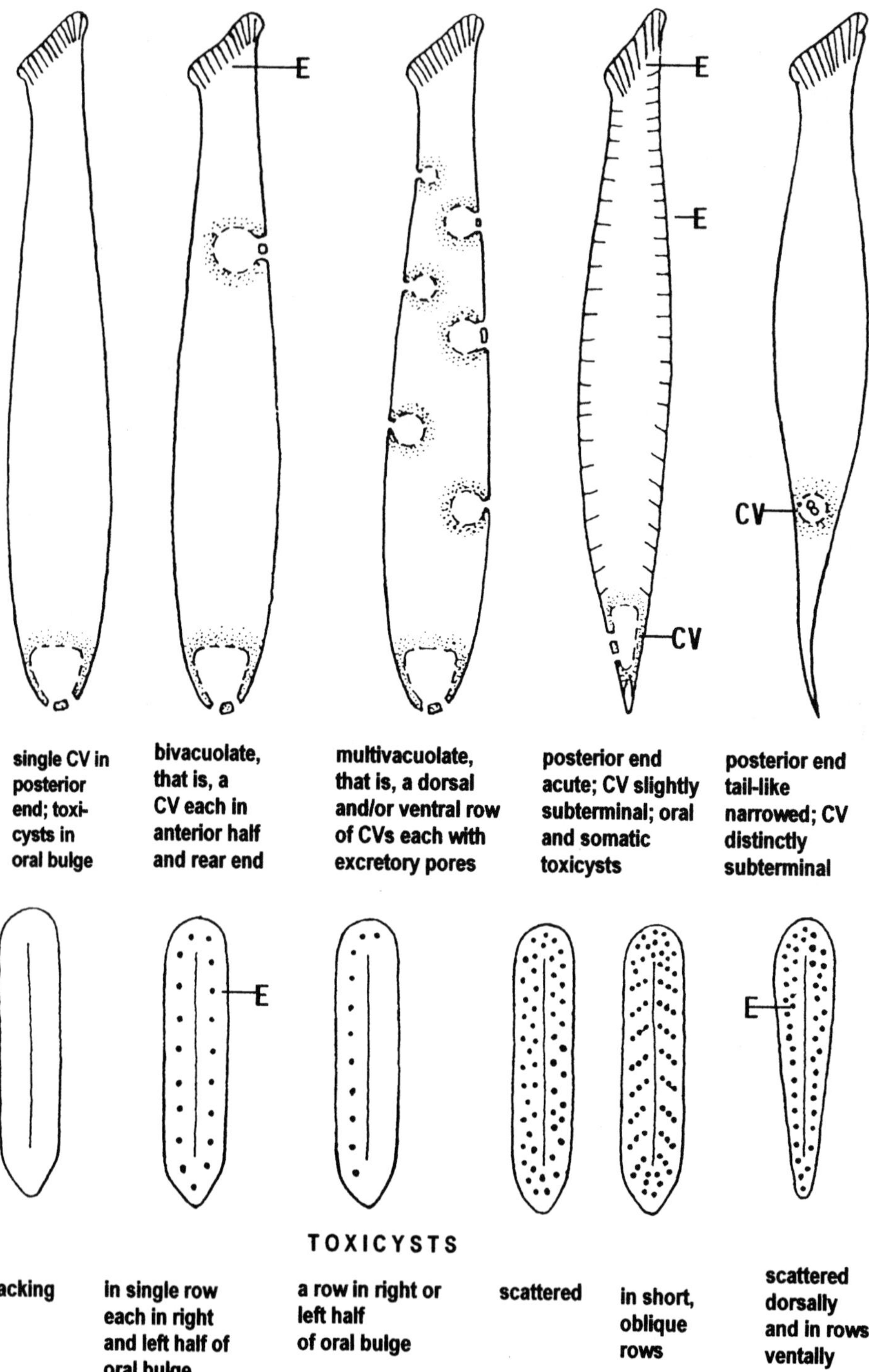

Fig. 5 Contractile vacuole (CV) and toxicyst (E) patterns and rare body shapes in spathidiid ciliates.

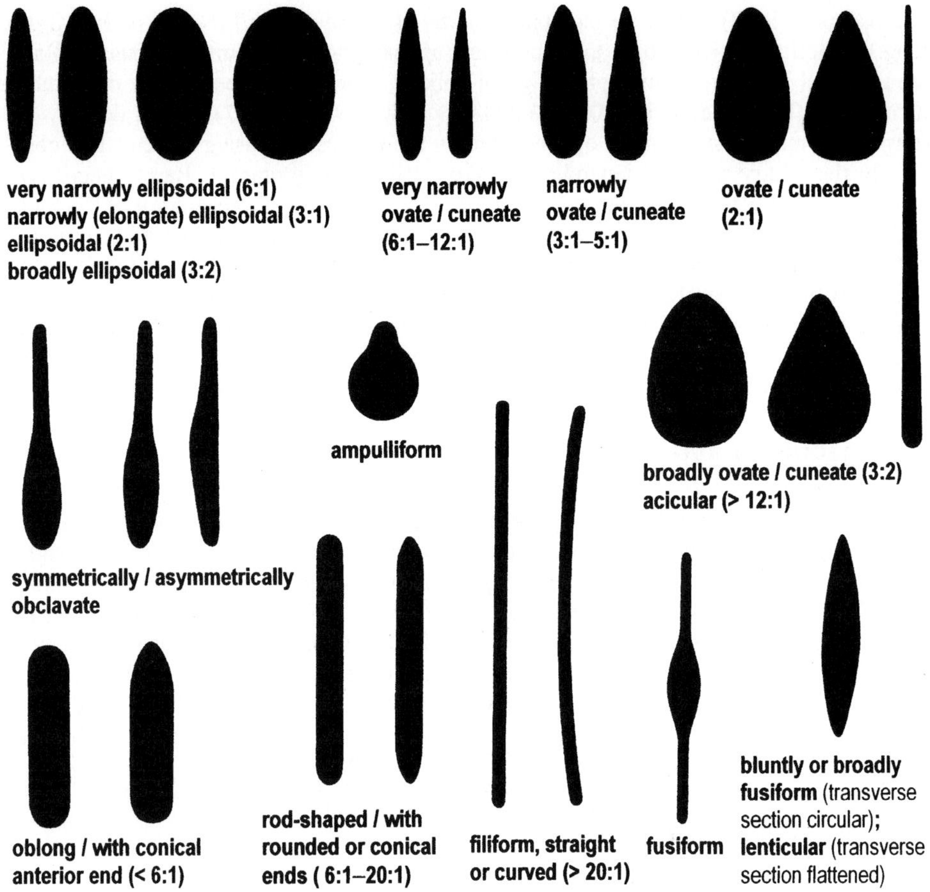

very narrowly ellipsoidal (6:1)
narrowly (elongate) ellipsoidal (3:1)
ellipsoidal (2:1)
broadly ellipsoidal (3:2)
very narrowly ovate / cuneate (6:1–12:1)
narrowly ovate / cuneate (3:1–5:1)
ovate / cuneate (2:1)
ampulliform
broadly ovate / cuneate (3:2)
acicular (> 12:1)
symmetrically / asymmetrically obclavate
oblong / with conical anterior end (< 6:1)
rod-shaped / with rounded or conical ends (6:1–20:1)
filiform, straight or curved (> 20:1)
fusiform
bluntly or broadly fusiform (transverse section circular); lenticular (transverse section flattened)

Fig. 6 Shapes of extrusomes in spathidiid ciliates: toxicysts. Length about 1–50 µm, often 5–20 µm.

shapes
loose
ordinary
dense
two size-types

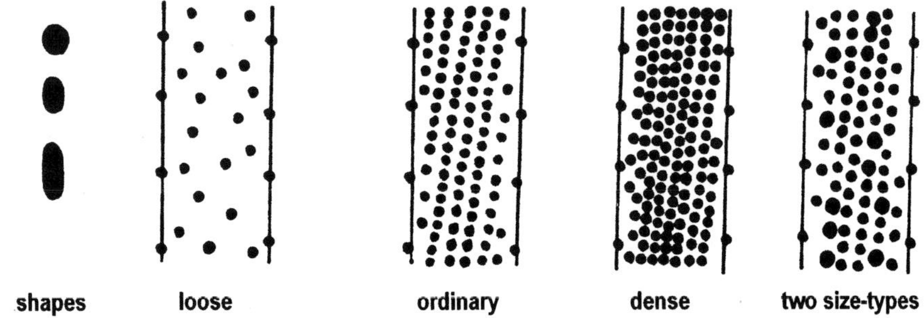

Fig. 7 Shapes and arrangement of extrusomes in spathidiid ciliates: mucocysts and related structures. Size about 0.2–2 µm.

microscope (Fig. 7). They are globular, broadly ellipsoidal, or ellipsoidal and ≤ 2 μm long. In the light microscope, the mucocysts appear granular ("cortical granules") and are arranged in rows following the slightly oblique course of the postciliary microtubular ribbons (FOISSNER et al. 2002, WILLIAMS et al. 1981; Fig. 7). When densely arranged, the mucocysts form an opaque sheet in the cortex, obscuring more or less completely the ciliary pattern, depending on their affinity to protargol. Rarely occur two types of mucocysts or mucocyst-like organelles, differing in size and/or refractivity and/or fine structure (FOISSNER et al. 2002, WILLIAMS et al. 1981). The mucocysts are not easily released. Methyl green-pyronin, which causes mucocyst extrusion in many ciliates (FOISSNER 1991), is usually ineffective. By chance, we were successful in obtaining some good SEM micrographs showing the release in several spathidiids (see volume II). These observations show that the "cortical granules" are indeed mucocysts and are extruded like those of other ciliates (HAUSMANN et al. 2003).

1·5 Cytoplasm and colour

All spathidiids are colourless. However, when packed with highly refractive food and/or lipid droplets and/or crystal-like inclusions, they appear dark or black under low brightfield magnification, for instance, *Arcuospathidium cooperi* (Fig. 56a) and *Rhinothrix porculus* (Fig. 149a–i). Some species are green due to ingested or symbiotic algae, for instance, *Spathidium chlorelligerum*. Crystal-like inclusions are frequent in some species, for instance, in *Apertospathula inermis* (Fig. 146a, d–f); in others, such inclusions are possibly partially digested paramylon grains from euglenoid prey. The food vacuoles may be large because the prey is sometimes ingested as a whole; soon, such vacuoles dissociate into several smaller ones with granular contents.

1·6 Movement

Basically, spathidiids glide and creep slowly to rapidly on and among the mud particles of limnetic and terrestrial habitats. However, when approaching the free water, they swim by rotation about the main body axis, whereby the oral region may be curved laterally and, in the long and slender species, swings more distinctly than the posterior, making the cells looking like swimming cones. The large, voluminous species usually glide majesticly, while the long or slender species show a worm-like behaviour wriggling like nematodes and slender hypotrichs. With very few exceptions (genus *Spathidiodes*), spathidiids are very flexible and can squeeze through small spaces. Pronounced contractility does not occur, but many species may show some shortening of the body, especially under mild coverslip pressure.

1·7 Somatic ciliature

1·7·1 The basic ciliary patterns (Fig. 8)

Spathidiids are asymmetric. Thus, both a ventral and dorsal side as well as a right and a

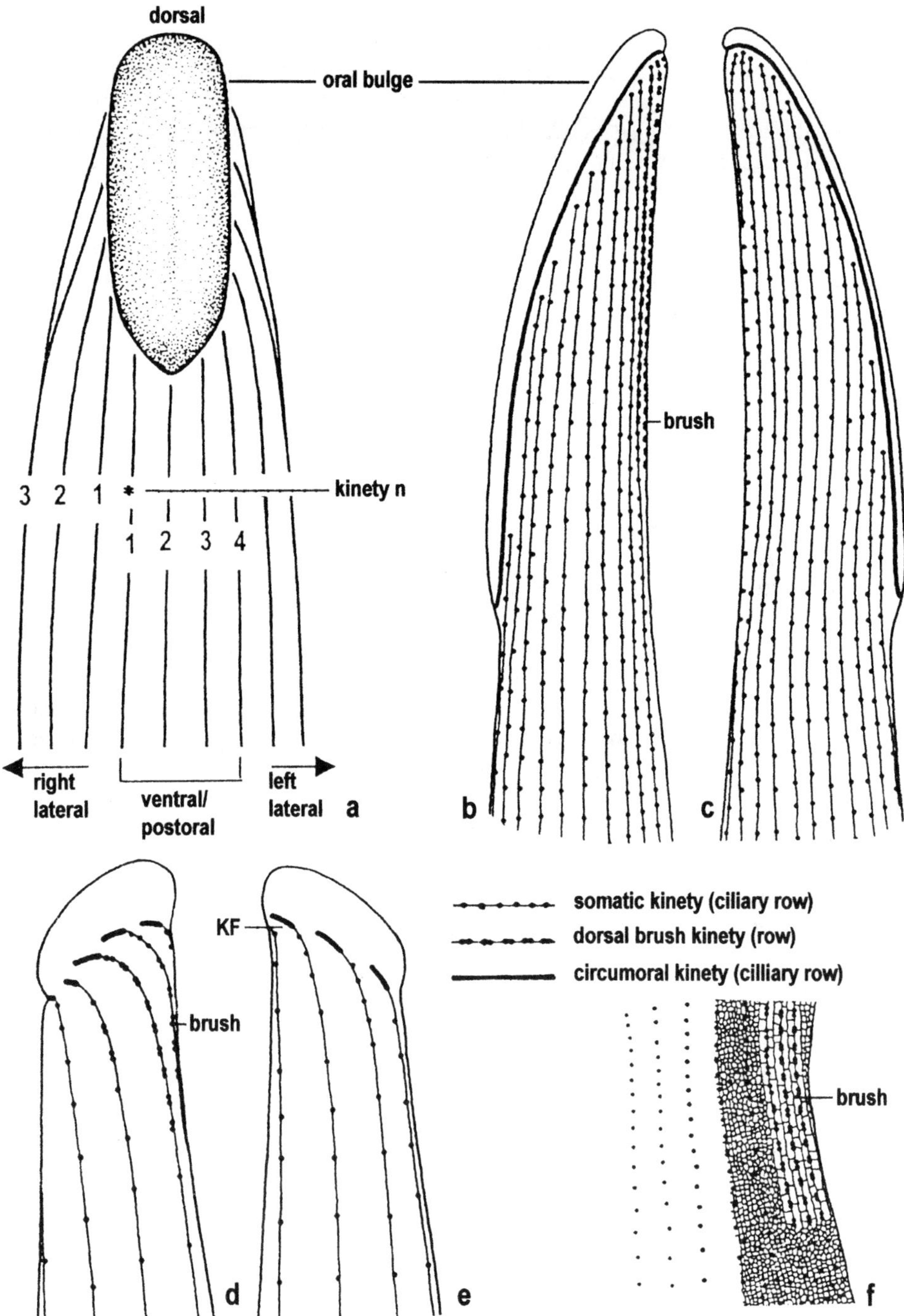

Fig. 8a–f Designation of ciliary rows (a), *Arcuospathidium* ciliary pattern (b, c), *Protospathidium* ciliary pattern (d, e), and silverline pattern (f) of spathidiid ciliates. For details, see text of general section. From FOISSNER (1980b, 1984) and original (a).

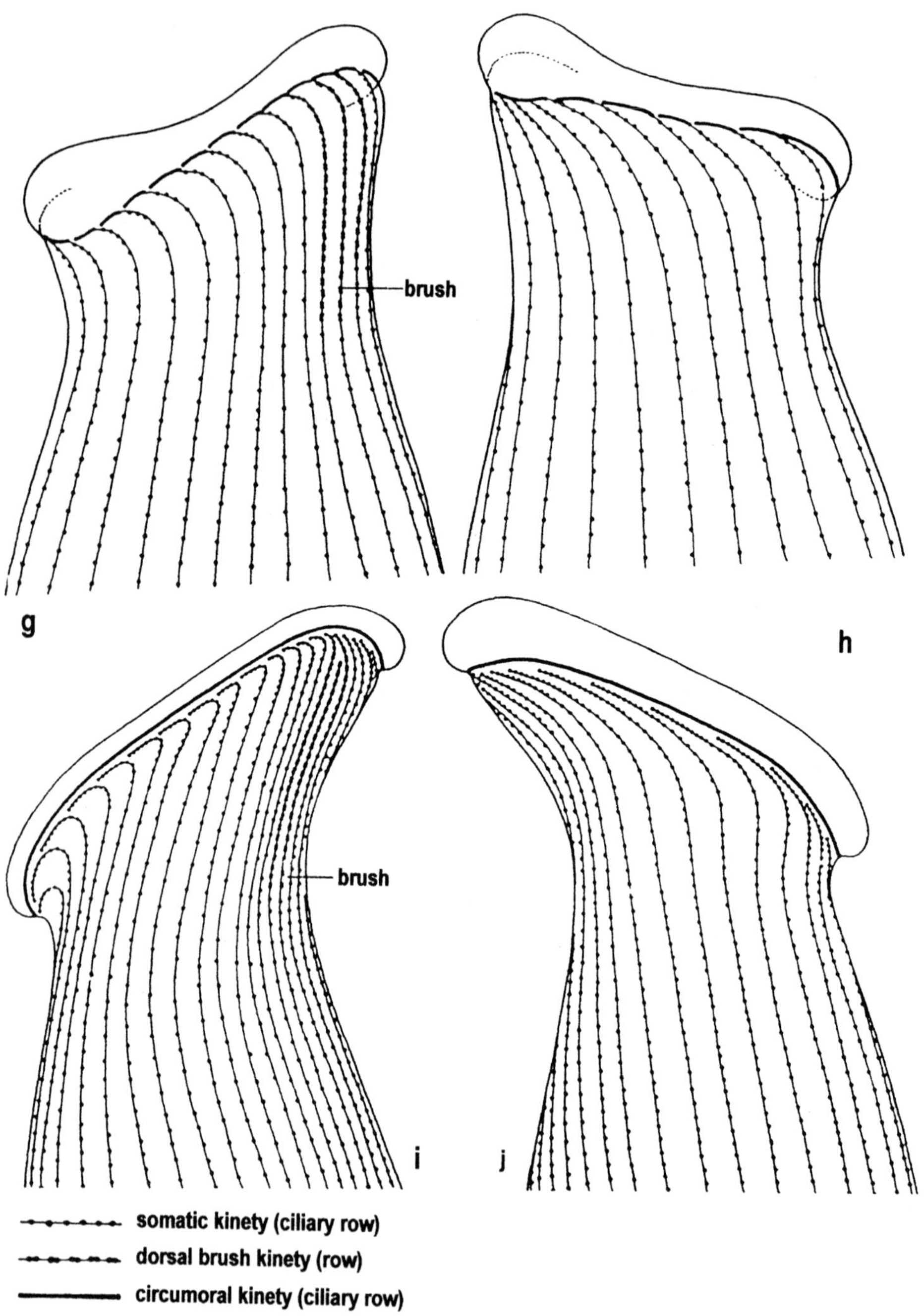

Fig. 8g–j *Spathidium* (g, h) and *Epispathidium* (i, j) ciliary pattern. For details, see text of general section. From FOISSNER (1984).

left side can be distinguished. However, the boundaries are fluent because all ciliary rows are alike. We suggest to designate the kineties as shown in figure 8a.

The spathidiids contained in this monograph have a simple, holotrichous (~ complete) ciliature composed of serially arranged monokinetids (= single, ciliated basal bodies). The monokinetids usually form equidistant rows extending from the circumoral kinety to the posterior body end; the rows may be slightly shortened both anteriorly and/or posteriorly and frequently contain some irregularities, such as small breaks or overlapping pieces; further, often some rows are strongly shortened anteriorly and/or posteriorly. Depending on row distance, we distinguish narrowly spaced rows (≤ 4 µm), ordinarily spaced rows (4–8 µm), and widely spaced rows (≥ 8 µm), as shown in figure 9.

The cilia within the rows are densely, ordinarily, or loosely spaced (Fig. 9). Usually, they are more densely spaced orally than postorally; rarely, vice versa. Frequently, the kineties contain some dikinetid-like kinetids, especially in the middle third. The posterior basal body of these pairs is ciliated, while the anterior is bare. Likely, the anterior kinetids are a reservoir for growing and/or dividing cells.

Based on FOISSNER (1984), we distinguish four basic ciliary patterns which, however, show some transitions and even may be obscured when the ciliary rows are distinctly shortened anteriorly and/or the species has less than 10 ciliary rows, for instance, in *Arcuospathidium namibiense*. Frequently, such species cannot be classified properly at the present state of knowledge. Originally, the patterns were used to define genera (FOISSNER 1984). In our system, the patterns define families, except of the *Epispathidium* pattern. The rank elevation will be discussed in the phylogenetic section of volume II. Briefly, it is due the discovery of new evolutionary lines (genera) having the same pattern but differing in other important features, for instance, the location of the dorsal brush in *Arcuospathidium* and *Cultellothrix* (XU & FOISSNER 2003).

Protospathidium pattern (Fig. 8d, e): The kinetofragments composing the circumoral kinety are separated by gaps one to four dikinetids wide and are connected to the ciliary rows. Thus, the circumoral kinety is discontinuous. The anterior region of the right side kineties is directed dorsally, while the anterior region of the left side kineties is directed ventrally.

Arcuospathidium pattern (Fig. 8b, c): The oral kinetofragments are aligned to a continuous circumoral kinety distinctly separate from the ciliary rows. The anterior end of the lateral somatic kineties is directed dorsally on **both** sides of the cell.

Spathidium pattern (Fig. 8g, h): The oral kinetofragments are usually connected to the ciliary rows from which they originated. Thus, minute discontinuities occur in the circumoral kinety. The anterior region of the right side kineties is directed dorsally, while the anterior region of the left side kineties is directed ventrally. The *Spathidium* pattern comprises several subtypes shown and discussed in the characterization of the genus.

Epispathidium pattern (Fig. 8i, j): The oral kinetofragments are aligned to a continuous circumoral kinety distinctly separate from the ciliary rows. The anterior region of the somatic kineties is densely ciliated and usually so distinctly curved dorsally (right side) or ventrally (left side) that the circumoral kinety is seemingly doubled, that is, the anterior region of the ciliary rows parallels the circumoral kinety.

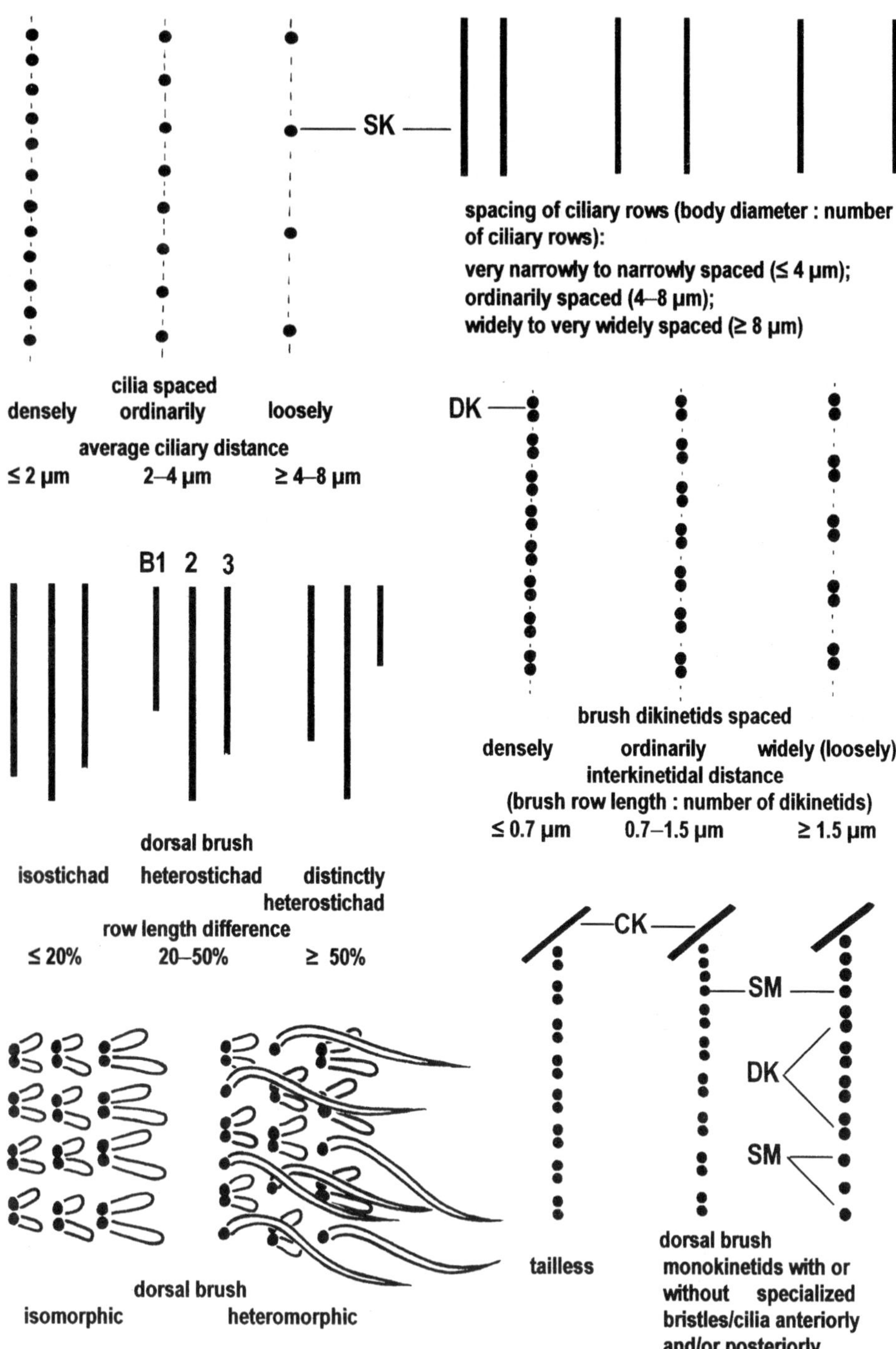

Fig. 9 Terminology of the somatic and dorsal brush ciliature. B1-3 – dorsal brush rows, CK – circumoral kinety, DK – brush dikinetids, SK – somatic ciliary rows, SM – specialized monokinetids.

1·7·2 The dorsal brush (Fig. 9)

The anterior region of some dorsal and/or left lateral ciliary rows is modified to the so-called "dorsal brush" or, simply, "brush". The area consists of specialized, narrowly spaced mono- and dikinetids with bristle-like cilia usually distinctly shorter than ordinary somatic cilia. The function of the brush is not known. The dorsal brush shows a considerable diversity and is thus of high significance in genus and species taxonomy.

Two features of the brush are of generic significance because they are not caused by simple space constraints. The first feature is the number of brush rows, which is highly stable: usually, there are three, but some species groups have only two (*Schmidingerophrya*) or more than three (*Semibryophyllum*). The second feature is the location of the brush: usually, the rows extend dorsally or dorsolaterally, while they are entirely or almost entirely displaced left laterally in *Cultellothrix* and *Latispathidium*.

A variety of brush features is used to characterize and distinguish species. Most details are shown in figure 9, and thus will be mentioned only briefly. The dorsal brush can be short (longest row ≤ 15% of body length in protargol preparations), ordinary (15–35%), or long (≥ 35% of body length). The individual brush rows may be of similar length (isostichad), of different length (heterostichad), or of very different length (distinctly heterostichad). Usually, the dorsal brush is isomorphic, that is, composed of bristles throughout; rarely, it is heteromorphic, that is, mixed with ordinary cilia (Fig. 9). The individual brush rows may commence and end with dikinetidal bristles. This is the tailless condition, which is rarely found, for instance, in most *Cultellothrix* species. Usually, the brush rows commence underneath the circumoral kinety with a short "anterior tail", that is, with a few ordinarily ciliated monokinetids. The posterior end of the brush rows is either dikinetidal, heteromorphic or, in the leftmost row 3, modified to an inconspicuous "posterior tail" composed of rather narrowly spaced monokinetids with up to 5 µm long bristles. The posterior tail of brush row 3 usually extends to near mid-body, rarely it is absent (some protospathidiids) or extends to near rear body end (some *Spathidium* species). Underneath the brush, the rows continue as ordinary somatic kineties and extend to the posterior end of the cell (Fig. 8b, d). Finally, the brush dikinetids may be densely, ordinarily, or widely spaced (Fig. 9).

During cell division, the brush bristles are generated as shown in figure 131f: a new bristle each is formed anteriorly of the parental cilia which gradually shorten to posterior brush bristles.

1·8 Oral apparatus

1·8·1 Structure (Fig. 10–12)

The location and structure of the spathidiid oral apparatus are basically as in other haptorid ciliates. However, details are sufficiently different to base two families on the oral apparatus, viz., the *Apertospathulidae* and the *Pharyngospathidiidae* (volume II). Likewise, the shape of the oral bulge varies considerably and is thus an important feature for species recognition (Fig. 10). Detailed electronmicroscopical investigations on the spathidiid oral bulge are lacking. However, the basic organization was described by

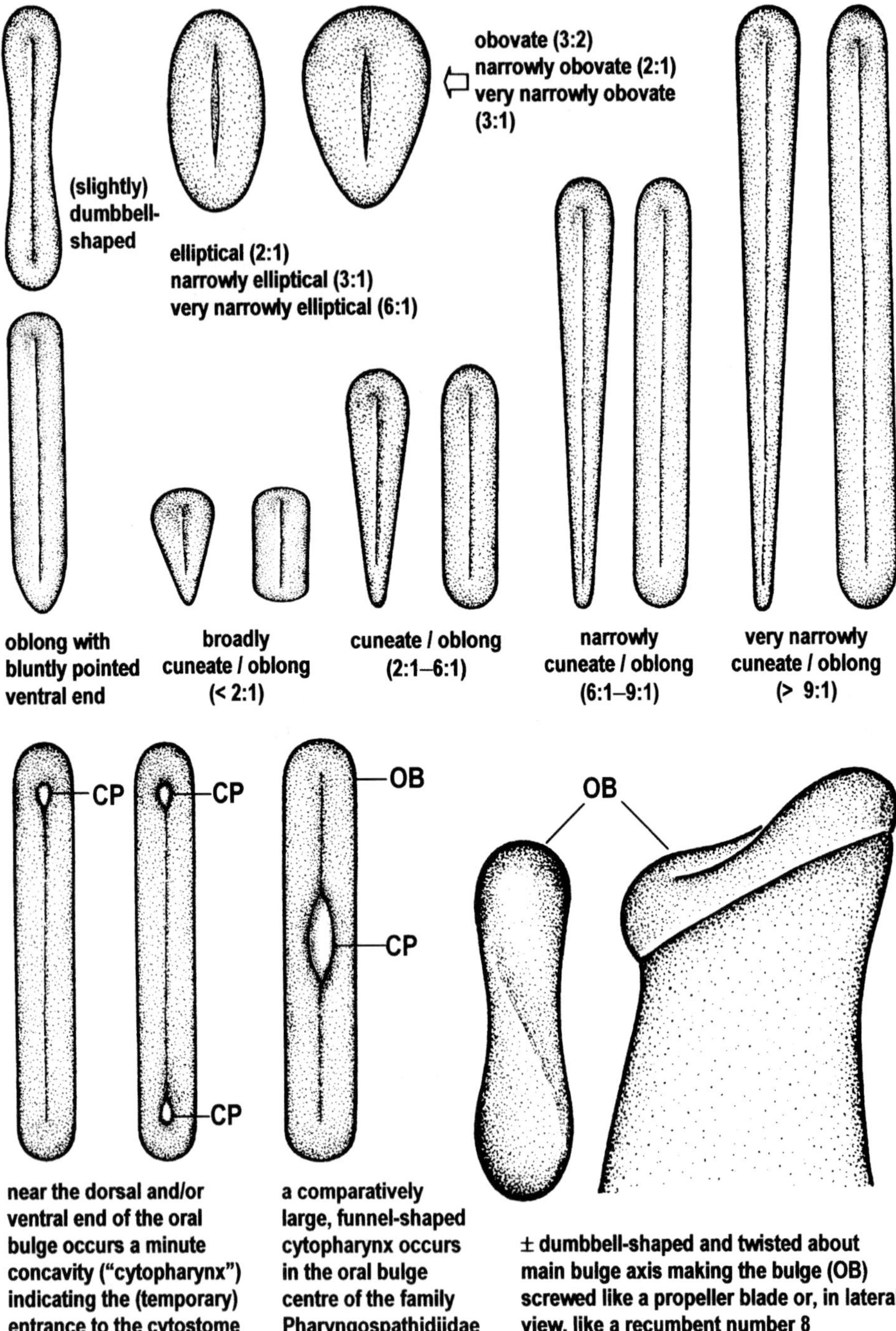

Fig. 10 Shapes of oral bulge and mouth patterns in spathidiid ciliates. CP – cytopharyngeal opening, OB – oral bulge.

BOHATIER et al. (1978). These findings and the light microscopical data collected during the preparation of the present monograph were used for the scheme shown in figure 12. Invariably, the spathidiid oral apparatus ("mouth") occupies the entire or almost the entire anterior body end. It consists of the oral bulge including the cytostome and the toxicysts; the circumoral kinety (ciliary row); and the oral basket.

The oral bulge is a more or less distinct, unciliated structure proximally bordered by the circumoral kinety. Frequently, the bulge is oblique (30°–60°) increasing the size of the oral area, especially in several arcuospathidiids, where it extends to or near to mid-body. We assume that this elongation is a derived feature. The shape (outline) of the oral bulge varies from elliptical to very narrowly cuneate/oblong (Fig. 10) and is roughly correlated with the familial classification: most protospathidiids have an elliptical or obovate bulge; in the spathidiids and pharyngospathidiids, the bulge is frequently oblong and twisted about the main axis making the bulge screwed like a propeller blade or, in lateral view, like a recumbent number 8; most arcuospathidiids have a cuneate to very narrowly cuneate oral bulge. The bulge cortex contains mucocysts ("cortical granules") and a fibre system sometimes distinctly impregnating with protargol and well recognizable in scanning electron micrographs (Fig. 119m, n, 135a–e, 138w, x). The fibre system consists of ribbons of microtubules corresponding to the transverse microtubule ribbons of the somatic basal bodies (Fig. 12). Thus, spathidiids have a rhabdos-type oral apparatus (CORLISS 1979). The microtubule ribbons originate from the bare basal bodies of the circumoral dikinetids and support/extend to the cytostome wall; depending on shape and size of the oral bulge, the fibres form a spiral (small, ± obovate bulge) or arrowhead-like (large, oblong bulge) pattern in the distal bulge surface.

The frontal (distal) surface of the oral bulge contains a flat groove which, basically, represents the cytostome, that is, the groove opens when prey is ingested. This is the simplest cytostome type and found in many species of various spathidiid families (Fig. 11a). The second cytostome type has a minute, obconical depression, which is lined by the transverse microtubule ribbons, near the dorsal (rarely also ventral) bulge end (Fig. 11b, 12). When feeding commences, the obconical depression opens, for instance, in *Arcuospathidium muscorum* (Fig. 14b). We call this minute depression "temporary cytostome" because it is open only during feeding. The temporary cytostome can be recognized in vivo, in the scanning electron microscope, and in protargol preparations (Fig. 60a, b, 114i, 118d, 119h, i, 139a, d). A third cytostome type is found in the new family Pharyngospathidiidae, described in the second volume. Here, the cytostome is a comparatively large, elliptical, permanent opening in the bulge centre. The wall of the permanent cytostome is lined by many transverse microtubule ribbons forming a conspicuous "inner oral basket" (Fig. 11c–e, 12).

The circumoral kinety marks the proximal margin of the oral bulge. Usually, it has the same shape as the oral bulge, rarely it is slightly different, for instance, an oblong, ventrally rounded oral bulge may be associated with a cuneate, ventrally tapering circumoral kinety. The circumoral kinety is ring-like closed in most genera. However, in the Apertospathulidae the ventral end is open because the left branch of the kinety is slightly to distinctly shortened. The circumoral kinety is composed of pairs of basal bodies (dikinetids) with gradually increasing distances from dorsal to ventral bulge end. The two basal bodies are arranged in a certain angle, and only the more posteriorly located one is ciliated (Fig. 12). Thus, the circumoral kinety consists of dikinetids forming a single ciliary row (Fig. 11a). The anterior basal body of the dikinetids is bare, but

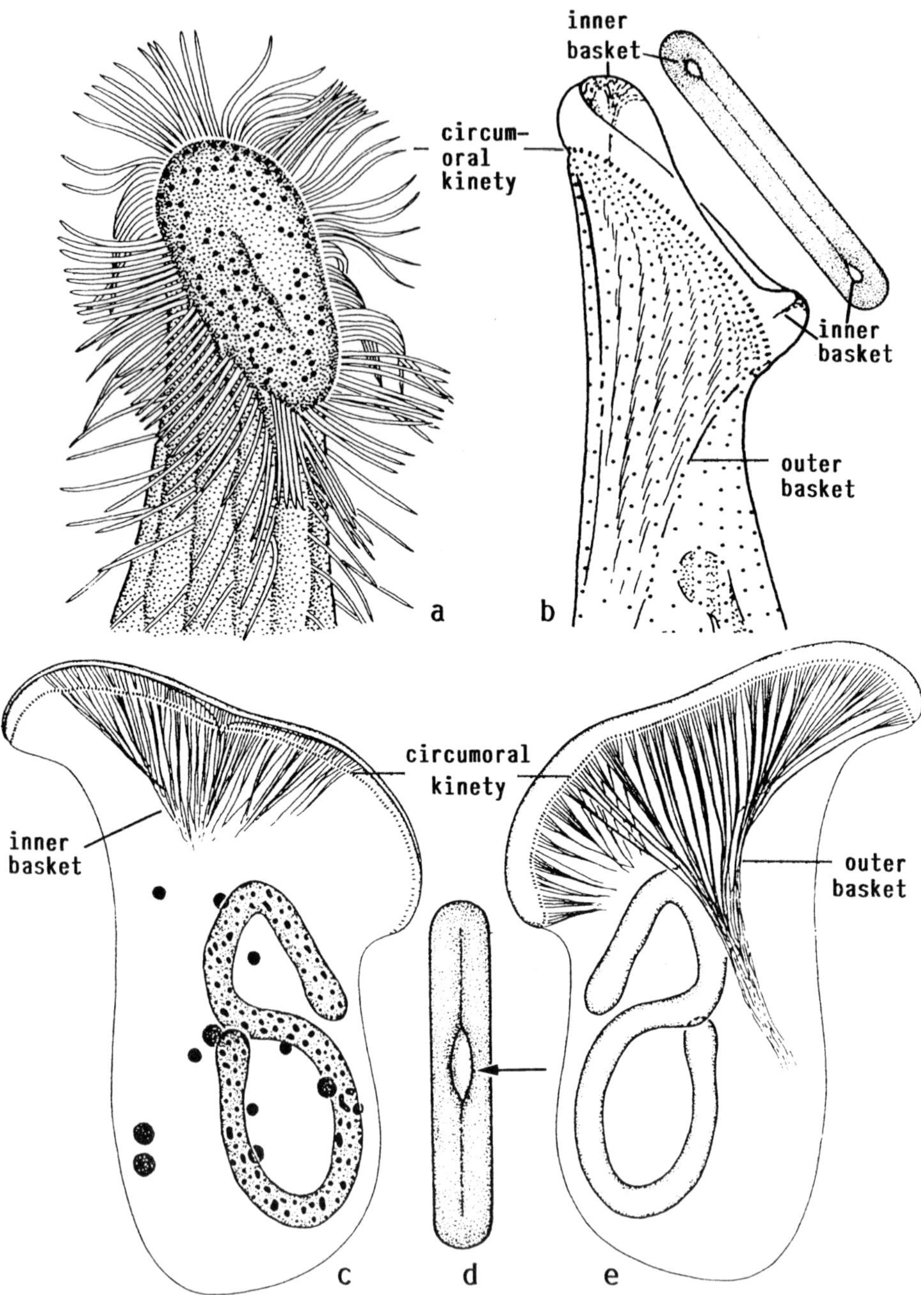

Fig. 11a–e Oral apparates of spathidiid ciliates. For details, see section 1.8 and next figure. **a)** Ventrolateral view of *Spathidium namibicola* (from FOISSNER et al. 2002). **b)** Right side and frontal view of *S. stammeri* (from FOISSNER 1987c). **c–e)** *Pharyngospathidium longichilum longichilum* (described in Vol. II) has a permanent cytostome (arrow) and a strongly developed inner and outer oral basket.

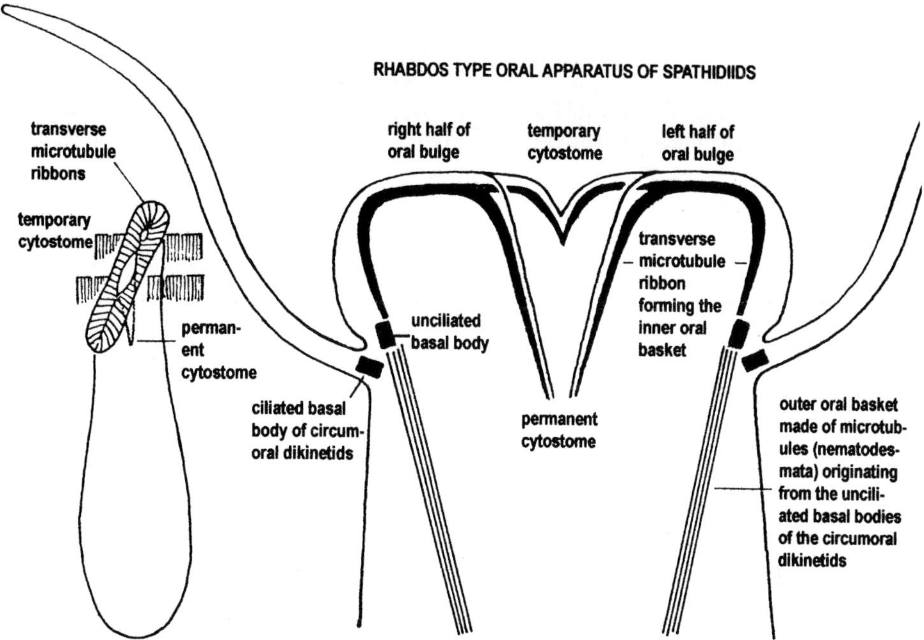

Fig. 12 Scheme of the rhabdos type oral apparatus of spathidiids. Based on our light microscopical data and the electronmicroscopical investigation of BOHATIER et al. (1978).

is associated with two important microtubule bundles: the oral basket rods (nematodesmata) originate from the proximal end of the basal bodies and constitute the (outer) oral basket, while the laterally originating transverse microtubule ribbons support the oral bulge and, if present, the pharyngeal opening, forming the inner oral basket (Fig. 11, 12).

1·8·2 Feeding (Fig. 13, 14)

Feeding of *Spathidium spathula* was described in detail by WOODRUFF & SPENCER (1921b, 1922). Unfortunately, these observations were overlooked by reviewers and disappeared from the modern literature (e.g. CALKINS & SUMMERS 1941, VERNI & GUALTIERI 1997). Thus, we provide an extended summary of their observations (Fig. 13a–h).

(i) *Spathidium spathula* readily paralyzes and swallows almost any small ciliate (e.g., *Colpidium*) with which it comes in contact, though flagellates of various kinds are immune.

(ii) *Spathidium* comes in contact with its prey solely through chance. They easily miss their prey. Nothing happens unless the quarry strikes the anterior end of the *Spathidium* nearly or quite in the center. Touching the edges produces no effect on either animal.

(iii) When the contact is made, the prey usually instantly becomes motionless and quickly shows vacuolization of the cytoplasm and disintegration of cilia. The stimulus afforded by the contact immediately stops the forward movements of the *Spathidium*, which gives a few rapidly repeated avoiding reactions that tend to keep it at or near the spot the capture was made. Meanwhile, the prey is moved around, apparently as a result of combined activity of the longer oral cilia and the gradually expanding edges of the oral bulge, until it may be conveniently be encircled. When the process of mouthing is successfully accomplished, nothing further is visible except the gradual sinking of the prey through the greatly expanded cytostome of the captor into the cytoplasm (Fig. 13a–h).

(iv) The whole process from contact to complete ingestion takes place in about thirty seconds, and the *Spathidium*, which usually sinks to the bottom during the process, within a few seconds more resumes its active foraging. Three such captures have been observed to occur within eight minutes and five over about an hour (Fig. 13a–h).

(v) Prey which has been paralyzed and has become removed from the oral region of the *Spathidium* is recovered in a majority of instances by a complex series of successively modified reactions, indicating "sensing at a distance". The factor involved in sensing is apparently a substance secreted by the *Spathidium* when the prey is paralyzed (authors: likely, these are the toxicysts which may form a long, reticulate mass).

Later, BALTES & WENZEL (1966) and WENZEL & BALTES (1967) investigated feeding and "sensing at a distance" in *Spathidium stammeri*. As concerns feeding in general, the data of WENZEL & BALTES (1967) largely match those of WOODRUFF & SPENCER (1922), while distance sensing was experimentally proven to depend on an organic acid – still effective after heating to 200°C – released by the prey. WENZEL & BALTES (1967) showed that *Spathidium* engulfes even Al_2O_3-particles, if they are impregnated with aqueous extracts of *Tetrahymena pyriformis* or meat extract.

Our own observations basically match those of WOODRUFF & SPENCER (1922) and those of related ciliates, e.g., *Homalozoon vermiculare* (KUHLMANN et al. 1980). Thus, we provide here only a brief overview (Fig. 14a–g) and refer to the individual species descriptions for details. Spathidiids are rapacious carnivores feeding on other protists, preferably ciliates and flagellates; some of the smaller species and those lacking extrusomes likely feed on bacteria, while the largest species may ingest small and middle-sized testate amoebae and small metazoans, mainly rotifers (Fig. 13i). The toxicysts play an important role in prey capture, as indicated by their great diversity (Fig. 6). The prey is ingested within a few minutes either whole (Fig. 14a, b, f, g) or as a mash when it is lysed outside (likely) by the toxicysts (Fig. 14e). The whole oral area is very extensible and thus can become funnel-shaped during the uptake of large prey (Fig. 14c), while the body usually becomes stouter and more or less deformed (Fig. 14b, e). When the prey touches the oral bulge, a bursiform or tube-like vacuole is generated at the site of the temporary or permanent cytostome (Fig. 14e–g). Then the opening widens more or less distinctly, depending on prey size, and the victim glides into the cell where a large vacuole has formed (Fig. 14b, e). WILLIAMS (1958) showed that axenic media (e.g., Osterhout´s solution) and the ciliate *Colpidium* are insufficient for growth of *Spathidium spathula*, which grows only in lettuce infusion with up to 3.2 divisions/day.

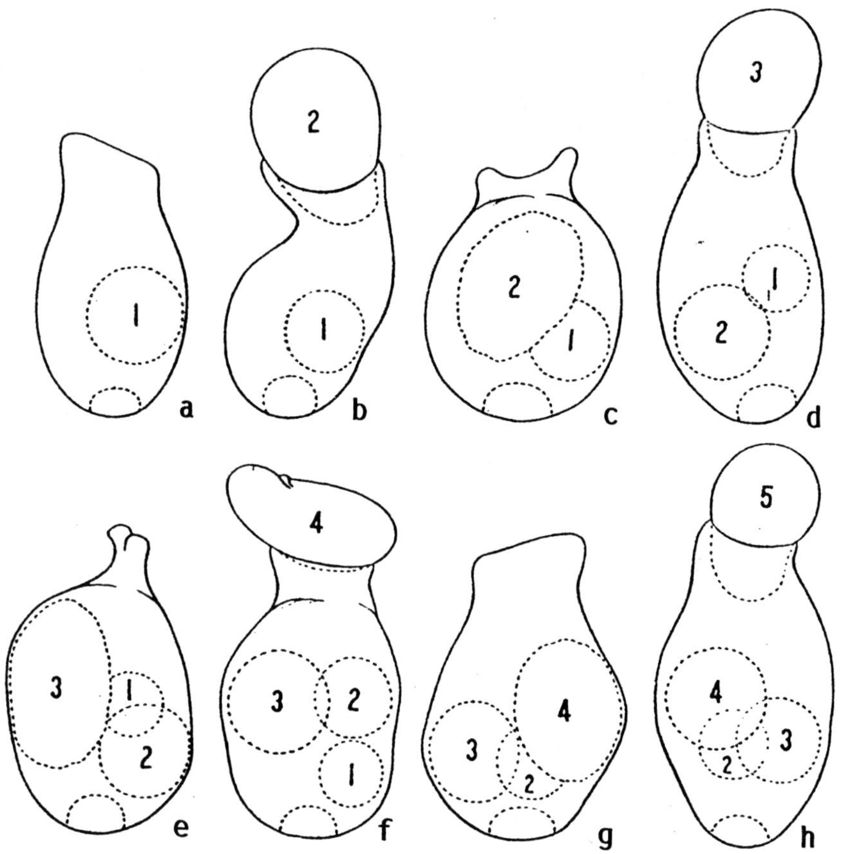

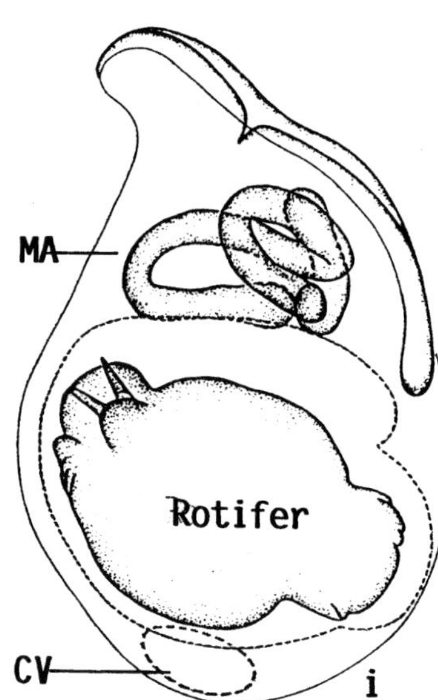

Fig. 13a–h Food uptake by a single specimen of *Spathidium spathula* over about an hour (from WOODRUFF & SPENCER 1922). **a)** Specimen when first observed. Contains a food vacuole (1) with a *Colpidium*. **b)** Seizing a second (2) *Colpidium*. **c)** Second *Colpidium* enclosed in food vacuole. **d)** Third (3) *Colpidium* engulfed. **e)** Third *Colpidium* in food vacuole. **f)** Seizure of fourth (4) *Colpidium*. **g)** Fourth *Colpidium* within vacuole; food vacuole with first *Colpidium* no longer discernible. **h)** Engulfing of fifth (5) *Colpidium*.

Fig. 13i *Pharyngospathidium longichilum longichilum* with an ingested rotifer (for details, see second volume). CV – contractile vacuole, MA – macronucleus.

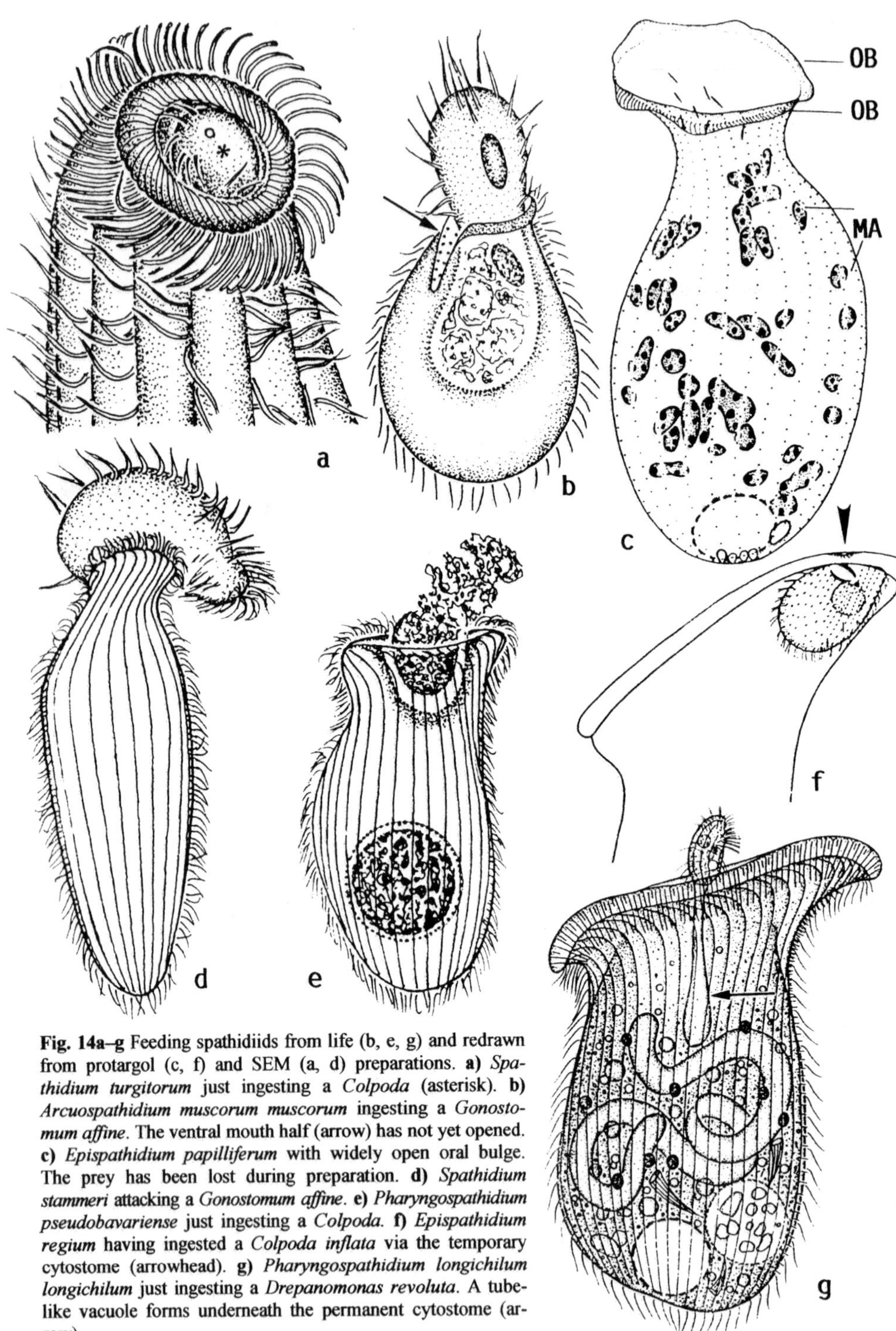

Fig. 14a–g Feeding spathidiids from life (b, e, g) and redrawn from protargol (c, f) and SEM (a, d) preparations. **a)** *Spathidium turgitorum* just ingesting a *Colpoda* (asterisk). **b)** *Arcuospathidium muscorum muscorum* ingesting a *Gonostomum affine*. The ventral mouth half (arrow) has not yet opened. **c)** *Epispathidium papilliferum* with widely open oral bulge. The prey has been lost during preparation. **d)** *Spathidium stammeri* attacking a *Gonostomum affine*. **e)** *Pharyngospathidium pseudobavariense* just ingesting a *Colpoda*. **f)** *Epispathidium regium* having ingested a *Colpoda inflata* via the temporary cytostome (arrowhead). **g)** *Pharyngospathidium longichilum longichilum* just ingesting a *Drepanomonas revoluta*. A tube-like vacuole forms underneath the permanent cytostome (arrow).

1·9 **Silverline pattern** (Fig. 8f)

Spathidiids and haptorids in general have a very narrowly meshed silverline pattern with polygonal meshes 0.5–2 µm in size; rarely the pattern is ordered in a linear fashion with meshes appearing quadrangular. In the area of the dorsal brush, the meshes are larger, assuming a platyophryid pattern, that is, the comparatively large meshes are divided by a median silverline extending between two brush rows (Fig. 8f). Considering this great uniformity, the silverline pattern is useful only for the classification of categories above the order, while useless for families, genera, and species.

2 **Life Cycle** (Fig. 15, 16)

The spathidiids have an ordinary ciliate life cycle, that is, the excysted theront feeds and becomes a trophont (Fig. 15). In nature, theronts and trophonts are rather similar, while overfed, large and ± distorted trophonts are common in pure cultures. The trophont divides and eventually encysts when environmental conditions become adverse (Fig. 15). Reproduction by budding has been observed in a culture-created species, that is, in *Spathidium polymorphum*, the descendant of *S. ascendens* (Fig. 16a). WENZEL (1955) emphasized that specimens originated from buds reproduced ordinarily.

The sexual life cycle is known only in one *Spathidium* species. The data show that it is as simple as the asexual cycle: the specimens divide several times and then conjugate. After conjugation they divide, provided sufficient food; rarely they encyst (Fig. 15).

Three to four generations are produced daily under good laboratory conditions, as in many other ciliates. However, division rate strongly decreased to one or less after 70–120 fissions in the cultures of WILLIAMS 1980 (Fig. 17, 18). In contrast, MOORE (1924a) and WOODRUFF & MOORE (1924) could keep cultures of *Spathidium* without conjugation or loss of viability for over 1,000 generations. Accordingly, they conclude that *Spathidium* can live indefinitely under entirely favourable environmental conditions without endomixis or conjugation.

2·1 **Division and stomatogenesis** (Fig. 17, 18; Table)

The data available indicate that division and stomatogenesis of the spathidiids match the mode known from other haptorid ciliates (FOISSNER 1996b). However, detailed data are available from only five species (see Table below): *Arcuospathidium cultriforme, A. muscorum, A. coemeterii, Protospathidium serpens,* and *Spathidium turgitorum*. Certainly, this is a small sample from the about 20 genera and 250 species described. None the less, the data show that cell and nuclear division occur according to textbook knowledge (HAUSMANN et al. 2003). Various small modifications occur, some of which might be genus and/or species-specific. See the individual species descriptions, for full documentation of the ontogenetic data available.

Briefly and using the terminology of FOISSNER (1996b), cell division occurs in the active (non-encysted) condition and is homothetogenic (homopolar). The macronucleus is homomeric and fuses to a globular mass in mid-dividers. Later, this mass extends,

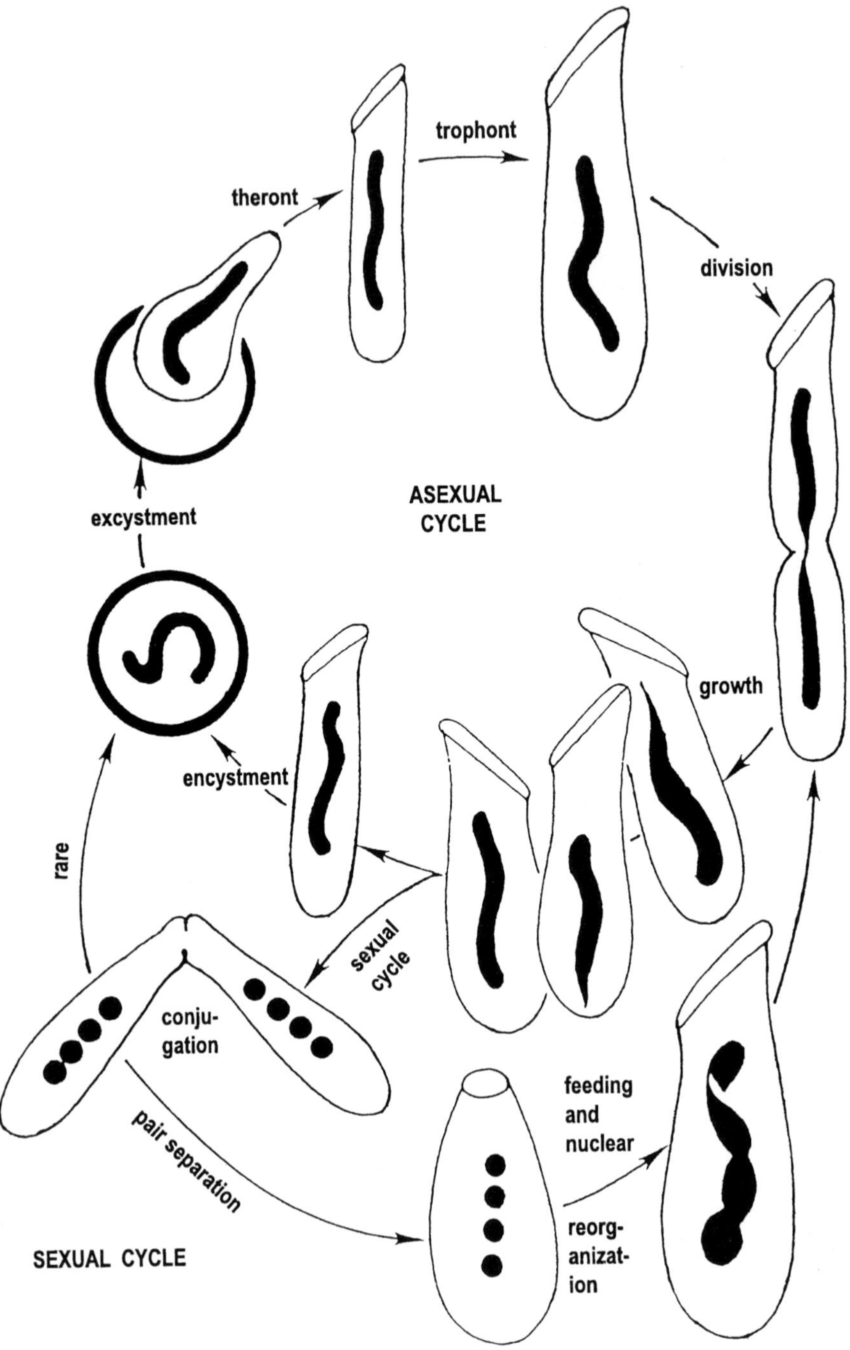

Fig. 15 Asexual and sexual life cycle of spathidiids. Based on MOODY (1912), MOORE (1924a, b), WOODRUFF & SPENCER (1924), and XU & FOISSNER (2004).

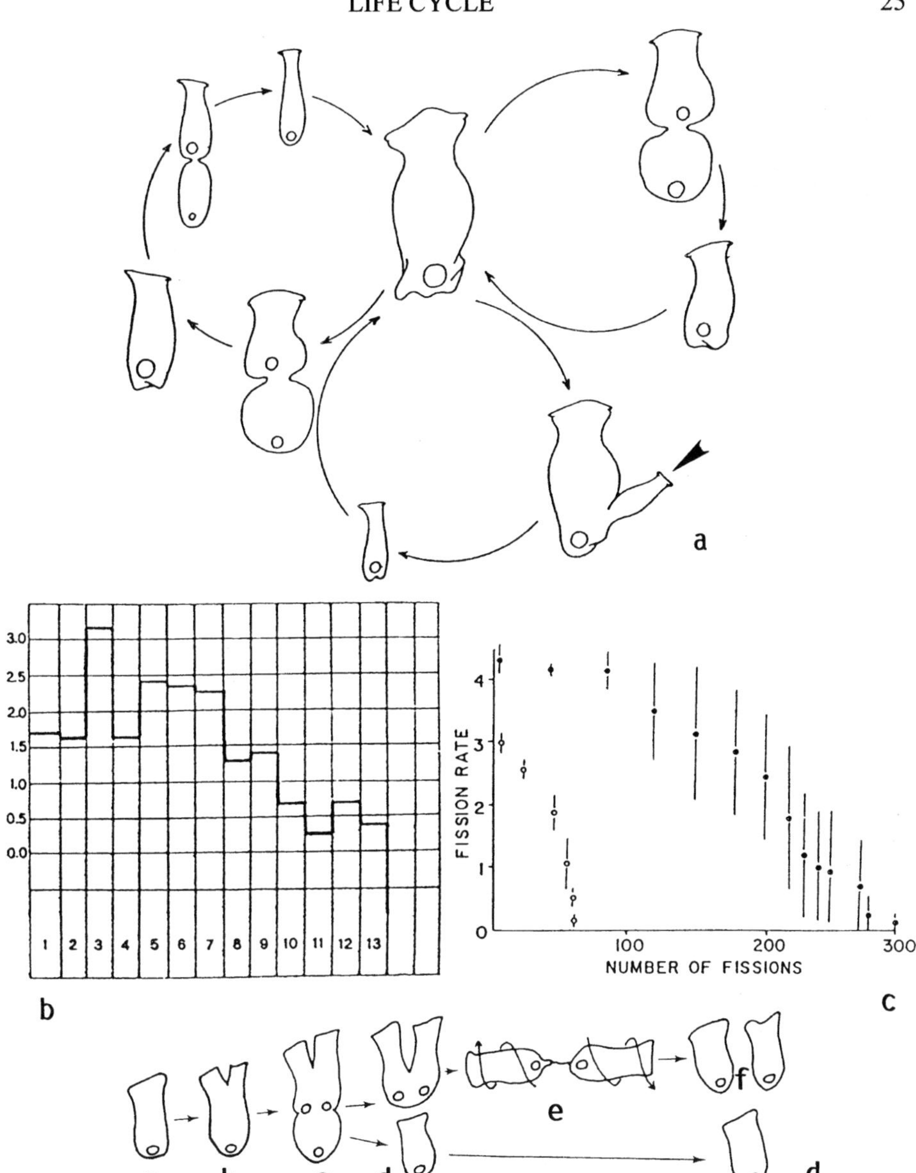

Fig. 16a Life cycle of *Spathidium polymorphum*, the descendant of *S. ascendens* (from WENZEL 1955). *Spathidium polymorphum* can reproduce by budding (arrowhead).

Fig. 16b *Spathidium spathula*, division rates averaged for ten-day periods between February 24, 1911 to July 7, 1911 (from MOODY 1912). Ordinates: average daily rate of division for three individuals. Abscissae: number of ten-day periods.

Fig. 16c *Spathidium spathula* (•, n = 30) and *S. muscicola* (○, n = 60), changes in the average daily rate of fission (± 2 S. E.) as a function of the age of clones (from WILLIAMS 1980).

Fig. 16d Fate of double specimens from an aged *S. spathula* clone (from WILLIAMS 1980). **a)** Cell with broadened oral bulge. **b)** Beginning a longitudinal division of the oral bulge. **c)** Continued longitudinal division of the anterior end together with the beginning of a transverse division. **d)** Completed transverse fission. **e)** Nearly completed longitudinal division of the proter. **f)** End result: 3 ciliates.

Ontogenetic comparison of spathidiids (from XU & FOISSNER 2005b)

Characteristics	*Arcuospathidium c. scalpriforme*	*A. muscorum*	*Cultellothrix coemeterii*	*Protospathidium serpens*	*Spathidium turgitorum*	*Homalozoon vermiculare*
Body becomes longer and more slender in early dividers	yes (distinctly)	?	yes	no (only stouter)	no	? (contractile)
Body distinctly inflated in fission area during middle stages	no	no	yes	yes	yes	no
Proter longer than opisthe	yes (ratio 1.5:1)	yes	yes	yes	yes	yes
Early dividers with indentation in fission area	yes	?	yes	yes	no?	yes
Blebs recognizable in fission area	yes	?	no	no	yes	yes
Macronucleus distinctly longer in early dividers	no	no	yes	no	no	no
Micronucleus greatly (≥ 3 times) enlarges in early middle dividers	no	no	yes	no	no	no
Dorsal brush row 1 generated in post-dividers	no	no	no	possibly	no	no
Dorsal brush row 2 produced distinctly earlier than rows 1 and 3	yes	?	no	no	no	no
Left side oral kinetofragments curve rightwards earlier than right ones	yes	yes	yes	no	no	?
Oral kinetofragments straight or slightly/distinctly curved	distinctly curved	distinctly curved	slightly curved	straight	slightly curved	distinctly curved
Individual oral kinetofragments separate or loosely aligned in very late dividers	aligned	aligned	aligned	separate	separate	separate
Circumoral kinety develops in simple or complex manner[1]	complex	complex	simple	simple	simple	simple
Shaping of oral bulge and circumoral ciliature completed in late dividers or post-dividers	late post-dividers	late post-dividers	late post-dividers	very late dividers	post-dividers	post-dividers
Division axis distinctly oblique?	yes	no	no	no	no	no
Anterior portion of opisthe's kineties strongly curved in mid-dividers	yes	yes	no	no	no	no
Shape of fission area in late dividers	clavate	clavate	roundish	roundish	roundish	roundish
References	XU & FOISSNER (2005b)	BERGER et al. (1983)	FOISSNER & LEI (2004)	this monograph	FOISSNER et al.(2002)	LEIPE et al. (1992)

[1] simple = by alignment of kinetofragments one after another; complex = by shifting and overlapping individual oral kinetofragments.

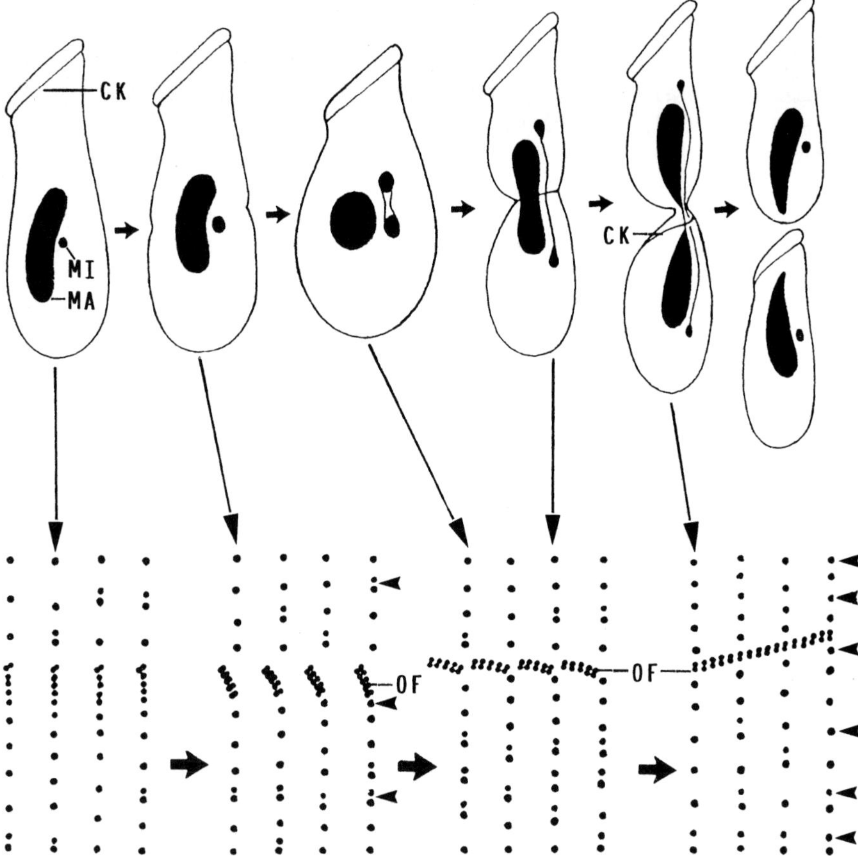

Fig. 17 Scheme of cell division and holotelokinetal stomatogenesis in spathidiid ciliates. Arrowheads mark newly produced basal bodies. CK – circumoral kinety, MA – macronucleus, MI – micronucleus, OF – oral kinetofragments composing the new opisthe circumoral kinety (CK).

divides, and forms the species-specific nuclear pattern (Fig. 17), showing complicated processes in the multinucleate species, for instance, *Spathidium turgitorum* (volume II and FOISSNER et al. 2002).

Stomatogenesis is holotelokinetal, that is, all ciliary rows produce kinetofragments assembling to the opisthe's circumoral kinety in very late dividers and post-dividers (Fig. 17). Several small variations occur (see Table above). The parental oral apparatus is not reorganized. Functional stomatogenesis is poorly understood, although XU & FOISSNER (2005b) suggested some models, based on their detailed study of the ontogenesis of *Arcuospathidium cultriforme*. Here, we cite literally their considerations; those related to evolution and phylogeny will be included in the general section of volume II.

Why is the division axis oblique in *Arcuospathidium cultriforme*? The long circumoral kinety (mouth) of *A. cultriforme* needs a lot of dikinetids. As there is no second (late) round of oral kinetid proliferation, the individual oral kinetofragments must be as long as possible to suffice for the new kinety. Basically, this is achieved by making the division axis oblique causing a larger perimeter of the fragment belt (Fig. 18a). Accordingly, blunt spathidiids (e.g., *A. muscorum*) and species with short mouth (e.g., *Spathidium turgitorum*) have a simple, transverse division axis (BERGER et al. 1983, FOISSNER et al. 2002). A further small space increase for dikinetids is achieved by curving the individual kinetofragments. A simple planimetric calculation shows that oblique division axis and fragment curving increase the useable perimeter by about 40%. Basically, however, curving of the oral kinetofragments is caused by the division blebs and independent of the space available for the growing kinetofragments (see below). Thus, it is unlikely that curving of the kinetofragments is a special adaptation to the restricted space available for the growing kinetofragments of *A. cultriforme* and other haptorids.

Why are the growing oral kinetofragments curved? The concave shape of the newly produced oral kinetofragments is a conspicuous and widespread feature in haptorids. In spathidiids and *Homalozoon* (LEIPE et al. 1992), the kinetofragments grow around the posterior and right margin of the division blebs, and thus assume the concave shape observed. There is little doubt that this applies also to other haptorids, for instance, *Dileptus* (GOLIŃSKA 1995).

Why become kinetofragments disordered in late dividers? In late and very late dividers of *A. cultriforme* (Fig. 77r, s) and *A. muscorum* (Fig. 65j–m), the opisthe's oral kinetofragments overlap more or less distinctly, some even becoming arranged one upon the other or side by side. At first glance, this appears as an effect of the general disorder caused by cell furrowing and the reduced space available for the fragment belt. However, the fragments never overlap in *Cultellothrix coemeterii*, a small species with comparatively short oral bulge (FOISSNER & LEI 2004). Thus, the overlap must have other reasons in long-mouthed species, likely the lack of a second round of oral dikinetid production which causes that the oral kinetofragments must be generated in full length. This, however, generates spatial constraints and disorder during furrowing.

How are the long mouth (oral bulge, circumoral kinety) and its steep slope obtained? Theoretically and in accordance with most data available, a spathidiid mouth could develop either by ingrowth or outgrowth. The ingrowth model suggests that mouth and slope are obtained by resorption of the anterior portion of the ventral ciliary rows or, in other words, the oral bulge grows into the ventral parental cortex and obtains, quasi automatically, the slope. The outgrowth model assumes that mouth and slope are obtained by some ventral growth of the forming oral area (Fig. 18b–d) and, in post-dividers, by a faster growth of the dorsal than ventral area (Fig. 18e, f).

None of the models can be rejected a priori because the monsters show that a new circumoral kinety can be generated between the ciliary rows, likely including the resorption of parental somatic kinetids (Fig. 79e, h, i). However, in *A. cultriforme* (XU & FOISSNER 2005b) and some other spathidiids (BERGER et al. 1983, FOISSNER et al. 2002), the ingrowth model is disproved by the lack of evidence for resorption of parental kinetids in the growing mouth area. Further, only half of the basal bodies found in morphostatic specimens occur in the oral area of early post-dividers, and the distances between the individual basal bodies are larger orally than postorally. To compensate for this deficit, the oral area should not extend into the parental ventral cortex and dissolve existing kinetids, but grow as a whole and concomitantly produce new kinetids. Thus, we favour the outgrowth model where length and slope of the oral bulge are obtained by unequal growth of the dorsal and ventral side (Fig. 18b–f).

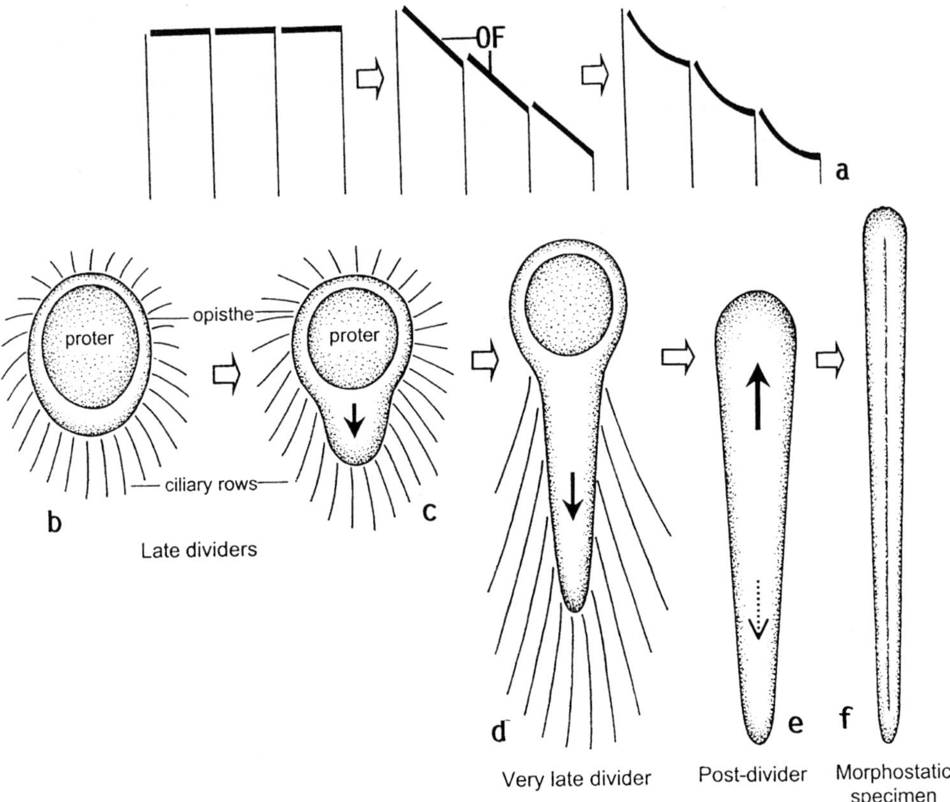

Fig 18a The long circumoral kinety of *A. cultriforme* needs a lot of dikinetids, that is, the individual oral kinetofragments (OF) must be as long as possible to suffice for the new circumoral kinety. This is achieved by making the division axis oblique and the fragments curved. Both provisions increase the useable perimeter by about 40%. From XU & FOISSNER (2005b).

Fig. 18b–f A model how the spathidiid oral bulge is produced. In *A. cultriforme* and several other spathidiids, both the oral bulge and ciliary pattern are generated mainly post-divisionally. Late and very late dividers perform growth mainly in the ventral fission area (b–d). After separation of the daughters, faster growth of the dorsal than ventral area makes the oral bulge more or less distinctly oblique (e, f). The shape of the oral bulge is genetically determined. From XU & FOISSNER (2005b).

As concerns *A. cultriforme* and other long-mouthed species, the oral area not only grows transversely in the fission region, but extends more or less far ventrally (posteriorly) in late dividers (Fig. 77t–x). We suggest this as a genus and/or species-specific process producing some space for the long oral kinetofragments because no growth can occur in the dorsal region where the daughters adhere.

The observations and models do not provide a reason for the dorsal curve of the ciliary rows in the mouth area of all spathidiids. This curve augments the oral ciliature and is thus likely related to the feeding process. At first glance, the curve seems caused by the sometimes distinct dorsal curve of the spathidiid oral area. However, such relationship is disproved by many species which have the oral area hardly curved dorsally, for instance, *Spathidium spathula* and *S. turgitorum* (FOISSNER 1984, FOISSNER et al.

2002). The lack of specific ontogenetic processes suggests that the dorsal curve is genetically fixed and belongs to the basic (plesiomorphic) equipment of the spathidiids.

How are the shape of the oral bulge and circumoral kinety obtained? When viewed ventrally, the spathidiid oral bulge may be obovate, oblong, or cuneate (Fig. 10). We did not find any indication that these shapes are caused by special ontogenetic constraints. Thus, they are genetically determined.

Was the ancestral dorsal brush two-rowed or three-rowed? The dorsal brush of *A. cultriforme* develops in a remarkable way not present or noticed in *A. muscorum* and *A. coemeterii* (BERGER et al. 1983, FOISSNER & LEI 2004): first, row 2 is generated, followed by rows 3 and 1 (Fig. 77k–o). In *Spathidium*, the brush rows develop concomitantly (see Table on page 26), while a certain population of *Protospathidium serpens* generates rows 2 and 3 during ontogenesis and row 1 in post-dividers (Fig. 52a–r). Further, most *Protospathidium* species have brush row 1 strongly shortened, and in some it is even lacking.

These observations and the proposed evolution of the spathidiids (Vol. II) suggest an ancestor with only two brush rows. We would not like to generalize this hypothesis, that is, apply it to other groups of haptorids because ontogenetic data are very sparse, the evolution of the haptorids is not known, and most species/genera have a three-rowed brush suggesting this as the plesiomorphic state.

What is the function of the division blebs? Division blebs are hemispherical protrusions underneath the prospective division furrow. They were first described in *Homalozoon vermiculare*, where they contain remnants (?) of microtubules and cortical granules (LEIPE et al. 1992). Later, division blebs were described also in *Spathidium turgitorum* (FOISSNER et al. 2002). We found them in *A. cultriforme* and several *Spathidium* species (Fig. 130b–d, Vol. II), suggesting a widespread occurrence in haptorid ciliates.

At first glance, the blebs appear as precursors of the prospective oral bulge. However, they are distinct only in early and middle dividers and disappear during late ontogenesis, long before oral bulge formation commences (FOISSNER et al. 2002, Vol. II). Thus, the function of the blebs remains obscure.

2·2 **Conjugation** (Fig. 19–26)

Specimens with macronuclear anlagen occur in most field populations (XU & FOISSNER 2004). Literature and the study of XU & FOISSNER (2004) suggest that such specimens are either exautogamonts (MOORE 1924a, b) or exconjugants. Nuclear reorganization during autogamy has not yet been investigated with modern techniques in spathidiids. The data from MOORE (1924a, b) suggest that autogamy occurs, like in several other ciliates (RAIKOV 1972), just before encystment, in the resting cyst, or during and after excystment. This is sustained by our field data that show a rather high percentage of transparent (non-feeding) specimens with two to six, usually four macronuclear pieces, while conjugating pairs were rare (XU & FOISSNER 2004). However, WILLIAMS (1989) did not observe autogamy and macronuclear reorganization in resting cysts of *S. spathula*. Further, he reported that 50–75 divisions occur between two conjugations.

WOODRUFF & SPENCER (1921a, c, 1924) provided pioneer data on the conjugation of *Spathidium*. Although the data are rather incomplete due to the lack of adequate staining

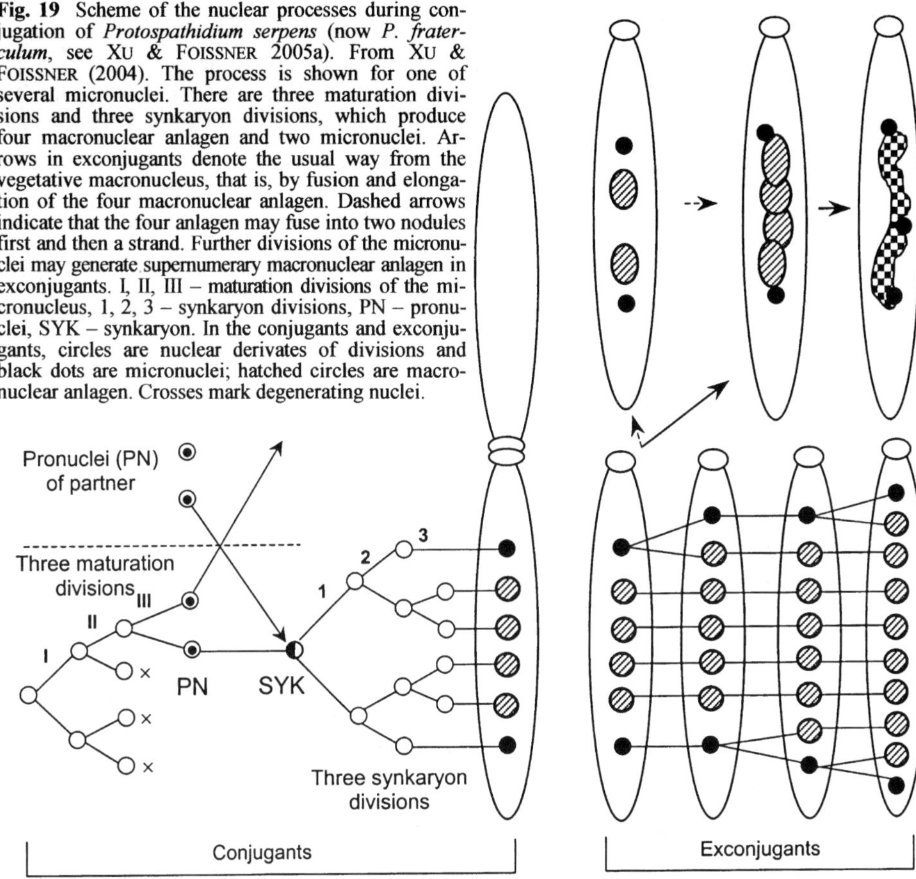

Fig. 19 Scheme of the nuclear processes during conjugation of *Protospathidium serpens* (now *P. fraterculum*, see XU & FOISSNER 2005a). From XU & FOISSNER (2004). The process is shown for one of several micronuclei. There are three maturation divisions and three synkaryon divisions, which produce four macronuclear anlagen and two micronuclei. Arrows in exconjugants denote the usual way from the vegetative macronucleus, that is, by fusion and elongation of the four macronuclear anlagen. Dashed arrows indicate that the four anlagen may fuse into two nodules first and then a strand. Further divisions of the micronuclei may generate supernumerary macronuclear anlagen in exconjugants. I, II, III – maturation divisions of the micronucleus, 1, 2, 3 – synkaryon divisions, PN – pronuclei, SYK – synkaryon. In the conjugants and exconjugants, circles are nuclear derivates of divisions and black dots are micronuclei; hatched circles are macronuclear anlagen. Crosses mark degenerating nuclei.

methods, those on body changes, duration, and fate of the exconjugants are still highly valuable and will thus reviewed here: "Animals ready to conjugate produce an effect on neighboring individuals which recalls the remarkable behaviour of *Spathidium* in relation to its paralyzed prey – the peristome regions of the Spathidia exhibiting a tendency to come in contact. Frequently, at such times, several individuals will become organically connected so that bizarre groups result. When normal peristomal contact has been established, the two animals shift and turn until the peristomal edges are exactly in apposition, and then endoplasmic fusion occurs. Usually this is accomplished within fifteen minutes. Then gradually the conjugants become smaller and more spherical, the maximum change in size and form being attained in about eight or ten hours, and maintained for about the same length of time when separation occurs. Thus at room temperature the animals remain paired from sixteen to twenty hours. The exconjugants retain for an hour or so the rounded form of conjugants, but their future history depends entirely upon the food supply. Unless Colpidia or similar ciliates are present, the Spathidia become greatly elongated, tiny, transparent cells and invariably die within twenty-four hours, except in the few cases where encystment takes place. When Colpidia are

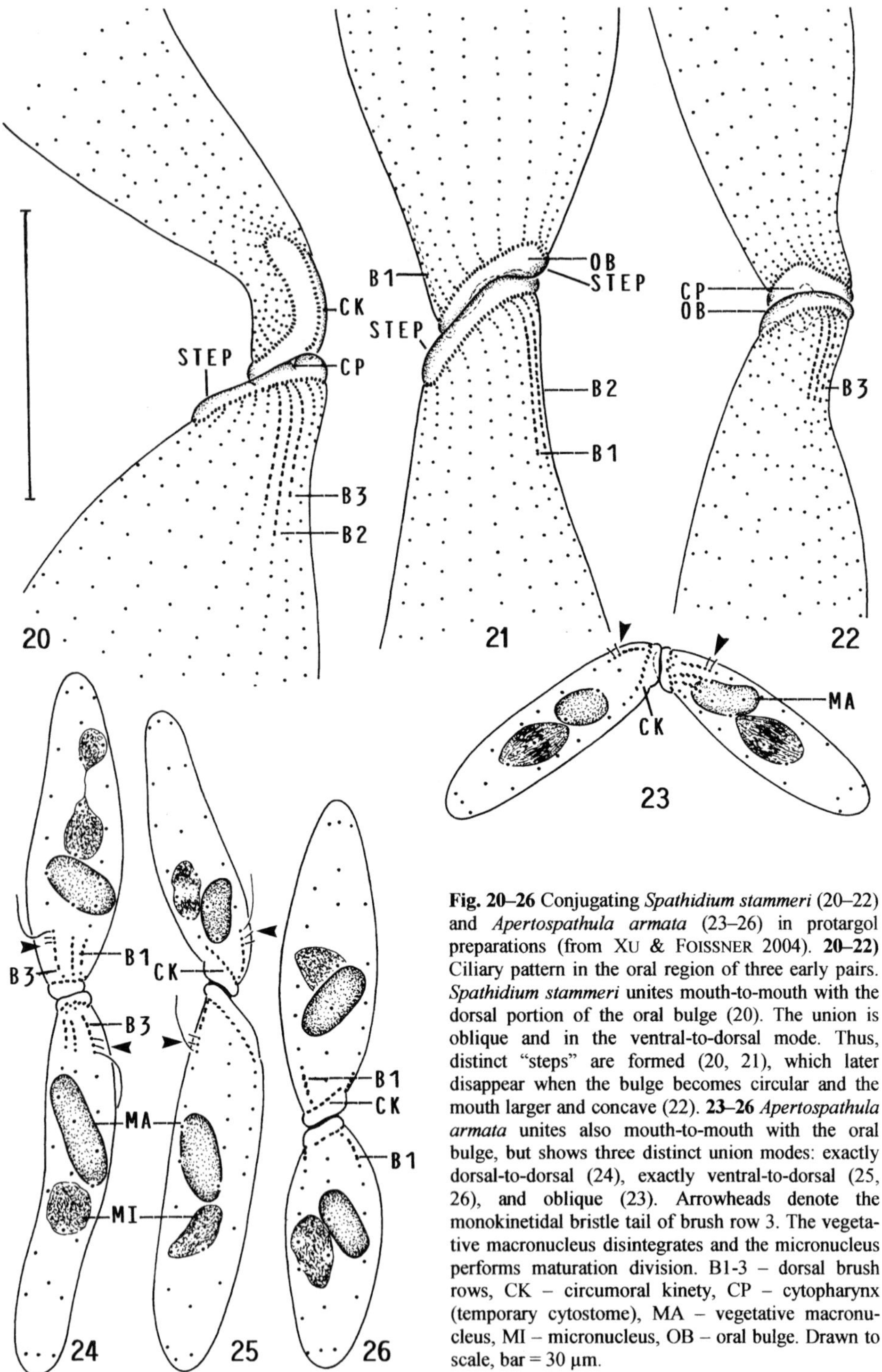

Fig. 20–26 Conjugating *Spathidium stammeri* (20–22) and *Apertospathula armata* (23–26) in protargol preparations (from XU & FOISSNER 2004). **20–22)** Ciliary pattern in the oral region of three early pairs. *Spathidium stammeri* unites mouth-to-mouth with the dorsal portion of the oral bulge (20). The union is oblique and in the ventral-to-dorsal mode. Thus, distinct "steps" are formed (20, 21), which later disappear when the bulge becomes circular and the mouth larger and concave (22). **23–26** *Apertospathula armata* unites also mouth-to-mouth with the oral bulge, but shows three distinct union modes: exactly dorsal-to-dorsal (24), exactly ventral-to-dorsal (25, 26), and oblique (23). Arrowheads denote the monokinetidal bristle tail of brush row 3. The vegetative macronucleus disintegrates and the micronucleus performs maturation division. B1-3 – dorsal brush rows, CK – circumoral kinety, CP – cytopharynx (temporary cytostome), MA – vegetative macronucleus, MI – micronucleus, OB – oral bulge. Drawn to scale, bar = 30 μm.

available, the Spathidia feed rapidly, become greatly distended with food, and divide usually within twenty-four hours".

Only very recently, XU & FOISSNER (2004) studied cell size and shape, nuclear changes, and the ciliary pattern during conjugation of *Protospathidium serpens*, using protargol impregnation and morphometry. Preliminary data were gathered from *Spathidium stammeri* and *Apertospathula armata*. Conjugation of *P. serpens* is temporary, isogamic, and without preconjugation divisions. Pair formation is heteropolar, and the partners unite obliquely with the oral bulge. The body becomes smaller and broader during conjugation, but no basic changes occur in the ciliary pattern. Conjugation and nuclear reconstruction follow the usual mode of ciliates. However, some peculiarities occur: only two of the four synkaryon derivatives of the second synkaryon division enter the third division and generate four macronuclear anlagen, which fuse to a single, long macronucleus strand. During conjugation, *S. stammeri* unites obliquely as *P. serpens*, while *A. armata* can pair dorsal-to-dorsal surface, ventral-to-dorsal surface, or obliquely as *P. serpens*. The nuclear processes of these three species are also rather different, showing a considerable diversity in union modes and nuclear events of spathidiids; *S. stammeri* even has preconjugation division.

WOODRUFF & SPENCER (1921a, c, 1924), WOODRUFF & MOORE (1924), and MOORE (1924a) performed also detailed experiments on the significance of conjugation in *Spathidium*. The data can be summarized as follows: (i) it has been possible to keep a culture of *Spathidium* under laboratory conditions without conjugation or loss of vitality for over 1,000 generations; (ii) *Spathidium* can live indefinitely under entirely favorable environmental conditions without endomixis or conjugation; (iii) conjugation in the majority of cases directly induces an immediate acceleration of the rate of reproduction; (iv) exconjugation cultures in the majority of cases outlive their parent (non-conjugant) cultures; (v) conjugation typically has a high survival value in the life of the organism; (vi) by permitting conjugation to occur successively at opportune periods, the rate of reproduction may be gradually 'built up' to a higher 'standard'; (vii) inbreeding to the F6 generation exhibited no deleterious effects.

These conclusions contrast results of a more recent study by WILLIAMS (1980): "Thirty newly excysted *Spathidium spathula* (MÜLLER) and 60 newly excysted *Spathidium muscicola* (KAHL) were selected as progenitors of clonal daily reisolation lines that omitted both conjugation and encystment. Daily division rates were determined for each clone either until it died or until the end of the 170 days of reisolation. Both species had reduced fission rates as they accumulated fissions in the absence of macronuclear reorganization. *Spathidium spathula* had a significant reduction of daily fission rate after 100–120 fissions and *S. muscicola* after 20–30 fissions (Fig. 16c). Older clones of both species contained a noticeable proportion of abnormal organisms (Fig. 16d). A significant increase (10.5 %) in daily fission rate occurred in aged sublines of *S. spathula* following conjugation (selfing) and its concomitant nuclear reorganization. *Spathidium muscicola* did not conjugate, but recently excysted sublines, compared to aged lines, had an increased daily fission rate".

In our opinion, this conflict is likely related to culture conditions. We made similar observations in a variety of ciliates: they grow well for some weeks or months, but then they decline; sometimes, they recover if the basal medium and/or food is changed.

2·3 Encystment, resting cysts, excystment (Fig. 27, 28)

Only MOORE (1924a, b), WENZEL (1959), BERGER et al. (1983), FOISSNER (1996a), FOISSNER et al. (2002), and XU & FOISSNER (2005a, b) provided light microscopical data on encystment, resting cysts, and excystment. Still, the most detailed data are those of MOORE (1924a, b), who studied more than fifty mass encystments in *Spathidium spathula*. Thus, her observations will be cited in full length. All authors agree that food depletion/addition is a main stimulus for encystment/excystment. This is supported by our observations: all resting cysts we studied during the preparation of the monograph were obtained by maintaining isolated specimens without food organisms in depression slides with a small quantity of Eau de Volvic or centrifuged soil percolate.

MOORE (1924b) described her observations as follows: "Individuals deprived of food become clear and transparent and assume a long slender form like that shown in figure 27a. As starvation continues, the preparatory stages of encystment set in. The body shortens and widens somewhat and the oral region becomes less plainly marked. Gradually only a slight indication of the anterior end remains, but the contractile vacuole at the opposite extremity is clearly visible. The cell becomes more and more rounded and finally forms a perfect sphere (Fig. 27b–d). As soon as this point is reached a thin membrane appears. In one individual carefully followed through the proceeding stages less than five minutes were required to complete the process – at 3:25 P.M., the cell was not quite spherical; at 3:30 it was a typical cyst (Fig. 27e).

At this stage the cyst is almost transparent, completely filled with light granular protoplasm; with a large persistent vacuole. Cysts of this kind are found on slides which twenty-four hours earlier contained normal individuals with few Colpidia. By the next day, however, a profound change has occurred. With the disappearance of the vacuole the cyst becomes contracted, its contents appear darker and more opaque, and the wall presents an irregular wrinkled appearance (Fig. 27g). An intermediate stage is occasionally found in which the protoplasm shrinks slightly away from the smooth wall, leaving a narrow space almost completely filled with granules (Fig. 27f). As the membrane contracts, however, the space is eliminated.

The opaque wrinkled condition apparently marks the most quiescent period of the whole process. Many encysted individuals degenerate at this point, but many after remaining in this state of inactivity for a shorter or longer time, usually from one to three days, begin to show signs of recovery. A small vacuole appears which gradually increases in size; the irregular contracted wall again smooths out, and the protoplasm once more transparent begins to rotate within the cyst wall (Fig. 27h–j). Signs of differentiation soon appear, and in less than an hour and a quarter a fully formed individual may be found rapidly revolving in the cyst (Fig. 27k–m). From the rounded portion containing the vacuole, in some cases, a narrow finger-like projection extends out, and as substance from the base gradually flows into this it becomes wider and more typical and finally assumes the proportion of a perfect organism, bent upon itself in conformity with the confining membrane. Occasionally division occurs within the cyst and two complete individuals are found intertwined about each other and constantly in motion.

Such fully differentiated individuals may move about in the cyst for hours without being able to escape. Unless the membrane is punctured from within or without,

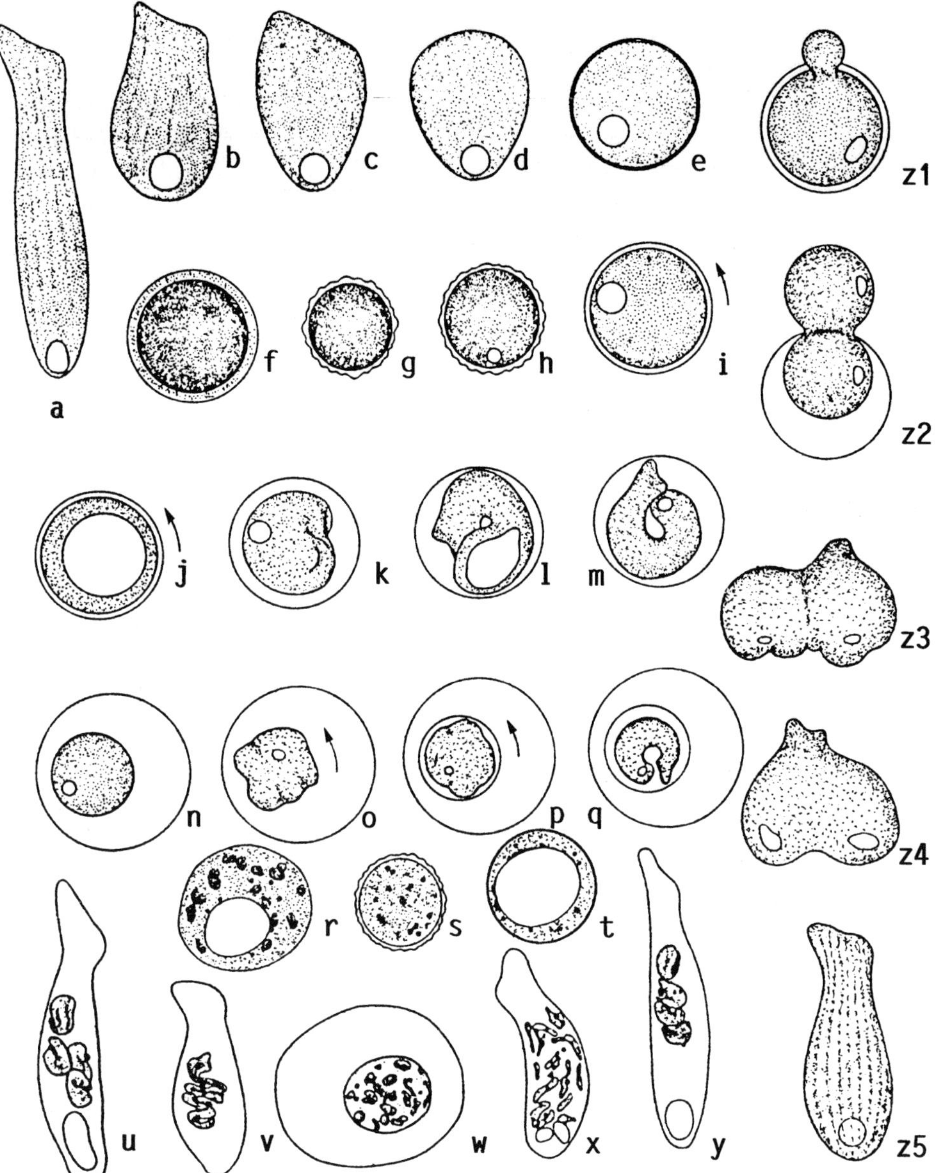

Fig. 27a–m *Spathidium spathula*, successive stages in encystment and excystment (from MOORE 1924b). **a)** Starved specimen prior to encystment. **b–d)** Stages of dedifferentiation. **e)** Early cyst with membrane present. **f, g)** Cyst maturation. **h–j)** Beginning of excystment. **k–m)** Stages in differentiation with fully formed specimen ready to escape (m).

Fig. 27n–q *Spathidium spathula*, reencystment with second cyst wall (p, q). From MOORE (1924b).

Fig. 27r–y *Spathidium spathula*, nuclear changes during encystment and excystment (from MOORE 1924b). **r–t)** Encystment: the macronuclear fragments degenerate to chromatin granules, while the micronuclei remain. **u, v, x, y)** Excysted specimens with macronuclear anlagen and micronuclei. **w)** Reencysted specimen with fragmented macronucleus.

Fig. 27z1–z5 *Spathidium spathula*, excystment without preceding differentiation (from MOORE 1924b). Figure (z5) shows the fully differentiated specimen one hour after excystment.

dedifferentiation finally sets in and re-encystment occurs. The active individual gradually slows down, rounds up into a ball, and appears as a compact homogeneous mass near the center of the original distended cyst which now resembles a 'halo'. After this second encystment a second process of differentiation may occur, even resulting in excystment. A new cyst membrane is not always formed, but where it does appear the resulting individual develops within the inner wall as in figure 27p–q. Cysts at this stage have also been found to return to the quiescent state; but whether or not a third differentiation may occur is not known. The extrusion of granules may occur also in the cysts with halos, but by the time differentiation is complete the space is in most cases absolutely clear.

Excystment. The process of excystment involves several curious phenomena. In the first place, it is difficult to understand why some fully formed individuals are able to break through the cyst wall while others find escape impossible. MOODY (1912) states that "by the addition of fresh medium it is often possible to recover *Spathidium* from the cysts". In the cultures which form the basis for the present study, however, no excysted individuals were observed for more than a year, although many actively revolving animals were seen during that time within the transparent membranes. Change of medium and the addition of food produced no evident effect. Drying, moreover, which CALKINS (1919) found necessary for inducing the excystment of *Uroleptus*, proved fatal in the case of *Spathidium*. No further signs of activity appeared in cysts deprived of water even for a short period, and disintegration rapidly took place.

Since reversal of morphogenetic processes may be repeated several times by the active protoplasm within the cyst, it seemed scarcely plausible to attribute the failure of excystment to any fundamental defect in the cell itself which might have been caused by the internal changes accompanying encystment. It appeared more probable that the difficulty lay in the nature of the cyst wall which proved impervious to rupture from within. To test this hypothesis, it was necessary only to puncture the cyst from without and liberate the enclosed individual. For this the point of a small sharp scalpel was used, and organisms partly or fully developed were recovered from the cysts. If only a small opening was made, the active individual soon worked its way through and continued swimming rapidly through the water. Due to its coiled position in the cyst, it presented somewhat the appearance of a corkscrew on first emerging and moved in an exaggerated spiral. Many forms obtained in this manner were rounded or irregular at the posterior end, and many were long and extremely thin, typically vermiform in shape. Within a few hours, however, normal individuals of minute size were formed in the majority of cases. These were isolated in separate depression slides, and division was found to occur in the larger forms about twenty-four hours later. The descendants of one of these were kept for fifteen days, during which time forty-two generations were produced. Obviously, the failure of such forms to excyst was in no way indicative of an inherent deficiency in the protoplasm itself, since vegetative existence was resumed as soon as the barrier was removed.

The first individuals were cut from cysts on January 29, 1922. Up to that time no independent excystment had been noted at all. Daily stock slides from pedigree cultures were then put aside for the accumulation of cysts in mass, and the operations were continued. Ten days later, two normal active forms were found on a slide which had contained only cysts on the previous day. Had they emerged without aid or had their cysts been pricked accidentally the day before? To determine this another slide containing cysts with fully differentiated individuals was set aside unmolested, to eliminate the possibility of external assistance. In twenty-four hours five excysted individuals were found, all of which divided within another day. After that, from time to time, occasional spontaneous excystment was found on slides containing individuals of widely diverse history. All were derived from the single pedigree series under cultivation at the time, but the actual conditions before encystment were as varied as possible. Except in one or two cases, only a few excysted individuals were obtained from a large number of cysts; but they represented cysts derived from ordinary vegetative forms, from the descendants of regenerated pieces, from recent exconjugants, and from individuals only a few generations removed from excystment.

No reason as yet can be assigned for the emergence of some individuals and the confinement of others. External conditions play no part, since cultures containing excysted forms were accorded no different treatment from many of those in which normal excystment never appeared. Moreover, in most cases a few individuals on a slide were able to penetrate the cyst wall, while the great majority which had passed through the same external changes could not effect an escape. In one instance only was the actual process of emergence observed. The actively moving individual, having succeeded in making a small hole in the cyst wall, pushed its way 'head first' through the opening in an instant and rapidly swam away. It is interesting to note that there is no collapse of the membrane after perforation in this manner, and clear transparent spheres with a single opening may be found scattered through the culture after liberation of the enclosed organisms.

One month later, without any previous indication of change, a totally different method of excystment was observed in individuals from the same pedigree culture. In this new mode of escape, more in accordance with that described by MOODY (1912), differentiation takes place chiefly outside the cyst wall. The cyst contents begin to rotate as a granular undifferentiated mass of protoplasm, containing only a small contractile vacuole. A portion is then extruded through a small opening in the membrane and appears as a bubble of protoplasm outside the cyst, yet still attached to the larger mass within (27z1). Gradually more and more of the inner substance passes to the outside until the stage shown in figure (27z2), is reached. After this point the process continues more rapidly, and in less than ten minutes excystment is complete. The new organism at this stage may be almost fully formed or more frequently merely an amorphous mass. Such an irregular individual as that shown in figure 27z3, immediately after excystment, is quickly transformed and appears as a perfect animal within an hour (Fig. 27z4, z5)".

MOORE (1924b) then continues with the description of nuclear reorganization (endomixis) during encystment and/or in the mature cyst and/or in excysting specimens (Fig. 27r–y). However, these data appear doubtful because MOORE (1924b) could stain the nuclear apparatus only after rupturing the cyst wall with a scalpel. Most of the data obtained by this crude method are likely artifacts. This is emphasized by the observations of WILLIAMS (1989) and XU & FOISSNER (2005a, b), who could not find nuclear reorganization in encysting and cystic *Spathidium spathula*, *Arcuospathidium cultriforme* and *Protospathidium fraterculum*. These authors studied protargol-impregnated encysting specimens and mature resting cysts. Their data largely agree with those of MOORE (1924b) cited above and show that these species likely produce kinetosome-resorbing (KS) cysts. Some peculiarities were discovered in *A. cultriforme*, one of the largest species of the group (XU & FOISSNER 2005b). The cyst wall of this species is

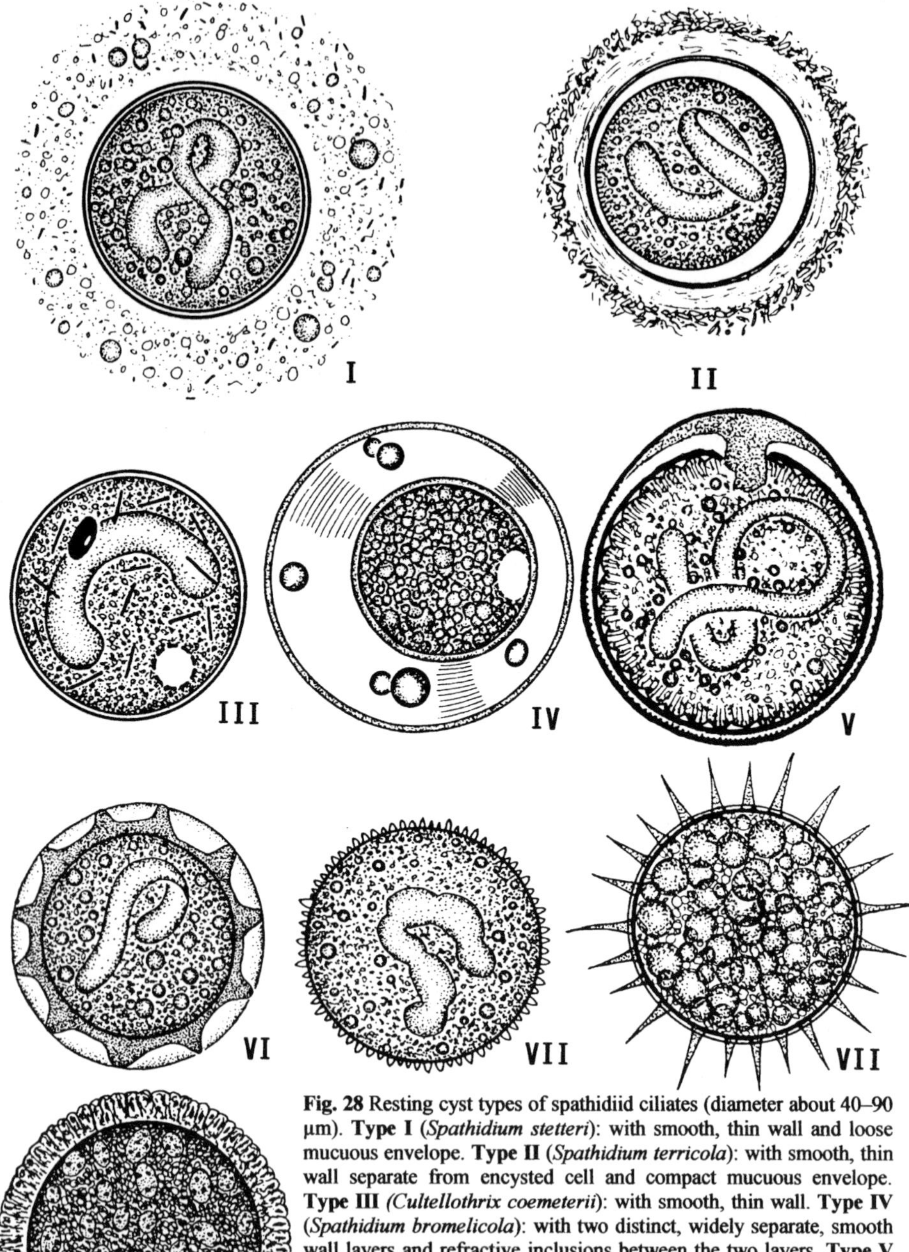

Fig. 28 Resting cyst types of spathidiid ciliates (diameter about 40–90 µm). **Type I** (*Spathidium stetteri*): with smooth, thin wall and loose mucuous envelope. **Type II** (*Spathidium terricola*): with smooth, thin wall separate from encysted cell and compact mucuous envelope. **Type III** *(Cultellothrix coemeterii*): with smooth, thin wall. **Type IV** (*Spathidium bromelicola*): with two distinct, widely separate, smooth wall layers and refractive inclusions between the two layers. **Type V** (*Epispathidium amphoriforme*): with smooth, rather thick wall and emergence pore. **Type VI** (*Arcuospathidium cultriforme cultriforme*): with very thick, roughly faceted wall. **Type VII** (*Spathidium namibicola*): with thin, aculeate wall. **Type VIII** (*Apospathidium atypicum*): with thin, spinose wall. **Type IX** (*Spathidium seppelti etoschense*): with curious, columnar wall.

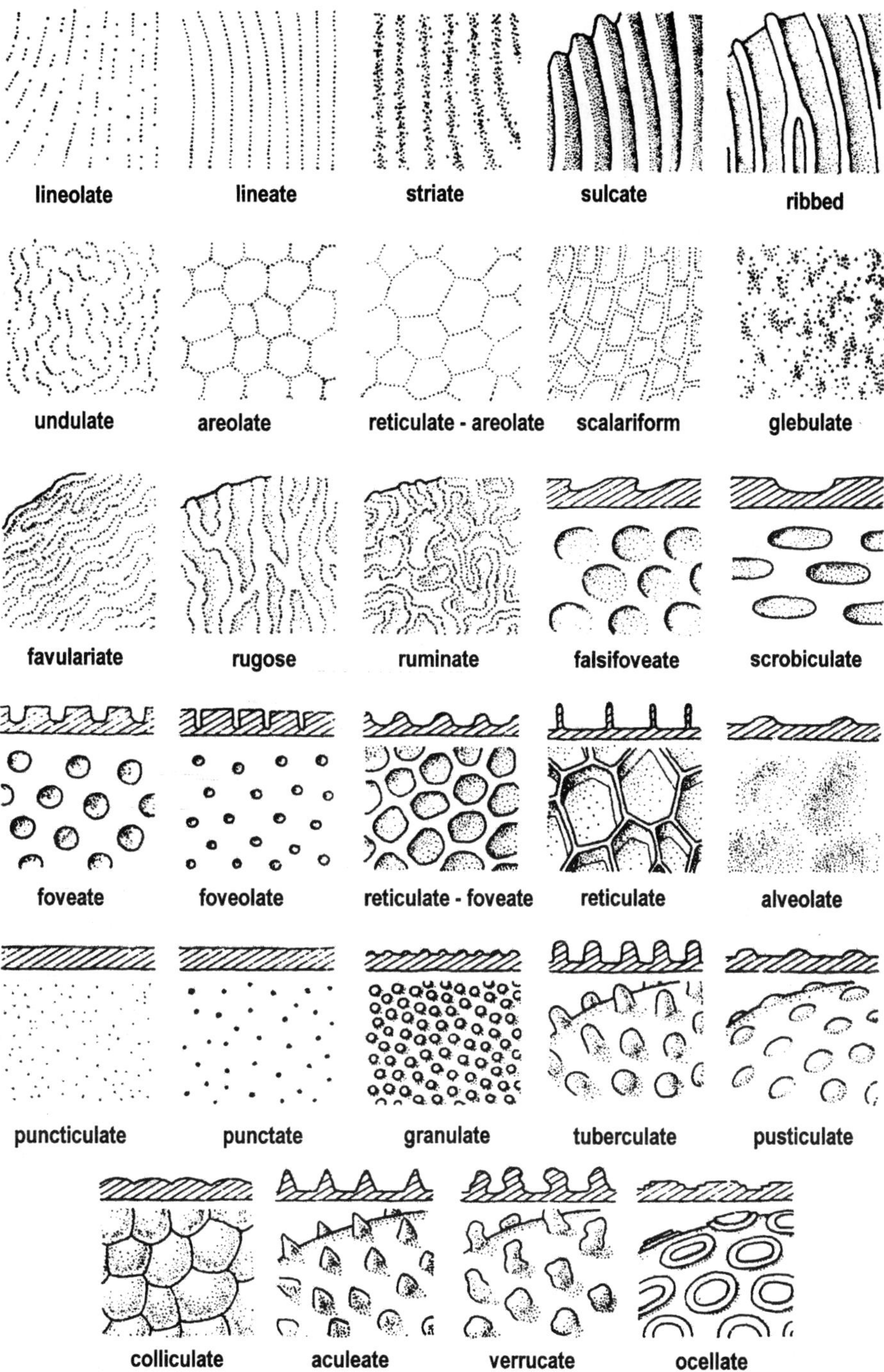

Fig. 29 Types of surfaces. From STEARN (1992), modified.

about 10 μm thick and its volume is almost thrice that of the cyst proper (170,430 μm^3 vs. 61,570 μm^3), while the macronucleus volume decreases from ~5,000 μm^3 to ~1000 μm^3.

Reproductive and exconjugant cysts have not been observed. However, species with similar interphase morphology may have quite different resting cysts (WENZEL 1959, XU & FOISSNER 2005a). Thus, resting cyst morphology should be included in the ordinary species description in future.

Few data are available on the morphological diversity of spathidiid resting cysts, although WENZEL (1959) and XU & FOISSNER (2000a) used wall details for distinguishing similar morphospecies. Thus, we investigated the cyst of several species during the preparation of this monograph. This showed a high and unexpected diversity, ranging from simple globules to cysts with conspicuous spines or some kind of escape opening. As yet, we could distinguish nine types (Fig. 28); likely, there are more. This suggests under-classification of the group. For instance, there are only 10 cyst types in the Colpodea, which are considered as a class (FOISSNER 1993).

2·4 Regeneration

MOODY (1912) and MOORE (1924c) showed that *Spathidium* possesses a considerable regeneration capacity. Unfortunately, these early experiments were not continued with modern staining methods. Thus, details of the ciliary pattern are not known.

The detailed study of MOORE (1924c) confirmed the preliminary results of MOODY (1912) and revealed that (i) all fragments which regenerate both morphologically and physiologically contain both macro- and micronuclear material; (ii) partial dedifferentiation of existing cortex structures precedes regeneration; (iii) complete restoration of body shape requires thirty minutes to one hour; (iv) several pieces which regenerated perfectly represent only 4% of the whole body volume; (v) complete regeneration is less frequent during conjugation than during vegetative life; (vi) pieces containing the synkaryon are capable of regenerating and dividing; (vii) fragments obtained after the beginning of micronuclear division and before the interchange of pronuclei fail to regenerate; (viii) conjugants cut soon after fusion or just prior to separation behave essentially like vegetative individuals; (ix) portions containing only resorbing macronuclear material and degenerating micronuclei may partially regain normal shape but subsequently dedifferentiate and never divide; (x) fragments obtained from starved individuals either fail to regenerate or become encysted after first passing through the processes of differentiation and dedifferentiation; (xi) it has been impossible to produce an amicronucleate race by artificial means.

3 Parasitism

No reports of parasitized spathidiids are available. Likewise, we never recognized parasitized specimens in our large collection. Further, we never saw parasitic nuclear bacteria.

4 Methods: Collecting, Culturing, Observing and Staining of Spathidiid Ciliates

4·1 Collecting

The methods for collecting and culturing spathidiid ciliates are treated only briefly as detailed descriptions are provided either in the introductions to the families or in the species descriptions. Furthermore, the general procedures as described, e. g., by the COMMITTEE ON CULTURES (1958) and LEE & SOLDO (1992) apply also to the spathidiids. The beginner may also consult the valuable booklet by FINLAY et al. (1988).

4·1·1 Limnetic and marine samples

Most described spathidiids are from limnetic or limnetic-terrestrial habitats, such as ditches and puddles with fallen leaves. Many have been discovered in "standing samples" with rotting material, such as algal masses and water plants. Few species each have been reported from activated sludge (AUGUSTIN & FOISSNER 1992, MADONI & GHETTI 1981b), the lake plankton (FOISSNER et al. 1999), streams and rivers (DETCHEVA 1983b, DUDICH 1967, FOISSNER 1997c), tank bromeliads (FOISSNER et al. 2003 and this monograph), and marine sites (KAHL 1930b). In all these habitats, spathidiids and other ciliates are sampled with conventional methods, that is, plankton nets (mesh-wide $\leq$ 10 µm), by scrapping the Aufwuchs, investigating mud and algal masses, and mats of cyanobacteria.

Our own experiences with limnetic and marine samples are frustrating. Although we saw several species, most were rare and did not grow in cultures. Never they became as abundant as in soil samples. Our very recent experiences suggest that several spathidiids appear late in decomposing limnetic samples. Thus, such samples should be controlled for at least one month, similar as the soil samples described below. A rich and almost inexplored habitat are bog ponds and puddles, where Dr. M. KREUTZ (pers. inf.) found about 30 species within a few years, but all in low numbers, except of *Rhinotrix porculus* (FOISSNER et al. 2005b). Our trials to cultivate some of these species failed. Obviously, the right methods for enriching spathidiids in limnetic and marine samples have not yet been discovered.

4·1·2 Moss and soil samples

Since KAHL (1930b) and WENZEL (1953), terrestrial mosses are known to be a rich source for spathidiids. Later, this knowledge was extended to soil (FOISSNER 1984, 1998, FOISSNER et al. 2005a). This is confirmed by the present monograph, where over 50 new species are described from soil collected worldwide. Recent experience suggests floodplain soils as a centre of spathidiid diversity, possibly "collecting" some or most of the limnetic species. The best way for growing these species is the non-flooded Petri dish method described by FOISSNER (1987a). In 2002, FOISSNER, AGATHA & BERGER updated the description, which is here quoted literally.

4·1·2·1 Sampling and sample processing

The material collected included mineral top soil (0–5 cm, rarely up to 10 cm depth) with fine plant roots, the humic layer, and the deciduous and/or grass litter from the soil surface. In soil with few organic materials and in sandy habitats, litter was sieved off the sand with an ordinary kitchen sieve, so that the final sample consisted of about 80% litter and 20% sand and gravel. Usually, 10 small subsamples were collected with a shovel from an area of about 100 m^2 and mixed to a composite sample. Bark samples were usually taken from one to three trees. The bark was collected with a knife, selecting for regions grown with mosses or lichens and/or containing some soil.

Generally, a "good" sample consists of 50% litter, humus and roots and 50% mineral soil. The litter and humus are very important because they release many nutrients when the sample is rewetted, stimulating growth of bacteria, fungi, flagellates, and amoebae, that is, the main food of ciliates. The nutrient increase obviously decouples microbiostasis, as explained in FOISSNER (1987a, 1997d).

All samples were air-dried for at least one month and then sealed in plastic bags. Such samples can be stored for years without any significant loss of species, provided they are from arid or temperate environments (FOISSNER 1997a, FOISSNER et al. 2002).

4·1·2·2 The non-flooded Petri dish method

a) Put the material in a Petri dish and loosely spread over the bottom of the dish in at least a 1 cm, better 2 to 3 cm thick layer. As concerns the present samples, sufficient material was available to fill a 2 cm high Petri dish 13 cm across or, rarely, a 3 cm high dish 18 cm in diameter. Basically, a large Petri dish (18 cm) is preferable because it provides more material for preparations.
b) Slightly over-saturate but do **not** flood the sample with distilled water. Water should be added to the sample until 5–20 ml will drain off when the Petri dish is tilted (45°) and the soil gently pressed with a finger. Complete saturation takes up to 12 hours, so check cultures after this time. Never flood the sample, that is, make an Aufguss ("infusion") because then only a few common species will develop. Further, the material should have been dry for at least one month.
c) Cover Petri dish and pinch a clip between bottom and lid for gas exchange. Generally, care must be taken that samples do not putrefy. This happens easily with saline material, soil containing animal excrements, and samples with very easily decomposable litter. If so, change the water in the sample and do not cover it for some days so that plenty of air is available; further, slightly under-saturate sample with water. Heavily saline soil (≥ 20‰) should be "washed", if no ciliates develop. Over-saturate the sample with water, as described above. After three days, remove the percolate and saturate again with water. Repeat two to four times, until ciliates develop.
d) A distinct succession occurs in the rewetted samples. Thus, they must be inspected on days 2, 6/7, 13/14, 21/22, and 30. Later inspections usually add only few species, likely because microbiostasis (ciliatostasis; see FOISSNER 1997d) increases and metazoan (rotifers, nematods) and protozoan (mainly heliozoans!) predators often became abundant. For inspection, the Petri dish is tilted some seconds and a rather large drop (~ 0.3 ml) of the drained water ("soil percolate") taken with a pipette and inspected for species; several such drops must be investigated from different sites of

the Petri dish, until the last drop adds but few species.

4·1·2·3 Collecting material for preparations

If a "difficult" species is noted, which happens in more than 70% of the samples, material for preparations must be collected. To obtain many specimens, the Petri dish is tilted (45°–60°) several times for a minute or so and the percolating water collected with a Pasteur pipette from several sites of the dish. If only little water (< 10 ml) drains from the sample and/or the species of interest is very rare, it should be sprinkled with 10–15 ml distilled water. This will cause an osmotic shock, detaching or rinsing many specimens from the soil particles and capillaries within about 10 min. Then, the procedure described above is repeated, that is, the Petri dish is tilted several times and the percolating soil water added to the first collection. Finally, the soil sample is again saturated with clean table water (e.g. Eau de Volvic) and stored for the next investigation. Certainly, these procedures strongly change the milieu, and thus a rather different ciliate community may develop after a week, possibly containing further "difficult" species. If so, the whole procedure is repeated, and so on.

Much care must be taken to keep the percolate clean of large (> 2 µm) soil particles, which would disturb the investigation of the preparations, while particles smaller than 2 µm hardly disturb, if not too numerous. To get clean material, note the following advice:

a) Usually, the percolating soil water which contains the organisms will be clean because the soil particles soon become stabilized by microbial activities, mainly by fungal hyphae and bacterial mucus. Thus, extreme care must be taken not to destroy the soil structure developed in the non-flooded Petri dish culture. Accordingly, the Petri dish must be handled gently and, if necessary, distilled water sprinkled softly on the surface. To increase percolation, mild finger pressure on the soil may be applied. Depending on the material sampled, the percolate has a light brown to orange colour (from lignins, humus colloids, etc.), which does not disturb the preparations (but see below).
b) The percolate is now gently shaken and large soil particles allowed to settle for about one minute. Then, the supernatant, which is now ready for preparations, is collected with a Pasteur pipette. Be careful not to lose bottom-dwellers. Occasionally, it may be helpful to sieve the percolate through a plankton net with 50–100 µm mesh-size or to concentrate it by mild centrifugation (max. 2000r/min for a few seconds), especially for preparations with expensive chemicals (osmium tetroxide in CHATTON-LWOFF silver nitrate impregnation).
c) If the sample is very saline (> 20 ‰), it may occur that no ciliates develop. Such samples can be "washed" every third day with table water, which decreases the salt concentration. Frequently, ciliates appear after the third or fourth wash, and often it are spathidiids that grow!

4·1·3 Cultivation

Most spathidiids are predators feeding on other ciliates and/or flagellates; cannibalism is not known. Thus, co-cultivation of prey is required. Culture methods have been described for some *Spathidium* species by MOODY (1912), MOORE (1924a, b), WENZEL

(1955), WILLIAMS (1980), and WOODRUFF & SPENCER (1924). Basically, the media used are simple, viz., yolk suspension and extracts of lettuce, hay, soil or beef providing growth of both, the Spathidia and the prey, usually *Colpidium* (now *Dexiostoma*) *campylum* and *Tetrahymena pyriformis*. Details will be provided in the individual species descriptions.

Our experiences with pure cultures of spathidiids and many other haptorids are frustrating: most clones either die or grow poorly dying after one or two weeks. However, fairly good results are occasionally obtained with semipure cultures set up with some ml of unfiltered soil percolate from the non-flooded Petri dish culture and Eau de Volvic (French table water) enriched with one to three crushed wheat grains to support growth of bacterivorous ciliates, which then serve as a food for the Spathidia contained. This simple method provided well-growing cultures of, for instance, *Protospathidium serpens* and *Arcuospathidium cultriforme*. Further, masses of spathidiids occasionally develop in the non-flooded Petri dish cultures, showing division, conjugation, and encystment.

4·3 Observing living ciliates

Many physical and chemical methods have been described for retarding the movement of ciliates in order to observe structural details (see FOISSNER 1991 for literature). Chemical immobilization (e.g., nickel sulphate) or physical slowing down by increasing the viscosity of the medium (e.g., methyl cellulose) are, in our experience, usually unsuitable. These procedures often change the shape of the cell or cause premortal alterations of various cell structures. The following simple method is therefore preferable: place about 0.5 ml of the raw sample on a slide and pick out (collect) the desired specimens with a micropipette under a compound microscope equipped with a low magnification (e.g., objective 4:1, eyepiece 10×). If specimens are large enough they can be picked out from a petri dish under a dissecting microscope. Working with micropipettes, the diameter of which must be adjusted to the size of the specimens, requires some training. Transfer the collected specimens, which are now in a very small drop of fluid, onto a slide. Apply small dabs of Vaseline (Petroleum jelly) to each of the four corners of a coverslip (or on the slide; it is useful to apply the jelly by an ordinary syringe with a thick needle). Place this coverslip on the droplet containing the ciliates. Press on the vaselined corners with a mounted needle until ciliates become slightly squeezed between slide and coverslip (Fig. 30a–d). As the pressure is increased the ciliates gradually become less mobile and more transparent. Hence, first the location of the main cell organelles (e.g., nuclear and oral apparatus, contractile vacuole) and then the details (e.g., extrusomes, micronucleus) can easily be observed under low (×100–400) and high (oil immersion objective) magnification.

The shape of the cells is of course altered by this procedure. Therefore, specimens taken directly from the sample with a large-bore (opening ~ 1 mm) pipette must first be investigated under low magnification (×100–400). Some species are too fragile to withstand handling with the micropipette and coverslip trapping without deterioration. Investigation with low magnification also requires some experience but it guarantees that undamaged cells are recorded. Video-microscopy is very useful at this point of investigation, especially for the registration of the swimming behaviour.

A compound microscope equipped with differential interference contrast is best for observing ciliates. If not available, use bright-field or phase-contrast; the latter is only useful for very flat species.

4·4 Staining procedures

Although there are numerous methods for staining ciliates, most of the older procedures (e.g., hematoxylin; see KIRBY 1950 for an excellent compilation of protocols) have been outdated by silver impregnation techniques and electron microscopy. Various silver stains are available, but all need some experience and are usually individually modified. However, familiarity with at least protargol impregnation and scanning electron microscopy (SEM) is an absolute prerequisite for studying spathidiid ciliates. These are thus described in great detail in order also to give even beginners a fair chance to obtain good slides.

Apart from silver impregnation, various other staining techniques are useful for taxonomic work with ciliates, especially the Feulgen nuclear reaction and supravital staining with methyl green-pyronin in order to reveal, respectively, the nuclear apparatus and the mucocysts.

Simple, viz. molecular formulae are given for the chemicals used, since usually only these are found in the catalogues of the suppliers (e.g., MERCK). In a laboratory manual it is thus convincing to use this style too, instead of the more correct constitutional or structural formulae.

Supravital staining with methyl green-pyronin. This simple method is an excellent technique for revealing the mucocysts of most ciliates (those of tetrahymenids, however, usually do not stain). Mucocysts are stained deeply and very selectively blue or red, and can be observed in various stages of swelling because the cells are not killed instantly. The nuclear apparatus is also stained.

Procedure

1. Pick out desired ciliates with a micropipette and place the small drop of fluid in the centre of a slide.
2. Add an equally sized drop of methyl green-pyronin and mix the two drops gently by swivelling the slide.
   Remarks: If ciliates were already mounted under the coverslip then add a drop of the dye at one edge of the coverslip and pass it through the preparation with a piece of filter paper placed at the other end of the coverslip.
3. Place a coverslip with vaselined corners on the preparation and squeeze specimens slightly.
   Remarks: Observe immediately. Cells die in the stain within some minutes. Mucocysts stain very quickly and many can be observed at various stages of swelling. To reveal the nuclear apparatus, cells should be fairly strongly squashed (= flattened). The preparation is temporary. After 5–10 minutes the cytoplasm often becomes heavily stained and obscures other details.

Reagents

1 g methyl green-pyronin (CHROMA-Gesellschaft, Küferstrasse 2, P.O. Box 1110, D–7316 Köngen)
ad 100 ml distilled water and filter
Remarks: This solution is very stable and can be used for years.

Protargol methods (protocol A in FOISSNER 1991 and recent experience). Protargol stains are indispensable for descriptive research on ciliates. Many protargol methods have been described, and none is perfect. Here, the variation which produces good results for spathidiids in our laboratory is communicated. This procedure works well with most ciliate and flagellate species (some, however, only rarely impregnate well, e.g., *Loxodes, Paramecium*) but requires at least 20 specimens. Contrary to the silver carbonate method, a single specimen cannot usually be handled successfully. Depending on the procedure used, protargol can reveal many cortical and internal structures, such as basal bodies, cilia, various fibrillar systems, and the nuclear apparatus. The silverlines, however, never impregnate. The shape of the cells is usually well preserved in permanent slides, which is an advantage for the investigation but makes photographic documentation difficult. However, micrographs as clear as those taken from wet silver carbonate impregnations can be obtained if the cells are scrapped off, selected with a micropipette, put on a new slide, pressed down with the coverslip, and photographed prior to fixing with sodium thiosulphate. A centrifuge may he used for step 2; staining jars (Fig. 30e) are necessary for steps 6–16. **The procedure is complicated and subject to many factors. Thus, be well organized and study the "Remarks" carefully.**

1. Fix organisms in Bouin's or Stieve's fluid for 10–30 minutes.
   Remarks: The fixation time has little influence on the quality of the preparation within the limits given. Ratio fixative: sample fluid should be at least 2:1. Pour ciliates into fixative using a wide-necked flask in order to bring organisms in contact with the fixative as quickly as possible. Both fixatives work well but may provide different results with certain organisms. Stieve's fluid may be supplemented with some drops of 2 % osmium tetroxide for better fixation of very fragile ciliates, e.g., the hypotrich *Urosoma*. This increases the stability of the cells but usually reduces their impregnability.
2. Concentrate by centrifugation and wash organisms 3–4 times in distilled water.
   Remarks: There are now two choices: either to continue with step 3 or transfer the material through 30–50–70 % alcohol into 70 % alcohol (ethanol) where it remains stable for several years. Transfer preserved material back through the graded alcohol series into distilled water prior to continuing with the next step. Impregnability of preserved material may be slightly different.
3. Clean 8 slides (or less if material is very scarce) per sample. The slides must be grease-free (clean with alcohol and flame).
   Remarks: Insufficiently cleaned slides may cause the albumen to detach. Mark slides on back if several samples are prepared together. We use staining jars with 8 sections so that we can work with 16 slides simultaneously by putting them back to back (Fig. 30e).
4. Put 1 small drop each of albumen-glycerol and concentrated organisms in the centre of a slide. Mix drops with a mounted needle and spread over the middle third.

Remarks: Use about equally sized drops of albumen-glycerol and suspended (in distilled water) organisms to facilitate spreading. The size of the drops should be adjusted so that the middle third of the slide is covered after spreading. Now remove sand, grains, etc. The thickness of the albumen layer should be equal to that of the organisms. Some thicker and thinner slides should, however, also be prepared because the thickness of the albumen layer may influence the quality of the preparation. Cells may dry out and/or shrink if the albumen layer is too thin; if it is too thick it may detach or the cells become impossible to study with the oil immersion objective.

5. Allow slides to dry for at least 2 hours or for 12 hours (overnight) at room temperature.
   Remarks: Slides may be allowed to dry for up to 24 hours but no longer if quality is to be maintained. Oven-dried (2 hours at 60° C) slides are usually also of poorer quality.
6. Place slides in a staining jar (Fig. 30e) filled with 95 % alcohol (ethanol) for 20–30 minutes. Place a staining jar with protargol solution into an oven (60° C).
   Remarks: Slides should not be transferred through an alcohol series into concentrated alcohol as this causes the albumen layer to detach! Decrease hardening time to 15–20 minutes if albumen is already rather old and/or not very sticky.
7. Rehydrate slides through 70 % alcohol and two distilled water steps for 5 minutes each.
8. Place slides in 0.2 % potassium permanganate solution. Remove first slide (or pair of slides) after 30 seconds and the others at 15 second intervals. Collect slides in a staining jar filled with distilled water.
   Remarks: Bleaching is by permanganate and oxalic acid (step 9). The procedure described above is necessary because each species has its optimum bleaching time. The sequence in which slides are treated should be recorded as the immersion time in oxalic acid must be proportional to that in the permanganate solution. The albumen layer containing the organisms should swell slightly in the permanganate solution and the surface should become uneven. If it remains smooth, the albumen is too sticky and this could decrease the quality of the impregnation. If the albumen swells strongly, it is possibly too weak (old) and liable to detach. Use fresh $KMnO_4$ solution for each series.
9. Quickly transfer slides to 2.5 % oxalic acid. Remove first slides (or pair of slides) after 60, 90, 120 and 160 seconds, the others at 20 second intervals. Collect slides in a staining jar filled with distilled water.
   Remarks: Same as for step 8! Albumen layer becomes smooth in oxalic acid.
10. Wash slides three times in distilled water for 3 minutes each.
11. Place slides in the warm (60° C) protargol solution and impregnate for 10–15 minutes at 60° C.
    Remarks: Protargol solution can be used only once.
12. Remove staining jar with the slides from the oven and allow to cool for 10 minutes at room temperature.
    Remarks: In the meantime organize six staining jars for developing the slides: distilled water – distilled water – fixative (sodium thiosulphate) – distilled water – 70 % alcohol – 100 % alcohol (ethanol).

13. Remove the first slide from the protargol solution and drop some developer on the albumen layer. Move slide gently to spread developer evenly. As soon as the albumen turns yellowish, pour off the developer, dip slide into the first two distilled water steps for about 2 seconds each and stop development by submerging the slide in the fixative (sodium thiosulphate), where it can be left for 1–5 minutes.
    Remarks: Now control impregnation with the compound microscope. The impregnation intensity is sufficient if the ciliary pattern is just recognizable. The permanent slide will be too dark if the ciliary pattern is distinct at this stage of the procedure! The intensity of the impregnation can be controlled by the concentration of the developer and the time of development. 5–10 seconds usually suffice for the diluted developer! Some species (e.g., most microthoracids) must be treated with undiluted developer. Development time increases with bleaching time. Therefore commence developing with those slides which were in the bleaching solutions for 60 and 120 seconds, respectively. The thinner the albumen layer, the quicker the development.
14. Fix slides in sodium thiosulphate for 2 min. Then wash in distilled water for about 3 min.
    Remarks: Do not wash too long; the albumen layer is very fragile and may detach!
15. Transfer slides to 70 % – 100 % – 100 % alcohol for 3–5 minutes each.
16. Clear by two 10 minute transfers through xylene.
17. Mount in synthetic neutral medium.
    Remarks: Do not dry slides between steps 16 and 17. The preparation is stable, provided step 14 is done correctly. The mounting medium should be rather viscous to avoid air-bubbles being formed when the solvent evaporates during drying.

Reagents

a) Bouin's fluid (prepare immediately before use; components can be stored)

15 parts	saturated, aqueous picric acid ($C_6H_3N_3O_7$; preparation: add an excess of picric crystals to, e.g., 1 litre of distilled water; shake solution several times within a week; some undissolved crystals should remain; filter before use)
5 parts	formalin (HCHO; commercial concentration, about 37 %)
1 part	glacial acetic acid (= concentrated acetic acid; $C_2H_4O_2$)

b) Stieve's fluid (slightly modified; prepare immediately before use; components can be stored)

38 ml	saturated, aqueous mercuric chloride (dissolve 60 g $HgCl_2$ in 1 litre of boiling distilled water)
10 ml	formalin (HCHO; commercial concentration, about 37 %)
3 ml	glacial acetic acid (= concentrated acetic acid; $C_2H_4O_2$)

c) Albumen-glycerol (2–6 month stability at 3°C)

15 ml	egg albumen
15 ml	concentrated (98–100 %) glycerol ($C_3H_8O_3$)

Pre-treatment of the egg albumen and preparation of the albumen-glycerol: Separate the white carefully from the yolk and embryo of three eggs (free range eggs are preferable to those from battery chickens, whose egg white is less stable and sticky).

Shake the white by hand (do not use a mixer!) for a minute in a narrow-mouthed 250 ml Erlenmeyer flask until a stiff white foam is formed. Allow the flask to stand for about 1 minute. Then pour the viscous rest of the egg white in a second Erlenmeyer flask and shake again until a stiff foam is formed. Repeat until most of the egg white is either stiff or becomes watery; usually 4–6 Erlenmeyer flasks of foam are obtained. Leave all flasks undisturbed for about 10 minutes and discard the watery albumen from the last flask. During this time a glycerol-like fluid percolates from the foam. This fluid is collected and used. Add an equal volume of concentrated glycerol and a small thymol crystal ($C_{10}H_{14}O$) for preservation to the mixture. Mix by shaking gently and pour mixture into a small flask. Leave undisturbed for two weeks. A whitish slime settles at the bottom of the flask. Decant the clear portion, discard slime and thymol crystal. A "good" albumen-glycerol drags a short thread when touched with a needle. The albumen is too thin (not sticky enough) or too old if this thread is not formed. Fresh albumen which is too thin may be concentrated by leaving it open for some weeks so that water can evaporate. If the albumen is too sticky, which may cause only one side of the organisms to impregnate well, it is diluted with distilled water or old, less sticky albumen to the appropriate consistency. The preparation of the albumen-glycerol must be undertaken with great care because much depends on its quality. Unfortunately, all commercial products which we have tried detach during impregnation.

d) 0.2 % potassium permanganate solution (stable for about 1 day)
   0.2 g potassium permanganate ($KMnO_4$) are dissolved in 100 ml distilled water
e) 2.5 % oxalic acid solution (stable for about 1 day)
   2.5 g oxalic acid ($C_2H_2O_4 \cdot 2\ H_2O$) are dissolved in 100 ml distilled water
f) 0.4–0.8 % protargol solution (stable for about 1 day)
   100 ml distilled water
   add 0.4–0.8 g protargol
   Remarks: Use light-brown "protargol for microscopy" presently available only by FLUKA. Some dark-brown, cheaper products do not work! Sprinkle powder on the surface of the water and allow to dissolve without stirring; use a wide-mouthed bottle for solving the protargol. Concentration of the protargol depends on its "strength", that is, on the silver contents.
g) Developer (mix in sequence indicated; sodium sulphite must be dissolved before hydroquinone is added)
   95 ml distilled water
   5 g sodium sulphite (Na_2SO_3)
   1 g hydroquinone ($C_6H_6O_2$)
   Remarks: This recipe yields the stock solution which is stable for some weeks and should be used undiluted for certain ciliates (step 13). Usually, however, it must be diluted with tap water in a ratio of 1:20 to 1:50 to avoid too rapid development and one-sided impregnation of the organisms. Freshly prepared developer is usually inadequate (the albumen turns greenish instead of brownish). The developer should thus be prepared from equal parts of fresh and old (slightly brownish) stock solutions. Take great care with the developer as its quality contributes highly to that of the slides. If the developer has lost its activity (which is not always indicated by a brown colour!) the silver is not or only insufficiently reduced and the organisms

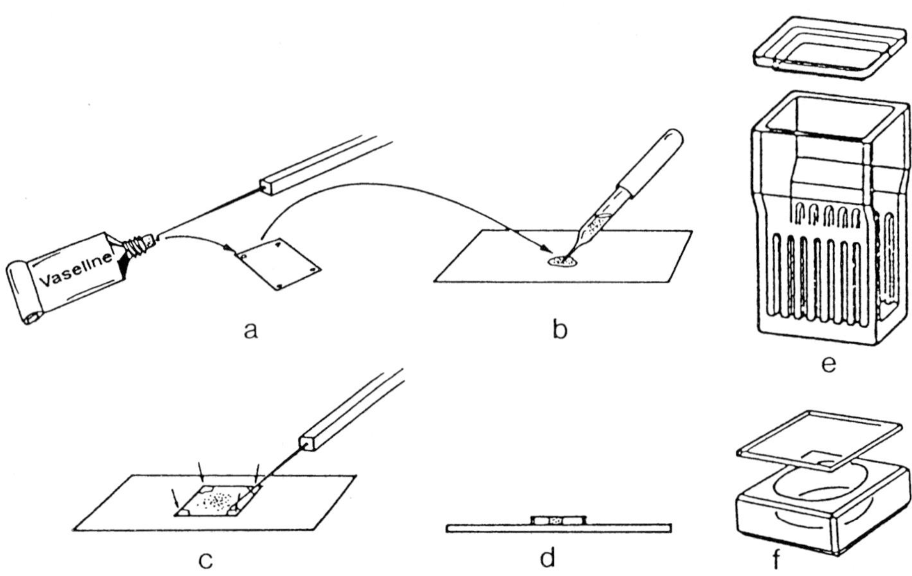

Fig. 30 a–f Life observation and staining of spathidiid ciliates (from FOISSNER 1991). **a–d)** Preparation of slides for observing living ciliates. **e)** Staining jar for 8 and 16 (back to back) slides, respectively. **f)** Watch-glass for cleaning of ciliates for scanning electron microscopy.

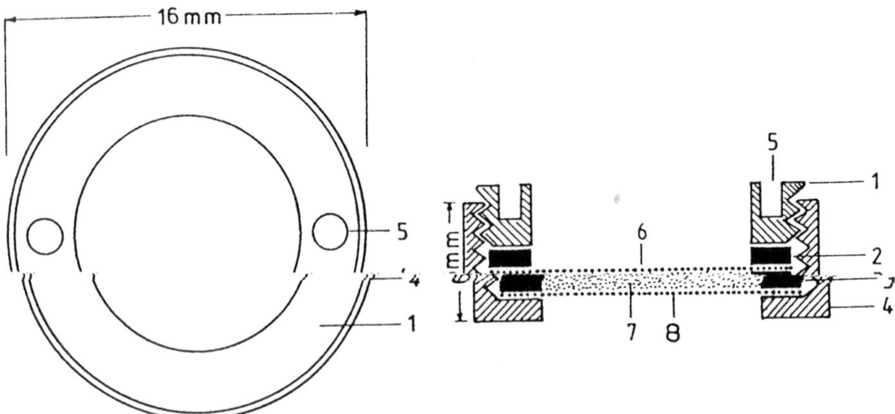

Fig. 31 Brass chamber for critical point-drying of protists (from FOISSNER 1991). 1 – threaded chamber lid; 2 – washer 2; 3 – washer 1; 4 – threaded chamber jacket; 5 – holes for forceps tips, used to screw lid into jacket; 6 – top steel net; 7 – sample; 8 – bottom steel net.

stain too faintly. A fresh developer should therefore be prepared for each "impregnation week" and some old developer kept. Fresh developer can be artificially aged by adding some sodium carbonate (Na_2CO_3). However, better results are obtained with air-aged solutions, that is, by a developer which has been kept uncovered for some days in a wide-mouthed bottle. It first turns yellowish, then light brown (most effective) and later dark brown and viscous (at this stage the developer has lost most of its activity but is still suitable for artificial ageing of fresh developer = 1:1 mixture mentioned above).

During the past five years, we obtained very good slides with the low-speed developer used by FRYD-VERSAVEL (pers. comm.). It is composed of 7 g boric acid, 1.5 g hydroquinone, 10 g sodium sulphite, and 75 ml acetone, all solved, one by one, in 420 ml distilled water. This developer is stable for some weeks and should be used only once. Pour developer into a staining jar and immerse slides, one by one, controlling impregnation intensity after 30–60 s. Usually, developing is finished within 1–5 min (if not, use the ordinary developer or double protargol concentration because slides should not be too long in the developer, as the albumen may detach). The further procedure is as described above (steps 14–17).

h) Fixative for impregnation (stable for several years)
25 g sodium thiosulphate ($Na_2S_2O_3 \cdot 5\ H_2O$) are dissolved in 1000 ml distilled water

Preparation for Scanning Electron Microscopy (SEM). Ciliate species cannot usually be identified solely by scanning electron microscopy because only a limited number of characters is revealed. However, SEM is useful for the beginner by allowing a three-dimensional view of the specimen and for the specialist in documenting details which are difficult to reveal with other methods. Only the method used by ourselves is described here. See textbooks for general SEM-techniques.

Procedure

1. Pour ciliates into Parducz' fixative and leave for about 30 minutes.
Remarks: Concentrate and clean material as thoroughly as possible (see step 2). Ratio of sample: fixative should be at least 1:1, better 1:2. Add some drops of 5 n HCl if fixative becomes milky when the material is added. Parducz' fluid preserves most ciliates very well. However, the cirri of the hypotrichous ciliates usually disintegrate into their component cilia. Hypotrichs should thus be fixed either in concentrated sublimate (dissolve 60 g $HgCl_2$ in 1 litre hot distilled water and allow to cool) or in a mixture composed of 4 parts concentrated sublimate and 1 part 2 % osmium tetroxide. A much better fixative for hypotrichs is that used by Barry Wicklow (pers. comm.): mix equal amounts of 2 % aqueous osmium tetroxide and 3 % glutaraldehyde and fix cells for 15–30 min. Wash in distilled water and proceed as described below (steps 2–7). Unfortunately, such material cannot be stored because crystals are formed. Thus, critical point drying must be followed immediately. The fixative preserves also many other ciliates well, although the metachronal ciliary waves are frequently not as distinct as with Parducz' fluid. We use this fixative now routinely for hypotrichs.
2. Wash ciliates at least 5 times with 0.05 M Na-cacodylate buffer.

Remarks: Ciliates in the buffer may be refrigerated for years. Washing must be done in a watch-glass (Fig. 30f) and a micropipette should be used to remove bacteria and organic debris. This cleaning of the material is essential but rather difficult and laborious, especially with small species (< 100 µm) and field material; thus cultures and/or pre-cleaned material (see below) should be used. The cleaning is performed as follows: Ciliates settle at the bottom of the fixation tube after 30 minutes (cp. step 1). Remove as much supernatant as possible with a pipette (do not centrifuge!). Then transfer the material to a watch-glass and allow to settle for about 5 minutes (use fume hood). Quickly remove most of the fixative with a micropipette under the dissecting microscope. Now wash the ciliates with the buffer by several passages through a large-bore (diameter about 1 mm) pipette. Bacteria and debris adhering to the ciliates are hereby mechanically removed. Again allow to settle, but control sedimentation with the dissecting microscope; remove supernatant containing bacteria and debris with a micropipette as soon as ciliates settle. This procedure must be repeated until the material is clean and all of the debris is removed. Use fractionated sedimentation if the sample contains several species differing in size and/or mass.

Field material: Larger species (> 100 µm) are picked out with a micropipette and sprinkled into the fixative. Several hundred specimens most be collected because loss of material may be considerable during the following steps. Small species can be prepared by this method only if abundant material is available. Accumulation can often be achieved by the following simple method: leave a freshly collected sample containing ample mud to stand for some hours at room temperature. Due to oxygen depletion the ciliates usually move to the surface where they can be skimmed off with a teaspoon.

3. Transfer cleaned ciliates with a small drop of buffer to the preparation chamber (Fig. 31).
   Remarks: Place the drop on the bottom plankton net of the chamber which is weighted with washer 1. The net must be dry to avoid spreading of the drop to the chamber margin and the washer. Place the top plankton net carefully on the drop, that is, on washer 1, using forceps. Weight top net with washer 2, close chamber with lid and immediately transfer into 30 % ethanol. The plankton net must have a mesh-size < 12 µm and can be used several times. It should fit exactly into the chamber, which is best achieved using an appropriate punch. Alternatively, metal grids with 10–20 µm mesh size, as used by soil scientists, can be used. They are stable for years.
4. Dehydrate chamber with ciliates in ethanol series (30–50–70–90–100–100 %) for 5 minutes each.
5. Dry chamber with ciliates in a critical-point drying apparatus.
   Remarks: We use CO_2 and change the alcohol at least 10 times. Amylacetate, as used previously (FOISSNER 1991), proved to be superfluous.
6. Open chamber and place ciliates on the SEM-stub. Use a mounted eyelash if ciliates do not fall from the net.
   Remarks: The dried ciliates usually form a lump at the bottom plankton net. This lump is carefully transferred (by holding the net over the stub and stripping off the ciliates with the eyelash) to the SEM-stub, where it is dispersed under the dissecting microscope with a mounted eyelash. The ciliates spread easily if cleaning and drying were sufficient. Be careful not to rewet the sample by your breath! Use a mouth protection fabric and a glass shield between the microscope and the sample.

Preparation of the SEM-stub: We use commercial aluminium SEM-stubs. To get a black, homogenous background the stub is covered with a Graphit-Tab available from several providers, e. g., the GRÖPL Company, Frauenhofnerstrasse 40, A-3430 Tulln, Austria (order no. G 3347 or G 3348, i. e., tabs with a diameter of 12 or 25 mm). Note that small species tend to sink into the graphit.

7. Sputter with gold. This is a very critical step! Use low (4 mA) sputter energy. Sputter about 10 times for 15 s each, with breaks of about 5 minutes to avoid heating. **Cilia become curled and denaturated if sputter energy is too high and/or the sample is slightly rewetted by your breath or perspiration!**

Reagents

a) Parducz' fixative (prepare immediately before use)
   4 ml aqueous 2 % osmium tetroxide (OsO_4)
   1 ml saturated, aqueous mercuric chloride ($HgCl_2$; preparation see protargol protocol)
b) 0.05 M Na-cacodylate buffer (can be stored for several months in the refrigerator; adjust to pH 7 with HCl)
   10.7 g dimethylarsinacid-sodium salt ($C_2H_6AsNaO_2 \cdot 3\ H_2O$) are dissolved in 1000 ml distilled water

B Systematic Section

1 How to Describe a Ciliate?

Guidelines for ciliate descriptions are not available. Thus, the chaos prevails. Some put the diagnosis ahead, others banish it to the end of the paper; some describe the "silver features" first, others put them behind the in vivo observations; some use prosa style, others prefer telegramese style, and so on. In our opinion, this bewildering situation, partially caused by inexperienced or very individual journal editors, can be ordered by looking beyond the narrow protozoological borders and adopting the style used by experienced taxonomists from other fields (MAYR 1975). Then, for instance, it becomes clear that telegramese style should be used and species description should commence with a concise "diagnosis" separating it from the nearest relatives = containing only or mainly those features needed to make a key.

Our species descriptions take into account the available knowledge, that is, follow some basic and established schemes: name of species, list of synonyms, nomenclature (if necessary), diagnosis, type locality, etymology, description s. str., occurrence and ecology, discussion/remarks. The "description s. str." is based on the sequence features are used during in vivo identification: size, shape, nuclear apparatus, contractile vacuole, extrusomes, cortex, cytoplasm, movement, somatic ciliary pattern, oral apparatus and ciliary pattern, resting cyst. Likewise, the characteristics of the individual features are usually described in a certain sequence: location, shape, size, peculiarities.

2 Species/Subspecies Concept

The species concept, of course, influences the number of species found and/or recognized as undescribed (LUCKOW, 1995, MCDADE, 1995, TURNER, 1999). We usually apply the phylogenetic species concept as defined by NIXON and WHEELER (1990): "A species is the smallest aggregation of populations (sexual) or lineages (asexual) diagnosable by a unique combination of character states in comparable individuals (semaphoronts)". Basically, this is a morphospecies concept which is, according to EHRENDORFER (1984) and FINLAY et al. (1996), as valid as any, and probably more pragmatic than any other; see EHRENDORFER (1984), LUCKOW (1995), MCDADE (1995), and TURNER (1999) for detailed discussion.

We do not consider ourselves as splitters, that is, we classify species as undescribed (new) only if populations can be separated from their nearest relatives by at least one distinct (non-morphometric) morphological feature, such as the presence/absence of caudal cirri or rod-shaped vs. fusiform extrusomes, or if quantitative differences, such as body size and/or number of ciliary rows, are really conspicuous (≥ 100%). Furthermore, we must have seen at least 10 individuals and studied the species in vivo and silver preparations, to provide reliable morphometrics, illustrations, and types.

MAYR (1963) defines a subspecies as "an aggregate of local populations of a species inhabiting a geographic subdivision of the range of the species, and differing taxonomically from other populations of the species". This concept, especially geographic isolation has been widely adopted, although there is still a lot of discussion (BÖHME 1978,

O'NEILL 1982, ROLÁN-ALVAREZ & ROLÁN 1995). Unfortunately, biogeography of protozoa is still in its infancy, and thus MAYR'S concept hardly can be applied. None the less, subspecies are useful also in protists, when used restrictively and as a simple taxonomic tool.

In the present monograph, we distinguish subspecies according to distinct morphometric differences in important features (e.g., number of ciliary rows) and/or qualitative (morphological) characters whose taxonomic value is still doubtful or not known. It is the last mentioned feature which makes the subspecies concept so useful: the name can be easily withdrawn if later research proves the features used to be unreliable, and the discoverer does not lose priority to "armchair" taxonomists if the subspecies later gets species rank (INTERNATIONAL COMMISSION ON ZOOLOGICAL NOMENCLATURE 1999). Furthermore, subspecies "collect" the infraspecific variation, that is, data which tend to be lost (ZUSI 1982), and enhance identification of species because of the broader concept; thus subspecies are especially useful for people and disciplines not specifically trained in taxonomy. In spite of the obvious advantages, protozoologists rarely used the subspecies/subgenus concept, although KAHL (1932) established some subgenera and varieties in ciliates and one third of the testate amoebae taxa are "variations" or "forms", most of which must be considered as subspecies according to the ICZN (FOISSNER & KORGANOVA 2000). Further, subspecies are common in extant and fossil foraminifera (BOLTOVSKOY 1954) and fossil tintinnids (BELOKRYS 1997). There is now a tendency to use them also in extant ciliates (FOISSNER et al. 2002, Song & WEI 1998).

A further main factor influencing the number of species recognized as undescribed is the treatment of literature data. Many of the old species descriptions lack type material and are poor compared with the present standard because the pioneers did not have the advantages of modern methods. Clearly, there is a tendency to disrespect the efforts of our predecessors and to establish new taxa with new methods. Our approach is to respect and re-interpret previous work and to neotypify species, provided that at least one main feature matches (see also chapter on neotypification in Vol. II).

3 How to Use the Monograph?

A few technical explanations are necessary for the proper use of the monograph.

(i) The original descriptions are either cited or are adapted to the style used in this monograph. In the latter case, great care was taken nothing to remove or to add. Further, we included all illustrations, even if poor, available for the individual species. With few exceptions, these are Xerox copies and thus show the species as it was originally published.

(ii) We quantified several features in three steps, for instance, the distance of the ciliary rows (Fig. 9): very narrowly to narrowly spaced ($\leq$ 4 µm), ordinarily spaced (4–8 µm), widely to very widely spaced ($\geq$ 8 µm). Usually, we left "ordinary" when the kinety distance was within the range given.

(iii) Usually, authors and dates are omitted from the names of the species in the Remarks section because this information is found in the individual species descriptions.

(iv) Magnifications. Scale bars are liked by reviewers and editors. However, usually they are circumstantial and superfluous when the size of the specimen or of a certain part of it is provided in the figure explanation. The old taxonomists knew of that. Image

KAHL with countless scale bars on the plates! Thus, we adopt the old style and give the size of the specimen in the figure explanation and scale bars only when appropriate.

The situation is more complex with micrographs. At first glance, scale bars appear indispensable, but when considered critically, they might also be superfluous or even misleading. What are the facts? A good description provides in vivo measurements of the main features, such as body length and width (first data set); a good description also provides detailed morphometrics from permanent silver preparations where, however, the specimens are more or less shrunken (second data set); and when the description is supplemented with SEM micrographs, a third data set is generated by scale bars because 20–60 % shrinkage is usual in SEM preparations. Now, we have three different sizes and scale bars for the same structure! Further bias is generated when specimens are more or less flattened by mild coverslip pressure to get detail-rich micrographs. All these problems and the fact that sizes are contained in the description and the tabulated morphometrics suggest that scale bars usually can be omitted from micrographs without loss of important information. Indeed, often they will be misleading and/or indicate higher precision than is actually present.

4 Key to Families

A simple family key is impossible for the spathidiids, that is, proper classification needs protargol impregnation. See figures 1, 4–12 for an explanation of the features used. An overall key to the species, independent of genus and family, will be provided in volume II.

1 Cytopharynx permanent, appearing as a comparatively distinct opening in the centre of the oral bulge in vivo and in preparations................ Pharyngospathidiidae

– Cytopharynx temporary, usually not recognizable in vivo; appears as a minute, fibrillar, obconical depression near dorsal bulge end in protargol preparations..... 2

2 Circumoral kinetofragments separated from each other by minute gaps 1–3 dikinetids wide (*Protospathidium* pattern). Most species small and/or narrow with oral bulge shorter than widest trunk region .... Protospathidiidae

– Circumoral kinetofragments arranged to a continuous circumoral kinety. Most species of ordinary to huge size with oral bulge at least as long as widest trunk region.. 3

3 Ciliary rows separated from circumoral kinety and curved dorsally on both sides of the cell .... Arcuospathidiidae

– Ciliary rows connected with circumoral kinety (*Spathidium* pattern) or their anterior ends so strongly curved that the circumoral kinety is seemingly doubled (*Epispathidium* pattern). Circumoral kinety closed or open ventrally.................... 4

4 Circumoral kinety open. Ciliary pattern often not unequivocally determinable....... .... Apertospathulidae

– Circumoral kinety closed. *Spathidium* or *Epispathidium* ciliary pattern ....Spathidiidae

Family *Protospathidiidae* nov. fam.

Diagnosis: Cylindroidal to very narrowly spatulate or obclavate Spathidiina with temporary cytostome. Circumoral kinetofragments separated from each other by minute gaps and attached to the somatic ciliary rows. Anterior end of right side ciliary rows directed dorsally, left side row ends directed ventrally. Oral bulge obovate or cuneate, small, that is, usually as long as or shorter than widest trunk region. Brush located dorsolaterally and very diverse.

Type genus: *Protospathidium* DRAGESCO & DRAGESCO-KERNÉIS, 1979.

Remarks: We discovered several new *Protospathidium* and *Protospathidium*-like spathidiids in the literature and in our unpublished material from soils globally – altogether 14 species. These species form two groups which are considered as distinct genera and are united in the new family Protospathidiidae. Likely, many further species remain to be discovered.

The most important feature of the species united in this family are the oral kinetofragments which are not aligned to a "continuous" circumoral kinety, but separated from each other by gaps one to four dikinetids wide. Unfortunately, the feature is of varying distinctness and occurs also in some species and/or specimens of other genera, for instance, in *Spathidium extensum* and cultivated specimens of *S. turgitorum*. On the other hand, separation of the circumoral kinetofragments may be indistinct, for instance, in the Antarctic population of *P. terricola*. The allocation of such species and populations remains doubtful and is usually based on additional features, such as the shape of the body and oral bulge. Most protospathidiids are very narrowly spatulate or cylindroidal (length:width ratio > 6:1) and 70–150 μm long, while others resemble small *Spathidium* or middle-sized *Arcuospathidium* species, especially *A. namibiense*. The nuclear pattern, the extrusomes, and the dorsal brush are highly diverse, and thus the most important features for species recognition. Brush row 1 is reduced to a few dikinetids in several species, and the dikinetids of row 3 are comparatively widely spaced. In sum, none of the family features is distinct in all species, but taken together, they define the generic home of most populations rather unambiguously.

As in the Spathidiidae and Arcuospathidiidae, part of the Protospathidiidae has the dorsal brush located laterally, especially *Edaphospathula paradoxa, E. espeletiae,* and *E. minor; Protospathidium serpens* and *P. muscicola* show at least a tendency. If such distinction is confirmed by further investigations, at least *Edaphospathula* should be split, like *Spathidium* and *Latispathidium*, respectively, *Arcuospathidium* and *Cultellothrix.*

Interestingly, most *Edaphospathula* (family Protospathidiidae) and *Latispathidium* (family Spathidiidae) species have ovate extrusomes and short left lateral kinetofragments usually composed of ≤ four dikinetids. Likely, these features evolved convergently, possibly driven by environmental constraints, such as body narrowing due to soil life.

Key to genera (requires protargol impregnation)

1 Extrusomes ≤ 4 μm long, basically ovate and massive. Right side oral kinetofragments composed of ≤ 3 dikinetids. Circumoral kinety usually cuneate .... *Edaphospathula*

– Extrusomes ≥ 4 μm long, basically rod-shaped and comparatively fine. Right side oral kinetofragments composed of ≥ 3 dikinetids. Circumoral kinety usually elliptical or obovate .... *Protospathidium*

Key to species of *Protospathidium* and *Edaphospathula*

The species of the two protospathidiid genera are often difficult to distinguish because some features are not easily recognized in these tiny organisms. Thus, reliable identification usually requires both careful in vivo observation (extrusomes, brush details!) and protargol impregnation (number of ciliary rows, details of dorsal brush and oral kinetofragments!).

1 Length usually ≥ 200 μm in vivo 2

– Length usually ≤ 170 μm in vivo 3

2 Size about 210 × 20 μm. Extrusomes fine and rod-shaped. On average 9 ciliary rows. Oral bulge hemispherical and about half as long as widest trunk region .... *P. namibicola*

– Size about 230 × 30 μm. Extrusomes bluntly fusiform. On average 13 ciliary rows. Oral bulge ordinarily convex occupying about 38% of maximum body width .... *P. arenicola*

3 Oral bulge extrusomes oblong or rod-shaped and 2–5 μm long 10

– Oral bulge extrusomes lacking or present and massive (ovate, fusiform etc.)........ 4

4 Extrusomes lacking .... *E. inermis*

– Extrusomes present, minute but massive, that is, bluntly fusiform, ovate or ampulliform and 1–2 × 0.8–1.5 μm in size 5

5 Macronucleus ellipsoidal to cylindroidal 6

– Macronucleus a long, tortuous strand or in many scattered nodules 8

6 Body bottle-shaped to elongate ovoidal, length:width ratio about 4:1 .... *E. minor*

– Body cylindroidal, vermiform, or elongate clavate, length:width ratio ≥ 7:1, on average 9–11:1 7

7 Size about 100 × 12 μm in vivo. Extrusomes ovate to ampulliform, 1.5 μm long. Dorsal brush occupies about 30% of body length; individual brush dikinetids widely spaced in all rows .... *E. brachycaryon*

– Size about 130 × 12 μm in vivo. Extrusomes bluntly fusiform, 3 μm long. Dorsal brush occupies about 15% of body length; individual brush dikinetids narrowly (ordinarily) spaced in all rows .... *E. gracilis*

8 Many scattered macronucleus nodules .... *E. fusioplites*

– Macronucleus a long, more or less tortuous and twisted strand 9

9 About 7 ciliary rows. Right side oral kinetofragments each composed of 3 dikinetids .... *E. paradoxa*

– About 11 ciliary rows. Right side oral kinetofragments each composed of 1 dikinetid .... *E. espeletiae*
10 A single, large macronucleus or many small, scattered macronucleus nodules ... 11
– Macronucleus a more or less nodulated, tortuous strand. Brush row 1 distinctly shorter than row 2 .... *P. serpens*
11 Single reniform, ellipsoidal, or dumbbell-shaped macronucleus 12
– Many small, scattered macronucleus nodules 13
12 Macronucleus reniform or ellipsoidal. 21 ciliary rows .... *P. terricola*
– Macronucleus reniform or dumbbell-shaped. 10 ciliary rows .... *P. vermiculus*
13 Slender (about 5–7:1). Dorsal bristles up to 4 µm long .... *P. muscicola*
– Very slender (≥ 10:1). Dorsal bristles up to 8 µm long .... *P. vermiforme*

Edaphospathula nov. gen.

Diagnosis: Protospathidiidae with right side oral kinetofragments composed of ≤ 3 dikinetids. Circumoral kinety broadly cuneate or obovate. Extrusomes ≤ 4 µm long, bascially ovate and massive (length:width ratio about 1.4–2.2:1).

Type species: *Protospathidium fusioplites* FOISSNER et al., 2005.

Etymology: Composite of the Greek substantive *edaphon* (soil organisms), the Latin noun *spatha* (spatula), and the diminutive suffix *ula*, meaning a small spatula living in soil. Feminine gender.

Remarks: The populations united in this genus can be considered either as *Spathidium* spp. with unusually widely spaced circumoral dikinetids or as protospathidiids with unusually small circumoral kinetofragments. We prefer the later interpretation because (i) the body is usually slender and the oral bulge small as in the type species of *Protospathidium*; (ii) *Spathidium* and *Arcuospathidium* species have closely spaced circumoral dikinetids (usually ≥ five between two kineties each), while protospathidiid circumoral kinetofragments are usually separated by distinct gaps and composed of ≤ five dikinetids; and (iii) several species have typical protospathidiid oral kinetofragments on the right side. The circumoral kinety is broadly cuneate, while the oral bulge is usually ellipsoidal or obovate, as in *Protospathidium*, a remarkable difference not yet fully understood.

Body shape and size and the number of ciliary rows are highly similar in all *Edaphospathula* species, except for *E. minor*. Thus, species are distinguished by the nuclear apparatus, the extrusomes, and the highly diverse dorsal brush. The extrusomes, though massive, are less than 3 µm long, and thus difficult to recognize and observe. None the less, they must be recorded very carefully because their shape and size are important features for species recognition. In most species, one to three dorsal ciliary rows project more or less distinctly onto the oral bulge, a rather characteristic but inconspicuous feature not (yet) included in the formal diagnosis.

Certainly, all these features are rather sophisticated, but present and correlated with each other in quite a lot of taxa. This suggests a specific evolutionary lineage and, hence, a distinct genus.

Edaphospathula fusioplites (FOISSNER, BERGER, XU & ZECHMEISTER-BOLTENSTERN, 2005) **nov. comb.** (Fig. 32a–x, 33a–e, 107a–l, 108k–n; Table 1, 2)

2005 *Protospathidium fusioplites* FOISSNER, BERGER, XU & ZECHMEISTER-BOLTENSTERN, Biodiv. Conserv., 14: 660 (Type slides with protargol-impregnated specimens from type and voucher localities are deposited in the Oberösterreichische Landesmuseum in Linz, Upper Austria.).

Diagnosis: Size about 110 × 12 µm in vivo. Cylindroidal to very narrowly obclavate with oblique to strongly oblique, obovate oral bulge about 2/3 as long as widest trunk region. Macronucleus in about 12 scattered nodules; multimicronucleate. Extrusomes ovate to broadly fusiform, 1.4–2 × 0.8–1.2 µm in size. On average 8 ciliary rows, 3 anteriorly differentiated to distinctly heterostichad dorsal brush occupying 17% of body length. Brush bristles up to 5 µm long, row 1 composed of an average of four dikinetids, row 2 of twelve, and row 3 of seven dikinetids followed by a monokinetidal bristle tail extending to mid-body. Oral kinetofragments each composed of 1 to 2 dikinetids.

Type locality: *Pinus nigra* forest soil in the Stampfltal near Vienna, Austria, E16°02' N47°53'.

Etymology: Composite of the Latin noun *fusus* (spindle) and the Greek noun *(h)oplites* (soldier ~ extrusome), referring to the fusiform to ovate extrusomes.

Description: We studied four populations of this species, but only the Austrian Stampfltal population was fully investigated. The North American and Kenyan specimens were also studied rather carefully (Table 2), while another Austrian population (Burgenland) and the South African population were routinely identified in vivo by the main features, viz., size and shape of body, oral bulge and extrusomes; the macronucleus pattern; and the number of ciliary rows. Although conspecificity is beyond reasonable doubt, the observations are kept separate, and the diagnosis and description contain only data from the Stampfltal population.

The populations contain rather many specimens with deviating macronucleus pattern most caused by exconjugant and post-divisional reorganization processes (see General Section), which obviously need long time in this species. Without knowledge of these peculiarities, the macronucleus patterns of *E. fusioplites* and *E. paradoxa* would appear highly variable (Table 1).

Size 90–140 × 10–15 µm in vivo, usually near 110 × 12 µm, as calculated from some in vivo measurements and the morphometric data (Table 2); length:width ratio 8.8–12.5:1, on average about 11:1 both in vivo and in protargol preparations. Shape cylindroidal to very narrowly obclavate, frequently slightly curved; anterior (oral) body end ordinarily to strongly slanted, posterior narrowly rounded, occasionally bluntly pointed, but never tail-like, widest in or slightly underneath mid-body (Fig. 32a, h–k, o, p, 107f–i, l). Macronucleus pattern difficult to recognize in vivo, basically nodular with individual nodules scattered in middle body third; however, half of the specimens show nuclear reorganization, that is, a mixture of nodules and short, moniliform pieces occurs

Table 1 Comparison of macronucleus pattern in several populations of *Edaphospathula fusioplites* and *E. paradoxa*

Species/population	Macronucleus pattern (proportion, %)				Number of specimens analysed
	Perfectly scattered nodules	Mixture of nodules and short strands	Perfect strand	Specimens with two or four nodules	
Edaphospathula fusioplites from Stampfltal (Austria)	45	40[1]	0	15	40
Edaphospathula fusioplites from Burgenland (Austria)	70	11[1]	3	16	74
Edaphospathula fusioplites from USA	70	30[1]	0	0	10
Edaphospathula paradoxa from Austria	0	27[2]	30	43	70

[1] Nodules and strands separate, but at nearly same plane, forming a discontinuous, tortuous strand.
[2] Nodules and strands still in line and often connected by a fine thread.

in about 40% of specimens; furthermore, 9% of specimens have four nodules and 6% only two (Fig. 32a, h–k, o, p, 107f–l, 108l–n; Table 2), that is, are likely exconjugants. Individual nodules globular to elongate ellipsoidal, on average 5 × 2.5 µm in protargol preparations, each usually containing some minute nucleoli. Micronuclei globular, most scattered among macronucleus nodules, some rather distant near anterior or posterior body end. Contractile vacuole in posterior body end, some excretory pores in pole area; several empty, acontractile (food?) vacuoles in posterior body half. Extrusomes in oral bulge and scattered in cytoplasm, broadly fusiform to indistinctly ovate, that is, both ends pointed or one end rather distinctly rounded; minute, that is, about 1.5–2 × 0.8 µm in size, but rather refractive and thus distinct in vivo, where they appear as conspicuous, bright dots when the oral bulge is viewed frontally; both, oral bulge and cytoplasmic extrusomes frequently impregnate rather intensely with protargol (Fig. 32a, b, e, k, o, s, u, 107a–e, j–l, 108k). Cortex very flexible, contains several rows of colourless, loosely arranged granules about 0.2 µm across between each two kineties. Cytoplasm colourless, contains few to many lipid droplets 1–5 µm across and, frequently, a subterminal vacuole with crystals and granular remnants causing a rather distinct inflation in protargol-prepared cells; occasionally specimens have a massive, large food vacuole containing amoeboid prey (Fig. 32a, l, o, 107h, i). Movement without peculiarities.

Cilia about 7 µm long in vivo, arranged in an average of eight equidistant, bipolar, ordinarily ciliated rows connected with inconspicuous oral kinetofragments (Fig. 32a, m–x, 107f–l; Table 2); one to three dorsal rows slightly projecting anteriorly, similar as in *E. minor* (Fig. 32m, r). Dorsal brush distinctly heterostichad and three-rowed occupying 17% of body length on average, a fourth row occurs in one out of 40 specimens analysed; all rows have some ordinary cilia anteriorly and continue as somatic kineties posteriorly. Brush rows of same structure anteriorly, that is, bristles rod-shaped and of nearly same length (5 µm) in vivo; posteriorly, bristle length decreases gradually with anterior bristles longer than posterior. Brush row 1 shorter than rows 2 and 3 comprising an average of only four dikinetids; middle row 2 longer than row 3, composed of twelve dikinetids on average; row 3 composed of an average of seven dikinetids much more widely spaced than those of rows 1 and 2, followed by a monokinetidal tail extending to mid-body with 1 µm long bristles (Fig. 32a, n–r, t–x, 107g, h, k, l; Table 2).

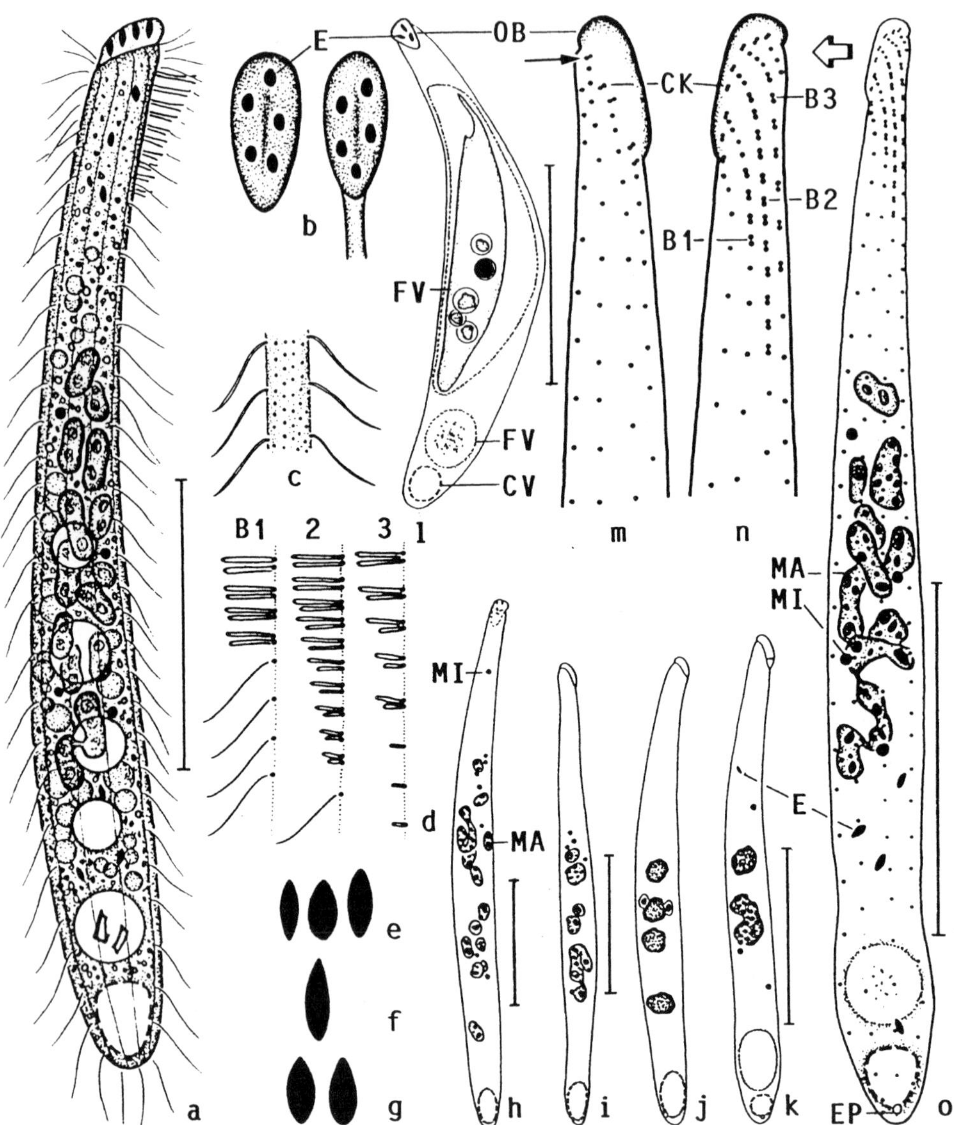

Fig 32a–o *Edaphospathula fusioplites* from life (a–g) and after protargol impregnation (h–o). From FOISSNER et al. (2005a). **a)** Left lateral view of a representative specimen, length 110 μm. **b)** Frontal views of oral bulge containing few, thick extrusomes. **c)** Surface view showing cortical granulation. **d)** Posterior portion of dorsal brush. **e–g)** Oral bulge extrusomes of specimens from Austria (1.5–2 × 0.8 μm in size), USA (2.5 × 0.8 μm), and the Republic of South Africa (2 × 1 μm). **h–k)** Variations of body shape and nucleus pattern. **l)** A specimen with a very large food vacuole containing amoeboid prey. **m–o)** Ciliary pattern of holotype specimen orientated slightly dorsolaterally, length 95 μm. Note individual oral kinetofragments composed of only one or two dikinetids, and the widely spaced dikinetids of brush row 3. Arrow marks projecting ciliary row. B1-3 – dorsal brush rows, CK – circumoral kinety, CV – contractile vacuole, E – extrusomes, EP – excretory pores, FV – food vacuole, MA – macronucleus nodules, MI – micronucleus, OB – oral bulge. Scale bars 30 μm.

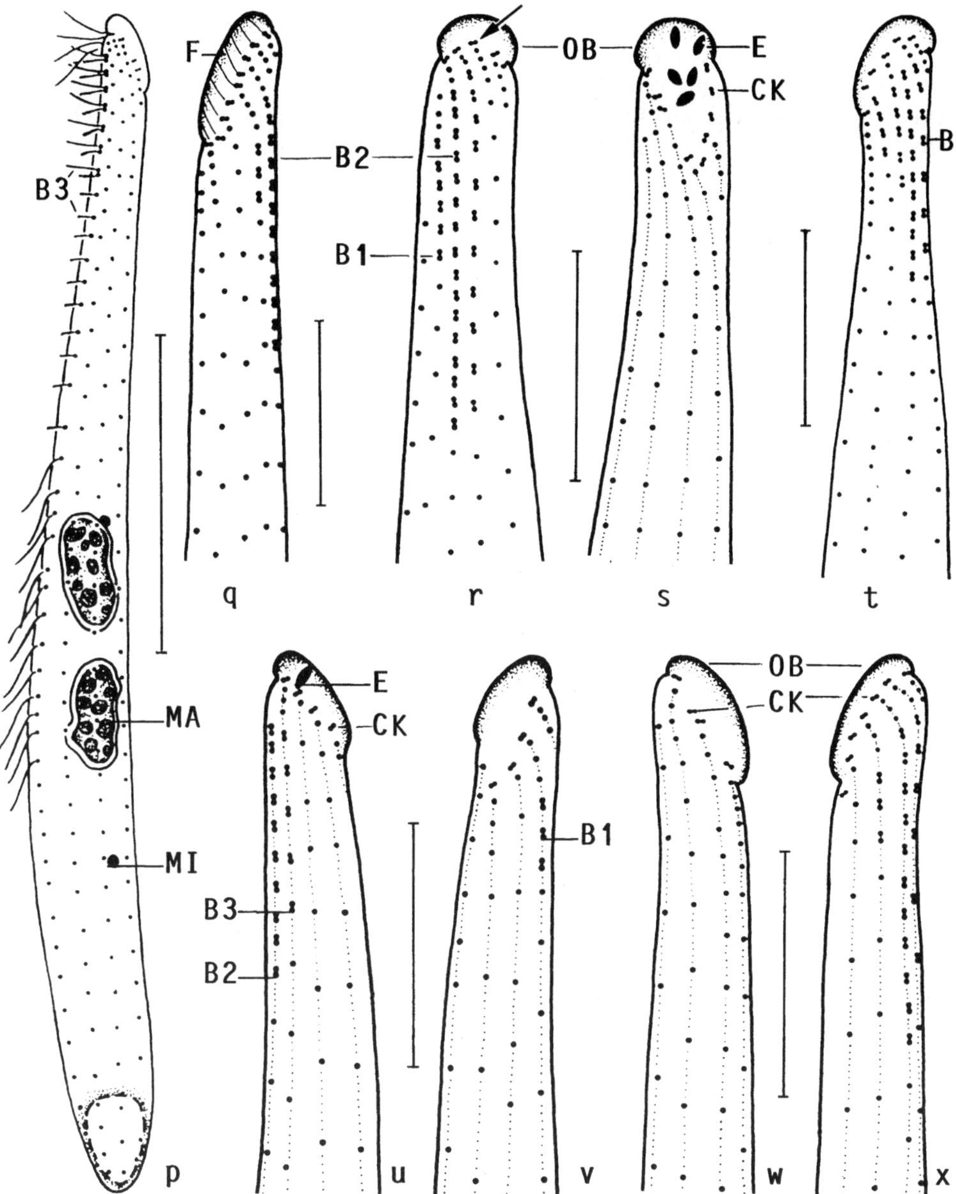

Fig. 32p–x *Edaphospathula fusioplites*, ciliary pattern after protargol impregnation (from FOISSNER et al. 2005a). **p, q)** Right side view and left anterior body portion of same specimen. Note monokinetidal tail of brush row 3 extending to mid-body. **r, s)** Dorsal and ventral anterior body portion of a specimen where the bulge extrusomes impregnated with protargol. Note the cuneate circumoral kinety. Arrow marks projecting ciliary rows. **t)** A specimen with four dorsal brush rows. **u, v)** Ciliary pattern of right and left anterior body portion of a specimen with only one oral dikinetid at anterior end of each somatic kinety. **w, x)** Ciliary pattern of right and left anterior body region of a specimen with oral kinetofragments composed of one or two dikinetids. B(1-3) – dorsal brush (rows), CK – circumoral kinety, CV – contractile vacuole, F – fibres, MA – macronucleus nodule, MI – micronucleus, OB – oral bulge. Scale bars 30 µm (p) and 10 µm (q–x).

Oral bulge rather distinct in vivo due to the compact, refractive extrusomes contained, basically, however, fairly inconspicuous because only about 3 µm high, hardly separate from body proper ventrally, and shorter than widest trunk region by about 1/3; slanted by 40°–70° and slightly convex in lateral view, while obovate when seen frontally (Fig. 32a, b, 107i–l). Individual oral kinetofragments inconspicuous because composed of only one to two dikinetids, forming a "circumoral kinety" comprising an average of only 13 dikinetids; in an extreme specimen, even only 8, that is, one each at end of kineties. Kinetofragments almost equidistantly spaced and thus inconspicuous on left side of cell, while separated from each other by gaps one to three dikinetids wide on right side. Each dikinetid associated with a single cilium and a fibre extending into oral bulge (Fig. 32m–x, 107f, i, j–l; Table 2). Oral basket rods (nematodesmata) not recognizable in vivo, also indistinct in over-impregnated cells; no oralized somatic monokinetids, as typical for the Fuscheriidae and Acropisthiidae (FOISSNER et al. 2002).

Observations on North American and African populations: The North American population matches the European specimens in all main features, for instance, body and extrusome shape, nucleus pattern, number and arrangement of ciliary rows, and the structure of the circumoral kinety and dorsal brush (Table 2). There are only a few minor differences: (i) body slightly smaller and stouter (length:width ratio 7.6 on average), likely because broad specimens with numerous food inclusions are more frequent; (ii) extrusomes slightly longer, viz., 2.5 × 0.8 µm (Fig. 32f); (iii) dikinetids of dorsal brush rather widely spaced not only in row 3 but also in row 1; (iv) brush bristles generally shorter, that is, only about 2 µm long in vivo.

The Kenyan specimens, which have broadly fusiform to ovate extrusomes, are slightly larger than those from Austria and the USA, increasing the total range of ciliary rows from 8–11 to 8–13 (Table 2). Further, they lack the slight anterior elongation of the dorsal ciliary rows (Fig. 33a–e). The South African specimens were identified in vivo, where they showed the same features as those of the other two populations. The extrusomes are even more conspicuous because they are slightly thicker (2 × 1 µm; Fig. 32g). The cortical granules are also minute, but more refractive and thus distinct in vivo. The dorsal brush is as in the American specimens.

Occurrence and ecology: As yet found at type locality (Pine forest soil from Austria, where it was moderately abundant), in a slightly saline inland soil rather near to the type locality (margin of Zicklacke, Burgenland), in the United States of America (rare in a grassland soil of Arizona between the towns of San Lucas and Caolinga; pH 6.2), in the Dominican Republic (rare in grassland soil from the Puerto Plata airport, north coast; pH 6.6; sample kindly provided by Dr. ALINE BERTHOLD); in South Africa (rare in soil from the bank of the Skeleton River in the Botanical Gardens of Kirstenbosch near Cape Town; pH 6.7), and in Kenya (Keekorok Lodge in the Massei Mara National Park, very hard grassland soil at bank of a river; pH 6.6). Thus, *E. fusioplites* is likely a cosmopolitan, possibly preferring acidic or circumneutral conditions. It is well adapted to soil life by the slender body.

Remarks: In vivo, *E. fusioplites* differs from all congeners by the scattered macronucleus nodules. *Protospathidium muscicola*, which has the same nuclear pattern, is much stouter (5–7:1 vs. 8–11:1) and has rod-shaped extrusomes.

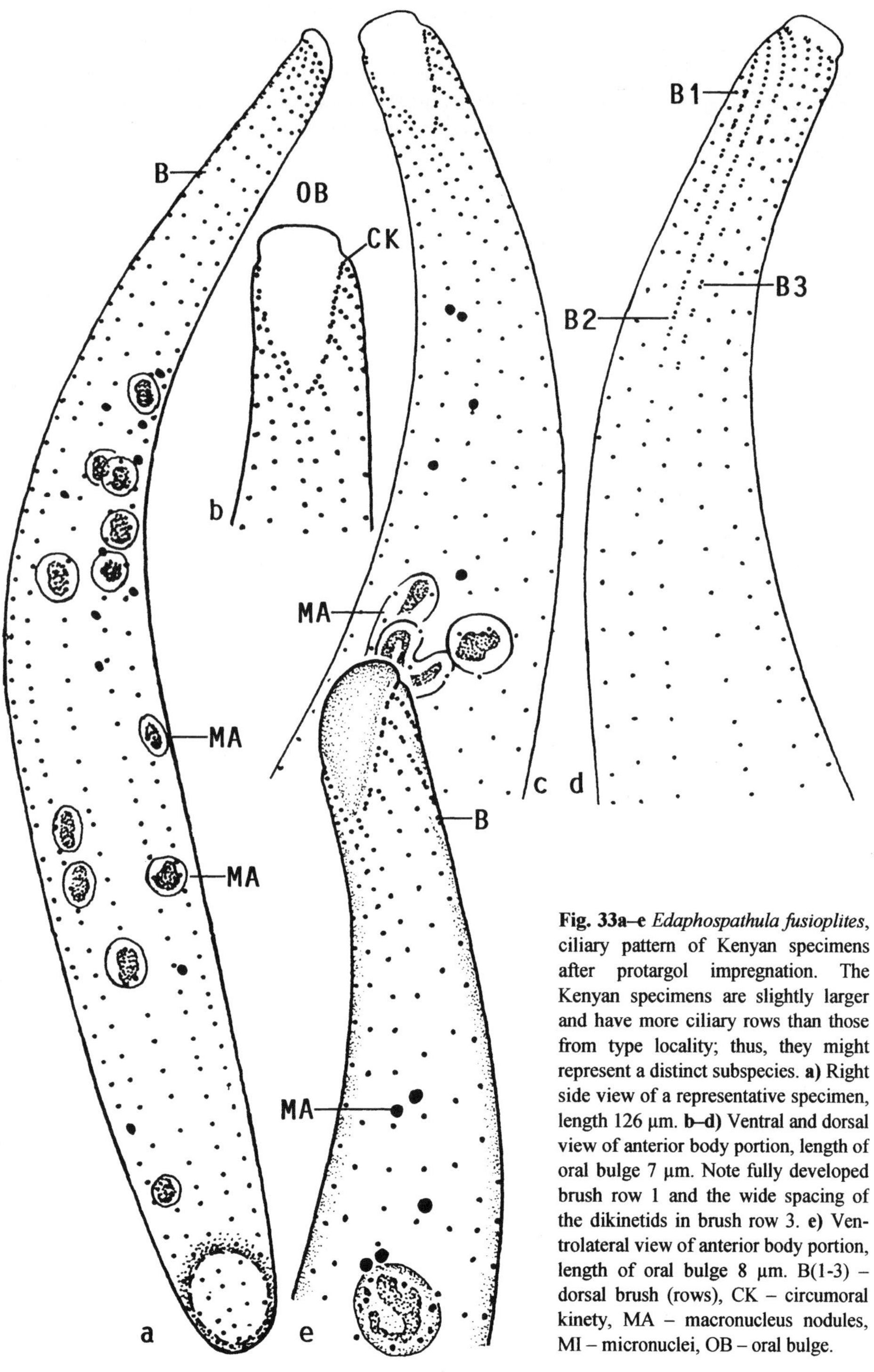

Fig. 33a–e *Edaphospathula fusioplites*, ciliary pattern of Kenyan specimens after protargol impregnation. The Kenyan specimens are slightly larger and have more ciliary rows than those from type locality; thus, they might represent a distinct subspecies. **a)** Right side view of a representative specimen, length 126 µm. **b–d)** Ventral and dorsal view of anterior body portion, length of oral bulge 7 µm. Note fully developed brush row 1 and the wide spacing of the dikinetids in brush row 3. **e)** Ventrolateral view of anterior body portion, length of oral bulge 8 µm. B(1-3) – dorsal brush (rows), CK – circumoral kinety, MA – macronucleus nodules, MI – micronuclei, OB – oral bulge.

At first glance, *E. fusioplites* highly resembles *Sikorops* in the arrangement and structure of the circumoral ciliature and the blunt extrusomes (FOISSNER 1999a, FOISSNER et al. 2002). However, *Sikorops* has nematodesmata (oral basket rods) originating from both the circumoral dikinetids and the anterior somatic monokinetids, and thus belongs to the family Acropisthiidae (FOISSNER et al. 2002).

Table 2 Morphometric data on *Edaphospathula fusioplites* from Austria (AU), the USA (US), and Kenya (KE), as well as on *E. paradoxa* from Austria (EP). From FOISSNER et al. (2005a)

Characteristics[1]	Pop[a]	$\overline{x}$	M	SD	SE	CV	Min	Max	n
Body, length	AU	97.6	96.0	14.2	3.1	14.5	75.0	130.0	21
	US	74.1	71.5	13.2	4.2	17.9	53.0	103.0	10
	KE	105.6	105.0	13.1	3.5	12.4	88.0	133.0	14
	EP	89.4	92.0	15.1	3.3	16.9	60.0	118.0	21
Body, width	AU	9.1	9.0	1.3	0.3	13.9	7.0	12.0	21
	US	10.0	10.0	1.6	0.5	16.1	6.0	12.0	10
	KE	13.6	13.5	2.9	0.8	21.4	9.0	18.0	14
	EP	9.9	9.0	2.0	0.4	19.9	7.0	15.0	21
Body length:width, ratio	AU	10.8	10.8	1.1	0.2	10.5	8.8	12.5	21
	US	7.6	7.3	1.8	0.6	24.1	4.8	11.5	10
	KE	8.0	8.2	1.4	0.4	17.8	5.6	10.6	14
	EP	9.3	8.9	1.6	0.4	17.7	5.7	12.0	21
Oral bulge, length	AU	6.9	7.0	0.8	0.2	12.0	5.0	8.0	21
	US	5.7	5.8	0.8	0.3	14.4	4.0	7.0	10
	KE	5.1	5.0	0.8	0.2	15.0	4.0	6.0	14
	EP	5.8	6.0	1.0	0.2	17.6	4.0	8.0	21
Oral bulge, width	AU	3.3	3.0	–	–	–	3.0	4.0	9
	US	2.8	3.0	–	–	–	2.5	3.0	3
	EP	2.9	2.5	0.8	0.4	28.3	2.0	4.0	5
Circumoral kinety to last dikinetid of brush row 1, distance	AU	7.5	8.0	1.1	0.2	14.4	6.0	10.0	21
	US	8.7	8.0	0.9	0.3	10.0	8.0	10.0	9
	KE	9.4	9.5	1.5	0.4	16.0	8.0	13.0	14
	EP	6.2	6.0	1.4	0.3	22.0	4.0	9.0	21
Circumoral kinety to last dikinetid of brush row 2, distance	AU	15.9	16.0	2.3	0.5	14.7	12.0	20.0	21
	US	12.1	12.0	1.9	0.6	15.7	10.0	16.0	9
	KE	16.8	16.5	2.5	0.7	15.2	14.0	21.0	14
	EP	17.9	18.0	2.0	0.4	11.2	13.0	22.0	21
Circumoral kinety to last dikinetid of brush row 3, distance	AU	13.5	13.0	2.1	0.5	15.3	10.0	18.0	21
	US	10.0	10.0	1.2	0.4	12.0	8.0	12.0	8
	KE	16.8	16.5	3.0	0.1	18.0	13.0	23.0	14
	EP	15.2	15.0	1.6	0.4	10.7	12.0	18.0	21
Anterior body end to first macronucleus nodule / macronucleus strand, distance	AU	29.2	30.0	6.2	1.4	21.4	15.0	39.0	21
	US	17.3	17.0	4.4	1.4	25.6	12.0	25.0	10
	EP	22.3	22.0	4.8	1.0	21.3	13.0	31.0	21
Nuclear figure, length	AU	45.4	40.0	15.8	3.4	34.8	24.0	92.0	21
	US	40.5	37.5	14.6	4.6	36.0	21.0	68.0	10
	EP	44.2	47.0	14.5	3.2	32.8	20.0	65.0	21
Macronucleus nodules, length	AU	5.4	5.0	1.4	0.3	24.9	4.0	9.0	21
	US	5.4	5.5	1.6	0.5	30.5	3.5	9.0	10
	KE	4.6	4.0	2.0	0.5	44.4	2.0	10.0	14
	EP	–	–	–	–	–	–	–	–
Macronucleus nodules or strand, width in mid	AU	2.9	2.5	0.8	0.2	27.2	2.0	4.0	21
	US	3.4	3.0	0.8	0.3	23.8	2.5	5.0	10

continued

Characteristics[1]	Pop[a]	$\overline{x}$	M	SD	SE	CV	Min	Max	n
	KE	3.0	3.0	0.5	0.1	15.5	2.5	4.0	14
	EP	3.0	3.0	0.6	0.1	19.3	2.0	5.0	21
Macronuclei, number	AU	11.6	13.0	3.4	0.7	29.5	5.0	15.0	21
	US	10.9	11.0	4.7	1.6	43.4	4.0	20.0	9
	KE	15.6	13.5	7.7	2.1	49.0	9.0	39.0	14
	EP	1.0	1.0	0.0	0.0	0.0	1.0	1.0	21
Micronuclei, across	AU	1.1	1.0	–	–	–	1.0	1.5	21
	US	1.1	1.0	–	–	–	1.0	1.5	8
	KE	1.2	1.2	–	–	–	1.0	1.3	13
	EP	1.8	2.0	–	–	–	1.5	2.0	9
Micronuclei, number	AU	4.5	4.0	1.4	0.3	31.7	2.0	9.0	21
	US	3.6	4.0	1.1	0.5	31.7	2.0	5.0	5
	KE	11.7	11.0	4.8	1.3	40.8	6.0	22.0	13
	EP	–	–	–	–	–	–	–	–
Circumoral dikinetids, number (*E. fusioplites*)	AU	12.4	13.0	1.2	0.3	9.4	8.0	13.0	21
Dikinetids in individual circumoral kinetofragments, number (*E. paradoxa*)	EP	3.1	3.0	0.6	0.1	20.2	2.0	4.0	21
Ciliary rows, number	AU	8.4	8.0	0.6	0.1	7.0	8.0	10.0	21
	US	10.3	10.0	0.7	0.2	6.6	9.0	11.0	10
	KE	11.1	11.0	0.9	0.2	7.8	10.0	13.0	14
	EP	7.1	7.0	–	–	–	7.0	8.0	21
Basal bodies in a right side ciliary row, number	AU	34.0	33.0	8.2	1.8	24.1	20.0	55.0	21
	US	44.3	42.0	9.2	3.1	20.7	32.0	60.0	9
	KE	52.8	52.5	11.6	3.1	22.0	32.0	74.0	14
	EP	41.3	39.0	11.1	2.4	26.8	26.0	63.0	21
Dorsal brush rows, number	AU	3.0	3.0	0.0	0.0	0.0	3.0	3.0	21
	US	3.0	3.0	0.0	0.0	0.0	3.0	3.0	9
	KE	3.1	3.0	–	–	–	3.0	4.0	14
	EP	3.0	3.0	0.0	0.0	0.0	3.0	3.0	21
Dikinetids in brush row 1, number	AU	4.2	4.0	1.1	0.2	26.8	3.0	7.0	21
	US	6.6	7.0	–	–	–	6.0	7.0	9
	EP	2.2	2.0	0.9	0.2	42.2	1.0	4.0	21
Dikinetids in brush row 2, number	AU	12.2	12.0	2.1	0.5	17.1	10.0	17.0	21
	US	12.1	12.0	1.4	0.5	11.3	10.0	15.0	9
	EP	17.2	17.0	3.0	0.7	17.7	12.0	24.0	21
Dikinetids in brush row 3, number	AU	7.0	7.0	1.4	0.3	20.7	5.0	11.0	21
	US	7.9	8.0	1.3	0.5	17.1	6.0	9.0	7
	EP	9.1	9.0	1.4	0.3	15.6	7.0	11.0	21

[1] Data based on mounted, protargol-impregnated (FOISSNER's method), and randomly selected specimens from non-flooded Petri dish cultures. Measurements in µm. CV – coefficient of variation in %, M – median, Max – maximum, Min – minimum, n – number of individuals investigated, SD – standard deviation, SE – standard error of arithmetic mean, $\overline{x}$ – arithmetic mean.

Edaphospathula paradoxa **nov. spec.** (Fig. 34a–v, 107 m –x ; Table 2)

D i a g n o s i s : Size about 110 × 12 µm in vivo. Cylindroidal to very narrowly obclavate with oblique, ovate oral bulge about 2/3 as long as widest trunk region. Macronucleus a nodulated, tortuous strand; multimicronucleate. Extrusomes ovate to ampulliform, 1.4–1.7 × 1–1.2 µm in size. On average 7 ciliary rows, 3 anteriorly differentiated to distinctly heterostichad dorsal brush occupying 20% of body length. Brush bristles up to 5 µm long, row 1 composed of an average of two dikinetids; row 2 of seventeen; and row 3 of nine dikinetids followed by a heteromorphic, monokinetidal bristle tail extending to mid-body. Oral kinetofragments each composed of an average of 3 dikinetids.

Type locality: *Pruno-Fraxinetum* floodplain forest soil from the Müllerboden near Vienna, Austria, E16°42' N48°.

Etymology: The Latin adjective *paradoxa* (curious) refers to the high percentage of specimens reorganizing the nuclear apparatus.

Description: Size 70–130 × 8–15 µm in vivo, usually about 110 × 12 µm, as calculated from some in vivo measurements and the morphometric data (Table 2); length:width ratio highly variable, viz., 5.7–12:1, on average 9:1 both in vivo and in protargol preparations. Shape cylindroidal to very slenderly ellipsoidal, frequently slightly curved; anterior (oral) body end ordinarily slanted, posterior narrowly rounded, widest in or slightly underneath mid-body, inconspicuously flattened in oral area (Fig. 34a–c, g, i, j, m, 107p–s). Macronucleus in middle quarters of cell, basically as in *P. serpens*, that is, a more or less distinctly nodulated and spiralized strand in 30% out of 70 specimens analysed (Fig. 34a, g, m, 107p, q); a mixture of short strands and/or nodules still in line and often connected by a fine thread in 27% of specimens (Fig. 107r); and 43% of the cells have two (12%) or four (31%) large, homogenously impregnated nodules (Fig. 107s), that is, are likely exconjugants; individual nodules with many small or a few large nucleoli up to 3 µm across. Most micronuclei along macronucleus strand, some rather distinct, exact number difficult to recognize due to many similarly sized and impregnated cytoplasmic inclusions; individual micronuclei globular to ellipsoidal, about 2 µm across (Fig. 34g, i, j, m). Contractile vacuole in posterior body end; excretory pores, interestingly, distinctly subterminal on dorsal side. Extrusomes in oral bulge and scattered in cytoplasm, ovate to broadly fusiform, minute, that is, 1.4–1.7 × 1–1.2 µm in size, but compact and thus distinct in vivo (Fig. 34a, d, 107n); oral extrusomes occasionally lightly impregnate with protargol; cytoplasmic extrusomes usually in minute vacuoles (Fig. 34e). Cortex very flexible, contains about five rows of narrowly spaced, colourless granules 0.5 µm across between each two kineties (Fig. 34e, f, 107m). Cytoplasm with few to many lipid droplets up to 5 µm across and, usually, many extrusomes, as described above. Glides rather rapidly on microscope slide wriggling like an eel.

Cilia about 6 µm long in vivo, arranged in an average of seven equidistant, bipolar, ordinarily ciliated rows connected with fairly distinct circumoral kinetofragments (Fig. 34a, g, m, n, 107o–s; Table 2); one to two dorsal rows, usually those left of brush, slightly elongated anteriorly, similar as in *E. minor* (Fig. 34r–t). Dorsal brush distinctly heterostichad and three-rowed occupying 20% of body length on average, fairly conspicuous because bristles of row 2 up to 5 µm high and very narrowly spaced, forming a membrane-like array in vivo; all rows with some ordinary cilia anteriorly and continuing as somatic kineties posteriorly (Fig. 34a, g, m, n, p, q, 107p, r, s; Table 2). Brush row 1 composed of an average of only two dikinetids each composed of a rod-shaped, about 1.5 µm long anterior bristle and an ordinary, 6 µm long posterior cilium. Brush row 2 composed of an average of seventeen narrowly spaced dikinetids, as described

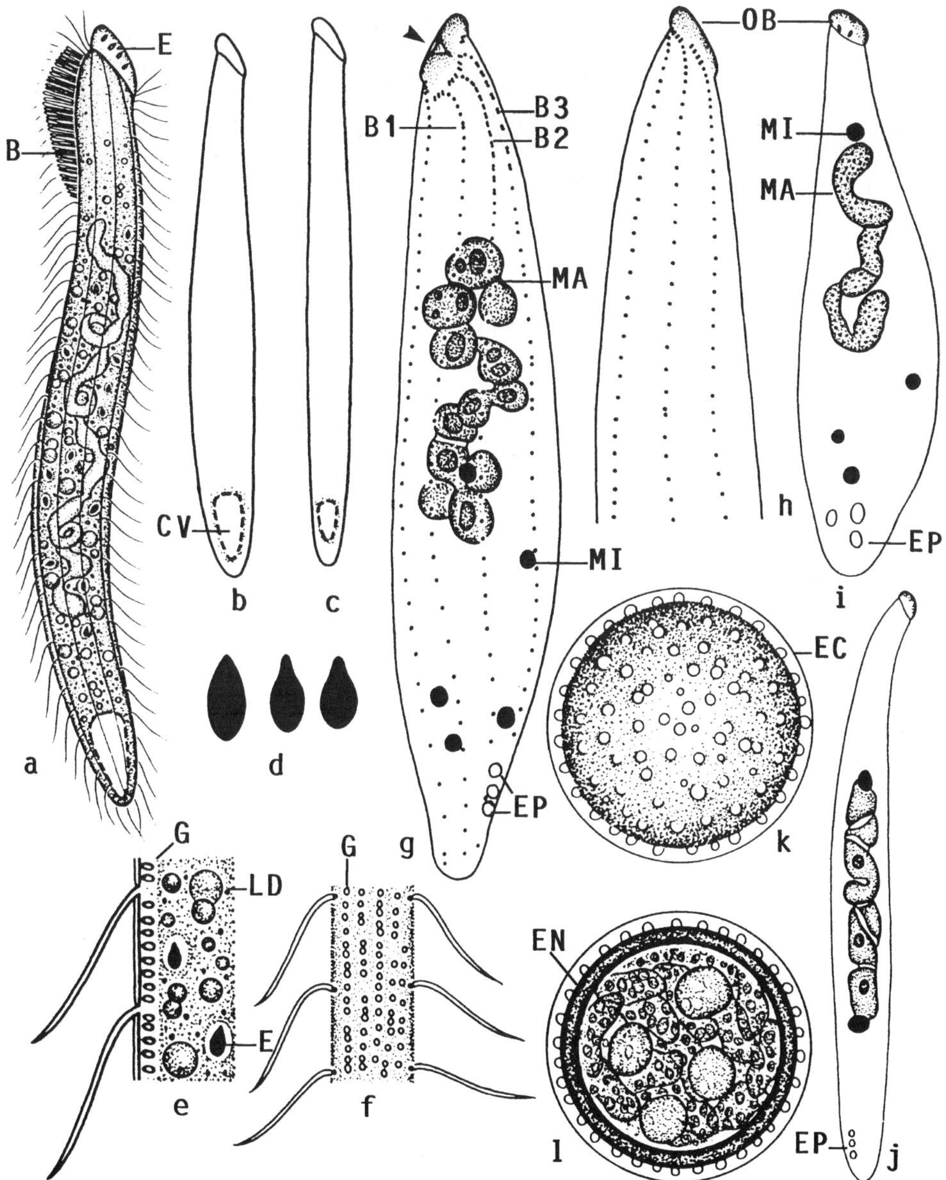

Fig. 34a–l *Edaphospathula paradoxa* nov. spec. from life (a–f, k, l) and after protargol impregnation (g–j). **a)** Right side view of a representative specimen with bristles of brush row 2 forming a membrane-like array, length 110 µm. **b, c)** Shape variants, length 90–100 µm. **d)** Oral bulge extrusomes, 1.4–1.7 × 1–1.2 µm. **e, f)** Optical section and surface view showing cortical granulation. **g, h)** Ciliary pattern of left and right side of a specimen with moniliform macronucleus, length 70 µm. Arrowhead marks cytostomial entrance. **i, j)** Exconjugants transforming four globular macronucleus anlagen into a long strand, length 65 µm and 67 µm. **k, l)** Surface view and optical section of a resting cyst 18 µm across. Note the tuberculate endocyst. B(1-3) – dorsal brush (rows), CV – contractile vacuole, E – extrusomes, EC – ectocyst, EN – endocyst, EP – excretory pores, G – cortical granules, LP – lipid droplets, MA – macronucleus, MI – micronuclei, OB – oral bulge.

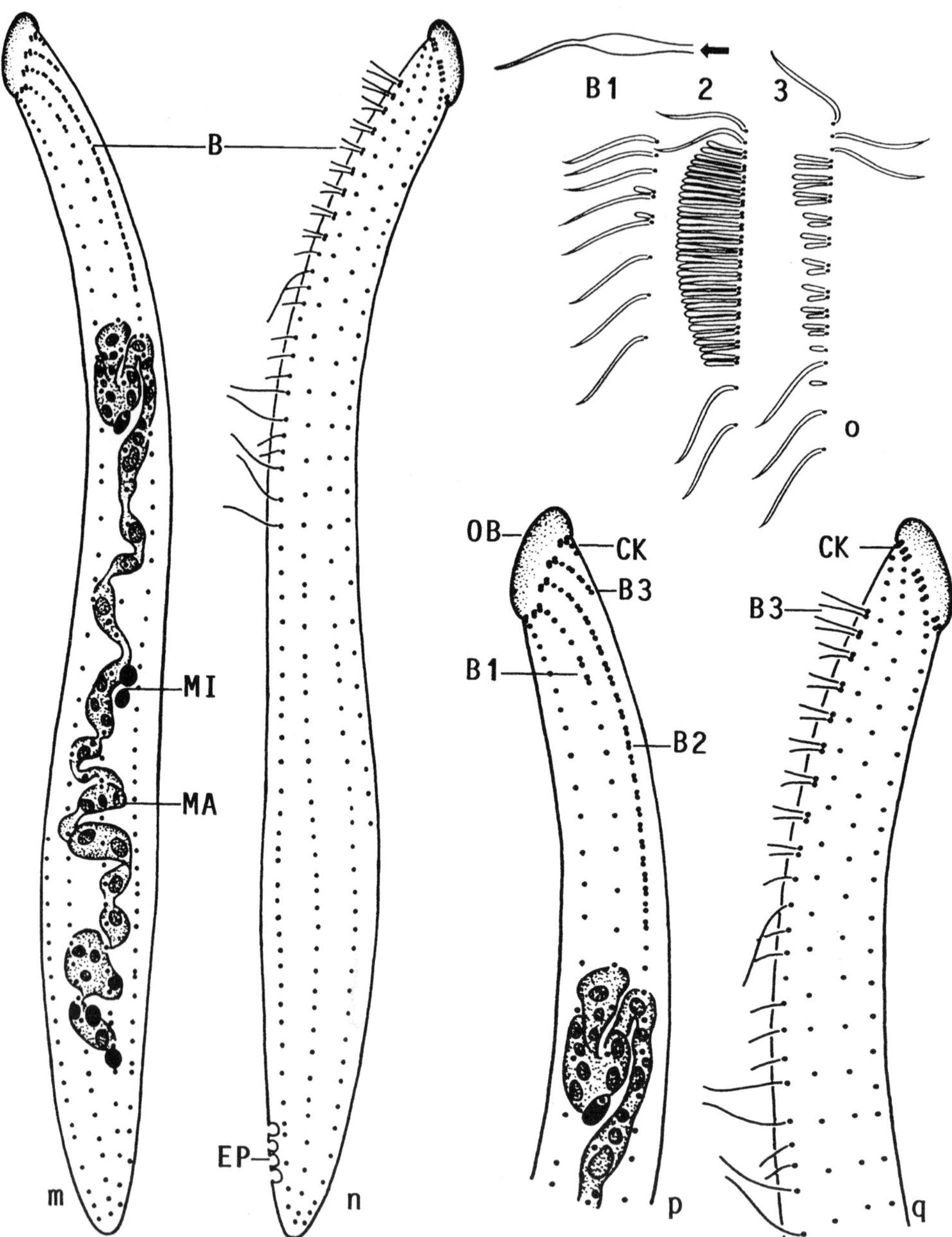

Fig. 34m–q *Edaphospathula paradoxa* nov. spec. after protargol impregnation (m, n, p, q) and from life (o). **m–q)** Ciliary pattern of left and right side and nuclear apparatus of holotype specimen, length 103 µm. The dorsal brush is highly complex (o–q): row 1 consists of only two dikinetids; row 2 has very narrowly spaced bristles up to 5 µm long; row 3 has comparatively widely spaced dikinetids with bristles up to 3 µm long, and a heteromorphic, monokinetidal tail, in which bristles and ordinary cilia irregularly alternate (q). Arrow in (o) marks an impaired bristle assuming a fusiform shape and becoming 8 µm long. Note the subterminal location of the excretory pores (n). B(1-3) – dorsal brush (rows), CK – circumoral kinety composed of small kinetofragments, EP – excretory pores, MA – macronucleus, MI – micronuclei, OB – oral bulge.

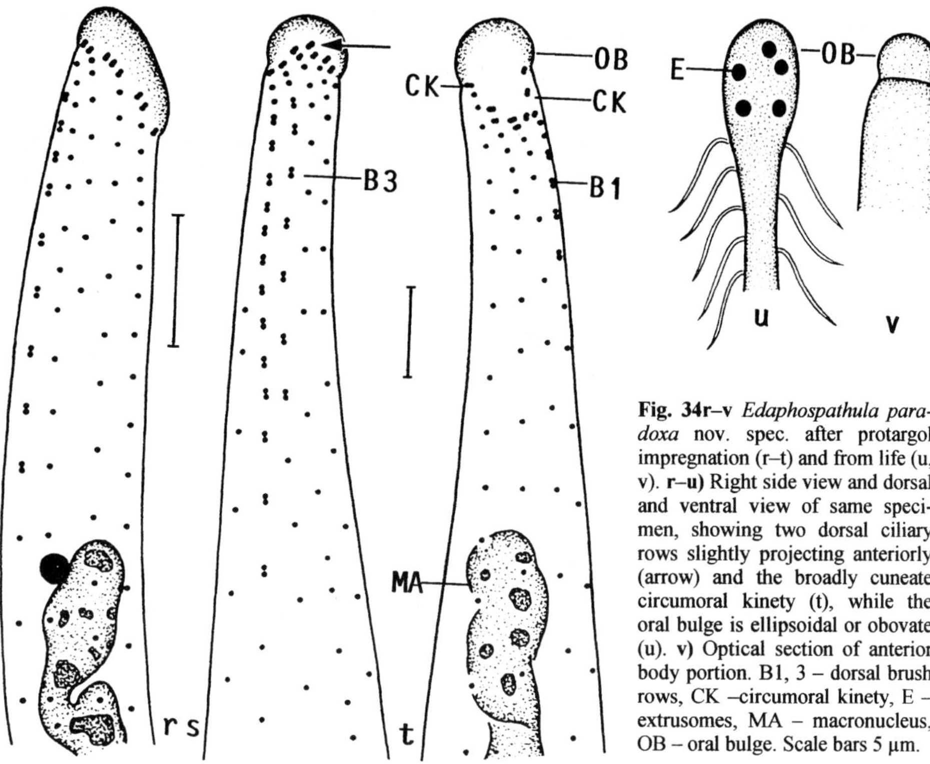

Fig. 34r–v *Edaphospathula paradoxa* nov. spec. after protargol impregnation (r–t) and from life (u, v). **r–u)** Right side view and dorsal and ventral view of same specimen, showing two dorsal ciliary rows slightly projecting anteriorly (arrow) and the broadly cuneate circumoral kinety (t), while the oral bulge is ellipsoidal or obovate (u). **v)** Optical section of anterior body portion. B1, 3 – dorsal brush rows, CK –circumoral kinety, E – extrusomes, MA – macronucleus, OB – oral bulge. Scale bars 5 µm.

above, bristles up to 5 µm long becoming gradually shorter anteriorly and posteriorly. Brush row 3 slightly shorter than row 2, composed of an average of nine comparatively widely spaced dikinetids with bristle length gradually decreasing from 3 µm anteriorly to 1.5–2 µm posteriorly; followed by a heteromorphic, monokinetidal tail extending to mid-body with 1.5 µm long bristles and ordinary cilia irregularly alternating (Fig. 34o).

Oral bulge rather distinct in vivo due to the compact, refractive extrusomes contained, basically, however, fairly inconspicuous because only about 3 µm high, hardly separate from body proper ventrally, and shorter than widest trunk region by about 1/3; slanted by approximately 45° and flat to slightly convex in lateral view, while ellipsoidal or obovate when seen frontally; bulge centre with minute, obconical depression (temporary cytostome) recognizable only in few, well-impregnated and oriented specimens (Fig. 34a, g–j, p, q, u, 107q; Table 2). Individual oral kinetofragments composed of an average of three dikinetids attached to ciliary rows and arranged in typical *Protospathidium* pattern, that is, separated from each other by gaps one to three dikinetids wide on right side; form broadly cuneate circumoral kinety (Fig. 34a, g, h, m, n, p, q, t, 107o–s; Table 2). Oral basket rods and/or oralized somatic monokinetids not recognizable.

Of ten specimens isolated, only one produced a colourless resting cyst 18 µm across, including the about 2.5 µm thick wall (Fig. 34k, l, 107t–x). Ectocyst and endocyst about 2 µm thick and compact, endocyst surface tuberculate, that is, studded with scattered, hyaline, 0.3–1.2 µm high hemispheres covered by a membrane-like structure. Cyst content granular with four large, pale macronuclear masses.

Occurrence and ecology: As yet found only at type locality, where it became rather abundant five weeks after rewetting the sample. See FOISSNER et al. (2005a) for details on site and soil. Briefly, it is nutrient-rich, calcareous clay with pH 7.2. The sample contained a highly diverse ciliate community composed of 120 species, of which several were undescribed, including the present species and *Semibryophyllum palustre*, described in volume II of the monograph.

Remarks: Easily confused with *E. fusioplites* (macronucleus a tortuous strand vs. in about 12 nodules) and *E. espeletiae* (see that species for separation).

Edaphospathula espeletiae (FOISSNER, 2000) **nov. comb.** (Fig. 35a–n, 108a–j; Table 3)

2000 *Sikorops espeletiae* FOISSNER, Stud. Neotrop. Fauna & Environm., 35: 56 (Type slides with protargol-impregnated specimens from type locality are deposited in the Oberösterreichische Landesmuseum in Linz, Upper Austria.).

Diagnosis: Size about 110 × 12 µm in vivo. Cylindroidal to very narrowly ellipsoidal with oblique, inconspicuous oral bulge about half as long as widest trunk region. Macronucleus filiform, tortuous; multimicronucleate. Extrusomes obovate, about 0.8 × 0.6 µm. On average 11 ciliary rows, 3 anteriorly differentiated to distinctly heterostichad dorsal brush occupying 20% of body length. Brush bristles up to 4 µm long, row 1 composed of an average of three dikinetids, row 2 of twenty, and row 3 of thirteen dikinetids followed by a heteromorphic, monokinetidal bristle tail extending to mid-body. Oral kinetofragments each composed of 1 to 2 dikinetids.

Type locality: Venezuela, Cordillera de Mérida, Páramo de Piedras Blancas about 2 km east of the Pico del Aquila (W70°48' N08°52'), on *Espeletia* leaves from dead, rotting *Espeletia* trunks.

Etymology: Genitive of genus *Espeletia*, the plant on which the species was found.

Description: Size in vivo 70 –140 × 8 – 16 µm, usually about 110 × 12 µm, length: width ratio also highly variable, viz., 5:1 – 11:1, on average 8.5:1 both in vivo and in protargol preparations (Table 3). Cylindroidal to very narrowly ellipsoidal, slightly narrowed towards both ends, posterior end tapered in about half of specimens (Fig. 35a, e, f, i, n, 108a, d). Macronucleus in central quarters of cell, filiform, helically coiled in about one third of specimens, in others coiled only in posterior half or tortuous. Usually three globular micronuclei attached to macronucleus in variable positions (Fig. 35a, e, h, n, 108a, d). Contractile vacuole in posterior end with about eight, slightly subterminally located, dorsolateral excretory pores (Fig. 35f). Extrusomes in oral bulge and cytoplasm, narrowly to ordinarily ovate, only 0.7 – 1 × 0.5 – 0.7 µm in size and thus difficult to recognize, do not impregnate with protargol (Fig. 35a, c–e). Cortex flexible, contains about six rows of minute (0.3 µm), colourless granules between each two ciliary rows (Fig. 35m). Cells colourless and hyaline, never dark by food and/or other inclusions, contain some lipid droplets 1 – 3 µm across. Swims rather rapidly by rotation about main body axis, the larger specimens appearing like swimming rods.

Cilia about 8 µm long in vivo, narrowly spaced in oral region of kineties, arranged in an average of eleven meridional and equidistant rows, except in ventral anterior area, where three kineties are close together and separated from the neighbouring ciliary rows by a more or less distinct gap (Fig. 35f, j, 108b; Table 3). Anterior end of kineties slightly curved, bears a single (possibly two in some dorsolateral kineties) dikinetid with the anterior (ciliated?) basal body slightly larger and more intensely impregnated than the (unciliated) posterior one; dikinetid lacking in ventralmost kinety in about half of specimens (Fig. 35i, l, 108b, e–g, i, j). Dorsal brush in anterior region of three dorsolateral kineties, distinctly heterostichad, consists of dikinetids having about 4 µm long, distally slightly inflated bristles associated with the anterior basal bodies and about 3 µm long, rod-shaped bristles with the posterior ones; usually some monokinetids or very narrowly spaced dikinetids at anterior end of each brush row (Fig. 35a, b, g, k, 108c, f, h). Brush row 1 inconspicuous because composed of only three dikinetids on average (Fig. 35k, f; Table 3). Brush row 2 longest because composed of twenty dikinetids on average (Fig. 35g, k, 108c). Brush row 3 shows several specializations: (i) dikinetids more widely spaced than in rows 1 and 2, rows 2 and 3 thus of almost same length, although row 2 has seven dikinetids more than row 3 on average (Table 3); (ii) a monokinetidal, posterior tail with shortened cilia; (iii) anterior basal body of dikinetids

Table 3 Morphometric data on *Edaphospathula espeletiae* (from FOISSNER 2000a)

Characteristics[1]	$\overline{x}$	M	SD	SE	CV	Min	Max	n
Body, length	102.6	105	16.4	3.6	16.0	66	148	21
Body, width underneath oral bulge	4.8	5	0.5	0.1	11.3	4	6	21
Body, maximum width	11.7	12	1.4	0.3	12.4	9	15	21
Macronucleus figure, length	45.6	45	10.2	2.2	22.4	30	75	21
Macronucleus, width	2.8	3	0.6	0.1	21.4	2	4	21
Macronuclei, number	1.0	1	0.0	0.0	0.0	1	1	21
Micronuclei, diameter	1.9	2	–	–	–	1.5	2.5	21
Micronuclei, number	3.1	3	1.0	0.2	30.7	1	5	21
Ciliary rows, number in mid-body	11.1	11	0.9	0.2	7.7	10	13	21
Basal bodies in a right side ciliary row, number	60.1	60	11.4	2.5	19.0	35	88	21
Dorsal brush rows, number	3.0	3	0.0	0.0	0.0	3	3	21
Circumoral kinety to end of brush row 1, distance	3.8	3	1.3	0.3	33.2	3	8	21
Circumoral kinety to end of brush row 2, distance	20.8	21	3.2	0.7	15.4	16	27	21
Circumoral kinety to end of brush row 3, distance	19.3	18	3.6	0.8	18.5	12	25	21
Dikinetids in brush kinety 1, number	3.2	3	11	0.3	35.2	1	5	21
Dikinetids in brush kinety 2, number	19.8	20	2.3	0.5	11.7	15	26	21
Dikinetids in brush kinety 3, number	13.4	13	2.0	0.4	15.0	10	18	21

[1] Data based on mounted, protargol-impregnated (FOISSNER´s method), and randomly selected specimens from a non-flooded Petri dish culture. Measurements in µm. CV – coefficient of variation in %, M – median, Max – maximum, Min – minimum, n – number of individuals investigated, SD – standard deviation, SE – standard deviation of mean, $\overline{x}$ – arithmetic mean.

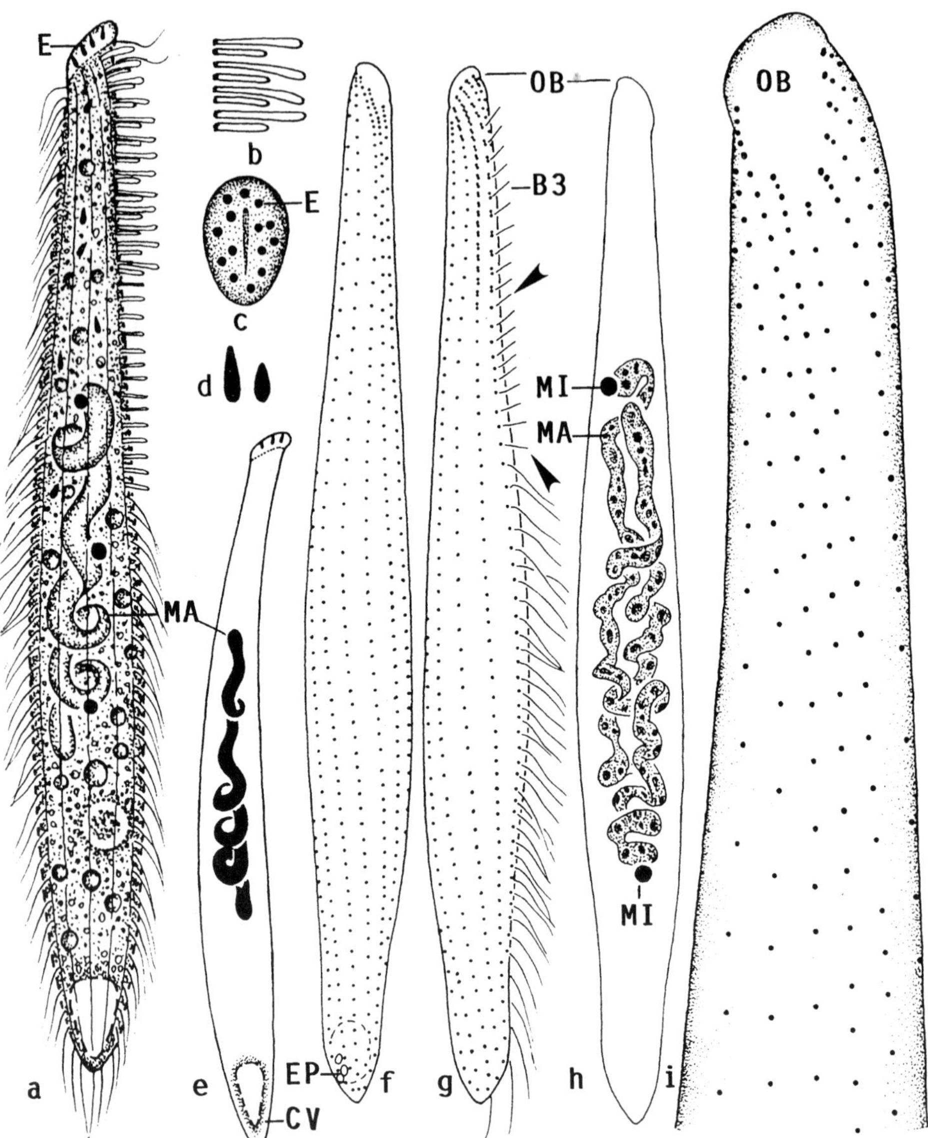

Fig. 35a–i *Edaphospathula espeletiae* from life (a–e) and after protargol impregnation (f–i). From FOISSNER (2000a). **a)** Left side view of a representative specimen, length 110 µm. **b)** Part of dorsal brush with up to 4 µm long bristles. **c)** Frontal view of oral bulge. The knobby extrusomes appear as refractive dots. **d)** Shape and size (0.7–1 × 0.5–0.7 µm) variability of oral bulge extrusomes. **e)** Shape variant with macronucleus coiled in posterior portion. **f–h)** Ciliary pattern of right and left side and nuclear apparatus of holotype specimen, length 110 µm. Note the highly specialized ciliary pattern of dorsal brush row 3 (arrowheads mark monokinetidal bristle tail; underneath the tail, the row has cilia of usual length. **i)** Ventral anterior end at high magnification, width at proximal end of oral bulge 5 µm. Note the inconspicuous dikinetid (anterior basal body slightly enlarged and more darkly impregnated than posterior one) at the anterior end of the ciliary rows (cp. Fig. 30j–l, 108b, e–j). CV – contractile vacuole, B(3) – dorsal brush (row), E – extrusomes in oral bulge, MA – macronucleus, MI – micronuclei, OB – oral bulge, EP – excretory pores of contractile vacuole.

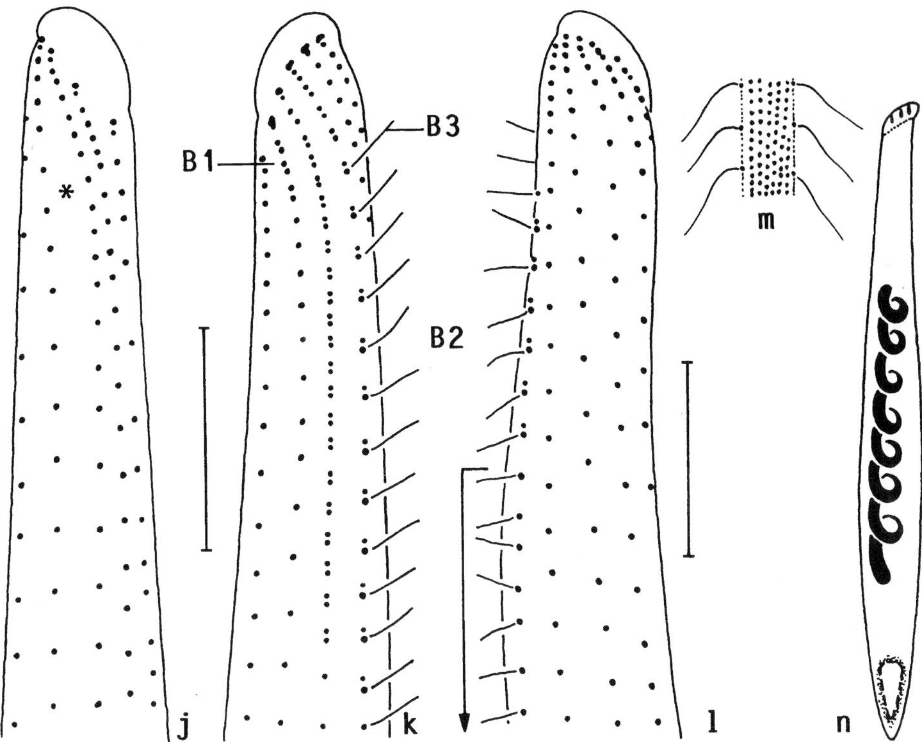

Fig. 35j–n *Edaphospathula espeletiae* from life (m, n) and after protargol impregnation (j – l). From FOISSNER (2000a). **j, k)** Ciliary pattern of right and left anterior portion of holotype specimen shown in figures (f, g). Note details of dorsal brush and indistinct circumoral dikinetid at anterior end of ciliary rows. Usually, the anterior basal body of the dikinetids is slightly larger and more darkly impregnated than the posterior one. Asterisk in figure (j) marks slightly increased distance between ventral and lateral ciliary rows. **l)** Dorsolateral view of anterior portion of another specimen. A heavily impregnated bristle emerges from the posterior basal body of the dikinetids composing brush row 3 and from its monokinetidal tail (arrowed bracket). **m)** Surface view showing cortical granulation. **n)** Shape variant with helically coiled macronucleus. B1-3 – dorsal brush rows. Scale bars 10 µm.

smaller and more faintly impregnated than posterior one; and (iv) posterior bristle of dikinetids and bristles of monokinetidal tail heavily impregnated (Fig. 35g, k, l, 108c, h; Table 3).

Oral bulge inconspicuous and indistinctly separate from body proper, obovate in frontal view, extrusomes contained appear as bright dots (Fig. 35a, c, e–i, 108a, b, e; Table 3). No distinct circumoral kinety because only a single dikinetid at anterior end of somatic kineties, as described above. Nematodesmata not recognizable, even in over-impregnated specimens (Fig. 35f, g, i–l, 108b, e, f, i, j).

Occurrence and ecology: As yet found only at type locality, a very special habitat occurring only in South America.

Remarks: FOISSNER remembers that classification of this species was difficult. Only when we had discovered *Edaphospathula fusioplites*, it became obvious that *Sikorops espeletiae* has the same organization, viz., oral kinetofragments composed of only one to three dikinetids. A close relationship of these populations is indicated also by distinct similarities in size and shape of the body and extrusomes. Thus, we transfer *Sikorops espeletiae* FOISSNER, 2000 to *Edaphospathula*.

Edaphospathula espeletiae differs from *E. fusioplites* by the macronucleus pattern (a long, tortuous strand vs. 12 scattered nodules), the number of kinetids in the somatic kineties (60 vs. 34–44), and the number of dikinetids composing dorsal brush rows 2 (20 vs. 12) and 3 (13 vs. 7). *Edaphospathula paradoxa* differs from *E. espeletiae* by the number of ciliary rows (7 vs. 11), the number of kinetids in the somatic kineties (41vs. 60), the number of dikinetids in brush row 3 (9 vs. 13) and, especially, the number of dikinetids composing the right side oral kinetofragments (usually 3 vs. 1); the last feature is not only important for distinguishing these species, but also for recognizing their generic home: *Edaphospathula*.

This comparison reveals that most differences separating these three taxa are rather sophisticated, and 20 years ago, they likely would not have been recognized as distinct species.

***Edaphospathula inermis* nov. spec.** (Fig. 36a–p; Table 4)

Diagnosis: Size about 90 × 10 µm in vivo. Cylindroidal with slightly narrowed ends and ordinarily oblique oral bulge about half as long as widest trunk region. Macronucleus usually a spiralized rod; usually bimicronucleate. No extrusomes. On average 10 ciliary rows, 3 anteriorly differentiated to distinctly heterostichad dorsal brush occupying 18% of body length. Brush bristles up to 4 µm long, row 1 composed of an average of one dikinetid, row 2 of fifteen, and row 3 of six dikinetids not followed by a monokinetidal bristle tail. Oral kinetofragments each composed of 1 to 2 dikinetids.

Type locality: Grassland soil from the surroundings of the Paiku-Tso lake, South Tibet, about 4700 m above sea level, E85°45' N28°45'.

Etymology: The Latin adjective *inermis* (unarmed) refers to the lacking extrusomes, a main feature of this species.

Description: Size 70–120 × 8–15 µm in vivo, usually near 90 × 10 µm, as calculated from some in vivo measurements and the morphometric data; length:width ratio 7–12.9:1, on average about 9:1 both in vivo and protargol preparations (Table 4). Shape cylindroidal with slightly narrowed ends, anterior quarter frequently slightly curved ventrally; anterior (oral) end ordinarily oblique, posterior usually narrowly rounded to bluntly pointed; widest in or underneath mid-body (Fig. 36a, b, e, h, k). Macronucleus in middle third of cell, usually a spiralized, more or less moniliform rod, rarely unspiralized, coiled, or highly tortuous; furthermore, five out of 50 specimens analysed have two or four large, homogenous nodules, that is, are likely exconjugants; contains many nucleoli up to 4 µm across. Usually two, rarely three micronuclei: one invariably at anterior end of macronucleus, other(s) in variable position near or attached to macronucleus; individual micronuclei globular to ellipsoidal, about 2 µm across in preparations

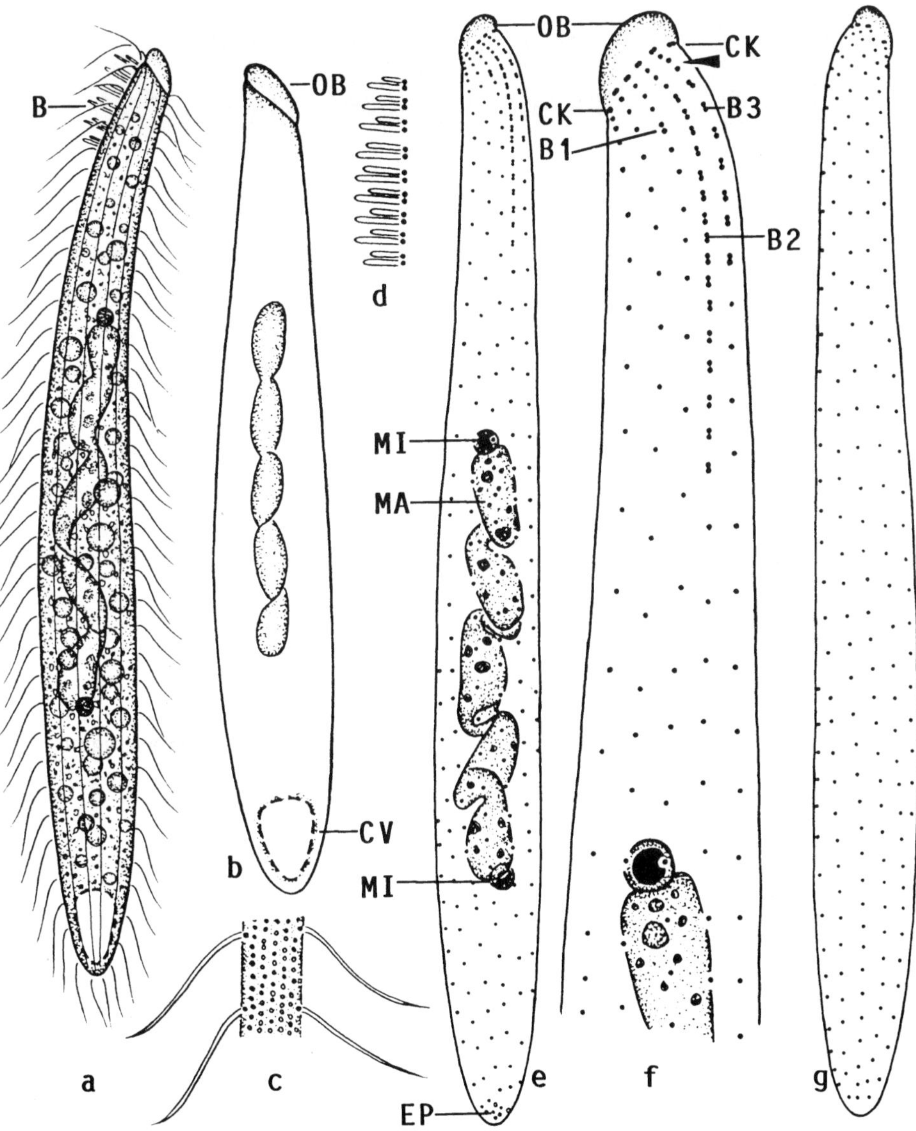

Fig. 36a–g *Edaphospathula inermis* nov. spec. from life (a–d) and after protargol impregnation (e–g). **a)** Right side view of a representative specimen, length 90 µm. Note lack of extrusomes and of a monokinetidal bristle tail in dorsal brush row 3, two outstanding, carefully checked features. **b)** Broad shape variant. **c)** Surface view showing inconspicuous cortical granulation. **d)** Anterior portion of a dorsal brush row. Note that the anterior bristle of the dikinetids is shorter than the posterior one, a specific feature recognizable also in some protargol-impregnated specimens (k, l). **e–g)** Ciliary pattern of left and right side and nuclear apparatus of holotype specimen. Note dominance of dorsal brush row 2, while row 1 consists of only one dikinetid. Arrowhead marks monokinetidal tail of ordinary cilia at anterior end of dorsal brush rows. B(1-3) – dorsal brush (rows), CK – circumoral kinety composed of kinetofragments each consisting of only one dikinetid, CV – contractile vacuole, EP – excretory pores of contractile vacuole, MA – macronucleus, MI – micronuclei, OB – oral bulge.

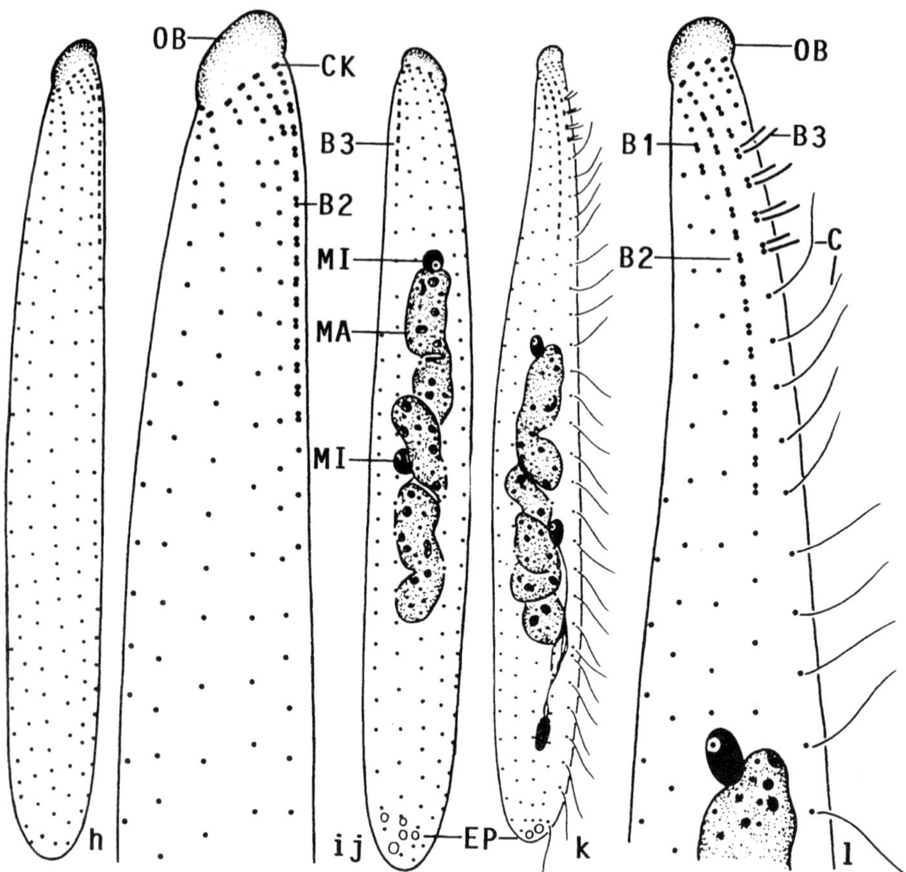

Fig. 36h–l *Edaphospathula inermis* nov. spec., somatic and oral ciliary pattern and nuclear apparatus after protargol impregnation. **h–j)** Left and right side view of a small specimen, length 67 µm. **k, l)** Dorsal view of a 80 µm long specimen with impregnated cilia, showing lack of a monokinetidal bristle tail in brush row 3 and that the anterior bristle of the dikinetids is shorter than the posterior bristle. Note that brush row 1 consists of only two dikinetids. B(1-3) – dorsal brush rows, C – ordinary somatic cilia, CK – circumoral kinety composed of kinetofragments each consisting of only one dikinetid, EP – excretory pores of contractile vacuole, MA – macronucleus, MI – micronuclei, OB – oral bulge.

(Fig. 36a, b, e, j, k). Contractile vacuole in posterior body end, some excretory pores in pole area. Extrusomes definitely not recognizable, even with interference contrast optics and in several, well-looking specimens. Cortex very flexible, contains about five rows of colourless, rather narrowly spaced granules circa 0.5 × 0.2 µm in size between each two ciliary rows. Cytoplasm colourless, contains many lipid droplets up to 3 µm across, thus likely predating on small ciliates and flagellates (Fig. 36a, b). Swims rather rapidly on microscope slide.

Somatic cilia about 8 µm long in vivo, arranged in an average of ten equidistant, bipolar, ordinarily ciliated rows anteriorly slightly curved and connected with inconspicuous oral kinetofragments each consisting of only one to two dikinetids; occasionally with small irregularities, such as minute breaks and/or supernumerary kinetids

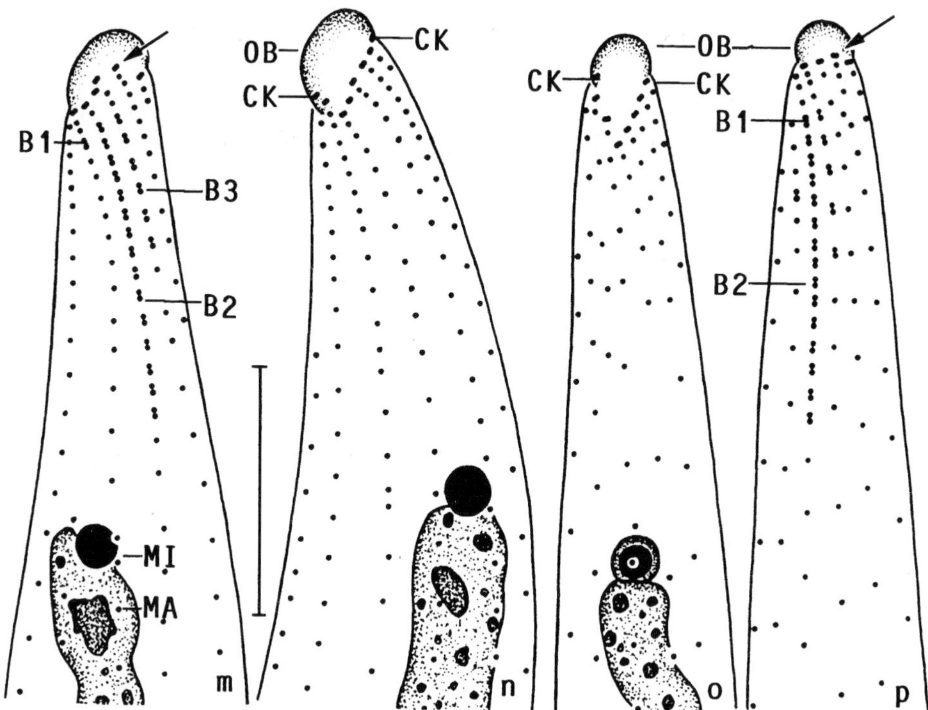

Fig. 36m–p *Edaphospathula inermis* nov. spec., somatic and oral ciliary pattern of anterior body region after protargol impregnation. **m)** Dorsal view showing dorsal brush and slightly projecting ciliary rows (arrow). **n)** Ventrolateral view showing broadly cuneate circumoral kinety. **o, p)** Ventral and dorsal view of same specimen, showing the cuneate circumoral kinety and the dorsal brush with row 1 consisting of only a single dikinetid. Arrow marks slightly projecting ciliary rows. B(1-3) – dorsal brush rows, CK – circumoral kinety composed of kinetofragments each consisting of only a single dikinetid, MA – macronucleus, MI – micronucleus, OB – oral bulge. Scale bar 10 µm for all figures.

outside rows (Fig. 36a, e, g, h, j, k; Table 4). Most dorsal kineties usually slightly elongated anteriorly, viz., extending onto dorsal surface of oral bulge, as in other members of the genus and recognizable only in few, well-oriented specimens (Fig. 36l, m, p). Dorsal brush distinctly heterostichad and three-rowed occupying about 18% of body length; all rows have one to several ordinary cilia anteriorly and continue as ordinary ciliary rows posteriorly. Brush bristles similar in all kineties, that is, rod-shaped with posterior bristle of dikinetids 3–4 µm long and anterior 2–3 µm. Brush row 1 very inconspicuous because composed of only 1–3 dikinetids; longest row 2 composed of an average of 15 rather narrowly spaced dikinetids; row 3 about half as long as row 2, composed of an average of only six comparatively widely spaced dikinetids, lacks a monokinetidal bristle tail, an unusual feature observed both in vivo and in some properly impregnated specimens (Fig. 36d, f, i, k, l, m, p; Table 4).

Oral bulge inconspicuous because indistinctly separate from body proper, hyaline due to the lack of extrusomes, only half as long as widest trunk region, and merely 2–3 µm high; slanted by about 45°, flat to distinctly convex, obovate when seen frontally (Fig. 36a, b, f, i, l, m–p; Table 4). Individual kinetofragments very indistinct because

Table 4 Morphometric data on *Edaphospathula inermis* (upper line) and *E. minor* (lower line; from FOISSNER et al. 2002)

Characteristics[1]	$\overline{x}$	M	SD	SE	CV	Min	Max	n
Body, length	82.7	80.0	10.4	2.3	12.6	66.0	103.0	21
	50.9	51.0	4.6	1.0	9.0	43.0	61.0	21
Body, width	8.8	9.0	1.4	0.3	16.0	7.0	13.0	21
	12.8	12.0	1.8	0.4	13.9	10.0	17.0	21
Body length:width, ratio	9.6	9.5	1.6	0.3	16.5	7.0	12.9	21
	4.0	4.1	0.6	0.1	15.9	3.0	5.5	21
Oral bulge, length (cell oriented laterally)	3.7	4.0	0.6	0.1	15.6	3.0	5.0	21
	3.4	3.5	–	–	–	3.0	4.0	21
Oral bulge, width (cell oriented ventrally)	2.7	3.0	0.3	0.1	12.6	2.0	3.0	11
	–	–	–	–	–	–	–	–
Oral bulge, height	1.9	2.0	0.2	–	10.6	1.5	2.0	21
	–	–	–	–	–	–	–	–
Circumoral kinety to last dikinetid of brush row 1, distance	2.7	3.0	0.8	0.2	29.8	2.0	5.0	21
	5.1	5.0	0.8	0.2	15.1	4.0	7.0	21
Circumoral kinety to last dikinetid of brush row 2, distance	14.6	14.0	1.7	0.4	11.5	11.0	17.0	21
	11.5	12.0	1.6	0.4	13.9	9.0	15.0	21
Circumoral kinety to last dikinetid of brush row 3, distance	8.2	8.0	0.9	0.2	10.7	7.0	10.0	21
	5.5	5.0	1.0	0.2	18.7	4.0	9.0	21
Anterior body end to nucleus	26.2	25.0	7.1	1.5	27.0	16.0	42.0	21
	25.5	25.0	3.0	0.7	11.7	20.0	31.0	21
Macronucleus figure, length	33.1	32.0	6.4	1.4	19.2	23.0	48.0	21
	12.1	12.0	1.4	0.3	11.6	10.0	14.0	21
Macronucleus, width	3.5	3.0	0.6	0.1	18.3	2.5	5.0	21
	5.7	6.0	0.6	0.1	10.2	5.0	7.0	21
Macronucleus, number	1.0	1.0	0.0	0.0	0.0	1.0	1.0	21
	1.0	1.0	0.0	0.0	0.0	1.0	1.0	21
Micronuclei, across	2.0	2.0	–	–	–	2.0	2.5	21
	3.0	3.0	–	–	–	2.5	3.5	21
Micronuclei, number	2.1	2.0	0.4	0.1	16.7	2.0	3.0	21
	1.0	1.0	0.0	0.0	0.0	1.0	1.0	21
Circumoral dikinetids, number	9.9	10.0	0.9	0.3	9.1	8.0	11.0	7
	–	–	–	–	–	–	–	–
Ciliary rows, number in mid-body	10.0	10.0	0.6	0.1	5.9	8.0	11.0	21
	7.1	7.0	–	–	–	7.0	8.0	21
Basal bodies in a right side ciliary row, number	41.0	40.0	9.0	2.0	21.9	24.0	64.0	21
	20.7	21.0	2.8	0.6	13.5	15.0	25.0	21
Dorsal brush rows, number	3.0	3.0	0.0	0.0	0.0	3.0	3.0	21
	3.0	3.0	0.0	0.0	0.0	3.0	3.0	21
Dikinetids in brush row 1, number	1.4	1.0	0.6	0.1	41.8	1.0	3.0	21
	3.8	4.0	0.8	0.2	20.4	3.0	6.0	21
Dikinetids in brush row 2, number	14.9	15.0	1.7	0.4	11.4	11.0	17.0	21
	10.0	10.0	1.3	0.3	12.7	8.0	13.0	21
Dikinetids in brush row 3, number	6.1	6.0	1.0	0.2	16.3	4.0	7.0	21
	4.7	5.0	0.9	0.2	19.6	3.0	7.0	21

[1] Data based on mounted, protargol-impregnated (FOISSNER's method), and randomly selected specimens from non-flooded Petri dish cultures. Measurements in µm. CV – coefficient of variation in %, M – median, Max – maximum, Min – minimum, n – number of individuals investigated, SD – standard deviation, SE – standard error of arithmetic mean, $\overline{x}$ – arithmetic mean.

composed of usually one, rarely two dikinetids, forming a cuneate circumoral kinety comprising an average of only 10 dikinetids. Nematodesmata not recognizable, even in over-impregnated specimens (Fig. 36f, i–p; Table 4).

Occurrence and ecology: As yet found only at type locality, where it was rare in the non-flooded Petri dish culture. The sample, kindly provided by Dr. NORBERT WINDING (Haus der Natur, Salzburg), was taken in a high mountain area covered with vegetation by only about 40%, viz., around the entrance to a colony of piping hares, and consisted of humic surface soil (0–7 cm) with pH 8.1 (in water).

Remarks: *Edaphospathula inermis* differs from all congeners and most other spathidiids by the lack of extrusomes, which we checked very carefully and is thus the main distinguishing feature. Otherwise it is highly similar to *E. espeletiae*, except for dorsal brush row 3, which lacks (vs. present) the monokinetidal bristle tail and is composed of only 6 (vs. 13) dikinetids. In vivo and preparations, *E. inermis* also highly resembles *P. serpens*, except for the extrusomes (absent vs. present), the number of dikinetids composing the oral kinetofragments (1–2 vs. 3–5), and dorsal brush row 2 (15 vs. usually 7–10).

Edaphospathula minor (FOISSNER, AGATHA & BERGER, 2002) **nov. comb.** (Fig. 37a–l; Table 4)

2002 *Sikorops minor* FOISSNER, AGATHA & BERGER, Denisia, 5: 209 (Type slides with protargol-impregnated specimens from type locality are deposited in the Oberösterreichische Landesmuseum in Linz, Upper Austria.).

Diagnosis: Size about 65 × 15 µm in vivo. Bottle-shaped to elongate ovoidal with ordinarily oblique, about 5 µm wide oral bulge. Macronucleus ellipsoidal; single micronucleus. Extrusomes bluntly fusiform, about 1.5 × 1 µm. On average 7 ciliary rows, 3 anteriorly differentiated to distinctly heterostichad dorsal brush occupying 23% of body length. Brush bristles up to 3 µm long, row 1 composed of an average of five dikinetids, row 2 of ten, and row 3 of five dikinetids followed by a monokinetidal bristle tail extending to mid-body.

Type locality: Highly saline, alkaline soil near the village of Himmafushi, North-Male Atoll, Maldives, E74° N3°.

Etymology: The Latin adjective *minor* (small) indicates that it is the smallest described species of the genus.

Description: Size 50–80 × 10–20 µm in vivo, usually about 65 × 15 µm; shrunken to 51 × 13 µm in protargol preparations showing a length:width ratio of 3–5.5:1, on average 4:1 both in vivo and in preparations (Table 4). Lateral view indistinctly spatulate, obclavate, bottle-shaped or almost cylindroidal, dorsal and ventral view obclavate to elongate ovoidal with posterior end moderately broadly rounded and bluntly pointed

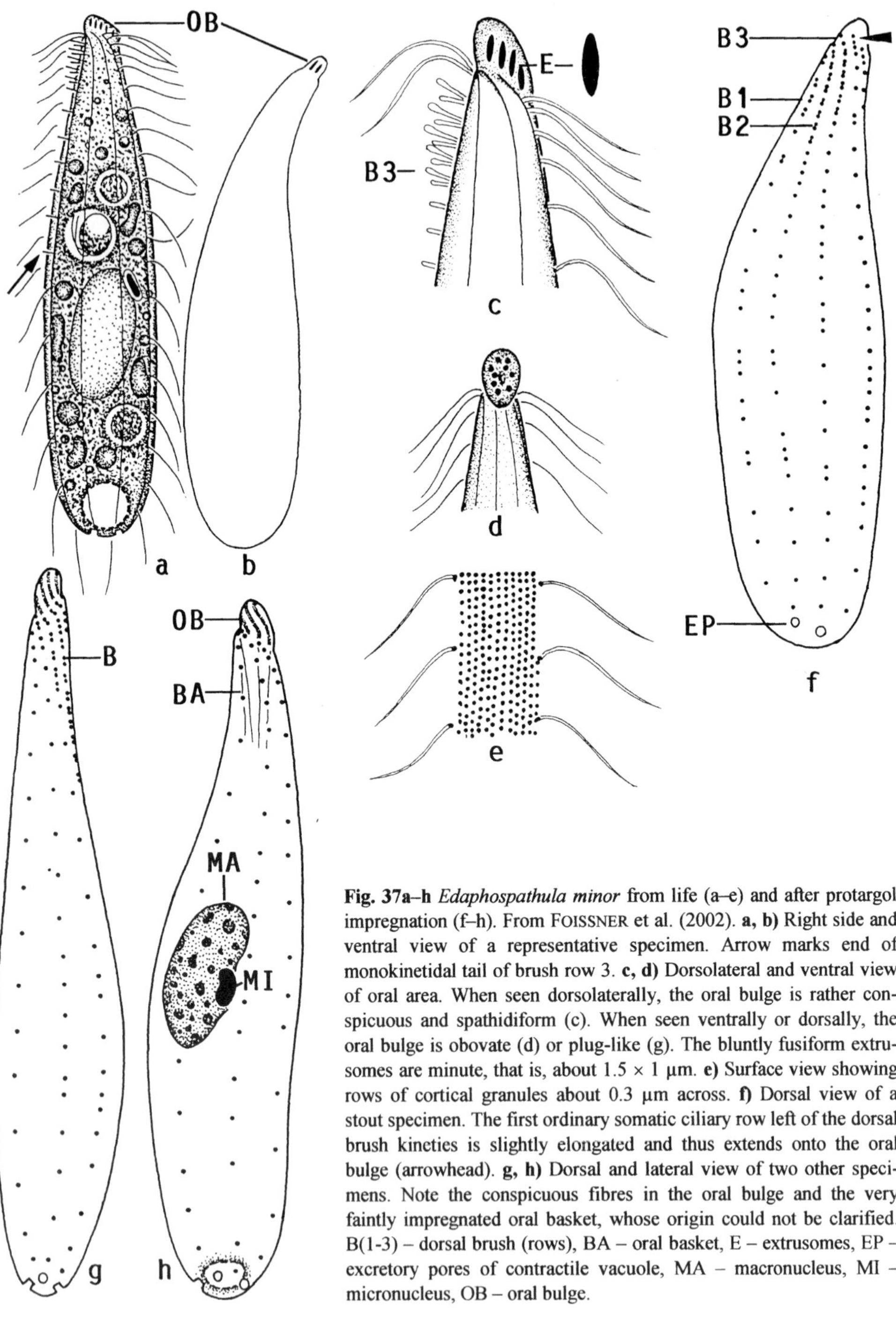

Fig. 37a–h *Edaphospathula minor* from life (a–e) and after protargol impregnation (f–h). From FOISSNER et al. (2002). **a, b)** Right side and ventral view of a representative specimen. Arrow marks end of monokinetidal tail of brush row 3. **c, d)** Dorsolateral and ventral view of oral area. When seen dorsolaterally, the oral bulge is rather conspicuous and spathidiform (c). When seen ventrally or dorsally, the oral bulge is obovate (d) or plug-like (g). The bluntly fusiform extrusomes are minute, that is, about 1.5 × 1 µm. **e)** Surface view showing rows of cortical granules about 0.3 µm across. **f)** Dorsal view of a stout specimen. The first ordinary somatic ciliary row left of the dorsal brush kineties is slightly elongated and thus extends onto the oral bulge (arrowhead). **g, h)** Dorsal and lateral view of two other specimens. Note the conspicuous fibres in the oral bulge and the very faintly impregnated oral basket, whose origin could not be clarified. B(1-3) – dorsal brush (rows), BA – oral basket, E – extrusomes, EP – excretory pores of contractile vacuole, MA – macronucleus, MI – micronucleus, OB – oral bulge.

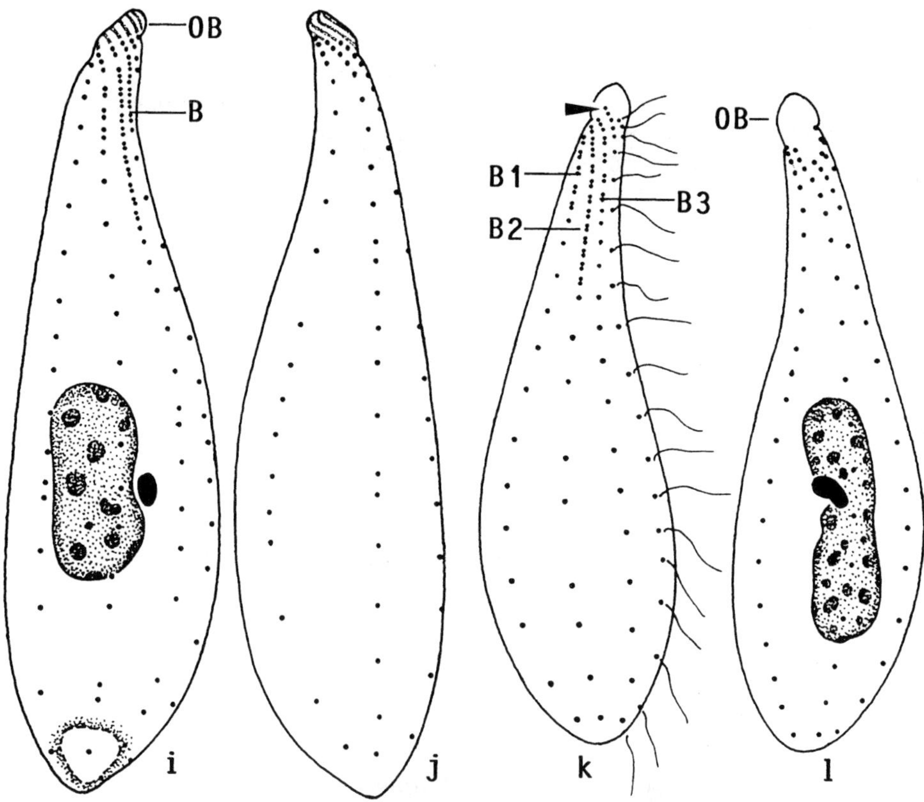

Fig. 37i–l *Edaphospathula minor*, oral and somatic ciliary pattern and nuclear apparatus after protargol impregnation (from FOISSNER et al. 2002). **i, j)** Left and right side view of holotype specimen. Note the low number (7) of ciliary rows and the distinct fibres in the oral bulge. **k, l)** Dorsal and ventral view of a small, obclavate specimen. The first ordinary somatic ciliary row (arrowhead) left of the dorsal brush kineties is slightly elongated and thus extends onto the oral bulge, which is conspicuously plug-shaped in this species, if it is viewed ventrally or dorsally. Brush row 2 is distinctly longer than rows 1 and 3. There is no distinct circumoral kinety recognizable, and the first basal body of each row is ciliated. B(1-3) – dorsal brush rows, OB – oral bulge.

anterior end flattened up to 2:1 (Fig. 37a–d, g, i, k). Macronucleus slightly below mid-body, broadly to elongate ellipsoidal, rarely distinctly reniform; nucleoli small and globular. Micronucleus attached to macronucleus at variable positions, discoidal because flattened about 2:1, broad view outline circular to broadly elliptical. Contractile vacuole in posterior body end, some excretory pores in pole area. Extrusomes only in oral bulge, bluntly fusiform, minute, that is, about 1.5 × 1 μm and thus difficult to recognize, do not impregnate with the protargol method used. Cortex very flexible, contains about 12 rows of colourless, refractive granules approximately 0.3 μm across between each two kineties. Cytoplasm colourless, in middle third often packed with food vacuoles as well as globular and irregular fat inclusions up to 6 μm across. Feeds on heterotrophic flagellates and likely also small ciliates, both rapidly digested because prey organisms could be not identified in the food vacuoles of protargol-impregnated cells. Swims slowly by rotation about main body axis.

Cilia about 8 µm long in vivo, densely spaced in narrowed oral region, in some specimens rather irregularly arranged due to interspersed, unciliated kinetids; arranged in an average of seven ordinarily spaced (6 µm), meridionally extending and anteriorly slightly curved rows with first row left of dorsal brush invariably slightly elongated and thus extending onto dorsal surface of oral bulge. Dorsal brush in anterior region of three dorsolateral kineties, distinctly heterostichad and of ordinary structure, that is, occupies 23% of body length and consists of ordinarily spaced dikinetids with up to 3 µm long bristles; two to three monokinetids or very closely spaced dikinetids at anterior end of individual brush rows. Brush rows 1 and 3 of almost same length each consisting of an average of four to five dikinetids, row 3 associated with a monokinetidal bristle tail extending to mid-body; row 2 about twice as long as rows 1 and 3, consists of ten dikinetids on average. All brush rows continue as ordinary somatic kineties posteriorly (Fig. 37a, f–l; Table 4).

Oral bulge inconspicuous because minute, that is, about 5 × 4 µm in vivo; basically spathidiform and slanted from dorsal to ventral when cell is viewed laterally, while flattened and thus projecting plug-like when viewed ventrally or dorsally. No distinct circumoral kinety, that is, ciliary rows simply end at base of oral bulge and the anteriormost basal body of each row is ciliated and associated with a conspicuous fibre spiralling to bulge centre. Oral basket rods recognizable only in over-impregnated specimens, rod origin remained unclear (Fig. 37a, d, f–l; Table 4).

Occurrence and ecology: The highly saline and alkaline (pH 8.6) sample from the type locality, kindly provided by Dr. WOLFGANG PETZ, contained much litter and some sandy soil collected under shrubs about 2 m inshore. In Namibia, *E. minor* occurred only at site (56), that is, in non-saline soil (FOISSNER et al. 2002). Found also in Cedar Creak, Utah, USA. Thus, *E. minor* is likely a euryhaline cosmopolitan.

Generic assignment and comparison with related species: Although oral dikinetids and oralized somatic kinetids could be not identified unequivocally, FOISSNER et al. (2002) classified this species in *Sikorops* because of the similar general organization, the blunt extrusomes, and the considerable similarity with *Sikorops namibiensis* FOISSNER et al., 2002. With the new knowledge available, it is likely that *Sikorops minus* belongs to *Edaphospathula*, especially because of the projecting ciliary rows on the dorsal side (Fig. 37f, k). However, we cannot exclude that the Maldivean species lacks oral dikinetids; if so, it would belong to the Enchelyina FOISSNER & FOISSNER, 1988a.

Generally, *E. minor* is an inconspicuous species with few distinct features. Thus, it is difficult to identify and to compare with older, often very incomplete descriptions of seemingly similar species. There are three *Spathidium* species in the literature which have some resemblance to *E. minor*: *Protospathidium vermiculus* (macronucleus oblong vs. ellipsoidal; several vs. single micronucleus; extrusomes rod-shaped and 3–4 µm long vs. bluntly fusiform and 1.5 µm long); *S. claviforme* (with distinct, dikinetidal circumoral kinety and 12 ciliary rows; FOISSNER 1987c); and *S. microstomum* (only 35–60 µm long and distinctly narrowed posteriorly providing the species with an entirely different shape). As concerns the congeners, only *E. brachycaryon* and *E. gracilis* have the same nuclear pattern; however, both are very slender (length:width ratio usually > 7:1 vs. 4:1) and are thus easily distinguished from the blunt *E. minor*. Furthermore, some *Lagynophrya* species look similar, but the sole species so far investigated in detail has a distinct, dikinetidal circumoral kinety (FOISSNER et al. 1999).

Edaphospathula brachycaryon **nov. spec.** (Fig. 38a–k, 109a–e; Table 5)

Diagnosis: Size about 100 × 12 μm in vivo. Elongate obclavate to cylindroidal with ordinarily oblique, obovate oral bulge about half as long as widest trunk region. Macronucleus oblong (~ 5:1); single micronucleus. Extrusomes ampulliform, circa 1.5 × 0.8 μm. On average 8 ciliary rows, 3 anteriorly differentiated to distinctly heterostichad dorsal brush with few, but unusually widely spaced (1.8 μm on average) dikinetids and thus occupying almost 1/3 of body length. Brush bristles up to 4 μm long, row 1 composed of an average of ten dikinetids, row 2 of fifteen, and row 3 of six dikinetids followed by a monokinetidal bristle tail extending to mid-body. Right side oral kinetofragments each usually composed of 1–2 dikinetids.

Type locality: Soil of reed-mace swamp at the bank of the Kanab Creek near the town of Kanab, Utah, USA, W112°30' N37°.

Etymology: Greek apposite noun composed of *brachy* (short) and *caryon* (nucleus), referring to the short macronucleus, a main feature of the species.

Description: Size 75–135 × 9–15 μm in vivo, usually near 100 × 12 μm, as calculated from some in vivo measurements and the morphometric data; length:width ratio 5.7–12.2:1, on average 9:1 both in vivo and in protargol preparations (Table 5). Shape elongate obclavate to cylindroidal with slightly narrowed ends, often somewhat curved; anterior (oral) end ordinarily oblique, posterior narrowly rounded; hardly flattened laterally (Fig. 38a, d, j, 109a–c). Macronucleus underneath mid-body, usually oblong and more or less distinctly curved, rarely semicircular or almost circular; contains many lobate and granular nucleoli. Invariably a single micronucleus attached to macronucleus in variable position (Fig. 38a, j, 109a–c; Table 5). Contractile vacuole in posterior body end, some excretory pores in terminal and subterminal pole area. Extrusomes attached with narrowed end to oral bulge, ampulliform, difficult to recognize because only about 1.5 × 0.8 μm in size (Fig. 38a–c); do not impregnate with the protargol method used. Cortex very flexible, contains about five rows of colourless, minute (≤ 0.3 μm) granules between each two ciliary rows. Cytoplasm colourless, contains some globular and irregular lipid inclusions and, occasionally, small oral baskets of microthoracid ciliates or euglenoid flagellates, obviously the preferred food. Movement without peculiarities.

Somatic cilia about 8 μm long in vivo, arranged in an average of eight equidistant, bipolar, ordinarily ciliated (average ciliary distance 3 μm in preparations) rows anteriorly slightly curved and connected with the inconspicuous oral kinetofragments; rarely with small irregularities, such as incomplete rows or an extra dikinetid right of brush row 1; no anteriorly projecting kineties (Fig. 38a, j, k, 109a, b, e; Table 5). Dorsal brush distinctly heterostichad, unusual due to the wide spacing of the dikinetids (about 1.8 μm in preparations, Table 5), making rows conspicuously long (row 2 on average 31% of body length) and distinctly different from those of *E. gracilis*, where the rows are short (10%) because composed of ordinarily spaced dikinetids; all rows

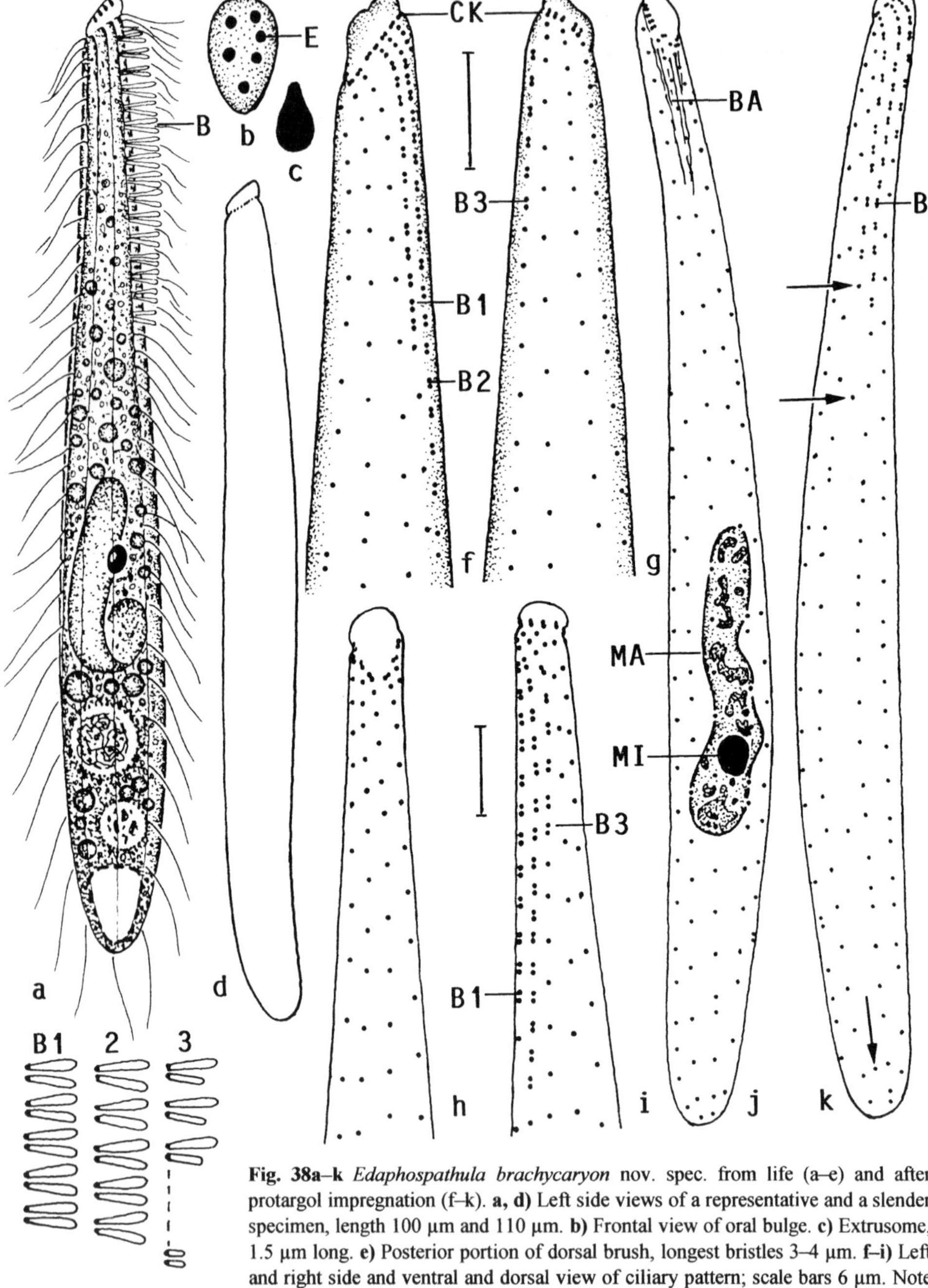

Fig. 38a–k *Edaphospathula brachycaryon* nov. spec. from life (a–e) and after protargol impregnation (f–k). **a, d)** Left side views of a representative and a slender specimen, length 100 μm and 110 μm. **b)** Frontal view of oral bulge. **c)** Extrusome, 1.5 μm long. **e)** Posterior portion of dorsal brush, longest bristles 3–4 μm. **f–i)** Left and right side and ventral and dorsal view of ciliary pattern; scale bars 6 μm. Note the widely spaced dikinetids of the dorsal brush. **j, k)** Holotype specimen, length 94 μm. Arrows mark incomplete kinety. B(1-3) – dorsal brush rows, BA – oral basket, CK – circumoral kinety, MA – macronucleus, MI – micronucleus, OB – oral bulge.

with some ordinary cilia anteriorly and continuing as somatic kineties posteriorly. Brush bristles similar in all rows, that is, slightly clavate and 3–4 µm long; anterior and posterior bristle of same length in row 1 and row 2 dikinetids, posterior bristle slightly shortened in row 3 dikinetids; monokinetidal tail of row 3 extends to mid-body (Fig. 38a, e–g, i, k, 109c–e; Table 5).

Oral bulge inconspicuous because minute, that is, only half as long as widest trunk region and 2–3 µm high; slanted by about 30°–50°, slightly convex or concave; obovate in frontal view, where extrusomes appear as bright dots (Fig. 38a, b, d, f, g, j). Individual oral kinetofragments indistinct because composed of usually only one, rarely two or three dikinetids forming a broadly cuneate, inconspicuous circumoral kinety. Nematodesmata rarely recognizable, originate from circumoral kinetofragments and form an inconspicuous oral basket (Fig. 38a, f–k, 109a, c, d; Table 5).

Occurrence and ecology: As yet found only at type locality, where it was rare in the non-flooded Petri dish culture (pH 7.2) set up with blackish soil mixed with many fine roots and some leaf litter. Whether *E. brachycaryon* is a limnetic or terrestrial species cannot be decided because of the swampy habitat; the slender shape indicates preference of soil environment.

Remarks: *Edaphospathula brachycaryon* is easily distinguished from most congeners and *Protospathidium* species by the simple nuclear apparatus and/or the slender shape and/or the ampulliform extrusomes. However, it is highly similar to *E. gracilis*, differing from that species by body size (about 100 × 12 µm vs. 160 × 13 µm in vivo), the oral bulge length (about 50% vs. 80% as long as widest trunk region), the extrusome shape and size (ampulliform and 1.5 µm long vs. bluntly fusiform and 3 µm long), and details of the dorsal brush (bristle pairs unusually widely vs. ordinarily spaced; special bristles absent vs. present in row 3). Further, *E. brachycaryon* and *E. gracilis* are almost indistinguishable from *Latispathidium similis* in vivo (Vol. II). Thus, and because likely further such species exist, identification should be checked in protargol preparations; in vivo observation is indispensable for the shape and size of the extrusomes.

Table 5 Morphometric data on *Edaphospathula gracilis* (upper line) and *E. brachycaryon* (lower line)

Characteristics[1]	$\overline{x}$	M	SD	SE	CV	Min	Max	n
Body, length	144.2	140.0	23.2	5.1	16.1	95.0	190.0	21
	89.4	85.0	14.3	3.3	16.0	69.0	122.0	19
Body, width	12.0	12.0	1.6	0.3	13.1	10.0	16.0	21
	10.2	10.0	1.6	0.4	15.5	8.0	14.0	19
Body length:width, ratio	12.2	12.3	2.2	0.5	18.2	8.2	15.5	21
	9.0	9.1	1.9	0.4	21.1	5.7	12.2	19
Oral bulge, length	9.3	9.0	1.4	0.3	15.3	7.0	12.0	21
	4.6	4.5	0.5	0.1	11.3	4.0	6.0	19
Oral bulge length:body width, ratio	0.8	0.8	0.1	–	16.2	0.5	1.0	21
	0.5	0.5	0.1	–	18.5	0.3	0.6	19
Oral bulge (circumoral kinety), width	3.5	3.5	0.5	0.2	14.3	3.0	4.0	7
	–	–	–	–	–	–	–	–
Oral bulge, height	2.1	2.0	0.4	0.1	17.9	1.5	3.0	21
	2.7	3.0	0.7	0.2	24.2	1.5	3.5	8
Circumoral kinety to last dikinetid of brush row 1, distance	12.0	12.0	2.0	0.4	16.6	8.0	16.0	21

continued

Characteristics[1]	$\overline{x}$	M	SD	SE	CV	Min	Max	n
	19.3	19.0	3.1	0.7	16.0	14.0	27.0	19
Circumoral kinety to last dikinetid of brush row 2, distance	13.4	14.0	2.1	0.5	15.6	9.0	17.0	21
	27.8	28.0	4.0	0.9	14.5	20.0	35.0	19
Circumoral kinety to last dikinetid of brush row 3, distance	9.3	9.0	1.6	0.3	16.7	6.0	13.0	21
	10.8	11.0	1.8	0.4	16.8	7.0	14.0	19
Anterior body end to macronucleus	57.5	60.0	11.8	2.6	20.5	31.0	80.0	21
	44.7	42.0	7.0	1.6	15.7	34.0	58.0	19
Macronucleus figure, length (*E. gracilis*)	25.9	25.0	6.5	1.4	25.1	13.0	42.0	21
Macronucleus, length (spread; *E. brachycaryon*)	21.9	20.0	4.9	1.1	22.3	15.0	30.0	19
Macronucleus, width in mid	6.3	6.0	1.0	0.2	15.3	5.0	8.0	21
	4.8	5.0	0.8	0.2	16.6	4.0	6.0	19
Macronucleus, number	1.0	1.0	0.0	0.0	0.0	1.0	1.0	21
	1.0	1.0	0.0	0.0	0.0	1.0	1.0	19
Micronucleus, length	3.0	3.0	0.3	0.1	9.1	2.5	4.0	21
	3.0	3.0	0.3	0.1	8.5	2.5	3.5	19
Micronucleus, width	2.2	2.0	0.6	0.1	25.3	1.5	3.0	21
	2.6	2.5	0.4	0.1	14.9	2.0	3.0	19
Micronucleus, number	1.0	1.0	0.0	0.0	0.0	1.0	1.0	21
	1.0	1.0	0.0	0.0	0.0	1.0	1.0	19
Dikinetids composing a right side circumoral kinetofragment, number	2.9	3.0	0.5	0.1	18.6	2.0	4.0	21
	1.5	1.0	–	–	–	1.0	3.0	19
Somatic ciliary rows, number	9.0	9.0	0.4	0.1	5.0	8.0	10.0	21
	8.3	8.0	0.5	0.1	5.5	8.0	9.0	19
Basal bodies in a right side ciliary row, number	41.0	40.0	9.0	2.0	21.9	24.0	64.0	21
	30.2	30.0	3.9	0.9	13.0	23.0	40.0	19
Dorsal brush rows, number	3.0	3.0	0.0	0.0	0.0	3.0	3.0	21
	3.0	3.0	0.0	0.0	0.0	1.0	1.0	19
Dikinetids in brush row 1, number	10.0	10.0	1.9	0.4	18.7	7.0	14.0	21
	10.2	10.0	1.2	0.3	11.5	8.0	12.0	19
Dikinetids in brush row 2, number	11.4	11.0	1.7	0.4	14.8	9.0	16.0	21
	15.4	15.0	1.4	0.3	9.3	13.0	18.0	19
Dikinetids in brush row 3, number	7.9	8.0	1.1	0.2	13.8	6.0	10.0	21
	5.8	6.0	1.0	0.2	17.8	4.0	8.0	19

[1] Data based on mounted, protargol-impregnated (FOISSNER's method), and randomly selected specimens from non-flooded Petri dish cultures; obviously inflated specimens were discarded. Measurements in µm. CV – coefficient of variation in %, M – median, Max – maximum, Min – minimum, n – number of individuals investigated, SD – standard deviation, SE – standard error of arithmetic mean, $\overline{x}$ – arithmetic mean.

***Edaphospathula gracilis* nov. spec.** (Fig. 39a–p, 109f–k; Table 5)

Diagnosis: Size about 160 × 13 µm in vivo. Cylindroidal with ordinarily oblique oral bulge about 80% as long as widest trunk region. Macronucleus oblong (~ 4:1); single micronucleus. Extrusomes bluntly fusiform, about 3 × 0.8 µm in size. On average 9 ciliary rows, 3 anteriorly differentiated to heterostichad dorsal brush with ordinarily spaced dikinetids occupying about 10% of body length. Brush bristles up to 4 µm long, row 1 composed of an average of ten dikinetids; row 2 of eleven; and row 3 of eight; both, rows 2 and 3 have a monokinetidal bristle tail extending to second third of body. Right side oral kinetofragments each composed of 2 to 4, usually 3 dikinetids.

Type locality: Artificial (?) soil from lawn of a hotel in the village of Sharm el Sheik, Sinai, Egypt, E34° N27°.

Etymology: The Latin adjective *gracilis* (gracile) refers to the slender body of the species.

Description: Size 110–210 × 10–15 µm in vivo, usually near 160 × 13 µm, as calculated from some in vivo measurements and the morphometric data; length:width ratio 8.2–15.5:1, usually near 12:1 both in vivo and protargol preparations (Table 5). Very narrowly spatulate to vermiform, usually cylindroidal with slightly narrowed ends, often somewhat curved anteriorly; anterior (oral) end ordinarily oblique, posterior narrowly rounded or bulbous due to the contractile vacuole contained; widest in or underneath mid-body, occasionally distinctly inflated in preparations; hardly flattened laterally (Fig. 39a, e–j, 109k). Macronucleus in mid-body and usually oblong (~ 4:1), more or less curved in seven out of 58 specimens analysed, and three individuals have a short, slightly nodulated and twisted strand; contains many nucleoli 1–2 µm across. Invariably a single micronucleus attached to mid-portion, rarely to anterior end of macronucleus (Fig. 39a, e–j). Contractile vacuole in posterior body end, some excretory pores in pole area. Oral bulge extrusomes comparatively distinct in vivo because bluntly fusiform and about 3 × 0.6–0.8 µm in size (Fig. 39a, b); do not impregnate with the protargol method used. Cortex flexible, contains very loosely spaced, colourless granules about 0.2 µm across between each two ciliary rows. Cytoplasm colourless, contains numerous lipid droplets up to 5 µm across and, frequently, many oral baskets of microthoracid prey ciliates (Fig. 39a). Appears as rather rapidly gliding or swimming rod on microscope slide.

Somatic cilia 8–10 µm long in vivo, arranged in an average of nine equidistant, bipolar, ordinarily ciliated rows (average ciliary distance 3.5 µm in preparations) anteriorly slightly curved and connected with fairly distinct circumoral kinetofragments; rarely with small irregularities, such as minute breaks and/or supernumerary kinetids outside rows (Fig. 39a, j–o; Table 5). Dorsal brush heterostichad and three-rowed occupying about 10% of body length; dikinetids ordinarily and equidistantly spaced in all rows (about 1.2 µm in preparations), which have some ordinary cilia anteriorly and continue as somatic kineties posteriorly; a short fourth row occurs right of row 1 in three out of 58 specimens analysed. Brush row 1 ends at same level as row 2, composed of an average of 10 dikinetids with up to 4 µm long, rod-shaped bristles. Brush row 2 composed of an average of 11 dikinetids, posterior bristles slightly inflated distally and increasing in length from 3 µm anteriorly to 4 µm posteriorly; rod-shaped anterior bristles 2–3 µm long; followed by a short, monokinetidal tail of 1 µm long bristles, an unusual pattern recognizable also in *Protospathidium muscicola*. Brush row 3 shorter by 31 % on average than rows 1 and 2, composed of of eight dikinetids followed by a monokinetidal tail extending to second third of body with 2 µm long bristles; anterior bristle of dikinetids minute, posterior about 3 µm long and distally slightly inflated in anterior third of row, while about 4 µm long and conspicuously inflated in posterior thirds of row (Fig. 39a, d, k, l, m, o, 109g–j; Table 5).

Oral bulge fairly distinct in vivo due to the rather refractive extrusomes contained; basically, however, inconspicuous because only about 80% as long as widest trunk region and merely 2–3 µm high; slanted by 20°– 40°, usually slightly concave, rarely

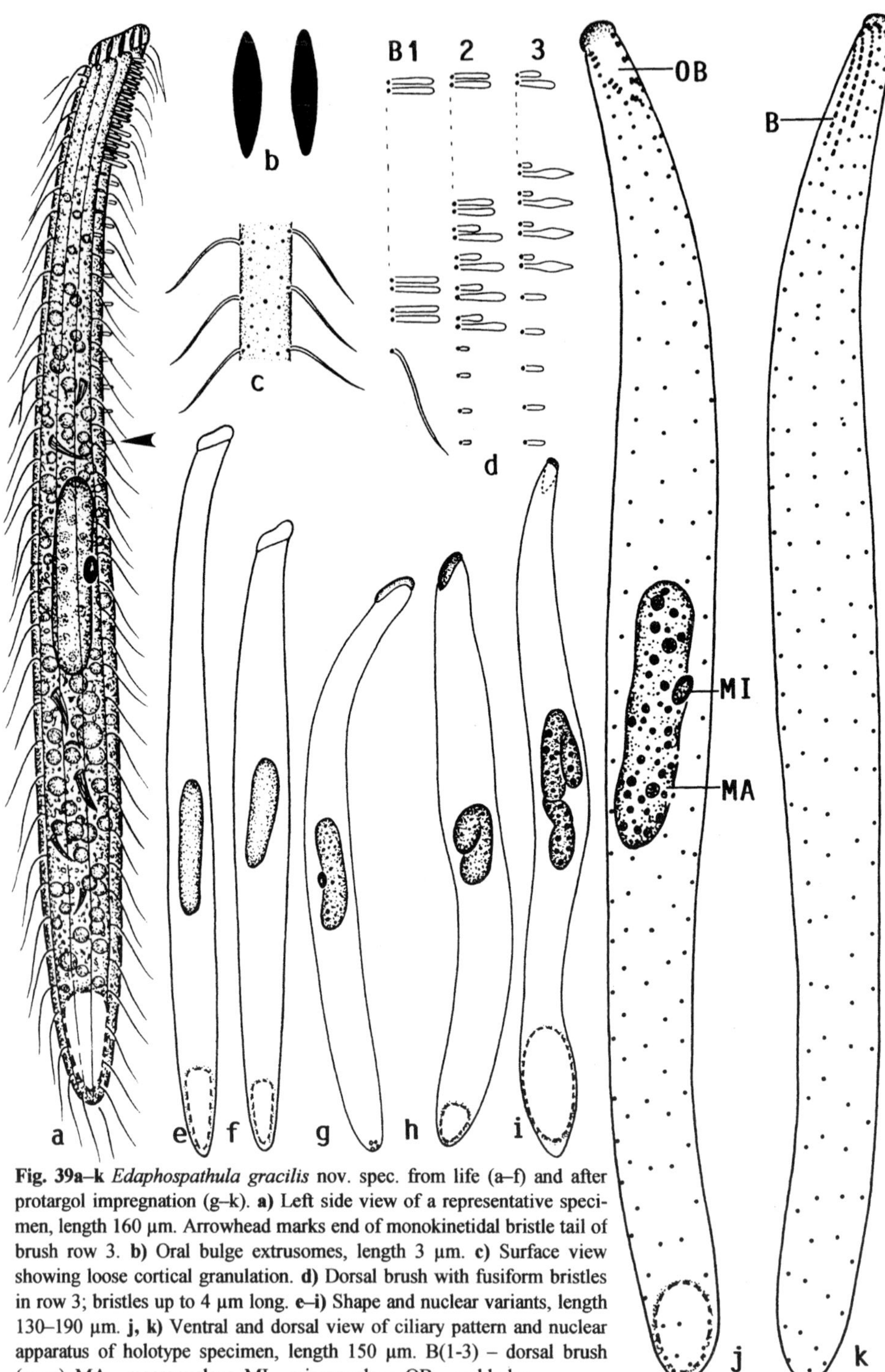

Fig. 39a–k *Edaphospathula gracilis* nov. spec. from life (a–f) and after protargol impregnation (g–k). **a)** Left side view of a representative specimen, length 160 µm. Arrowhead marks end of monokinetidal bristle tail of brush row 3. **b)** Oral bulge extrusomes, length 3 µm. **c)** Surface view showing loose cortical granulation. **d)** Dorsal brush with fusiform bristles in row 3; bristles up to 4 µm long. **e–i)** Shape and nuclear variants, length 130–190 µm. **j, k)** Ventral and dorsal view of ciliary pattern and nuclear apparatus of holotype specimen, length 150 µm. B(1-3) – dorsal brush (rows), MA – macronucleus, MI – micronucleus, OB – oral bulge.

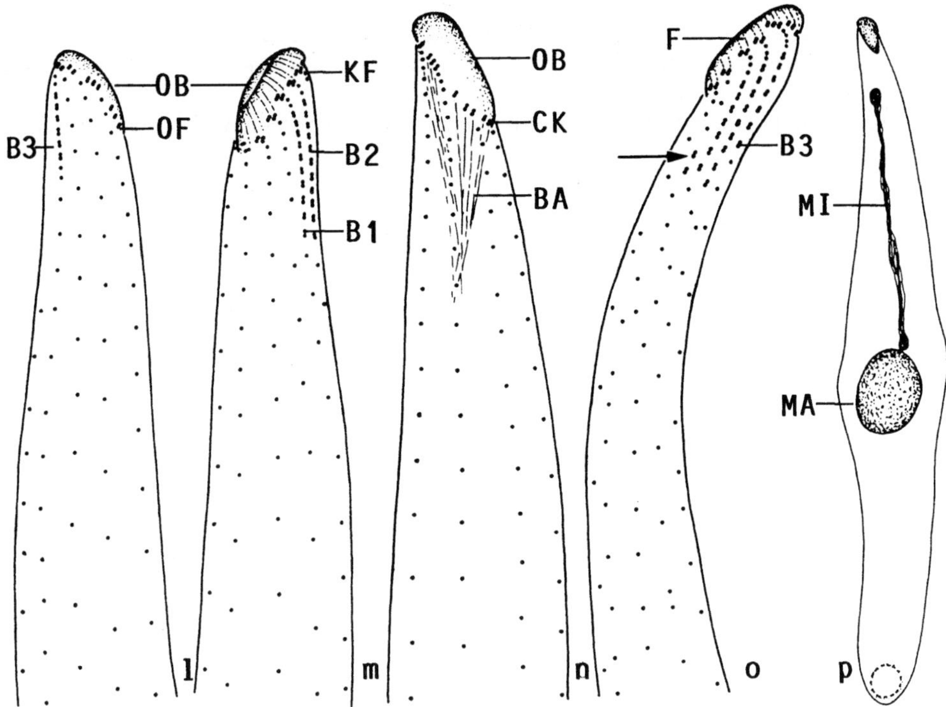

Fig. 39l–p *Edaphospathula gracilis* nov. spec., oral and somatic ciliary pattern after protargol impregnation. **l, m)** Right and left side view of same specimen, showing that the dorsal brush is located dorsally, length of oral bulge 9 µm. **n, o)** Ventrolateral and dorsolateral view, length of oral bulge 10 µm and 12 µm. Arrow marks a short, fourth dorsal brush row. **p)** Nuclear pattern of a middle divider, length 137 µm. B(1-3) – dorsal brush rows, BA – oral basket, CK – circumoral kinety, F – fibres, MA – macronucleus, MI – micronucleus, OB – oral bulge, OF – oral kinetofragments.

distinctly convex (Fig. 39a, e–h, l–o, 109f, k; Table 5). Individual oral kinetofragments fairly distinct because usually composed of two or three, rarely four, on average three dikinetids forming a cuneate circumoral kinety; each dikinetid associated with a cilium, a faintly impregnated nematodesma, and two fibres extending into oral bulge centre. Oral basket recognizable only in some properly impregnated specimens, composed of fine nematodesma bundles forming an about 20 µm long cone (Fig. 39j, l–o, 109f–j; Table 5).

Occurrence and ecology: As yet found only at type locality, where it became abundant in the non-flooded Petri dish culture. The sample, kindly provided by Dr. WOLFGANG PETZ, likely consisted mainly of artificial soil (compost or composted activated sludge) because it was taken from an only about 5 cm thick layer above the sandy ground. The material was black and "fat", hardly contained any litter, had pH 8.2 in water, and was sown with lawn grasses a few days before the sample was taken.

Remarks: The generic allocation of this species is not entirely clear because the right side oral kinetofragments are composed of three dikinetids (*Protospathidium* ≥ three dikinetids, *Edaphospathula* ≤ three dikinetids). However, *E. gracilis* is highly similar to *E. brachycaryon*, which has typical *Edaphospathula* kinetofragments; hence, the Egyptian population is assigned to the same genus.

Edaphospathula gracilis differs from *E. brachycaryon* mainly by the extrusomes (bluntly fusiform and 3 μm long vs. ampulliform and 1.5 μm long) and the dorsal brush: although both have similar numbers of dikinetids, the rows of *E. brachycaryon* are twice as long as those of *E. gracilis* because the dikinetids of *E. brachycaryon* are unusually widely spaced (Table 5). This difference is so conspicuous that it is recognizable also in vivo. Further differences and separation from similar species, see discussion of *E. brachycaryon*.

Protospathidium DRAGESCO & DRAGESCO-KERNÉIS, 1979

1979 *Protospathidium muscicola* gen. n., sp. n. DRAGESCO & DRAGESCO-KERNÉIS, Acta Protozool., 18: 411 – Type by monotypy (see also next entry; formal diagnosis not provided).

1984 *Protospathidium* DRAGESCO & DRAGESCO-KERNÉIS, 1979 – FOISSNER, Stapfia, 12: 66 (formal diagnosis of the genus and fixation of type species: *Protospathidium muscicola* DRAGESCO & DRAGESCO-KERNÉIS, 1979).

1986 *Protospathidium* DRAGESCO & DRAGESCO-KERNÉIS, 1979 – DRAGESCO & DRAGESCO-KERNÉIS, Faune trop., 26: 155 (some sort of formal diagnosis, obviously according to FOISSNER 1984).

Diagnosis: Protospathidiidae with right side oral kinetofragments composed of ≥ 3 dikinetids. Circumoral kinety elliptical to obovate. Extrusomes usually ≥ 4 µm long, comparatively fine and rod-shaped (length:width ratio about 5–12:1).

Type species: *Protospathidium muscicola* DRAGESCO & DRAGESCO-KERNÉIS, 1979.

Etymology: Not given in original description. Composite of the Greek *proto* (the first) and the generic name *Spathidium* (small spatula), obviously referring to the hypothesis that this kind of spathidiids represents the plesiomorphic state of the group. Neuter gender.

Remarks: See family Protospathidiidae for key to species. The features separating *Protospathidium* and *Edaphospathula* are not easily recognizable. Thus, careful in vivo observation and protargol impregnation are required for reliable identification. Further details, see discussions of family Protospathidiidae and genus *Edaphospathula.*

Protospathidium terricola FOISSNER, 1998 (Fig. 40a–t; Table 6)

1997 *Protospathidium terricola* FOISSNER, in press – PETZ & FOISSNER, Polar Rec., 33: 310 (description of an Antarctic population; voucher slides with protargol-impregnated specimens deposited at the locality described below).

1998 *Protospathidium terricola* FOISSNER, Europ. J. Protistol., 34: 218 (Type slides with protargol-impregnated specimens from type locality are deposited in the Oberösterreichische Landesmuseum in Linz, Upper Austria.).

Nomenclature: Although FOISSNER's paper was submitted earlier than that of PETZ & FOISSNER, it appeared slightly later. However, FOISSNER (1998) has priority because this is definitely stated in the paper by PETZ & FOISSNER (1997) and no other priorities are violated.

Diagnosis (of type population): Size about 90 × 25 µm in vivo. Narrowly spatulate with conspicuous, oblique oral bulge shorter than widest trunk region by about 35%. Macronucleus elongate reniform; single micronucleus. Extrusomes rod-shaped, about 5 µm long. On average 21 ciliary rows, 3 anteriorly modified to moderately distinct, heterostichad dorsal brush occupying 22% of body length; brush row 1 not reduced, only slightly shorter than row 2. Right side oral kinetofragments each composed of 4–5 dikinetids.

Type locality: Grassland soil from Mt. Kenya near the Lodge "The Ark" in the Mount Kenya National Park, Kenya, equatorial Africa, E37° N0°.

Etymology: The compound Latin noun *terricola* (living in soil) refers to the habitat the species was discovered.

Description (of Kenyan type population; Fig. 40a–i; Table 6): Size in vivo 70–100 × 20–30 µm. Elongate bursiform, usually looking like a swimming sausage, anterior portion frequently slightly curved, middle and posterior portion more or less distinctly inflated, depending on nutrition state (Fig. 40a, c, e). Unflattened, except for slightly compressed anterior third. Macronucleus in mid-body, oblong to elongate reniform, rarely slenderly ellipsoidal, contains many tiny nucleoli. Micronucleus globular, attached to macronucleus, but not in fixed position. Contractile vacuole in posterior end. Extrusomes mainly in oral bulge, rod-shaped, about 5 µm long (Fig. 40a–c). Cortex flexible, contains about four rows of minute, colourless granules between each two kineties (Fig. 40d). Cytoplasm usually packed with lipid droplets 0.5–3 µm across. Swims moderately rapidly by rotation about longitudinal body axis.

Cilia 10–12 µm long in vivo, rather unevenly spaced, arranged in an average of 21 rows. Somatic kineties evenly spaced, extend meridionally, anterior ends conspicuously curved and composed of 3–6 distinct dikinetids associated with long, fine nematodesmata. Dorsal brush of usual structure, i.e., composed of narrowly spaced dikinetids having short, slightly clavate cilia; at anterior end of each brush kinety 2–6 monokinetids (Fig. 40a, e, f, i).

Oral bulge conspicuous, although indistinctly separate from body proper, because comparatively high and refractive due to the extrusomes contained. Bulge centre conically depressed, that is, with pronounced temporary cytostome. Circumoral kinety discontinuous because dikinetidal fragments separated by gaps one to two dikinetids wide and adhering to their respective somatic kineties (Fig. 40a, e–i).

Description (of Antarctic, Whitney Point population; Fig. 40j–t; Table 6): Size highly variable, in vivo about 134 × 31 µm (Table 6). Shape dependent on nutrition state, usually narrowly spatulate with oblique oral portion distinctly narrowed and curved dorsally (Fig. 40j, p–s); cross-section circular, oral portion laterally slightly flattened. Macronucleus in living specimens about 30–38 × 12–22 µm, basically reniform but rather variable in detail (Fig. 40j, l, p–r): sausage-like (28%, n=115), C-shaped (25%), ellipsoidal (24%), or helically wound (23%); contains spherical to ellipsoidal nucleoli 1.5–4 µm across. Micronucleus globular to slightly ellipsoidal, adjacent to macronucleus. Contractile vacuole in posterior end, with up to nine excretory pores. Oral bulge extrusomes rod-shaped with rounded ends, 4–8 µm, usually about 5 µm long in vivo; fusiform extrusomes, likely developmental stages, occur in the cytoplasm. Cortex flexible and colourless, with four to five, rarely up to eight rows of minute, pale granules between each two ciliary rows; individual granules about 0.2 µm across, stain reddish with methyl green-pyronin but are not extruded. Cytoplasm with numerous small, colourless granules; some lipid droplets up to 4 µm across; and food vacuoles containing bright green globules (algae?) up to 13 µm across, many greenish spherical to ellipsoidal (up to 6 × 3 µm) cyanobacteria, flagellates, and ciliates (for example,

Table 6 Morphometric data on *Protospathidium terricola* from Kenya, Africa (KE; from FOISSNER 1998) and Antarctica (AN; from PETZ & FOISSNER 1997)

Characteristics[1]	Pop	$\overline{x}$	M	SD	SE	CV	Min	Max	n
Body, length in vivo	AN	133.8	132.5	25.8	6.9	19.3	88.0	175.0	14
Body, width in vivo	AN	30.8	31.0	3.5	0.4	11.4	25.0	35.0	12
Body, length	KE	80.7	82.0	9.6	3.0	11.9	65.0	93.0	10
	AN	101.0	91.0	23.6	4.2	23.3	68.0	150.0	31
Body, width	KE	22.5	22.0	3.6	1.1	16.1	16.0	28.0	10
	AN	32.3	33.0	5.6	1.0	17.3	23.0	44.0	31
Body length:width, ratio	KE	3.6	3.7	0.5	0.2	14.2	2.7	4.2	10
Oral bulge, length	KE	14.3	15.0	1.7	0.6	11.6	12.0	17.0	10
	AN	17.2	17.0	3.5	0.6	20.3	10.0	26.0	31
Oral bulge, height	AN	3.9	4.0	1.0	0.2	24.9	2.0	6.0	30
Anterior end to macronucleus, distance	AN	59.5	60.0	15.2	2.7	25.6	31.0	87.0	31
Macronucleus, length	KE	28.1	26.5	4.2	1.3	15.0	22.0	35.0	10
	AN	33.6	33.0	4.7	0.8	13.8	24.0	43.0	31
Macronucleus, width	KE	7.6	8.0	0.8	0.3	11.1	6.0	9.0	10
	AN	13.4	13.0	2.3	0.2	16.8	8.5	21.0	31
Micronucleus, largest diameter	KE	3.5	3.5	–	–	–	3.0	4.0	10
	AN	3.1	3.0	0.5	0.2	15.9	2.5	4.0	6
Extrusomes, length	AN	5.8	6.0	0.8	0.1	13.6	4.0	7.0	31
Macronucleus, number	KE	1.0	1.0	0.0	0.0	0.0	1.0	1.0	10
	AN	1.0	1.0	0.0	0.0	0.0	1.0	1.0	30
Micronucleus, number	KE	1.0	1.0	0.0	0.0	0.0	1.0	1.0	10
	AN	1.0	1.0	0.0	0.0	0.0	1.0	1.0	5
Circumoral kinety to end of brush row 1, distance	KE	15.0	15.0	2.6	0.9	17.0	11.0	18.0	9
	AN	15.2	14.0	3.8	0.7	25.1	9.0	23.0	31
Circumoral kinety to end of brush row 2, distance	KE	18.0	18.0	2.1	0.7	11.5	15.0	21.0	9
	AN	18.7	19.0	3.9	0.7	20.8	12.0	30.0	31
Circumoral kinety to end of brush row 3, distance	KE	10.8	10.0	1.7	0.6	15.9	8.0	14.0	9
	AN	9.9	9.0	2.6	0.5	26.0	6.0	15.5	28
Brush kinety 3, total length, i.e., with tail	AN	52.8	48.0	10.7	1.9	20.3	34.0	72.0	31
Ciliary rows, number	KE	21.3	21.0	1.6	0.5	7.7	19.0	25.0	10
	AN	17.4	17.0	1.2	0.2	6.6	15.0	19.0	31
Basal bodies in a right side ciliary row, number	KE	42.2	39.5	7.3	2.3	17.3	35.0	55.0	10
	AN	42.3	43.0	6.6	1.3	15.7	31.0	60.0	24
Dorsal brush rows, number	KE	3.0	3.0	0.0	0.0	0.0	3.0	3.0	10
	AN	3.0	3.0	–	–	–	3.0	4.0	31
Dikinetids in brush row 1, number	AN	14.3	14.0	2.9	0.5	20.3	9.0	20.0	31
Dikinetids in brush row 2, number	AN	20.3	20.0	3.6	0.6	17.6	15.0	28.0	31
Dikinetids in brush row 3, number	AN	8.9	9.0	1.6	0.3	17.7	5.0	12.0	31

[1] Data on Kenyan population based on mounted, protargol-impregnated (FOISSNER´s method), and randomly selected specimens from a non-flooded Petri dish culture. Data on Antarctic population based, if not stated otherwise, on protargol-impregnated (WILBERT´s method), mounted, and randomly selected specimens from a non-flooded Petri dish culture. Measurements in µm. CV – coefficient of variation in %, M – median, Max – maximum, Min – minimum, n – number of individuals investigated, SD – standard deviation, SE – standard error of arithmetic mean, $\overline{x}$ – arithmetic mean.

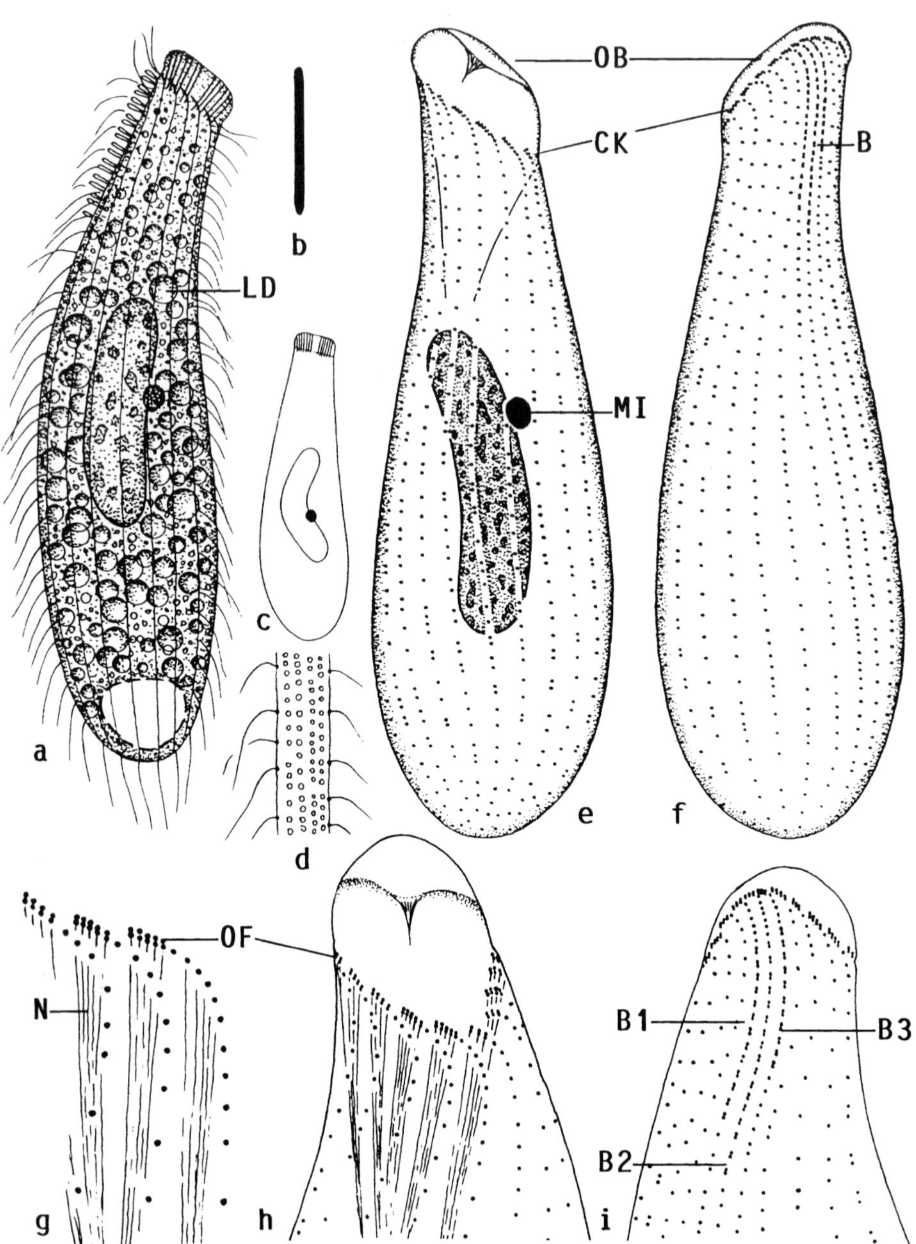

Fig. 40a–i *Protospathidium terricola*, Kenyan type population from life (a–d) and after protargol impregnation (e–i). From FOISSNER (1998). **a)** Right side view of a representative specimen packed with lipid droplets, length 90 µm. **b)** Extrusome, length 5 µm. **c)** Shape variant. **d)** Surface view showing cortical granulation. **e, f)** Ciliary pattern of right and left side and nuclear apparatus of holotype specimen, length 90 µm. **g)** Oral kinetofragments at high magnification. **h, i)** Ciliary pattern of ventral and dorsal side, length 33 µm. B(1-3) – dorsal brush (rows), CK – circumoral kinety, LD – lipid droplets, MI – micronucleus, N – nematodesmata, OB – oral bulge, OF – oral kinetofragments.

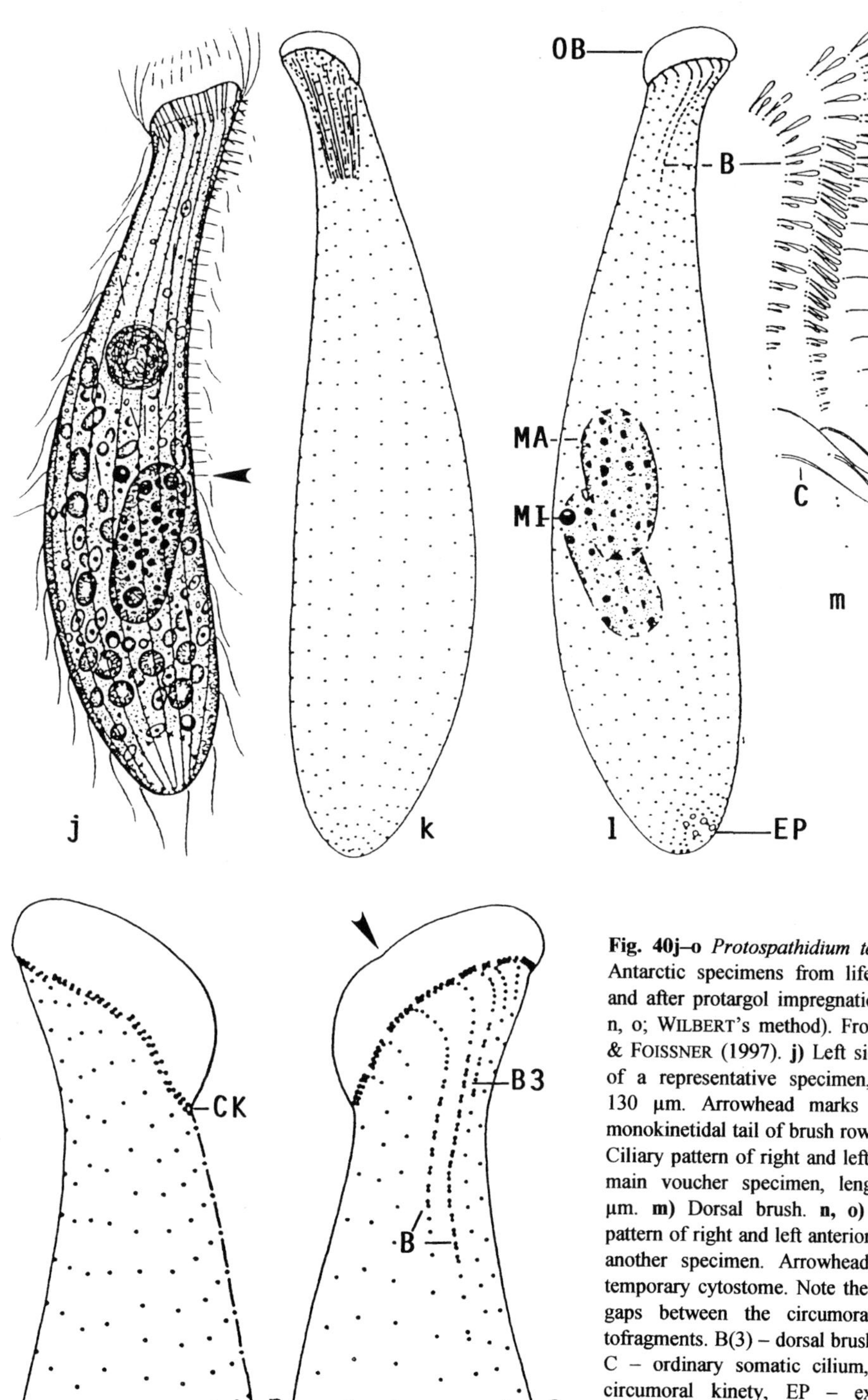

Fig. 40j–o *Protospathidium terricola*, Antarctic specimens from life (j, m) and after protargol impregnation (k, l, n, o; WILBERT's method). From PETZ & FOISSNER (1997). **j)** Left side view of a representative specimen, length 130 µm. Arrowhead marks end of monokinetidal tail of brush row 3. **k, l)** Ciliary pattern of right and left side of main voucher specimen, length 135 µm. **m)** Dorsal brush. **n, o)** Ciliary pattern of right and left anterior side of another specimen. Arrowhead marks temporary cytostome. Note the minute gaps between the circumoral kinetofragments. B(3) – dorsal brush (row), C – ordinary somatic cilium, CK – circumoral kinety, EP – excretory pores, MA – macronucleus, MI – micronucleus, OB – oral bulge.

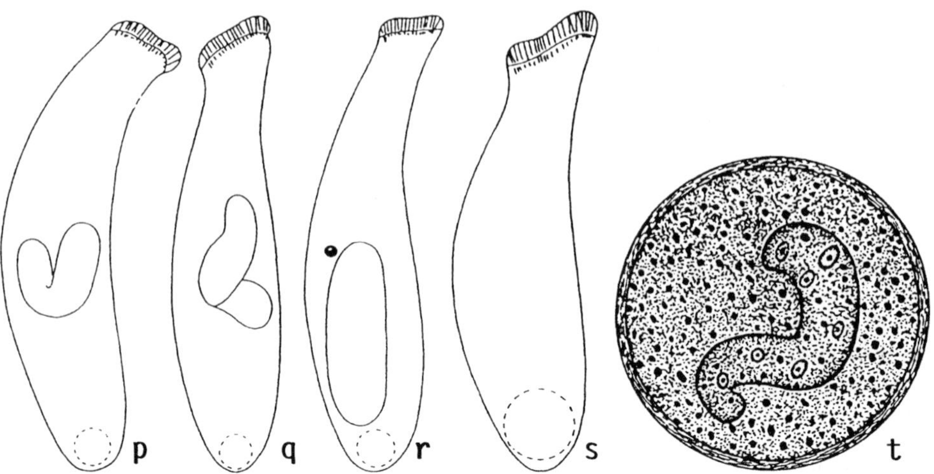

Fig. 40p–t *Protospathidium terricola*, Antarctic specimens from life. From PETZ & FOISSNER (1997). **p–s)** Variability of shape of body and macronucleus. **t)** Resting cyst, 50 μm across.

Odontochlamys wisconsinensis). Rotates about main body axis with neck usually curved dorsally when swimming; glides between and on soil particles, often recoiling and changing direction.

Dorsal brush three-rowed and heterostichad, rows 1 and 2 almost of same length and mainly composed of paired bristles (Table 6); row 3 consists of a short, dikinetidal anterior portion and a long, monokinetidal tail with about 2.5 μm long bristles terminating near mid-body; brush rows with distinct anterior tail, posterior ends often with one to six apparently unciliated basal bodies before continuing backwards as ordinary somatic kineties (Fig. 40m, n). Bristles of brush dikinetids highly differentiated, clavate, anterior cilium usually longer (about 2–4 μm) than posterior (approximately 1–2.5 μm), in rear portion of brush gradually shortened; two to eight 6–8 μm long and thus conspicuous bristles in anterior portion of row 3 (Fig. 40m).

Oral bulge conspicuous, though only about half as long as widest trunk region, because about 5 μm high and packed with refractive extrusomes; ordinarily slanted, centre with small, conical depression. Circumoral kinety obovate in frontal view, discontinuous, that is, composed of dikinetidal fragments adhering to anterior ends of somatic kineties and separated from each other by minute gaps often difficult to recognize, especially on right side (Fig. 40n, o).

Specimens may encyst rapidly, that is, the first cyst appeared 6 min after transfer from the raw culture to Eau de Volvic; after 30 min all (about 10) individuals were encysted. One-week-old cysts 49–52 μm in diameter ($\bar{x}$ 50, n 4), wall about 1.5 μm thick, smooth, vitreous. Cytoplasm with many about 1 μm-sized granules and tortuous to C-shaped macronucleus containing globular nucleoli (Fig. 40t).

Occurrence and ecology: To date found in equatorial Africa (type population) and, probably (see below), in Antarctica, Wilkes Land, where it occurred on 28 January 1994 in the 0–2 cm layer of algal ornithogenic soil, Whitney Point, Clark Peninsula; on 16 December 1993 in the 0–3 cm soil layer, Beall Island, Windmill Islands;

and on 2 December 1993 in the 0–3 cm soil layer, north coast of Shirley Island, Windmill Islands. In Antarctica, *P. terricola* occurred in fresh as well as in air-dried and remoistened soil samples with pH 5.1–5.5 and at 8– 20°C.

Remarks: The type population has clearly separated oral kinetofragments and thus belongs to *Protospathidium* (Fig. 40e–h). *Protospathidium terricola* differs from most congeners by the simple, reniform macronucleus, a stable feature easy to recognize even in vivo. *Protospathidium vermiculus*, which has a similar macronucleus, is smaller (65 × 15 μm vs. 90 × 25 μm) and has only 10 (vs. 21) ciliary rows. In vivo, *P. terricola* is easily confused with *Spathidium claviforme* (19–25 vs. 10–13 ciliary rows) and *S. vermiforme*, which is smaller (length 70–100 μm vs. 40–60 μm) and has only about 10 ciliary rows (vs. 19–25) and a more cylindroidal body shape (vs. elongate bursiform). Thus, identification should be checked in protargol preparations, where the separated oral kinetofragments are recognizable.

We kept separate the descriptions of the Kenyan and Antarctic populations because they are probably not conspecific. Basically, they match well because the few significant morphometric differences are likely caused by the different preparation methods used (Table 6), that is, PETZ's prepared specimens (WILBERT's method) are obviously inflated as shown by the length:width ratio in live (4.3:1) and prepared cells (3.1:1). Thus, the Kenyan specimens, which have a length:width ratio of about 3.6:1 both in vivo and preparations, match well, and only a single, main difference remains, viz., the oral kinetofragments, which are indistinctly separated in the Antarctic specimens (Fig. 40n, o). With the new knowledge available since PETZ's study (FOISSNER et al. 2002, present monograph), it seems possible that the Antarctic population is closely related to *Spathidium claviforme*. This is emphasized by the large size and low number of ciliary rows, both approaching that species.

Protospathidium vermiculus (KAHL, 1926) **nov. comb.** (Fig. 41a–p, 110 a–i; Table 7)

1926 *Spathidium vermiculus* KAHL, Arch. Protistenk., 55: 269.
1930 *Spathidium vermiculus* KAHL, 1926 – KAHL, Tierwelt Dtl., 18: 156 (revision).
1930 *Spathidium cucumis* spec. n. BAUMEISTER – KAHL, Tierwelt Dtl., 18: 159 (species described by BAUMEISTER in KAHL; new synonym).
1962 *Spathidium cucumis* – BAUMEISTER, Planktonkunde: 65 (with original figure, Fig. 41d!).
1962 *Spathidium cucumis* KAHL, 1953 (BAUMEISTER) – VUXANOVICI, Studii Cerc. Biol., 14: 203 (wrong dating; brief redescription).
1962 *Spathidium cucumis* BAUMEISTER, 1930 – DINGFELDER, Arch. Protistenk., 105: 554 (misidentification; see *Spathidium piliforme*).
1980 *Spathidium vermiculus* KAHL, 1926 – FOISSNER, Verh. Zool.-bot. Ges. Wien, 188/119: 104 (brief redescription).

Synonymy: To make our decisions transparent, they will be discussed in detail. They are based, to a large extent, on the thorough investigation of a population very likely representing *S. vermiculus*.

The identity of the two species synonymized is difficult to clarify because small spathidiids generally pose problems and the original descriptions are meagre and burdened

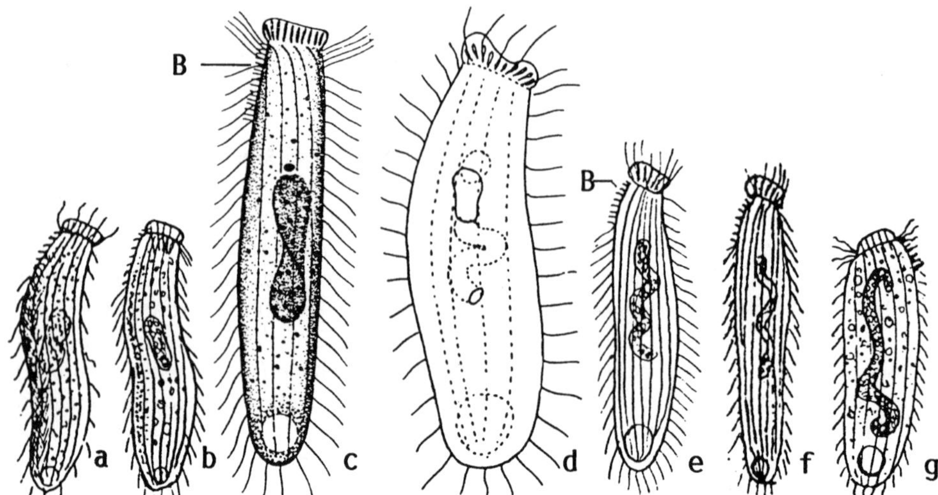

Fig. 41a–g *Protospathidium vermiculus* (a–c) and its proposed synonym, *P. cucumis* (d–g) from life. **a)** From KAHL (1926), length 40 µm (actually about 50 µm; calibration error, see text). **b)** From KAHL (1930a), length 50 µm. **c)** From FOISSNER (1980b), length 53 µm. **d)** From BAUMEISTER (1962), size 59 × 13 µm. **e)** From KAHL (1930b), according to BAUMEISTER, length 60 µm. **f)** From KAHL (1943), length 60 µm. **g)** From VUXANOVICI (1962a), length 80 µm. B – dorsal brush.

with the following uncertainties: (i) KAHL (1926, 1930b) provides two different sizes for *S. vermiculus*, viz., 40 µm and 50 µm, likely due to a calibration error in (1926); (ii) the figures in KAHL (1926, 1930b) differ considerably, especially the length of the extrusomes: in the (1926) figure (Fig. 41a) and description, they do not extend into the body proper, while they do so in the (1930b) figure (Fig. 41b), where they are about twice as long as the height of the oral bulge; (iii) KAHL (1930b) states: "oral bulge as wide as broadest trunk region", but in the figures it occupies only 61–67% (Fig. 41a, b), just as in the neotype population from Iceland (Table 7); (iv) the macronucleus is reniform according to the description, but ellipsoidal in the illustration (Fig. 41b).

Spathidium cucumis also poses severe problems: (i) the figure shown by KAHL (1930b) differs considerably from that provided by BAUMEISTER (1962), though KAHL (1930b) states "according to BAUMEISTER" in the figure explanation (Fig. 41d, e); (ii) the figure changed again in KAHL (1943), where the cell is very narrowly ellipsoidal and the oral bulge even more pronounced (Fig. 41f).

Original description of *S. vermiculus* by KAHL (1926): "Length 40 µm, very flexible and metabolic moving worm-like in the mud; parallel-sided, rarely inflated by food inclusions. Extrusomes 3.3 µm long (KAHL 1930b), do not extend into the hyaline, strongly granulated cytoplasm, that is, are about as long as the height of the comparatively wide, button-shaped oral bulge. Dorsal brush distinct. Striation and ciliature fine, not narrow. Macronucleus reniform. One cell with two heads (Fig. 41a)".

Redescription of *S. vermiculus* by FOISSNER (1980b): "Length 50–60 µm in vivo. Body very slender and worm-like flexible, about 2:1 flattened, soon becomes inflated under coverslip. Oral bulge about 3 µm high, sigmoidal, contains slightly comma-shaped extrusomes not extending into the cytoplasm. About 10 ciliary rows with circa 25 cilia each, ciliation thus loose; rows anteriorly curved and more densely and longer ciliated (a re-evaluation of the original notes suggest a protospathidiid pattern, not yet known when the paper was submitted). Dorsal brush three-rowed and with short bristles, occupies only one quarter of body length. Silverlines form very narrow, linearly oriented meshes. Contractile vacuole in rear end, surrounded by many granules. Roots in mud. Differs from KAHL's *S. vermiculus* only by the shape of the macronucleus (Fig. 41c)".

Original description of *S. cucumis* by KAHL (1930b): "Small (60–70 µm), slender, cucumber-shaped, unflattened ciliate (from liquid manure) with high oral bulge projecting dorsally and ventrally. Size 59 × 13 µm. Slightly narrowed anteriorly. Oral bulge button-shaped and projecting, contains short trichocysts. Dorsal brush bristles short. Ciliary rows ordinarily spaced and ciliated (Fig. 41e)". BAUMEISTER (1962) adds "nucleus spiral" and a figure obviously not known to KAHL and later authors (Fig. 41d).

Redescription of *S. cucumis* by VUXANOVICI (1962a): "Similar to the original description. Cortex with 6–7 ciliary rows per side; cytoplasm transparent, contains some minute algae 1–2 µm across. Nucleus a long, tortuous strand. Oral bulge about half as wide as broadest trunk region (Fig. 41g)".

When the original descriptions and figures are compared and the discrepancies shown above are added, a pronounced similarity of *S. vermiculus* and *S. cucumis* cannot be denied. Both largely agree in body shape and size; the button-shaped and thus conspicuous oral bulge; the minute extrusomes; and the short macronucleus, whose shape details are highly variable, also in the neotype population. Thus, there is not a single, important feature separating the two species unequivocally, suggesting synonymy at the present state of knowledge.

Type material and neotypification: No type material is available. Neotypification is necessary to erase all the problems detailed above and to stabilize the species (FOISSNER 2002). We suggest to fix the Iceland population, described below, as a neotype because (i) it matches the original descriptions of *S. vermiculus* and *S. cucumis*; (ii) it is from a similar habitat (moorland) as KAHL's *S. vermiculus*; and (iii) it is from the same main biogeographic region (holarctic).

Diagnosis (includes all information known, emphasizing the new, solid data from the Iceland population): Size about 65 × 15 µm in vivo. Narrowly spatulate with oblique, obovate, button-shaped oral bulge shorter than widest trunk region by 30–40%. Macronucleus circa 20 µm long, oblong, reniform, dumbbell-shaped, or spiralized; usually bimicronucleate. Extrusomes rod-shaped, fine, about as long (3–4 µm) as height of oral bulge. On average about 10 ciliary rows, 3 anteriorly differentiated to inconspicuous, almost isostichad dorsal brush occupying one quarter of body length and comprising up to 3 µm long bristles: brush row 1 composed of an average of 10 dikinetids, row 2 of 12, and row 3 of 9 dikinetids not followed by a monokinetidal bristle tail. Right side oral kinetofragments composed of 5 dikinetids on average.

Type and neotype localities: The type locality is a shallow road drain in the Eppendorf moorland near Hamburg, Germany, E10° N53°30'. The neotype locality is a mire in the surroundings of the village of Thingvellier, about 45 km east of the town of Reykjavik, SW-Iceland, W21°10' N64°15'.

Etymology: Not given in original description. The Latin noun *vermiculus* (little worm) obviously refers to the worm-like behaviour of the species. *Cucumis*, the name of the junior synonym, refers to the cucumber-like shape.

Description of Iceland neotype population: Size 50–75 × 10–20 µm in vivo, usually about 65 × 15 µm, as calculated from some in vivo measurements and protargol preparations, assuming a shrinkage of 10%; length:width ratio 2.8–5.4:1, on average near 4.5:1 both in vivo and prepared specimens (Table 7); rather fragile soon becoming inflated under mild coverslip pressure. Narrowly spatulate with oblique anterior (oral) body end about 60% as long as widest trunk region usually found in second third of cell, posterior end narrowly to moderately broadly rounded; ventral and dorsal aspect rather conspicuous due to the comparatively large, almost hemispherical oral bulge (Fig. 41h–j, l, n–p, 110a–e); unflattened and not contractile. Macronucleus slightly underneath mid-body, contains many globular to lobate nucleoli, about 20 × 5 µm in size and basically oblong in vivo, in detail, however, highly variable: more or less distinctly dumbbell-shaped in 50 specimens (out of 77 cells investigated); oblong in 8 specimens; reniform in 7 individuals; spiral or tortuous in 4 specimens each; and in one to four nodules in 4 specimens (likely exconjugants). Two to three globular to broadly ellipsoidal micronuclei in variable positions adjacent to, rarely rather distant from macronucleus (Fig. 41h, j, m–p, 110a–e; Table 7). Contractile vacuole in rear body end, two to four excretory pores in dorsal pole area. Extrusomes studded in oral bulge and scattered in cytoplasm, in vivo rod-shaped, fine, and about 3 µm long; anterior end impregnates heavily, appearing as a distinct granule (Fig. 41o, p, 110a, b, f, h, i). Cortex very flexible and distinctly furrowed by ciliary rows; cortical granules not recognizable, that is, either lacking or very inconspicuous. Cytoplasm colourless and hyaline, contains some lipid droplets up to 5 µm across. Feeds on heterotrophic flagellates and small ciliates digested in food vacuoles up to 20 µm across. Swims rather slowly rotating about main body axis.

Somatic cilia 8 µm long in vivo, arranged in 10, rarely 11 bipolar, equidistant, rather widely spaced rows slightly more densely ciliated anteriorly than posteriorly, especially in first and/or second row of right (ventral) side. Kineties ordinarily ciliated (average ciliary distance 2.3 µm) and attached to oral kinetofragments, anterior portion distinctly curved dorsally on right side, while slightly curved ventrally on left, and abutting on oral kinetofragments at right angles on dorsal side (Fig. 41h, l–n, 110f–i; Table 7). Dorsal brush three-rowed and almost isostichad, inconspicuous because longest row 2 occupies only 21% of body length and bristles in vivo merely 2–3 µm long and only slightly inflated distally; rows of similar length and continuing as ordinary somatic kineties posteriorly, dikinetids of row 3 slightly wider spaced than those of rows 1 and 2; row 3 without monokinetidal bristle tail, an unusual feature; anterior tail of rows lacking or indistinct, that is, composed of only one to two cilia (Fig. 41h, m, n, 110g; Table 7).

Oral bulge slightly convex and obliquely truncate by 30° to 45° in lateral view, conspicuous in vivo and protargol preparations, though shorter than widest trunk region by 30–40 %, because about 4 µm high and slightly projecting laterally and dorsally; thus, almost hemispherical when viewed dorsally and obovate when viewed ventrally (Fig. 41h, i, l–p, 110a, g, h; Table 7). Circumoral kinety of similar or same shape as oral bulge, that is, obovate and composed of dikinetidal kinetofragments separated from

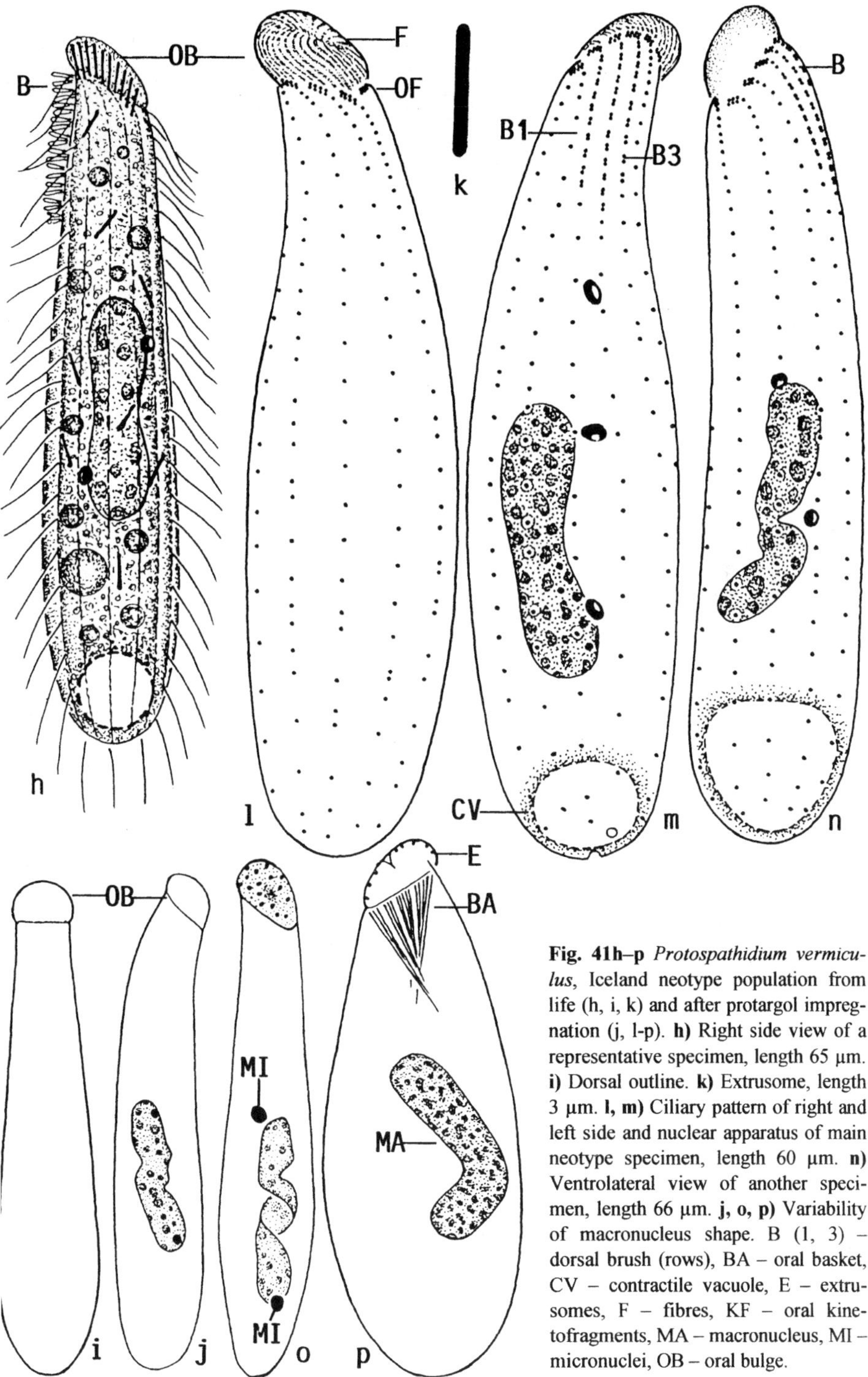

Fig. 41h–p *Protospathidium vermiculus*, Iceland neotype population from life (h, i, k) and after protargol impregnation (j, l-p). **h)** Right side view of a representative specimen, length 65 µm. **i)** Dorsal outline. **k)** Extrusome, length 3 µm. **l, m)** Ciliary pattern of right and left side and nuclear apparatus of main neotype specimen, length 60 µm. **n)** Ventrolateral view of another specimen, length 66 µm. **j, o, p)** Variability of macronucleus shape. B (1, 3) – dorsal brush (rows), BA – oral basket, CV – contractile vacuole, E – extrusomes, F – fibres, KF – oral kinetofragments, MA – macronucleus, MI – micronuclei, OB – oral bulge.

Table 7 Morphometric data on *Protospathidium vermiculus* from Iceland

Characteristics[1]	$\bar{x}$	M	SD	SE	CV	Min	Max	n
Body, length	57.6	58.5	4.9	1.4	8.5	49.0	66.0	12
Body, width	13.4	13.0	2.6	0.8	19.5	10.0	20.0	12
Body length:width, ratio	4.4	4.5	0.6	0.2	14.2	2.8	5.4	12
Oral bulge, length	7.5	8.0	0.7	0.2	9.0	6.0	8.0	12
Oral bulge, height	3.5	3.5	–	–	–	3.0	4.0	12
Macronucleus figure, length	18.7	19.0	2.7	0.8	14.7	14.0	24.0	12
Macronucleus, maximum width	4.3	4.0	1.0	0.3	22.7	3.0	6.0	12
Micronuclei, largest diameter	1.7	1.6	–	–	–	1.5	2.0	12
Macronucleus, number	1.0	1.0	0.0	0.0	0.0	1.0	1.0	12
Micronuclei, number	2.4	2.0	–	–	–	2.0	3.0	12
Circumoral kinety to end of brush row 1, distance	9.5	10.0	1.6	0.5	16.5	7.0	12.0	12
Circumoral kinety to end of brush row 2, distance	12.2	12.0	1.3	0.4	10.4	10.0	14.0	12
Circumoral kinety to end of brush row 3, distance	9.6	10.0	1.0	0.3	10.4	8.0	11.0	12
Ciliary rows, number	10.1	10.0	–	–	–	10.0	11.0	12
Basal bodies in a right side ciliary row, number	25.4	25.0	3.3	1.0	13.0	21.0	31.0	12
Dorsal brush rows, number	3.0	3.0	0.0	0.0	0.0	3.0	3.0	12
Dikinetids in brush row 1, number	9.6	10.0	1.0	0.3	10.4	7.0	13.0	12
Dikinetids in brush row 2, number	11.9	12.0	0.9	0.3	7.6	11.0	14.0	12
Dikinetids in brush row 3, number	8.8	9.0	1.3	0.4	15.1	6.0	11.0	12
Right side oral kinetofragments, number of dikinetids	4.8	5.0	0.7	0.2	14.9	4.0	6.0	12

[1] Data based on mounted, protargol-impregnated (FOISSNER's method), randomly selected specimens from a non-flooded Petri dish culture. Measurements in µm. CV – coefficient of variation in %, M – median, Max – maximum, Min – minimum, n – number of individuals investigated, SD – standard deviation, SE – standard error of arithmetic mean, $\bar{x}$ – arithmetic mean.

each other by gaps two to three dikinetids wide. Individual kinetofragments composed of an average of five dikinetids, each associated with a 8–10 µm long cilium and a comparatively distinct fibre extending spirally to the bulge centre, where a minute, obconical depression (temporary cytostome) is sometimes recognizable. Oral basket distinct in protargol preparations, obconical and 10–20 µm long, composed of cuneate nematodesma bundles originating from the oral kinetofragments (Fig. 41h, l–p, 110d–i; Table 7).

Occurrence and ecology: KAHL (1926) discovered *P. vermiculus* in a shallow road drain of the Eppendorf moorland in the surroundings of Hamburg, Germany. It became abundant when the drain was filled with leaf litter. Later, he found some specimens in similar habitats and classified the species as mesosaprobic (KAHL 1930b). The redescription of FOISSNER (1980b) is based on two small populations from temporary pasture ponds in the Austrian Central Alps (Grossglockner area), about 1500–2500 m above sea-level (FOISSNER 1980a). The Iceland neotype population occurred in a non-flooded Petri dish culture (pH 4.5) of dried moss and grass/shrub roots (*Carex rostrata, Eriophorum angustifolium, Comarum palustre*) from a mire in southwest Iceland (see type localities). The species became abundant two weeks after rewetting the sample and was associated with a variety of limnetic and terrestrial ciliates. *Protospathidium vermiculus* has been recorded also by several other authors, all from Europe, except of GONG (1986), who reported it from Lake Donghu, Wuhan, China: few specimens with a length of about 55 µm in a sample with rewetted *Calmus* leaves from Bavaria (WENZEL 1953); rather frequent, but not abundant, in the mud of the Hamburg Harbour, Germany (BARTSCH & HARTWIG 1984); in a slightly acidic (pH 5.5–6.9) thermal lake (11°C) in Slovakia (MATIS & STRAKOVÁ-STRIEŠKOVÁ 1991); and in the alpha-mesosaprobic mud of one of twelve sampling stations in the Stirone river, northern Italy (MADONI & GHETTI 1981a).

The type locality of *S. cucumis* is unknown, but likely in Bavaria, where BAUMEISTER lived and worked. It was abundant in a drain heavily polluted with liquid manure. VUXANOVICI (1962a) rediscovered some specimens in lake Fundeni near Bucarest, Rumania. DETCHEVA (1981, 1983a) reported *S. cucumis* from alpha-mesosaprobic mud of one out of eighteen sampling stations in the Maritza river system, Bulgaria.

Remarks: The Iceland population matches perfectly both, *S. vermiculus* and *S. cucumis*. All variations observed in these species occur also in the Iceland specimens. Thus, identification is beyond reasonable doubt and the population used as a neotype for *S. vermiculus*. Concomitantly, the species is transferred to *Protospathidium* because its oral ciliary pattern matches this genus.

Protospathidium vermiculus differs from *P. terricola* by body size (65 × 15 µm vs. 90 × 25 µm in vivo) and number of micronuclei (2 to 3 vs. 1) and ciliary rows (10 vs. 17–21). In vivo, *P. vermiculus* is easily confused with *P. serpens* which, however, is larger (65 × 15 µm vs. 90 × 15 µm in vivo); has a much longer (20 µm vs. 35–50 µm), tortuous and nodulated macronucleus (vs. basically oblong); and possesses a distinctly reduced brush row 1 (on average composed of 9 vs. 2–6 dikinetids). For distinction from *Edaphospathula minor,* see that species.

Protospathidium namibicola FOISSNER, AGATHA & BERGER, 2002 (Fig. 42a–n, 111a–c, e; Table 8)

2002 *Protospathidium namibicola* FOISSNER, AGATHA & BERGER, Denisia, 5: 305 (Type slides with protargol-impregnated specimens from type locality are deposited in the Oberösterreichische Landesmuseum in Linz, Upper Austria.).

Diagnosis: Size about 210 × 20 µm. Cylindroidal to elongate obclavate with conspicuous, hemispherical oral bulge about half as long as widest trunk region. Macronucleus a circa 130 µm long, tortuous, flattened strand. Extrusomes fine and rod-shaped, about 5 × 0.5 µm in size. On average 9 ciliary rows, 3 anteriorly differentiated to inconspicuous, heterostichad dorsal brush occupying 10% of body length and comprising up to 4 µm long bristles: brush row 1 composed of an average of 13 dikinetids, row 2 of 16, and row 3 of 10 dikinetids followed by a monokinetidal bristle tail. Right side oral kinetofragments usually composed of 4 dikinetids each.

Type locality: Dune soil (sand) in the Central Namib Escarpment, north of the village of Solitaire, Namibia, E16° S23°50'.

Etymology: The compound Latin noun *namibicola* (inhabiting the Namib Desert) refers to the habitat the species was discovered.

Description: Size 150–280 × 10–30 µm in vivo, usually about 210 × 20 µm; length:width ratio also highly variable, viz., 6.2–19.2:1, on average near 11:1 both in vivo and preparations (Table 8); unflattened and not contractile. Shape basically elongate obclavate gradually broadening posteriorly, widest region in or below mid-body, both ends narrowly rounded (Fig. 42a, d, i, 111e); slenderest specimens rod-shaped, broadest ordinarily obclavate (Fig. 42f, g); more or less distinctly inflated immediately after ingesting large prey ciliates. Macronucleus invariably an about 130 µm long, tortuous strand usually commencing in second third of body and extending to near body end; flattened ribbon-like and/or more or less distinctly spiralized in about 70% of specimens; nucleoli globular, small; condenses to a globular mass in middle dividers. Many ellipsoidal micronuclei along macronucleus, difficult to identify because of similarly sized and impregnated cytoplasmic inclusions (Fig. 42a, f, g, j, 111e; Table 8). Contractile vacuole in rear body end, on average five excretory pores on left posterior pole area. Extrusomes packed in oral bulge and scattered in cytoplasm, posterior third occasionally impregnates with protargol; individual extrusomes fine and rod-shaped with rounded ends, about 5 µm long (Fig. 42a, c, h). Cortex very flexible, contains about eight rows of minute (≤ 0.4 µm), colourless granules between each two ciliary rows (Fig. 42h). Cytoplasm colourless, contains few to many lipid droplets up to 5 µm across and remnants of prey ciliates, such as oral baskets and cytoplasmic crystals. Feeds on small (*Leptopharynx costatus*, *Pseudochilodonopsis mutabilis*) and medium-sized (*Gonostomum affine*) ciliates ingested whole because still identifiable in early food vacuoles; some specimens contained five to ten, only partially digested prey ciliates, showing that *P. namibicola* is an effective predator. Glides and swims slowly, often appearing vermiform due to the strong flexibility and curved anterior body half (Fig. 42d).

Cilia about 10 µm long in vivo, arranged in an average of nine equidistant, mostly bipolar rows connected with the circumoral kinetofragments and rather loosely ciliated in cylindroidal neck region (Fig. 42a, i, j; Table 8). Dorsal brush three-rowed and heterostichad, inconspicuous because occupying only 10% of body length and bristles merely up to 4 µm long; brush region soft because occasionally slightly inflated in protargol preparations. Anterior tail of brush rows conspicuous, composed of closely spaced, ordinary cilia, continue as somatic kineties posteriorly; row 1 on average composed of 13 dikinetids, row 2 of 17, and row 3 of 10 dikinetids followed by a monokinetidal bristle tail; brush dikinetids with a 3–4 µm long, distally slightly inflated anterior bristle and a 2–3 µm long posterior bristle (Fig. 42a, b, l, n, 111c; Table 8).

Oral bulge conspicuous in vivo, although only half as wide as broadest trunk region, because hemispherical and up to 6 µm high, appears as a distinct knob at low and middle magnifications (Fig. 42a, d, g, 111a, b); obovate in frontal view and bright due to the extrusomes contained (Fig. 42c, m). Circumoral kinety at base of oral bulge, composed of dikinetidal kinetofragments attached to the somatic ciliary rows and separated from each other by gaps one to three dikinetids wide; individual kinetofragments composed of three to six dikinetids with zigzagging basal bodies, dikinetidal organization

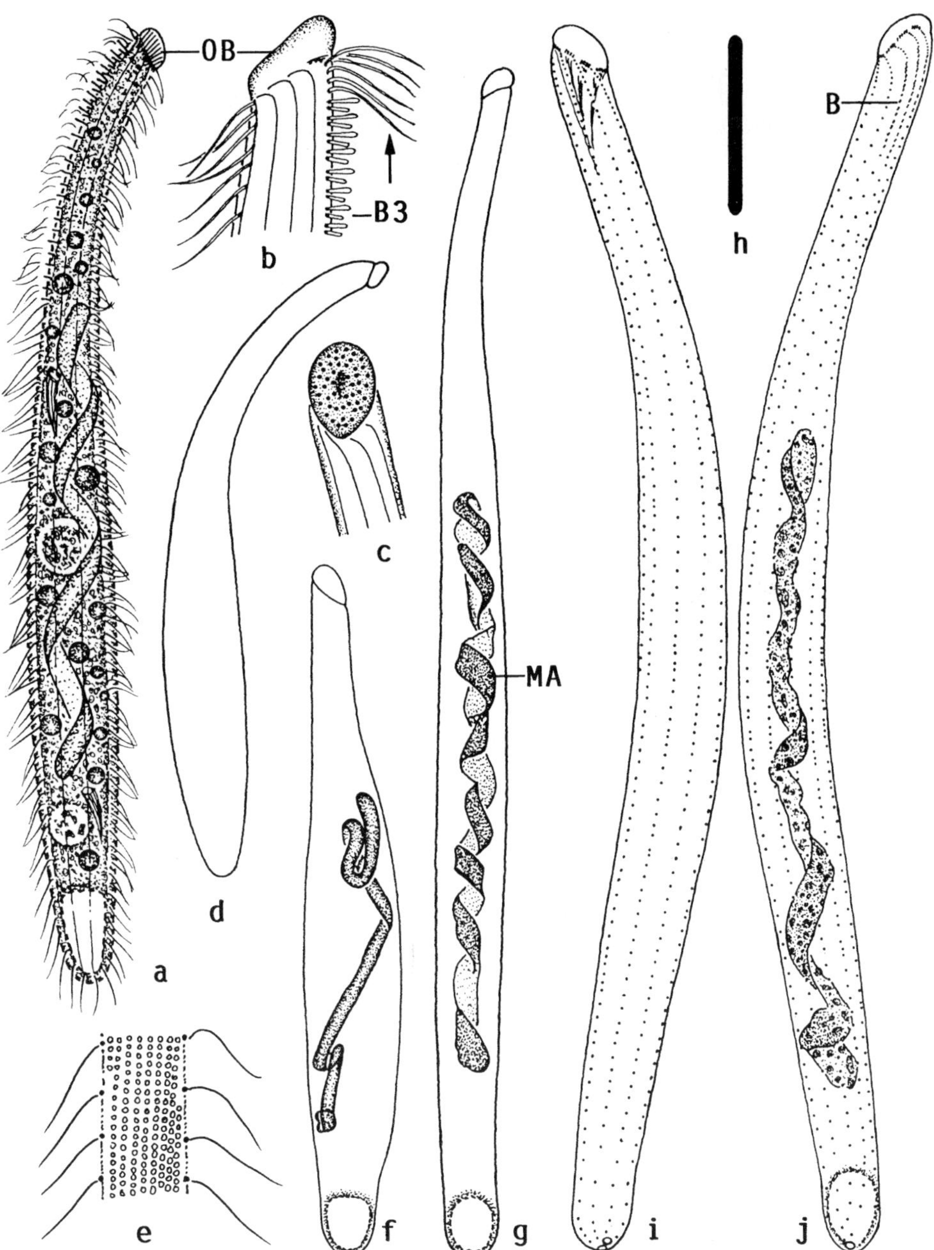

Fig. 42a–j *Protospathidium namibicola* from life (a–e, h) and after protargol impregnation (f, g, i, j). From FOISSNER et al. (2002). **a)** Right side view of a representative specimen, length 210 µm. **b)** Left anterior portion, showing monokinetidal anterior tail (arrow) of brush row 3. **c)** Frontal view of oral bulge. **d)** Shape variant. e) Surface view showing cortical granulation. **f, g)** One of the broadest, respectively, slenderest specimens found, length 122 and 152 µm. Note the flattened, spiralized macronucleus. **h)** Extrusome, length 5 µm. **i, j)** Ciliary pattern of right and left side and nuclear apparatus of holotype specimen, length 200 µm (for details, see next plate). B(3) – dorsal brush (row), MA – macronucleus, OB – oral bulge.

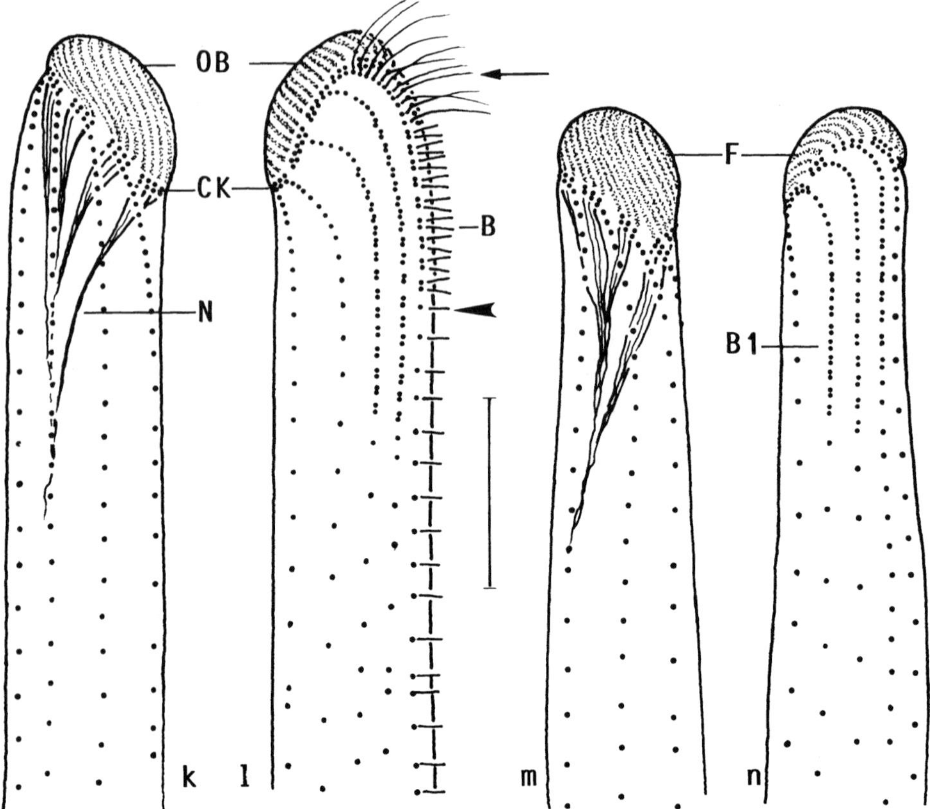

Fig. 42k–n *Protospathidium namibicola*, somatic and oral infraciliature of right (k, m) and left (l, n) anterior body portion after protargol impregnation; figures (k, l) show the same cell as figures (i, j), that is, the holotype specimen (from FOISSNER et al. 2002). Arrow marks the long monokinetidal tail at anterior end of brush rows, while the arrowhead denotes the monokinetidal posterior bristle tail of brush row 3. Note the hemispherical oral bulge, which contains distinct fibres (F) extending spirally to the bulge centre. Nematodesmata (oral basket rods, N) originate only from the circumoral kinetofragments (CK). B (1) – dorsal brush (row), CK – circumoral kinety (fragments), F – fibres, N – nematodesmata, OB – oral bulge. Scale bar 10 µm.

clearly recognizable only in strongly impregnated cells. Each dikinetid associated with a rather thick fibre spiralling to bulge centre, and a short nematodesma contributing to the comparatively (body size!) inconspicuous oral basket (Fig. 42a, i–n, 111a–c; Table 8).

Occurrence and ecology: To date found only at type locality, where it was abundant, indicating a high prey density not only in the non-flooded Petri dish culture, but also in the natural environment. *Protospathidium namibicola* is well adapted to the sandy habitat by its slender, flexible body.

Remarks: *Protospathidium namibicola* is a conspicuous species differing from its supposed nearest relative, *P. arenicola,* by body size (188 × 18 µm vs. 227 × 25 µm in protargol preparations), the shorter (134 µm vs. 274 µm) and more tortuous macronucleus, the number of ciliary rows (9 vs. 14), the shape of the oral bulge (hemispherical vs. ordinarily discoidal) and, especially, the shape of the extrusomes (fine rods vs. very narrowly ellipsoidal and distinctly asymmetric; Fig. 112u, y, z). It is the last mentioned feature, present also in the USA specimens, which convinced us that these populations represent distinct species. With respect to the morphometrics mentioned above, the USA specimens are more or less distinctly in between those from Namibia and Australia.

Protospathidium namibicola and *P. arenicola* have a clear identity and thus cannot be confused with *P. muscicola* or *P. vermiforme*, which are smaller (< 150 µm on average) and have scattered macronucleus nodules and a distinctly reduced brush row 1; furthermore, *P. vermiforme* is often tail-like posteriorly. Generally, however, care must be taken not to confuse *P. namibicola* with other slender spathidiids, for instance, *Arcuospathidium namibiense* (brush with up to 15 µm long bristles, scattered macronucleus nodules), members of the *Spathidium procerum* group (oral bulge distinctly longer, *Spathidium* ciliary pattern; volume II), and *Spathidium bonneti* (scattered macronucleus nodules, brush row 1 reduced to a single dikinetid; volume II). Furthermore, there are some slender *Enchelyodon* species, for instance, *E. armatides* FOISSNER et al. (2002), which look fairly similar due to the knobby oral bulge.

Table 8 Morphometric data on *Protospathidium namibicola* (PN; from FOISSNER et al. 2002) and *P. arenicola* from the USA (PAU) and Australia (PAA)

Characteristics[1]	Species	$\overline{x}$	M	SD	SE	CV	Min	Max	n
Body, length	PN	188.3	192.0	26.8	5.9	14.3	136.0	250.0	21
	PAU	203.0	205.0	29.0	6.3	14.3	154.0	255.0	21
	PAA	227.6	215.0	27.1	5.9	11.9	180.0	280.0	21
Body, width	PN	17.6	18.0	4.3	0.9	24.4	9.0	24.0	21
	PAU	27.2	26.0	6.6	1.4	24.2	16.0	46.0	21
	PAA	25.0	25.0	7.1	1.6	35.0	16.0	44.0	21
Body length:width, ratio	PN	11.4	10.4	3.4	0.8	30.2	6.2	19.2	21
	PAU	7.9	7.4	2.5	0.5	31.2	4.3	14.1	21
	PAA	9.7	9.0	2.6	0.6	26.7	4.7	14.9	21
Oral bulge, length	PN	9.3	10.0	1.0	0.2	10.4	7.0	11.0	21
	PAU	10.2	10.2	0.9	0.2	9.2	8.0	12.0	21
	PAA	12.9	13.0	1.8	0.4	14.2	8.0	16.0	21
Oral bulge, height	PN	4.4	4.0	1.0	0.2	23.2	3.0	6.0	21
	PAU	3.6	3.5	0.4	0.1	10.8	3.0	4.0	21
	PAA	3.4	3.5	–	–	–	3.0	4.0	21
Body width:oral bulge length, ratio	PN	1.9	1.8	0.4	0.1	23.7	1.1	2.8	21
	PAU	2.7	2.4	0.7	0.2	27.2	1.6	4.6	21
	PAA	1.9	1.8	0.5	0.1	24.7	1.3	3.1	21
Anterior body end to macronucleus, distance	PN	60.7	60.0	13.3	2.9	22.0	44.0	95.0	21
	PAU	61.3	57.0	12.8	2.8	20.9	45.0	90.0	21
	PAA	61.3	58.0	10.1	2.2	16.5	46.0	83.0	21
Macronucleus figure, length	PN	97.9	96.0	20.8	4.5	21.3	60.0	136.0	21
	PAU	105.2	115.0	27.0	5.9	25.7	40.0	146.0	21
	PAA	129.3	125.0	26.0	5.7	20.1	78.0	180.0	21
Macronucleus, length (spread; approximate)	PN	134.5	140.0	–	–	–	90.0	180.0	21
	PAU	211.0	200.0	–	–	–	100.0	360.0	21
	PAA	273.8	250.0	–	–	–	160.0	500.0	21
Macronucleus, width	PN	4.7	5.0	1.0	0.2	21.8	3.0	7.0	21
	PAU	4.6	5.0	0.7	0.2	14.8	3.0	6.0	21

continued

Characteristics[1]	Species	$\overline{x}$	M	SD	SE	CV	Min	Max	n
	PAA	3.8	4.0	0.9	0.2	24.4	3.0	7.0	21
Circumoral kinety to end of brush row 1, distance	PN	15.6	15.0	2.3	0.5	14.6	12.0	21.0	21
	PAU	20.1	20.0	4.4	1.0	21.7	13.0	27.0	21
	PAA	25.1	25.0	3.6	0.8	14.5	18.0	32.0	21
Circumoral kinety to end of brush row 2, distance	PN	18.5	18.0	2.6	0.6	13.9	14.0	24.0	21
	PAU	21.1	22.0	3.9	0.9	18.3	15.0	30.0	21
	PAA	25.8	26.0	4.3	0.9	16.6	20.0	36.0	21
Circumoral kinety to end of brush row 3, distance	PN	12.0	12.0	1.6	0.4	13.6	9.0	17.0	21
	PAU	13.2	13.0	2.4	0.5	18.0	9.0	18.0	21
	PAA	16.0	16.0	1.6	0.4	10.0	13.0	18.0	21
Ciliary rows, number	PN	9.4	10.0	1.2	0.3	12.8	7.0	11.0	21
	PAU	12.6	13.0	1.3	0.3	9.9	11.0	15.0	21
	PAA	14.2	14.0	0.9	0.2	6.6	12.0	16.0	21
Ciliated kinetids in a ventral ciliary row, number	PN	80.9	78.0	13.9	3.0	17.1	63.0	115.0	21
	PAU	111.5	110.0	18.4	4.0	16.5	80.0	155.0	21
	PAA	107.2	100.0	26.6	5.8	24.9	70.0	158.0	21
Dorsal brush rows, number	PN	3.0	3.0	0.0	0.0	0.0	3.0	3.0	21
	PAU	3.0	3.0	0.0	0.0	0.0	3.0	3.0	21
	PAA	3.0	3.0	0.0	0.0	0.0	3.0	3.0	21
Dikinetids in brush row 1, number	PN	13.0	13.0	2.9	0.6	21.9	8.0	18.0	21
	PAU	13.1	14.0	2.8	0.6	21.6	5.0	18.0	21
	PAA	18.2	18.0	3.3	0.7	18.4	12.0	24.0	21
Dikinetids in brush row 2, number	PN	15.9	17.0	2.7	0.6	17.2	11.0	21.0	21
	PAU	14.3	14.0	2.2	0.5	15.2	9.0	18.0	21
	PAA	18.9	19.0	4.0	0.9	21.5	12.0	27.0	21
Dikinetids in brush row 3, number	PN	9.7	10.0	1.8	0.4	18.5	7.0	13.0	21
	PAU	9.3	9.0	1.4	0.3	14.9	7.0	12.0	21
	PAA	10.9	11.0	2.2	0.5	20.1	7.0	16.0	21
Excretory pores, number	PN	5.3	5.0	1.8	0.4	33.2	3.0	8.0	21
	PAU	6.7	7.0	1.1	0.4	25.4	4.0	10.0	21
	PAA	5.6	5.0	2.0	0.4	35.8	3.0	10.0	21
Dikinetids composing a right side oral kinetofragment, number	PN				mostly 4.0				
	PAU	3.9	4.0	0.8	0.2	21.3	3.0	6.0	21
	PAA	3.7	4.0	–	–	–	3.0	5.0	21

[1] Data based on mounted, protargol-impregnated (FOISSNER's method), and randomly selected specimens from non-flooded Petri dish cultures. Measurements in µm. CV – coefficient of variation in %, M – median, Max – maximum, Min – minimum, n – number of individuals investigated, SD – standard deviation, SE – standard error of arithmetic mean, $\overline{x}$ – arithmetic mean.

Protospathidium arenicola **nov. spec.** (Fig. 43a–m, 44a–n, 111d, g, h, 112a–z; Table 8)

D i a g n o s i s : Size about 250 × 27 µm in vivo. Cylindroidal to elongate obclavate with ordinary (discoidal) oral bulge about half as long as widest trunk region. Macronucleus an about 270 µm long, highly tortuous, flattened strand. Extrusomes narrowly ellipsoidal and asymmetric, about 5 × 1 µm in size. On average 14 ciliary rows, 3 anteriorly differentiated to heterostichad dorsal brush occupying 11% of body length and comprising up to 4 µm long bristles: brush row 1 composed of an average of 13–18

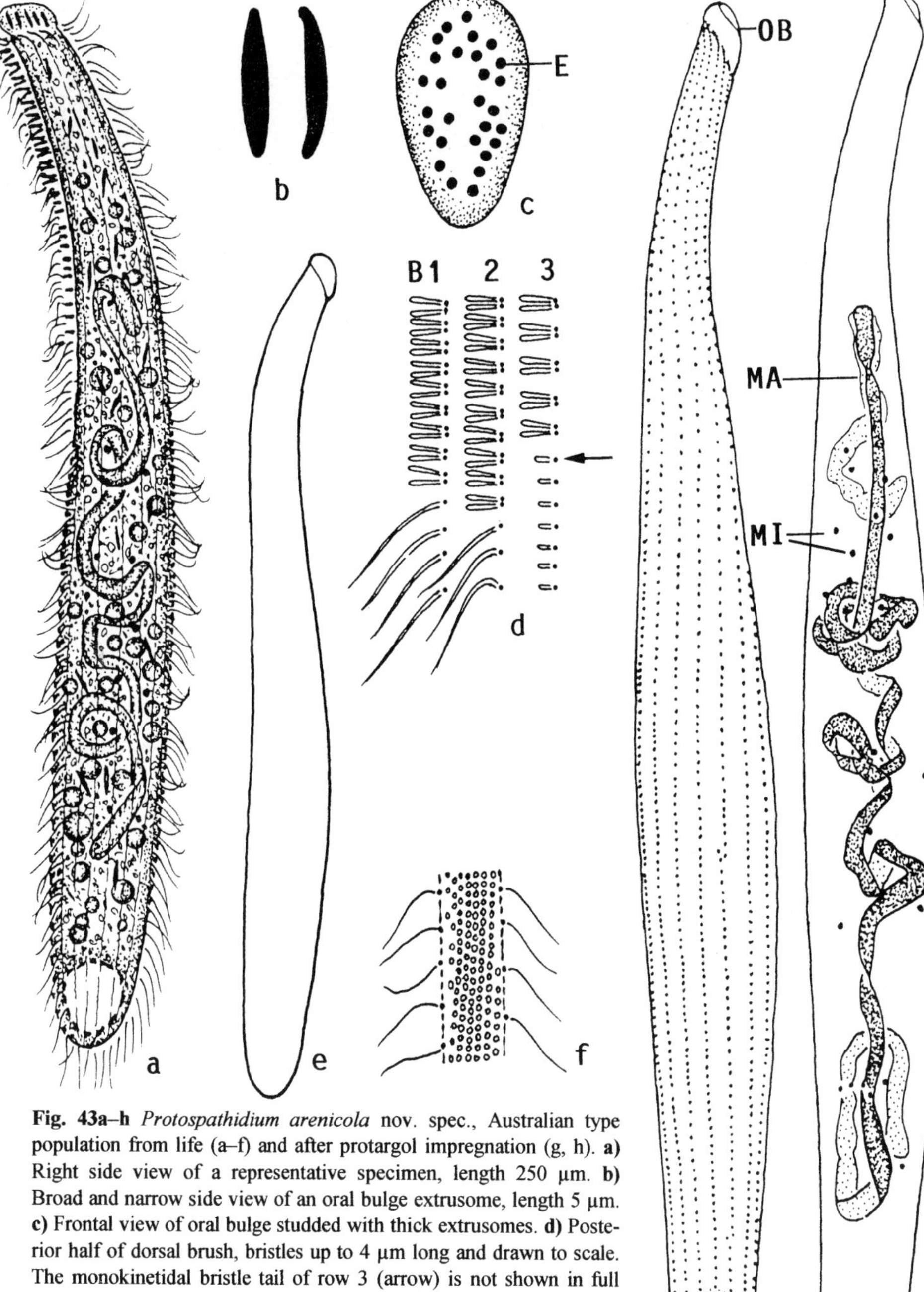

Fig. 43a–h *Protospathidium arenicola* nov. spec., Australian type population from life (a–f) and after protargol impregnation (g, h). **a)** Right side view of a representative specimen, length 250 µm. **b)** Broad and narrow side view of an oral bulge extrusome, length 5 µm. **c)** Frontal view of oral bulge studded with thick extrusomes. **d)** Posterior half of dorsal brush, bristles up to 4 µm long and drawn to scale. The monokinetidal bristle tail of row 3 (arrow) is not shown in full length. **e)** Slender shape variant. **f)** Surface view showing cortical granulation. **g, h)** Right side ciliary pattern and nuclear apparatus of holotype specimen, length 245 µm. For details, see following figures. Note the ribbon-like flattened macronucleus. B(1-3) – dorsal brush (rows), E – extrusomes, MA – macronucleus, MI – micronuclei, OB – oral bulge.

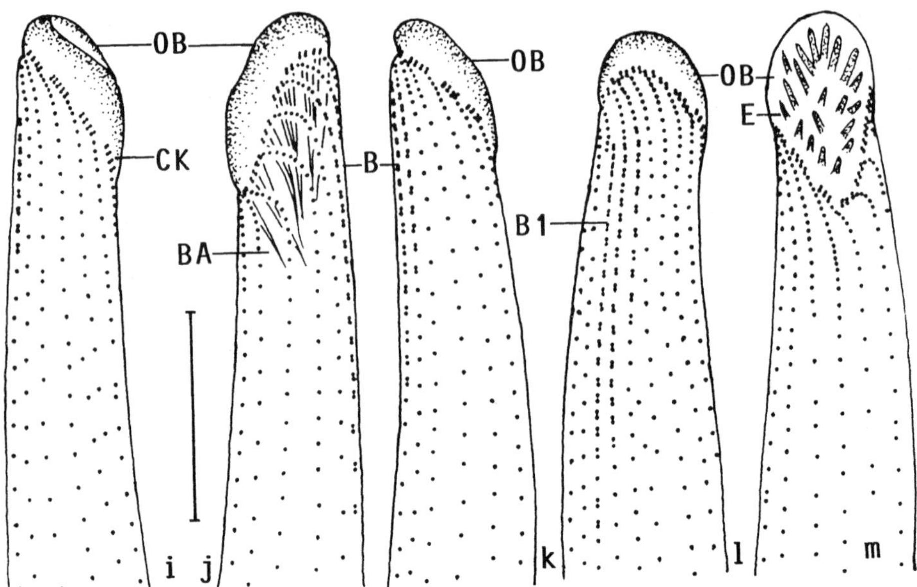

Fig. 43i–m *Protospathidium arenicola* nov. spec., Australian type population after protargol impregnation. **i, j)** Right and left side ciliary pattern in anterior portion of holotype specimen (overviews, see figures 43g, h). **k)** Right side ciliary pattern of another specimen. **l, m)** Dorsal and ventral ciliary pattern and extrusomes in anterior body portion. Scale bar 15 µm for all figures. B(1) – dorsal brush (row), BA – oral basket, CK – circumoral kinetofragments (kinety), E – extrusomes, OB – oral bulge.

dikinetids, row 2 of 14–19, and row 3 of 9–11 dikinetids followed by a monokinetidal bristle tail. Right side oral kinetofragments usually composed of 4 dikinetids each.

Type locality: Forest soil in the surroundings of Alice Springs, that is, a hill beside the road to the Ayers Rock, Australia, E133° S24°.

Etymology: The compound Latin noun *arenicola* (living in sand) refers to the habitat the species was discovered.

Description: Size 180–300 × 20–40 µm in vivo, usually near 250 × 27 µm; length:width ratio also highly variable, viz., 4.7–14.9:1, on average about 9:1 both in vivo and in protargol preparations (Table 8); unflattened and not contractile. Very narrowly spatulate to cylindroidal and often slightly obclavate, widest region in or below mid-body, both ends narrowly rounded, anterior half usually slightly curved; slenderest specimens rod-shaped, broadest very narrowly obclavate (Fig. 43a, b, g, 112q, v; Table 8). Macronucleus an about 270 µm long, highly tortuous, more or less ribbon-like flattened strand commencing in second quarter of cell and extending to near body end, occasionally distinctly spiralized or broken into two long pieces; nucleoli globular, small. On average 25 globular micronuclei attached to or near to macronucleus (Fig.

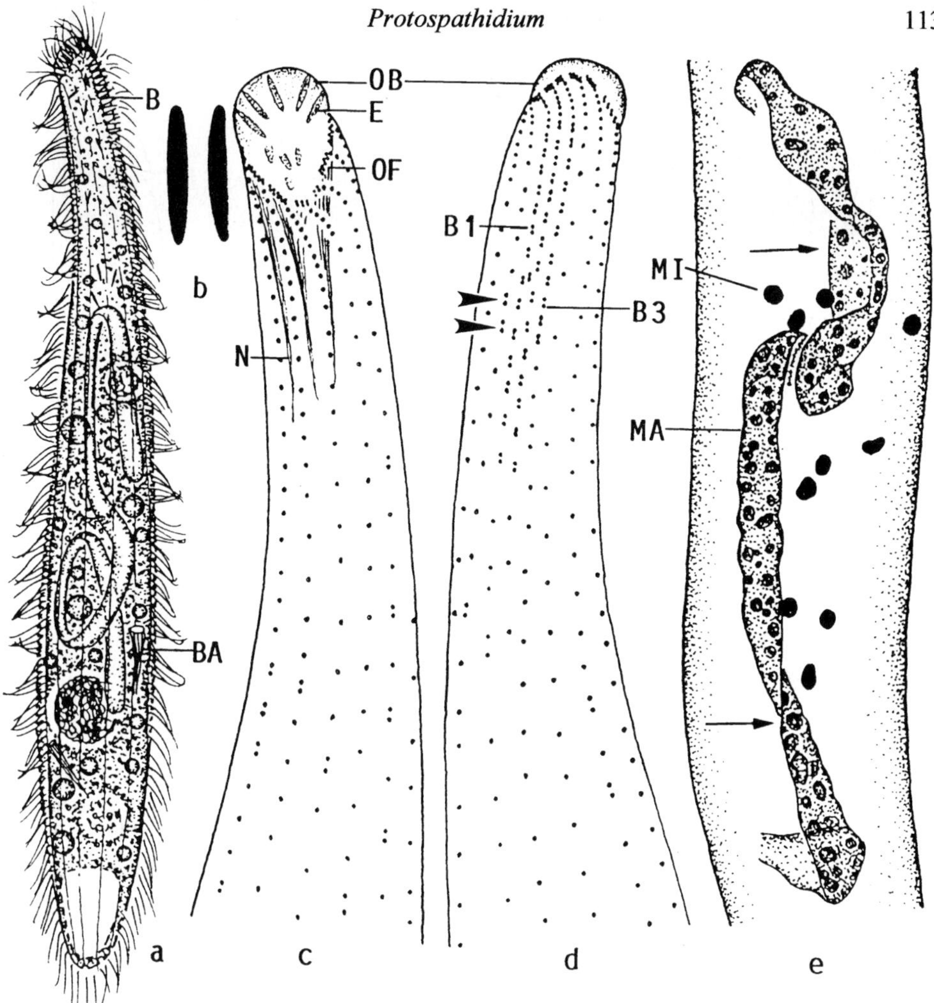

Fig. 44a–e *Protospathidium arenicola* nov. spec., USA specimens from life (a, b) and after protargol impregnation (c–e). **a)** Left side view of a representative specimen, length 230 µm. Note the long, flattened macronucleus and oral baskets (BA) of microthoracid ciliate prey. **b)** Oral bulge extrusome seen from the symmetrical and asymmetrical side, length 4 µm. **c, d)** Ventral and dorsal view of ciliary pattern in anterior body region, length of oral bulge 9 µm. The oral bulge extrusomes are recognizable because they are highly refractive and faintly impregnated. The nematodesmata form distinct bundles originating from the individual circumoral kinetofragments. This is a typical feature of *Protospathidium*. Arrowheads mark some dikinetids right of brush row 1. **e)** Posterior part of nuclear apparatus of the specimen shown in figure (j), width of cell 19 µm. Arrows mark regions where the macronucleus flattening is recognizable. B(1-3) – dorsal brush (rows), BA – oral basket of microthoracid ciliate prey, E – extrusomes, MA – macronucleus, MI – micronucleus, N – nematodesma bundle, OB – oral bulge, OF – circumoral kinetofragments.

43a, h, 112q, v; Table 8). Contractile vacuole in rear body end, on average six excretory pores scattered in pole area; no second contractile vacuole in anterior body half. Extrusomes studded in oral bulge and scattered in cytoplasm, impregnate brownish with protargol sometimes showing a dark globule in posterior end; those in cytoplasm often narrowly ovate in protargol preparations; oral bulge extrusomes in vivo about 4–5 × 0.8–1.1 µm in size and asymmetrical, that is, narrowly ellipsoidal in broad side view

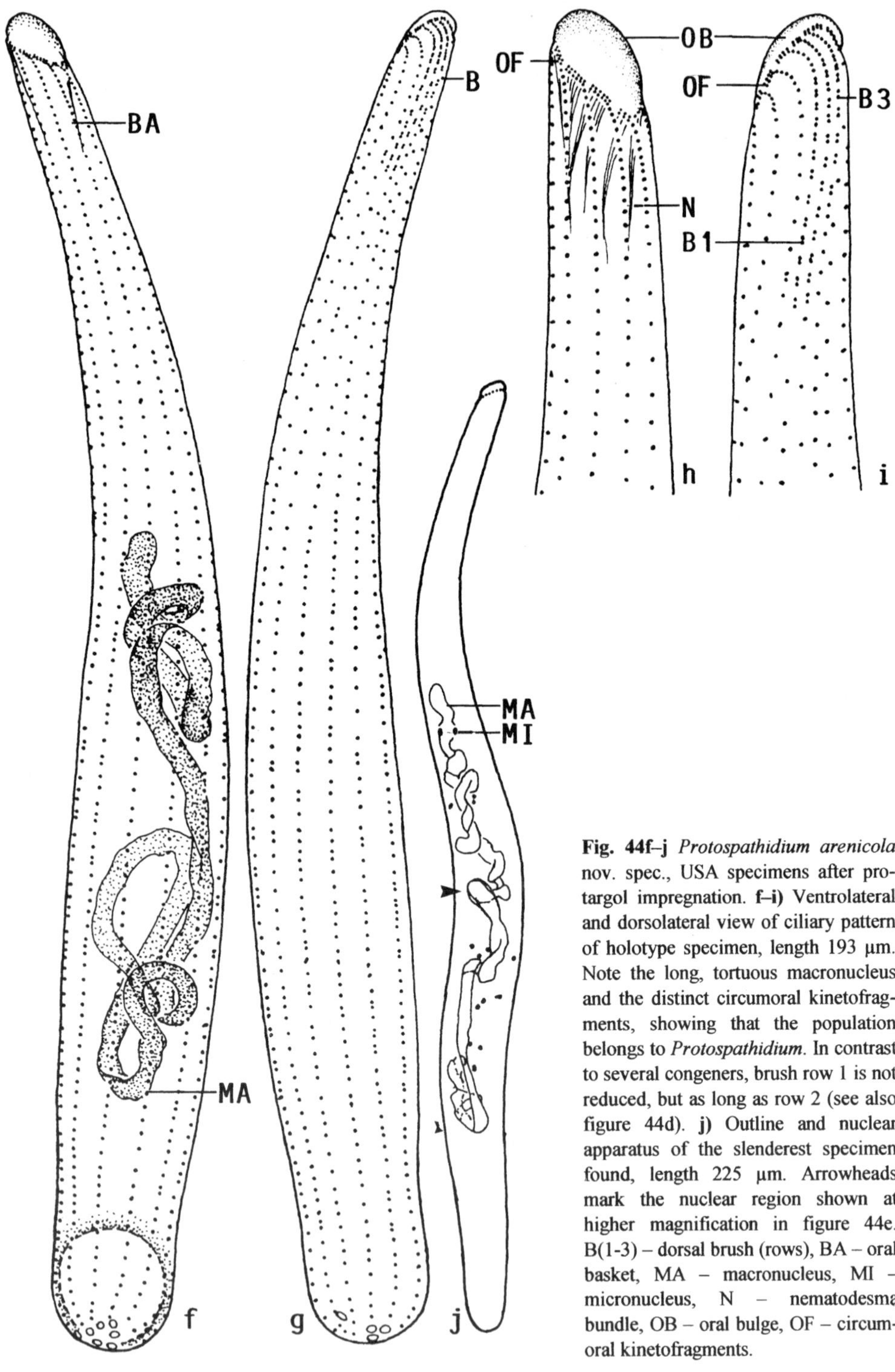

Fig. 44f–j *Protospathidium arenicola* nov. spec., USA specimens after protargol impregnation. **f–i)** Ventrolateral and dorsolateral view of ciliary pattern of holotype specimen, length 193 µm. Note the long, tortuous macronucleus and the distinct circumoral kinetofragments, showing that the population belongs to *Protospathidium*. In contrast to several congeners, brush row 1 is not reduced, but as long as row 2 (see also figure 44d). **j)** Outline and nuclear apparatus of the slenderest specimen found, length 225 µm. Arrowheads mark the nuclear region shown at higher magnification in figure 44e. B(1-3) – dorsal brush (rows), BA – oral basket, MA – macronucleus, MI – micronucleus, N – nematodesma bundle, OB – oral bulge, OF – circumoral kinetofragments.

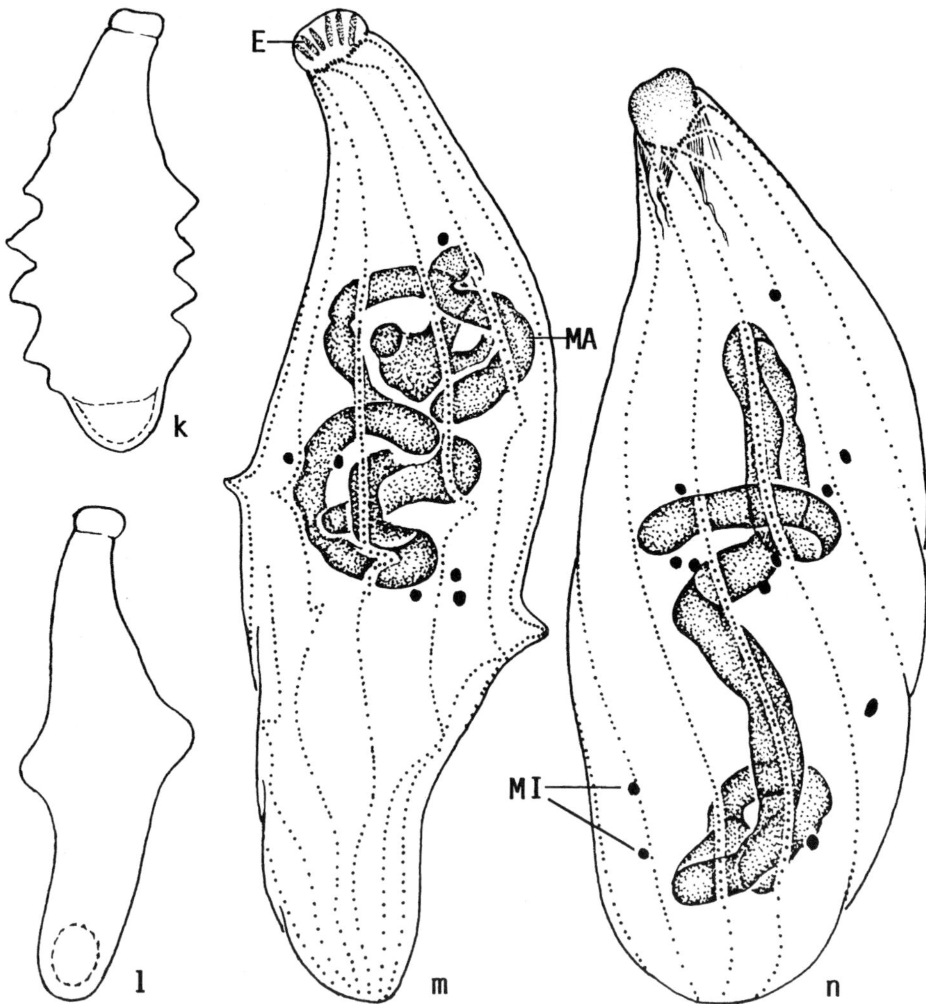

Fig. 44k–n *Protospathidium arenicola* nov. spec., USA specimens from life (k, l) and after protargol impregnation (m, n). These figures are from a very early stage of culture growth, where most specimens are distinctly wrinkled forming conspicuous, conical processes (k–m). Such specimens have more or less spiral ciliary rows and can extend slowly (n), showing that spathidiids have some sort of contractility. Length 115 µm (m), 112 µm (n). E – extrusomes, MA – macronucleus, MI – micronuclei.

and indistinctly C-shaped in narrow side view (Fig. 43a–c, m, 112s–u, y, z). Cortex very flexible, contains about seven rows of minute (0.3 µm), colourless granules between each two ciliary rows. Cytoplasm colourless, contains few to many lipid droplets up to 5 µm across and remnants of prey, e.g., oral baskets of microthoracid ciliates or euglenids. Swims and glides slowly moving to and fro the anterior body third.

Cilia about 8 µm long in vivo, arranged in an average of 14 equidistant, rather densely ciliated rows connected with circumoral kinetofragments; frequently, one or two rows slightly shortened anteriorly. Dorsal brush heterostichad and inconspicuous because occupying only 11% of body length and bristles merely up to 4 µm long in vivo. All rows of similar structure and with conspicuous anterior tail of closely spaced, ordinary cilia; continue as somatic kineties posteriorly; frequently with one or two small breaks; posterior tail of row 3 extending to second body third (Fig. 43a, g–m, 112r, s, w, x; Table 8).

Oral bulge fairly conspicuous in vivo, although only half as wide as broadest trunk region, because rather distinctly separate from body proper, 4–5 µm high, and bright due to the refractive extrusomes contained; obovate in frontal view with extrusomes appearing as brilliant dots. Circumoral kinety not obovate as oral bulge, but cuneate; composed of kinetofragments separated from each other by gaps one to three dikinetids wide; individual kinetofragments each composed of three to five, on average four zigzagging dikinetids associated with short nematodesmata forming a comparatively (body length!) inconspicuous oral basket (Fig. 43a, g–m, 112r, s, w, x; Table 8).

USA population: We found *P. arenicola* also in an inland sand dune near the town of Saint-Anthony, Idaho, W112° N43°. The dune was almost completely covered with a thin layer of moss and shrubs. The sample consisted of leaf litter, moss, and sand, and had pH 6.8 in water.

The Idaho specimens are more similar to *P. arenicola* than *P. namibicola* because they have the same extrusome and oral bulge shape, while most morphometrics are in between, for instance, the number of ciliary rows: 9 in *P. namibicola*, 13 in the Idaho specimens, and 14 in the Australian population. For details, see figures 44a–j, 111d, g, h, 112a–p, and Table 8.

See explanations to figures 44k–n and the remarks section in *Spathidium contractile* (volume II), for the description of a curious phenomenon observed in a very young culture, where specimens were distinctly wrinkled and contracted.

Occurrence and ecology: *Protospathidium arenicola* occurred at several sites in Australia, indicating wide distribution on this continent. All these sites, the USA locality, and the type locality of *P. namibicola* are very sandy, suggesting that these species prefer such habitats. We did not find them in European inland sand dunes, indicating restricted distribution. The long, slender body suggests that both evolved in terrestrial environments (FOISSNER 1987a). In the non-flooded Petri dish cultures, abundance was high at the Australian type locality, while low in the Idaho sample.

Remarks: For comparison with similar species, see *P. namibicola*.

Protospathidium vermiforme FOISSNER, AGATHA & BERGER, 2002 (Fig. 45a–j)

1981 *Protospathidium bonneti* nov. comb. (BUITKAMP, 1977) – FOISSNER, Zool. Jb. Syst., 108: 271 (misidentification).

1986 *Protospathidium bonneti* (BUITKAMP, 1977) – DRAGESCO & DRAGESCO-KERNEIS, Faune trop., 26: 156 (revision).

2002 *Protospathidium vermiforme* FOISSNER, AGATHA & BERGER, Denisia, 5: 310 (Type slides with protargol-impregnated specimens are deposited in the Oberösterreichische Landesmuseum in Linz, Upper Austria, under the name *Protospathidium bonneti*.).

Diagnosis: Size 80–160 × 5–13 µm in vivo. Vermiform with distinct, oblique oral bulge slightly shorter than widest trunk region. About 15 macronucleus nodules and 8 ciliary rows, 3 anteriorly differentiated to conspicuous dorsal brush with up to 8 µm long bristles. Oral kinetofragments each composed of 3–4 dikinetids.

Type locality: Grassland soil from the surroundings of the "Wallack-Haus" at the Grossglockner-Hochalpenstrasse, Carinthia, Austria, E12°50' N47°.

Etymology: The Latin adjective *vermiforme* refers to the worm-like body shape.

Description: Size 80–160 × 5–13 µm in vivo, usually near 120 × 10 µm, slightly to distinctly flattened laterally. Vermiform, anterior (oral) body end obliquely truncate and slightly shorter than broadest trunk region, widest in middle third, posterior portion tail-like narrowed or narrowly rounded; both body ends fragile, that is, often dissolve under coverslip, tail possibly sometimes lacking. Nuclear apparatus in middle body third, composed of about 15 macronucleus nodules circa 4 × 2.7 µm in size and, likely, several micronuclei; nodules occasionally incompletely separated, producing more or less moniliform pieces. Contractile vacuole subterminal in base of tail or terminal if tail is lacking or lost, bulges body end when filled completely (Fig. 45a–c, e). Extrusomes not studied, possibly rod-shaped. Cortex fragile, bright. Cytoplasm colourless, contains some refractive granules. Moves slowly, worm-like, and thus easily confused with small nematodes at low magnification.

Somatic cilia about 8 µm long in vivo, arranged in about eight bipolar and ordinarily spaced rows more densely ciliated anteriorly than postorally, especially first row on right side having about ten very narrowly spaced cilia anteriorly, similar as in some congeners. Individual ciliary rows composed of about 30 cilia, that is, rather loosely ciliated and attached to circumoral kinetofragments in typical *Protospathidium* pattern (Fig. 45a, b, f, i). Dorsal brush three-rowed, not yet studied in detail in vivo, where it is rather conspicuous due to some about 8 µm long bristles anteriorly, similar as in *Arcuospathidium namibiense*; most bristles, however, only 2 µm long; all rows with a short tail of ordinary cilia anteriorly, especially distinct in row 1 composed of only three dikinetids. Row 2 longer than rows 1 and 3, composed of about 15 narrowly spaced dikinetids. Row 3 slightly shorter than row 2, composed of about 10 comparatively widely spaced dikinetids (Fig. 45a, b, g, h, j).

Oral bulge distinct in vivo because about 4 µm high and somewhat knobby, slightly twisted along main body axis and likely obovate in frontal view. Individual oral kinetofragments composed of three to four dikinetids. Nematodesmata neither recognizable in vivo nor protargol preparations (Fig. 45a, b, d, f–j).

Occurrence and ecology: The FOISSNER group misidentified this and similar species as *Spathidium bonneti* for a long time (see below and FOISSNER et al. 2002). Meanwhile, it is clear that there are several very slender, middle-sized spathidiids belonging to even different genera. Thus, all former records are doubtful and not mentioned here. Only the records from sites near to the locality are reliable, that is, those from the Grossglockner and Gastein region (FOISSNER 1981b, FOISSNER & PEER 1985). These

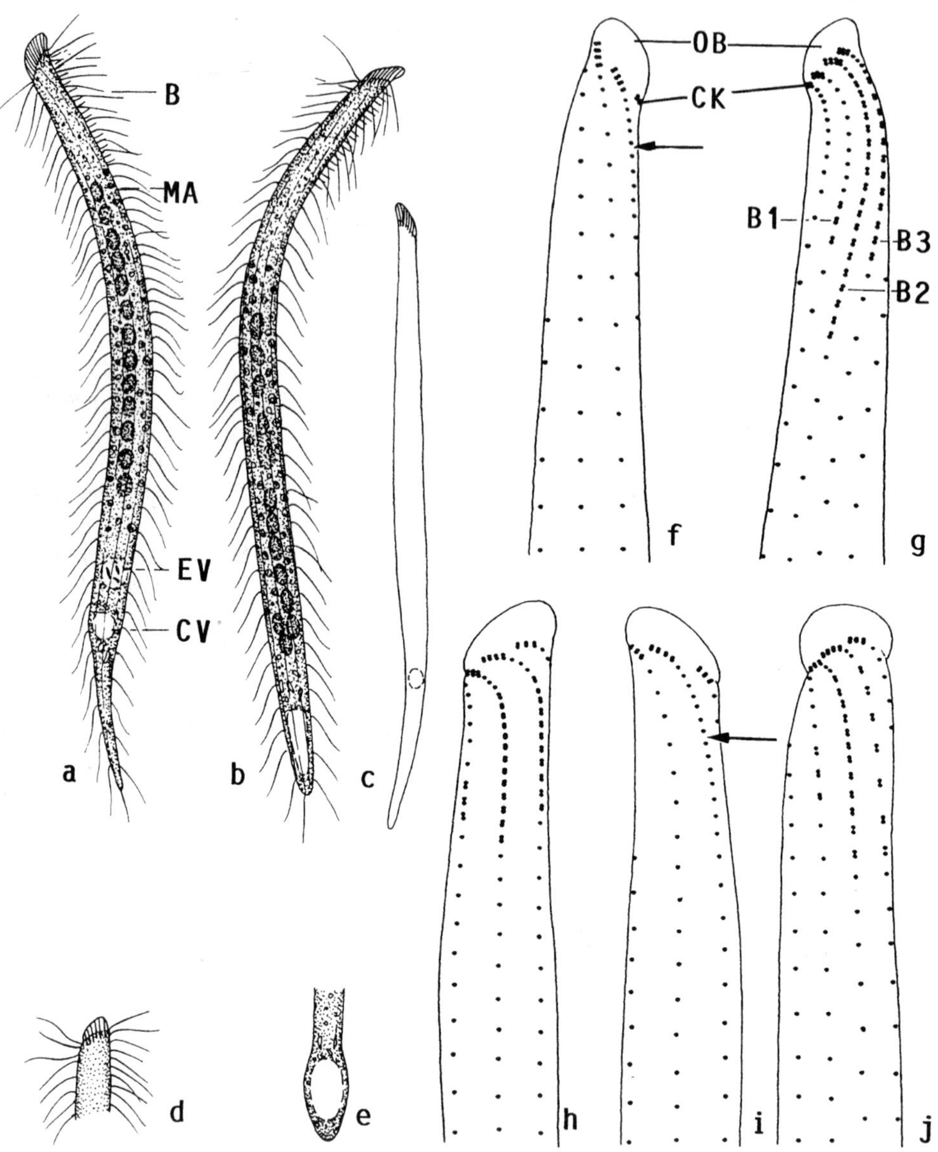

Fig. 45a–j *Protospathidium vermiforme* (from FOISSNER 1981a) from life (a–e) and after protargol impregnation (f–j). **a, b)** Left side views of representative specimens, length 125 µm and 110 µm. **c)** A very slender specimen, length 150 µm. **d)** Dorsolateral view of slightly spiraled anterior body region. **e)** Posterior body end when contractile vacuole is filled. **f–j)** Ciliary pattern in anterior right (f, i), left (g, h), and dorsal (j) body region; figure pairs (f, g; h, i) show both sides of two specimens, length 27 µm, 46 µm, 31 µm. Arrows mark condensation of cilia in anterior portion of a ventrolateral kinety. The protospathidiid oral kinetofragments are very distinct. Brush row 1 consists of three dikinetids far away from the circumoral kinety. B (1-3) – dorsal brush (rows), CK – circumoral kinetofragments, CV – contractile vacuole, EV – egestion vacuole, MA – macronucleus, OB – oral bulge.

data show that *P. vermiforme* is rather common in alpine grassland soils of Austria. Further, it is likely restricted to terrestrial habitats because similar species never have been reported from limnetic environments.

Remarks: FOISSNER (1981a) redescribed *Spathidium bonneti* BUITKAMP, 1977a from soil of the Austrian Central Alps and transferred it to *Protospathidium*, assuming that BUITKAMP overlooked the protospathidiid oral kinetofragments in his population from the Ivory Coast. However, DRAGESCO & DRAGESCO-KERNÉIS (1986) doubted FOISSNER's identification and suggested that the Austrian population is a new species and BUITKAMP's species the representative of a new genus. As concerns the species, we fully agree with DRAGESCO & DRAGESCO-KERNÉIS because we learned that there are quite a lot of slender spathidiids in soil, some of which have, like BUITKAMP's species, a continuous circumoral kinety, for instance, *Arcuospathidium namibiense*. Comparing the figures of *Protospathidium vermiforme* (Fig. 45a–c) with that of *Spathidium bonneti* (Fig. 62o), FOISSNER's misidentification appears understandable. Such mistakes are simply caused by our great ignorance of ciliate diversity outside the holarctic region. Whether BUITKAMP's species belongs to *Spathidium* or *Arcuospathidium* needs further investigation.

Clearly, *P. vermiforme* needs redescription from type locality to obtain detailed information on extrusomes and morphometrics. In vivo, *P. vermiforme* is indistinguishable from *Arcuospathidium namibiense* and *Spathidium bonneti*. Thus, protargol impregnation is required to reveal the structure of the circumoral kinety: composed of a continuous line of dikinetids in *Spathidium* and *Arcuospathidium* vs. short, dikinetidal fragments attached to the ciliary rows in *Protospathidium* (Fig. 45f–j). As concerns the congeners, only *P. serpens* and *P. muscicola* are similar to *P. vermiforme*. However, *P. serpens* is distinctly stouter (< 6:1 vs. > 10:1) and has a tortuous, irregularly nodulated macronucleus; likewise, *P. muscicola* is stouter and has the macronucleus nodules scattered.

Protospathidium muscicola DRAGESCO & DRAGESCO-KERNÉIS, 1979 (Fig. 46a–u, 47a–m, 48a–k, 113a–k; Table 9)

1979 *Protospathidium muscicola* gen. n., sp. n. DRAGESCO & DRAGESCO-KERNÉIS, Acta Protozool., 18: 411.

1984 *Protospathidium serpens* (KAHL, 1930) – BERGER, FOISSNER & ADAM, Zool. Jb. Syst., 111: 355 (misidentification; mainly description of ontogenesis; voucher slides with protargol-impregnated specimens of this Austrian population are deposited in the Oberösterreichische Landesmuseum in Linz, Upper Austria).

1986 *Protospathidium muscicola* DRAGESCO & DRAGESCO-KERNÉIS, 1979 – DRAGESCO & DRAGESCO-KERNÉIS, Faune trop., 26: 156 (comparison with *Spathidium bonneti*, as redescribed by FOISSNER 1981a; two figures rather different from those in DRAGESCO & DRAGESCO-KERNÉIS 1979, although designated as "d'après DRAGESCO & DRAGESCO-KERNÉIS 1979").

Type material: DRAGESCO & DRAGESCO-KERNÉIS will deposit type slides with protargol-impregnated specimens in the Oberösterreichische Landesmuseum in Linz, Upper Austria (Prof. Dr. JEAN DRAGESCO, pers. comm.). Voucher slides with protargol-impregnated Venezuelan and Botswanan specimens, described here, are deposited at the same locality.

Diagnosis (based on literature and newly investigated populations from Venezuela and Botswana): Size about 80–110 × 10–20 µm in vivo. Narrowly spatulate to obclavate with oblique, obovate oral bulge about half as long as widest trunk region. Usually 14–23 ellipsoidal, scattered macronucleus nodules and several micronuclei. Extrusomes oblong to very narrowly ovate, 3–4 µm long. On average 9–11 ciliary rows, 3 anteriorly differentiated to distinctly heterostichad dorsal brush occupying 17–22% of body length and comprising up to 4 µm long bristles: brush row 1 composed of only 2–4 dikinetids, row 2 of 12–15, and row 3 of 9–10 dikinetids followed by some monokinetidal bristles. Right side oral kinetofragments each comprising an average of 4–6 dikinetids. Type VII resting cysts with rounded spines 0.5–2 µm high.

Type locality: Not mentioned in original description. On request, Prof. Dr. JEAN DRAGESCO informed us that he discovered *P. muscicola* in moss from the University garden of Cotonou, Benin, E2°20' N6°15'.

Etymology: Not given in original description. The compound Latin noun *muscicola* (inhabiting moss) obviously refers to the habitat the species was discovered.

Description: Size and length:width ratio of individual populations rather variable in protargol preparations (Table 9): 52–110 × 10–18 µm, $\bar{x}$ 75.2 × 13 µm, n ? (African type; DRAGESCO & DRAGESCO-KERNÉIS 1979); 50–82 × 7–14 µm, $\bar{x}$ 64.8 × 9.6 µm (Austrian population; BERGER et al. 1984; Table 9); 74–110 × 15–23 µm, $\bar{x}$ 94.9 × 19 µm (Venezuelan population; Table 9). Taking into account some preparation shrinkage and the Botswanan specimens described below, an usual in vivo size of 80–110 × 10–20 µm and a length:width ratio of 5–7:1 can be estimated, with largest specimens up to 140 µm long (DRAGESCO & DRAGESCO-KERNÉIS 1979). Shape narrowly spatulate to very narrowly ellipsoidal, or slightly obclavate and more or less curved (most Venezuelan specimens); anterior (oral) body end obliquely truncate by 25° to 45° and almost half as long as broadest trunk region, widest in or underneath mid-body, posterior end narrowly to moderately broadly rounded; slightly flattened laterally, especially in hyaline anterior body portion (Fig. 46a, b, k, q, s–u, 47a, h, 48a–d, 113a–d). Nuclear apparatus composed of scattered macronucleus nodules and several minute, globular micronuclei not clearly recognizable in type population, though DRAGESCO & DRAGESCO-KERNÉIS (1979) marked a large nodule as a micronucleus (Fig. 46c). Number of macronucleus nodules rather variable (Fig. 46a, c, k, q, 47a, i, m, 48a–d, 113a–d; Table 9): 10–20 in type population, 15–30 in Austrian specimens (BERGER et al. 1984), and 9–21 in Venezuelan population, where 41 out of 60 specimens analysed have perfectly scattered nodules (including five early dividers, which show the true pattern best), 18 cells have a mixture of scattered nodules and one or a few short strands comprising nodules usually connected by a fine strand, one specimen (possibly a late post-divider) has six rather long pieces (Fig. 46t), and one cell (possibly a monster and/or post-divider, as indicated by the imperfectly shaped oral bulge and circumoral kinety) has a long, tortuous strand (Fig. 46s); nucleoli scattered and 1–2 µm across. Contractile vacuole in posterior body end, several excretory pores in dorsal pole area (Fig. 46a, k, r, t, 47a, l). Extrusomes studded in oral bulge and scattered in cytoplasm. Oral bulge extrusomes asymmetrically oblong and 2.5–3 × 0.3–0.4 µm in size in populations from Botswana and the Republic of South Africa (Fig. 46, upper right corner); very narrowly

Table 9 Morphometric data on a Venezuelan (upper line; original) and an Austrian (lower line; from BERGER et al. 1984) population of *Protospathidium muscicola*

Characteristics[1]	$\overline{x}$	M	SD	SE	CV	Min	Max	n
Body, length	94.9	98.0	11.1	2.4	11.7	74.0	110.0	21
	64.8	67.6	9.3	1.8	14.4	50.0	82.0	28
Body, width	19.0	19.0	2.1	0.5	11.0	15.0	23.0	21
	9.6	9.5	1.6	0.3	16.3	7.0	14.0	28
Body length:width, ratio	5.1	4.9	0.8	0.2	16.8	3.7	6.8	21
	7.0	6.8	1.8	0.3	25.0	3.6	10.2	28
Oral bulge, length (cells oriented laterally)	8.7	9.0	1.2	0.3	13.6	7.0	11.0	21
	5.5	5.3	0.6	0.1	11.1	4.5	7.0	28
Oral bulge, width (cells oriented ventrally)	7.5	7.0	1.0	0.2	13.0	6.0	9.0	21
	–	–	–	–	–	–	–	–
Oral bulge, height	2.9	3.0	0.3	0.1	10.9	2.5	3.5	8
	2.3	2.3	0.5	0.1	22.8	1.5	3.0	28
Circumoral kinety to last dikinetid of brush row 1, distance	7.0	7.0	1.5	0.3	20.8	5.0	10.0	21
	6.3	6.0	0.9	0.2	14.2	4.0	8.0	28
Circumoral kinety to last dikinetid of brush row 2, distance	21.0	21.0	3.1	0.7	15.0	13.0	27.0	21
	11.0	11.0	1.3	0.2	11.8	9.0	13.0	28
Circumoral kinety to last dikinetid of brush row 3, distance	17.1	17.0	3.1	0.7	17.9	10.0	22.0	21
	11.1	11.0	2.1	0.4	18.9	7.0	15.0	28
Anterior body end to first macronucleus nodule, distance	29.9	30.0	4.2	0.9	14.0	22.0	40.0	21
	–	–	–	–	–	–	–	–
Macronucleus figure, length	47.2	48.0	7.5	1.7	15.9	36.0	60.0	21
	–	–	–	–	–	–	–	–
Macronucleus nodules, length	6.1	5.5	2.2	0.5	36.1	3.0	11.0	21
	–	–	–	–	–	–	–	–
Macronucleus nodules, width	4.0	4.0	0.8	0.2	19.1	2.5	6.0	21
	2.7	2.5	0.7	0.1	24.8	2.0	4.0	28
Macronucleus nodules, number	14.1	14.0	2.9	0.6	20.6	9.0	21.0	21
	–	–	–	–	–	15.0	30.0	28
Micronuclei, across	1.4	1.5	0.2	0.0	15.8	1.0	1.5	21
	–	–	–	–	–	–	–	–
Micronuclei, number	6.0	6.0	1.9	0.4	30.7	4.0	9.0	21
	–	–	–	–	–	–	–	–
Ciliary rows, number	10.5	10.0	1.1	0.2	10.3	9.0	12.0	21
	8.8	9.0	0.8	0.1	9.0	7.0	10.0	28
Ciliated kinetids in a right side kinety, number	37.5	37.0	6.0	1.3	15.9	26.0	51.0	21
	33.8	33.5	4.5	0.8	13.2	26.0	42.0	28
Dorsal brush rows, number	3.0	3.0	0.0	0.0	0.0	3.0	3.0	21
	3.0	3.0	0.0	0.0	0.0	3.0	3.0	28
Dikinetids in brush row 1, number	4.1	4.0	1.0	0.2	24.5	2.0	6.0	21
	4.6	4.0	1.1	0.2	24.3	3.0	7.0	28
Dikinetids in brush row 2, number	14.9	15.0	2.4	0.5	16.2	9.0	19.0	21
	11.7	12.0	1.9	0.4	15.9	7.0	14.0	28
Dikinetids in brush row 3, number	9.7	10.0	2.0	0.4	20.4	6.0	14.0	21
	8.8	9.0	1.2	0.2	14.2	7.0	11.0	28
Dikinetids composing a right side oral kinetofragment, number	4.1	4.0	0.6	0.1	15.3	3.0	6.0	21
	5.5	5.5	–	–	–	4.0	7.0	4

[1] Data based on mounted, protargol-impregnated (FOISSNER's method), and randomly selected specimens from non-flooded Petri dish cultures. Measurements in µm. CV – coefficient of variation in %, M – median, Max – maximum, Min – minimum, n – number of individuals investigated, SD – standard deviation, SE – standard error of arithmetic mean, $\overline{x}$ – arithmetic mean.

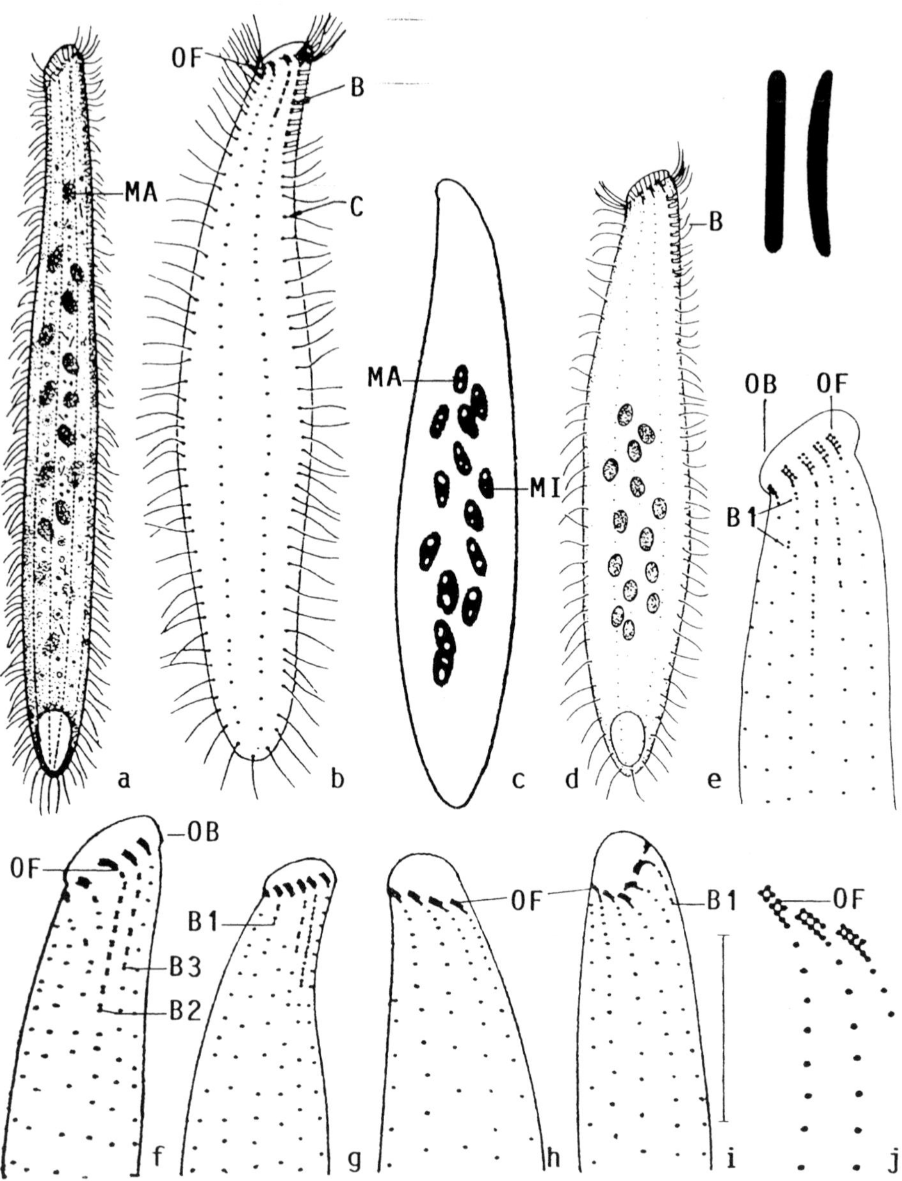

Fig. 46a–j *Protospathidium muscicola*, African type population from life (a), after protargol impregnation (b, d–j), and after Feulgen reaction (c). From DRAGESCO & DRAGESCO-KERNÉIS (1979, a–c, f–j; 1986, d, e). **a, d)** Left side overviews, length 90 µm, 80 µm. **b)** Ciliary pattern of left side, length 90 µm. **c)** Nuclear apparatus, length 62 µm. **e–g)** Left side views of anterior body portion. **h, i)** Right side and ventral view of anterior body portion. **j)** The oral kinetofragments are dikinetidal, separated from each other, and attached to the ciliary rows. Upper right corner: asymmetric oral bulge extrusome of a specimen from the Republic of South Africa, length 3 µm (original). B(1-3) – dorsal brush (rows), C – somatic cilia, MA – macronucleus nodules, MI – micronucleus, OB – oral bulge, OF – oral kinetofragments. Scale bar 20 µm for (f–i).

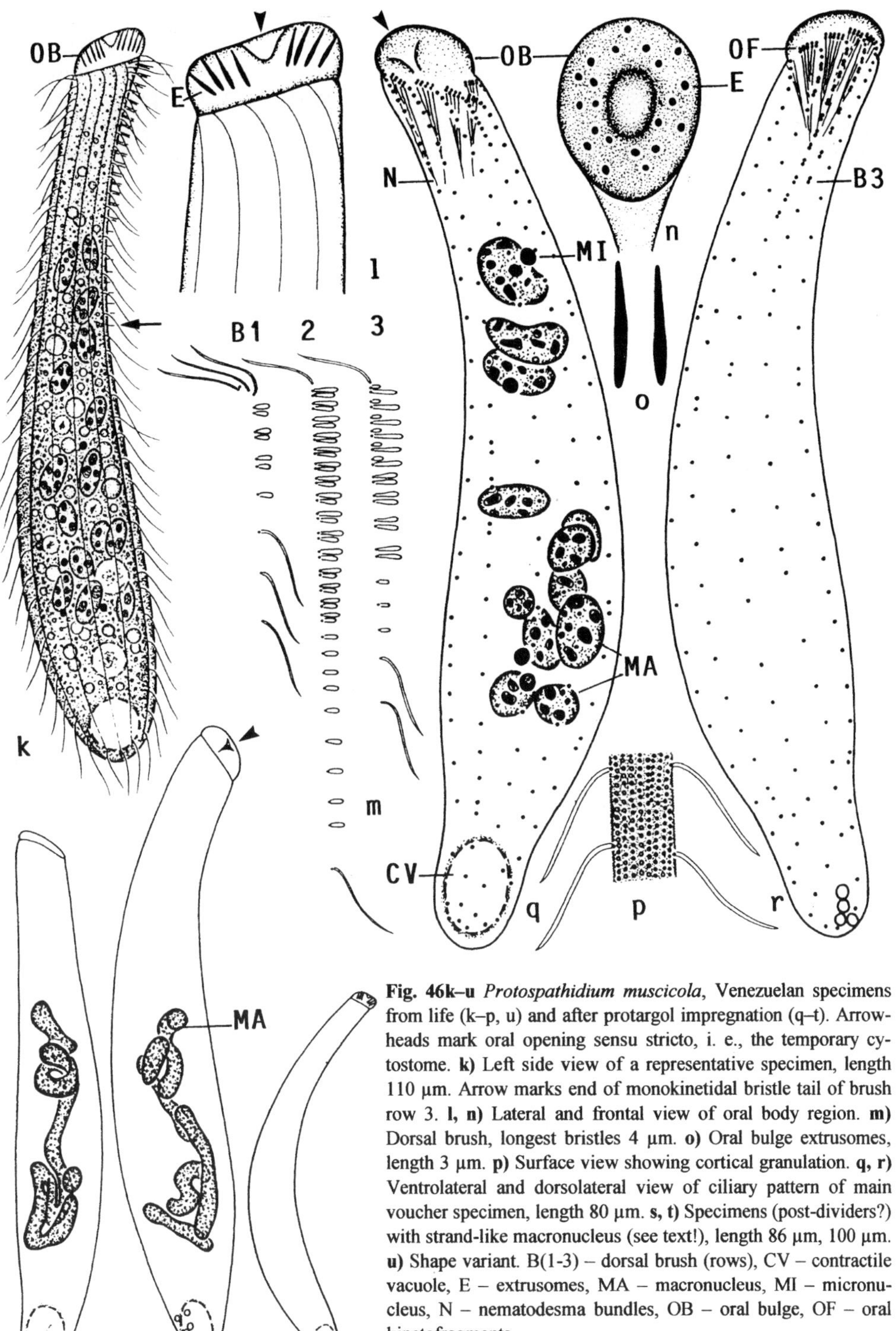

Fig. 46k–u *Protospathidium muscicola*, Venezuelan specimens from life (k–p, u) and after protargol impregnation (q–t). Arrowheads mark oral opening sensu stricto, i. e., the temporary cytostome. **k)** Left side view of a representative specimen, length 110 µm. Arrow marks end of monokinetidal bristle tail of brush row 3. **l, n)** Lateral and frontal view of oral body region. **m)** Dorsal brush, longest bristles 4 µm. **o)** Oral bulge extrusomes, length 3 µm. **p)** Surface view showing cortical granulation. **q, r)** Ventrolateral and dorsolateral view of ciliary pattern of main voucher specimen, length 80 µm. **s, t)** Specimens (post-dividers?) with strand-like macronucleus (see text!), length 86 µm, 100 µm. **u)** Shape variant. B(1-3) – dorsal brush (rows), CV – contractile vacuole, E – extrusomes, MA – macronucleus, MI – micronucleus, N – nematodesma bundles, OB – oral bulge, OF – oral kinetofragments.

ovate and about 3 × 0.5 µm in Venezuelan cells (Fig. 46a, k, l, o, 47c); do not impregnate with the protargol method used. Cortex flexible, contains up to eight rows of minute (≤ 0.4 µm), colourless granules between each two ciliary rows. Cytoplasm colourless, contains few to many lipid droplets 1–4 µm across and up to 5 µm-sized food vacuoles with remnants of bacteria and heterotrophic flagellates. Glides and swims rather rapidly.

Somatic cilia 4.5 µm long in protargol-impregnated (?) African specimens, while 9 µm (in vivo) in Venezuelan and Botswanan individuals, arranged in an average of nine to twelve, usually bipolar and ordinarily spaced rows more densely ciliated anteriorly than posteriorly. Ciliary rows rather loosely ciliated in Venezuelan specimens and attached to oral kinetofragments, anterior region slightly curved dorsally on right side, while abutting on oral kinetofragments at right angles on dorsal and left side (Fig. 46a, b, f–k, q, r, 47a, h, i, 113a–d; Table 9). Dorsal brush three-rowed and distinctly heterostichad due to the strongly shortened row 1, inconspicuous because longest row 2 occupying only about 17–22 % of body length and bristles merely up to 4 µm long in vivo; basically similar in the three populations studied, detailed data, however, available only for Venezuelan and Botswanan specimens; all rows possess a short tail of 1–3 ordinary cilia anteriorly, and the Venezuelan specimens have a monokinetidal tail of 1–2 µm long bristles posteriorly, an unusual pattern compared to congeners (where detailed data are, however, sparse) and other spathidiids which have associated a monokinetidal bristle tail only with row 3, as does the Botswanan population; all rows continue posteriorly as ordinary somatic kineties with bristles and ordinary cilia occasionally mixed in transition zone. Brush row 1 composed of an average of only 2–4 dikinetids with 2 µm long, slightly inflated bristles; brush row 2 distinctly longer than row 1 and usually slightly longer than row 3, composed of an average of 10–15 ordinarily spaced dikinetids with bristle length gradually decreasing from 3.5 µm anteriorly to 2 µm posteriorly, and the slightly inflated anterior bristle of the dikinetids a bit longer than the rod-shaped posterior bristle; brush row 3 usually slightly shorter than row 2, composed of an average of 7–10 comparatively widely spaced dikinetids with bristle length gradually decreasing from 4 µm anteriorly to 1.5 µm posteriorly, and anterior bristle of dikinetids distinctly shortened in anterior half of row (Fig. 46a, b, d–i, k, m, r, 47a, d, i, j, l, 48b, d, h; Table 9). Dorsal brush of Botswanan specimens rather different (Fig. 47d), specifically, rows 1 and 2 lack a monokinetidal bristle tail posteriorly. On the other hand, both populations have dikinetids comparatively widely spaced in row 3 (Fig. 46r, 472j, l).

Oral bulge rather conspicuous, especially in Austrian specimens, though only about half as long as widest trunk region, because 2.5–4 µm high in vivo and somewhat knobby, that is, projecting laterally and thus obovate to almost circular in frontal view; obliquely truncate by 25° to 45°, surface flat to slightly convex, centre conically depressed and lined by fine fibres recognizable after protargol impregnation (Fig. 46a, f–i, k, l, n, q, r, t, u, 47a, b, h–m, 48a, d, e, f, h; Table 9). Circumoral kinety of same shape as oral bulge, composed of dikinetidal kinetofragments obliquely attached to the respective ciliary rows and separated from each other by gaps one to three dikinetids wide. Individual kinetofragments on average composed of four to six dikinetids each associated with an about 9 µm long cilium and a fine, short nematodesma. Oral basket inconspicuous, recognizable only after protargol impregnation, composed of cuneate, about 20 µm long nematodesma bundles originating from the circumoral kinetofragments (Fig. 46b, d–j, q, r, 47h–m, 48a, d, e, g, h; Table 9).

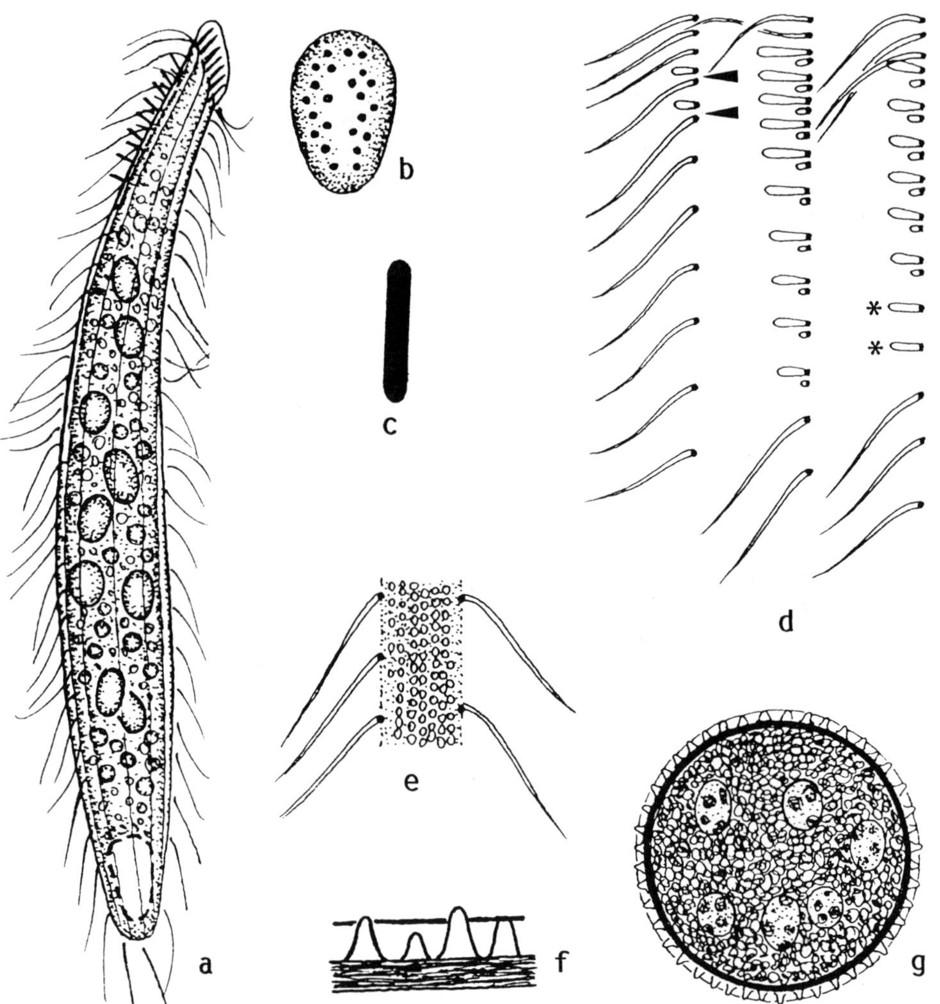

Fig. 47a–g *Protospathidium muscicola*, Botswanan specimens from life. **a)** Right side view of a representative, swimming specimen, length 110 µm. **b)** Frontal view of oral bulge studded with extrusomes. **c)** Oral bulge extrusome, 3 × 0.4 µm. **d)** Dorsal brush (bristles and cilia drawn to scale, longest cilia 8 µm). Row 1 consists of two dikinetids only (arrowheads), and the posterior basal body of the dikinetids bears an ordinary cilium. Row 3 has a short monokinetidal tail composed of two bristles (asterisks). **e)** Surface view showing densely arranged, pale cortical granules < 0.5 µm in size. **f, g)** Resting cyst (20 µm across) and detail of wall.

Recently, we found *P. muscicola* also in Africa, viz., in floodplain soil of the Zambezi, Botswana (Fig. 47a–m). This population matches the original description in body shape and size (very narrowly spatulate, about 80–120 × 10–15 µm in vivo), the scattered macronucleus nodules, the number of ciliary rows (9–12) and, especially, in the short, oblong extrusomes (2.5–3 × 0.3–0.4 µm in five specimens investigated). The complex dorsal brush is shown in figure 47d and differs considerably from that of the Venezuelan specimens (Fig. 46m). This and the slightly different extrusomes indicate

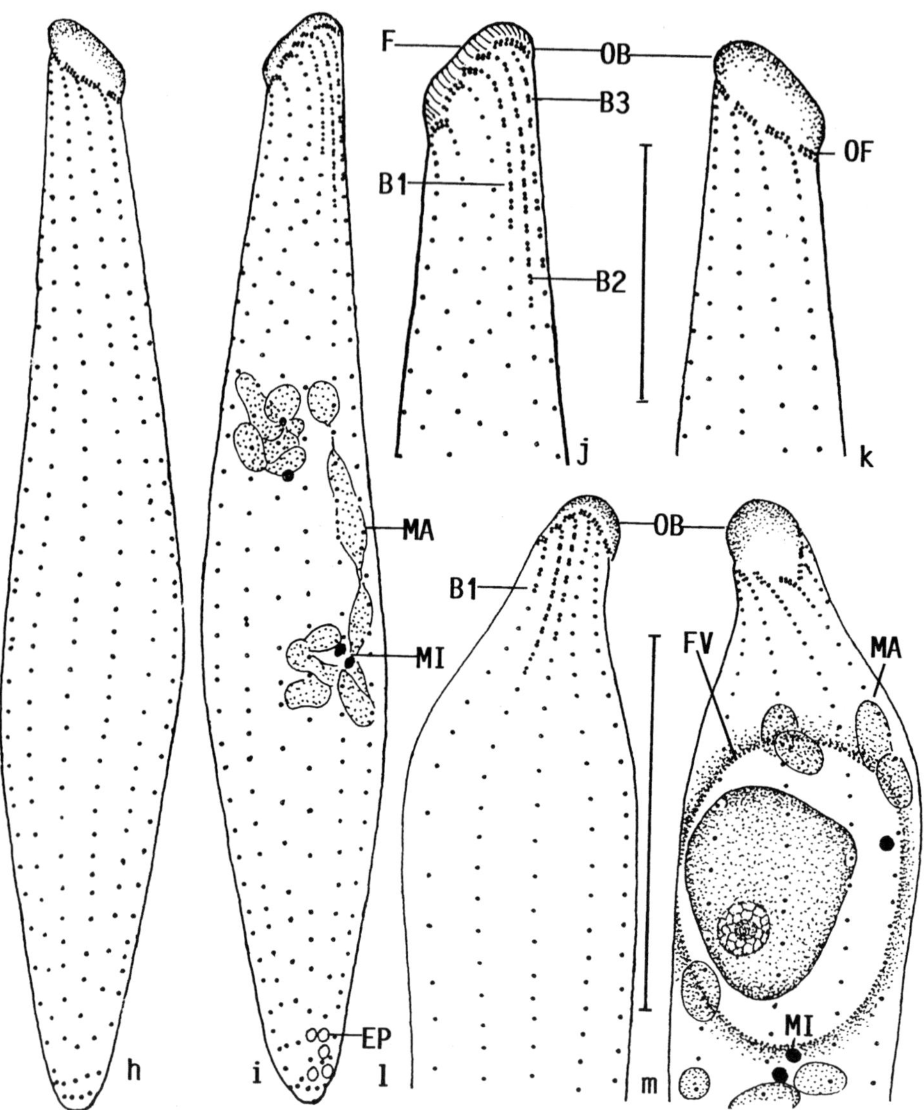

Fig. 47h–m *Protospathidium muscicola*, Botswanan specimens after protargol impregnation. Note that the type locality is also in Africa, viz., in Benin. **h–k)** Right and left side views of a specimen with long brush row 1 and an irregularly moniliform macronucleus, indicating that it is a late post-divider; length 120 µm. Over 90% of the specimens of this population have the usual macronucleus pattern, that is, about 20 scattered nodules, as shown in figure (m). **l, m)** Dorsal and ventral view of anterior half of a specimen having just captured a naked amoeba in a large food vacuole. This specimen has an ordinary brush row 1 (composed of few dikinetids, see also figure 47d) and macronucleus pattern (scattered nodules). B 1-3 – dorsal brush rows, EP – excretory pores of the contractile vacuole, F – fibres, FV – food vacuole, MA – macronucleus (nodules), MI – micronuclei, OB – oral bulge, OF – oral kinetofragments. Scale bars 20 µm.

some biogeographic specialization.

The resting cysts of the Botswanan specimens are unique in having countless conical processes which are 1–2 µm high and rounded distally (Fig. 47f, g, 113i–k). The cysts are colourless and have an average diameter of 22 µm (19–28 µm, n8). The thin wall consists of a 0.8–1 µm thick dense layer and, separated by the about 1 µm wide spine layer, a thin membrane likely covering the processes. The macronucleus nodules do not fuse. The resting cyst of a population from the Galapagos Islands are very similar to those from Botswana; however, the spines are only 0.5–1 µm high and thus easily overlooked (use a magnification ≥ ×600!).

Ontogenesis (Fig. 48a–k; after BERGER et al. 1984): Division commences with a significant ($P < 0.001$) growth of the cells from an average of 65 µm to 100 µm and a proliferation of basal bodies in mid-body, especially in the dorsal brush rows. No changes are recognizable in the nuclear apparatus (Fig. 48d, e). By further proliferation of basal bodies originate the dorsal brush dikinetids and short, dikinetidal oral kinetofragments, some of which are slightly curved to the right; concomitantly, the ciliary rows divide in mid-body, that is, above the newly formed oral kinetofragments, and half of the macronucleus nodules each become located in the proter and opisthe (Fig. 48e, f). Later, however, the macronucleus nodules fuse to a globular mass in mid-body. Figures 48g–k show a late divider with distinct division furrow and thus fully separated ciliary rows. Both daughters are about 62 µm long, that is, have already the average size of interphase specimens (Table 9). The dorsal brush and the oral kinetofragments are fully developed, but hardly curved rightwards, showing that shaping of the circumoral kinety occurs only in very late dividers or post-dividers. The globular nuclear mass of middividers has extended to a long rod and the micronuclei almost finished division, but are still connected by a long, fine strand. No changes are recognizable in the parental oral structures.

Occurrence and ecology: DRAGESCO & DRAGESCO-KERNÉIS (1979) discovered *P. muscicola* in moss from the university garden of Contonou, Benin. Great numbers developed at 30°C, but the culture soon declined. The Botswanan population described above, was rather abundant in the non-flooded Petri dish culture at room temperature (~ 20°) and stayed for three weeks. Another population was found in the Republic of South Africa. BERGER et al. (1984) recorded *P. muscicola* from the upper (0–5 cm) litter and soil layer of an alpine beech forest in the surroundings of the Brunnalm, Hagengebirge, Salzburg, 1360 m above sea-level. In Venezuela, *P. muscicola* occurred in a circumneutral (pH 7.3) mixture of moss and saline soil from the Morrocay National Park. Very recently, we found *P. muscicola* also in a soil sample from the Galapagos Islands. These data indicate a cosmopolitan distribution of *P. muscicola* (but see below); it is, however, a rare species not occurring, for instance, in 73 samples from Namibia (FOISSNER et al. 2002).

Remarks: The original description is perfect, except of detailed morphometrics. Later, DRAGESCO & DRAGESCO-KERNÉIS (1986) provided revised (?)/schematic (?) figures (Fig. 46d, e) rather different from those contained in the original publication (Fig. 46a–c, f–j). BERGER et al. (1984) obviously misidentified the Austrian population as *P. serpens*, redescribed by FOISSNER (1981a); they mentioned the different nuclear

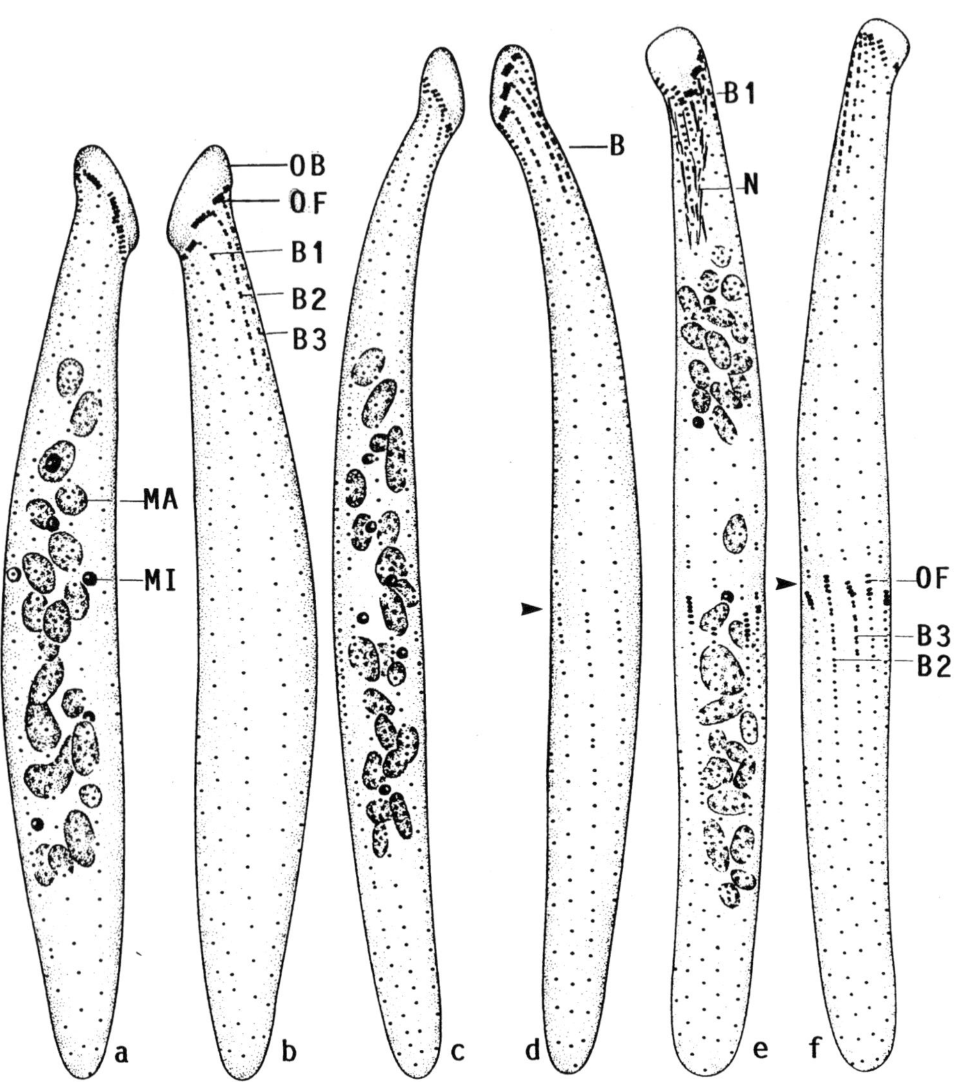

Fig. 48a–f *Protospathidium muscicola*, oral and somatic ciliary pattern and nuclear apparatus of Austrian specimens after protargol impregnation (from BERGER et al. 1984). **a, b)** Ciliary pattern of right and left side of a representative, morphostatic specimen, length 70 μm. Note the conspicuous oral bulge and the macronucleus nodules. **c, d)** Right and left side view of a very early divider (length 100 μm) commencing basal body proliferation underneath the prospective division furrow marked by an arrowhead. The nuclear apparatus is still unchanged. Generally, dividers are significantly longer and more slender than morphostatic specimens. **e, f)** Ventral and dorsal view of an early divider (length 100 μm) having produced oral kinetofragments and a new dorsal brush underneath the prospective division furrow marked by an arrowhead. The macronucleus nodules and the micronuclei form a cluster each in proter and opisthe. However, later they fuse to an ellipsoidal mass in mid-body. B(1-3) – dorsal brush (rows), MA – macronucleus nodules, MI – micronucleus, N – nematodesmata, OB – oral bulge, OF – oral kinetofragments.

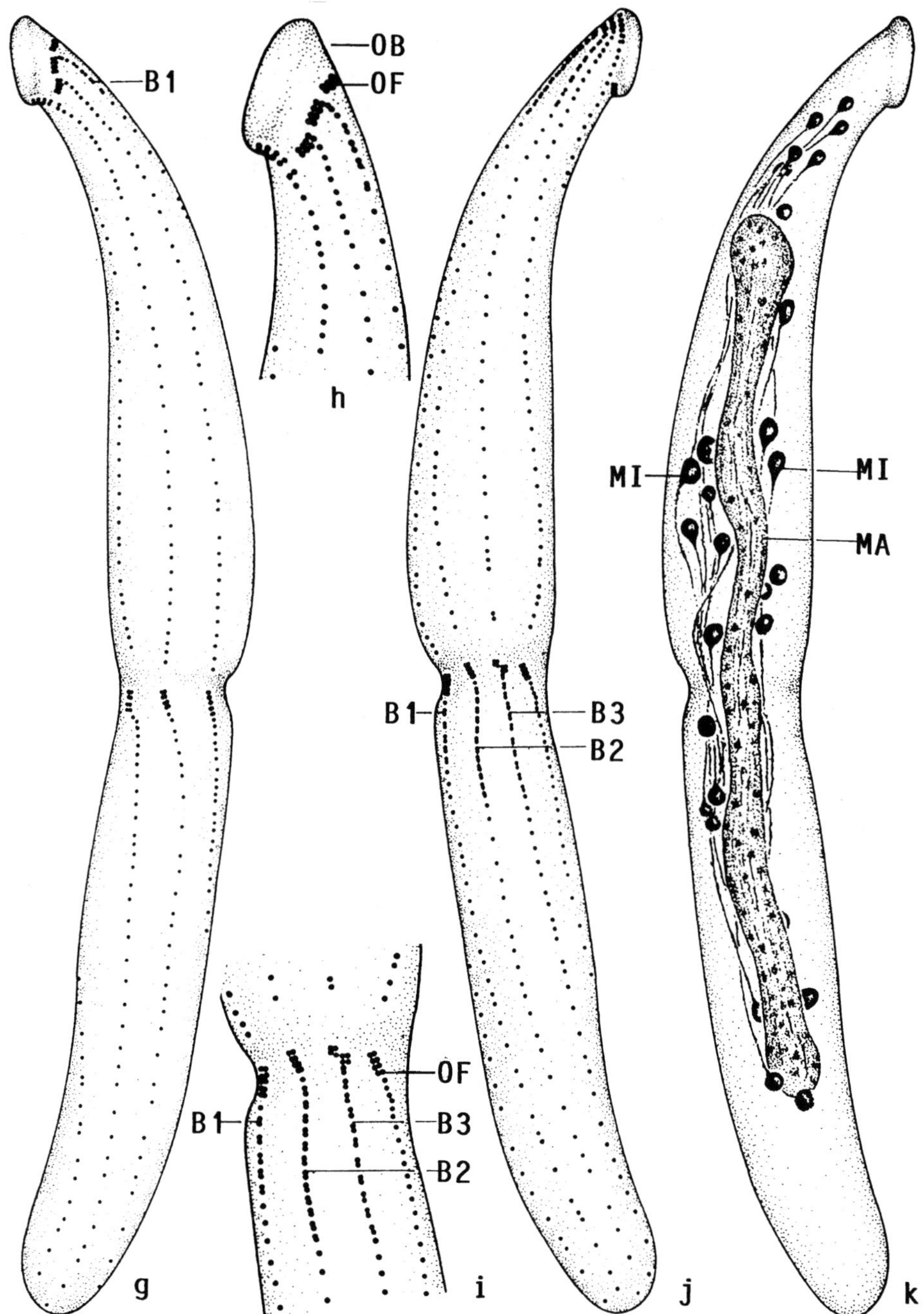

Fig. 48g–k *Protospathidium muscicola*, ciliary pattern and nuclear apparatus of a late divider from Austrian population, length 123 µm (from BERGER et al. 1984). Note division furrow, many dividing micronuclei, and the newly formed oral kinetofragments most composed of four dikinetids (i); the parental oral structures do not change (h). B(1-3) – dorsal brush (rows), MA – elongating macronucleus, MI – micronuclei, OB – oral bulge, OF – oral kinetofragments.

pattern, but did not discuss *P. muscicola* which they likely overlooked. The Austrian and Venezuelan populations match the African specimens in all main characteristics. Thus, conspecificity is reasonable and the species diagnosed with the three descriptions available. However, the very narrowly ovate extrusomes of the Venezuelan specimens (vs. oblong in two African populations), the dorsal brush peculiarities, and the different length of the cyst spines indicate some biogeographic specialization. Indeed, *P. muscicola* has a "brother" with tubercular resting cysts, recently discovered in tank bromeliads from Jamaika.

Protospathidium muscicola differs from most congeners by the scattered macronucleus nodules, except of *P. vermiforme* (about 80–110 × 10–20 µm vs. 80–160 × 5–13 µm, that is, distinctly stouter, viz., 6.3:1 vs. 13:1; brush bristles up to 4 µm vs. 8 µm long) and *Edaphospathula fusioplites* (about 80–110 × 10–20 µm vs. 110 × 12 µm, that is, distinctly stouter, viz., 6.3:1 vs. 9:1; extrusomes oblong and 3–4 µm long vs. ovate and 1.5 µm long; oral kinetofragments composed of 4–6 vs. 1–3 dikinetids). Specimens with incompletely fragmented macronucleus are rather easily confused with *P. serpens* (Fig. 46s, t, 47i).

Protospathidium serpens (KAHL, 1930) FOISSNER, 1981 (Fig. 49a–v, 50a–q, 51a–v, 52a–r, 53a–s, 54a–r, 113l–n, 114a–s, 115a–m; Tables 10–12)

Note added lately: Very recently, XU & FOISSNER (2005a) split this species, using the structure of the resting cyst as a main feature: with conspicuous spines in *P. serpens* (Fig. 51o) and smooth in *P. fraterculum* (Fig. 55p). *Protospathidium fraterculum* comprises the Antarctic populations studied by FOISSNER (1996c) and PETZ & FOISSNER (1997) as well as the South African population investigated by XU & FOISSNER (2005a). All other populations stay with *P. serpens* because their resting cysts are unknown. Further, XU & FOISSNER (2005a) studied an Austrian population of *P. serpens* and fixed it as a neotype. Likely, *P. fraterculum* is composed of two (sub)species, differing in brush row 3 which possesses (Antarctic populations) or lacks (South African population) a monokinetidal bristle tail. Unfortunately, it was too late to include all these new data in the present monograph.

1930 *Spathidium serpens* KAHL, Tierwelt Dtl., 18: 158.

1981 *Protospathidium serpens* nov. comb. (KAHL, 1930–35) – FOISSNER, Zool. Jb. Syst., 108: 274 (slides with protargol-impregnated specimens deposited in the Oberösterreichische Landesmuseum in Linz, Upper Austria).

1984 *Protospathidium serpens* (KAHL, 1930) – BERGER, FOISSNER & ADAM, Zool. Jb. Syst., 111: 355 (misidentification, see *P. muscicola*).

1996 *Protospathidium serpens* (KAHL, 1930) FOISSNER, 1981 – FOISSNER, Acta Protozool., 35: 106 (description of an Antarctic population; voucher slides with protargol-impregnated specimens deposited as described above).

1997 *Protospathidium serpens* (KAHL, 1930) FOISSNER, 1981 – PETZ & FOISSNER, Polar Rec., 33: 308 (detailed morphometry and brief description of two Antarctic populations; voucher slides with protargol-impregnated specimens deposited as described above).

2004 *Protospathidium serpens* (KAHL, 1930) FOISSNER, 1981 – XU & FOISSNER, J. Eukaryot. Microbiol., 51: 605 (detailed description of conjugation of a species later recognized to be *P. fraterculum*; see next entry).

2005 *Protospathidium serpens* (KAHL, 1930) – XU & FOISSNER, J. Eukaryot. Microbiol., 52: 298 (neotypification with an Austrian population; neotype slides deposited as described above).

Type material: No original type material is available. However, the identity was threatened by the newly established *P. fraterculum* XU & FOISSNER, 2005a. Thus, XU & FOISSNER (2005a) investigated an Austrian population of *P. serpens* and fixed it as a neotype.

Diagnosis (includes all populations known): Size about 90 × 15 µm in vivo. Cylindroidal with oblique, obovate, rather distinct oral bulge about half as long as widest trunk region. Macronucleus a tortuous, more or less nodulated strand; several micronuclei. Extrusomes rod-shaped, 2–5 µm long. On average 9–13 ciliary rows, 2 to 4, usually 3 anteriorly differentiated to dorsal brush occupying 11–15% of body length. Brush bristles up to 4 µm long, row 1 usually composed of 2–4 dikinetids, row 2 of 7–10, and row 3 of 4–7 dikinetids; monokinetidal bristle tail of row 3 distinct or lacking. Right side oral kinetofragments composed of an average of 3–5 dikinetids. Type VIII resting cyst with spines longer than 3 µm.

Type and neotype locality: Shallow rod drain in the surroundings of Hamburg, E10° N53°30'. The neotype locality is meadow soil under a temporary puddle in the town of Salzburg, Austria, E13°40' N47°47'.

Etymology: Not given in the original description. The Latin adjective *serpens* (serpentine) obviously refers to the serpentine swimming behaviour observed by KAHL (1930b).

Description: Size 40–180 × 8–25 µm in vivo, usually about 80–100 × 15–20 µm, as calculated from in vivo measurements and protargol preparations of five populations; length:width ratio also highly variable, viz., 3.5–9:1, with an overall average of about 6:1 in vivo and 4.8:1 in protargol preparations (Tables 10, 11). General appearance very narrowly spatulate to ellipsoidal or obclavate because slimy, inconspicuously flattened laterally, and oblique anterior (oral) body end only about half as long as broadest trunk region usually in third quarter of cell, posterior end narrowly to moderately broadly rounded, never tail-like elongated; often slightly, rarely distinctly curved (concave) ventrally or slightly sigmoidal, Namibian specimens up to 30% contractile and with amoeboid shape changes when rooting in the mud of the Petri dish (Fig. 49a, b, d, h, j, m, n, p, 50a–d, h–m, 51a, g–k, 114a–c, o–s, 115c). Nuclear apparatus in central quarters of cell. Macronucleus a more or less tortuous, nodulated strand, as stated in the original description (Fig. 49a) and evident from very early dividers, which show the true pattern best (Fig. 52a, c, e); in detail, however, highly variable in all populations. FOISSNER (1996a) observed three main patterns in 79 Antarctic specimens: nodular strand (48 cells, Fig. 49j, n), cylindroidal (27 cells, Fig. 49o), ellipsoidal (4 cells, Fig. 49q). PETZ & FOISSNER (1997) reported the following patterns in 92 Antarctic specimens (Fig. 50a–d): nodular strand (60%), rather widely wound (38%), cylindroidal (2%); in another Antarctic population, they observed the following pattern in 106 specimens: narrowly tortuous strand (58%), nodular strand (37%), cylindroidal (5%), U-shaped (2%), ellipsoidal (1%). A similar variability is evident in the Namibian

Table 10 Morphometric data on various populations of *Protospathidium serpens*

Characteristics[1]	Pop[2]	$\overline{x}$	M	SD	SE	CV	Min	Max	n
Body, length	NAM	106.4	104.0	29.7	9.0	27.9	78.0	170.0	11
	ANF	78.8	77.0	9.0	2.3	11.5	63.0	95.0	15
	AP1[3]	82.9	86.0	14.9	–	18.0	48.0	102.0	20
	AP2	72.2	70.0	17.9	–	24.8	38.0	114.0	31
Body, width	NAM	18.7	20.0	4.1	1.2	21.8	13.0	23.0	11
	ANF	18.5	18.0	2.5	0.7	13.7	13.0	22.0	15
	AP1[3]	20.7	20.0	5.0	–	24.3	11.0	30.0	20
	AP2	14.9	14.0	2.8	–	16.0	11.0	22.0	31
Body length:width, ratio	NAM	5.8	5.3	1.5	0.4	25.2	3.6	9.0	11
	ANF	4.3	4.2	0.5	0.1	11.3	3.5	5.4	15
Oral bulge, length	ANF	7.6	8.0	1.0	0.3	13.0	6.0	9.0	15
	AP1	9.0	9.0	1.1	–	11.8	6.0	10.0	16
	AP2	7.9	8.0	1.3	–	16.3	5.0	11.0	31
Oral bulge, height	AP1	3.1	3.0	–	–	23.0	2.0	4.0	16
	AP2	2.5	2.5	–	–	19.2	1.5	3.0	31
Anterior body end to macronucleus, distance	NAM	29.8	30.0	9.9	3.0	33.2	18.0	50.0	11
	ANF	27.9	25.0	6.9	1.8	24.9	20.0	46.0	15
	AP1	28.2	29.5	8.7	–	30.8	15.0	50.0	20
	AP2	22.1	22.0	5.0	–	22.7	10.0	34.0	31
Macronucleus figure, length	NAM	45.7	44.0	19.5	5.9	42.7	22.0	90.0	11
	ANF	31.3	30.0	6.1	1.6	19.6	20.0	42.0	15
	AP1	35.9	35.0	11.4	–	31.9	17.0	58.0	20
	AP2	34.2	34.0	6.5	–	19.1	20.0	48.0	31
Macronucleus, width	NAM	4.5	4.0	–	–	–	4.0	5.0	11
	ANF	5.1	5.0	1.2	0.3	24.3	4.0	8.0	15
	AP1	6.7	7.0	1.7	–	25.4	4.0	9.0	20
	AP2	5.8	6.0	1.0	–	17.7	4.0	8.0	31
Micronuclei, largest diameter	NAM	2.3	2.5	–	–	–	1.5	2.8	5
	ANF	1.7	1.8	–	–	–	1.5	2.0	15
	AP1	2.1	2.0	–	–	–	1.5	3.0	28
	AP2	2.7	2.5	–	–	–	2.0	3.5	31
Micronuclei, number	NAM	5.3	5.5	1.0	0.4	20.0	4.0	7.0	8
	ANF	3.3	3.0	1.2	0.3	35.6	2.0	5.0	15
	AP1	2.9	3.0	1.1	–	36.9	1.0	5.0	20
	AP2	2.1	2.0	0.6	–	30.0	1.0	4.0	32
Circumoral kinety to end of brush row 1, distance	NAM	6.4	7.0	3.1	0.9	48.3	2.0	10.0	11
	ANF	4.3	4.0	1.2	0.3	27.2	3.0	7.0	15
	AP1	3.6	3.5	0.6	–	16.4	3.0	5.0	21
	AP2	4.6	4.3	1.6	–	34.7	2.5	8.5	26
Circumoral kinety to end of brush row 2, distance	NAM	13.2	13.0	4.1	1.2	30.7	5.0	20.0	11
	ANF	10.5	11.0	2.1	0.6	20.3	7.0	14.0	15
	AP1	9.3	9.5	1.2	–	13.4	7.0	12.0	22
	AP2	9.0	9.0	2.4	–	26.4	4.5	12.5	31
Circumoral kinety to end of brush row 3, distance	NAM	9.3	9.0	2.8	0.9	30.5	3.0	13.0	11
	ANF	9.4	10.0	1.6	0.4	16.5	7.0	11.0	15
	AP1	8.8	8.0	2.0	–	22.3	6.0	13.0	15
	AP2	7.1	7.0	2.4	–	34.2	3.5	12.0	28
Ciliary rows, number	NAM	9.0	9.0	1.6	0.5	17.2	7.0	12.0	11
	ANF	11.6	11.0	0.7	0.2	6.4	11.0	13.0	15
	AP1	13.0	13.0	1.2	–	9.0	11.0	16.0	31
	AP2	11.1	11.0	0.8	–	7.4	10.0	12.0	32
Basal bodies in a right side ciliary row, number	NAM	56.4	58.0	19.6	5.9	34.8	25.0	84.0	11
	ANF	30.7	30.0	5.6	1.4	18.2	20.0	40.0	15

continued

Characteristics[1]	Pop[2]	$\overline{x}$	M	SD	SE	CV	Min	Max	n
	AP1	36.2	37.0	6.3	–	17.3	24.0	50.0	17
	AP2	22.4	22.0	5.1	–	22.8	12.0	34.0	30
Dorsal brush rows, number	NAM	2.6	3.0	–	–	–	2.0	3.0	30
	ANF	3.5	4.0	–	–	–	3.0	4.0	15
	AP1	3.1	3.0	–	–	–	3.0	4.0	22
	AP2	3.0	3.0	–	–	–	3.0	4.0	32
Dikinetids in brush row 1, number	NAM	6.1	7.0	2.4	0.7	39.2	2.0	9.0	11
	ANF[4]	3.0	3.0	0.8	0.2	25.2	2.0	4.0	15
	AP1	3.4	3.0	1.0	–	27.5	1.0	5.0	23
	AP2	2.9	3.0	1.0	–	36.5	1.0	5.0	28
Dikinetids in brush row 2, number	NAM	14.6	14.0	3.6	1.1	24.9	8.0	20.0	11
	ANF	8.5	8.0	1.5	0.4	17.2	7.0	12.0	15
	AP1	9.2	9.0	1.5	–	16.3	6.0	12.0	29
	AP2	6.9	7.0	1.6	–	23.5	4.0	10.0	31
Dikinetids in brush row 3, number	NAM	7.6	8.0	1.0	0.3	13.5	6.0	10.0	11
	ANF	5.2	5.0	0.7	0.2	13.0	4.0	7.0	15
	AP1	6.6	7.0	1.4	–	21.2	4.0	9.0	31
	AP2	4.5	4.5	1.2	–	25.9	3.0	7.0	30
Dikinetids in a right side oral kinetofragment, number	NAM	5.1	5.0	1.1	0.3	22.3	4.0	8.0	11

[1] Data based protargol-impregnated, randomly selected specimens (preparation and culture methods, see footnote 2). Measurements in µm. CV – coefficient of variation in %, M – median, Max – maximum, Min – minimum, n – number of individuals investigated, SD – standard deviation, SE – standard error of arithmetic mean, $\overline{x}$ – arithmetic mean.

[2] Populations: NAM – Namibian site 16 (see FOISSNER et al. 2002 for detailed site description) specimens from a semipure culture impregnated with FOISSNER's protargol protocol (new, unpublished data); ANF – Antarctica, Signey Island specimens from a non-flooded Petri dish culture impregnated with FOISSNER's protargol protocol (from FOISSNER 1996a) ; AP1 – Antarctica, Clark Peninsula specimens from a non-flooded Petri dish culture impregnated with WILBERT's protargol method (from PETZ & FOISSNER 1997) ; AP2 – Antarctica, Windmill Islands specimens from a non-flooded Petri dish culture impregnated with WILBERT's protargol protocol (from PETZ & FOISSNER 1997).

[3] In vivo measurements: length 68.1 µm, M 66.5, SD 20.9, CV 30.7, Min 43, Max 117, n 12; width 12.8 µm, M 12, SD 4.9, CV 38.2, Min 8, Max 25, n 11.

[4] About half of the Antarctic specimens have four brush rows, likely an additional row at right side of brush, as indicated by the low number of dikinetids ($\overline{x}$ 1.9). To make data of this compilation comparable, we used values from rows 2, 3, and 4 in FOISSNER's publication.

specimens (Fig. 114o–s, 115g–i). One to six, usually two to three micronuclei 2–3 µm across in variable position adjacent to or in shallow macronuclear depressions; sometimes rather distant from macronucleus, in an Antarctic population often a micronucleus each near anterior and posterior end of macronucleus (Fig. 49h, n, o, q, 50a–d, m, 51n, p, 52a, 114c, o–s, 115c, g–i; Tables 10, 11). Contractile vacuole in rear body end, three to six excretory pores in dorsal pole area (Fig. 49a, d, i, j, p, 51a, n, p). Extrusomes arranged in two to three indistinct circles around central depression of oral bulge and scattered in cytoplasm, do not impregnate with FOISSNER's and WILBERT's protargol methods. Oral bulge extrusomes oblong to rod-shaped and fine in seven populations, while very narrowly ellipsoidal and slightly curved in Peloponnese and Venezuelan specimens; in vivo 2.5–5 µm long in Austrian and Antarctic specimens, while only 2–2.5 µm in Venezuelan, North American and Peloponnese cells (Fig. 49a, d, k, l, 50g, n, 51b, 114h, 115e, j, k). Cortex bright and very flexible, about 0.7 µm thick in Austrian

Table 11 Morphometric data on neotype specimens of *Protospathidium serpens* from Austria (upper line) and *P. fraterculum* from South Africa (lower line). From XU & FOISSNER (2005a)

Characteristics[1]	$\overline{X}$	M	SD	CV	Min	Max	n
Body, length	90.3	92.0	7.2	7.9	77.0	105.0	22
	68.9	67.0	7.6	11.0	58.0	83.0	22
Body, width	18.4	19.0	3.5	18.8	10.0	24.0	22
	13.2	13.5	2.7	20.4	8.0	19.0	22
Body length:width, ratio	5.1	4.9	1.3	25.2	3.5	9.8	22
	5.4	5.4	1.0	18.7	3.7	7.3	22
Oral bulge, length	8.4	8.0	1.1	13.6	7.0	11.0	22
	6.8	7.0	0.9	13.3	5.0	8.0	22
Oral bulge, height	3.8	4.0	0.6	15.0	3.0	5.0	22
	2.6	2.5	–	–	2.0	3.0	22
Oral bulge length:body width, ratio	0.5	0.5	0.1	21.3	0.3	0.8	22
	0.5	0.5	0.1	19.5	0.3	0.8	22
Circumoral kinety to last dikinetid of brush row 1, distance	5.7	6.0	0.9	16.3	4.0	8.0	22
	6.9	7.0	1.1	15.8	5.0	9.0	22
Circumoral kinety to last dikinetid of brush row 2, distance	14.0	14.0	1.9	13.5	10.0	19.0	22
	10.3	10.0	1.6	15.4	7.0	13.0	22
Circumoral kinety to last dikinetid of brush row 3, distance	11.7	12.0	1.8	15.2	8.0	15.0	22
	9.3	9.0	1.5	15.7	6.0	12.0	22
Anterior body end to macronucleus, distance	27.6	26.0	8.7	31.4	12.0	54.0	22
	21.8	20.0	5.9	26.9	15.0	36.0	22
Macronucleus figure, length	32.9	33.0	7.3	22.2	16.0	48.0	22
	26.7	26.0	6.0	22.5	14.0	40.0	22
Macronucleus, length (spread)	46.2	45.0	–	–	32.0	60.0	22
	40.5	40.0	–	–	27.0	55.0	22
Macronucleus, width in middle third	4.4	4.0	0.8	18.1	3.0	6.0	22
	3.7	4.0	0.6	17.6	3.0	5.0	22
Macronucleus, number	1.0	1.0	0.0	0.0	1.0	1.0	22
	1.0	1.0	0.0	0.0	1.0	1.0	22
Micronuclei, across	1.9	2.0	0.3	18.2	1.5	3.0	22
	2.3	2.3	–	–	2.0	3.0	22
Micronuclei, number	4.2	4.0	1.2	28.2	2.0	7.0	22
	2.8	3.0	0.8	29.3	2.0	5.0	22
Ciliary rows, number	10.2	10.0	1.1	10.8	9.0	12.0	22
	12.3	12.0	0.6	5.1	11.0	14.0	22
Basal bodies in a right side ciliary row, number	31.0	31.0	6.0	19.2	21.0	42.0	22
	41.1	40.5	6.7	16.2	32.0	56.0	22
Dorsal brush rows, number	3.0	3.0	0.0	0.0	3.0	3.0	22
	3.0	3.0	0.0	0.0	3.0	3.0	22
Dikinetids in brush row 1, number	3.4	3.0	1.0	28.3	2.0	6.0	22
	6.8	7.0	1.0	14.4	5.0	9.0	22
Dikinetids in brush row 2, number	12.4	12.0	2.0	16.5	8.0	17.0	22
	13.0	12.0	2.0	15.2	10.0	16.0	22
Dikinetids in brush row 3, number	7.5	7.0	1.4	18.2	5.0	11.0	22
	9.5	9.0	1.6	16.6	7.0	13.0	22
Circumoral dikinetids in a right side kinetofragment, number	3.8	4.0	0.7	19.2	3.0	5.0	22
	3.6	4.0	0.7	18.1	3.0	5.0	22
Resting cysts of *P. fraterculum*, across	23.4	23.0	3.2	13.6	20.0	28.0	11
Macronucleus, length (spread)	30.8	30.0	3.8	12.4	25.0	36.0	11
Macronucleus, width (middle)	4.2	4.0	0.6	14.4	3.0	5.0	11
Micronuclei, across	2.3	2.0	0.4	18.0	2.0	3.0	11
Micronuclei, number	2.8	3.0	0.6	21.4	2.0	4.0	11

[1] Data on *P. serpens* are based on mounted, protargol-impregnated (FOISSNER's method), and randomly selected specimens from a non-flooded Petri dish culture. Data on *P. fraterculum* are based on the same preparation method, but well-looking specimens were selected. Measurements in µm. CV – coefficient of variation in %, M – median, Max – maximum, Min – minimum, n – number of specimens investigated, SD – standard deviation, $\overline{X}$ – arithmetic mean.

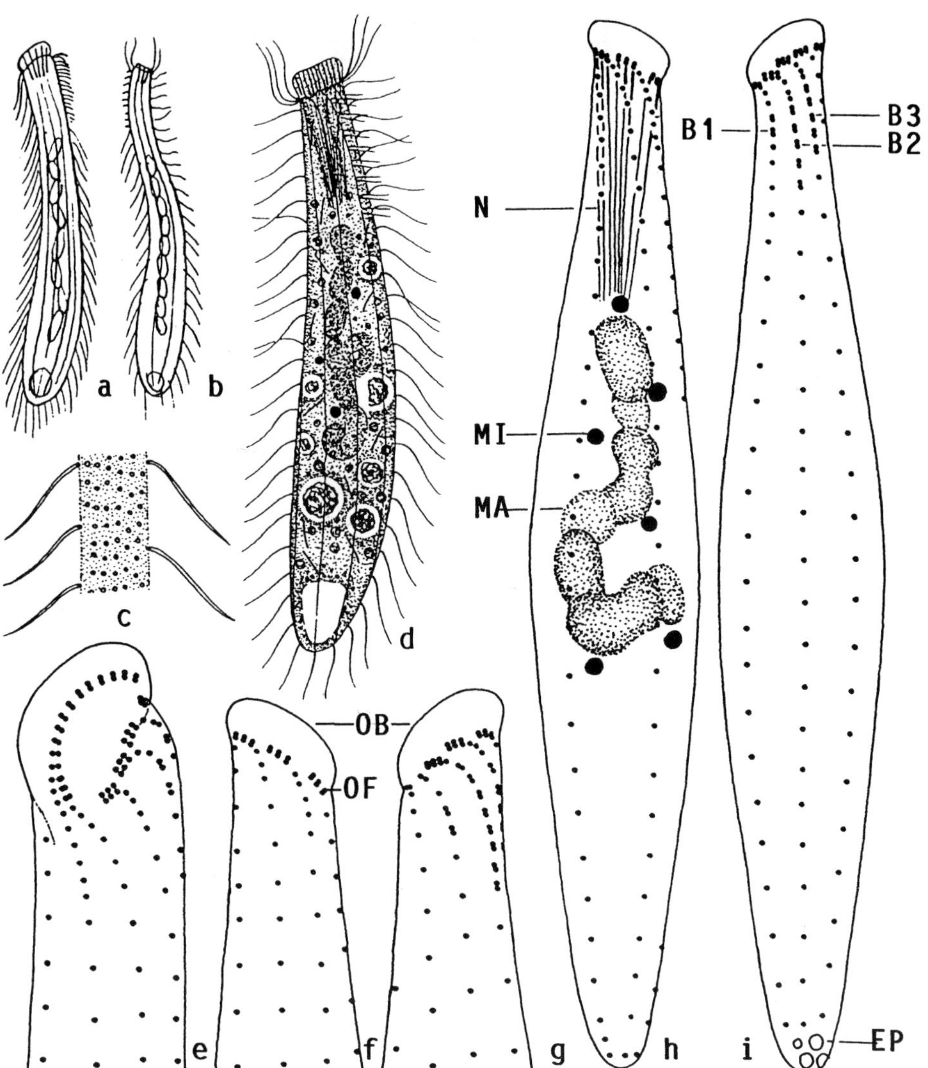

Fig. 49a–i *Protospathidium serpens* from life (a–d) and after protargol impregnation (e–i). Figures (a, b) from KAHL (1930b), (d–i) from FOISSNER (1981a), (c) original. **a)** German type specimen from a shallow road drain, length 90 µm. **b)** A similar, unnamed species with very flat oral bulge and distinct extrusomes, length 100 µm. **c)** Surface view showing the minute (~ 0.2 µm) and loosely arranged cortical granules of a Namibian specimen. **d)** Left side view of a representative specimen from alpine soil of Austria, length 90 µm. **e–i)** Slightly schematized figures of the ciliary and nuclear pattern of the Austrian population studied by FOISSNER (1981a), length 60 µm. The oral kinetofragments are distinctly separated and usually composed of three dikinetids. The excretory pores of the contractile vacuole are on the dorsal pole area. The dorsal brush consists of three rows; however, row 1 is usually composed of only two dikinetids. B1-3 – dorsal brush rows, EP – excretory pores, MA – macronucleus, MI – micronucleus, N – nematodesmata (oral basket rods), OB – oral bulge, OF – oral kinetofragments.

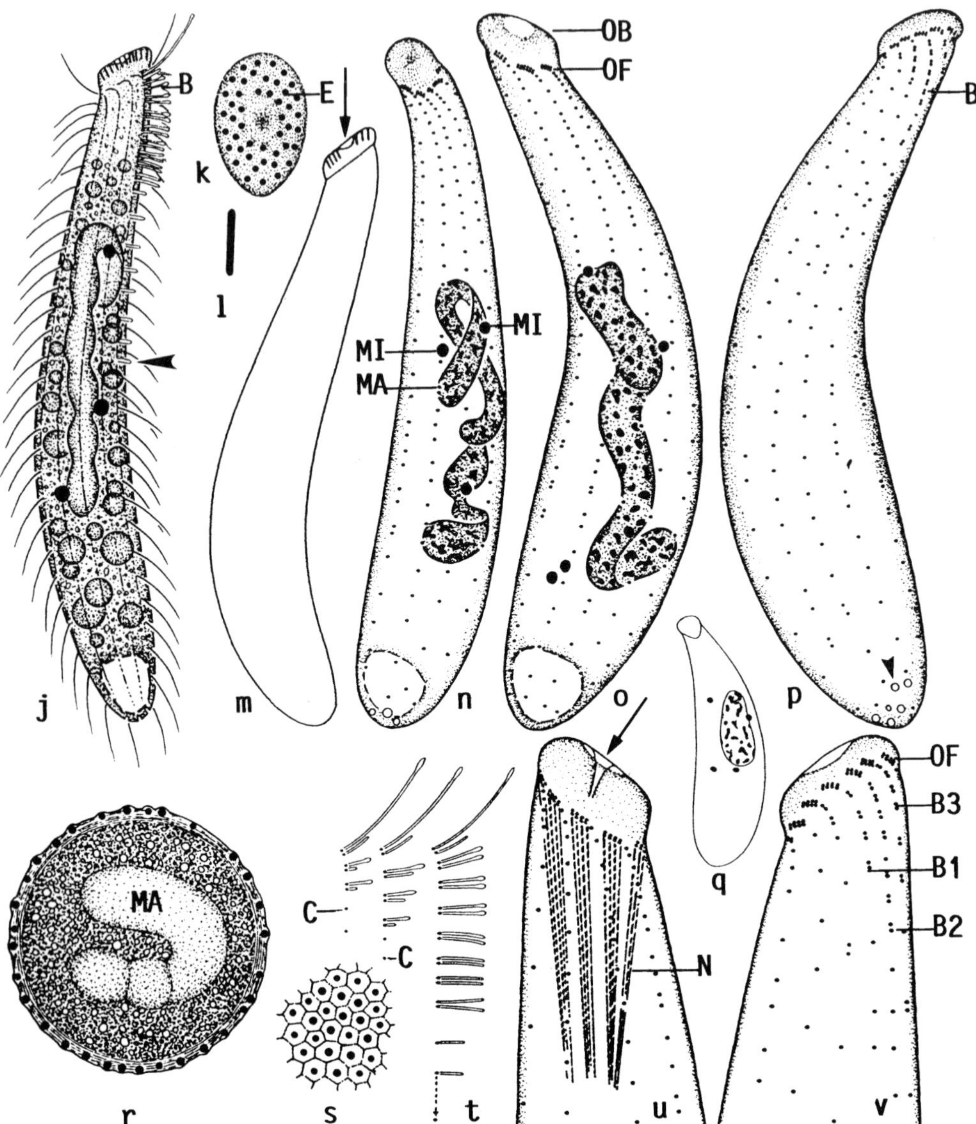

Fig. 49j–v *Protospathidium serpens* (now *P. fraterculum* XU & Foissner, 2005a) from life (j–m, r–t) and after protargol impregnation (n–q, u, v). Antarctic population studied by FOISSNER (1996a). **j)** Left side view of a representative specimen, length 80 µm. Arrowhead marks posterior end of monokinetidal bristle tail of brush row 3. **k)** Frontal view of oral bulge studded with extrusomes. **l)** Oral bulge extrusome, length 3 µm. **m)** Stout specimen with mouth entrance marked by arrow. **n)** Ciliary pattern and nuclear apparatus of a ventrally oriented specimen, length 84 µm. **o, p)** Ciliary pattern of main voucher specimen, length 87 µm. Arrowhead marks excretory pores. **q)** Specimen with ellipsoidal macronucleus, length 63 µm. **r, s)** Optical section and surface view of resting cyst, diameter 30 µm. **t)** Dorsal brush according to in vivo observations. **u, v)** Infraciliature of anterior right and left side, length of oral bulge 9 µm. Arrow marks temporary cytostome lined by thin fibres. Note some bristle dikinetids right of brush row 1, forming an indistinct fourth row. B(1-3) – dorsal brush (rows), C – basal bodies of ordinary cilia, E – extrusomes, MA – macronucleus, MI – micronucleus, N – nematodesmata, OB – oral bulge, OF – oral kinetofragments.

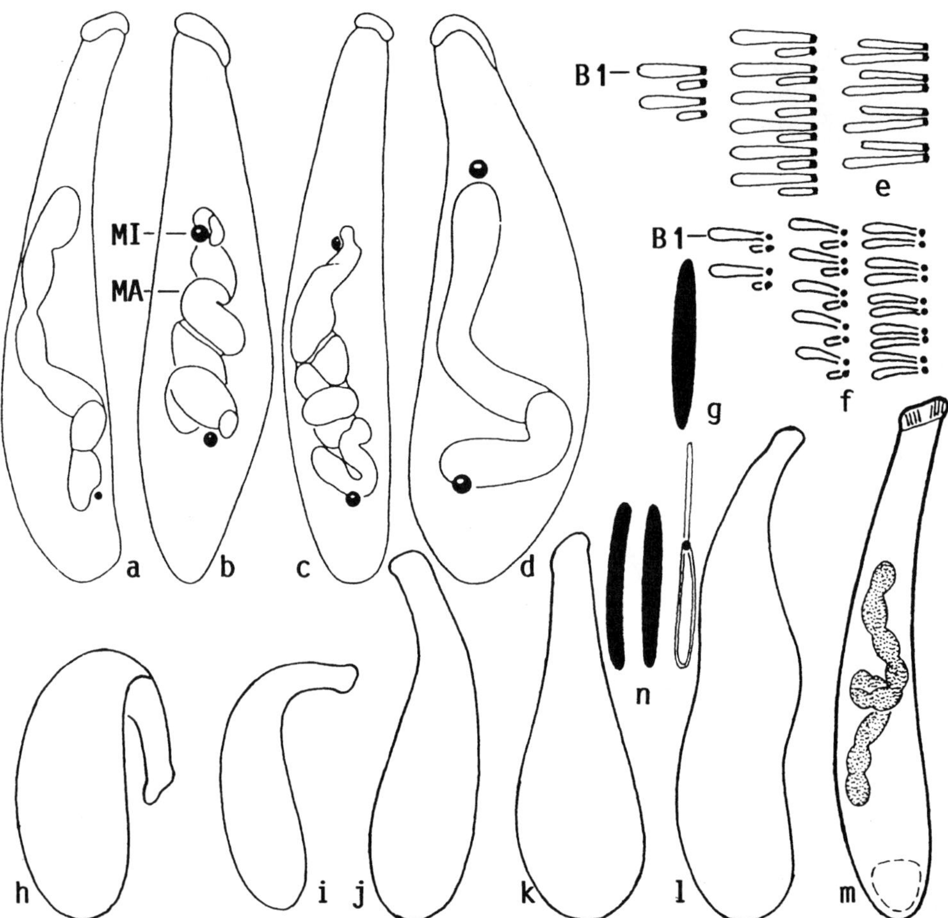

Fig. 50a–n *Protospathidium serpens* after protargol impregnation (a–d) and from life (e–n). **a–d)** Variability of macronucleus in two Antarctic populations, length 65–105 µm (from PETZ & FOISSNER 1977). **e, f)** Middle portion of dorsal brush of a Namibian and a Venezuelan specimen, longest bristles 3 µm. **g)** Extrusome of a Greek specimen, length 2–2.5 µm. **h–m)** Shape variability of cultivated specimens from Namibia rooting in the mud of the Petri dish (h–l) and swimming in the medium (m). **n)** Two sides of a resting extrusome and an exploded one from a Venezuelan specimen, length 2.5–3 µm. B1 – dorsal brush row 1.

specimens (Fig. 51e, f), while 1–1.5 µm thick and jelly-like in Australian and Venezuelan cells; contains about five rows of minute (0.3–0.6 × 0.3–0.5 µm) granules between each two ciliary rows; granules loosely to ordinarily spaced in African and Austrian specimens (Fig. 49c, 51e, f), while densely arranged in Australian and Venezuelan cells (Fig. 115m). Cytoplasm rather hyaline, contains colourless lipid droplets 1–5 µm across and 3–20 µm-sized food vacuoles with bacteria, heterotrophic flagellates, and small ciliates. Glides slowly and serpentinously or rapidly on microscope slides and among soil particles, rotates about main body axis when swimming.

Somatic cilia 7–12 µm long in vivo, arranged in an average of nine to thirteen bipolar and ordinarily spaced rows more densely ciliated anteriorly than posteriorly,

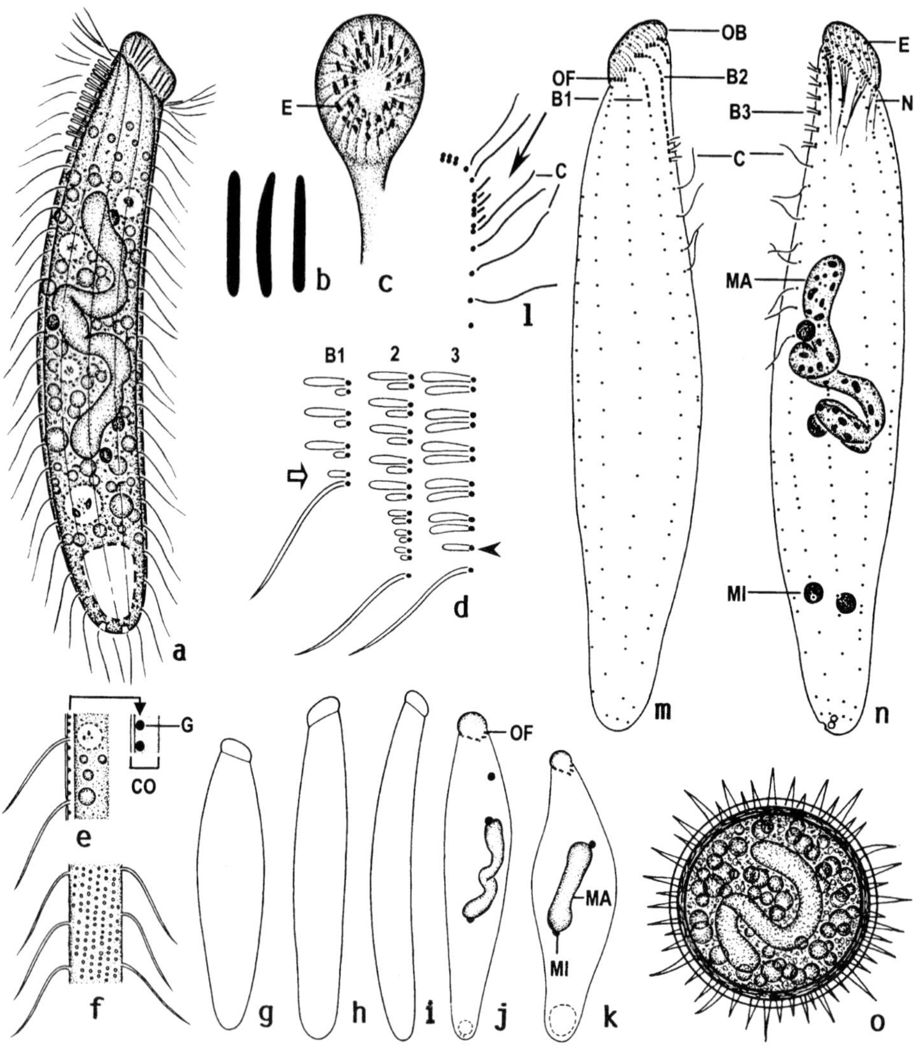

Fig. 51a–o *Protospathidium serpens*, Austrian neotype specimens from life (a–i, o) and after protargol impregnation (j–n). From XU & FOISSNER (2005a). **a)** Right side view of a representative specimen, length 100 µm. **b)** Basically, the oral bulge extrusomes are rod-shaped and about 2.5–3 × 0.3 µm in size. **c)** Frontal view of oral bulge studded with extrusomes arranged in two to three rough circles around the indistinct central bulge depression, that is, the temporary cytostome. **d)** Dorsal brush. The last dikinetid of row 1 bears a 1 µm long bristle (blank arrow) and a 7 µm long ordinary cilium. A monokinetidal bristle tail is lacking in row 3; rarely, a single bristle is recognizable (arrowhead). **e, f)** Optical section and surface view showing the cortical granules (G) embedded in an about 0.7 µm thick, jelly layer (cp. Fig. 115m). **g–k)** Variability of body shape and nuclear pattern. **l–n)** Ciliary pattern of left and right side and nuclear apparatus of main neotype specimen, length 95 µm. Note lack of a monokinetidal bristle tail in brush row 3, a main difference to the Antarctic population of *P. fraterculum* (Fig. 51u, v, 115a) **o)** Type VIII resting cyst with an about 1.5 µm thick, spiny wall, diameter 18 µm without spines. B 1-3 – dorsal brush rows, C – somatic cilia, CO – cortex, E – extrusomes, G – cortical granules, MA – macronucleus, MI – micronuclei, N – nematodesmata, OB – oral bulge, OF – oral kinetofragments.

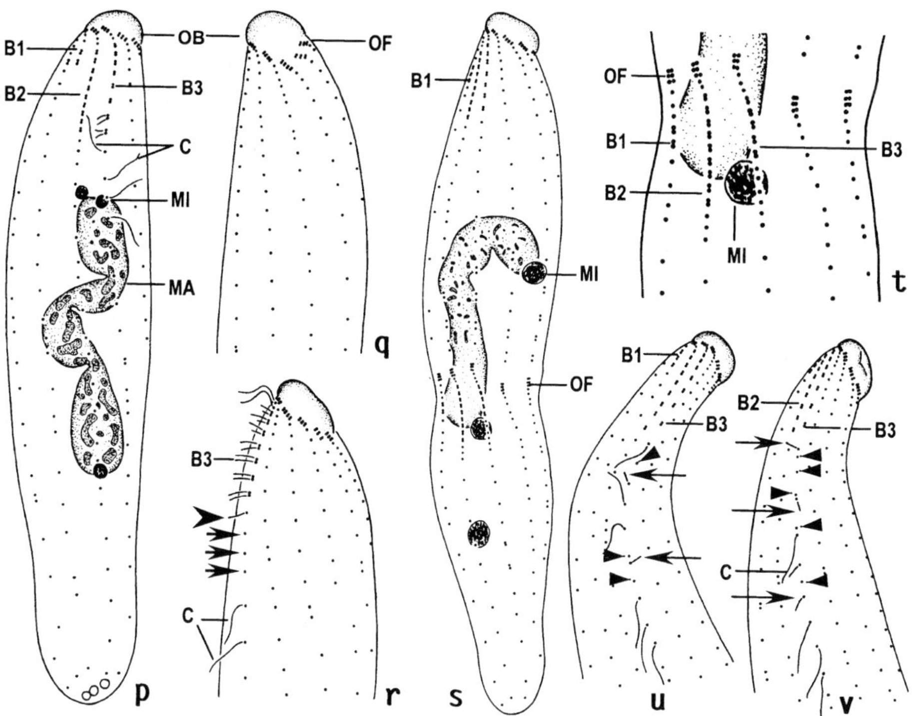

Fig. 51p–v *Protospathidium serpens,* Austrian neotype specimens (p–t) and *P. fraterculum* (u, v) after protargol impregnation. From XU & FOISSNER (2005a). **p, q)** Dorsal and ventral view of same specimen, length 90 µm. Note lack of a monokinetidal bristle tail at posterior end of brush row 3. **r)** Right side anterior portion of a specimen with brush row 3 followed by a single bristle (arrowhead) and three bare basal bodies (short arrows). **s, t)** An early divider, as indicated by the condensing macronucleus and the inflated micronuclei. The three brush rows (B1–3) of the opisthe are not generated concomitantly, but row 1 is formed later than rows 2 and 3. **u, v)** Dorsal views of Antarctic *P. fraterculum* specimens investigated by FOISSNER (1996a). The monokinetidal tail of brush row 3 extends to mid-body and is composed of bristles (arrows) mixed with ordinary cilia and bare basal bodies (arrowheads). B 1-3 – dorsal brush rows, C – ordinary somatic cilia, MA – macronucleus, MI – micronuclei, OB – oral bulge, OF – oral kinetofragments.

especially in first and/or second row of right (ventral) side. Kineties ordinarily ciliated (average ciliary distance 2–3 µm) and attached to oral kinetofragments, anterior portion slightly curved dorsally on right side, while abutting on oral kinetofragments at right angles on left and dorsal side (Fig. 49a, d, h–j, o, p, u, v, 51a, m, n, p, 114a–h; Tables 10, 11). Dorsal brush basically three-rowed, in almost half of the Namibian specimens only two-rowed, while four-rowed in half of the Antarctic population investigated by FOISSNER (1996a); inconspicuous because row 1 reduced to a few dikinetids, longest row 2 occupies only about 12% of body length, and bristles merely up to 4 µm long in vivo (Fig. 49a, d, e, g, i, p, t, v, 51a, d, m, n, p, r, 113m, n, 114a, c, e–g, j, 115a–c, l; Tables 10, 11). Brush details rather similar in several populations. Namibian site (70) specimens (unpubl. data, Fig. 50e; FOISSNER et al. 2002): row 1, anterior bristles of dikinetids 2.5 µm long, posterior 1 µm, both clavate; row 2, anterior bristle of dikinetids

3 µm long, posterior 1.5 µm, both clavate; row 3, anterior bristle of dikinetids 2.5 µm long, posterior 3 µm, both rod-shaped. Peloponnese specimens (unpubl. data): row 1, anterior bristles of dikinetids 3 µm long, posterior 1.5 µm, both clavate; row 2 as row 1; row 3, anterior and posterior bristle of dikinetids 2 µm long and rod-shaped. Venezuelan specimens (unpubl. data, Fig. 50f, 115l): row 1, anterior bristles of dikinetids 3 µm long and clavate, posterior 0.5 µm long; row 2 as row 1, but posterior bristles 1 µm long; row 3, both bristles 3 µm long and clavate. Marion Island specimens (FOISSNER 1996a; Fig. 49j, p, t, v, 51u, v, 115a, c): row 1, anterior bristles of dikinetids 2 µm long and inflated, posterior 0.5 µm; row 2, anterior bristles of dikinetids 2–3 µm long, posterior 0.5–2 µm, both clavate; row 3, anterior and posterior bristles of dikinetids 3.5 µm long and rod-shaped, with monokinetidal tail of 2 µm long bristles extending to mid-body; anterior bristle of first dikinetid of each row elongated to 8–10 µm and distally inflated, a special feature definitely lacking in, at least, populations from Namibian site (70), Venezuela, and the Republic of South Africa. Austrian neotype specimens (Fig. 51a, d, l–n, p, r, 113m, n, 114a, c, e–g, j; Table 12): brush located slightly dorsolaterally; row 1 composed of 2–6, usually only 3 dikinetids, the last pair with an 1 µm long anterior bristle and a 7 µm long ordinary cilium posteriorly, anterior bristle of other dikinetids about 2 µm long and slightly inflated in mid, posterior bristle knob-like because only about 0.5 µm long. Brush row 2 longest, composed of 8–17, on average 12 ordinarily spaced dikinetids, bristles similar to those of row 1 in anterior portion, but posterior bristles up to 1 µm long; length of bristles sharply decreases posteriorly. Brush row 3 slightly shorter than row 2, composed of 5–11, on average 7 comparatively widely spaced dikinetids; anterior and posterior bristle of dikinetids each up to 3 µm long and sometimes V-like spread, slightly curved, and clavate; monokinetidal bristle tail lacking or reduced to a single, 1.5 µm long bristle neighboured the next ordinary cilium, as shown by many properly-impregnated specimens and the scanning electron micrographs. These data suggest the following general pattern of the dorsal brush of *P. serpens*: (i) three-rowed, but frequently with a high proportion of two- or four-rowed specimens; (ii) inconspicuous because longest brush row 2 occupies merely about 12% of body length and bristles usually only up to 4 µm long; (iii) none or few ordinary cilia between brush rows and oral kinetofragments; (iv) row 1 reduced usually consisting of only two to four dikinetids with anterior bristles longer than posterior ones and slightly clavate; row 2 is the longest usually consisting of seven to twelve dikinetids with anterior bristles longer than posterior ones and slightly clavate; row 3 slightly shorter than row 2 usually consisting of four to seven comparatively widely spaced dikinetids with rod-shaped bristles of either same length or with the anterior bristles shorter than the posterior ones, which contrasts rows 1 and 2, where the anterior bristles are longer than the posterior ones; (iv) monokinetidal bristle tail of brush row 3 present or lacking, depending on population.

Oral bulge rather conspicuous, though only about half as long as widest trunk region, because packed with refractive extrusomes, 3–5 µm high in vivo, and somewhat knobby, that is, projecting laterally and thus obovate in frontal view; obliquely truncate by 30° to 45°, surface flat and with spiral fibre pattern converging in the distinctly depressed bulge centre lined by fine fibres sometimes recognizable in protargol preparations (Fig. 49a, d, g–k, o, p, u, v, 51a, c, m, n, p–r, 113l, 114b, d, e, h–j, 115d–f; Tables 10, 11). Circumoral kinety of similar or same shape as oral bulge, composed of dikinetidal kinetofragments obliquely attached to the respective ciliary rows and separated from each other by gaps two to four dikinetids wide. Individual kinetofragments on average composed of three to five dikinetids, depending on population, each associated with a 8–10 μm long cilium and a fine nematodesma. Oral basket inconspicuous, recognizable only after protargol impregnation, composed of cuneate, 20–30 μm long nematodesma bundles originating from the circumoral kinetofragments (Fig. 49e–i, o, p, u, v, 51l–n, p–r, 113l–n, 114d–j, 115a, b, d, j; Tables 10, 11).

The Antarctic specimens encysted within 15 min when transferred from the non-flooded Petri dish culture to Eau de Volvic (PETZ & FOISSNER 1997). Resting cysts 25–32 μm across, usually near 30 μm, brownish, wall about 2 μm thick, highly refractive, contains conspicuous, compact granules causing cyst surface to become tubercular and, respectively, honey-combed in lateral and surface view (FOISSNER 1996a, PETZ & FOISSNER 1997). Cytoplasm finely granulated, macronucleus tortuous (Fig. 49r, s).

Of 50 Austrian (Salzburg) specimens, about half encysted within 48h. Four-days-old resting cysts spherical ($\bar{x}$ 17.6 μm, M 18, SD 1.1, CV 6.2, Min 16, Max 20, n 12), colourless, and with 1.5 μm thick, spiny wall; spines up to 3.5 μm long and flexible, likely originate from inner wall. Nucleus tortuous. Cytoplasm packed with lipid droplets up to 3 μm across (Fig. 51o, 114k–n).

Obviously, the Salzburg cysts are entirely different from those of the Antarctic specimens studied by FOISSNER (1996a), strongly suggesting different species! We checked the original notes from the Antarctic specimens. This showed that the data are likely correct, although no micrographs were made. Thus, XU & FOISSNER (2005a) classified the Antarctic populations as a new species, *P. fraterculum*, which is sustained by a South African population also having a smooth cyst wall.

Morphology and ontogenesis of a Namibian population (Fig. 52a–r): Division was studied in specimens from Namibian site 16 (see FOISSNER et al. 2002 for site description). Cultures were established in Eau de Volvic enriched with some drops of percolate from the non-flooded Petri dish culture and three wheat grains to stimulate growth of bacteria and prey protozoa.

The cultivated specimens show a high variability in all features, even in the number of ciliary rows (Table 10), and especially in the macronucleus pattern, ranging, like in the Antarctic populations, from a single, ellipsoidal nodule to a very long, tortuous, more or less nodulated strand sometimes broken in two or three pieces (Fig. 114o–s, 115g–i). Further, specimens show up to 30% contractility and amoeboid changes of body shape when rooting in the mud on the bottom of the Petri dish (Fig. 50h–l). Some of the variability is caused by dying post-conjugants and monstrous individuals being either very small (< 50 μm) or very large (> 150 μm). Variability drastically decreases if only very early dividers, that is, "healthy" specimens are analyzed. They all have an ordinary size (length $\bar{x}$ 105 μm, SD 11.5, SE 2.7, CV 10.9, Min 85, Max 132, n 18) and a long, slightly tortuous and nodulated macronucleus, which is thus the true pattern in this species (Fig. 52a, c, 114o, p, 115g).

In contrast to *P. muscicola*, very early dividers of *P. serpens* have the same length (105 μm, see above) as morphostatic specimens (Table 10). Division commences with the production of basal bodies slightly underneath mid-body in those kineties which bear the dorsal brush in proter and opisthe. Usually, even an oblique dikinetid, obviously belonging to the prospective oral kinetofragments, is recognizable at the anterior end of each developing brush kinety. No changes are recognizable in the

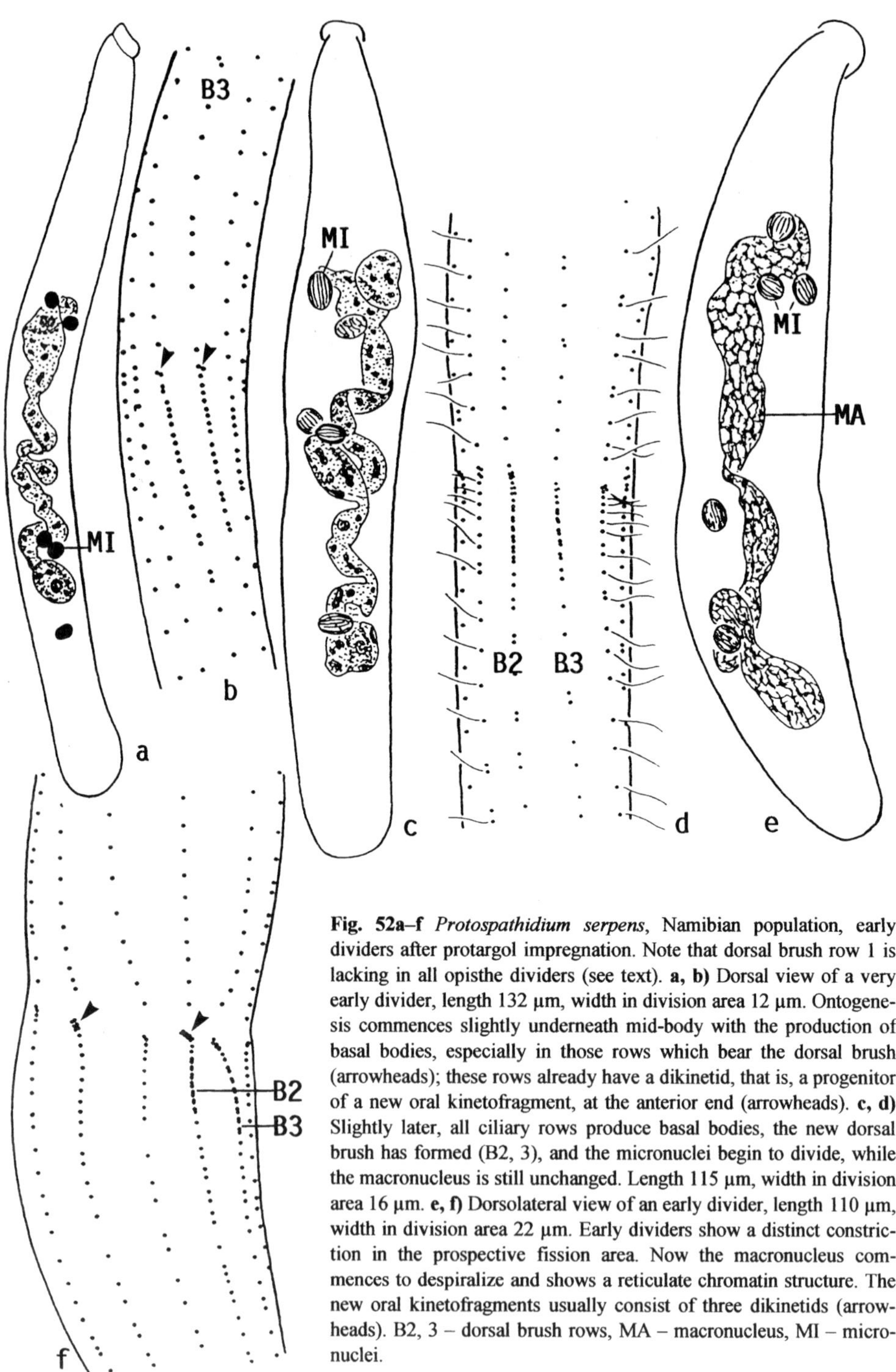

Fig. 52a–f *Protospathidium serpens*, Namibian population, early dividers after protargol impregnation. Note that dorsal brush row 1 is lacking in all opisthe dividers (see text). **a, b)** Dorsal view of a very early divider, length 132 µm, width in division area 12 µm. Ontogenesis commences slightly underneath mid-body with the production of basal bodies, especially in those rows which bear the dorsal brush (arrowheads); these rows already have a dikinetid, that is, a progenitor of a new oral kinetofragment, at the anterior end (arrowheads). **c, d)** Slightly later, all ciliary rows produce basal bodies, the new dorsal brush has formed (B2, 3), and the micronuclei begin to divide, while the macronucleus is still unchanged. Length 115 µm, width in division area 16 µm. **e, f)** Dorsolateral view of an early divider, length 110 µm, width in division area 22 µm. Early dividers show a distinct constriction in the prospective fission area. Now the macronucleus commences to despiralize and shows a reticulate chromatin structure. The new oral kinetofragments usually consist of three dikinetids (arrowheads). B2, 3 – dorsal brush rows, MA – macronucleus, MI – micronuclei.

nuclear apparatus (Fig. 52a, b). Somewhat later, a surprising change of body shape occurs, viz., a slight indentation in the prospective fission area (Fig. 52c, d). This has been observed also in *Cultellothrix coemeterii*, and might have been overlooked in previous studies. Basal body proliferation occurs now in all ciliary rows, as obvious from narrowly spaced granules, of which only the posterior one is ciliated. Dikinetids are formed in the opisthe brush kineties, and minute oral kinetofragments, each usually composed of two dikinetids, occur at the anterior end of the broken ciliary rows. The macronucleus appears unchanged, while the micronuclei double their size from about 2.5 µm to 4–5 µm becoming ellipsoidal and fibrous (Fig. 52c, d). Slightly later, when the now somewhat inflated and despiralized macronucleus shows a fibrous, reticular structure and the oral kinetofragments consist of usually three dikinetids, the indentation in the prospective fission area becomes as distinct as in early late dividers (Fig. 52e, f). The opisthe dorsal brush is completed, that is, composed of dikinetids and some ordinary monokinetids between the anterior end of the rows and the newly produced oral kinetofragments (Fig. 52f). However, brush row 1, which consists of two to nine dikinetids and is lacking in almost half of the morphostatic cells (Table 10), is lacking in all early and late dividers checked, even in specimens, in which the proter has three rows (Fig. 52b, d, p). This indicates that it is produced in post-dividers (cp. *Arcuospathidium cultriforme*, where dividers show that the brush rows are not produced concomitantly). Alternatively, the fast-growing population could have developed in direction of specimens with two rows only, as indicated by the high percentage of two-rowed morphostatic specimens.

Early mid-dividers have an ellipsoidal to slightly fusiform outline, that is, loose the central indentation and become more or less inflated in mid-body, where the condensing, fibrous macronucleus mass and the inflated micronuclei accumulate (Fig. 52g, h). The opisthe oral kinetofragments now consist of two to five dikinetids, that is, are almost completed. In mid-dividers (Fig. 52i–k), the newly formed, straight oral kinetofragments curve slightly rightwards and consist of two to ten, usually four to six dikinetids, as in morphostatic cells (Table 10), and intense proliferation of basal bodies occurs underneath the fragments, showing that the main portion of new basal bodies is produced here, that is, not along the whole kineties (Fig. 52h, k). The macronucleus eventually condenses to a large globule with fibrous content, and the micronuclei commence fission division, that is, become distinctly fusiform and show many longitudinally oriented fibres, likely bundles of microtubules distributing the chromosomes, which are sometimes recognizable as minute, more heavily impregnated rods. Soon, however, the micronuclei divide and move apart, assuming a compact structure and remaining connected by a fine, argyrophilic thread (Fig. 52i, j).

In early late dividers (Fig. 52l, m), the globular macronucleus mass extends to a short rod and the micronuclei grow slightly and show a more loose consistency. The newly produced somatic basal bodies have fully developed cilia and become distributed in the kineties. The new oral kinetofragments rarely assume a slightly concave shape and curve distinctly to the right. Late dividers (Fig. 52n, o) have a similar length as morphostatic specimens and show a distinct indentation in mid-body. A hemispherical, bare protuberance, viz., the precursor of the oral bulge, occurs at the anterior end of the opisthe, while newly formed excretory pores of the contractile vacuole are recognizable at the posterior end of the proter. The macronucleus mass extends to a long, slightly tortuous rod, which becomes divided in the fission area and shows a fibrous structure

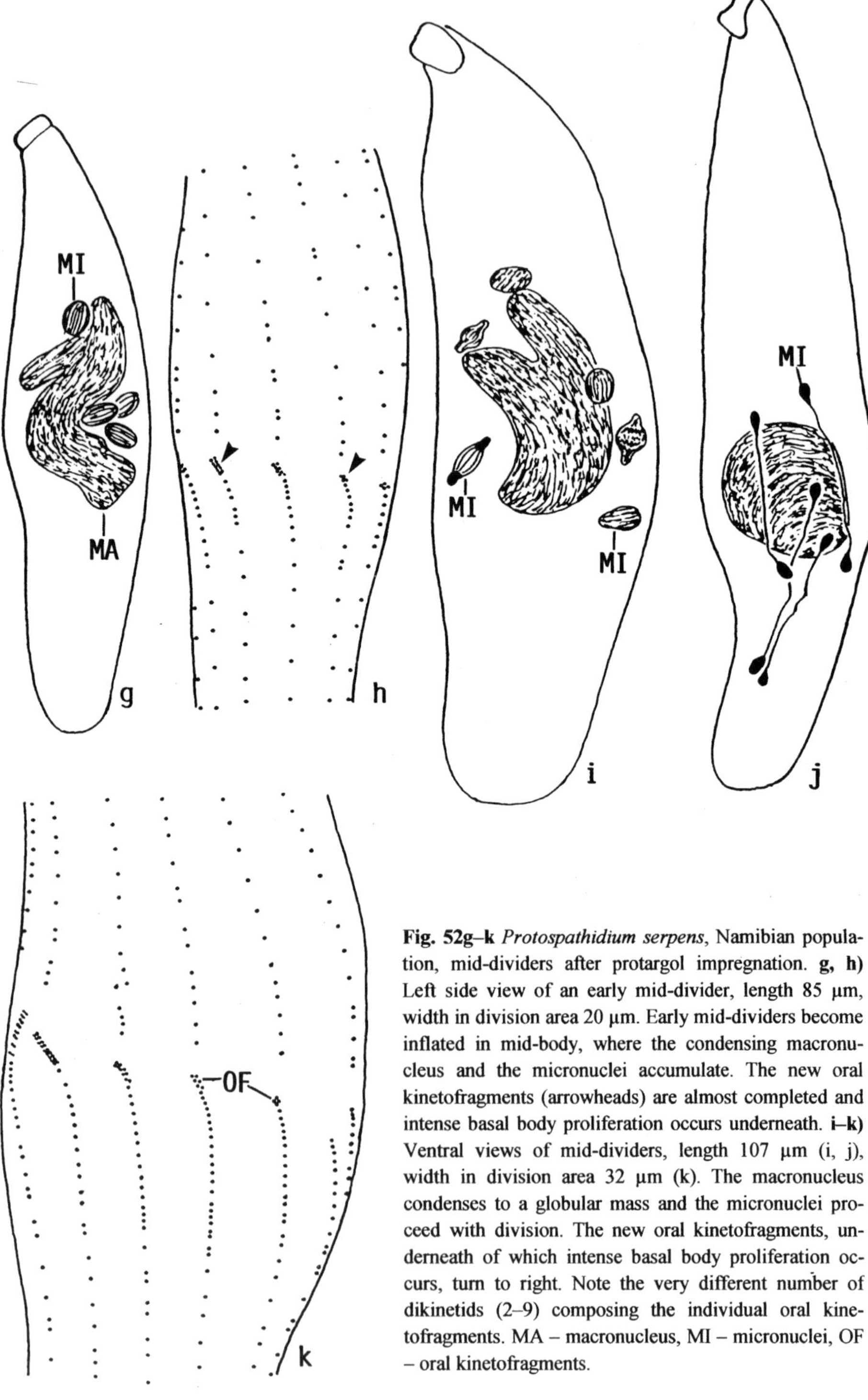

Fig. 52g–k *Protospathidium serpens*, Namibian population, mid-dividers after protargol impregnation. **g, h)** Left side view of an early mid-divider, length 85 µm, width in division area 20 µm. Early mid-dividers become inflated in mid-body, where the condensing macronucleus and the micronuclei accumulate. The new oral kinetofragments (arrowheads) are almost completed and intense basal body proliferation occurs underneath. **i–k)** Ventral views of mid-dividers, length 107 µm (i, j), width in division area 32 µm (k). The macronucleus condenses to a globular mass and the micronuclei proceed with division. The new oral kinetofragments, underneath of which intense basal body proliferation occurs, turn to right. Note the very different number of dikinetids (2–9) composing the individual oral kinetofragments. MA – macronucleus, MI – micronuclei, OF – oral kinetofragments.

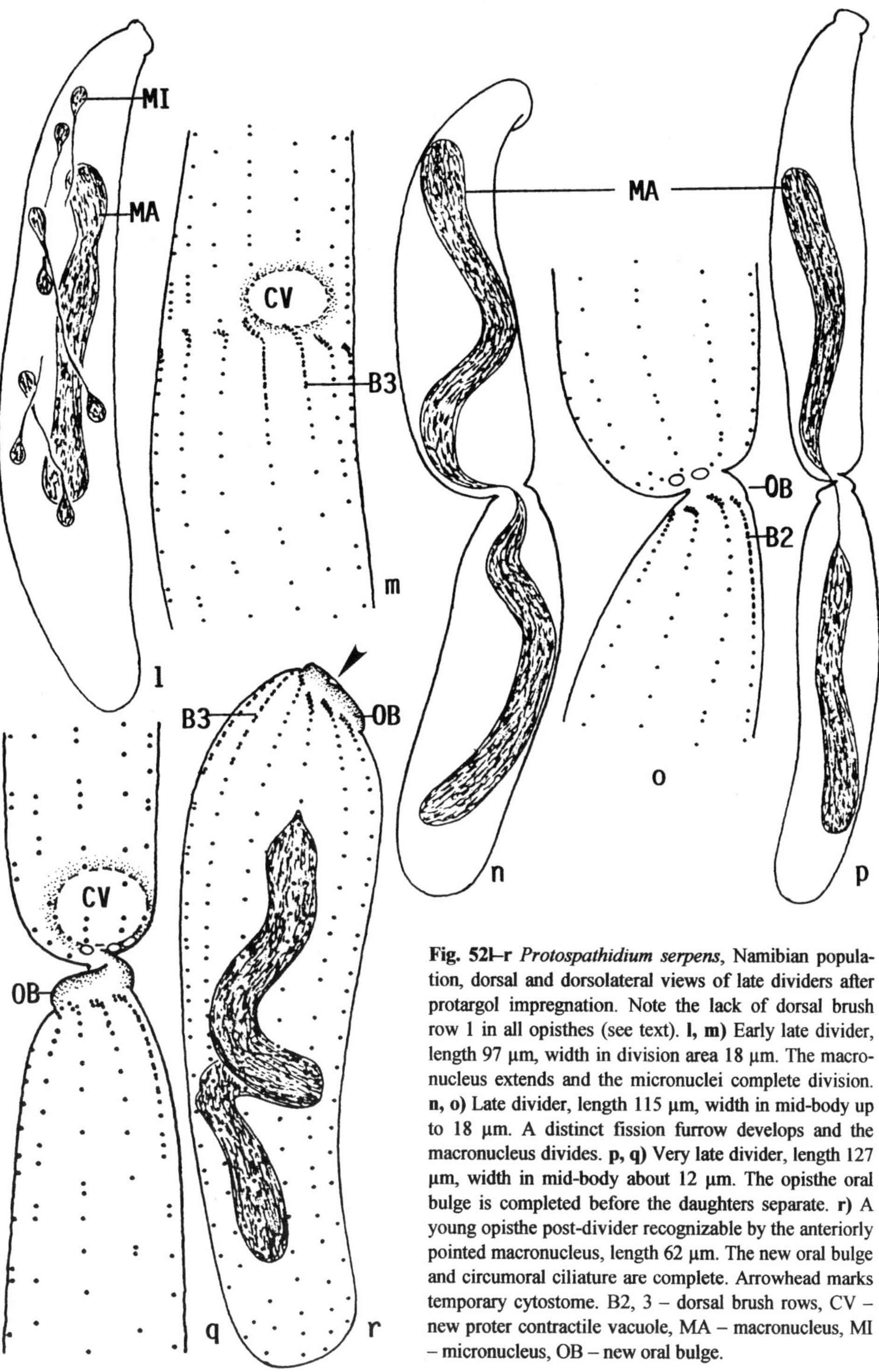

Fig. 52l–r *Protospathidium serpens*, Namibian population, dorsal and dorsolateral views of late dividers after protargol impregnation. Note the lack of dorsal brush row 1 in all opisthes (see text). **l, m)** Early late divider, length 97 μm, width in division area 18 μm. The macronucleus extends and the micronuclei complete division. **n, o)** Late divider, length 115 μm, width in mid-body up to 18 μm. A distinct fission furrow develops and the macronucleus divides. **p, q)** Very late divider, length 127 μm, width in mid-body about 12 μm. The opisthe oral bulge is completed before the daughters separate. **r)** A young opisthe post-divider recognizable by the anteriorly pointed macronucleus, length 62 μm. The new oral bulge and circumoral ciliature are complete. Arrowhead marks temporary cytostome. B2, 3 – dorsal brush rows, CV – new proter contractile vacuole, MA – macronucleus, MI – micronucleus, OB – new oral bulge.

with many minute condensations, likely developing nucleoli (Fig. 52n, o). This macronucleus structure is recognizable also in early post-dividers. The shape of the daughters of very late dividers is already fairly similar to that of small-sized interphase specimens (Fig. 52p, q). The macronucleus divided in two rod-shaped pieces each about two thirds as long as the body and more or less distinctly pointed in the fission area. The micronuclei also completed division, but could not be studied in detail because of many similar-sized and impregnated food inclusions. The oral kinetofragments assume their definite position and arrangement, but are still slightly wider spaced than in morphostatic specimens. The oral bulge is almost fully developed and connected to the proter with the dorsal half (Fig. 52p, q). Thus, early opisthe post-dividers are usually recognizable only by the pointed anterior end of the macronucleus (Fig. 52r). They are stouter than typical morphostatic cells, but have fully developed the circumoral ciliature and the oral bulge, including the temporary cytostome.

The parental oral apparatus and dorsal brush do not show any changes during division, as in the other species investigated so far. We could not follow the development of the oral basket, which impregnated too faintly. Likewise, no blebs are recognizable in the division area, in contrast to *Spathidium spathula* and *S. turgitorum*; possibly they are too minute in this small species or did not withstand the preparation procedures.

Ontogenetic comparison: Division of *P. serpens* is highly similar to that of other spathidiids, but shows some features possibly unique to the genus or, at least, different to those known from *Spathidium* and *Arcuospathidium* (see Table on page 26).

(i) Early dividers show a transient indentation in the prospective fission area. This was observed also in the Austrian population studied by XU & FOISSNER (2005a). A reinvestigation of the Antarctic population studied by FOISSNER (1996a) showed this indentation too, but it was less distinct than in the Namibian and Austrian specimens.

(ii) The Austrian specimens studied by XU & FOISSNER (2005a) likely have division blebs not recognizable in the Namibian dividers.

(iii) The oral kinetofragments remain straight, that is, are not curved as in *Spathidium*, *Arcuospathidium*, and *Cultellothrix*.

(iv) All brush rows are generated in early opisthe dividers in the above mentioned Austrian and Antarctic specimens, while the Namibian specimens generate brush row 1 probably post-divisionally. Brush row 1 develops slightly later than rows 2 and 3 in the Austrian and Antarctic specimens.

(v) Shaping of the oral bulge and circumoral kinety occurs distinctly earlier in *Protospathidium* than in *Spathidium* and *Arcuospathidium*, viz., in very late dividers, respectively, in early post-dividers.

Conjugation: Spontaneous mass conjugation occurred in a stationary semipure culture of *P. serpens* from the Republic of South Africa. This event was studied by XU & FOISSNER (2004) in protargol-impregnated specimens. Later, the species was recognized to be *P. fraterculum* (XU & FOISSNER 2005a). We provide an extended summary of the observations and the full set of figures. An abstract is given in chapter 2.2. Basically, conjugation and nuclear reconstruction follow the usual mode of ciliates. However, some peculiarities occur: only two of the four synkaryon derivatives of the second synkaryon division enter the third division and generate four macronuclear anlagen, which then fuse to the species-specific strand (Fig. 53d–s, 54a–n). No basic changes

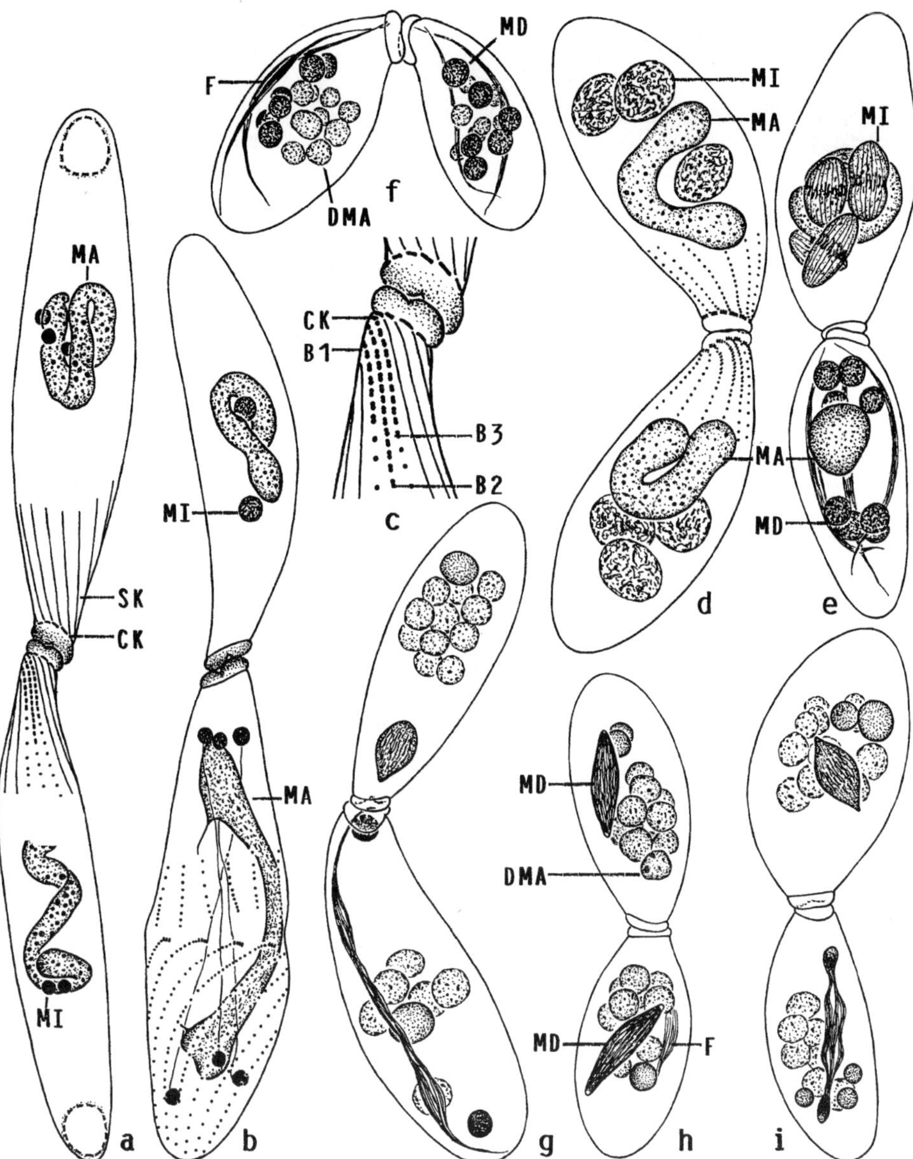

Fig. 53a–i *Protospathidium fraterculum*, protargol-impregnated early and maturating conjugants, drawn to scale (from XU & FOISSNER 2004). **a, c)** Very early conjugant, length 142 µm. The macronucleus commences to condense and the micronuclei are in interphase to early prophase. **b)** A deviating, early conjugant with one partner performing asexual division, length 125 µm. **d)** Prophase of the first maturation division, length 85 µm. **e)** Metaphase in the anterior and telophase in the posterior partner of the second maturation division, length 82 µm. The macronucleus condensed to a globular mass. **f)** Very late telophase of the second maturation division, length 45 µm. The macronucleus disintegrates into many globules. **g–i)** Metaphase to telophase of the third maturation division, length 90 µm, 66 µm, 73 µm. B2, 3 – dorsal brush rows, CK – circumoral kinety, DMA – disintegrating vegetative macronucleus, F – rest of division spindle, MA – vegetative macronucleus, MD – maturation derivatives, MI – micronuclei, SK – somatic kineties.

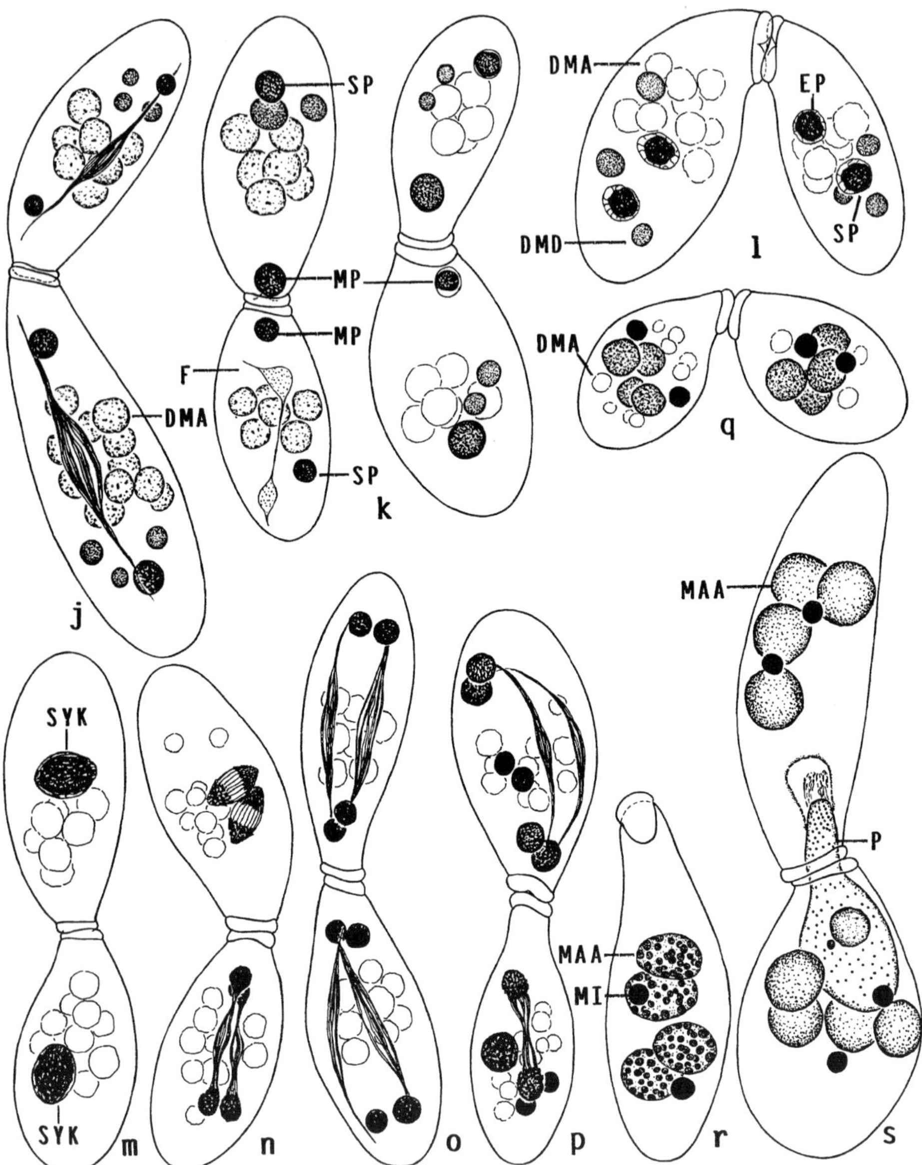

Fig. 53j–s *Protospathidium fraterculum*, protargol-impregnated, maturating and synkaryon developing conjugants and exconjugants, drawn to scale (from XU & FOISSNER 2004). **j)** Late telophase of the third maturation division, length 85 µm. **k)** Pronuclei formation, length 70 µm. **l)** The migratory pronucleus approaches the partner's stationary pronucleus in prophase, length 45 µm. **m)** Synkaryon formation, length 70 µm. The synkaryon is broadly ellipsoidal in the prometaphase of the first division. **n, o)** Anaphase to telophase of the second synkaryon division, length 75 µm. **q)** A very late conjugant, length 46 µm. The pairs separate after this stage, and thus early exconjugants have, typically, four macronucleus anlagen and two micronuclei. **r)** A very early exconjugant, length 50 µm. **s)** Bilateral reconjugation, length 95 µm. DMA – disintegrating vegetative macronucleus, DMD – degenerating maturation derivative, EP – exchanged pronucleus, F – rest of division spindle, MP – migratory pronucleus, SP – stationary pronucleus, SYK – synkaryon.

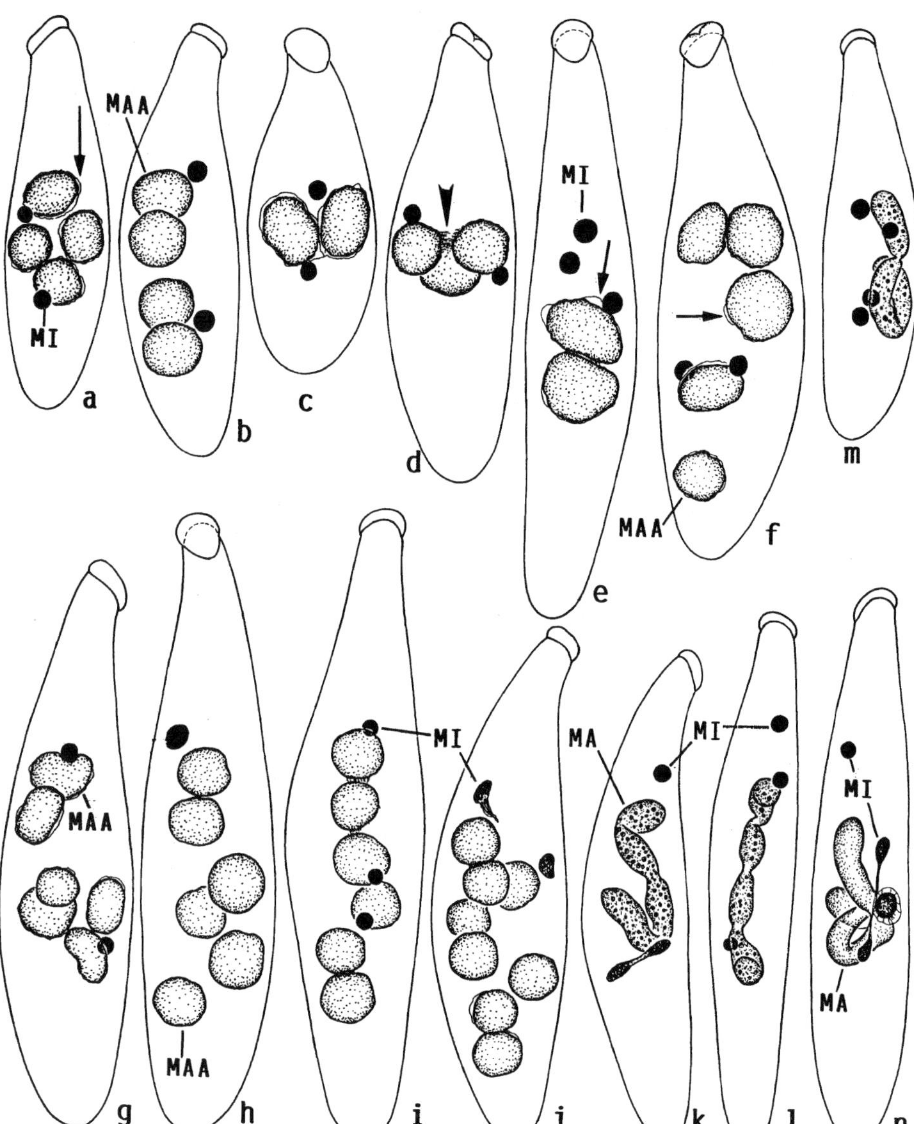

Fig. 54a–n *Protospathidium fraterculum*, exconjugant nuclear reconstruction after protargol impregnation, drawn to scale (from Xu & Foissner 2004). **a, b)** Two exconjugants each with the typical four macronuclear anlagen and two micronuclei, length 53 µm, 58 µm. Note the wrinkled nuclear membrane (arrow). **c–e)** Exconjugants with two or three macronuclear anlagen, which are generated by stepwise fusion of the four macronuclear anlagen, length 47 µm, 62 µm, 80 µm. Arrowhead marks some curious fibres between anlagen. **f–j)** Exconjugants with five to eight macronuclear anlagen and one to three micronuclei, length 73 µm, 77 µm, 83 µm, 85 µm, 70 µm. Five, six or eight macronucleus anlagen, instead of the usual four, are likely generated by one or both progenitor micronuclei. **k–n)** Very late exconjugants with the macronucleus strand composed of four more or less distinct, ellipsoidal nodules, that is, the former four macronucleus anlagen, length 67 µm, 77 µm, 55 µm, 75µm. MA – vegetative macronucleus, MAA macronucleus anlagen, MI – micronuclei.

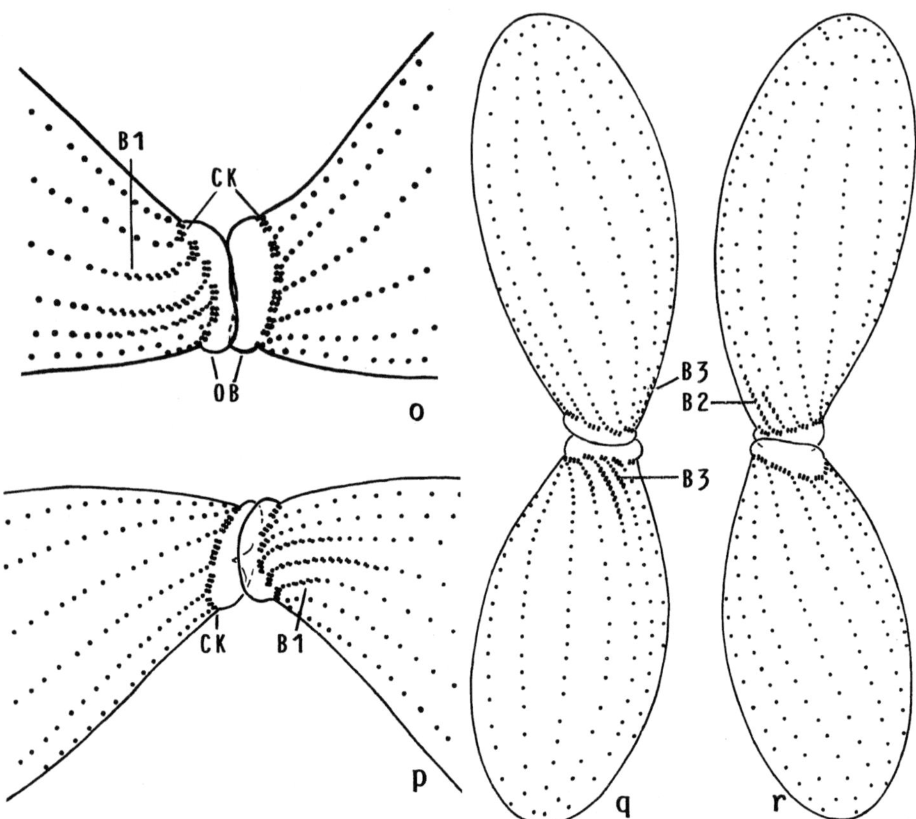

Fig. 54o–r *Protospathidium fraterculum,* ciliary pattern of the protargol-impregnated conjugants shown in figures 53d and 53h (from XU & FOISSNER 2004). Pair formation is heteropolar, and the partners unite with the oral bulge obliquely. Thus, the dorsal brush is fully recognizable only in one cell. However, it is impossible to determine whether union is basically dorsal-to-dorsal or ventral-to-dorsal because the bulge is minute and roundish. Conjugation does not cause basic changes in the ciliary pattern, though some adjustments occur, for instance, the dorsal brush dikinetids become obliquely arranged due to the distinct body shortening. B1-3 – dorsal brush rows, CK – circumoral kinety, OB – oral bulge.

occur in the ciliary pattern (Fig. 53a, c, 54 o–r). There is no preconjugation division.

Pair formation is heteropolar and all pairs were united in such a way that the dorsal brush was *fully* recognizable only in one cell (Fig. 53a, 54o–r). Thus the partners united obliquely. The conjugants may form a rod-like or strongly arched pair. Among a total of 15 early conjugants, 47 % were rod-shaped or slightly arched, while 53 % were arched by 30° to almost 180° (Fig. 53a–d). The same proportion was recognizable in 34 late conjugants. Thus, there is no correlation between pair shape and progress through conjugation. The nuclear apparatus is hardly changed during pairing. Two deviating cases were observed: in the first pair, one of the partners performed asexual cell and nuclear division, while the micronuclei of the other partner were in an early prophase stage (Fig. 53b); in the second pair, one partner had two condensed macronucleus masses, likely due to a split vegetative nucleus.

There are three maturation divisions (Fig. 53d–h). They were associated with distinct changes in body size, body shape, and the appearance of the nuclear apparatus. The body distinctly shortened and broadened from 69 × 13 µm to 40 × 19 µm, that is, became ellipsoidal on average. Detailed morphometrics showed that conjugation is likely isogamic, though one of the partners is occasionally shorter by up to one third. The food vacuoles disappeared before the formation of the pronuclei. During the formation of the pronuclei and the synkaryon, the mouth opening enlarged. The migratory pronucleus approached the partner's stationary pronucleus in prophase state (Fig. 53l), and the synkaryon was thus formed by fusion of the pronuclei (Fig. 53m).

The three synkaryon divisions were associated with further body diminution, which reached a maximum in late conjugants with four macronuclear anlagen. Of the four synkaryon derivatives, only two entered the third division and produced four macronuclear anlagen; the two non-dividing derivatives condensed to micronuclei about 3 µm across (Fig. 53m–q). The conjugants separated after the third synkaryon division. Very early exconjugants were recognizable by their small size and the very typical nuclear apparatus, viz., four macronuclear anlagen and two micronuclei (Fig. 53q, r, 54a, b). Further, they were able to feed. Vegetative body size was reached only in late exconjugants with fusing macronucleus anlagen. Bilateral reconjugation was observed in one pair (Fig. 53s). The cells were full of food vacuoles and one partner showed a just captured ciliate.

Occurrence and ecology: The following compilation cannot take into account the recent split of the species by XU & Foissner (2005a) because data on resting cysts are usually lacking. Probably, populations with smooth cyst wall (= *P. fraterculum*) occur mainly in the southern hemisphere. Basically, *Protospathidium serpens*-like populations are the most frequent protospathidiids and have been reported from all main biogeographic regions (FOISSNER 1998). However, they are infrequent and rarely develop high abundances in the non-flooded Petri dish cultures, but they can be rather easily cultivated in Eau de Volvic (French table water) enriched with some drops of the soil percolate and a few crushed wheat grains to stimulate growth of small ciliates and flagellates, the preferred food. *Protospathidium serpens*-like populations occur in a huge variety of terrestrial and semiterrestrial habitats including terrestrial mosses, soils of high mountains and from Antarctica, highly saline soils from the Etosha Pan in Namibia, and mud from rock-pools and puddles. Thus, *P. serpens* is euryoecious and likely cosmopolitan preferring, however, terrestrial habitats.

Holarctic records: The type locality is in the surroundings of Hamburg, Germany, where KAHL (1930b) discovered *P. serpens* in a shallow road drain, a semiterrestrial habitat. Further records from Germany: in two samples of dry, terrestrial moss from Bavaria, length of specimens 60–90 µm (WENZEL 1953); in evolved coastal dune soil of northern Germany at pH 6.7–6.9 (GORALCZYK & VERHOEVEN 1999); in beech forest soil near Göttingen (FOISSNER 2000b). Many records are known from Austria (FOISSNER & FOISSNER 1988b): in soil (pH 4.2–6.4) from six alpine sites of Styria (Grossglockner area) 879 m to 2310 m above sea level (FOISSNER 1981b); in soil (pH 4.2–6.4) from five alpine sites of Salzburg (Gastein area) 1780 m to 1950 m above sea-level (FOISSNER & PEER 1985); in conventionally and biologically farmed grassland soil of Styria (FOISSNER et al. 1990); in soil from three out of four sites in Tyrol, contaminated with heavy metals (POHLA et al. 1994); with 10–30% frequency in soils (pH 6.9–7.5) from wheat fields, lowland forests, and beech forests of Lower Austria and Salzburg

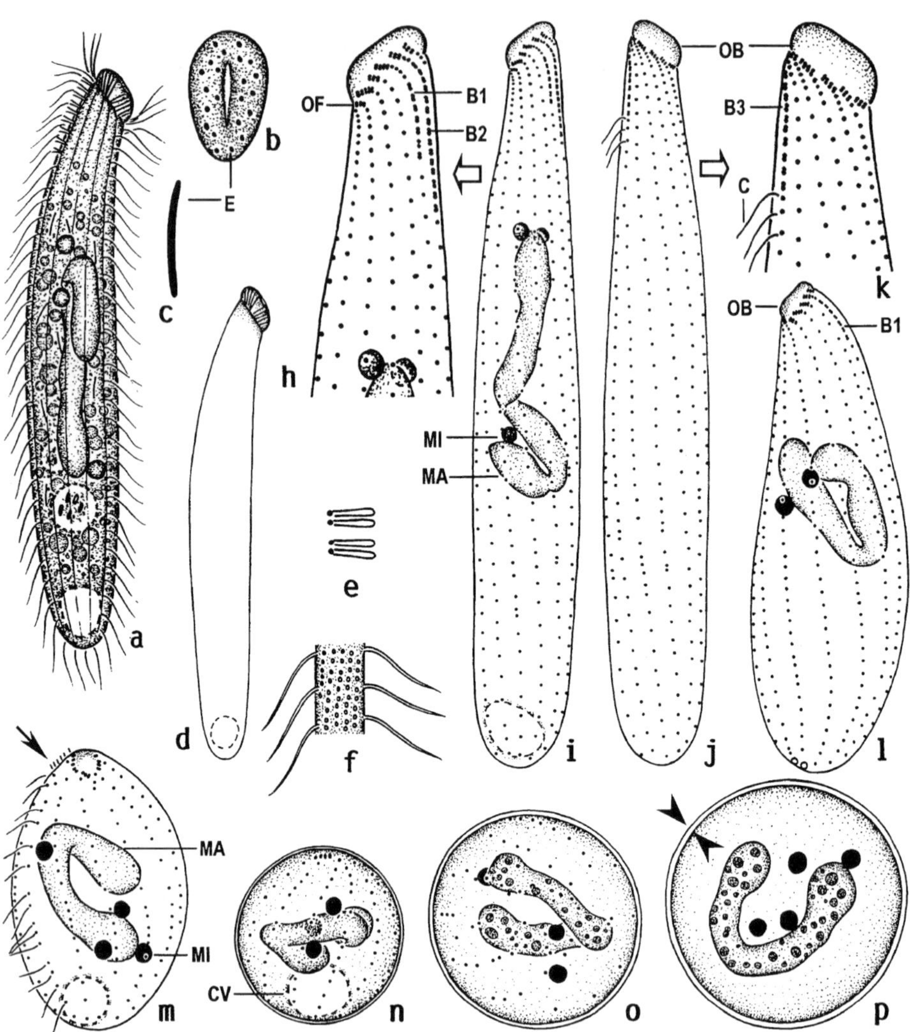

Fig. 55a–p *Protospathidium fraterculum* from life (a–f) and after protargol impregnation (g–p). From XU & FOISSNER (2005a). **a)** Right side view of a representative specimen, length 90 µm. **b)** Frontal view of oral bulge studded with extrusomes. **c)** Extrusomes are rod-shaped, slightly curved, and about 4 µm long. **d)** A long and slender specimen, length 115 µm. **e)** Dorsal brush bristles are up to 3 µm long. **f)** Surface view showing cortical granulation. **h–k)** Ciliary pattern of left and right side and nuclear apparatus of holotype specimen, length 85 µm. Brush row 1 is composed of an average of seven dikinetids, and row 3 lacks a monokinetidal bristle tail, a main difference to the Antarctic populations which have a tail and thus likely represent a different subspecies or species. **m–p)** Encystment (for details, see XU & FOISSNER 2005a). The smooth wall of the mature cyst (p) is the sole feature distinguishing *P. fraterculum* from *P. serpens* which has a spiny cyst wall (Fig. 51o). The arrow indicates the highly reduced dorsal brush bristles. Opposed arrowheads mark the smooth, about 1 µm thick wall of the resting cyst. Note that the nuclear apparatus does not reorganize. See Table 12 for morphometrics. B1-3 – dorsal brush rows, C – somatic cilia, CV – contractile vacuole, E – extrusomes, MA – macronucleus, MI – micronuclei, OB – oral bulge, OF – oral kinetofragments.

Table 12 Comparison of main features of *Protospathidium serpens* and *P. fraterculum* (from Xu & Foissner 2005a). Data marked with (*) are from protargol preparations; those in brackets are means.

Characteristics	*Protospathidium serpens* complex			*P. serpens* neotype	*P. fraterculum*			
Source	Kahl (1930b)	Foissner (1981a)	Xu & Foissner (2005a)	Xu & Foissner (2005a)	Xu & Foissner (2005a)	Foissner (1996a)	Petz & Foissner (1997)	
Locality	Germany	Austria	Namibia	Austria	South Africa	Antarctica	Pop 1 Antarctica	Pop 2
Body, length (µm; in vivo)	70–90	70–90	~ 120	70–130	90–120	70–100	43–117 (68)	?
Body, width (µm; in vivo)	?	10–15	~ 20	12–25	15–20	12–18	8–25 (13)	?
Body, length:width ratio (in vivo)	6.7	6.8	6	5.5	6	4.3*	5.3	4.8*
Oral bulge, length (µm)*	?	~ 8	?	7–11 (8.4)*	5–8 (6.8)*	6–9 (7.6)*	6–10 (9)*	5–11 (7.9)*
Macronucleus nodulated	yes	yes	yes	yes	yes	yes	yes	yes
Extrusomes oblong or rod-shaped	yes	yes	yes	yes	yes	yes	yes	yes
Ciliary rows, number	~ 10	7–9 (8.3)*	7–12 (9)*	9–12 (10.2)*	11–14 (12.3)*	11–13 (11.6)*	11–16 (13)*	10–12 (11.1)*
Brush rows, number	?	3	2–3 (2.6)*	3*	3*	3–4 (3.5)*	3–4 (3.1)*	3–4 (3)*
Dikinetids in brush row 1, number	?	~ 2[1]	2–9 (6.1)*	2–6 (3.4)*	5–9 (6.8)*	2–4 (3)*	1–5 (3.4)*	1–5 (2.9)*
Dikinetids in brush row 2, number	?	6–8[1]	8–20 (14.6)*	8–17 (12.4)*	10–16 (13)*	7–12 (8.5)*	6–12 (9.2)*	4–10 (6.9)*
Dikinetids in brush row 3, number	?	4–6[1]	6–10 (7.6)*	5–11 (7.5)*	7–13 (9.5)*	4–7 (5.2)*	4–9 (6.6)*	3–7 (4.5)*
Brush dikinetids, total number	?	~ 15[2]	~ 28*	~ 22*	~ 29*	~ 17*	~ 19*	~ 14*
Brush row 3 with a distinct tail of monokinetidal bristles	?	?	no	no	no	yes	yes	yes
Resting cysts spiny or smooth	?	?	?	spiny	smooth	smooth	smooth	smooth

[1] According to description

[2] Counted from the two specimens figured.

(FOISSNER et al. 1985, 2005a). Other records: Peloponnese, Greece and North America (FOISSNER, unpubl.); rare in agricultural soils of Slovakia (TIRJAKOVÁ 1988); in limnetic habitats of the Ukraine, without any data, misidentification thus cannot be excluded (BABKO & KOVALCHUCK 1992).

Paleotropic records: In soil (pH 6) from a marshy site near the Sheldrick waterfalls in the Shimba Hills of Kenya (FOISSNER 1999a); in 6 out of 73 soil samples from Namibia, viz., in mud (pH 6.2) from puddles and rock-pools and in saline (15‰, pH6.2–8.4) soils from the coastal area and the Etosha Pan (FOISSNER et al. 2002).

Neotropic and Australic records (details unpublished; but see figures 115j–m): FOISSNER (1998).

Antarctic records, all likely belonging to *P. fraterculum* (XU & FOISSNER 2005a): *Deschampsia antarctica* grass sward from Signey Island (FOISSNER 1996a); cold-desert (fellfield) soil samples from two sites in the surroundings of Casey Station, Wilkes Land, East Antarctica (PETZ & FOISSNER 1997); in *Cotula plumosa* moss and lava gravel with little vegetation (pH 6.3) from Marion Island, southern ocean (FOISSNER 1996c). Note that Table 3 in SUDZUKI (1979) contains *P. serpens*, but only as a global list of terrestrial ciliates not yet reported from Antarctica!

R e m a r k s : Although the overall appearance of the populations investigated is very similar, they differ considerably in details, such as the dorsal brush, the length of the extrusomes, and the number of ciliary rows (Table 12). Thus, *P. serpens* could be a complex of species, which is emphasized by the different resting cysts, the main reason for the recent split of the species by XU & FOISSNER (2005a). On the other hand, it shows high variability in pure cultures, suggesting that most populations belong to the same, rather variable species. This is sustained by the knowledge that some of the differences have little weight, for instance, the length of the extrusomes, or might be caused by misobservations, for instance, details of the dorsal brush. Thus, the rather different number of ciliary rows remains as a main difference which, however, should also not be over-emphasized because the extremes overlap in most populations (Table 12). FOISSNER (1996a) could not exclude synonymy of *P. serpens* and *P. muscicola* due to the high variability of the macronucleus. With the new knowledge available from dividing specimens, it becomes obvious that they have a different macronucleus pattern (a nodulated, tortuous strand vs. many scattered nodules), and are thus distinct species.

Protospathidium serpens is the most frequent species of the genus and fairly easily identified by the following combination of characters: body cylindroidal and in vivo about 60–120 × 10–20 µm in size; macronucleus a nodulated, tortuous strand; extrusomes rod-shaped and fine, 2–5 µm long; about 10 ciliary rows; oral bulge rather distinct, button-shaped; resting cyst with conspicuous spines. However, identifications should be checked in protargol preparations because of rather similar species in other genera, for instance, *Spathidium rusticanum* and *Armatospathula periarmata*.

Family Arcuospathidiidae nov. fam.

Diagnosis: Spathidiina with temporary cytostome and somatic ciliary rows anteriorly curved dorsally on both sides of cell and distinctly separate from circumoral kinety. Brush located dorsally, dorsolaterally, or on left side. Extrusomes restricted to oral bulge or in both oral bulge and somatic cortex. Oral bulge usually as long as or longer than widest trunk region, cuneate to very elongate elliptical; ventral end of circumoral kinety closed.

Type genus: *Arcuospathidium* FOISSNER, 1984.

Remarks: There are at least six genera (*Arcuospathidium*, *Cultellothrix*, *Armatospathula*, *Apertospathula*, *Longispatha*, *Rhinothrix*) which have the ciliary rows distinctly separate from the circumoral kinety and curved dorsally on both sides of the cell. These are distinct features because most other spathidiid genera have the left side ciliary rows attached to the circumoral kinety and more or less distinctly curved ventrally. Thus, we united these genera in two distinct families, the Arcuospathidiidae nov. fam. and the Apertospathulidae FOISSNER, XU and KREUTZ, 2005, which differ in the oral bulge and circumoral kinety. Likely, further genera remain to be assigned or discovered, for instance, part of the *Semibryophyllum* and *Cultellothrix* species and the genus *Spathidiodes*, which likely also belongs to the Arcuospathidiidae, as indicated by the shape of the oral bulge.

Unfortunately, a main family character, that is, the closed circumoral kinety can be determined only in protargol preparations. Generic separation occurs mainly according to the location of the dorsal brush and the presence/absence of somatic cortical extrusomes.

We assign to the new family four genera (one new) and 31 species/subspecies, of which 9 are described here for the first time. Although this is a considerable gain, we are convinced that many (most?) species are still undiscovered and several of the old *Spathidium* species belong to this family. Our suggestion is based on the observation that we continuously discover new species in this group, especially in the terrestrial-limnetic transition zone, that is, in mud and soil from ephemeral puddles, floodplains, and green river beds, for instance, the Chobe River in Botswana, Africa. Many of the species are difficult to investigate and easily overseen because they are small (< 100 µm) and rare, that is, do not grow to large numbers in non-flooded Petri dish cultures.

Key to genera (requires silver impregnation, see above!)

1	With body (somatic cortical) extrusomes	*Armatospathula*
–	Without body extrusomes	2
2	Cortex comparatively rigid and hyaline (bright)	*Spathidiella*
–	Cortex flexible and of usual appearance	3
3	Dorsal brush mainly on dorsal surface of cell	*Arcuospathidium*
–	Dorsal brush mainly on left side of cell	*Cultellothrix*

Arcuospathidium FOISSNER, 1984

1984 *Arcuospathidium* FOISSNER, Stapfia, 12: 74 – Type species (by original designation): *Spathidium cultriforme* PENARD, 1922.

Improved diagnosis: More or less distinctly knife-shaped Arcuospathidiidae with brush located dorsally or dorsolaterally; individual brush rows with anterior tail of ordinary cilia.

Etymology: Not given in original description. The name is a composite of the Latin noun *arcus* (arc, curve) and the generic name *Spathidium* (small spatula), referring to the often long and curved oral bulge and the similarity to the genus *Spathidium*.

Remarks: Although we "cleaned" the genus by transferring some uncommon species to new genera (*Cultellothrix*, *Apertospathula*, *Armatospathula*), it still contains a variety of organization types (species with and without extrusomes, with one or several contractile vacuoles, with a single macronucleus or many macronucleus nodules....), indicating polyphyly. Further, the dorsal brush is highly diverse, ranging from lack of row 1 (*A. namibiense namibiense*) to three rows with almost same length (*A. multinucleatum*). Frequently, rows 1 and 2 have similar length, while row 3 is shortened; and usually row 2 has more dikinetids than rows 1 and 3. Invariably, brush dikinetids are oriented in main kinety axis, even if they are very densely spaced, as in row 2 of *A. pachyoplites*. However, further splitting of *Arcuospathidium* would be premature at the present state of knowledge.

In our monograph, *Arcuospathidium* comprises 17 species of which 4 are new to science. Most species were discovered rather recently, indicating a high proportion of undescribed species.

Key to species (careful live observation usually sufficient)

1 With two contractile vacuoles, one in anterior body half and another in posterior body end .... *A. bulli*
– With single contractile vacuole in posterior body end 2
2 With extrusomes 4
– Without extrusomes 3
3 Macronucleus slenderly reniform .... *A. cooperi*
– Two rather narrowly spaced macronucleus nodules .... *A. vermiforme*
4 Macronucleus ellipsoidal or oblong........ 5
– Macronucleus a cylindroidal or tortuous strand or in many scattered nodules........ 6
5 Extrusomes rod-shaped, 3–5 μm long; 6–9 ciliary rows .... *A. pelobium*
– Extrusomes cuneate, about 4 × 1 μm; circa 14 ciliary rows .... *A. deforme*
6 Macronucleus a cylindroidal or tortuous strand 10
– Many scattered macronucleus nodules 7
7 Vermiform, length:width ratio ≥10:1. Extrusomes about 5 μm long and thick or only 1–2 μm long and thus difficult to recognize 8
– Narrowly to very narrowly spatulate, length:width ratio 5–10:1. Extrusomes rod-shaped, 4–5 μm long .... *A. multinucleatum*

8 Extrusomes conspicuous, that is, about 5 × 1 µm. More than 10 ciliary rows .... *A. virungense*
– Extrusomes only 1–2 µm long. Less than 8 ciliary rows. Dorsal brush with some 12–18 µm long bristles anteriorly 9
9 Extrusomes oblong. Two dorsal brush rows .... *A. namibiense namibiense*
– Extrusomes narrowly to broadly ovate. Three brush rows .. *A. namibiense tristicha*
10 Size usually 80–130 × 25–40 µm, length:width ratio often ≤ 4:1. Extrusomes rod-shaped to obclavate, 4–6 µm long. Oral bulge extends to near mid-body .... *A. muscorum*
a) Extrusomes rod-shaped to very narrowly cuneate. Usually 14–18 ciliary rows.... .... *A. muscorum muscorum*
b) Extrusomes symmetrically to asymmetrically obclavate. About 12 ciliary rows.. .... *A. muscorum rhopaloplites*
– Length usually ≥ 150 µm, very narrowly spatulate to vermiform with length:width ratio > 6:1 on average 11
11 Oral bulge ≤ 25% of body length. Less than 12 ciliary rows. Extrusomes narrowly ovate or obovate 12
– Oral bulge ≥ 30% of body length. More than 15 ciliary rows. Extrusomes basically rod-shaped or pencil-shaped 13
12 Extrusomes narrowly ovate, about 7 × 1.4 µm.............................. *A. pachyoplites*
– Extrusomes very narrowly obovate (i.e. attached to oral bulge with broad end!), about 5 × 1 µm .... *A. vlassaki*
13 Size about 200 × 18 µm in vivo with oral bulge occupying approximately 1/3 of body length. About 17 ciliary rows. Dorsal brush conspicuous, that is, with up to 10 µm long bristles .... *A. lorjeae*
– Size about 230–280 × 30–40 µm with oral bulge occupying 1/3 to 2/3 of body length. More than 25 ciliary rows. Dorsal brush inconspicuous, that is, bristles ≤ 5 µm long 14
14 Oral bulge almost half body length. Oral bulge extrusomes scattered, that is, form two or more rough rows in each bulge half .... *A. cultriforme scalpriforme*
– Oral bulge about 1/3 or 2/3 of body half. Extrusomes form a rough row in each bulge half 15
15 Oral bulge about 1/3 of body length .... *A. cultriforme cultriforme*
– Oral bulge about 2/3 of body length .... *A. cultriforme megastoma*

Arcuospathidium cooperi FOISSNER, 1996 (Fig. 56a–i, 116a–f; Table 13)

1996 *Arcuospathidium cooperi* FOISSNER, Biol. Fertil. Soils, 23: 288 (Type slides with protargol-impregnated specimens from type locality are deposited in the Oberösterreichische Landesmuseum in Linz, Upper Austria.).

Diagnosis: Size about 130 × 20 µm in vivo. Narrowly to very narrowly spatulate with strongly oblique, very narrowly cuneate oral bulge approximately 1.6 times longer than widest trunk region. Macronucleus oblong; single micronucleus. No extrusomes. On average 11 ciliary rows, 3 anteriorly differentiated to heterostichad dorsal brush

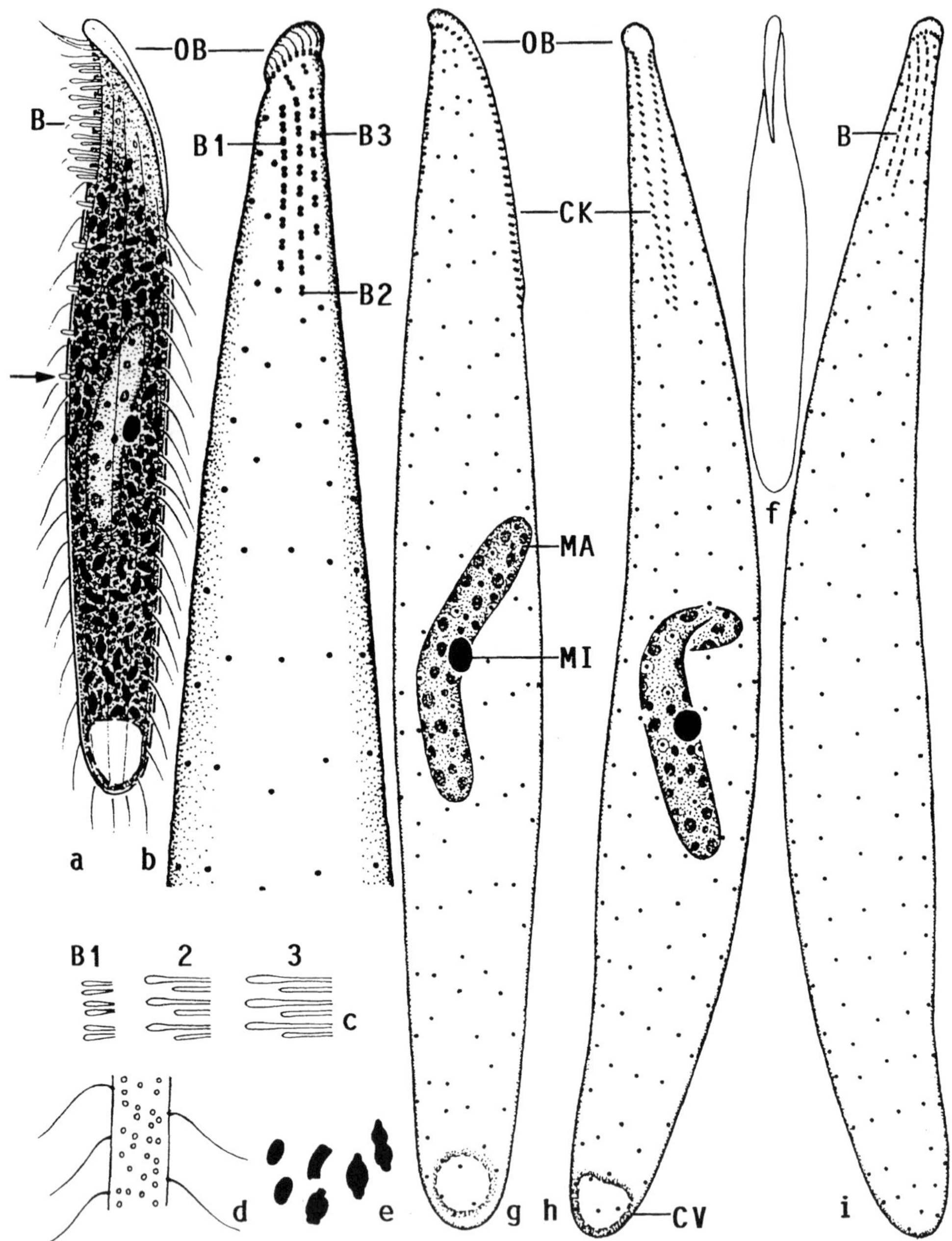

Fig. 56a–i *Arcuospathidium cooperi* from life (a, c–f) and after protargol impregnation (b, g–i). From FOISSNER (1996c). **a, e)** Right side view of a representative specimen packed with highly refractive inclusions (e), length 120 µm. Arrow marks end of bristle tail of brush row 3. **b, c)** Dorsal view of anterior body portion (length 48 µm) showing details of dorsal brush with bristles up to 5 µm long. **d)** Cortical granulation. **f)** Ventral view. **g)** Ciliary pattern of right side, length 133 µm. **h, i)** Ciliary pattern of ventral and dorsal side and nuclear apparatus of holotype specimen, length 132 µm. B(1-3) – dorsal brush (rows), CK – circumoral kinety, CV – contractile vacuole, MA – macronucleus, MI – micronucleus, OB – oral bulge.

occupying 12% of body length. Brush bristles up to 5 µm long: row 1 composed of an average of 14 dikinetids, row 2 of 15, and row 3 of 10 dikinetids followed by a monokinetidal bristle tail extending to mid-body.

Type locality: Moss and soil from Marion Island, Prince Edward Islands, Southern Indian Ocean, E37°51' S46°52'.

Dedication: Named after Dr John COOPER (University of Cape Town, Republic of South Africa), who provided the sample containing the species.

Description: Size 90–160 × 18–24 µm in vivo, usually near 130 × 20 µm, as calculated from some in vivo measurements and the morphometric data; length:width ratio 4.4–8.1:1, on average 6.5:1 both in vivo and protargol preparations (Table 13). Narrowly to very narrowly spatulate, flattened only in oral region, widest usually underneath proximal end of distinctly slanted oral bulge; neck indistinct (Fig. 56a, g, f, 116a). Macronucleus near centre of body, oblong, frequently slightly curved, 30 × 6 µm on average in protargol preparations, contains many globular nucleoli. Micronucleus globular, attached to macronucleus. Single contractile vacuole in posterior end (Fig. 56a, g, h, 116a; Table 13). No extrusomes recognizable in oral bulge and cytoplasm, not even with interference contrast optics and in an Austrian population found very recently. Cortex flexible, colourless, contains loose rows of pale granules 0.5–1 µm across. Cytoplasm conspicuously dark at x100 due to innumerable, highly refractive, crystalline (?) inclusions causing macronucleus and contractile vacuole to stand out as bright blisters; individual crystals 3–5 µm long, of various shape, do not dissolve in water when cell is disrupted (Fig. 56a, e). Movement serpentine and slow.

Somatic cilia 8–10 µm long in vivo, loosely spaced, arranged in an average of 11 equidistant, bipolar rows anteriorly slightly curved dorsally and usually distinctly separate from circumoral kinety. Dorsal brush heterostichad and composed of three, rarely four rows of clavate bristles, very short occupying only 12% of body length; usually, all rows have a minute tail of ordinary cilia anteriorly. Brush row 1 composed of an average of 14 dikinetids with 2 µm long bristles. Longest row 2 composed of an average of 15 dikinetids with 3–4 long anterior and 2 µm long posterior bristles. Row 3 shorter by an average of 24% than rows 1 and 2, but with monokinetidal tail extending to mid-body with 2 µm long bristles; composed of an average of 10 dikinetids with 4–5 µm long anterior and 3–4 µm long posterior bristles (Fig. 56a–c, g–i, 116 a, c; Table 13).

Oral bulge inconspicuous because hyaline, narrow, only up to 3 µm high, and gradually merging into body proper ventrally; on average 1.6 times as long as widest trunk region, distinctly slanted by about 70°, moderately convex in lateral view and very narrowly cuneate in frontal view; temporary cytostome neither recognizable in protargol preparations nor scanning micrographs. Circumoral kinety of same shape as oral bulge, distinctly separate from ciliary rows, composed of comparatively widely spaced dikinetids, of which only one basal body bears a 8–10 µm long cilium. Nematodesmata not recognizable (Fig. 56a, f–h, 116a, b, d–f; Table 13).

Austrian specimens with type I resting cysts about 30 µm across; slime cover circa 5 µm high, wall about 1 µm thick; refractive inclusions partially maintained, cyst thus dark at low bright field magnification.

Table 13 Morphometric data on *Arcuospathidium cooperi* (upper line) and *Arcuospathidium vermiforme* (lower line). From FOISSNER (1996c, 1984)

Characteristics[1]	$\overline{x}$	M	SD	SE	CV	Min	Max	n
Body, length	120.1	118.0	18.4	5.6	15.3	87.0	153.0	11
	141.0	150.0	28.3	6.9	20.1	70.0	185.0	17
Body, width in lateral view	18.5	18.0	1.0	0.3	5.6	17.0	20.0	11
	13.9	14.0	1.6	0.4	11.6	11.0	18.0	17
Body length:width, ratio	6.5	6.4	1.0	0.3	15.1	4.4	8.1	11
	10.2	10.3	2.2	0.5	21.7	5.7	13.8	17
Oral bulge, length	29.3	30.0	3.2	1.0	10.8	24.0	35.0	11
	26.7	27.0	3.6	0.9	13.5	17.0	34.0	17
Oral bulge length: body width, ratio	1.6	1.6	0.2	0.1	11.4	1.2	1.8	11
	1.9	1.9	0.3	0.1	14.2	1.4	2.5	17
Circumoral kinety to last dikinetid of brush row 1, distance	13.5	14.0	1.7	0.5	12.6	10.0	16.0	11
	14.2	14.0	3.2	0.8	22.9	7.0	20.0	17
Circumoral kinety to last dikinetid of brush row 2, distance	14.8	16.0	1.9	0.6	12.8	12.0	17.0	11
	14.0	14.0	3.5	0.8	24.9	7.0	20.0	17
Circumoral kinety to last dikinetid of brush row 3, distance	10.2	10.0	1.4	0.4	13.7	7.0	12.0	11
	8.9	9.0	1.9	0.5	21.6	6.0	12.0	17
Anterior body end to macronucleus, distance	48.4	50.0	11.2	3.4	23.2	32.0	69.0	11
	74.1	76.0	21.4	5.2	28.9	34.0	107.0	17
Macronucleus nodules, length	30.5	30.0	3.6	1.1	11.9	22.0	35.0	11
	15.7	15.0	3.3	0.8	21.1	10.0	24.0	17
Macronucleus nodules, width	5.8	6.0	0.8	0.2	12.9	5.0	7.0	11
	7.6	7.0	1.1	0.3	14.0	6.0	10.0	17
Macronuclei, number	1.0	1.0	0.0	0.0	0.0	1.0	1.0	100
	1.9	2.0	0.3	0.8	17.7	1.0	2.0	17
Micronucleus, length	3.8	4.0	–	–	–	3.0	4.0	11
	4.3	4.2	–	–	–	4.0	5.6	17
Micronucleus, width	2.7	3.0	–	–	–	2.0	3.0	11
	2.4	2.2	–	–	–	1.6	3.0	17
Micronucleus, number	1.0	1.0	0.0	0.0	0.0	1.0	1.0	100
	0.9	1.0	–	–	–	0.0	1.0	17
Ciliary rows, number	11.2	11.0	0.9	0.3	7.8	10.0	13.0	11
	11.3	11.0	1.0	0.2	8.7	10.0	14.0	17
Kinetids in a right side ciliary row, number	25.5	25.0	4.6	1.4	17.9	20.0	35.0	11
	31.3	31.0	6.0	1.5	19.3	14.0	40.0	17
Dorsal brush rows, number	3.1	3.0	–	–	–	3.0	4.0	11
	3.0	3.0	0.0	0.0	0.0	3.0	3.0	17
Dikinetids in brush row 1, number	13.5	14.0	1.7	0.5	12.6	10.0	16.0	11
	10.1	10.0	1.9	0.5	19.1	7.0	14.0	17
Dikinetids in brush row 2, number	14.8	16.0	1.9	0.6	12.8	12.0	17.0	11
	10.5	11.0	2.1	0.5	20.3	6.0	15.0	17
Dikinetids in brush row 3, number	10.2	10.0	1.4	0.4	13.7	7.0	12.0	11
	6.6	7.0	1.1	0.3	16.2	4.0	9.0	17

[1] Data based on mounted, protargol-impregnated (FOISSNER's method), and randomly selected specimens from non-flooded Petri dish cultures. Measurements in µm. CV – coefficient of variation in %, M – median, Max – maximum, Min – minimum, n – number of individuals investigated, SD – standard deviation, SE – standard error of arithmetic mean, $\overline{x}$ – arithmetic mean.

Occurrence and ecology: As yet found at type locality, where it was abundant in the non-flooded Petri dish culture, and in a soil sample from Upper Austria, where it was rare. Likely, *A. cooperi* is not a predator because it lacks extrusomes. It is well adapted to the soil habit by the slender and flexible body.

Remarks: Within the genus *Arcuospathidium*, only *A. cooperi* and *A. vermiforme* lack extrusomes. The absence of extrusomes, at least in the light microscope, was very carefully checked in several specimens and with optimum technical equipment. Certainly, this is a curious, negative feature found, however, also in some *Apertospathula*, *Spathidium* and *Protospathidium* species. Size, shape, ciliary pattern, and terrestrial habitat of *A. cooperi* highly resemble *A. vermiforme* which, however, has two macronucleus nodules. This combination of features, viz., the lack of extrusomes and an oblong macronucleus, distinguishes *A. cooperi* from all other species of the order.

Arcuospathidium vermiforme FOISSNER, 1984 (Fig. 57a–m, 116g–k; Table 13)

1984 *Arcuospathidium vermiforme* FOISSNER, Stapfia 12: 79 (partim, that is, population I, while population II belongs to *A. pelobium*. Type slides with protargol-impregnated specimens from type locality are deposited in the Oberösterreichische Landesmuseum in Linz, Upper Austria.).

Diagnosis: Size about 160 × 16 µm in vivo. Cylindroidal with strongly oblique, very narrowly cuneate oral bulge approximately twice as long as widest trunk region. Two macronucleus nodules and a single micronucleus in variable position. No extrusomes. On average 11 ciliary rows, 3 anteriorly differentiated to heterostichad dorsal brush occupying 10% of body length. Brush bristles up to 2 µm long (5 µm and in complex pattern in another population): row 1 composed of an average of 10 dikinetids, row 2 of 11, and row 3 of 7 dikinetids likely followed by a monokinetidal bristle tail.

Type locality: Grassland soil from the village of Schaming, Eugendorf near Salzburg, Austria, E13°7' N47°52'.

Etymology: Not given in original description. *Vermiforme* (worm-shaped) obviously refers to the slender shape of the species.

Description: Size 80–200 × 10–20 µm in vivo, usually near 160 × 16 µm, as calculated from some in vivo measurements and the morphometric data; length:width ratio 5.7–13.8:1, on average about 10:1 both in vivo and protargol preparations (Table 13). Cylindroidal to vermiform with middle third slightly to distinctly inflated, depending on nutrition state, anterior (oral) end strongly oblique, flattened only in oral region, frequently somewhat twisted about main body axis; neck indistinct (Fig. 57a, e–k); minute and stout, likely malformed individuals rather frequent; fragile, often dying on microscope slide becoming shorter and broader. Nuclear apparatus usually underneath mid-body and basically composed of two narrowly spaced macronucleus nodules and

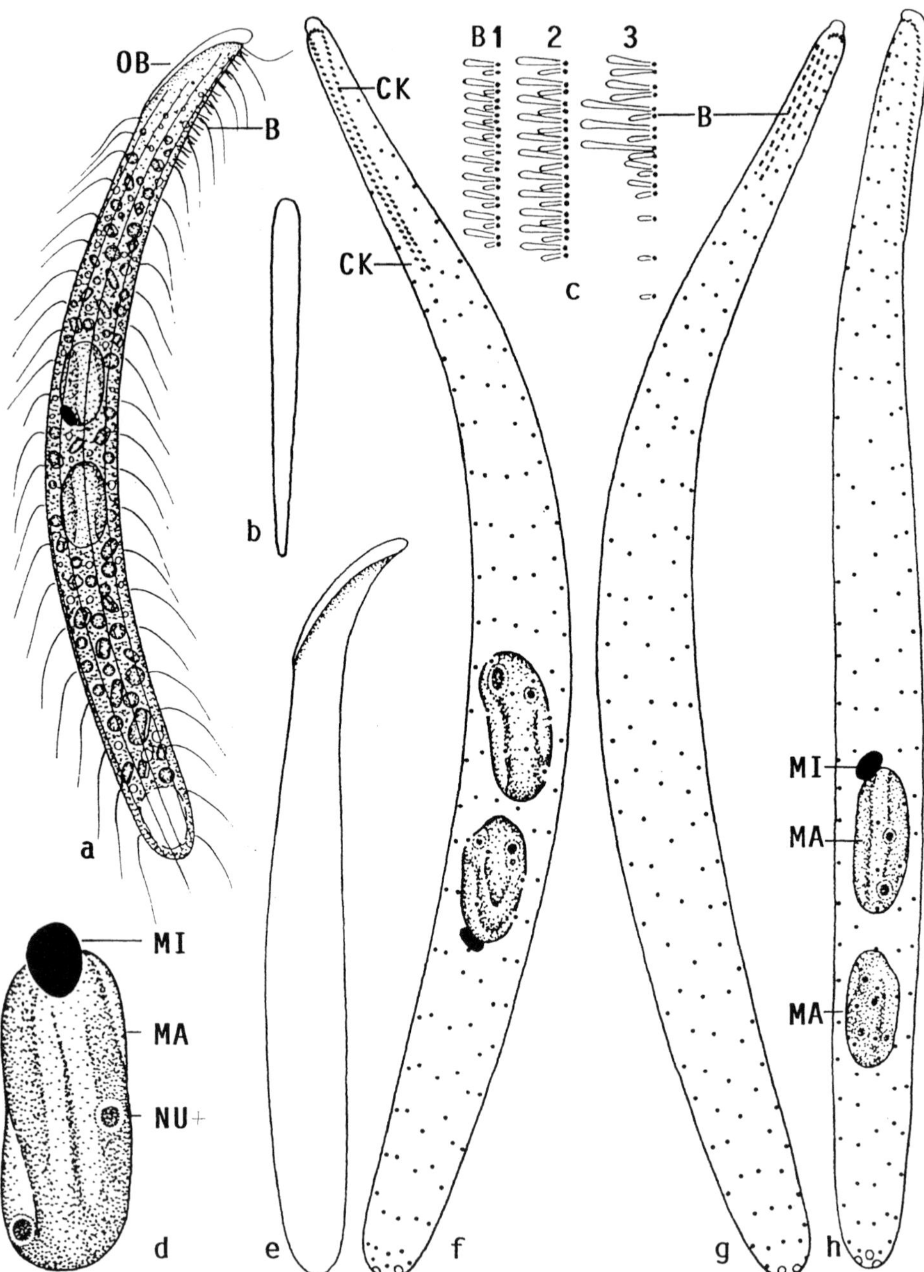

Fig. 57a–h *Arcuospathidium vermiforme* from life (a–e) and after protargol impregnation (f–h). From FOISSNER (1984). **a)** Left side view of a representative specimen, length 160 µm. **b)** Frontal view of oral bulge. **c)** Dorsal brush of a Carinthian specimen, longest bristles 5 µm. **d)** Macronucleus nodule. **e)** Rare shape variant. **f, g)** Ciliary pattern of ventral and dorsal side and nuclear apparatus of holotype specimen, length 157 µm. **h)** Right side ciliary pattern, length 154 µm. B(1-3) – dorsal brush (rows), CK – circumoral kinety, MA – macronucleus nodules, MI – micronucleus, NU – nucleolus, OB – oral bulge.

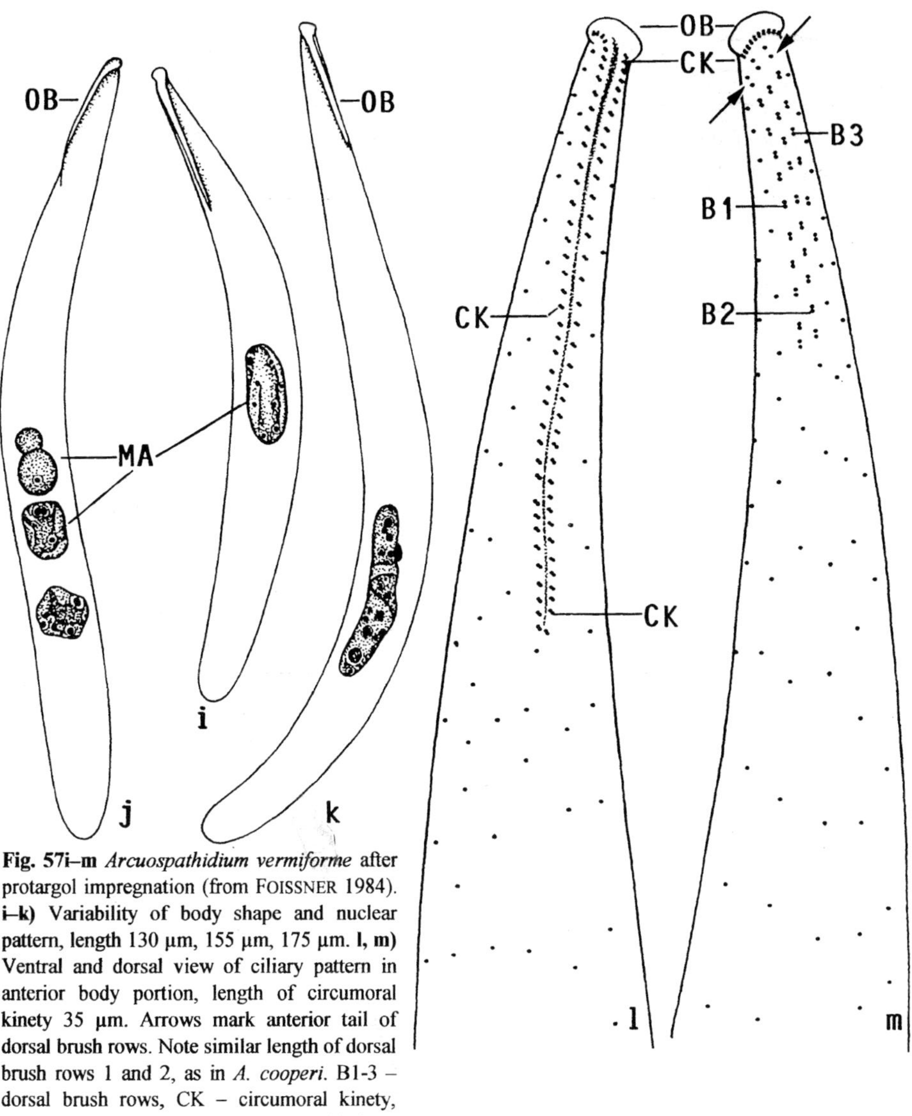

Fig. 57i–m *Arcuospathidium vermiforme* after protargol impregnation (from FOISSNER 1984). **i–k)** Variability of body shape and nuclear pattern, length 130 µm, 155 µm, 175 µm. **l, m)** Ventral and dorsal view of ciliary pattern in anterior body portion, length of circumoral kinety 35 µm. Arrows mark anterior tail of dorsal brush rows. Note similar length of dorsal brush rows 1 and 2, as in *A. cooperi*. B1-3 – dorsal brush rows, CK – circumoral kinety, MA – macronucleus nodules, OB – oral bulge.

an ellipsoidal micronucleus frequently attached to the anterior nodule; micronucleus lacking in 1 of 17 specimens analysed. Macronucleus nodules ellipsoidal, rarely globular, in vivo very hyaline and with irregular, longitudinal furrows; contain only few, minute nucleoli heavily impregnating with protargol. Of 311 randomly selected, well-looking specimens, 277 (89.1%) have two ellipsoidal to globular macronucleus nodules; 26 (8.4%) possess an oblong nucleus; 7 (2.2%) have a single, globular nodule; and 1 specimen has four globular nodules (Fig. 57a, d, f, h–k, 116k; Table 13). Likely, most variability is caused by exconjugants and post-dividers. Contractile vacuole in rear body end, some excretory pores in pole area. Lack of extrusomes checked in several specimens each from type locality and Carinthia. Cortex thin, bright, contains rows of minute granules. Cytoplasm colourless, usually packed, especially in Carinthian specimens (116j), with square crystals and irregular organic inclusions up to 5 µm across. Swims and creeps slowly and serpentinously.

Somatic cilia about 10 µm long in vivo, loosely spaced (average ciliary distance 4.5 µm; Table 13), arranged in an average of 11 equidistant, bipolar rows anteriorly slightly curved dorsally and usually distinctly separate from circumoral kinety. Dorsal brush heterostichad and composed of three rows of clavate bristles, very short, that is, occupies only 10% of body length; usually, all rows possess a minute tail of ordinary cilia anteriorly (Fig. 57m). Brush rows 1 and 2 of similar length, each composed of about 10 dikinetids; row 3 shortened by an average of 38%, composed of an average of seven dikinetids, likely has a monokinetidal bristle tail. Bristles of type population not studied in detail, only about 2 µm long; those of Carinthian specimens up to 5 µm long and structured as shown in figure 57c (Fig. 57a, f–h, l, m, 116i; Table 13).

Oral bulge inconspicuous because hyaline, narrow, only up to 2 µm high at dorsal end, and gradually merging into body proper ventrally; on average about twice as long as widest trunk region, distinctly slanted by about 70°, moderately convex, and very narrowly cuneate in frontal view; temporary cytostome not recognizable. Circumoral kinety very narrow, of same shape as oral bulge, distinctly separate from ciliary rows; composed of comparatively widely spaced dikinetids each associated with an about 10 µm long cilium and a fine nematodesma contributing to the inconspicuous oral basket (Fig. 57a, b, e–l, 116g–i; Table 13).

Occurrence and ecology: As yet found only in Austria, viz., in Salzburg and Carinthia; very abundant in the non-flooded Petri dish culture with soil from type locality. In Carinthia, *A. vermiforme* occurred in a non-flooded Petri dish culture with mud and soil from a dry, temporary puddle in the Austrian Central Alps almost 2000 m above sea level (puddle 1 in FOISSNER 1980a). Likely, *A. vermiforme* is not a predator because it lacks extrusomes. It is well adapted to the soil environment by the slender, flexible body.

Remarks: See *A. cooperi*! This is a highly characteristic species unlikely to have been overlooked in any other of the 1000 soil samples investigated by the senior author. Thus, there is a fair probability that it is endemic to Europe.

Arcuospathidium pelobium **nov. spec.** (Fig. 58a–s; Table 14)

1984 *Arcuospathidium vermiforme* population II (slides with protargol-impregnated specimens are deposited, under the name *Arcuospathidium vermiforme*, in the Oberösterreichische Landesmuseum in Linz, Upper Austria).

Diagnosis (based on type and Austrian population): Size about 50–100 × 10–12 µm in vivo. Claviform to cylindroidal with oblique, cuneate oral bulge about as long as widest trunk region. Macronucleus ellipsoidal. Oral bulge extrusomes fine and rod-shaped, 3–5 µm long. On average 7–9 ciliary rows, 3 anteriorly differentiated to heterostichad dorsal brush occupying about 14% of body length. Brush bristles up to 4 µm

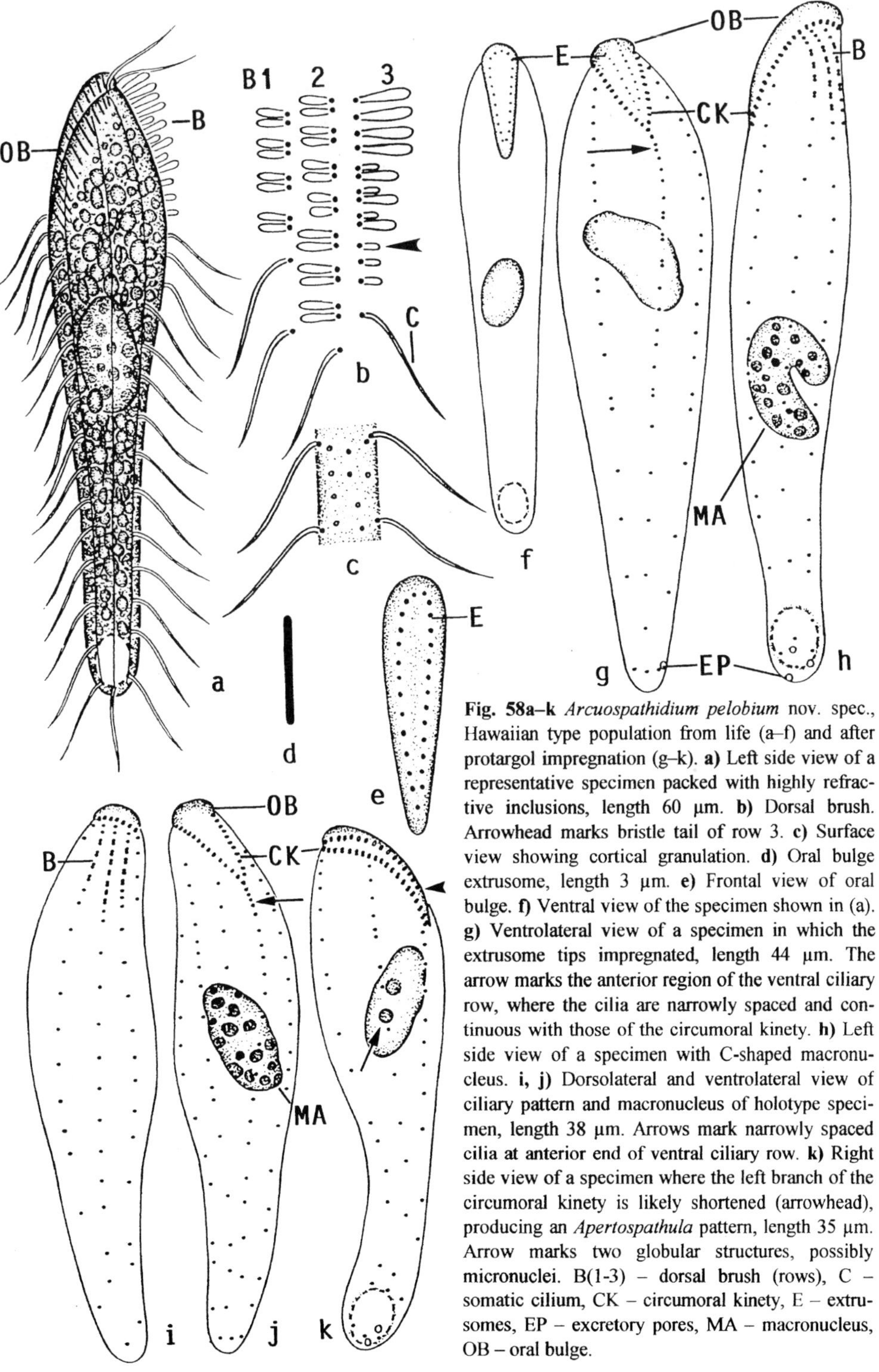

Fig. 58a–k *Arcuospathidium pelobium* nov. spec., Hawaiian type population from life (a–f) and after protargol impregnation (g–k). **a)** Left side view of a representative specimen packed with highly refractive inclusions, length 60 µm. **b)** Dorsal brush. Arrowhead marks bristle tail of row 3. **c)** Surface view showing cortical granulation. **d)** Oral bulge extrusome, length 3 µm. **e)** Frontal view of oral bulge. **f)** Ventral view of the specimen shown in (a). **g)** Ventrolateral view of a specimen in which the extrusome tips impregnated, length 44 µm. The arrow marks the anterior region of the ventral ciliary row, where the cilia are narrowly spaced and continuous with those of the circumoral kinety. **h)** Left side view of a specimen with C-shaped macronucleus. **i, j)** Dorsolateral and ventrolateral view of ciliary pattern and macronucleus of holotype specimen, length 38 µm. Arrows mark narrowly spaced cilia at anterior end of ventral ciliary row. **k)** Right side view of a specimen where the left branch of the circumoral kinety is likely shortened (arrowhead), producing an *Apertospathula* pattern, length 35 µm. Arrow marks two globular structures, possibly micronuclei. B(1-3) – dorsal brush (rows), C – somatic cilium, CK – circumoral kinety, E – extrusomes, EP – excretory pores, MA – macronucleus, OB – oral bulge.

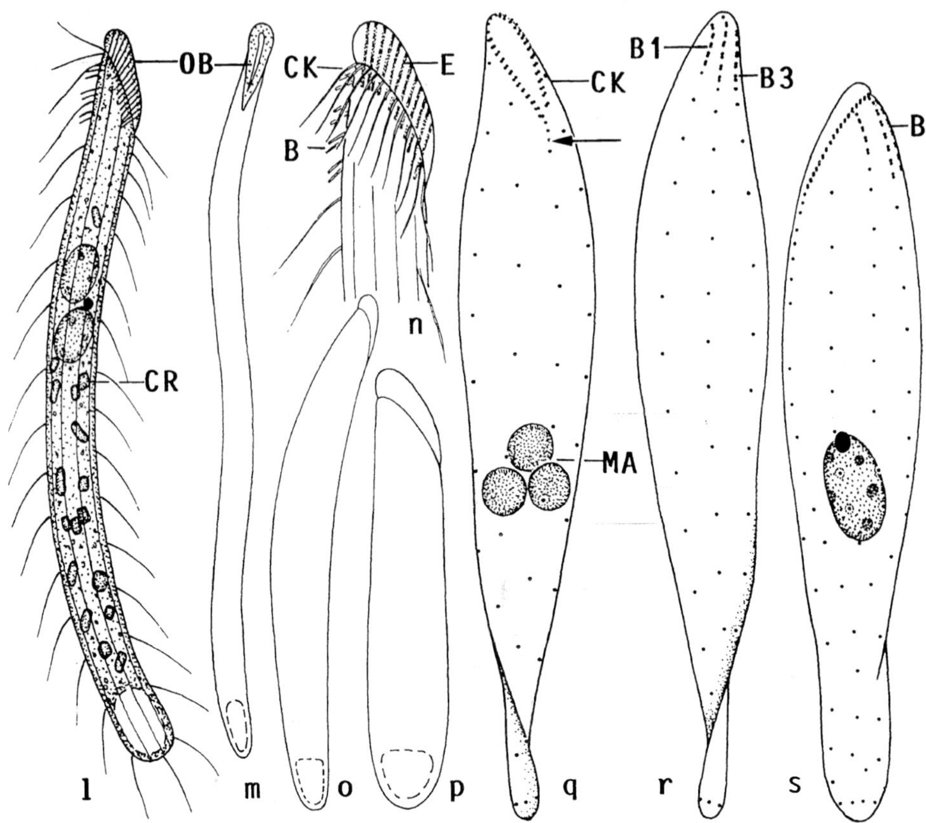

Fig. 58l–s *Arcuospathidium pelobium* nov. spec., Austrian specimens from life (l–p) and after protargol impregnation (q–s). From FOISSNER (1984). **l, m)** Right side and ventral view of a specimen with two macronucleus nodules, length 90 µm. **n)** Oral body portion at higher magnification. **o, p)** Shape and size variants, length 80 µm, 63 µm. **q, r)** Ventrolateral and dorsolateral view of ciliary pattern and nuclear apparatus of main voucher specimen, length 67 µm. Arrow marks condensation of cilia at anterior end of ventral row. **s)** Ciliary pattern of left side of a specimen with typical nuclear pattern, length 70 µm. B(1, 3) – dorsal brush (rows), CK – cilia of circumoral kinety, CR – crystals (?), E – extrusomes, MA – macronucleus, OB – oral bulge.

long: row 1 composed of an average of 4 dikinetids, row 2 of 6, and row 3 of 4 dikinetids followed by a short monokinetidal bristle tail.

Type locality: Mud from a temporary stream in Hawaii, North Kohala, W155°50' N20°04'.

Etymology: The Greek *pelobia* (living in mud) refers to the habitat the species was discovered, viz., in floodplain soil.

Description (type population; Fig. 58a–k): Size 40–80 × 10–15 µm, usually about 55 × 12 µm in vivo, unflattened; length:width ratio 3.3–5.6:1, on average about 5:1

Table 14 Morphometric data on *Arcuospathidium pelobium* from Hawaii (upper line) and Austria (lower line; from FOISSNER 1984)

Characteristics[1]	$\overline{x}$	M	SD	SE	CV	Min	Max	n
Body, length	44.9	44.0	8.1	1.8	17.9	34.0	62.0	19
	57.7	56.5	10.2	3.2	17.6	43.0	69.0	10
Body, width	10.2	10.0	1.5	0.4	15.1	8.0	13.0	19
	10.8	11.0	1.4	0.5	13.2	9.0	13.0	10
Body length:width, ratio	4.5	4.2	0.7	0.2	16.0	3.3	5.6	19
	5.4	5.6	1.0	0.3	17.7	3.8	7.4	10
Oral bulge (circumoral kinety), length	10.1	10.0	1.9	0.4	18.5	7.0	15.0	18
	7.7	7.0	1.4	0.4	17.9	6.0	10.0	10
Oral bulge, height	1.6	1.5	–	–	–	1.5	2.0	7
				not measured				
Oral bulge length:body width, ratio	1.0	1.0	0.2	0.1	24.6	0.6	1.7	19
	0.7	0.7	0.1	0.1	16.9	0.6	0.9	10
Circumoral kinety to last dikinetid of brush row 1, distance	4.2	4.0	–	–	–	4.0	5.0	12
				not measured				
Circumoral kinety to last dikinetid of brush row 2, distance	6.6	6.0	1.8	0.5	27.8	5.0	12.0	12
	7.3	7.2	0.7	0.2	9.1	6.0	8.0	10
Circumoral kinety to last dikinetid of brush row 3, distance	3.9	4.0	–	–	–	3.0	4.0	12
				not measured				
Anterior body end to macronucleus, distance	14.8	16.0	4.0	0.9	27.4	5.0	20.0	19
				not measured				
Macronucleus, length	8.0	8.0	2.0	0.5	24.7	4.0	11.0	19
	7.2	8.0	3.4	1.1	46.9	2.7	12.0	10
Macronucleus, width	4.1	4.0	0.7	0.2	17.4	3.0	6.0	19
	5.1	5.3	1.9	0.6	37.2	2.0	8.0	10
Macronucleus, number	1.0	1.0	0.0	0.0	0.0	1.0	1.0	19
	1.8	1.0	1.1	0.4	63.1	1.0	4.0	10
Ciliary rows, number	6.9	7.0	0.4	0.1	6.2	6.0	8.0	17
	8.8	8.0	1.1	0.4	12.9	8.0	11.0	10
Basal bodies in a right side ciliary row, number	17.7	18.0	3.1	0.8	17.8	12.0	24.0	14
				not counted				
Dorsal brush rows, number	3.0	3.0	0.0	0.0	0.0	3.0	3.0	17
	3.0	3.0	0.0	0.0	0.0	3.0	3.0	10
Dikinetids in brush row 1, number	3.9	4.0	–	–	–	3.0	4.0	12
				not counted				
Dikinetids in brush row 2, number	6.4	6.0	1.0	0.3	15.5	5.0	8.0	12
				not counted				
Dikinetids in brush row 3, number	3.8	4.0	–	–	–	3.0	4.0	12
				not counted				
Circumoral kinetids, number	28.3	28.0	–	–	–	27.0	30.0	3
				not counted				

[1] Hawaiian data based on 19 out of 22 specimens found in the preparations (FOISSNER's method) from material of the non-flooded Petri dish culture. The Austrian population is also from a non-flooded Petri dish culture and contains all well-impregnated specimens found in the preparations (FOISSNER's method). Measurements in µm. CV – coefficient of variation in %, M – median, Max – maximum, Min – minimum, n – number of individuals investigated, SD – standard deviation, SE – standard error of arithmetic mean, $\overline{x}$ – arithmetic mean.

both in vivo and in protargol preparations (Table 14). Shape more or less distinctly clavate, rarely bluntly fusiform or elongate ellipsoidal, rear third frequently wrinkled, that is, with irregular folds and crests; anterior (oral) end slanted by about 45°, posterior narrowly rounded and bluntly bulbous when contractile vacuole is filled; neck indistinct (Fig. 58a, f–k). Macronucleus in mid-body, often ellipsoidal, rarely C-shaped or globular, contains many nucleoli 1–2 µm across.Micronucleus(ei) not unequivocally recognizable. Contractile vacuole in rear body end, several excretory pores in dorsal and lateral pole area. Extrusomes in a row along margin of oral bulge, that is, form a cuneate pattern sometimes recognizable also in protargol preparations due to the impregnated anterior end of the extrusomes (Fig 58f, g). Individual extrusomes rod-shaped with rounded ends, fine, about 3 µm long (Fig. 58d). Cortex thin and flexible, contains very loosely spaced, colourless granules about 0.2 µm across. Cytoplasm usually packed with discoidal, highly refractive structures (crystals?) 2–4 × 1–1.5 µm in size (Fig. 58a). Food vacuoles not recognizable. Movement without peculiarities.

Somatic cilia about 8 µm long in vivo, arranged in an average of seven equidistant, mostly bipolar and rather loosely ciliated rows with anterior end distinctly separate from circumoral kinety, but only indistinctly curved dorsally, *Arcuospathidium* ciliary pattern thus indistinct (Fig. 58a, g–k). Dorsal brush heterostichad and three-rowed, rather conspicuous, though occupying only 15% of body length, because bristles up to 4 µm long in vivo. All brush rows have one or several monokinetids anteriorly and continue as ordinary ciliary rows posteriorly. Brush row 1 composed of an average of four dikinetids with 2.5 µm long bristles. Brush row 2 composed of an average of six dikinetids with bristles 2.5 µm long anteriorly and 3 µm posteriorly. Brush row 3 composed of an average of four dikinetids with bristles gradually decreasing in length from 4 µm anteriorly to 2 µm posteriorly, long anterior bristles indistinctly dikinetidal; monokinetidal tail inconspicuous because consisting of only three to four bristles (Fig. 58a, b, h, i).

Oral bulge about as long as widest trunk region, slightly convex, slanted by 40°–60°, dorsal portion higher than ventral gradually merging into body proper; in frontal view cuneate and conspicuous due to the extrusome row (Fig. 58a, e; Table 14). Circumoral kinety of similar shape as oral bulge, continuous, composed of about 28 narrowly spaced dikinetids; right end usually connected with a ventral somatic kinety having some narrowly spaced cilia anteriorly, as in several congeners. Circumoral kinety likely open in one out of 22 specimens analysed, that is, right branch longer than left by about four dikinetids, as in *Apertospathula*. Oral basket rods not recognizable in vivo and protargol preparations (Fig. 58a, g–k).

Description of Austrian population (Fig. 58l–s): Size 60–140 × 6–15 µm in vivo, usually near 100 × 10 µm, according to six specimens measured in vivo; protargol-impregnated cells distinctly smaller, viz., 58 × 11 µm on average (Table 14), likely due to strong shrinkage and some inflation during preparation, as indicated by the different length:width ratios (about 10:1 vs. 5.4:1). Spatulate to cylindroidal, in preparations also slightly clavate, laterally flattened up to 2:1, anterior (oral) end distinctly slanted, posterior rounded or inflated by contractile vacuole; neck indistinct; very fragile, specimens frequently die on the slide becoming shorter and broader (Fig. 58l, m, o, p). Nuclear pattern likely consisting of an ellipsoidal, very hyaline macronucleus and an ordinary, globular micronucleus, as indicated by the median value; however, specimens with two to four macronucleus nodules, likely post-conjugants, are frequent (Table 14). Contractile vacuole in posterior body end. Oral bulge extrusomes rod-shaped and about 5 µm long, do not impregnate with protargol (Fig. 58n, s); arrangement not studied.

Cortex thin and fragile, frequently with longitudinal folds, making specimens wrinkled. Cytoplasm hyaline and colourless, in posterior half sometimes packed with cubic, yellowish crystals with a size of about 2 × 1 µm, making cells dark (refractive). Swims and creeps slowly and serpentinuously.

Somatic cilia about 8 µm long in vivo, arranged in an average of nine equidistant, bipolar rows whose anterior pattern is hardly recognizable due to the loose spacing of the cilia, except at anterior end of ventral row, where some cilia are very narrowly spaced, as in several congeners. Dorsal brush at anterior end of three dorsal kineties, inconspicuous because occupying only 13% of body length and individual rows composed of less than 10 dikinetids with short, rod-shaped bristles (Fig. 58l, n, q–s; Table 14).

Oral bulge of ordinary appearance in vivo, while almost invisible in protargol preparations; slanted by 40°–60° and about as long as trunk width in vivo, while occupying only two thirds of body width in the prepared, inflated specimens; up to 4 µm high at dorsal end, while gradually merging into body surface ventrally, flat to slightly convex in lateral view, cuneate in frontal view; temporary cytostome not recognizable, bulge midline, however, occasionally slightly depressed (Fig. 58l–n; Table 14). Circumoral kinety cuneate as oral bulge, distinctly separate from ciliary rows, composed of narrowly spaced dikinetids each associated with an about 8 µm long cilium; nematodesmata not recognizable (Fig. 58n, q, s).

Occurrence and ecology: Both, the type and the Austrian population were found in semiterrestrial environments, viz., the mud of a temporary stream, respectively, in floodplain soil. Thus, this species likely occurs also in, or even prefers limnetic conditions. See FOISSNER et al. (1985) for a detailed description of the Austrian site (Vogelsang floodplain), where *A. pelobium* was frequent but never abundant in the non-flooded Petri dish cultures. The Hawaiian stream was dry when the sample was taken and inhabited by another curious ciliate, viz., *Idiocolpoda pelobia* FOISSNER.

Remarks: The Hawaiian *A. pelobium* matches the Austrian specimens in all main features, except of the in vivo shape. Thus, conspecificity is likely.

FOISSNER (1984) classified this species as "population II of *A. vermiforme*", but with some reservation. With today's knowledge, it is obvious that this population is different not only from *A. vermiforme* (without extrusomes, very different morphometrics), but also from all other described *Spathidium* taxa. Thus, we assign it to *A. pelobium*.

The general appearance of *A. pelobium* highly resembles *Apertospathula inermis* (extrusomes lacking) and *A. armata* (extrusomes ellipsoidal and 1.5 µm long vs. rod-shaped and 3–5 µm long) which, however, have much shorter brush bristles (1 µm vs. up to 4 µm). A reinvestigation of the protargol preparations from the Austrian specimens confirmed that they have a closed circumoral kinety, and thus belong to *Arcuospathidium*.

Great care must be taken in identifying these inconspicuous, small species; likely, there are many! As usual, the best features are the nuclear pattern and the shape and size of the extrusomes.

Arcuospathidium deforme **nov. spec.** (Fig. 59a–i, 118 a, b; Table 14)

Diagnosis: Size 130 × 20 µm in vivo. Narrowly spatulate, oblong, or clavate with oblique, cuneate oral bulge slightly longer than widest trunk region. Macronucleus oblong; 2 micronuclei. Oral bulge extrusomes narrowly cuneate and asymmetric, about 4 × 1 µm. On average 14 ciliary rows, 3 anteriorly differentiated to heterostichad dorsal brush occupying 16% of body length. Brush bristles in complex pattern and up to 7 µm long: row 1 composed of an average of 16 dikinetids, row 2 of 18, and row 3 of 9 dikinetids followed by a short monokinetidal bristle tail.

Type locality: Soil from a green portion ("green river bed") of the Chobe River near the Muchenje Safari Lodge, Botswana, E24°40' S18°.

Etymology: The Latin adverb *deforme* (misshapen) refers to the body shape which changes from oblong to clavate when the specimens are disturbed.

Description: Size 70–150 × 15–25 µm in vivo, rarely up to 180 µm long, usually about 130 × 20 µm, as calculated from some in vivo measurements and the morphometric data; length:width ratio 3.4–7.8:1 in preparations, on average near 6:1 both in vivo and in prepared specimens (Table 15). Narrowly spatulate to cylindroidal in fresh preparations, that is, when collected from the non-flooded Petri dish culture, soon becoming more or less distinctly clavate, even without coverslip (Fig. 59 a, b); anterior (oral) end oblique, posterior rounded; slightly flattened in oral area; neck indistinct (Fig. 59 a, c, h, i). Macronucleus usually in middle body third, oblong to cylindroidal, rarely U-like curved; nucleoli globular, small, and numerous. Usually two globular, foamy micronuclei attached to mid-macronucleus, often close together and only faintly impregnated (Fig. 59 a, e, h, i; Table 15). Contractile vacuole in posterior body end. Extrusomes form a row each in right and left half of oral bulge and are scattered in anterior half of cytoplasm, impregnate rather intensely with the protargol method used; asymmetrically cuneate and about 4 × 1 µm in size both in vivo and preparations, where the oral bulge contains an average of 14 extrusomes (Fig. 59a, c, e, 118a, b; Table 15). Cortex very flexible, no cortical granules recognizable. Cytoplasm colourless and packed with 1 µm-sized granules and lipid droplets up to 5 µm across. Swims rapidly by rotation about main body axis.

Cilia about 12 µm long in vivo, arranged in an average of 14 equidistantly spaced, loosely ciliated rows distinctly shortened anteriorly, except of rows on dorsal side; ciliary pattern thus indeterminable, but likely arcuospathidiid because shortened rows are common in this family. Dorsal brush exactly on dorsal side of cell, occupies 16% of body length on average, three-rowed (rarely occurs a fourth row) and heterostichad with row 3 extending along a flat ridge, occasionally an ordinary cilium between individual brush rows and circumoral kinety; all rows continue as ordinary somatic kineties to posterior body end. Brush row 1 composed of an average of 16 dikinetids, row 2 of 18, and row 3 of nine dikinetids followed by about six monokinetidal, 1.5 µm long bristles. All brush bristles rod-shaped and as thick as ordinary cilia, bristle length, however, highly different forming complex pattern (Fig. 59 a, d–f, 118a; Table 15).

Oral bulge ordinarily slanted, conspicuous because 4–5 µm high and about 7 µm wide at dorsal end, studded with refractive extrusomes (Fig. 118a), and slightly longer

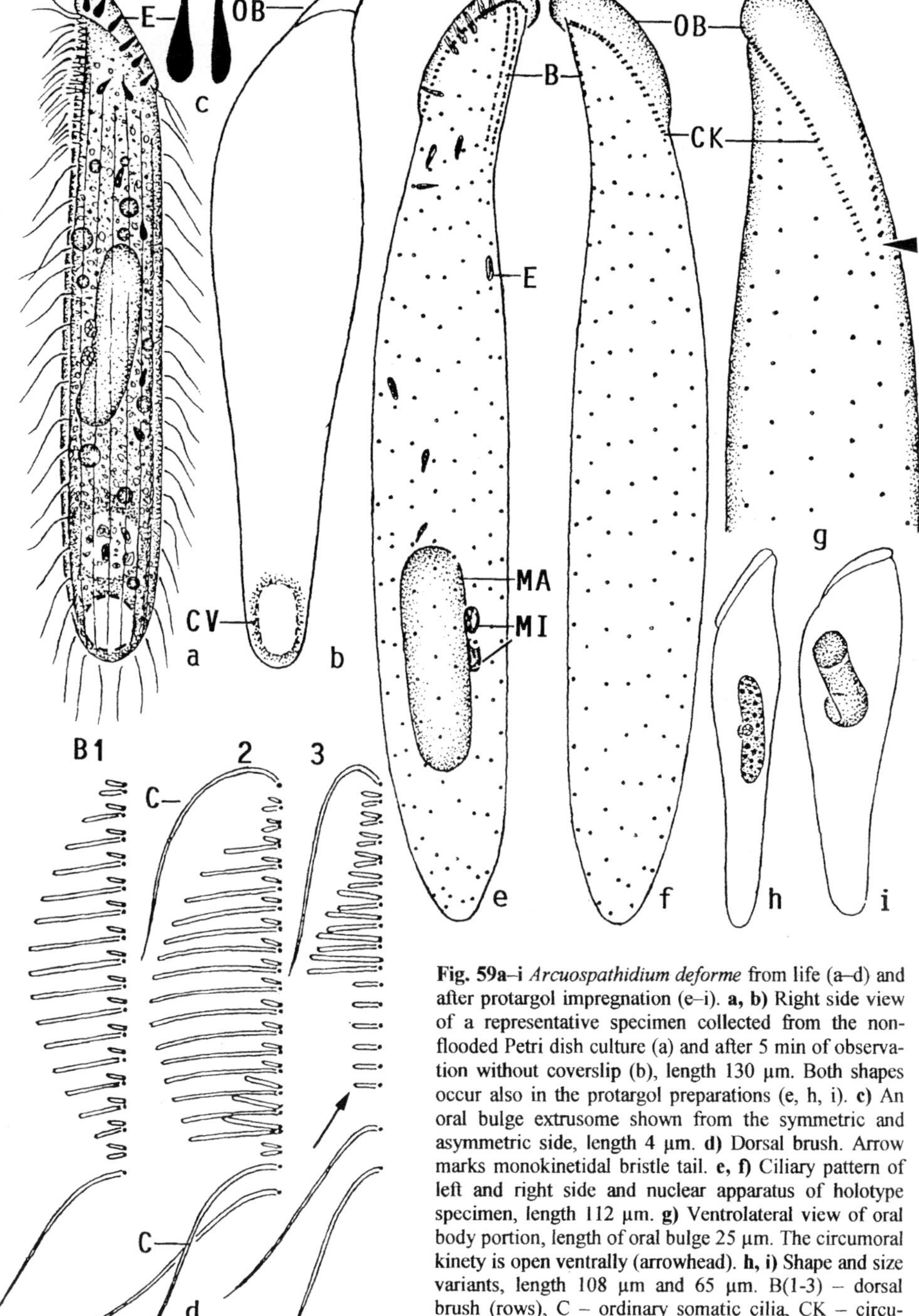

Fig. 59a–i *Arcuospathidium deforme* from life (a–d) and after protargol impregnation (e–i). **a, b)** Right side view of a representative specimen collected from the non-flooded Petri dish culture (a) and after 5 min of observation without coverslip (b), length 130 µm. Both shapes occur also in the protargol preparations (e, h, i). **c)** An oral bulge extrusome shown from the symmetric and asymmetric side, length 4 µm. **d)** Dorsal brush. Arrow marks monokinetidal bristle tail. **e, f)** Ciliary pattern of left and right side and nuclear apparatus of holotype specimen, length 112 µm. **g)** Ventrolateral view of oral body portion, length of oral bulge 25 µm. The circumoral kinety is open ventrally (arrowhead). **h, i)** Shape and size variants, length 108 µm and 65 µm. B(1-3) – dorsal brush (rows), C – ordinary somatic cilia, CK – circumoral kinety, E – extrusomes, MA – macronucleus, MI – micronuclei, OB – oral bulge.

Table 15 Morphometric data on *Arcuospathidium deforme*

Characteristics[1]	$\overline{x}$	M	SD	SE	CV	Min	Max	n
Body, length	117.5	115.0	20.8	5.8	17.7	65.0	160.0	13
Body, width	19.9	19.0	2.4	0.7	12.0	16.0	24.0	13
Body length: width, ratio	6.0	6.0	1.2	0.3	20.1	3.4	7.8	13
Oral bulge (circumoral kinety), length	23.5	23.0	3.6	1.0	15.1	16.0	24.0	13
Body width: oral bulge length, ratio	0.9	0.9	0.2	0.1	19.8	0.7	1.3	13
Oral bulge, width at dorsal end	6.0	6.0	0.7	0.2	11.1	5.0	7.0	10
Oral bulge, height	3.6	4.0	–	–	–	3.0	4.0	7
Oral bulge extrusomes, length	3.7	4.0	–	–	–	3.0	4.0	13
Oral bulge extrusomes, number	13.6	14.0	4.6	1.3	34.0	5.0	22.0	13
Circumoral kinety to end of brush row 1, distance	18.1	18.0	3.6	1.0	19.6	9.0	24.0	13
Circumoral kinety to end of brush row 2, distance	19.2	19.0	3.6	1.0	18.8	10.0	25.0	13
Circumoral kinety to end of brush row 3, distance	10.2	10.0	1.7	0.5	16.5	6.0	13.0	13
Anterior body end to macronucleus, distance	47.0	48.0	13.1	3.6	27.9	15.0	67.0	13
Macronucleus, length	33.2	32.0	6.8	1.9	20.4	21.0	46.0	13
Macronucleus, width	7.1	7.0	0.8	0.2	10.7	6.0	8.0	13
Macronucleus, number	1.0	1.0	0.0	0.0	0.0	1.0	1.0	13
Micronuclei, length	4.3	4.0	–	–	–	4.0	5.0	12
Micronuclei, width	3.9	4.0	0.5	0.1	12.5	3.0	5.0	12
Micronuclei, number	2.0	2.0	0.0	0.0	0.0	2.0	2.0	10
Ciliary rows, number	14.1	14.0	1.1	0.3	7.9	13.0	16.0	13
Kinetids in a right side ciliary row, number	22.6	23.0	5.3	1.5	23.2	14.0	35.0	13
Dorsal brush rows, number	3.1	3.0	–	–	–	3.0	4.0	13
Dikinetids in brush row 1, number	15.8	16.0	2.2	0.6	14.0	11.0	20.2	13
Dikinetids in brush row 2, number	17.9	18.0	2.1	0.6	11.9	13.0	22.0	13
Dikinetids in brush row 3, number	9.2	9.0	1.1	0.3	11.8	7.0	11.0	13

[1] Data based on mounted and protargol-impregnated (FOISSNER's method) cells from a non-flooded Petri dish culture. All specimens found were used. Measurements in µm. CV – coefficient of variation in %, M – median, Max – maximum, Min – minimum, n – number of individuals investigated, SD – standard deviation, SE – standard error of arithmetic mean, $\overline{x}$ – arithmetic mean.

than widest trunk region on average; slightly to distinctly convex in lateral view, cuneate in frontal view. Circumoral kinety of same shape as oral bulge, ventral end rather distinctly opened in some specimens (Fig. 59g), composed of >50 dikinetids each associated with a fine nematodesma contributing to the pronounced, but faintly impregnated oral basket (Fig. 59a, e–g; Table 15).

Occurrence and ecology: See *Apertospathula similis*, which was discovered in the same sample, where it was as rare as that species. The shape changes in the aerob preparations (Fig. 59 a, b) indicate that *A. deforme* is microaerobic or anaerobic, although symbiotic bacteria are not recognizable.

Remarks: The generic classification of this population is difficult because the ciliary pattern is not recognizable (see above) and the circumoral kinety is rather distinctly opened in some specimens (Fig. 59g), as typical for *Apertospathula*. However, both branches of the circumoral kinety end at the same level, while the right branch is longer than the left in *Apertospathula*. Thus, we classify the species in *Arcuospathidium*.

The narrowly cuneate extrusomes and the curious dorsal brush distinguish *A. deforme* from all described congeners. However, there are several species from other families and genera, which look quite similar, e.g., *Protospathidium serpens, Longispatha* spp., *Clutellothrix* spp., and *Armatospathula costaricana*. Thus, identification requires matching of all characteristics given in the diagnosis and should be checked in protargol preparations.

Arcuospathidium multinucleatum FOISSNER, 1999 (Fig. 60a–m; Table 16)

1999 *Arcuospathidium multinucleatum* FOISSNER, Biodiv. Conserv., 8: 330 (Type and voucher slides with protargol-impregnated specimens from type and other localities are deposited in the Oberösterreichische Landesmuseum in Linz, Upper Austria.).

Diagnosis (contains only type population): Size about 140 × 20 µm in vivo. Narrowly to very narrowly spatulate with oblique to strongly oblique, narrowly cuneate oral bulge approximately 1.6 times longer than widest trunk region. On average 47 macronucleus nodules and several micronuclei. Extrusomes rod-shaped, 4–5 µm long. On average 15 ciliary rows, 3 anteriorly differentiated to isostichad dorsal brush occupying 18% of body length. About 20 dikinetids with bristles up to 4 µm long in each brush row, monokinetidal bristle tail of row 3 extends to mid-body.

Type locality: Forest soil near the village of Limuru, about 25 km NE of the town of Nairobi, Kenya, Equatorial Africa, E36°50' S1°.

Etymology: The Latin adjective *multinucleatum* (many nuclei) refers to the main feature of the species, viz., the numerous macronucleus nodules.

Description: Size 100–200 × 15–40 µm in vivo, usually 130–170 × 20–30 µm, depending on population. Narrowly to very narrowly spatulate, length:width ratio 5:1–10:1, likewise depending on specimen and population (Fig. 60a, d, l, m; Table 16). Oral area distinctly flattened laterally and separated from cylindroidal postoral portion by narrowed neck, often conspicuously convex and occasionally curved laterally, providing specimens with an inconspicuous twist. Cells very flexible and rather fragile, those from Australian population contractile by up to 50% under coverslip pressure. Macronucleus nodules scattered throughout trunk, ellipsoidal, number highly variable, depending on specimen and population (Fig. 60a, g; Table 16); rarely, nodules obtain a moniliform configuration, i.e., are connected by fine strands. Micronuclei scattered among

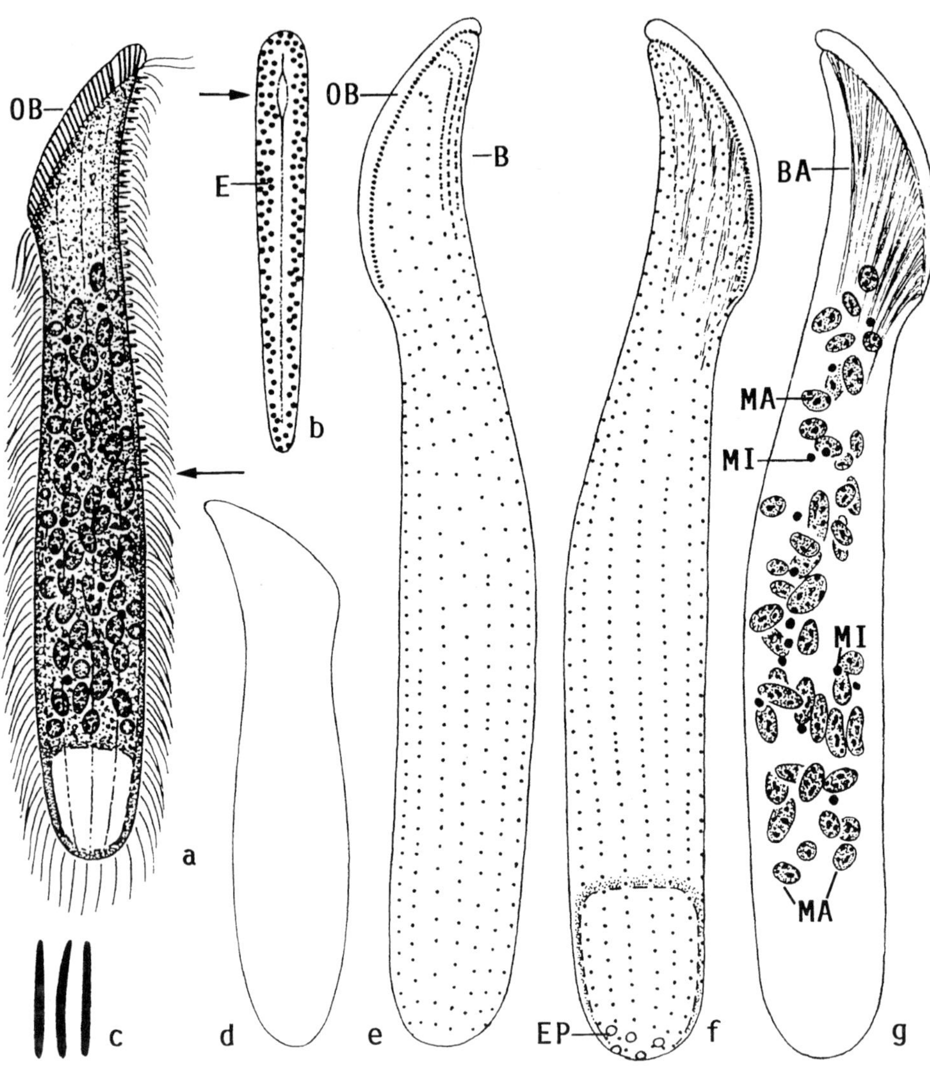

Fig. 60a–g *Arcuospathidium multinucleatum* from life (a–d) and after protargol impregnation (e–g). From FOISSNER (1999a). **a)** Left lateral view of a representative specimen from type population, length 140 µm. Note brush row 3, which extends with minute, monokinetidal bristles to mid-body (arrow). **b)** Frontal view of oral bulge, which is packed with extrusomes. Arrow marks minute conical depression near dorsal bulge end, viz., the temporary cytostome. **c)** Extrusomes are rod-shaped to indistinctly ellipsoidal, slightly curved, and 4–5 µm long. **d)** Diagram of a representative specimen from the Simba Hills population. These specimens are, on average (Table 16) stouter than those from type locality (Fig. 60a, e). **e–g)** Ciliary pattern of left and right side as well as nuclear apparatus and oral basket of holotype specimen, length 122 µm. B – dorsal brush, BA – oral basket, E – extrusomes, EP – excretory pores of contractile vacuole, MA – macronucleus nodules, MI – micronuclei.

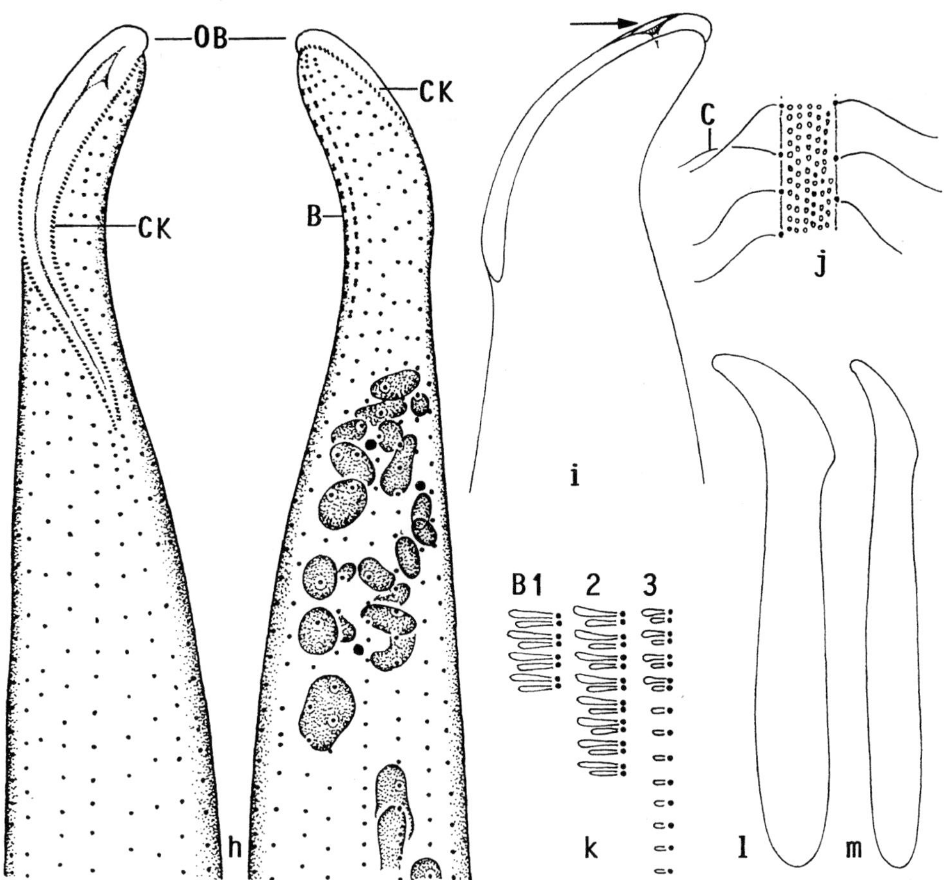

Fig. 60h–m *Arcuospathidium multinucleatum* from life (j–m) and after protargol impregnation (h, i). From FOISSNER (1999a). **h)** Ciliary pattern of anterior ventral and dorsal side of a specimen from type population. The oral bulge is distinctly cuneate and 32 µm long. **i)** Anterior portion of a specimen from another population, showing the minute conical depression near the dorsal end of the oral bulge (arrow), viz., the temporary cytostome. **j)** Surface view showing cortical granulation. **k)** Posterior portion of dorsal brush of a Venezuelan specimen; bristles up to 4 µm long. **l, m)** Diagrams of representative individuals from Australia. Specimens of this population are, on average (Table 16), more slender than those from type locality (Fig. 60a). B (1-3) – dorsal brush (rows), C – cilia, CK – circumoral kinety, OB – oral bulge.

macronucleus nodules, globular and heavily impregnated with protargol. Contractile vacuole in posterior body end, several excretory pores in pole area. Extrusomes in vivo very similar in all populations, rod-shaped to very narrowly ellipsoidal or cuneate, 4–5 µm long, studded in oral bulge and scattered in cytoplasm (Fig. 60c); about 3 µm long and more or less heavily impregnated in protargol slides. Cortex very flexible, contains about five rows of colourless granules < 1 µm across between each two ciliary rows (Fig. 60j), pellicle of Australian specimens partially covered by 2–3 µm long bacterial rods. Cytoplasm, except for flat and hyaline oral portion, more or less densely packed with 1 – 7 µm-sized, lipid droplets, depending on nutrition state. Very likely feeds

Table 16 Morphometric data on three populations (Pop) of *Arcuospathidium multinucleatum*: KT (Kenya, type), KS (Kenya, Shimba Hills), AU (Australia). From FOISSNER (1999a)

Characteristics[1]	Pop	$\overline{x}$	M	SD	CV	Min	Max	n
Body, length	KT	124.1	124.5	10.7	8.6	103.0	138.0	10
	KS	141.0	140.0	15.2	10.8	120.0	165.0	10
	AU	156.4	159.5	25.4	16.3	118.0	192.0	10
Body, width	KT	16.7	17.0	1.2	6.9	15.0	18.0	10
	KS	29.3	30.0	4.9	16.6	20.0	35.0	10
	AU	15.6	15.5	2.4	15.5	12.0	19.0	10
Body length:width, ratio	KT	7.4	7.5	0.7	8.8	6.1	8.4	10
	KS	4.9	4.8	0.7	14.9	4.0	6.0	10
	AU	10.2	10.2	1.8	17.8	7.2	13.3	10
Oral bulge (circumoral kinety), length	KT	27.2	28.0	3.4	12.5	21.0	31.0	10
	KS	41.6	42.5	5.4	12.9	35.0	50.0	10
	AU	37.0	35.0	7.7	20.9	26.0	47.0	10
Oral bulge length:body width, ratio	KT	1.6	1.6	0.2	10.8	1.3	1.9	10
	KS	1.5	1.4	0.2	14.6	1.2	1.8	10
	AU	2.4	2.5	0.6	24.9	1.5	3.5	10
Macronculei, number	KT	46.9	45.0	11.2	23.8	30.0	65.0	10
	KS	39.6	38.0	10.2	25.8	24.0	55.0	10
	AU	32.6	33.0	9.7	29.9	14.0	44.0	10
Macronucleus nodules, length	KT	5.5	5.3	2.3	41.7	3.0	11.0	10
	KS	8.9	10.0	2.3	25.6	5.0	12.0	10
	AU	6.5	6.0	1.4	22.1	5.0	9.0	10
Macronucleus nodules, width	KT	2.7	2.7	0.6	22.2	2.0	4.0	10
	KS	4.4	4.0	1.1	24.4	3.0	7.0	10
	AU	4.3	4.0	1.2	27.0	3.0	6.0	10
Micronuclei, number	KT	12.3	11.5	2.7	21.7	8.0	17.0	10
	KS	12.0	13.0	3.3	27.8	5.0	17.0	10
	AU	8.8	8.5	2.6	30.2	5.0	12.0	10
Micronuclei, largest diameter	KT	1.6	1.6	0.1	7.2	1.4	1.8	10
	KS	2.0	2.0	0.0	0.0	2.0	2.0	10
	AU	1.8	2.0	0.2	12.9	1.3	2.0	10
Ciliary rows, number	KT	14.6	15.0	1.7	11.7	11.0	17.0	10
	KS	17.5	18.0	1.4	8.2	15.0	19.0	10
	AU	12.8	12.0	1.1	8.9	12.0	15.0	10
Basal bodies in a right side ciliary row, number	KT	60.3	57.5	15.1	25.1	40.0	90.0	10
	KS	57.3	57.5	7.4	12.9	45.0	68.0	10
	AU	50.7	46.0	13.9	27.5	34.0	72.0	10
Dorsal brush rows, number	KT	3.1	3.0	–	–	3.0	4.0	10
	KS	3.0	3.0	–	–	3.0	3.0	10
	AU	3.1	3.0	–	–	3.0	4.0	10
Circumoral kinety to last dikinetid of brush row 1, distance	KT	21.4	21.5	2.9	10.8	17.0	25.0	10
	KS	29.5	30.0	4.4	14.9	22.0	35.0	10
	AU	20.4	19.5	3.9	19.4	15.0	30.0	10
Circumoral kinety to last dikinetid of brush row 2, distance	KT	21.8	22.0	3.1	14.3	15.0	25.0	10
	KS	31.6	30.5	4.4	13.9	25.0	40.0	10
	AU	24.2	23.5	3.8	15.6	19.0	32.0	10
Circumoral kinety to last dikinetid of brush row 3, distance	KT	18.1	18.0	3.0	16.4	11.0	22.0	10
	KS	25.7	25.0	5.8	22.5	17.0	35.0	10
	AU	23.3	22.5	3.9	16.7	19.0	31.0	10

[1] Data based on mounted, protargol-impregnated (Foissner's method), and randomly selected specimens from non-flooded Petri dish cultures. Measurements in µm. CV – coefficient of variation in %, M – median, Max – maximum, Min – minimum, n – number of individuals investigated, SD – standard deviation, $\overline{x}$ – arithmetic mean.

mainly on ciliates. Swims rather clumsily but creeps versatilely on soil particles, showing great flexibility.

Somatic cilia about 10 µm long, ordinarily spaced, arranged in 12–18 longitudinal rows anteriorly curved dorsally and distinctly separate from circumoral kinety; monokinetidal anterior tail of dorsal brush kineties curved ventrally, occasionally abutting to circumoral kinety (Fig. 60e, f, h). Dorsal brush isostichad, occupies 16%–24% of body length, depending on population (Table 16), consists of three (rarely four) rows of almost same length; bristles 2–4 µm long and slightly clavate; row 3 continues with about 1 µm long, monokinetidal bristles to mid-body (Fig. 60a, e, h, k).

Oral bulge oblique to strongly oblique and narrowly cuneate, on average 1.6 times longer than widest trunk region, bright due to the many extrusomes contained. Temporary cytostome near dorsal end of oral bulge, more distinct in Australian than African specimens, in vivo occasionally recognizable as a minute notch in bulge outline and as a tiny opening in frontal view; distinct in excellently prepared specimens due to the fine fibres lining the cytostome wall (Fig. 60a, b, h, i). Circumoral kinety of same shape as oral bulge, that is, narrowly cuneate and composed of densely spaced dikinetids each associated with a cilium and a very fine nematodesma; oral basket thus inconspicuous (Fig. 60g).

Occurrence and ecology: As yet found at type locality (forest in the Escarpment Mountains near Nairobi, mixture of leaf litter and red soil, pH 6.6), in the Shimba Hills Nature Reserve (Kenya), in Australia (Green Island near Cairns, mixture of litter and brownish soil from the Palm-girdle about 30 m inshore, pH 7.1), and in Venezuela (0–5 cm very sandy Savannah soil near Puerto Ayacucho, pH 6.0). These data indicate that *A. multinucleatum* is very likely a cosmopolitan, possibly preferring circumneutral conditions.

Remarks: FOISSNER (1999a) studied three populations of this species in detail, and several others cursorily (see occurrence and ecology section). They agree in the main features (size, slender shape, many macronucleus nodules, short extrusomes, 10–20 ciliary rows, isostichad dorsal brush), but differ in some morphometric characteristics (Table 16), which might be considered by some colleagues to be of significance for splitting the species into several subspecies. Thus, the diagnosis contains only the type population. Differing features from the other populations are mentioned, if important, in the description and figure explanations.

Arcuospathidium multinucleatum is a typical member of the genus and easily identified by the numerous, scattered macronucleus nodules. This feature is found also in the *A. namibiense* group which, however, comprises very slender species with short oral bulge about as long as widest trunk region.

Arcuospathidium namibiense FOISSNER, AGATHA & BERGER, 2002

2002 *Arcuospathidium namibiense* FOISSNER, AGATHA & BERGER, Denisia, 5: 283.

Improved diagnosis: Size about 160 × 10 µm in vivo. Vermiform with inconspicuous, strongly oblique, cuneate to obovate oral bulge occupying about 85% of body width. Approximately 20 scattered macronucleus nodules and 2 conspicuous, narrowly ellipsoidal to lenticular micronuclei. Extrusomes oblong or narrowly to broadly ovate/cuneate, 0.8–2 µm long. Usually 4–6 ciliary rows, 2 or 3 anteriorly differentiated to heterostichad dorsal brush occupying 6–10% of body length on average. Brush conspicuous due to some 15–20 µm long bristles at anterior end of row 3. Row 1 lacking or consisting of few bristles.

Etymology: Named after the country discovered, viz., Namibia, Southwest Africa.

Remarks: The diagnosis has been adjusted to the style used in this book and improved because the ovate extrusomes were erroneously described as "oblanceolate" by FOISSNER et al. (2002).

FOISSNER et al. (2002) split *A. namibiense* into two subspecies, differing mainly in the shape of the extrusomes (oblong vs. ovate), but also in details of the brush (two vs. three rows) and the number of ciliary rows (four vs. five). Recent observations on an Austrian population suggest a third type with 40–90 macronucleus nodules. However, we hesitate to formally recognize it as third subspecies because the number of macronucleus nodules is often highly variable; on the other hand, we do not change the original diagnosis (about 20 macronucleus nodules), but prefer to await further data. Generally, this type of *Arcuospathidium* needs a broad species/subspecies concept because the decisive features are difficult to recognize.

Arcuospathidium namibiense has so few ciliary rows that their arrangement in the oral area, which is crucial for the generic assignment, is difficult to follow. The dorsal brush occupies the left side, suggesting classification in *Latispathidium* FOISSNER et al. (2005a), a genus not yet known when FOISSNER et al. (2002) described *A. namibiense*. However, *Latispathidium* has a *Spathidium* ciliary pattern, while that of *A. namibiense* resembles *Arcuospathidium*. *Arcuospathidium namibiense* has conspicuously elongated brush bristles at anterior end of row 3. However, this feature is found also in some typical *Spathidium* and *Latispathidium* species, and is thus likely not an indicator of relationship or a useful generic character. Possibly, *A. namibiense* is misclassified, but assigning it to *Latispathidium* would contaminate this genus.

Arcuospathidium namibiense differs, inter alia, by the scattered macronucleus nodules from most described congeners, especially *A. vermiforme* (two macronucleus nodules), *A. cooperi* (single, oblong macronucleus), and *A. vlassaki* (single, rod-shaped macronucleus). Only *A. multinucleatum* and *A. virungense* have the same nuclear pattern as *A. namibiense*. The former is easily distinguished by body shape (14:1 vs. 7:1 in preparations) and number of ciliary rows (4–6 vs. 11–19), while the later has thick, 4–5 µm long extrusomes and > 10 ciliary rows.

In vivo, *A. namibiense* may be similar to *Spathidium procerum*. However, this species is a "real" *Spathidium*, has up to 10 µm long, rod-shaped extrusomes, lacks elongated dorsal bristles, and possesses 10 ciliary rows. As concerns the elongated dorsal bristles, *A. namibiense* resembles *Spathidium falciforme*, which, however, is only 40–60 µm long.

Arcuospathidium namibiense is easily confused with *Edaphospathula fusioplites* (with vs. without distinctly elongated anterior brush bristles), *Protospathidium vermiforme* (protospathidiid ciliary pattern; check by protargol impregnation!) and, especially, *Spathidium bonneti*. However, a more detailed comparison reveals significant differences, although the description of *S. bonneti* is brief and lacks detailed morphometrics and data about the extrusomes. The most important difference concerns the conspicuously elongated (15–20 µm!) and thus highly characteristic bristles at top of brush row 3. These bristles and their basal bodies are so distinct in vivo and protargol preparations that one cannot assume that BUITKAMP (1977a) overlooked them. A further main difference is the body's length:width ratio, which is 30:1 in *S. bonneti*, according to BUITKAMP's description and illustration (Fig. 62o). In *A. namibiense namibiense* and *A. namibiense tristicha,* only two of 42 specimens analyzed have a ratio of 24:1, while the average values are much lower, viz., 13.7:1 and 15.1:1 (Table 17). The third main difference concerns the location of the single dikinetid composing brush row 1 (Fig. 62n, q): distinctly underneath the circumoral kinety in *S. bonneti,* while close to the circumoral kinety in *A. namibiense tristicha.* Finally, the location of the contractile vacuole (distinctly subterminal vs. terminal) and body shape (distinctly vs. slightly narrowed posteriorly) are slightly different. FOISSNER et al. (2002) discussed the problem with Dr. BUITKAMP, whose slides of *Spathidium bonneti* are, unfortunately, completely bleached. In his letter, BUITKAMP stated: (i) My prepared specimens have a similar size as yours, but I remember that they were distinctly narrower in vivo; (ii) With interference contrast, I could recognize in two specimens at least one 15 µm long dorsal bristle, which I obviously overlooked previously, but the whole dorsal brush pattern seems to be distinctly different in *S. bonneti* and *A. namibiense*; (iii) I believe that *S. bonneti* and *A. namibiense* are different species.

Arcuospathidium namibiense namibiense FOISSNER, AGATHA & BERGER, 2002 (Fig. 61a–j; Table 17)

2002 *Arcuospathidium namibiense namibiense* FOISSNER, AGATHA & BERGER, Denisia, 5: 285 (Type slides with protargol-impregnated specimens from type locality are deposited in the Oberösterreichische Landesmuseum in Linz, Upper Austria.).

Diagnosis: Extrusomes oblong. 4–5 ciliary rows, 2 anteriorly modified to brush.

Type locality: Soil from *Aloe dichotoma* forest near the Gariganus Guest Farm, Namibia, E18°25' S26°30'.

Description: Size highly variable, in vivo 110–210 × 8–15 µm, usually about 160 × 12 µm; length:width ratio also highly variable, that is, 9:1 – 24:1, on average 14:1 in preparations (Table 17). Cylindroidal to vermiform, frequently strongly curved or sigmoidal and slightly twisted about main body axis, rather distinctly flattened laterally, neck indistinct (Fig. 61a, h); acontractile and rather fragile, specimens thus sometimes distorted and/or inflated in protargol preparations. Nuclear apparatus in middle quarters of cell on average (Fig. 61a, h). Macronucleus nodules basically scattered with a tendency to form a cluster each in second and third quarter of cell; rarely in rough series or in single accumulation near body centre; individual nodules globular to narrowly ellipsoidal, on average ellipsoidal, usually contain one or two comparatively large nucleoli, rarely several small ones; unite to a globular mass in mid-dividers. Usually two micronuclei, one each in anterior and posterior third of nuclear figure; individual micronuclei

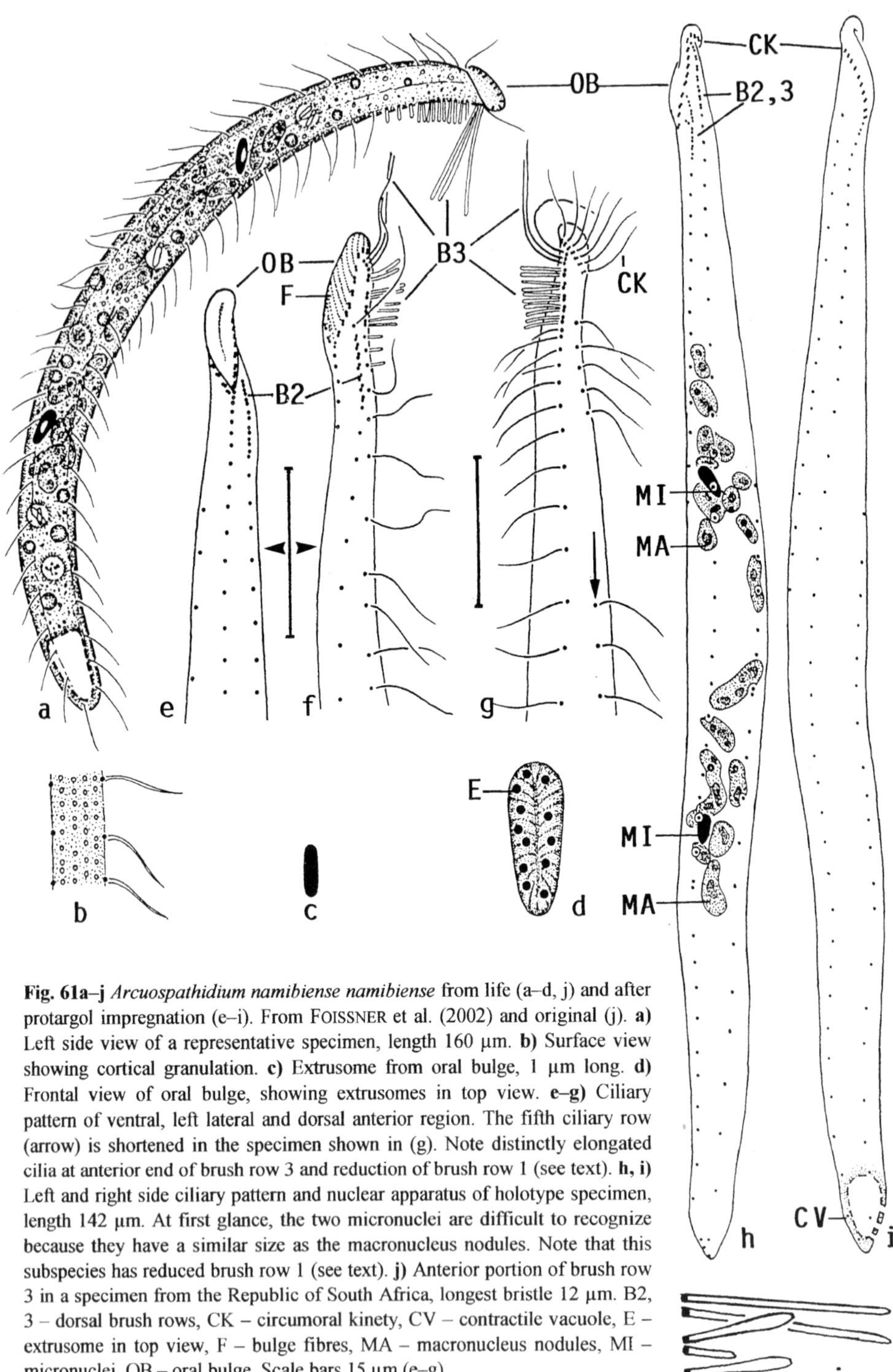

Fig. 61a–j *Arcuospathidium namibiense namibiense* from life (a–d, j) and after protargol impregnation (e–i). From FOISSNER et al. (2002) and original (j). **a)** Left side view of a representative specimen, length 160 μm. **b)** Surface view showing cortical granulation. **c)** Extrusome from oral bulge, 1 μm long. **d)** Frontal view of oral bulge, showing extrusomes in top view. **e–g)** Ciliary pattern of ventral, left lateral and dorsal anterior region. The fifth ciliary row (arrow) is shortened in the specimen shown in (g). Note distinctly elongated cilia at anterior end of brush row 3 and reduction of brush row 1 (see text). **h, i)** Left and right side ciliary pattern and nuclear apparatus of holotype specimen, length 142 μm. At first glance, the two micronuclei are difficult to recognize because they have a similar size as the macronucleus nodules. Note that this subspecies has reduced brush row 1 (see text). **j)** Anterior portion of brush row 3 in a specimen from the Republic of South Africa, longest bristle 12 μm. B2, 3 – dorsal brush rows, CK – circumoral kinety, CV – contractile vacuole, E – extrusome in top view, F – bulge fibres, MA – macronucleus nodules, MI – micronuclei, OB – oral bulge. Scale bars 15 μm (e–g).

conspicuous because narrowly ellipsoidal, bluntly fusiform, ovate, or slenderly ovate and about 4 × 2 µm in size, that is, almost as large as macronucleus nodules; micronuclei lacking or not clearly recognizable in about one third of specimens due to similarly sized and shaped cell inclusions. Contractile vacuole in rear body end, some subterminal excretory pores on ventral side. Extrusomes in oral bulge, oblong, extremely minute, that is, about 1–1.2 µm long; do not impregnate with the protargol method used (Fig. 61a, c). Cortex very flexible, thin, contains inconspicuous rows of loosely spaced, colourless granules less than 0.5 µm across. Cytoplasm hyaline, contains some bright lipid droplets 1–2 µm across and about 5 µm-sized food vacuoles each with one or two bacterial rods. Slowly gliding and wriggling like an eel, showing great flexibility when creeping among soil particles.

Cilia about 7 µm long in vivo, invariably arranged in four or five, usually four slightly spiralling, equidistant rows composed of loosely spaced (3–6 µm apart) cilia and bare kinetids above some ciliated ones. All rows abut on circumoral kinety, except for one or two (in specimens with five rows) distinctly shortened right side rows. Brush conspicuous, although occupying only 10% of body length, because of the long bristles easily recognizable even at low magnification (≥ ×100). Brush row 1 is lacking, compared with *A. namibiense tristicha*, where it is present but consists of only one or few dikinetids. To make morphometrics comparable, we designate the first (right) brush row of *A. namibiense namibiense* as row 2. Dorsal brush row 2 composed of an average of seven dikinetids bearing about 5 µm long bristles, usually separated from circumoral kinety by a single monokinetid, continues as ordinary somatic ciliary row posteriorly. Brush row 3 anteriorly with two to four narrowly spaced, about 15 µm long and thus highly conspicuous, rod-shaped bristles obliquely spread posteriorly in gliding and swimming specimens; long bristles followed by an average of five dikinetids with about 5 µm long bristles (seen also in a population from the Republic of South Africa; Fig. 61j); then row 3 continues with a few 2–3 µm long, monokinetidal bristles or ordinary cilia to rear body end (Fig. 61a, e–i; Table 17).

Oral bulge about 3 µm high in vivo, minute because slightly narrower than widest postoral body region and occupying only 6% of body length on average; slanted by 40°–70° to main body axis, obovate to slenderly obovate or cuneate in frontal view, contains the minute extrusomes described above. Circumoral kinety composed of comparatively widely spaced dikinetids, each bearing a single cilium and a fibre extending to bulge centre. Oral basket (rods) not recognizable in vivo or protargol preparations, not even in over-stained specimens (Fig. 61a, d, e–i; Table 17).

Table 17 Morphometric data on *Arcuospathidium namibiense namibiense* (upper line) and *A. namibiense tristicha* (lower line). From FOISSNER et al. (2002)

Characteristics [1]	$\overline{x}$	M	SD	SE	CV	Min	Max	n
Body, length	147.7	143.0	27.4	6.0	18.5	100.0	190.0	21
	138.9	140.0	16.7	3.7	12.0	103.0	170.0	21
Body, width	10.8	10.0	1.9	0.4	17.6	7.0	15.0	21
	9.1	9.0	1.5	0.3	16.9	6.0	11.0	21
Body length:width, ratio	13.7	14.0	4.1	0.9	29.3	8.0	24.0	21
	15.2	15.0	3.6	0.8	23.0	11.0	24.0	21
Oral bulge (circumoral kinety), length	8.1	8.0	0.8	0.2	10.0	6.0	9.0	21

continued

Characteristics [1]	$\overline{x}$	M	SD	SE	CV	Min	Max	n
	8.6	8.0	1.1	0.2	13.0	7.0	12.0	21
Oral bulge, height	2.4	3.0	–	–	–	2.0	3.0	21
	1.9	2.0	–	–	–	1.0	3.0	21
Circumoral kinety to last dikinetid of brush row 1, distance			lacking					21
	2.2	2.0	0.5	0.1	24.1	2.0	4.0	21
Circumoral kinety to last dikinetid of brush row 2, distance	9.0	8.0	2.5	0.6	27.9	5.0	14.0	21
	14.7	15.0	2.0	0.4	13.5	11.0	18.0	21
Circumoral kinety to last dikinetid of brush row 3, distance	6.3	6.0	1.3	0.3	20.2	4.0	8.0	21
	8.2	8.0	1.4	0.3	17.1	6.0	11.0	21
Anterior body end to first macronuclear nodule, distance	35.4	37.0	10.0	2.2	28.2	11.0	50.0	21
	50.4	50.0	9.9	2.2	19.7	36.0	70.0	21
Nuclear figure, length	73.2	67.0	16.3	3.6	22.3	49.0	97.0	21
	61.2	62.0	12.4	2.7	20.2	36.0	84.0	21
Macronuclear nodules, length	6.2	6.0	1.9	0.4	30.4	3.0	12.0	21
	5.5	5.0	1.7	0.4	31.4	4.0	10.0	21
Macronuclear nodules, width	2.9	3.0	0.6	0.1	21.8	2.0	4.0	21
	3.0	3.0	0.9	0.2	30.7	1.0	5.0	21
Macronuclear nodules, number	19.2	20.0	6.8	1.5	35.6	6.0	34.0	21
	25.3	22.0	7.0	1.5	27.7	16.0	42.0	21
Micronuclei, length	4.2	4.0	0.6	0.1	14.4	3.0	5.0	21
	5.1	5.0	0.8	0.2	15.4	4.0	7.0	21
Micronuclei, width	1.8	2.0	–	–	–	1.0	3.0	21
	1.8	2.0	–	–	–	1.0	2.0	21
Micronuclei, number	2.2	2.0	0.5	0.1	23.4	2.0	4.0	21
	2.0	2.0	0.0	0.0	0.0	2.0	2.0	21
Somatic kineties, number	4.1	4.0				4.0	5.0	21
	5.1	5.0	–	–	–	5.0	6.0	21
Ciliated kinetids in a right lateral kinety, number	35.8	35.0	7.5	1.6	21.0	23.0	54.0	21
	52.2	52.0	8.9	2.0	17.1	37.0	68.0	21
Dorsal brush rows, number [2]	2.0	2.0	0.0	0.0	0.0	2.0	2.0	21
	3.0	3.0	0.0	0.0	0.0	3.0	3.0	21
Dikinetids in brush row 1, number			lacking					21
	1.3	1.0	0.6	0.1	43.6	1.0	3.0	21
Dikinetids in brush row 2, number	7.0	7.0	1.6	0.4	23.0	2.0	9.0	21
	15.1	15.0	2.1	0.5	13.7	11.0	19.0	21
Dikinetids in brush row 3, number [3]	5.5	5.0	1.0	0.2	17.8	4.0	7.0	21
	7.3	7.0	1.1	0.2	14.5	4.0	9.0	21
Distinctly elongated brush bristles, number	3.3	4.0	0.9	0.2	25.8	2.0	4.0	17
	2.8	3.0	0.6	0.1	21.4	2.0	4.0	21
Circumoral kinetids, number	19.8	20.0	2.3	0.5	11.4	14.0	25.0	21
	30.9	30.0	4.2	0.9	13.4	25.0	40.0	21
Excretory pores, number	3.7	3.0	1.7	0.5	44.8	1.0	8.0	13
	5.8	6.0	2.2	0.7	37.1	3.0	10.0	10

[1] Data based on mounted, protargol-impregnated (FOISSNER's method), and selected (distinctly inflated or distorted specimens excluded in *A. namibiense namibiense*) or randomly selected (*A. namibiense tristicha*) specimens from non-flooded Petri dish cultures. Measurements in µm. CV – coefficient of variation in %, M – median, Max –maximum, Min – minimum, SD – standard deviation, SE – standard error of arithmetic mean, $\overline{x}$ – arithmetic mean.

[2] First row lacking in 1 out of 25 specimens of *A. namibiense tristicha*.

[3] Elongated bristles at anterior end not included because dikinetidal organization not clearly recognizable.

Occurrence and ecology: To date found at type locality, where it was rather abundant in the non-flooded Petri dish culture, and in a soil sample from the Republic of South Africa, where it was rare. This subspecies is obviously much rarer than *A. namibiense tristicha* and possibly even confined to Africa. It is well adapted to the soil environment by the slender, highly flexible body.

Remarks: See introduction to species.

Arcuospathidium namibiense tristicha FOISSNER, AGATHA & BERGER, 2002 (Fig. 62a–m, p–x; 117a–p; Table 17)

2002 *Arcuospathidium namibiense tristicha* FOISSNER, AGATHA & BERGER, Denisia, 5: 288 (Type slides with protargol-impregnated specimens from type locality and the Austrian voucher locality are deposited in the Oberösterreischische Landesmuseum in Linz, Upper Austria.).

Improved Diagnosis: Extrusomes narrowly to broadly ovate/cuneate. 5–6 ciliary rows, 3 anteriorly modified to dorsal brush.

Type locality: Bark from a *Maytenus oleoides* tree (Celastraceae) in the botanical garden of Cape Town, Republic of South Africa, E18°25' S33°53'.

Etymology: Composite of the Greek words *tri* (three) and *sticha* (row), referring to the three dorsal brush rows.

Description (of type population): Size highly variable, that is, 110–190 × 6–12 µm in vivo, usually about 150 × 9 µm; length:width ratio also highly variable, that is, 11–24:1, on average 15.2:1 in preparations (Table 17). Cylindroidal to vermiform, rarely coiled, usually strongly curved or sigmoidal and distinctly twisted about main body axis, flattened only in anterior region, neck indistinct; when gliding, anterior body half often straight, while posterior third more or less distinctly curved, providing cells with a characteristic L-shaped appearance (Fig. 62a–c, r, t–v). Nuclear apparatus in middle and anterior half of last third of cell on average (Fig. 62a, l). Macronucleus nodules basically scattered with a tendency to form a cluster each in anterior and posterior end of nuclear figure; rarely in series or distinct clusters in anterior and posterior body half; individual nodules globular to oblong, on average broadly ellipsoidal, usually contain some small nucleoli. Invariably two micronuclei, one each in anterior and posterior third of nuclear figure; individual micronuclei conspicuous because narrowly ellipsoidal or fusiform and about 5 × 2 µm in size, that is, almost as long as macronuclear nodules. Contractile vacuole in rear end, some subterminal excretory pores on ventral side. Extrusomes in both sides of oral bulge, narrowly to broadly ovate/cuneate, extremely minute, that is, about 1–1.2 × 0.5 µm; impregnate with the protargol method used and then resemble basal bodies (Fig. 62a, d, g, j, s; 17b–d). Cortex very flexible, thin, contains five to six rows of colourless, minute (< 0.5 µm) granules between each two ciliary rows. Cytoplasm hyaline, contains some bright lipid droplets 1–3 µm across and up to 12 µm-sized food vacuoles, once with an almost intact *Cyrtolophosis mucicola*. Glides rather rapidly on microscope slide, wriggling like an eel when creeping among soil particles.

Cilia about 7 µm long in vivo, invariably arranged in five to six, often five distinctly

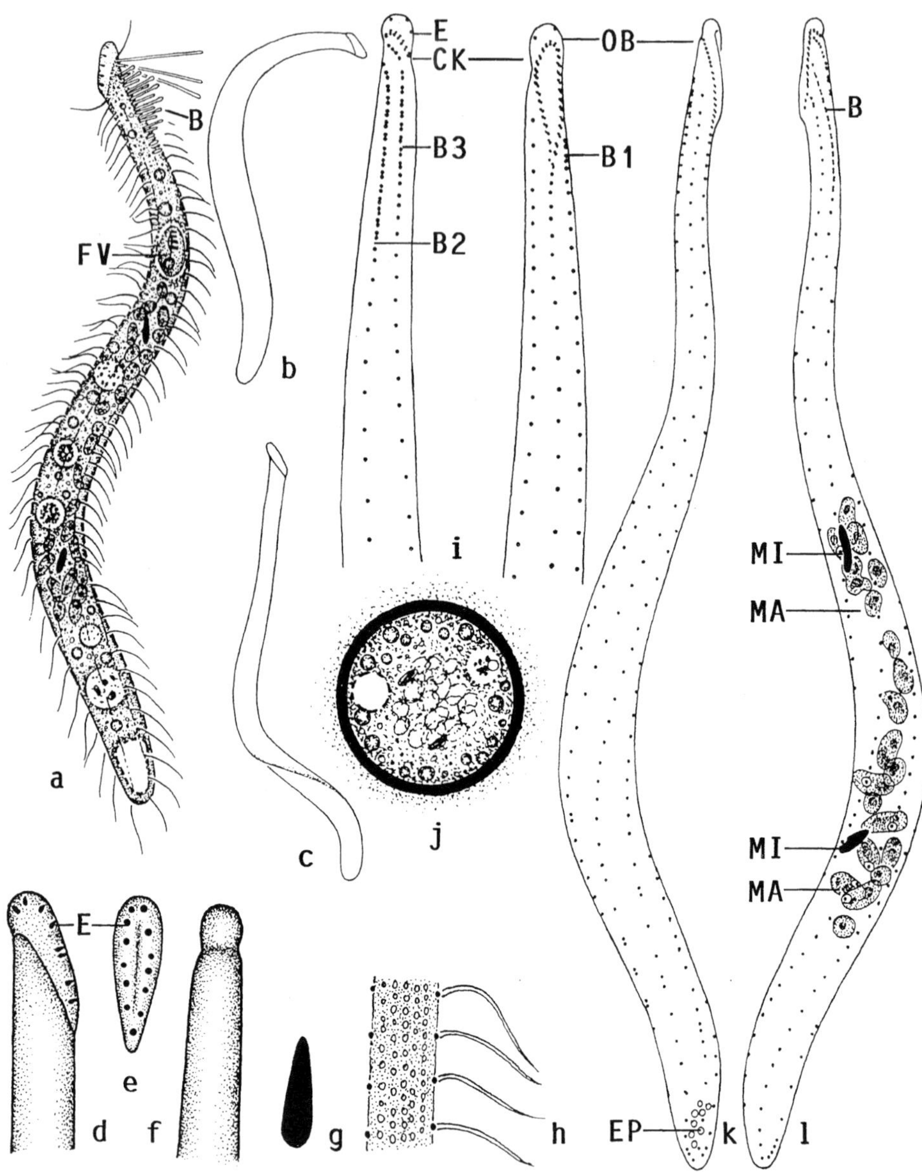

Fig. 62a–l *Arcuospathidium namibiense tristicha*, Namibian type specimens (a–i, k, l) and resting cyst of an Austrian specimen (j) from life (a–h, j) and after protargol impregnation (i, k, l). From FOISSNER et al. (2002) and original (j). **a)** Left side view of a representative specimen, length 150 µm. **b, c)** Shape variants. **d–f)** Right lateral, frontal, and dorsal view of oral area. **g)** Narrowly cuneate oral bulge extrusome, length 1 µm. **h)** Surface view showing cortical granulation. **i)** Ciliary pattern of dorsal and ventral side in oral body portion, length of oral bulge 11 µm. Brush details, see figures 62m, q. **j)** Fully developed (?) resting cyst, 23 µm across. **k, l)** Right and left side ciliary pattern and nuclear apparatus of holotype specimen, length 132 µm. B(1-3) – dorsal brush (rows), CK – circumoral kinety, E – extrusomes, EP – excretory pores of contractile vacuole, FV – food vacuole, MA – macronucleus nodules, MI – micronuclei, OB – oral bulge.

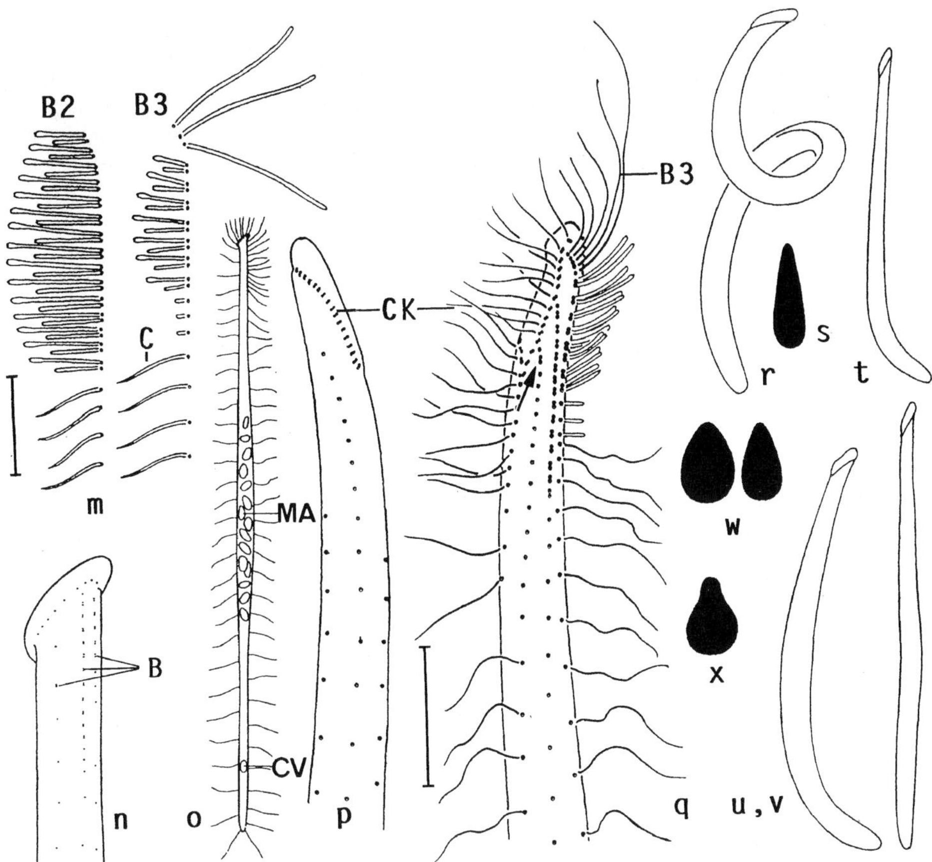

Fig. 62m–x *Arcuospathidium namibiense tristicha* (m, p–x; from FOISSNER et al. 2002 and originals) and *Spathidium bonneti* (n, o; from BUITKAMP 1977a), Namibian (m, p, q), Ivory Coast (n, o), Benin (r–v), Austrian (w), and Tibetan (x) specimens from life (m, o, r–x) and after protargol impregnation (n, p, q). **m)** Dorsal brush rows 2 and 3 of a very carefully studied specimen. Cilia drawn to scale (10 µm). Note that the somatic cilia are shorter than most brush bristles. **n, o)** *Spathidium bonneti*, 150 × 5 µm, is more slender than *A. namibiense* (30:1 vs. 14–16:1) and lacks elongated bristles at top of brush row 3. Furthermore, the single dikinetid composing brush row 1 is much farther underneath the circumoral kinety than in *A. namibiense* (q, arrow). **p, q)** Ciliary pattern of anterior right and left side. Arrow marks the single dikinetid composing brush row 1; it is much more close to the circumoral kinety than in *S. bonneti* (n). **r, t, u, v)** Shape variants, those shown in (t, u) are most typical. **s, w, x)** Depending on population, the extrusomes of *A. namibiense tristicha* are 1–2 µm long and narrowly to broadly ovate/cuneate, rarely even ampulliform (x). When exploded, they become drumstick-shaped and 3–5 µm long (Fig. 117i, j). B(2, 3) – dorsal brush (rows), C – somatic cilium, CK – circumoral kinety, CV – contractile vacuole, MA – macronucleus nodules. Scale bars 10 µm (m, p, q).

spiralling, equidistant, ordinarily ciliated rows abuting on circumoral kinety, except of one or two (in specimens with six rows) right side rows distinctly shortened anteriorly (Fig. 62a, i–l, p, q; Table 17). Brush conspicuous, although occupying only 11% of body length, due to the long bristles already recognizable at low magnification (≥ x100). Brush row 1 consists of one to three, usually only one dikinetid separated from circumoral kinety by a single monokinetid; lacking in one out of 25 specimens investigated.

Brush row 2 about 15 µm long in vivo, separated from circumoral kinety by a single monokinetid, continues as ordinary somatic kinety posteriorly, composed of an average of 15 ordinarily spaced dikinetids with up to 10 µm long bristles gradually decreasing in length at both row ends; anterior bristle of dikinetids longer than posterior and slightly inflated distally. Brush row 3 about 10 µm long in vivo, anteriorly with some very closely spaced, 15–20 µm long and thus highly conspicuous, rod-shaped bristles obliquely spread backwards in gliding and swimming specimens (Fig. 117a, b); composed of an average of seven dikinetids with up to 6 µm long bristles gradually decreasing in length at both row ends; anterior bristles of dikinetids longer than posterior and slightly inflated distally; continues with some 2–3 µm long, monokinetidal bristles before extending as an ordinary somatic kinety backwards (Fig. 62a, i–l, m, q; Table 17).

Oral bulge up to 4 µm high in vivo and thus rather conspicuous, although slightly narrower than widest postoral region and occupying only about 6% of body length, slanted by 40–60° to main body axis, narrowly obovate in frontal view, contains the minute extrusomes described above. Circumoral kinety composed of rather closely spaced dikinetids each bearing an about 7 µm long cilium. Oral basket (rods) not recognizable in vivo or protargol preparations, not even in over-stained specimens (Fig. 62a, d–f, i–l, p, q, 117b; Table 17).

Observations on other populations: *A. namibiense tristicha* is rather frequent and in vivo best identified by the minute, narrowly to broadly ovate/cuneate extrusomes observed in more than 20 populations from soils globally. A small population from Benin (in soil from the farm of the Agricultural Faculty at the National University in the town of Abomey; kindly provided by Prof. Jean DRAGESCO) was studied in some detail (Fig. 62r–v). It matches the Namibian type specimens well, especially in having long (about 13 µm as opposed to 7 µm long somatic cilia) bristles at anterior end of brush row 3 and minute, cuneate extrusomes which, however, did not impregnate with protargol. The macronucleus is more frequently indistinctly moniliform than in the Namibian specimens. Brush row 1 is lacking in about 20% of specimens and usually consists of three to six dikinetids in the others, while rows 2 and 3 have only about half the number of dikinetids found in the Namibian specimens. The Tibetan and Brazilian specimens perfectly match those of the type population, except of the extrusomes which are ampulliform (Fig. 62x), while those of a Venezuelan population are ovate as in the type specimens and double their length when exploded, showing that are typical toxicysts (117i, j).

Arcuospathidium namibiense tristicha occurs also in the holarctic biogeographical region, viz., in the USA and Austria. The USA specimens (13–35, on average 23 macronucleus nodules, n 6) are very similar to those of the South African population, while the Austrian specimens have a significantly higher number of macronucleus nodules (40–90, $\bar{x}$ 64, n 9), indicating that they might represent a distinct subspecies, as discussed in the introduction to the species. On the other hand, all other features are as in *A. namibiense namibiense* and/or *A. namibiense tristicha* (Fig. 62w, 117k–p). Brief description of the Austrian population from spruce forest soil of the Neuwald area in Styria (for site details, see FOISSNER et al. 2005a): (i) shape and size as in type specimens, viz., 130–180 × 8–12 µm (n 5), on average 150 × 10µm (117k, l); (ii) central body quarters packed with an average of 64 macronucleus nodules sometimes forming short, moniliform pieces (117m, n); (iii) micronuclei in vivo as conspicuous as in type specimens, but not impregnated with protargol; (iv) 5–15 thick, ovate to broadly ovate extrusomes in oral bulge, individual extrusomes 0.8–2 × 0.6–0.8 µm in vivo (Fig. 62w, 117n, o); (v) moves like small worms, that is, wriggles slowly about main body axis, while curved or L-shaped when gliding on microscope slide (Fig. 62r, t, u); (vi) somatic ciliature as in Namibian specimens, cilia 7 µm long in vivo and arranged in five to six rows; (vii) anterior brush bristles 12–15 µm long and slightly thicker than the ordinary somatic cilia, followed by pairs of 5–8 µm long bristles with anterior (posterior in type!) bristle of a pair slightly shorter than posterior (117l, n, o); (viii) brush row 1 composed of one to three dikinetids with 4–5 µm long bristles; (ix) oral bulge minute, oblique (≥ 45%), and cuneate to obovate in frontal view, circa 2 µm high at dorsal end (Fig. 117l, m); (x) about 20 circumoral dikinetids, which matches not *A. namibiense tristicha* (circa 30) but *A. namibiense namibiense*; (xi) few of the specimens isolated formed fully developed (?) resting cysts about 23 µm across and with an 0.8–1 µm thick, smooth wall covered by a thin slime layer (Fig. 62j, 117p).

These observations confirm FOISSNER et al. (2002) in distinguishing the two subspecies mainly according to the shape of the extrusomes. Refined molecular data will probably show further (sub)species.

Occurrence and ecology: The data mentioned above suggest a global distribution and wide ecological range of *A. namibiense tristicha*, although we cannot exclude that it is a complex of (sub)species. It occurs, for instance, in savannah soils of Namibia (FOISSNER et al. 2002) and in various forest soils of Austria (FOISSNER 2004b, FOISSNER et al. 2005a). *Arcuospathidium namibiense tristicha*, which is well adapted to the soil environment by the worm-like body, became abundant in the non-flooded Petri dish culture from type locality; usually, however, numbers are low.

Remarks: See introduction to species.

Arcuospathidium muscorum (DRAGESCO & DRAGESCO-KERNÉIS, 1979) FOISSNER, 1984

Diagnosis (includes two subspecies and six populations from three biogeographical regions): Size about 80–130 × 25–40 µm in vivo. Spatulate with oblique, very narrowly cuneate oral bulge extending in anterior body third to half. Macronucleus a moderately long, tortuous strand; single micronucleus. Extrusomes rod-shaped to narrowly obcuneate or obclaviform, 4–6 × 0.3–0.8 µm in size. Usually 12–18 ciliary rows, 3 anteriorly differentiated to heterostichad dorsal brush occupying about 14% of body length: brush rows 1 and 2 of similar length, row 3 shortened and with long monokinetidal tail.

Etymology: Not mentioned in the original description. The Latin *muscorum* obviously refers to the habitat (moss) the species was discovered.

Remarks: We split this cosmopolitan species into two subspecies, according to the shape of the extrusomes. They are rod-shaped or, rarely, slightly obcuneate in many

Holarctic, African, and South American populations, while more or less distinctly obclavate in several Australian populations, suggesting some biogeographic differentiation.

Arcuospathidium muscorum differs from the congeners with long, tortuous macronucleus mainly by the single (vs. many) micronucleus and the small (field specimens usually ≤ 100 µm vs. > 100 µm), stout (≤ 4:1 vs. ≥ 5:1) body. All described *Apertospathula* and *Cultellothrix* species have not only a different ciliary pattern, but also another macronucleus type, except of *C. velhoi* and *C. lionotiformis* which are, inter alia, larger (> 100 µm vs. < 100 µm) and knife-shaped (vs. spatulate). Small specimens of *Spathidium contractile* may be very similar to *A. muscorum*, but have a different ciliary pattern, several (vs. one) micronuclei, and two (vs. one) types of oral bulge extrusomes. Small specimens and/or cells with short macronucleus resemble *Cultellothrix coemeterii* which, however, has a distinctly shorter oral bulge (34–48% vs. 25% of body length).

Arcuospathidium muscorum muscorum (DRAGESCO & DRAGESCO-KERNÉIS, 1979) FOISSNER, 1984 **nov. stat.** (Fig. 63a–t, 64a–q, 65a–m, 118c, d, 119a–q; Tables 18, 19)

1979 *Spathidium muscorum* DRAGESCO & DRAGESCO-KERNÉIS, Acta Protozool., 18:408.

1981 *Spathidium muscorum* DRAGESCO & DRAGESCO-KERNÉIS, 1979 – FOISSNER, Zool. Jb. Syst., 108: 279 (redescription).

1983 *Spathidium muscorum* DRAGESCO & DRAGESCO-KERNÉIS, 1979 – BERGER, FOISSNER & ADAM, J. Protozool., 30: 532 (brief redescription and ontogenesis).

1984 *Arcuospathidium muscorum* (DRAGESCO & DRAGESCO-KERNÉIS, 1979) nov. comb. – FOISSNER, Stapfia, 12: 74.

1998 *Arcuospathidium muscorum* (DRAGESCO & DRAGESCO-KERNÉIS, 1979) FOISSNER, 1984 – FOISSNER, Europ. J. Protistol., 34: 199 (biogeography).

2000 *Arcuospathidium muscorum* (DRAGESCO & DRAGESCO-KERNÉIS, 1979) FOISSNER, 1984 – FOISSNER, Stud. Neotrop. Fauna & Environm., 35: 60 (description of a Venezuelan population).

Type material: Not mentioned in the original description; probably, it will be deposited at the Oberösterreichische Landesmuseum in Linz, Upper Austria. Voucher slides with protargol-impregnated specimens from all Austrian populations and the Venezuelan population are deposited at the locality mentioned before.

Diagnosis: Extrusomes rod-shaped to very narrowly obcuneate. Usually 14–18 ciliary rows.

Type locality: Not mentioned in the original description. On request, Prof. Dr. JEAN DRAGESCO informed us that he discovered *A. muscorum* in moss from the University garden of Cotonou, Benin, E2°20' N6°15'.

Description: All known and some new data are put together because morphological conspecificity of the populations is beyond reasonable doubt. As usual, cultivated specimens tend to be larger.

Table 18 Morphometric data on three populations of *Arcuospathidium muscorum muscorum*. AF – Austrian Central Alps, Guttal grassland, ~1900m a.s.l (from FOISSNER 1981a); AB – Austrian Central Alps, Gastein grassland, ~2000m a.s.l. (from BERGER et al. 1983); VE – Venezuela, Cordillera de Mérida, ~ 4200m a.s.l. (from FOISSNER 2000a).

Characteristics[1]	Pop	$\bar{x}$	M	SD	SE	CV	Min	Max	n
Body, length	AF	66.9	64.0	12.3	3.3	18.4	48.0	90.0	15
	AB	93.0	95.0	8.3	1.7	8.9	78.0	105.0	25
	VE	86.5	83.0	11.1	3.0	12.8	71.0	105.0	14
Body, width at proximal end of circumoral kinety	AF	11.9	12.0	2.0	0.5	17.1	9.0	17.0	15
	AB	27.7	27.0	5.3	1.1	19.2	18.0	37.0	25
	VE	15.1	15.0	3.9	1.0	25.8	9.0	23.0	14
Body, maximum postoral width	AF	17.4	16.0	3.9	1.0	22.5	13.0	26.0	15
	AB	40.3	40.0	5.5	1.1	13.6	29.0	50.0	25
	VE	25.4	26.0	4.0	1.1	15.7	19.0	35.0	14
Body length:width, ratio	AF	3.9	4.0	0.5	0.1	13.9	2.8	4.9	15
	AB	2.4	2.4	0.4	0.1	17.0	1.8	3.2	25
	VE	3.5	3.4	0.6	0.2	16.1	2.3	4.4	14
Oral bulge (cord of circumoral kinety), length	AF	27.7	25.0	5.6	1.4	20.1	21.0	40.0	15
	AB	42.7	46.0	12.9	2.6	30.2	31.0	55.0	25
	VE	30.9	32.0	4.1	1.1	13.3	23.0	38.0	14
Circumoral kinety to last dikinetid of brush row 2, distance	AF	9.5	9.0	2.2	0.6	22.6	6.0	19.0	15
	AB	12.5	13.0	1.7	0.3	13.4	9.0	16.0	25
	VE	11.4	12.0	1.7	0.5	15.2	8.0	14.0	14
Circumoral kinety to last dikinetid of brush row 3, distance	AF	6.0	6.0	1.6	0.4	26.7	4.0	9.0	15
	AB	9.2	9.0	1.4	0.3	15.0	7.0	12.0	25
	VE	7.7	8.0	1.8	0.5	23.0	5.0	11.0	14
Macronucleus figure, length[2]	AF	24.3	23.0	9.1	2.4	37.6	17.0	52.0	15
Macronucleus, width	AF	5.4	5.0	1.9	0.5	35.2	3.0	9.0	15
	AB	6.2	6.0	0.9	0.2	15.1	5.0	9.0	25
	VE	4.4	5.0	0.9	0.3	21.2	3.0	6.0	14
Micronucleus, largest diameter	AF	5.1	5.3	0.6	0.2	11.0	4.0	6.0	9
Macronucleus, number[2]	AF	1.0	1.0	0.0	0.0	0.0	1.0	1.0	15
	AB	1.0	1.0	0.0	0.0	0.0	1.0	1.0	25
	VE	1.0	1.0	0.0	0.0	0.0	1.0	1.0	14
Micronucleus, number	AF	1.0	1.0	0.0	0.0	0.0	1.0	1.0	15
	AB	1.0	1.0	0.0	0.0	0.0	1.0	1.0	25
	VE	1.0	1.0	0.0	0.0	0.0	1.0	1.0	14
Ciliary rows, number	AF	14.2	14.0	1.9	0.5	13.4	12.0	18.0	15
	AB	21.1	21.0	2.8	0.6	13.1	16.0	27.0	25
	VE	13.3	14.0	1.5	0.4	10.8	11.0	15.0	14
Dorsal brush rows, number[3]	AF	3.3	3.0	–	–	–	3.0	4.0	15
	AB	3.2	3.0	–	–	–	3.0	4.0	25
	VE	3.1	3.0	–	–	–	3.0	4.0	14

[1] Data based on mounted, protargol-impregnated (FOISSNER's method), and randomly selected specimens from non-flooded Petri dish cultures (AF, VE) and a pure culture (AB). Measurements in µm. CV – coefficient of variation in %, M – median, Max – maximum, Min – minimum, n – number of individuals investigated, SD – standard deviation, SE – standard error of arithmetic mean, $\bar{x}$ – arithmetic mean.

[2] Specimens with unusual pattern (e.g., 2 to 4 nodules, likely post-conjugates or just excysted cells) excluded.

[3] Fourth row usually right of row 1 and consisting of only few dikinetids (Fig. 64o).

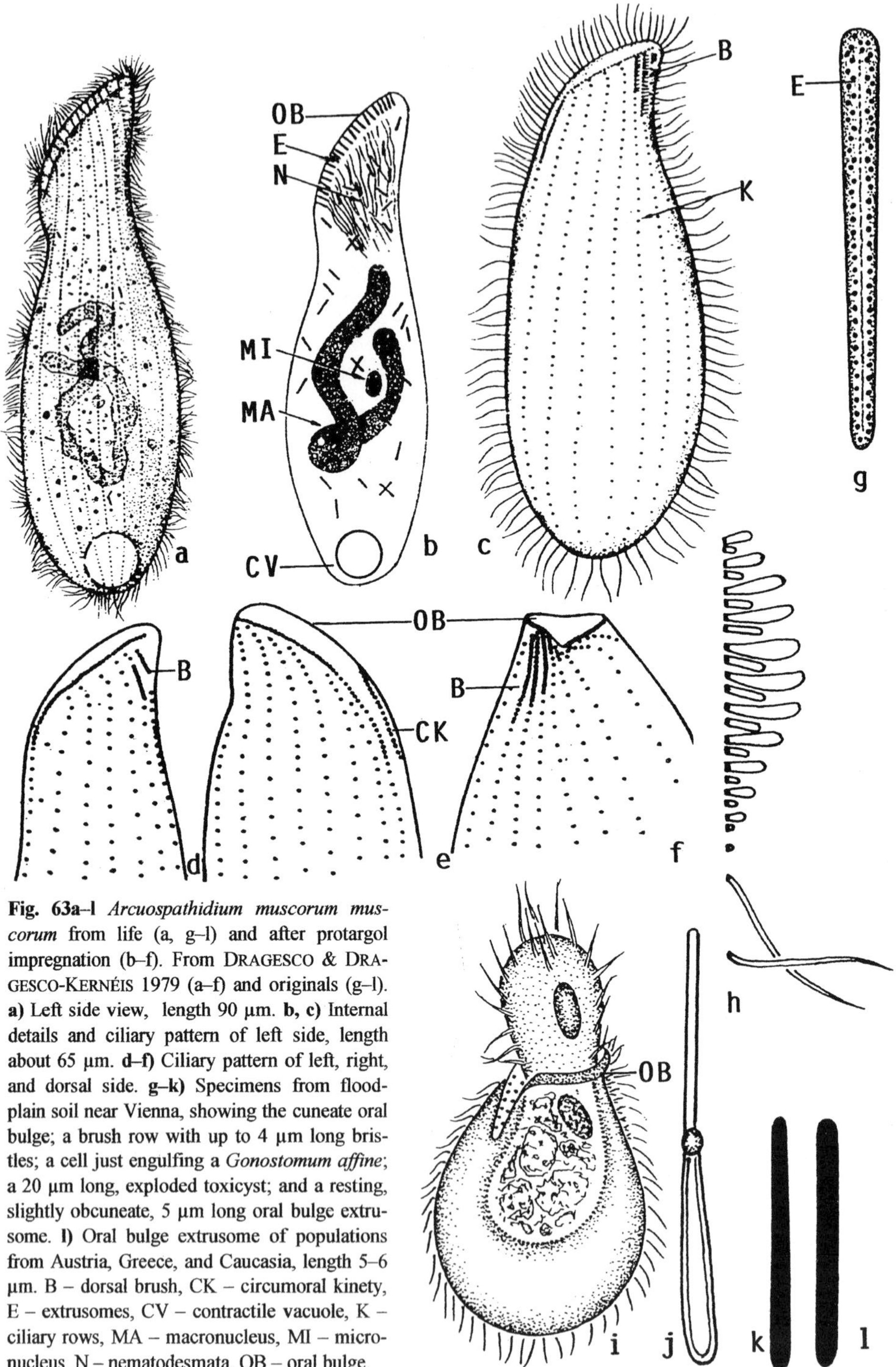

Fig. 63a–l *Arcuospathidium muscorum muscorum* from life (a, g–l) and after protargol impregnation (b–f). From Dragesco & Dragesco-Kernéis 1979 (a–f) and originals (g–l). **a)** Left side view, length 90 µm. **b, c)** Internal details and ciliary pattern of left side, length about 65 µm. **d–f)** Ciliary pattern of left, right, and dorsal side. **g–k)** Specimens from floodplain soil near Vienna, showing the cuneate oral bulge; a brush row with up to 4 µm long bristles; a cell just engulfing a *Gonostomum affine*; a 20 µm long, exploded toxicyst; and a resting, slightly obcuneate, 5 µm long oral bulge extrusome. **l)** Oral bulge extrusome of populations from Austria, Greece, and Caucasia, length 5–6 µm. B – dorsal brush, CK – circumoral kinety, E – extrusomes, CV – contractile vacuole, K – ciliary rows, MA – macronucleus, MI – micronucleus, N – nematodesmata, OB – oral bulge.

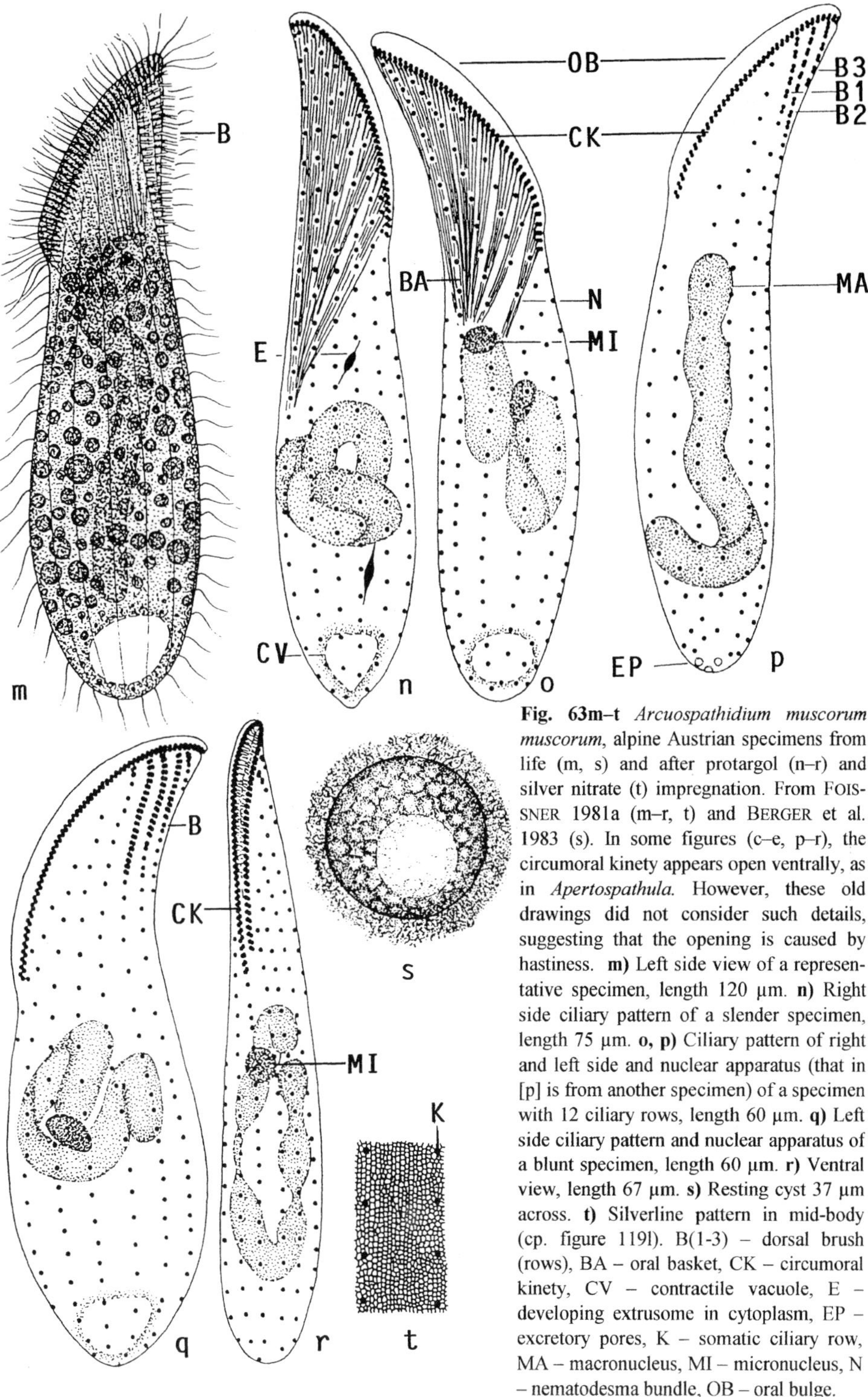

Fig. 63m–t *Arcuospathidium muscorum muscorum*, alpine Austrian specimens from life (m, s) and after protargol (n–r) and silver nitrate (t) impregnation. From FOISSNER 1981a (m–r, t) and BERGER et al. 1983 (s). In some figures (c–e, p–r), the circumoral kinety appears open ventrally, as in *Apertospathula*. However, these old drawings did not consider such details, suggesting that the opening is caused by hastiness. **m)** Left side view of a representative specimen, length 120 µm. **n)** Right side ciliary pattern of a slender specimen, length 75 µm. **o, p)** Ciliary pattern of right and left side and nuclear apparatus (that in [p] is from another specimen) of a specimen with 12 ciliary rows, length 60 µm. **q)** Left side ciliary pattern and nuclear apparatus of a blunt specimen, length 60 µm. **r)** Ventral view, length 67 µm. **s)** Resting cyst 37 µm across. **t)** Silverline pattern in mid-body (cp. figure 119l). B(1-3) – dorsal brush (rows), BA – oral basket, CK – circumoral kinety, CV – contractile vacuole, E – developing extrusome in cytoplasm, EP – excretory pores, K – somatic ciliary row, MA – macronucleus, MI – micronucleus, N – nematodesma bundle, OB – oral bulge.

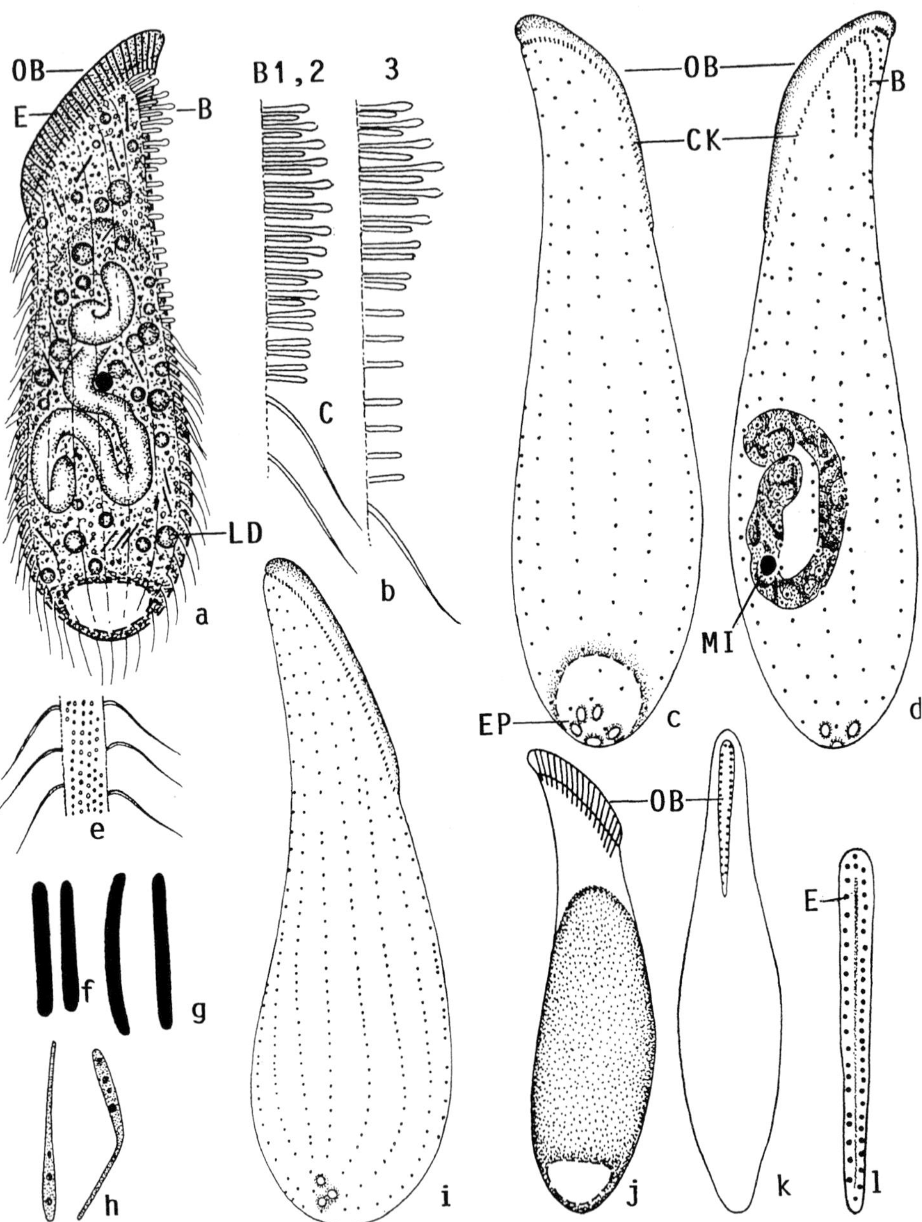

Fig. 64a–l *Arcuospathidium muscorum muscorum*, Venezuelan specimens from life (a, b, e–h, j–l) and after protargol impregnation (c, d, i). From FOISSNER (2000a). **a)** Left side view of a representative specimen, length 90 µm. **b)** Dorsal brush. Arrowhead marks monokinetidal bristle tail of row 3. **c, d, i)** Ciliary pattern of right and left side of a slender and a moderately broad specimen, length 75 µm and 85 µm. **e)** Surface view showing cortical granulation. **f, g)** Oral bulge extrusomes from two populations, length 4 µm and 5–6 µm. **h)** Incompletely (?) exploded toxicysts are 8 µm long. **j, k)** Right side and ventral view of same specimen. **l)** Frontal view of oral bulge. B(1-3) – dorsal brush (rows), C – ordinary somatic cilium, CK – circumoral kinety, E – extrusomes, EP – excretory pores, LD – lipid droplet, MI – micronucleus, OB – oral bulge.

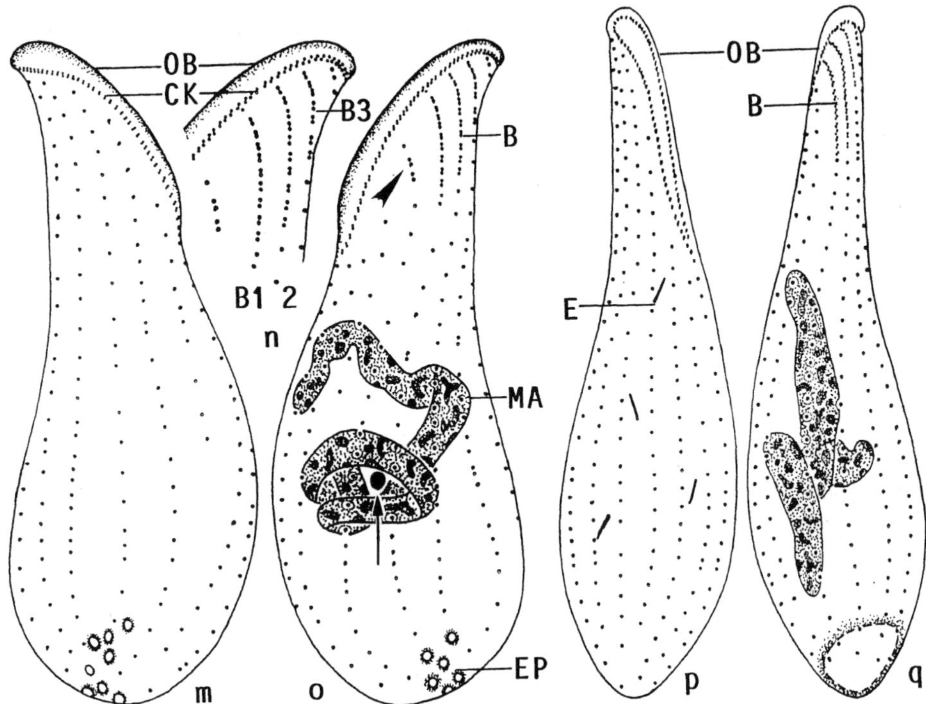

Fig. 64m–q *Arcuospathidium muscorum muscorum*, Venezuelan specimens after protargol impregnation. From FOISSNER (2000a). **m–o)** Right and left side view of ciliary pattern and nuclear apparatus of main voucher specimen, length 70 µm. Arrow denotes the micronucleus. Arrowhead marks minute, additional brush row composed of only two dikinetids. **p, q)** Ventrolateral and dorsolateral view of ciliary pattern and macronucleus, length 72 µm. The brush is dorsolaterally located. B(1-3) – dorsal brush (rows), CK – circumoral kinety, E – extrusome, EP – excretory pores of contractile vacuole, MA – macronucleus, OB – oral bulge.

Size 40–140 × 25–50 µm in vivo, usually 80–130 × 25–40 µm, depending on population and material (from field or cultures); frequently, very small (length 40–60 µm) and rather large (120–140 µm) individuals occur together, even in field. Length:width ratio 1.8–4.9:1, on average 2.4–3.9:1 both in vivo and protargol preparations; rather fragile, specimens thus tend to shrink considerably in protargol preparations (Tables 18, 19). Shape highly variable, spatulate to narrowly spatulate with anterior (oral) outline conspicuously scalpriform, that is, distinctly slanted and ventrally separated from body proper by the rather distinct neck; dorsal outline straight to slightly sigmoidal; posterior end narrowly to broadly rounded; laterally flattened up to 2:1, in oral area up to 4:1; very flexible and somewhat contractile under mild coverslip pressure (Fig. 63a–c, m–r, 64a, c, j, k, m, 65a, 118a, b, 119a–c, p). Nuclear apparatus in middle third of cell. Macronucleus usually a highly tortuous, more or less moniliform strand often forming a globular mass in mid-body; nucleoli globular, minute, numerous. Invariably a single, ellipsoidal micronucleus in variable position near macronucleus, rather large, that is, 3 to 5 µm long. Post-conjugates (?) with two large, ellipsoidal macronucleus nodules rather common in the population investigated by FOISSNER 1981a (Fig. 63a, b, m–r, 64a, d, o, q, 65a, 118a, 119c, j). Contractile vacuole in posterior end, about 10 excretory pores in and near centre of pole (Fig. 63a, m–o, 64a, c, d, o, p, 118a). Cytopyge adjacent to contractile vacuole; fecal mass about 5 µm across, composed of a slimy matrix containing some refractive inclusions.

Only one type of extrusomes in oral bulge, form a row each in right and left bulge half of Venezuelan specimens, while two rough rows are recognizable in the dorsal bulge half of Austrian cells from Müllerboden and Salzburg (Fig. 63g, 64l); impregnate strongly with silver carbonate, but not with the protargol method used, except of certain, fusiform to indistinctly cuneate developmental stages in the cytoplasm (Fig. 63n, 64p). Individual extrusomes carefully studied in four populations from Austria (FOISSNER 1981a; Müllerboden, Fig. 63k; Salzburg and Tyrol, Fig. 63l), two populations from Venezuela (Fig. 64f, g), and a population each from Greece and Caucasia (Fig. 63l); very similar in all specimens, that is, rod-shaped to indistinctly obcuneate with rounded ends, straight to slightly curved, and 4–6 × 0.3–0.5 µm in size (Fig. 63k, l, 64f, g). Thus, we believe that DRAGESCO & DRAGESCO-KERNÉIS (1979), who described the extrusomes as fine, only 2µm long rods, based the observations on protargol-impregnated specimens, where often only a certain part of the extrusomes impregnates, likely the toxin-containing portion (Fig. 63a, b). Exploded extrusomes of typical toxicyst structure, anterior half thinner and brighter than posterior, which may contain small globules, possibly toxin droplets (Fig. 63j, 64h, 119q); about 8 µm long in Venezuelan specimens, while 20–25 µm in Müllerboden cells from Austria, a remarkable difference considering the almost identical resting length (4 µm vs. 5 µm).

Cortex very flexible, rather distinctly furrowed by ciliary rows, contains five to ten rows of minute (≤ 0.3 µm), colourless granules between each two kineties (Fig. 64e). Silverline pattern well recognizable in KLEIN-FOISSNER silver nitrate preparations, finely reticular with meshes 0.3–0.5 µm across, forming 15–20 mesh lines between ciliary rows (Fig. 63t, 119l); meshes slightly enlarged in brush area. Cells usually finely granular and hyaline in oral portion, while trunk dark at low magnification (≤ x100) when crammed with food inclusions and lipid droplets up to 7 µm across (Fig. 63a, m, 64a, j). In the non-flooded Petri dish cultures, mainly feeds on medium-sized ciliates, such as *Colpoda cucullus* and *Gonostomum affine*, in pure cultures also on yeast and *Tetrahymena pyriformis* (BERGER et al. 1983). When prey touches the oral bulge, it opens in midline beginning near the dorsal bulge end, where the temporary cytostome is located. Then, the prey glides slowly into the cell, whereby it fragmentates and is collected in a large food vacuole (Fig. 63j, 119a, h, i, k, m, o). Movement without peculiarities, that is, glides rather rapidly on microscope slide and soil particles.

Cilia 8–10 µm long in vivo, arranged in an average of 13–21, usually 14–18 equidistant, meridionally extending rows each with about 20–40 cilia more narrowly spaced in middle body quarters than anteriorly and posteriorly (Tables 18, 19). Ciliary rows usually distinctly separate from circumoral kinety, only occasionally one or a few rows with some narrowly spaced cilia close to the circumoral kinety. *Arcuospathidium* ciliary pattern indistinct because anterior end of left side ciliary rows usually slightly curved ventrally in dorsal half of cell; anterior region of kineties of left ventral half often distinctly shortened and slightly curved dorsally, as typical for *Arcuospathidium* (Fig. 63a, c–e, m–r, 64a, c, d, m, o, 65a, 119a–g, p). Dorsal brush in anterior region of three, rarely four dorsolateral kineties, occupies only 11–14% of body length (Table 18), composed of dikinetids often so narrowly spaced that they cannot be distinguished individually and from monokinetidal anterior tail cilia (119f, g); thus, FOISSNER (1981a) and BERGER et al. (1983) oriented the dikinetids transversely for graphical reasons (Fig. 63c, d, f, m, p, q, 64a, d, o, q, 65a, 119e–g). Details of brush rather similar in the more carefully investigated specimens from Austria, Greece, and Venezuela (Fig. 63h, 64b, 119e–g, p; Table 18): row 1 composed of 6–10 dikinetids and slightly shorter than row 2, composed of 9–13 dikinetids; row 3 shorter than rows 1 and 2 composed of 8–12 dikinetids followed by a monokinetidal tail extending to mid-body with 2 μm long, stiff bristles and continuing posteriorly with a mixture of bristles and ordinary cilia (Fig. 119p); longest anterior bristle of pairs 2–5 μm high, posterior 2–3 μm, depending on population and specimen; anterior bristle of pairs gradually shortened to both ends of rows, producing a rather conspicuous, convex pattern; anterior bristle of pairs gradually inflated distally (Austria; Fig. 63h, 119e–g), rod-shaped (Greece), or slightly clavate (Venezuela; Fig. 64b).

Oral bulge conspicuous because (i) occupying an average of 36% to 46% of body length, respectively, about as long or slightly longer than widest trunk region, depending on population and specimen (Table 18), and (ii) scalpriform, that is, slightly to distinctly convex and conspicuously slanted with ventral end gradually merging into body proper (Fig. 63a–e, m–q, 64a, c, d, i, j, m, 65a, 119a–e, p). Bulge very narrowly cuneate in frontal view, surface ornamented by (i) a distinct furrow in midline, where the bulge opens during feeding; (ii) a comparatively inconspicuous temporary cytostome recognizable only in SEM micrographs as a whirl near dorsal end of bulge; and (iii) by an arrowhead-like pattern of ridges produced by microtubule bundles originating from the circumoral dikinetids (Fig. 63g, r, 64k, l, p, q, 118c, d, 119a, d, h, i, k, m, n, o). Circumoral kinety of same shape as oral bulge, composed of dikinetids more narrowly spaced in dorsal than ventral half. Individual circumoral kinetids associated with a metachronally beating cilium and an up to 20 μm long nematodesma. Oral basket conspicuous and rather strongly impregnated with the protargol method used, except for Venezuelan specimens (Fig. 63b–e, m–r, 64m–q, 65a, 118c, d, 119a–k, n).

Type I resting cyst studied by BERGER et al. (1983) in cultivated specimens from the Austrian Central Alps (Fig. 63s). Cysts globular with an average diameter of 36.7 μm (M 36, SD 3.0, SE 0.6, CV 8.2, Min 32, Max 43, n 25); covered by an up to 5 μm thick mucous layer containing yeast cells, bacteria, and debris. Cyst wall smooth, colourless, and only circa 1 μm thick.

Ontogenesis (Fig. 65a–m): Division was studied by BERGER et al. (1983) in a cultivated population from the Austrian Central Alps. It is very similar to that described for *A. cultriforme*, however, with some variation (Table on p. 26). No changes are recognizable in the proter daughter.

As usual, ontogenesis commences with the proliferation of basal bodies on the widest part of the cell, viz., in mid-body. The rows that bear the dorsal brush proliferate a little earlier and more strongly than the others (Fig. 65c). Soon, the opisthe dorsal brush and dikinetidal oral kinetofragments are formed (Fig. 65d, e). The macronucleus assumes a curious, ring-like shape and still has globular nucleoli (Fig. 65d–f). Excretory pores appear above the prospective division furrow.

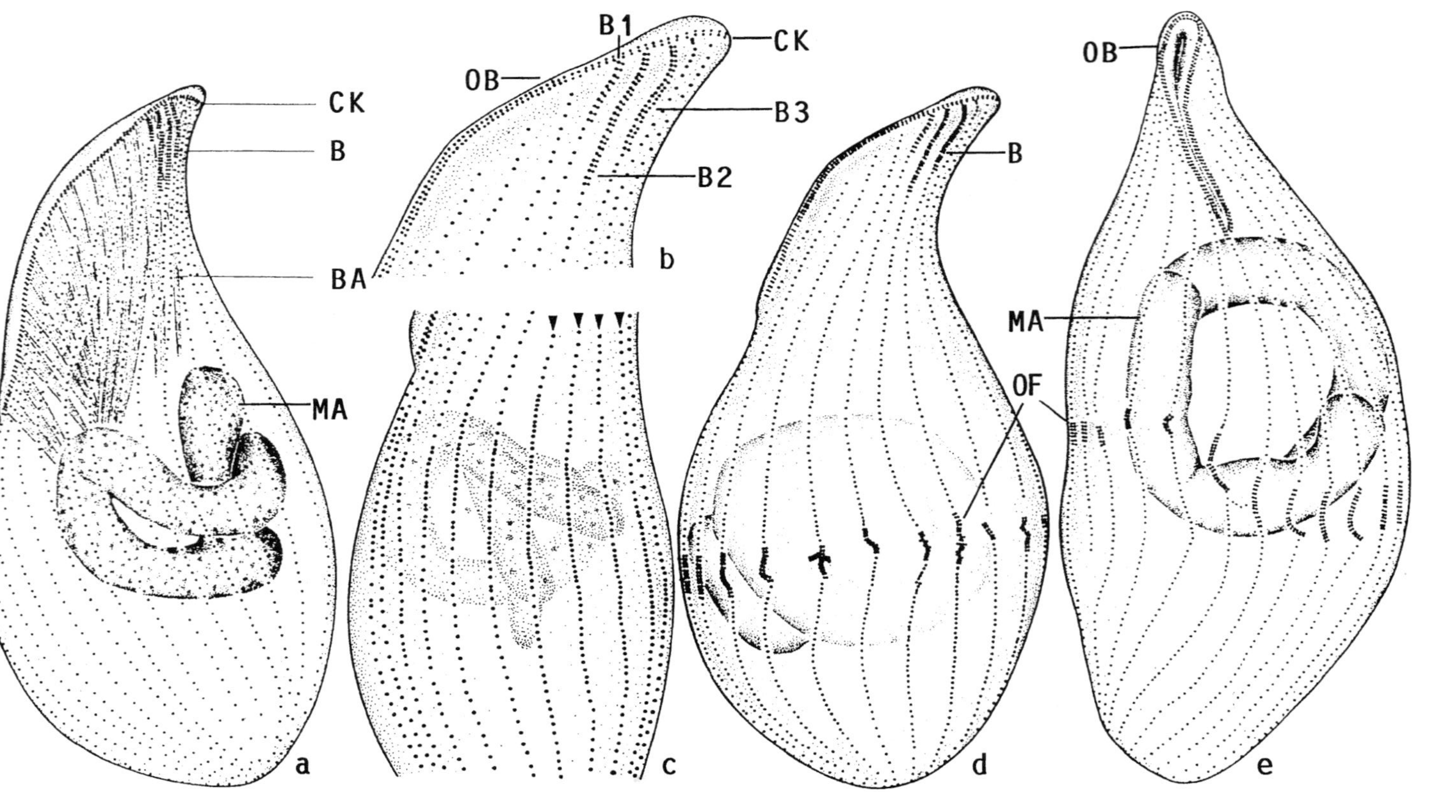

Fig. 65a–e *Arcuospathidium muscorum muscorum*, interphase (a, b) and divisional (c–e) morphology of the cultivated Austrian population studied by BERGER et al. (1983) after protargol impregnation. In these and the following figures, the longitudinally arranged brush dikinetids are transversely oriented because they are so narrowly spaced that they would appear as a thick line when drawn ordinarily. **a, b)** Left side view of a representative specimen, length 93 µm. **c)** Left side view of a very early divider with intense basal body production in mid-body. Arrowheads mark rows bearing the dorsal brush in the proter. **d, e)** Lateral and ventral view of early dividers, length 91 µm and 103 µm. B(1-3) – dorsal brush (rows), BA – oral basket, CK – circumoral kinety, MA – macronucleus, OB – oral bulge, OF – opisthe oral kinetofragments.

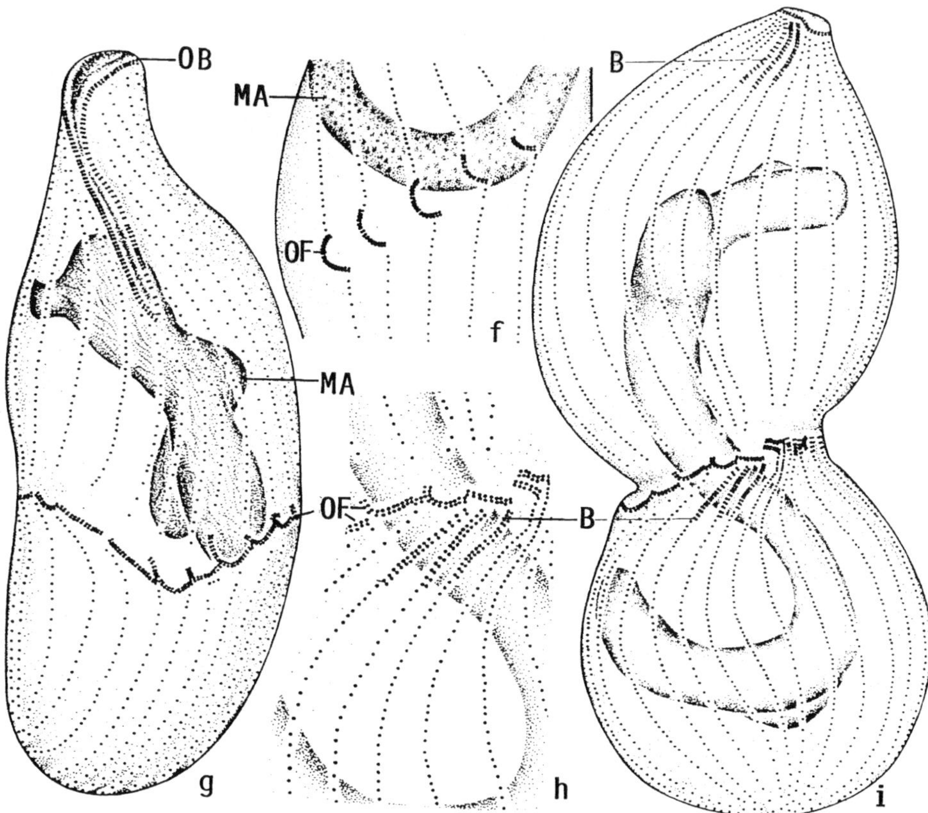

Fig. 65f–i *Arcuospathidium muscorum muscorum*, dividers of the cultivated Austrian population studied by BERGER et al. (1983) after protargol impregnation. **f)** Mid-region of an early divider showing curved opisthe oral kinetofragments, width 40 µm. **g)** Ventral view of a middle divider, length 95 µm. Note the condensing macronucleus and the opisthe's oral kinetofragments, which are now arranged transversely. **h, i)** Dorsal view of a middle to late divider with growing macronucleus, length 105 µm. The opisthe brush is fully developed and the newly formed oral kinetofragments consist of narrowly spaced dikinetids. B – dorsal brush, MA – macronucleus, OB – oral bulge, OF – opisthe oral kinetofragments.

In early middle dividers, the newly produced oral kinetofragments become concave, detach from the ciliary rows from which they were generated, and turn to the right, forming a conspicuous, slightly irregular row underneath mid-body. The individual kinetofragments are closely spaced, but still recognizable and overlap more or less distinctly. The macronucleus dissolves the nucleoli and condenses to an ellipsoidal, fibrous mass (Fig. 65f–h).

Early late dividers show a distinct division furrow, a complete opisthe dorsal brush, and a circular arrangement of the opisthe's oral kinetofragments. The macronucleus elongates to a sigmoidal rod extending in both daughters (Fig. 65h, i). In late and very late dividers, the opisthe's oral area grows ventrally, assuming a clavate shape because the daughters adhere dorsally. The new oral kinetofragments commence to fuse and arrange along the forming oral bulge, whereby they overlap distinctly, especially in the ventral mouth half, producing the curious, irregular pattern typical for *Arcuospathidium*

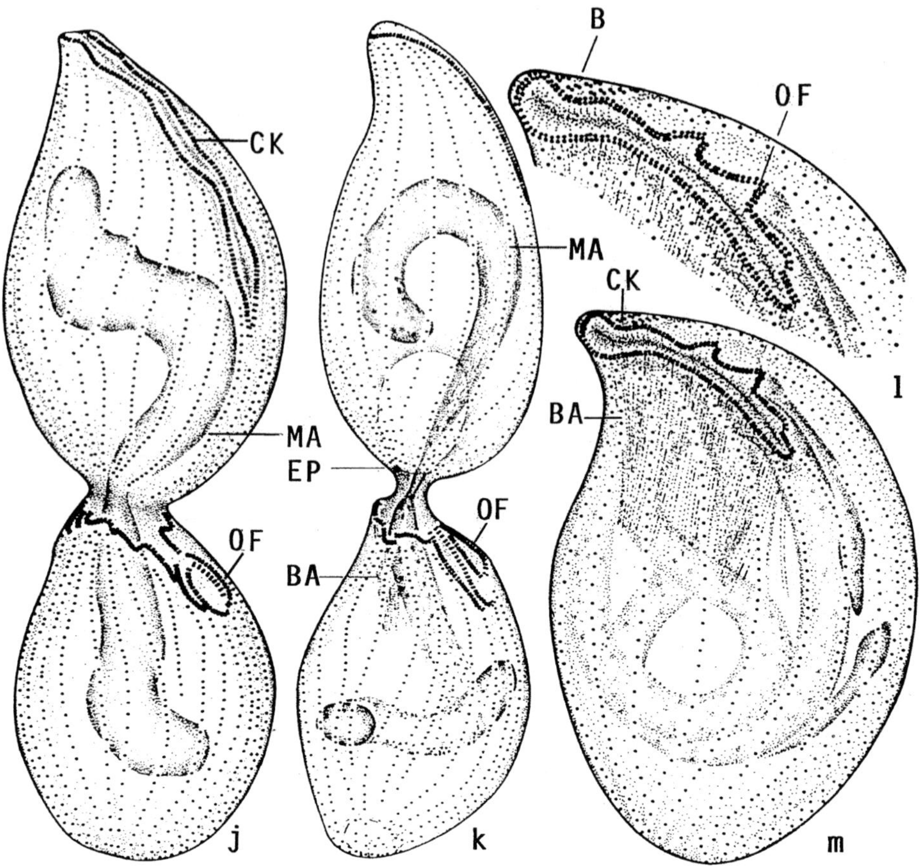

Fig. 65j–m *Arcuospathidium muscorum muscorum*, dividers of cultivated Austrian population studied by BERGER et al. (1983) after protargol impregnation. **j, k)** Ventrolateral view of ciliary pattern of a late and a very late divider, length 110 µm and 128 µm. Note the complex arrangement of the oral kinetofragments forming the opisthe's circumoral kinety. **l, m)** Ventrolateral view of ciliary pattern of a opisthe post-divider, length 55 µm. The left branch of the new circumoral kinety is still rather irregular. B – dorsal brush, BA – forming oral basket, CK – circumoral kinety, EP – newly formed excretory pores, MA – macronucleus, OF – oral kinetofragments forming the opisthe's circumoral kinety.

(Fig. 65j, k). Oral basket rods and spherical nucleoli develop just before the daughters separate (Fig. 65k). The differentiation of the opisthe's oral apparatus is completed after cytokinesis. It is remarkable that the left part of the circumoral kinety is always considerably delayed in its development. The filial products are small and broadly ellipsoidal (Fig. 65l, m).

Occurrence and ecology: *Arcuospathidium muscorum mucorum* has been recorded from most main biogeographic regions, except of Australia and Antarctica (FOISSNER 1998). However, the distribution is not uniform because this species is much more frequent in the holarctic region than in Africa and South America. This is obvious from the following data: FOISSNER (1981a) and FOISSNER & PEER (1985) recorded *A. muscorum muscorum* from 10 out of 20 sites in the Austrian Central Alps; FOISSNER et al. (1985, 2005a) observed it in 11 out of 19 sites from Lower Austria and Vienna; and FOISSNER (2000a) found it in 3 out of 20 sites in Germany; in Africa (Kenya, Namibia), *A. muscorum muscorum* occurred in only 9 out of 82 samples (FOISSNER 1999a, FOISSNER et al. 2002); it is also rare in South America (FOISSNER 1997a, 2000a, and unpubl.), and did not occur in 37 samples from Australia (BLATTERER & FOISSNER 1988, FOISSNER 1997a). Further records from Europe and Asia: Austria and Germany (FOISSNER 2000b, 2004b, FOISSNER & FOISSNER 1988b, LEHLE 1989, POHLA et al. 1994), Slovakia (TIRJAKOVÁ 1988), Italy (VERNI & ERRA 1994), Greece (Fig. 63l), Caucasia (Fig. 63l), and the Kuril Islands (ALEKPEROV 1989).

BERGER et al. (1983) could cultivate *A. muscorum muscorum* in Eau de Volvic with yeast and a species of the *Tetrahymena pyriformis* complex added as food. Likewise, DRAGESCO & DRAGESCO-KERNÉIS obtained rapidly growing cultures at temperatures around 30°C. This easy cultivation matches the broad range of terrestrial habitats the species has been found: coniferous and deciduous forest soils; high mountain and lowland soils; soil from savannahs, grasslands, and cultivated fields; crust soil in the Cen tral Namib Desert; and mud and soil from puddles, rock-pools, and trees. In spite of this, *A. muscorum muscorum* is likely not ubiquitous because it is usually found in less than half of the samples, even in Europe, where it is common. However, the factors regulating its occurrence are not known. Likely, *A. muscorum muscorum* is not confined to terrestrial environments, but occurs also in typical limnetic habitats, as suggested by the records from floodplain soil (FOISSNER et al. 2005a) and mud from road puddles and stream rock-pools (FOISSNER et al. 2002).

Remarks: *Arcuospathidium muscorum muscorum* is a widespread and rather frequent species. Thus, its late discovery is surprising; however, our efforts to synonymize *A. muscorum* with a previously described species failed. The generic allocation might be questioned because it seemingly lacks ordinary cilia at the anterior end of the dorsal brush rows, as *Cultellothrix* spp. However, scanning electron micrographs (Fig. 119e–g) show the anterior tails unequivocally, suggesting that they were overlooked in silver preparations due to the narrow spacing of the brush kinetids.

Table 19 Comparison of main features in four populations of *Arcuospathidium muscorum muscorum* and one population of *A. muscorum rhopaloplites* (for details, see Tables 18, 20)

Populations[1]	Body, length			Body, width			Oral bulge, length			Kineties, number		
	$\overline{x}$	Min	Max	$\overline{x}$	Min	Max	$\overline{x}$	Min	Max	$\overline{x}$	Min	Max
Type from Benin, Africa, n?	57	37	86	22	10	54	?	?	?	17?	16	18
Austrian population (AF, n 15)	67	48	90	18	13	26	28	21	40	14	12	18[2]
Austrian population (AB, n 25)	93	78	105	40	29	50	45	31	55	21	16	27
Venezuelan population (VE, n 14)	86	71	105	25	19	35	31	23	38	13	11	15[3]
A. muscorum rhopaloplites	91	63	108	33	24	40	30	20	40	12	11	13

[1] Data based on mounted, protargol-impregnated specimens (for details and literature, see Table 18). Max – maximum, Min – minimum, $\overline{x}$ – arithmetic mean.

[2] 15–18, on average 16 rows in six specimens from another site of this area (FOISSNER, unpubl.).

[3] Similar numbers occur in an USA population.

The diagnosis includes all four populations investigated because conspecificity is beyond reasonable doubt. As usual, the cultivated specimens are rather large (Tables 18, 19), likely showing the upper range of size and kinety number, while the non-flooded Petri dish populations from Austria and Venezuela probably represent the ordinary and lower range (Tables 18, 19). FOISSNER (2000a) supposed that the Venezuelan population could be a distinct subspecies because it has an average of only 13 ciliary rows. However, this matches the high mountain specimens from Austria (Tables 18, 19) and a population from an inland sand dune in the USA, contained in the type slides of *Spathidium paraclaviforme* (described in Vol. II) and *Protospathidium arenicola.*

***Arcuospathidium muscorum rhopaloplites* nov. sspec.** (Fig. 66a–l; Table 20)

Diagnosis: Extrusomes symmetrically to asymmetrically obclavate. About 12 ciliary rows.

Type locality: Forest soil in the surroundings of Alice Springs, that is, on a hill beside the road to the Ayers Rock, Australia, E133° S24°.

Etymology: Composite of the Greek nouns *rhopalos* (club) and *(h)oplites* (soldier), referring to the obclavate extrusomes.

Description: This subspecies is highly similar to *A. muscorum muscorum*, which has been described in detail above (Table 19). Thus, we shall provide only a brief description, emphasizing features not recognizable in the morphometrics (Table 20) and figures 66a–l.

Prepared specimens stouter (2.8:1) than live cells (4:1), likely due to inequal shrinkage. Spatulate with oblique to strongly oblique oral bulge occupying an average of 34% of body length (Table 20). Shape frequently as shown in figure 66j, that is, with straight dorsal margin and a distinct concavity at ventral end of oral bulge. Macronucleus glomerate in nearly half of specimens, about 45 μm long when spread, strand often partially flattened up to 2:1. Micronucleus attached to macronucleus, ellipsoidal, likely lacking in about half of specimens (Fig. 66a, h, k, l). Cortical granules either very small or lacking. Extrusomes studied in three populations, studded in oral bulge and scattered in cytoplasm, arranged in one to three rough rows in both bulge halves (Fig. 66b, c); asymmetrically obclavate, rarely narrowly obcuneate, about 5 × 0.6–0.8 μm in size (Fig. 66d–f); do not impregnate with the protargol method used, except for certain, about 3 μm long, narrowly cuneate cytoplasmic developmental stages (Fig. 66l). Feeds on heterotrophic flagellates and small ciliates, e.g., *Protocyclidium terricola*.

Ciliary rows widely spaced, arranged in typical *Arcuospathidium* pattern. Dorsal brush as in *A. muscorum muscorum*, anterior tails composed of only one to two cilia, rarely lacking in one or two of the three rows (Fig. 66a, g, h, I, k). Oral bulge and circumoral kinety narrowly cuneate to almost oblong, temporary cytostome near dorsal end; oral basket distinct (Fig. 66a–c, g–i, l).

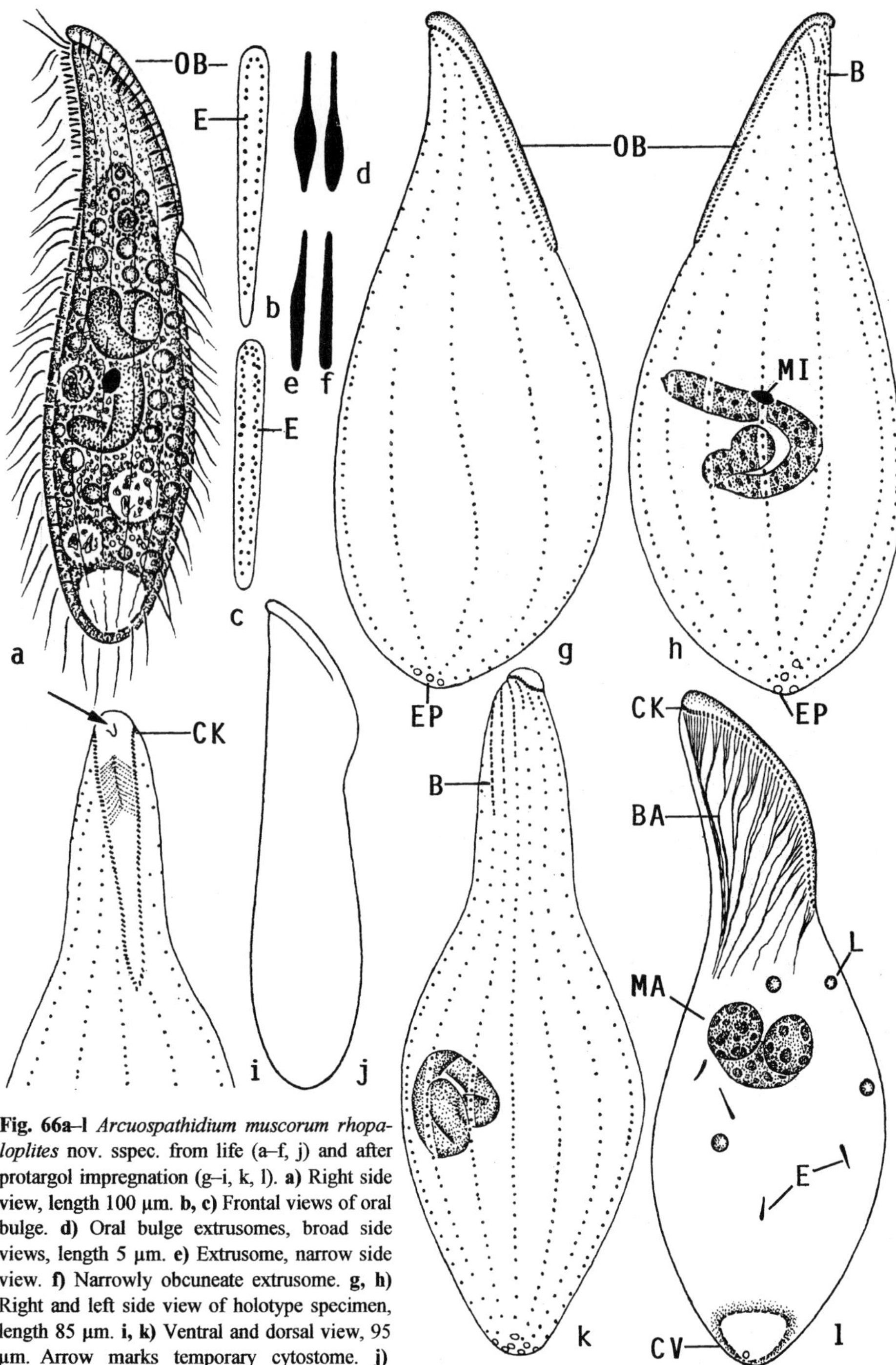

Fig. 66a–l *Arcuospathidium muscorum rhopaloplites* nov. sspec. from life (a–f, j) and after protargol impregnation (g–i, k, l). **a)** Right side view, length 100 µm. **b, c)** Frontal views of oral bulge. **d)** Oral bulge extrusomes, broad side views, length 5 µm. **e)** Extrusome, narrow side view. **f)** Narrowly obcuneate extrusome. **g, h)** Right and left side view of holotype specimen, length 85 µm. **i, k)** Ventral and dorsal view, 95 µm. Arrow marks temporary cytostome. **j)** Shape variant. **l)** Main organelles, length 85 µm. B – dorsal brush, BA – oral basket, CK – circumoral kinety, E – extrusomes, EP – excretory pores, L – lipid droplet, MA – macronucleus, MI – micronucleus, OB – oral bulge.

Table 20 Morphometric data on *Arcuospathidium muscorum rhopaloplites*

Characteristics[1]	$\overline{x}$	M	SD	SE	CV	Min	Max	n
Body, length	90.6	95.0	11.8	2.7	13.0	63.0	108.0	19
Body, maximum postoral width	32.9	34.0	3.7	0.9	11.3	24.0	40.0	19
Body length:width, ratio	2.8	2.8	0.4	0.1	14.3	2.1	3.5	19
Oral bulge, length[2]	30.4	30.0	5.1	1.2	16.7	20.0	40.0	19
Body length:oral bulge length, ratio	3.0	3.0	0.4	0.1	12.4	2.3	3.6	19
Oral bulge, width[2]	3.8	4.0	0.9	0.3	24.6	3.0	6.0	13
Oral bulge, height near dorsal end[2]	2.1	2.0	–	–	–	1.5	3.0	10
Anterior body end to macronucleus, distance	42.2	42.0	9.4	2.2	22.3	19.0	60.0	19
Circumoral kinety to end of brush row 1, distance	11.6	11.0	2.4	0.6	20.7	8.0	16.0	19
Circumoral kinety to end of brush row 2, distance	11.3	11.0	1.5	0.4	13.5	9.0	14.0	19
Circumoral kinety to end of brush row 3, distance	7.7	8.0	1.6	0.4	21.3	3.0	10.0	19
Macronucleus figure, length	22.7	22.0	7.1	1.6	31.4	14.0	36.0	19
Macronucleus, length (spread)	46.0	45.0	–	–	–	25.0	68.0	19
Macronucleus, width	4.8	5.0	0.7	0.2	14.2	4.0	6.0	19
Macronuclei, number	1.0	1.0	0.0	0.0	0.0	1.0	1.0	19
Micronucleus, length	3.6	3.5	–	–	–	3.0	4.0	7
Micronucleus, width	2.3	2.5	–	–	–	2.0	2.5	7
Micronuclei, number[3]	1.0	1.0	0.0	0.0	0.0	1.0	1.0	7
Ciliary rows, total number	12.1	12.0	0.5	0.1	3.8	11.0	13.0	19
Dorsal brush rows, number	3.0	3.0	0.0	0.0	0.0	3.0	3.0	19
Kinetids in a right side ciliary row, number	44.4	45.0	8.4	1.9	18.9	30.0	64.0	19
Dikinetids in brush row 1	12.8	13.0	2.2	0.5	16.8	9.0	16.0	19
Dikinetids in brush row 2	12.5	12.0	1.1	0.3	8.6	11.0	14.0	19
Dikinetids in brush row 3	8.5	9.0	1.7	0.4	20.6	5.0	12.0	19
Excretory pores, number	8.0	8.0	1.5	0.4	19.1	6.0	11.0	19

[1] Data based on mounted, protargol-impregnated (FOISSNER's method), and randomly selected specimens from a non-flooded Petri dish culture. Measurements in µm. CV – coefficient of variation in %, M – median, Max – maximum, Min – minimum, n – number of individuals investigated, SD – standard deviation, SE – standard error of arithmetic mean, $\overline{x}$ – arithmetic mean.

[2] Measured as distances from circumoral kinety.

[3] Possibly lacking in many specimens.

Occurrence and ecology: As yet found only in Australia, where *A. muscorum rhopaloplites* is as common as *A. muscorum muscorum* in other biogeographic regions. Whether or not the latter occurs in Australia at all, needs to be investigated.

The *Arcuospathidium vlassaki* group

The three species collected in this group look very similar at first glance because they are slender and large (~ 150–200 μm) and have a comparatively short (1 to 2 times body width) oral bulge with thick extrusomes. However, on more detailed investigation, considerable differences become obvious, suggesting species status for all populations (Table 21).

The generic classification of the three *Arcuospathidium* species is difficult because they are so loosely ciliated on the left side that the ciliary pattern cannot be determined unequivocally. The overall similarity with *S. patinum* and *S. halophilum* indicates a close relationship with *Spathidium*.

Tabelle 21 Comparison of several similar *Arcuospathidium* and *Spathidium* species

Characteristics[1]	*A. vlassaki*	*A. pachyoplites*	*A. virungense*	*S. patinum*[2]	*S. halophilum*[2]
Length:width, ratio ($\overline{x}$)[1]	9.9:1	9.0:1	10.8:1	13.7	10.3
Macronucleus[1]	rod-shaped	tortuous strand	scattered nodules	moniliform	moniliform
Micronuclei[1]	fusiform	fusiform	fusiform	fusiform	ellipsoidal
Oral bulge length:body width, ratio[1]	1.4:1	1.9:1	2.1:1	2.1:1	1.4:1
Extrusomes attached with broad end?	yes	no	no	no	yes
Extrusomes form 1 or 2 rows in oral bulge	1	2	1 (2)	1	2
Dikinetids very narrowly spaced in brush row 2[1]	yes	yes	yes	yes	no
Length of brush row 2:number of dikinetids, ratio[1]	~1:1	~1:1	~1.1:1	~1:1	~1.3:1

[1] Based upon protargol-impregnated specimens; others in vivo.
[2] Described in volume II.

Arcuospathidium vlassaki FOISSNER, 2000 (Fig. 67a–r, 120a–y, 121a–c, 122a, m; Tables 21, 22)

2000 *Arcuospathidium vlassaki* FOISSNER, Biol. Fertil. Soils, 30: 470 (Type slides with protargol-impregnated specimens from type locality are deposited in the Oberösterreichische Landesmuseum in Linz, Upper Austria.).

Diagnosis (of type population): Size about 200 × 18 μm in vivo. Cylindroidal with inconspicuous, strongly oblique, narrowly cuneate oral bulge approximately 1.4 times longer than widest trunk region. Macronucleus rod-shaped, 83 μm long on average. Micronuclei numerous and bluntly fusiform. Extrusomes about 5 × 1 μm in size, very narrowly obovate, that is, attached with broad end to only left half of oral bulge. On average 11 ciliary rows, 3 anteriorly differentiated to heterostichad dorsal brush occupying 15% of body length. Brush bristles up to 5 μm long: row 1 composed of an average of 16 dikinetids, row 2 of 24, and row 3 of 17 dikinetids followed by a monokinetidal bristle tail extending to mid-body.

Type locality: Highly saline soil from the margin (*Suaeda articulata* zone) of the Etosha Pan, Namibia, E16° S19°.

Dedication: Named in honor of Prof. Dr. K. VLASSAK (Katholieke Universiteit Leuven, Belgium), on occasion of his 65th birthday.

Description: Size 100–260 × 15–25 µm, usually about 200 × 18 µm in vivo; length:width ratio also very variable, viz., 5–19:1, on average near 10:1 both in vivo and protargol preparations (Table 22). Saudi Arabian specimens slightly larger and also highly variable, viz., 190–300 × 15–30 µm, usually near 250 × 20 µm in vivo (n 9). Narrowly spatulate to vermiform, flattened only in oblique oral region, usually widest in mid-body, largest specimens often distinctly curved (Fig. 67a, d, i, 120a, b, j). Saudi Arabian specimens usually more distinctly cylindroidal, that is, hardly widened in mid-body, frequently worm-like curved and more or less distinctly twisted about main body axis; oral area often slightly axe-like separated from body proper, that is, with rather distinct neck usually lacking in Namibian specimens (Fig. 120j, k, l, t, u). Macronucleus in middle third of cell, rod-shaped, occasionally tortuous; contains many small and large nucleoli. Micronuclei near macronucleus and scattered in cytoplasm, conspicuous both in interphase (Fig. 67a, h, i) and dividing (Fig. 67m–o) specimens because bluntly fusiform, and thus easily confused with extrusomes. Single contractile vacuole in posterior body end. Extrusome features highly characteristic and identical in Namibian and Saudi Arabian specimens; individual toxicysts **attached with broad end** to **only left half** of oral bulge (Fig. 67a, b) and scattered in cytoplasm, in vivo very narrowly obovate, slightly curved, compact, and 5–6 × 1 µm in size (Fig. 67e, 120d–i, m); become asymmetrically lageniform before explosion in Saudi Arabian specimens (Fig. 120m–p); usually do not impregnate with protargol. Somatic and oral bulge cortex highly flexible, contain narrowly spaced rows of colourless granules ≤ 0.2 µm across (Fig. 120b, g). Cytoplasm colourless, without crystals, well-fed specimens dark at low magnification (≤ x100) due to many lipid droplets 1–5 µm across and 15 µm wide vacuoles with prey remnants, i.e., mainly crystals from hypotrichs (Fig. 67a). Feeds on various ciliates, such as *Vorticella astyliformis*, *Pseudocohnilembus binucleatus*, and hypotrichs; overfed specimens heavily deformed. Swims and creeps slowly and serpentinously.

Somatic cilia rather widely and irregularly spaced, 8–10 µm long in vivo, except for oral area where occasionally some 3–5 µm long bristles alternate with ordinary cilia (Fig. 67c); arranged in an average of 11 bipolar, equidistant rows anteriorly fairly distinctly separate from circumoral kinety and slightly curved dorsally, except for dorsal rows frequently inconspicuously curved ventrally (Fig. 67a, i, j, k, p–r, 120q, 121c). Dorsal brush at anterior end of three dorsolateral kineties, inconspicuous because occupying only 15% of body length and bristles merely up to 3 µm long (5 µm in Saudi Arabian specimens) in vivo; usually, rows have an anterior tail of one to three ordinary cilia. Brush rows 1 and 3 about as long as oral bulge each composed of an average of 16–17 dikinetids; row 2 slightly longer than oral bulge and composed of 24 densely spaced dikinetids on average; row 3 with a monokinetidal bristle tail extending to

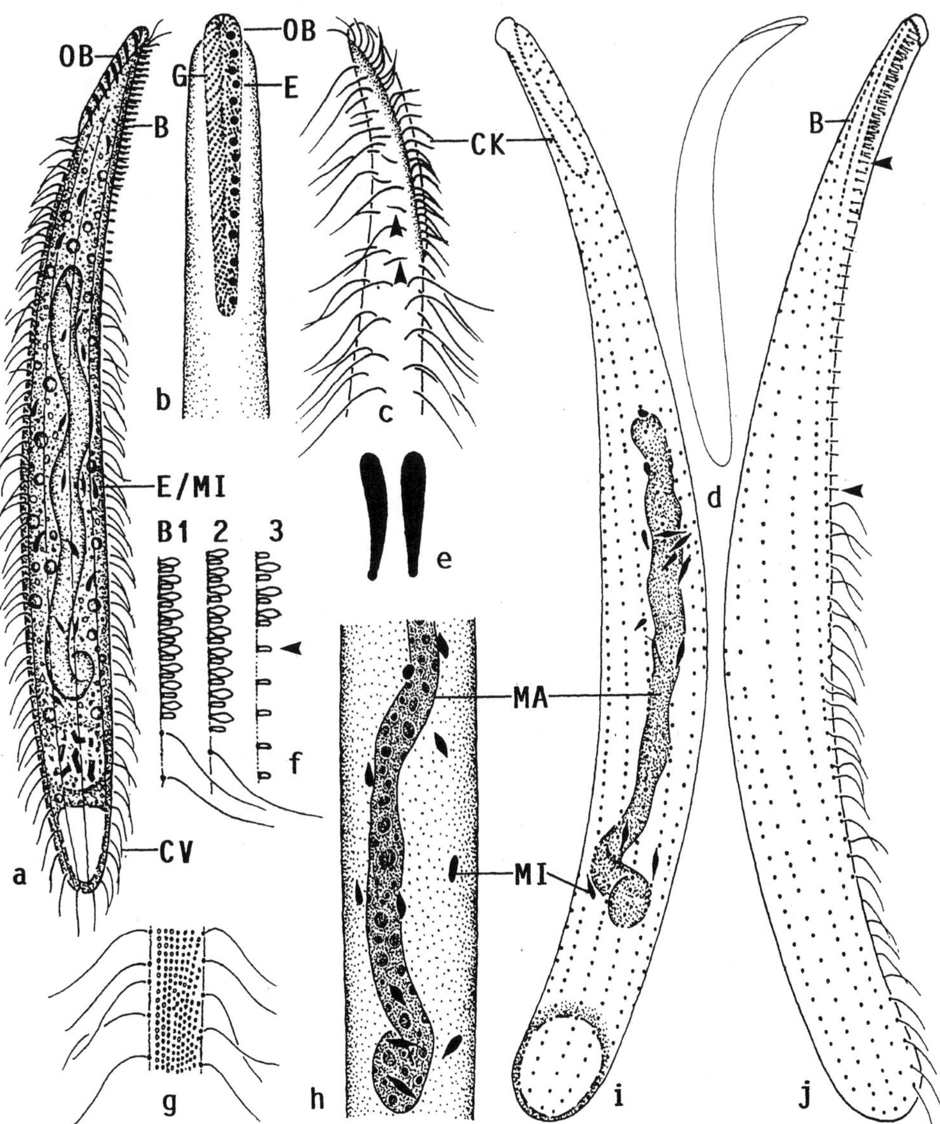

Fig. 67a–j *Arcuospathidium vlassaki* from life (a–g) and after protargol impregnation (h–j). From FOISSNER (2000c). **a)** Left side view of a representative specimen, length 200 µm. Extrusomes and micronuclei have a similar size and shape and are thus difficult to distinguish. **b)** Frontal view of oral bulge, which contains extrusomes only in the left half. **c)** Short bristles (arrowheads) irregularly alternate with ordinary cilia in the oral portion. **d)** Slender shape variant. **e)** Two views of the same extrusome, length 5 µm. **f)** Posterior region of dorsal brush; row 3 has a monokinetidal bristle tail (arrowhead) extending to mid-body (j). **g)** Surface view showing cortical granulation. **h)** Nuclear apparatus, width of cell 16 µm. Note bluntly fusiform (more or less narrowly ovate when viewed obliquely) to ellipsoidal micronuclei. **i, j)** Ciliary pattern of ventral and dorsal side and nuclear apparatus of holotype specimen, length 160 µm. Arrowheads mark bristle tail of brush row 3. CV – contractile vacuole, CK – circumoral kinety, B(1-3) – dorsal brush (rows), E – extrusomes, G – cortical granules, MA – macronucleus, MI – micronuclei, OB – oral bulge.

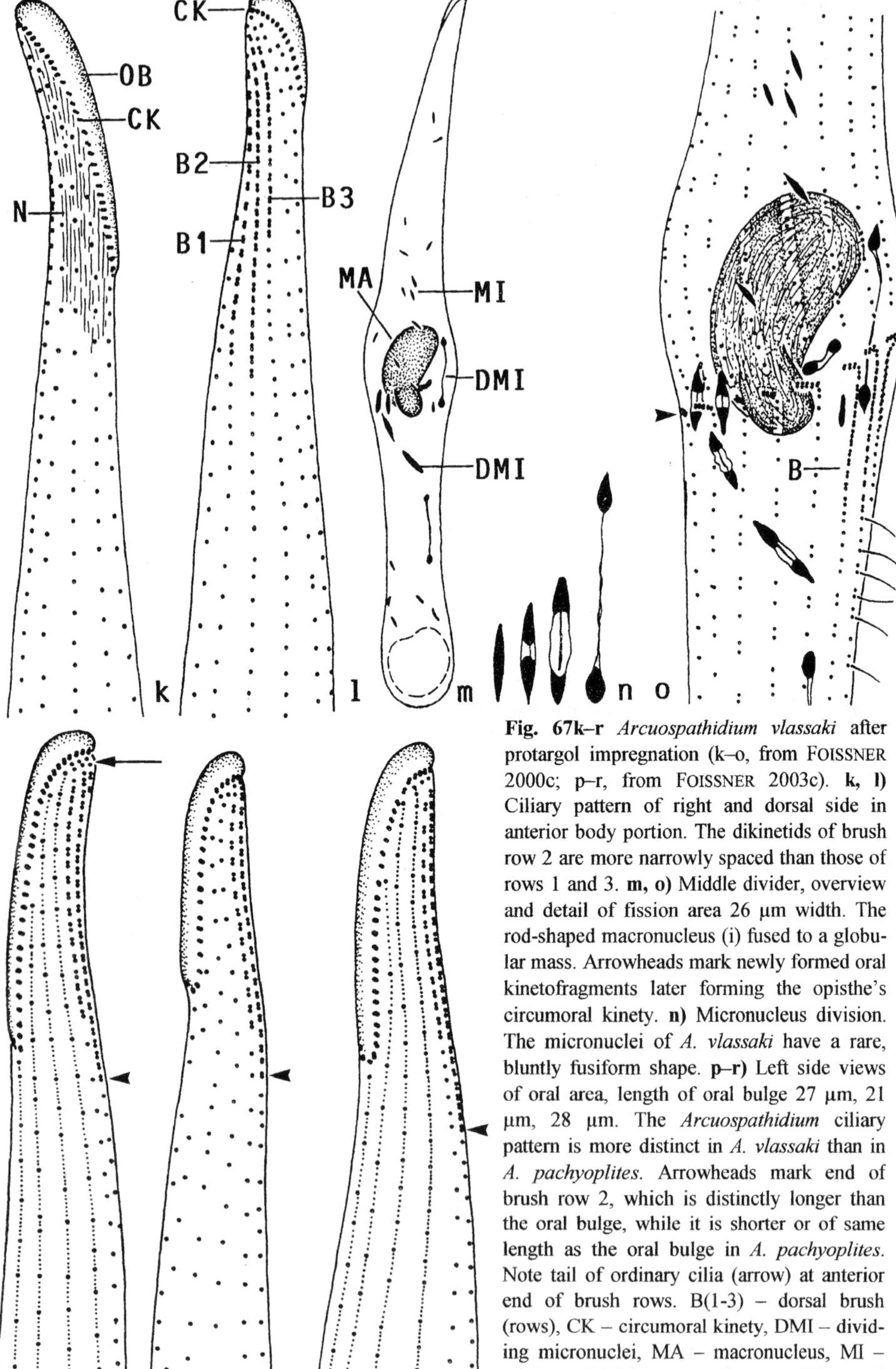

Fig. 67k–r *Arcuospathidium vlassaki* after protargol impregnation (k–o, from FOISSNER 2000c; p–r, from FOISSNER 2003c). **k, l)** Ciliary pattern of right and dorsal side in anterior body portion. The dikinetids of brush row 2 are more narrowly spaced than those of rows 1 and 3. **m, o)** Middle divider, overview and detail of fission area 26 µm width. The rod-shaped macronucleus (i) fused to a globular mass. Arrowheads mark newly formed oral kinetofragments later forming the opisthe's circumoral kinety. **n)** Micronucleus division. The micronuclei of *A. vlassaki* have a rare, bluntly fusiform shape. **p–r)** Left side views of oral area, length of oral bulge 27 µm, 21 µm, 28 µm. The *Arcuospathidium* ciliary pattern is more distinct in *A. vlassaki* than in *A. pachyoplites*. Arrowheads mark end of brush row 2, which is distinctly longer than the oral bulge, while it is shorter or of same length as the oral bulge in *A. pachyoplites*. Note tail of ordinary cilia (arrow) at anterior end of brush rows. B(1-3) – dorsal brush (rows), CK – circumoral kinety, DMI – dividing micronuclei, MA – macronucleus, MI – micro-nuclei, N – nematodesmata, OB – oral bulge.

second quarter of body (Fig. 67a, j, l, p–r, 120c, v, w, 121c; Table 22); length relations of brush rows not maintained in scanning preparations, where specimens are usually strongly shrunken (cp. Fig. 120v, w with 120x, y,121c). Anterior bristle of dikinetids clavate and 2.5–3 µm long, posterior bristle rod-shaped and about 1 µm long (Fig. 67f). Brush of Saudi Arabian specimens more complicated according to in vivo observations and scanning electron micrographs (Fig. 120v–y, 121c): anterior and posterior bristle of row 1 dikinetids slightly clavate and about 2 µm long each; dikinetids of row 2 with rod-shaped, 1.5 µm long anterior bristle and clavate, 3 µm long posterior bristle; dikinetids of row 3 with rod-shaped, 4 µm long anterior bristle and clavate, 5 µm long posterior bristle; bristle length decreases to ≤ 1 µm at posterior end of rows, where occasionally bristles and ordinary cilia irregularly alternate for a short distance. The scanning micrographs basically confirm the in vivo observations, but show that the posterior bristles are possibly shorter than the anterior ones, as in the Namibian specimens.

Oral bulge inconspicuous in vivo, protargol preparations, and scanning micrographs because narrow and only up to 3 µm high at dorsal end, containing few extrusomes only in left bulge half, and occupying merely 13% of body length (Fig. 67a, c, k, p–r, 120a, r, s); oral area of Saudi Arabian specimens more conspicuous because slightly axe-shaped and thus more clearly separated from body proper ventrally (Fig. 120k, l, q, t, u). Oral bulge on average 1.4 times (about 2 times in Saudi Arabian specimens) as long as widest trunk region, distinctly slanted by 60°–80°, flat to moderately convex in lateral view, narrowly cuneate when seen frontally; temporary cytostome neither recognizable in vivo nor protargol preparations and scanning micrographs (Fig. 67a–c, d, k, p–r, 120a–c, k, l, q–u; Table 22). Circumoral kinety narrow, not cuneate as oral bulge, but narrowly oblong to slightly dumbbell-shaped, fairly distinctly separate from ciliary rows, composed of comparatively widely spaced dikinetids each associated with an only about 4 µm (8 µm in Saudi Arabian specimens) long cilium and a very fine nematodesma contributing to the inconspicuous, lightly impregnated oral basket (Fig. 67a, i, k, p–r, 120r, s).

Occurrence and ecology: In Namibia, *A. vlassaki* occurred only at type locality, although several samples were taken in the region (FOISSNER et al. 2002). This indicates a very patchy distribution. The species could be cultivated in soil percolate by adding some cracked, sterilized wheat grains to stimulate the growth of bacterivorous ciliates (mainly *Pseudocohnilembus binucleatus*), which served as food for *A. vlassaki*. *Arcuospathidium vlassaki* and the food ciliates co-existed and divided readily in this culture. In Saudi Arabia, *A. vlassaki* also occurred in a highly saline area (120‰, pH 7.5 in water), viz., the Al Hasan oasis near the Arabian Gulf. The sample consisted of surface crusts, litter, and soil (mainly yellow-red sand) and grass roots from the upper 0–5 cm.

Remarks: Both, *A. vlassaki* and *A. pachyoplites* are very variable, as shown by the high coefficients of variation (Table 22). However, high variability of body size and related features is quite common in tiny, vermiform species, and thus must not be overinterpreted. In any case, we are quite sure not to have mixed two similar species.

This species is rather similar to *A. pachyoplites*, differing mainly in the location (only in left bulge half vs. in left and right bulge half) and attachment (with broad end vs. acute end) of the extrusomes (further details, see *A. pachyoplites*). Although these are sophisticated features recognizable only in vivo, they are highly important because found in widely separated populations.

Table 22 Morphometric data on *Arcuospathidium vlassaki* (upper line) and *Arcuospathidium pachyoplites* (lower line). From FOISSNER (2000c, 2003c)

Characteristics[1]	$\overline{x}$	M	SD	SE	CV	Min	Max	n
Body, length	162.3	160.0	43.0	9.4	26.5	94.0	260.0	21
	158.9	160.0	25.5	5.6	16.1	115.0	210.0	21
Body, width in lateral view	16.4	16.0	2.6	0.6	16.2	13.0	25.0	21
	19.3	18.0	6.1	1.3	31.6	12.0	17.5	21
Body length:width, ratio	9.9	9.5	3.1	0.7	31.8	4.7	18.6	21
	9.0	8.2	3.3	0.7	36.7	4.8	17.5	21
Oral bulge, length	21.8	21.0	5.2	1.1	24.1	13.0	37.0	21
	34.4	35.0	6.2	1.4	18.0	21.0	44.0	21
Oral bulge, width	–	–	–	–	–	–	–	–
	4.6	4.5	0.7	0.1	14.6	3.0	5.5	21
Oral bulge length:body width, ratio	1.4	1.4	0.4	0.1	32.0	0.7	2.6	21
	1.9	1.9	0.6	0.1	33.4	0.8	3.3	21
Circumoral kinety to last dikinetid of brush row 1, distance	20.8	20.0	6.3	1.4	30.4	13.0	35.0	21
	17.2	18.0	3.2	0.7	18.5	11.0	23.0	21
Circumoral kinety to last dikinetid of brush row 2, distance	25.0	25.0	5.4	1.2	21.7	16.0	40.0	21
	27.2	27.0	4.8	1.0	17.6	18.0	38.0	21
Circumoral kinety to last dikinetid of brush row 3, distance	17.7	17.0	3.7	0.8	20.9	12.0	27.0	21
	17.0	18.0	3.8	0.8	21.2	12.0	27.0	21
Macronucleus, length (spread and thus approximate)	82.9	80.0	–	–	–	40.0	160.0	21
	111.3	100.0	–	–	–	80.0	150.0	21
Macronucleus, width	5.1	5.0	0.6	0.1	11.2	4.0	6.0	21
	4.2	4.0	0.8	0.2	18.3	3.0	5.0	21
Macronucleus, number	1.0	1.0	0.0	0.0	0.0	1.0	1.0	21
	1.0	1.0	0.0	0.0	0.0	1.0	1.0	21
Micronucleus, length	3.6	4.0	0.6	0.1	16.7	3.0	5.0	21
	–	–	–	–	–	–	–	–
Micronucleus, width	1.0	1.0	–	–	–	~1	–	21
	–	–	–	–	–	–	–	–
Micronuclei, number	11.6	12.0	2.7	0.6	22.1	7.0	17.0	21
	–	–	–	–	–	–	–	–
Ciliary rows, number	11.6	11.0	1.5	0.3	12.9	9.0	15.0	21
	10.0	10.0	0.9	0.2	8.7	9.0	12.0	21
Kinetids in a right side ciliary row, number	64.7	65.0	16.2	3.5	25.1	38.0	100.0	21
	56.2	56.0	10.8	2.4	19.2	36.0	78.0	21
Dorsal brush rows, number	3.0	3.0	0.0	0.0	0.0	3.0	3.0	21
	3.0	3.0	0.0	0.0	0.0	3.0	3.0	21
Dikinetids in brush row 1, number	15.7	15.0	4.1	0.9	26.0	12.0	30.0	21
	11.4	11.0	2.2	0.5	18.9	7.0	15.0	21
Dikinetids in brush row 2, number	24.4	24.0	4.6	1.0	18.9	19.0	35.0	21
	28.0	28.0	5.7	1.2	20.4	18.0	38.0	21
Dikinetids in brush row 3, number	16.8	17.0	3.5	0.8	21.0	10.0	27.0	21
	13.3	13.0	2.0	0.4	14.7	11.0	19.0	21

[1] Data based on mounted, protargol-impregnated (FOISSNER's method), and randomly selected specimens from a pure culture (*A. vlassaki*) and a non-flooded Petri dish culture (*A. pachyoplites*). Measurements in µm. CV – coefficient of variation in %, M – median, Max – maximum, Min – minimum, n – number of individuals investigated, SD – standard deviation, SE – standard error of arithmetic mean, $\overline{x}$ – arithmetic mean.

Arcuospathidium pachyoplites FOISSNER, 2003 (Fig. 68a–p, 122a–k, n–v; Tables 21, 22)

2003 *Arcuospathidium pachyoplites* FOISSNER, Acta Protozool., 42: 146 (Type slides with protargol-impregnated specimens from type locality are deposited in the Oberösterreichische Landesmuseum in Linz, Upper Austria.).

Diagnosis: Size about 170 × 20 µm in vivo. Cylindroidal with strongly oblique, narrowly cuneate oral bulge about twice as long as widest trunk region. Macronucleus tortuous, figure formed about 80 µm long; several bluntly fusiform micronuclei. Extrusomes conspicuous because narrowly ovate and 7 × 1.4 µm in size, scattered in both halves of oral bulge and attached to bulge cortex with narrowed anterior end. On average 10 ciliary rows, 3 anteriorly differentiated to heterostichad dorsal brush occupying 17% of body length. Brush bristles up to 4 µm long: row 1 composed of an average of 11 dikinetids, row 2 of 28, and row 3 of 13 dikinetids followed by a short monokinetidal bristle tail.

Type locality: Saline, very sandy coastal soil from the surroundings of the village of Choroni, Henry Pittier National Park, north coast of Venezuela, South America, W67°45' N10°15'.

Etymology: Apposite noun composed of the Greek words *pachy* (thick) and (*h*)*oplites* (soldier ~ extrusome), referring to the conspicuous extrusomes.

Description: Size 120–230 × 15–40 µm in vivo, usually near 170 × 20 µm, as calculated from some in vivo measurements and the morphometric data (Table 22). Narrowly spatulate to vermiform with an average length:width ratio of 9:1, but "handle" much longer than "blade", that is 4.7:1, flattened only in oral region; oral bulge bluntly pointed anteriorly and rather distinctly separated from narrowed neck, strongly oblique, that is, almost in parallel with main body axis; posterior end narrowly rounded, in preparations occasionally almost globular when the contractile vacuole is filled (Fig. 68a, f, g, l, 122n, o). Macronucleus extending in posterior two thirds of body, basically a long, slightly tortuous strand with ends frequently coiled, spiralized and/or inflated; occasionally in two long pieces. Nucleoli small, globular and numerous, rarely reticulate (Fig. 68a, f–h, l, 122a, n, o). Probably, 5–10 micronuclei often not unequivocally distinguishable from extrusomes in vivo and protargol preparations; usually fusiform and about 3 × 1–1.5 µm in impregnated specimens (Fig. 68g, h). Contractile vacuole in rear body end, several excretory pores in pole area; definitely no second contractile vacuole in anterior body half. Extrusomes accumulated in both sides of oral bulge and scattered in cytoplasm, attached to oral bulge cortex with pointed anterior end; basically narrowly to very narrowly ovate, but with several modifications within and between specimens, as shown in figure 68e; 6–8 × 1–1.7 µm in size and compact, that is, rather long, thick, and highly refractive, making them very conspicuous at even low magnification (x100; Fig. 122a, c) and in silver preparations, where they appear as strongly refractive inclusions (Fig. 122u, v); mature (bulge) extrusomes never impregnate with the protargol method used, while a certain cytoplasmic developmental stage impregnates

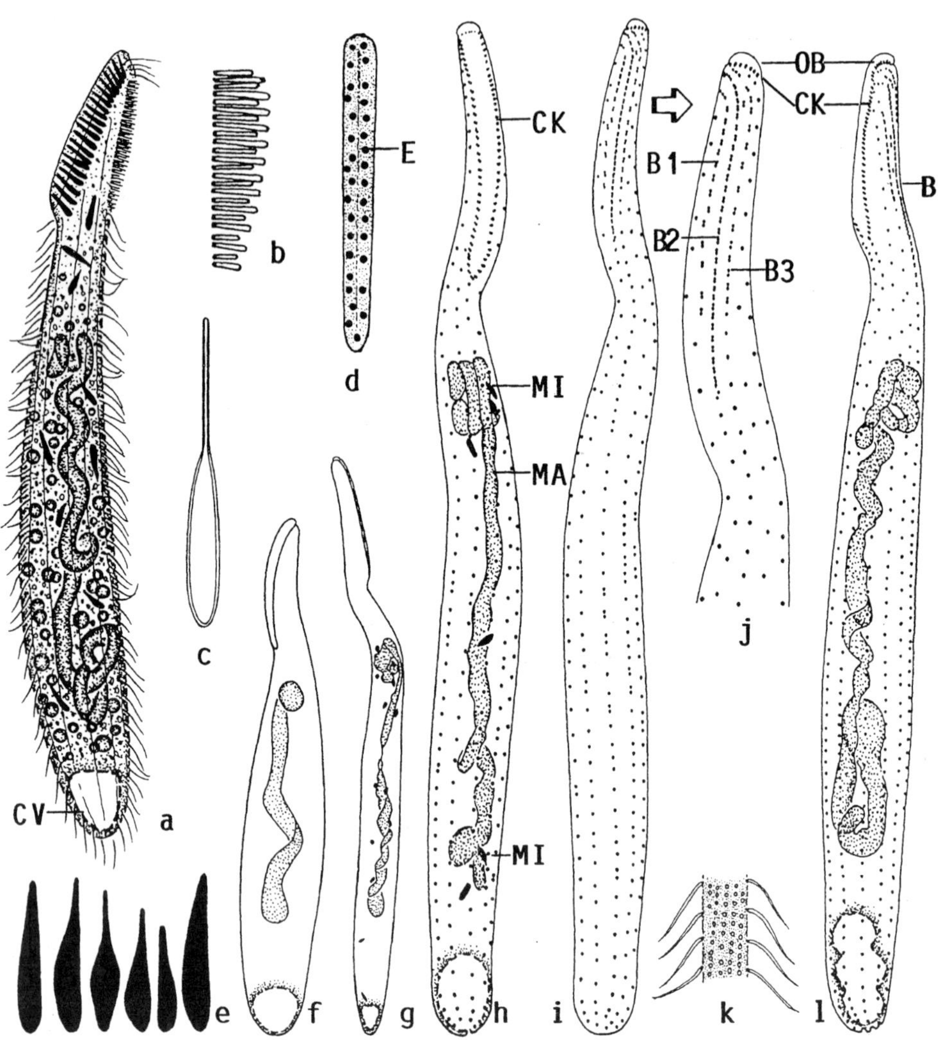

Fig. 68a–l *Arcuospathidium pachyoplites* from life (a–e, k) and after protargol impregnation (f–j, l). From FOISSNER (2003c). **a)** Left side view of a representative specimen, length 170 μm. The oral bulge is very conspicuous due to the massive extrusomes contained. **b)** Posterior end of brush row 2, longest bristles 4 μm. **c)** Exploded toxicyst, length 20 μm. **d)** Frontal view of the indistinctly cuneate oral bulge studded with extrusomes. **e)** Oral bulge extrusomes, length 6–8 μm. **f, g)** A broad and a slender specimen; note variability of macronucleus. **h–j)** Ciliary pattern of ventral and dorsal side and nuclear apparatus of holotype specimen, length 162 μm. Note the oblong circumoral kinety and the dorsal brush dikinetids, which are much more closely spaced in middle row 2 than in rows 1 and 3. **k)** Surface view showing cortical granulation. **l)** Left side view of another specimen with rather distorted dorsal brush and typical macronucleus with coiled and inflated ends, length 168 μm. B(1-3) – dorsal brush (rows), CK – circumoral kinety, CV – contractile vacuole, E – extrusomes, MA – macronucleus, MI – micronuclei, OB – oral bulge.

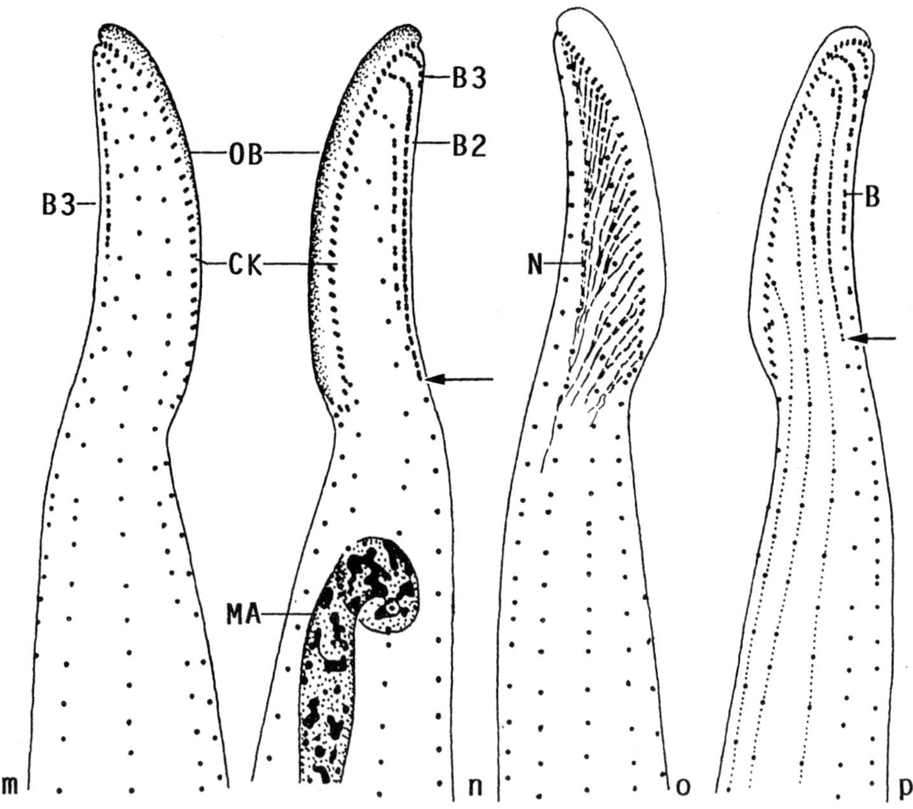

Fig. 68m–p *Arcuospathidium pachyoplites*, right and left side view of oral body portion of two specimens after protargol impregnation, length of oral bulge 34 µm, 30 µm (from FOISSNER 2003c). Dotted lines connect individual basal bodies of ciliary rows. These figures show: the rather long and very steep oral bulge; the dorsal brush dikinetids much more closely spaced in middle row 2 than in rows 1 and 3; that brush row 2 (arrows) is of about same length as the oral bulge; and, finally, the indistinct *Spathidium* pattern produced by the left side kineties, whose first basal body is very near to the circumoral kinety. B(2, 3) – dorsal brush (rows), CK – circumoral kinety, MA – macronucleus, N – nematodesmata, OB – oral bulge.

brownish, like the micronuclei (see above). Exploded extrusomes about 20 µm long and of typical toxicyst structure (Fig. 68a, c, e, 122a, c, e–k, u, v). Cortex very flexible, contains about five rows of minute granules approximately 0.2 µm across between each two ciliary rows. Cytoplasm colourless, usually with many lipid droplets 0.5–5 µm across; some specimens with a large food vacuole containing massive prey, likely ciliates. Movement conspicuously slow and serpentine, glides on microscope slide and among soil particles, showing great flexibility.

Cilia about 8 µm long in vivo, arranged in an average of 10 equidistant, meridional, ordinarily ciliated rows abutting on circumoral kinety in acute angles more distinct at right than left side, where basal bodies are widely spaced, producing an intermediate *Arcuospathidium-Spathidium* pattern (Fig. 68m–p, 122p–r, and remarks below). Dorsal brush not as stable as in many congeners, but often with small irregularities, such as minute breaks within rows, supernumerary dikinetids outside rows, or even some extra bristles forming a short fourth row (Fig. 68t); basically, however, three-rowed, heterostichad, and inconspicuous because shorter than oral bulge, occupying only 17% of body length and bristles merely up to 4 μm long, decreasing to 2 μm at end of rows; all rows with one or few ordinary cilia anteriorly and continuing as somatic kineties posteriorly; in vivo, bristles slightly inflated distally and anterior bristle of dikinetids shorter than posterior. Brush rows 1 and 3 of similar length and with comparatively widely spaced dikinetids; row 2 distinctly longer than rows 1 and 3 and with dikinetids so narrowly spaced (≤ 1 μm) that they are difficult to illustrate; row 3 with some minute, about 1 μm long, monokinetidal bristles forming a short tail extending to second body third (Fig. 68a, h–j, l–p, 122d, p, r, s; Table 22).

Oral bulge very conspicuous due to the large and highly refractive extrusomes contained and the steep slope almost in parallel with main body axis (Fig. 68a, f, g, l, m, o, 122a–c, o, p, r; Table 22); basically, however, of ordinary size and shape, that is, about twice as long as widest trunk region, moderately convex, and dorsally slightly higher than ventrally; oblong to narrowly cuneate and studded with ex-trusomes in frontal view (Fig. 68d). Circumoral kinety of same shape as oral bulge, composed of comparatively widely and frequently slightly irregularly arranged dikinetids, each associated with a cilium and a fine basket rod recognizable only in over-impregnated specimens (Fig. 68h, l–p, 122b, p, q, u, v).

Occurrence and ecology: As yet found only at type locality, that is, 10–20 m inshore the beach of the village of Choroni, Henry Pittier National Park, north coast of Venezuela. The sample consisted of very sandy coastal soil up to 10 cm depth and the mouldy top leave litter from shrubs, Cactaceae, and grasses. The rewetted mixture had 10‰ salinity and pH 6.7. The species become abundant one week after rewetting the sample. Prey is obviously digested rapidly because only few specimens with prey remnants were found in the protargol slides.

Remarks: This species does not fix well, as is often the case with saline material. Some specimens look rather distorted and/or inflated by large food inclusions and/or insufficient preservation. Furthermore, the slides contain some very small specimens (post-dividers?) and cells with a distinctly shorter oral bulge, possibly belonging to another species. All these poorly preserved and unusual specimens, roughly 10% of the population, were excluded from description and morphometry by FOISSNER (2003c).

At first glance, this species appears as a typical *Arcuospathidium* because the anterior ends of the ciliary rows are seemingly curved dorsally at both sides of the oral bulge, a diagnostic feature of the genus. However, closer investigation shows that most left side rows commence with a basal body very near to the circumoral dikinetids, giving the rows a *Spathidium*-like pattern (Fig. 68n, p). On the other hand, body shape and the steep, oblong oral bulge argue for a classification in *Arcuospathidium*, as does a comparison with the supposed nearest relative, *A. vlassaki*, where the *Arcuospathidium* ciliary pattern is more distinct (Fig. 67p–r).

Of the five species compared in Table 21, *A. pachyoplites* is most similar to *A. vlassaki*. However, a more detailed comparison reveals differences significant at species level, most related to the extrusomes emphasizing the need of thorough in vivo data: narrowly ovate with acute end attached to the oral bulge vs. very narrowly obovate attached with broad end; scattered in entire oral bulge vs. forming a single row in left bulge half; 7 × 1.4 µm and thus highly conspicuous vs. 5 × 1 µm and of ordinary appearance (Fig. 122h–m). Although there is some variation in these features, the differences are obvious, stable, and conspicuous. The second main distinguishing feature concerns the relative length of the oral bulge and dorsal brush: the longest brush row 2 is of same length or shorter than the oral bulge in *A. pachyoplites* (Fig. 68h, i, n, p, 122r), while brush row 2 is of same length or longer than the oral bulge in *A. vlassaki* (Fig. 67p–r), the average values being 34.4 µm and 27.2 µm vs. 21.8 µm and 25 µm (Table 21). Further, minor differences in sum supporting separation of the South American and African populations: dikinetids very narrowly spaced (≤ 1 µm) only in brush row 2 vs. rows 2 and 3; body width:oral bulge length ratio 0.56 vs. 0.75, that is, bulge relatively longer in *A. pachyoplites* than in *A. vlassaki*; *Arcuospathidium* ciliary pattern less distinct in the former than the latter, as proved by the reinvestigation of the African type population. Observations on other populations are needed to prove whether these inconspicuous differences are stable or within the range of natural variability of the taxa concerned.

Arcuospathidium virungense nov. spec. (Fig. 69a–m, 142e–g; Tables 21, 22)

Diagnosis: Size about 170 × 15 µm in vivo. Cylindroidal with strongly oblique, very narrowly cuneate oral bulge about twice as long as widest trunk region. On average 23 ellipsoidal macronucleus nodules and 9 bluntly fusiform micronuclei. Extrusomes conspicuous because narrowly ovate and 4–5 µm long; attached to oral bulge cortex with narrowed anterior end. On average 10 ciliary rows, 3 anteriorly differentiated to heterostichad dorsal brush occupying about 18% of body length. Brush bristles up to 7 µm long: row 1 composed of an average of 8 dikinetids, row 2 of 25, and row 3 of 9 dikinetids followed by a monokinetidal bristle tail extending to mid-body.

Type locality: Soil and moss from the rain forest in the Virunga National Park, Rwanda, equatorial Africa, E30° S2°, about 2500 m above sea level.

Etymology: *Virungense* is an adjective derived from Virunga, that is, the area the species was discovered.

Description: Size 150–200 × 12–20 µm in vivo, usually about 170 × 15 µm, as calculated from some in vivo measurements and the morphometric data (Table 23). Cylindroidal with rather distinct neck and an average length:width ratio near 11:1, flattened only in oral region; anterior (oral) end strongly oblique, posterior rounded (Fig. 69a, f, i, k). Nuclear apparatus in middle quarters of cell, composed of an average of 23 scattered, ellipsoidal macronucleus nodules and 9 conspicuously fusiform micronuclei deeply impregnated with the protargol method used (Fig. 69a, c, f, 142 e, f; Table 23). Macronucleus pattern rather variable, likely due to post-dividers with a strand-like or moniliform nucleus. Of 122 specimens analysed, 89 have scattered nodules, 26 have a moniliform strand and/or several short pieces, and 7 specimens have a tortuous strand. Early dividers have scattered nodules, showing that this is the true

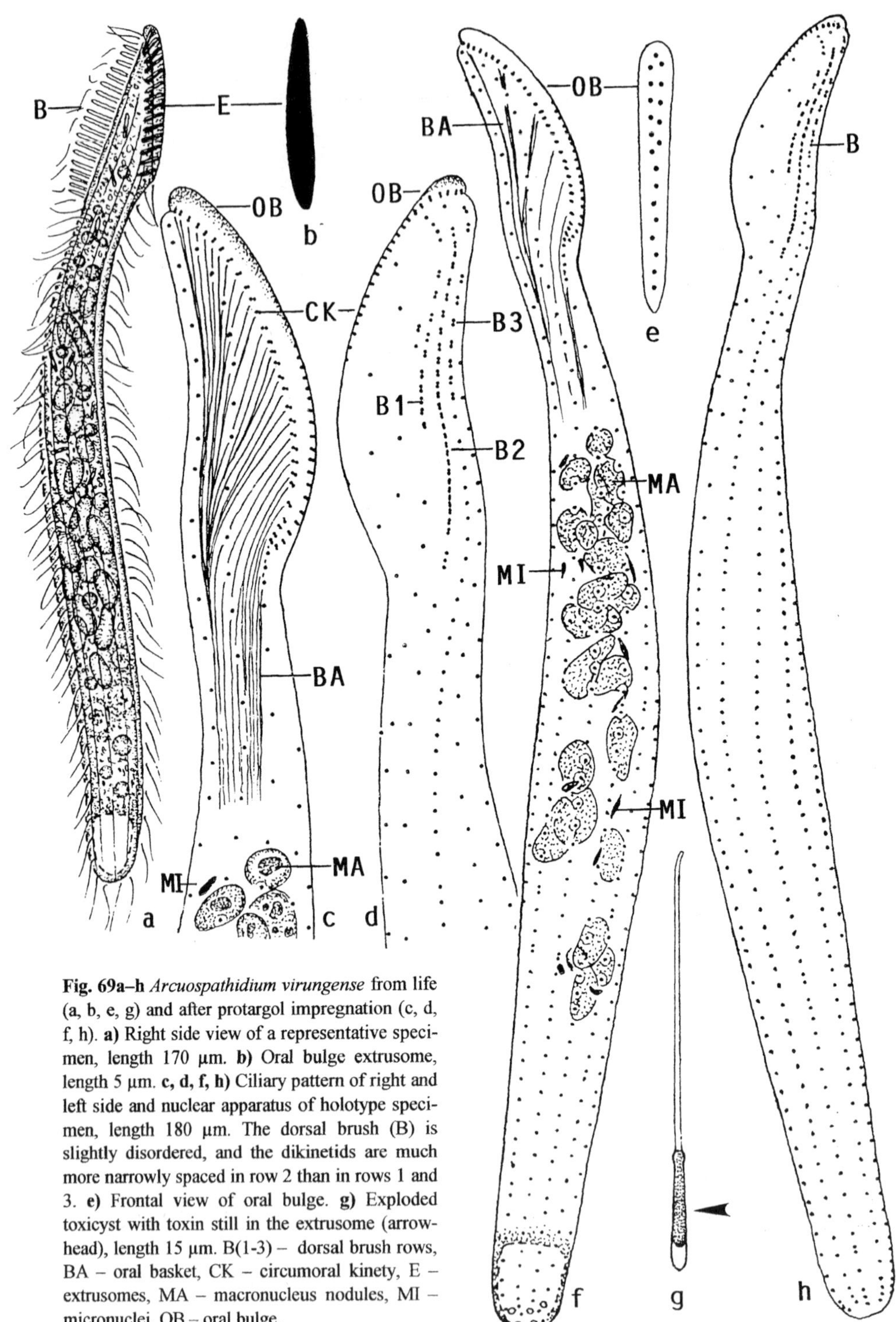

Fig. 69a–h *Arcuospathidium virungense* from life (a, b, e, g) and after protargol impregnation (c, d, f, h). **a)** Right side view of a representative specimen, length 170 µm. **b)** Oral bulge extrusome, length 5 µm. **c, d, f, h)** Ciliary pattern of right and left side and nuclear apparatus of holotype specimen, length 180 µm. The dorsal brush (B) is slightly disordered, and the dikinetids are much more narrowly spaced in row 2 than in rows 1 and 3. **e)** Frontal view of oral bulge. **g)** Exploded toxicyst with toxin still in the extrusome (arrowhead), length 15 µm. B(1-3) – dorsal brush rows, BA – oral basket, CK – circumoral kinety, E – extrusomes, MA – macronucleus nodules, MI – micronuclei, OB – oral bulge.

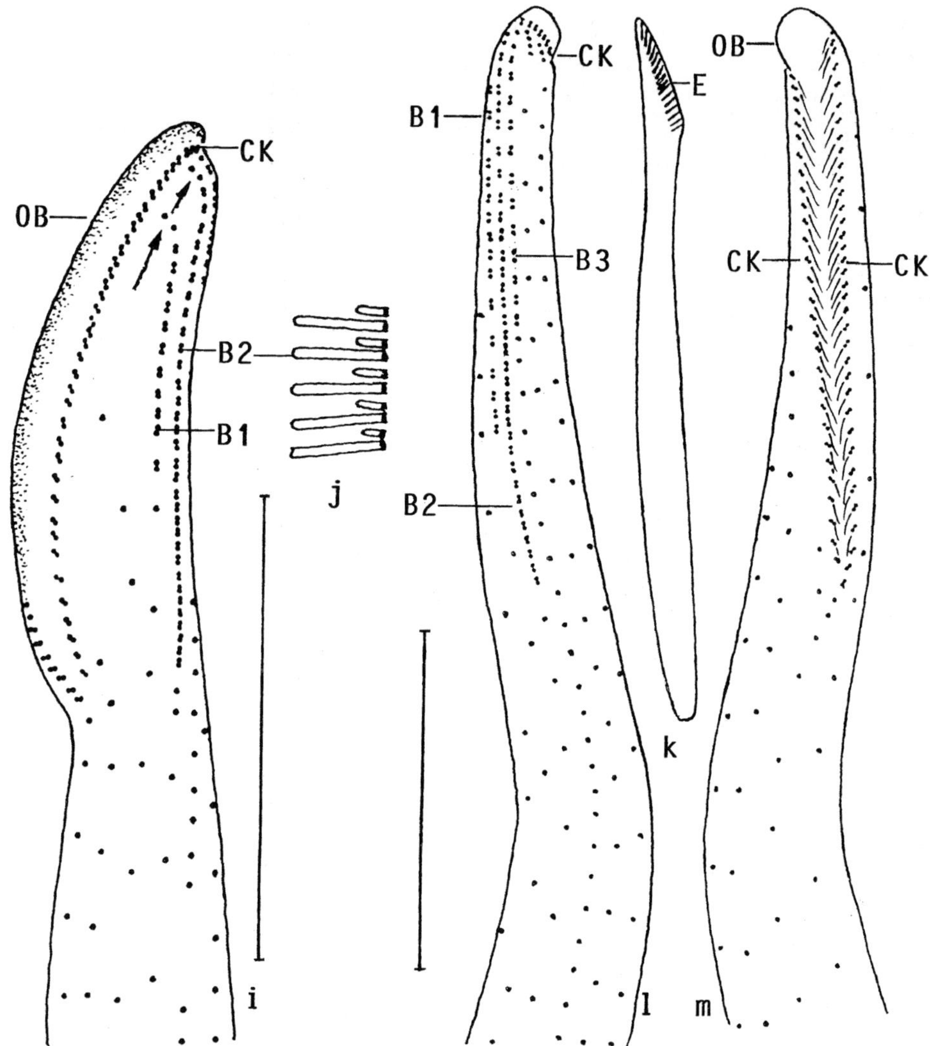

Fig. 69i–m *Arcuospathidium virungense* after protargol impregnation (i, l, m) and from life (j, k). **i, j)** The left oral body portion is so sparsely ciliated that the arcuospathidiid ciliary pattern cannot be recognized unequivocally. Arrows mark anterior tail of ordinary cilia in the dorsal brush rows. Brush row 2 has up to 7 µm long bristles (j) and the dikinetids are very narrowly spaced in the posterior half. **k)** Shape variant. **l, m)** Dorsal and ventral view of oral body portion showing the narrowly cuneate circumoral kinety and the very narrowly spaced dikinetids in the posterior half of brush row 2. Brush row 3 continues posteriorly with rather narrowly spaced, monokinetidal bristles (basal bodies). B1-3 – dorsal brush rows, CK – circumoral kinety, OB – oral bulge. Scale bars 25 µm.

interphase pattern, which is confirmed by the high percentage (73%) of this pattern in morphostatic cells. Contractile vacuole in rear body end, several excretory pores in pole area; no second contractile vacuole in anterior body half. Extrusomes scattered in cytoplasm and accumulated in oral bulge, forming two rows in broadened dorsal third and a

Table 23 Morphometric data on *Arcuospathidium virungense*

Characteristics[1]	$\overline{x}$	M	SD	SE	CV	Min	Max	n
Body, length	161.4	162.5	14.2	4.1	8.8	140.0	185.0	12
Body, width	15.3	14.5	2.6	0.8	17.0	12.0	20.0	12
Body length:width, ratio	10.8	10.5	1.6	0.5	14.8	8.5	13.3	12
Oral bulge, length	31.8	31.0	4.8	1.3	15.0	25.0	45.0	13
Oral bulge length:body width, ratio	2.1	2.0	0.4	0.1	17.6	1.5	2.8	12
Anterior body end to end of brush row 1, distance	17.8	17.0	5.1	1.4	28.7	8.0	26.0	13
Anterior body end to end of brush row 2, distance	29.6	28.0	5.1	1.4	17.1	25.0	45.0	13
Anterior body end to end of brush row 3, distance	17.8	16.0	5.3	1.5	29.7	14.0	34.0	13
Macronucleus figure, length	68.0	63.5	13.2	3.8	19.4	53.0	90.0	12
Macronucleus nodules, length	6.9	7.0	2.4	0.7	34.1	4.0	12.0	12
Macronucleus nodules, width	3.8	3.5	1.0	0.3	25.4	3.0	6.0	12
Macronucleus nodules, number	22.8	23.5	5.8	1.7	25.7	13.0	30.0	12
Micronuclei, length	3.7	4.0	0.6	0.2	16.8	3.0	5.0	12
Micronuclei, width	1.0	1.0	0.0	0.0	0.0	1.0	1.0	12
Micronuclei, number	8.8	8.5	2.3	0.7	26.1	5.0	12.0	12
Ciliary rows, number	10.3	10.0	1.2	0.3	11.2	8.0	12.0	12
Kinetids in a right side ciliary row, number	51.7	47.5	12.1	3.5	23.5	37.0	70.0	12
Dorsal brush rows, number	3.0	3.0	0.0	0.0	0.0	3.0	3.0	12
Dikinetids in brush row 1, number	7.8	8.0	2.1	0.6	26.6	5.0	12.0	13
Dikinetids in brush row 2, number	25.6	25.0	3.0	0.8	11.6	22.0	33.0	13
Dikinetids in brush row 3, number	9.3	9.0	1.8	0.5	19.8	7.0	14.0	13

[1] Data based on mounted, protargol-impregnated (FOISSNER's method), and randomly selected specimens from a non-flooded Petri dish culture. Measurements in µm. CV – coefficient of variation in %, M – median, Max – maximum, Min – minimum, n – number of individuals investigated, SD – standard deviation, SE – standard error of arithmetic mean, $\overline{x}$ – arithmetic mean.

single row in narrow ventral thirds, attached to bulge cortex with acute anterior end, impregnate faintly with protargol; narrowly ovate and slightly curved, conspicuous because strongly refractive and thick, that is, about 5 × 1 µm in vivo (Fig. 69a, b, 142g). Exploded extrusomes about 15 µm long and of typical toxicyst structure (Fig. 15g). Cortex flexible, contains several rows of minute, colourless granules between each two kineties. Cytoplasm colourless, usually with rather many lipid droplets up to 4 µm across. Food vacuoles about 5 µm across, contain bacteria. Movement not noted.

Cilia about 8 µm long in vivo, arranged in an average of 10 equidistant, meridional rows with cilia distinctly loosened in oral and neck areas; ciliary pattern thus not determinable (Fig. 69c, d, f, h, i; Table 23). Dorsal brush three-rowed, heterostichad and conspicuous, though occupying only 18% of body length, due to the up to 7 µm long bristles of row 2; individual rows with one to three ordinary cilia anteriorly and continuing as somatic kineties posteriorly, frequently with small irregularities, such as minute breaks or some extra dikinetids forming a short fourth row. Rows 1 and 3 of similar length, each composed of 8–9 ordinarily spaced dikinetids; row 3 with monokinetidal tail of 2 µm long bristles comparatively narrowly spaced and extending to mid-body. Row 2 distinctly longer than rows 1 and 3, composed of an average of 26 dikinetids very narrowly spaced in posterior half; anterior bristle of dikinetids 2 µm long, posterior bristle conspicuous because increasing in length from about 4 µm at ends of row to 7 µm in mid-row (Fig. 69a, d, h–j, l; Table 23).

Oral bulge rather conspicuous due to the steep slope and the thick, highly refractive extrusomes contained; basically, however of ordinary size and shape, that is, about twice as long as body width, moderately convex, and narrowly to very narrowly cuneate in frontal view. Circumoral kinety of similar shape as oral bulge, composed of comparatively widely spaced dikinetids, each associated with a cilium and a well impregnated nematodesma; oral basket thus distinct extending to second third of cell (Fig. 69a, c, f, h, i, k, m; Table 23).

Occurrence and ecology: As yet found only at type locality. In the non-flooded Petri dish culture, the species became rather abundant two weeks after rewetting. We thank Prof. E. STÜBER (Salzburg, Haus der Natur) for providing the sample.

Remarks: This new species is highly similar to *A. vlassaki* from saline soil of Namibia, and to *A. pachyoplites* discovered in saline coastal soil of Venezuela (Table 21). *Arcuospathidium virungense* differs from these species mainly by the macronucleus pattern: many nodules vs. a long strand. Minor differences are the length of the brush bristles (7 µm vs. ≤5 µm), the number of dikinetids in brush rows 1 and 3 (double in *A. vlassaki*), the mode of extrusome attachment (with broad end in *A. vlassaki*) and, especially, the habitat (saline vs. ordinary soil). Possibly, these three species are closely related Gondwanan relicts.

Arcuospathidium bulli FOISSNER, 2000 (Fig. 70a–n, 123g–o; Table 24)

2000 *Arcuospathidium bulli* FOISSNER, Biol. Fertil. Soils, 30: 473 (Type slides with protargol-impregnated specimens from type locality are deposited in the Oberösterreichische Landesmuseum in Linz, Upper Austria.).

Diagnosis: Size about 260 × 25 µm. Cylindroidal with conspicuous, strongly oblique oral bulge occupying approximately 27% of body length. Macronucleus very long and tortuous. Micronuclei numerous and globular. Two contractile vacuoles, one in anterior body half, the other in rear end. Extrusomes oblong, 4 × 1 µm, attached to left bulge half. On average 22 ciliary rows, 3 anteriorly differentiated to heterostichad dorsal brush occupying 25% of body length. Brush bristles up to 4 µm long: row 1 composed of an average of 48 dikinetids, row 2 of 50, and row 3 of 26 dikinetids followed by a monokinetidal bristle tail extending to second quarter of body.

Type locality: Savannah soil near the village of Gabiro, Virunga National Park, Rwanda, equatorial Africa, E30° S2°.

Dedication: Named in honour of Prof. ALAN T. BULL (Kent University, UK), chief editor of *Biodiversity and Conservation*, acknowledging his open mind on taxonomy.

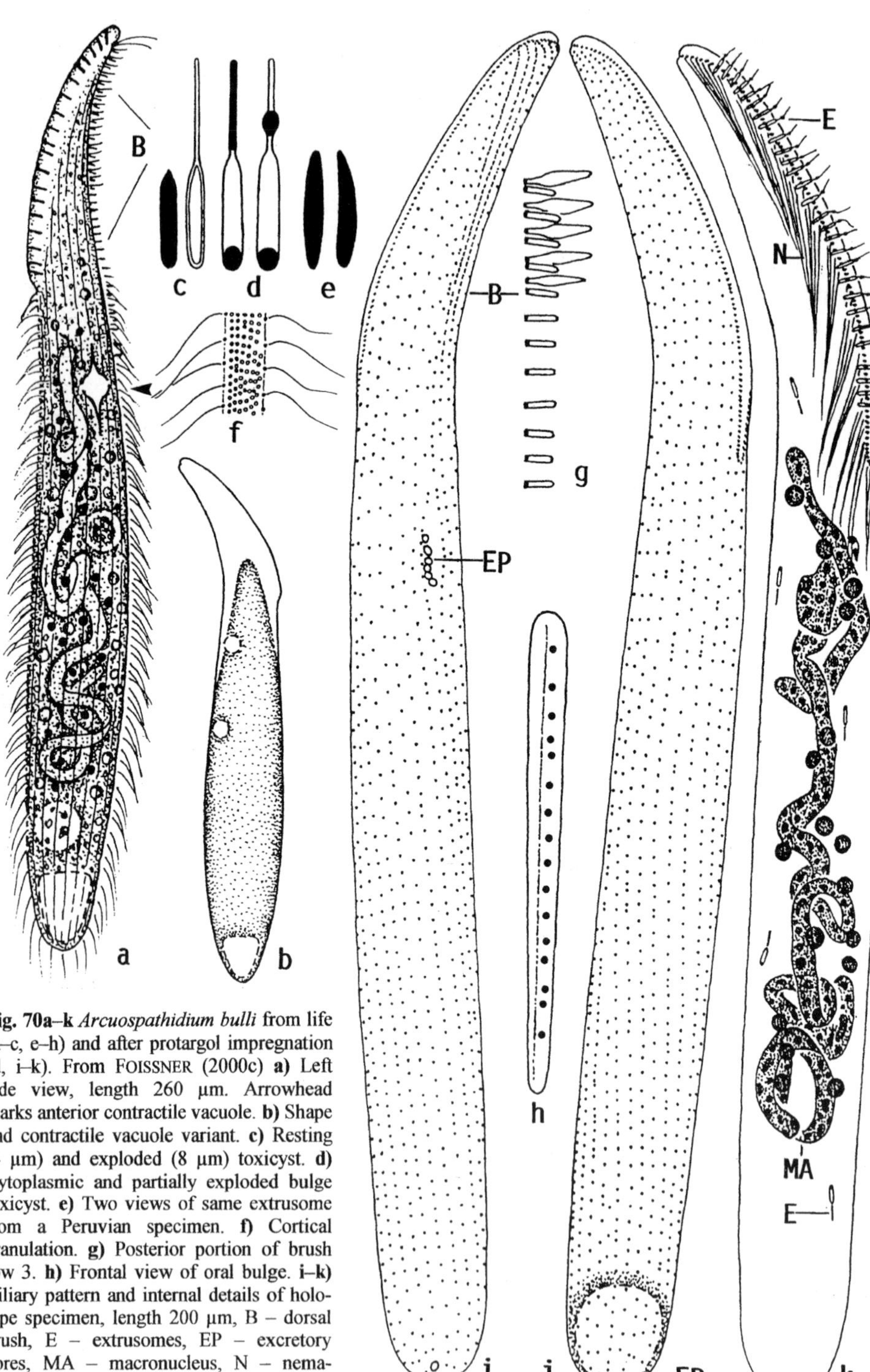

Fig. 70a–k *Arcuospathidium bulli* from life (a–c, e–h) and after protargol impregnation (d, i–k). From FOISSNER (2000c) **a)** Left side view, length 260 µm. Arrowhead marks anterior contractile vacuole. **b)** Shape and contractile vacuole variant. **c)** Resting (4 µm) and exploded (8 µm) toxicyst. **d)** Cytoplasmic and partially exploded bulge toxicyst. **e)** Two views of same extrusome from a Peruvian specimen. **f)** Cortical granulation. **g)** Posterior portion of brush row 3. **h)** Frontal view of oral bulge. **i–k)** Ciliary pattern and internal details of holotype specimen, length 200 µm, B – dorsal brush, E – extrusomes, EP – excretory pores, MA – macronucleus, N – nematodesma bundles.

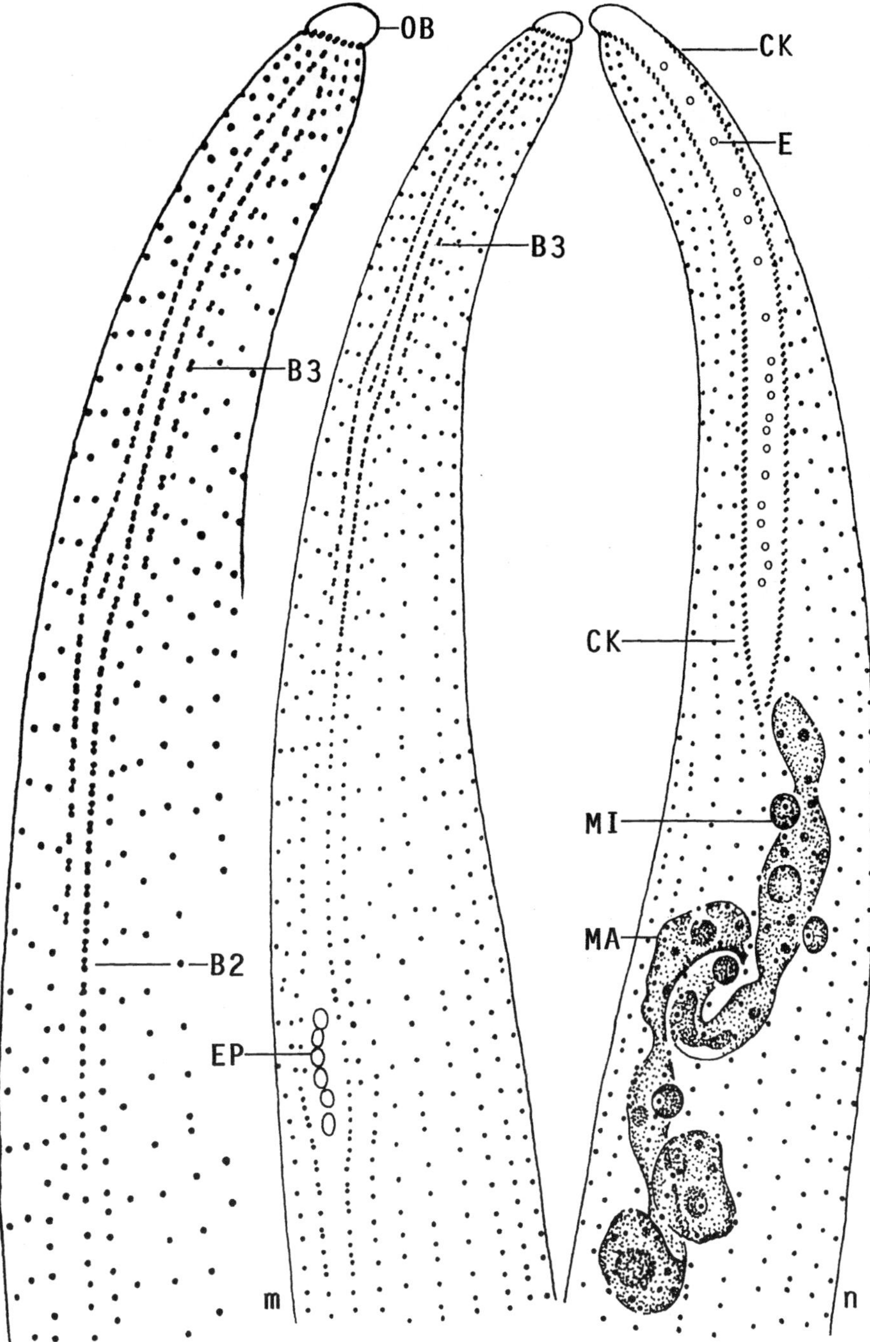

Fig. 70l–n *Arcuospathidium bulli*, ciliary pattern of anterior dorsal and ventral half after protargol impregnation, length 100 µm. From FOISSNER (2000c). B(2, 3) – dorsal brush rows, CK – circumoral kinety, E – extrusome, EP – pores of anterior contractile vacuole, MA – macronucleus, MI – micronucleus, OB – oral bulge.

Description: Size highly variable, in vivo 190–300 × 20–40 μm, usually about 260 × 25 μm, contracts up to 30% under slight coverslip pressure; length:width ratio also very variable, 7:1 – 15:1, on average 10:1. Very narrowly spatulate to rod-shaped, flattened only in obliquely truncated oral region, usually widest in mid-body (Fig. 70a, b, k; Table 24). Macronucleus in middle third of cell, long and tortuous, occasionally composed of two to three long pieces; contains many small and large nucleoli. Micronuclei near macronucleus, globular, not as compact as is usual (Fig. 70a, k). One to three, usually one contractile vacuole in anterior body half and a second in rear end, both with several excretory pores; anterior vacuoles with short collecting canals (Fig. 70a, b, i, m). Extrusomes attached with pointed end to only left half of oral bulge and scattered in cytoplasm (Fig. 70a, h, k, 123j–i); in vivo oblong with conical anterior end, about 4 × 1 μm in size, very compact and thus appearing as bright dots when viewed frontally (Fig. 70c, h, 123g–i, m); exploded toxicysts in vivo club-shaped and about 8 μm long, hyaline with double-contoured wall (Fig. 70c); cytoplasmic toxicysts with darkly impregnated anterior rod and a globule in posterior end of shaft (Fig. 70d, 123l); partially extruded bulge toxicysts with an intensively impregnated globule each in proximal half of shaft and posterior end of capsule (Fig. 70d, 123k). Somatic and oral bulge cortex very flexible, contain narrowly spaced rows of granules about 0.5 μm across (Fig. 70f). Cytoplasm colourless, without crystals, well-fed specimens dark at low magnification (≤ x100) due to many lipid droplets 1–5 μm across. Feeds mainly on rotifers and ciliates. Glides slowly among soil particles and on microscope slides.

Somatic cilia rather widely and irregularly spaced, about 10 μm long, form longitudinal rows anteriorly curved dorsally on both sides of oral bulge and distinctly separate from circumoral kinety (Fig. 70a, i, j). Dorsal brush at anterior end of three dorsal kineties, rows 1 and 2 each about 50 μm long, row 3 shorter but continuing with monokinetidal, 2 μm long bristles to second quarter of cell; brush dikinetids narrowly spaced except in row 3, anterior bristle about 4 μm long and lanceolate, posterior bristle 2 μm long and rod-shaped (Fig. 70a, g, i, l, m, 123k; Table 24).

Oral bulge distinctly oblique and slightly convex, very narrowly cuneate, conspicuous because about 70 μm long and left half filled with thick, bright extrusomes (Fig. 70a, b, k, n, 123h, j, k); occupies about 27% of body length, respectively, about three times as long as widest trunk region. Circumoral kinety at base of oral bulge, oblong with pointed ventral end, composed of comparatively widely spaced dikinetids (Fig. 70i, j, n, 123j, n, o). Nematodesmata (oral basket rods) distinct, form long, obliquely extending bundles producing conspicuous oral basket (Fig. 70k, 123j).

Occurrence and ecology: Found in Rwanda (tropical Africa, type locality) and, with the same characteristics, in a sandy, slightly saline coastal soil (pH 7.5) of Phuket Island, Thailand. A population from soil in the surroundings of Lake Titicaca (Peru, South America) possessed bluntly fusiform, slightly curved extrusomes (Fig. 70e), indicating a distinct subspecies. This is sustained by the specimens from Thailand, whose extrusomes were exactly like those from the type population. However, at the present state of knowledge, such difference appears insufficient for a distinct taxon.

Remarks: As yet, *A. bulli* is the sole member of the family having a contractile vacuole not only in rear end, as usual, but also in anterior half. This might be an indication for a distinct evolutionary lineage, that is, for a generic separation. Basically,

Table 24 Morphometric data on *Arcuospathidium bulli* (from FOISSNER 2000c)

Characteristics[1]	$\overline{x}$	M	SD	SE	CV	Min	Max	n
Body, length	223.0	200.0	47.8	14.4	21.4	156.0	295.0	11
Body, width	21.6	21.0	3.2	1.0	15.0	17.0	28.0	11
Body length:width, ratio	10.4	10.0	2.2	0.7	20.9	7.4	15.3	11
Oral bulge, length[2]	62.0	60.0	9.9	3.0	16.0	50.0	80.0	11
Body length:oral bulge length, ratio	3.6	3.6	0.5	0.1	12.8	2.9	4.4	11
Macronucleus figure, length	121.4	103.0	40.1	12.1	33.0	75.0	200.0	11
Macronucleus, width	4.0	4.0	0.9	0.3	22.4	3.0	5.0	11
Micronuclei, length	3.2	3.0	–	–	–	3.0	4.0	11
Micronucleus, width	3.1	3.0	–	–	–	–	–	11
Brush row 1, length (from circumoral kinety to last dikinetid)	47.9	49.0	9.5	2.9	19.9	35.0	60.0	11
Brush row 2, length (from circumoral kinety to last dikinetid)	55.7	54.0	11.2	3.4	20.1	42.0	80.0	11
Brush row 3, length (exclusive monokinetidal tail)	40.6	35.0	10.0	3.0	24.7	30.0	60.0	11
Ciliary rows, number	22.7	22.0	1.7	0.5	6.8	21.0	26.0	11
Basal bodies, number in a right side kinety[3]	101.7	95.0	21.9	6.6	21.5	75.0	145.0	11
Brush rows, number	3.0	3.0	0.0	0.0	0.0	3.0	3.0	11
Brush row 1, number of dikinetids	48.0	50.0	8.1	2.4	16.8	29.0	60.0	11
Brush row 2, number of dikinetids	49.6	50.0	7.2	2.2	14.6	36.0	60.0	11
Brush row 3, number of dikinetids	26.4	27.0	3.6	1.1	13.8	20.0	32.0	11
Macronucleus, number	1.0	1.0	0.0	0.0	0.0	1.0	1.0	11
Micronuclei, number	21.8	20.0	10.9	3.3	50.2	7.0	40.0	11
Anterior contractile vacuole, distance to anterior cell end	81.8	71.0	19.3	5.8	23.6	65.0	130.0	11
Anterior contractile vacuole, number of pores	5.8	5.0	1.7	0.5	28.6	3.0	8.0	11

[1] Data based on mounted, protargol-impregnated (FOISSNER's method), and randomly selected specimens from a non-flooded Petri dish culture. Measurements in µm. CV – coefficient of variation in %, M – median, Max – maximum, Min – minimum, n – number of individuals investigated, SD – standard deviation, SE – standard error of arithmetic mean, $\overline{x}$ – arithmetic mean.

[2] Measured as length of cord from distal to proximal end of circumoral kinety.

[3] Inclusive unciliated granules in line with ciliated ones.

however, *A. bulli* appears to be a member of the *Arcuospathidium cultriforme*-group, as indicated by distinct similarities in body size and shape; the macronucleus pattern; the shape and location of the extrusomes; and the long dorsal brush. Formerly, the increased number of contractile vacuoles posed problems with the generic allocation (FOISSNER 2000c). Now, with the refined diagnosis of *Supraspathidium*, the population is a clear member of *Arcuospathidium*.

Arcuospathidium bulli highly resembles members of the *A. cultriforme*-group, except for the anterior contractile vacuole(s). *Spathidium faurefremieti* is also rather similar to *A. bulli* and even has anterior contractile vacuoles (see Vol. II). However, the ciliary rows are arranged in pronounced *Spathidium* pattern. In vivo, these two species can be distinguished by the oral bulge (about 3 vs. 1.5 times as long as widest trunk region) and the extrusomes (thick and with conical anterior end vs. simple, thin rods).

The *Arcuospathidium cultriforme* group

FOISSNER et al. (2002) and XU & FOISSNER (2005b) provided convincing evidences for classifying the following taxa into a special assemblage, viz., the *Arcuospathidium cultriforme* group: *A. cultriforme cultriforme*, *A. cultriforme scalpriforme*, *A. cultriforme megastoma*, and *A. lorjeae*. These taxa agree in important details of their body organization, such as size and shape, oral bulge, nuclear pattern, and the oblong extrusomes. Likely, they developed from a *A. cultriforme*-like ancestor by either elongating the oral bulge (*A. cultriforme scalpriforme*, *A. cultriforme megastoma*) or reducing significantly body width and number of ciliary rows (*A. lorjeae*). Possibly, *A. bulli* also belongs to this group, though the second, anterior contractile vacuole indicates a specific lineage. The *Arcuospathidium cultriforme* group might even represent a distinct (sub)genus, as indicated by the special resting cysts.

Arcuospathidium cultriforme (PENARD, 1922) FOISSNER, 1984

1922 *Spathidium cultriforme* PENARD, Infusoires, p. 25.
1984 *Arcuospathidium cultriforme* (PENARD, 1922) nov. comb. – FOISSNER, Stapfia, 12: 78 (designation as type species of the genus *Arcuospathidium*).
2002 *Arcuospathidium cultriforme cultriforme* (PENARD, 1922) FOISSNER, 1984 nov. stat. – FOISSNER, AGATHA & BERGER, Denisia, 5: 299 (improved diagnosis and ranking as a subspecies).
2005 *Arcuospathidium cultriforme* (PENARD, 1922) FOISSNER, 1984 – XU & FOISSNER, Protistology, 4: 8 (improved diagnosis and monographic revision).

Diagnosis (includes three subspecies): Size 150–450 × 20–90 μm in vivo, usually near 240 × 40 μm. Narrowly to very narrowly knife-shaped with oblique to strongly oblique, very narrowly oblong to cuneate oral bulge occupying 1/3, 1/2 or 2/3 of body length. Macronucleus an indistinctly nodulated, tortuous strand, rarely occur two long pieces, many small nodules or a mixture of nodules and short strands; multimicronucleate. Extrusomes arranged in a row each in right and left half of oral bulge or scattered, basically oblong with a size of 4–8 × 0.7–1.1 μm in vivo. On average 25–37 ciliary rows, 3 anteriorly differentiated to heterostichad dorsal brush occupying 23–36%, usually about 28% of body length; brush bristles up to 5 μm long in vivo. Type VI resting cyst with thick, faceted wall.

Etymology: Not given in original description. Possibly, *cultriforme* (knife-shaped) refers to both, the body ("corps allongé en forme de couteau ...") and the exploded toxicysts ("... couteau suédois fermé ... mais le couteau s'est ouvert").

Remarks: The diagnosis is based on FOISSNER et al. (2002) and XU & FOISSNER (2005b), who distinguished three subspecies according to the length of the oral bulge and the arrangement of the extrusomes: *A. cultriforme cultriforme* (mouth about 1/3 of body length, a row of extrusomes each in right and left bulge half), *A. cultriforme scalpriforme* (mouth about 1/2 of body length, bulge extrusomes scattered), and *A. cultriforme megastoma* (mouth about 2/3 of body length, a row of extrusomes each in right and left bulge half).

XU & FOISSNER (2005b) carefully revised these taxa and added many new data. They considered neotypification as inappropriate because the original descriptions are sufficiently detailed for recognition and the status of the taxa has not yet settled. However, if there ever should be a combined morphological and gene sequence study of *A. cultriforme* and *A. scalpriforme*, these populations should be fixed as neotypes because, basically, their identity is threatened by *A. cultriforme megastoma* and populations with more or less different macronucleus pattern (XU & FOISSNER 2005b).

FOISSNER et al. (2002) distinguished *A. cultriforme cultriforme* and *A. cultriforme scalpriforme* only by the length of the oral bulge. XU & FOISSNER (2005b) added two features, viz., the arrangement (in rows vs. scattered) of the extrusomes and the brush bristles of row 3 (anterior bristle shorter than posterior in *A. cultriforme scalpriforme*, while vice versa in *A. cultriforme cultriforme*, *A. lorjeae*, and *A. bulli*; Fig. 128a, c, 129d, 134c, f). However, accurate data are available only from a few populations each. Thus, we cannot yet be entirely confident in these features, but they are good starters for falsification. The first feature is included in the diagnoses because it has been seen in at least three populations each, while brush bristle details have been reliably documented (by SEM) in only one population each.

Taken together, further studies should painstakingly document the following features: ratio of body and oral bulge length; the macronucleus pattern; the arrangement of the extrusomes (in rows or scattered); shape and size of the extrusomes; the arrangement of the brush dikinetids (straight or oblique); the number of ciliary rows; and brush row 3 (anterior or posterior bristle of pairs shortened). Certainly, gene sequence data would be very helpful.

Biogeography: As concerns *A. cultriforme cultriforme* and *A. cultriforme scalpriforme*, both appear to be cosmopolites (FOISSNER 1998). However, the data of XU & FOISSNER (2005b) cast doubts on this view because both South American populations of *A. cultriforme cultriforme* have thinner extrusomes and obliquely arranged bristle dikinetids (Fig. 71s, t, 124c, 127d–f, 129a–c), suggesting distinct (sub)species, especially if gene sequence data would be in the same line. The extrusome differences become impressive, if the micrographs are compared (Fig. 124c, h–k, 127d–f). Interestingly, the South American populations of *A. cultriforme cultriforme* have very similar extrusome shapes as the *A. cultriforme scalpriforme* populations from South Africa and Brazil (compare Fig. 71s, t and 75l, n). Likewise, the rather different averages in the number of ciliary rows (Table 25) and the different nuclear patterns suggest that the complex consists of more (sub)species that presently recognized.

Arcuospathidium cultriforme cultriforme (PENARD, 1922) FOISSNER, 1984 (Fig. 71a–z, 72a–s, 73a–w, 74a–i, 124–129; Table 25)

1922 *Spathidium cultriforme* PENARD, Infusoires, p. 25.
1930 *Spathidium cultriforme* PENARD – KAHL, Arch. Protistenk., 70: 383 (brief redescription; likely a misidentification).
1930 *Spathidium cultriforme* PENARD, 1922 – KAHL, Tierw. Dtl., 18: 164 (revision).
1975 *Spathidium cultriforme* PENARD, 1922 – FRYD-VERSAVEL, IFTODE & DRAGESCO, Prostistologica, 11: 511–514 (brief redescription from protargol-impregnated, Italian specimens).

1984 *Arcuospathidium cultriforme* (PENARD, 1922) nov. comb. – FOISSNER, Stapfia, 12: 78 (redescription from life and after protargol impregnation of Austrian specimens from Burgenland).

2002 *Arcuospathidium cultriforme cultriforme* (PENARD, 1922) FOISSNER, 1984 nov. stat. – FOISSNER, AGATHA & BERGER, Denisia, 5: 299 (improved diagnosis and ranking as a subspecies).

2005 *Arcuospathidium cultriforme cultriforme* (PENARD, 1922) FOISSNER, 1984 – XU & FOISSNER, Protistology, in press (species monograph; voucher slides of the populations mentioned in Table 25 are deposited in the Oberösterreichische Landesmuseum in Linz (LI), Upper Austria).

Diagnosis: Mouth (oral bulge) about one third of body length. Extrusomes in a row each right and left of oral bulge midline, rarely only a single row in left half.

Type locality: Wall mosses in Switzerland (Chemin de la Montagne, sur un vieux mur.).

Description: Size considerably variable within and between populations, average values ranging from 192 × 28 µm to 306 × 45 µm and variation coefficients of up to 22% (Table 25). Taking into account the in vivo measurements and 10–20% preparation shrinkage, in vivo ranges of 150–450 × 20–90 µm have been calculated, with an overall average of 240 × 40 µm, which is close to PENARD's data (200–260 × 40–50 µm). Length:width ratio highly variable within populations (CV 21–37%), while rather constant between populations, ranging from 5.7:1 to 7.2:1, with the average of 6.4:1 matching the in vivo ratio of about 6:1. Contractile by up to one third of body length, especially in oral region; contracts and extends slowly, size changes thus difficult to recognize.

Usually narrowly to very narrowly knife-shaped or lanceolate (3:1–12:1, on average about 6:1) due to the long and strongly slanted oral bulge (40°–80°, on average 65°). Oral portion more or less distinctly curved dorsally and bluntly pointed anteriorly; neck of ordinary distinctness; trunk cylindroidal, strongly inflated after ingestion of rotifers or large ciliates; posterior end rounded or bluntly pointed after systole of contractile vacuole; laterally flattened up to 2:1 in hyaline oral region, while only slightly flattened or unflattened postorally; sometimes twisted about main body axis, especially when just collected from soil habitat, that is, the non-flooded Petri dish culture (Fig. 71a, g, z, 72a–c, g, h, m, o, 73a–c, h–p, 74e, 126a, c, d, 127a, b, 131d).

Nuclear apparatus in middle quarters of cell (Fig. 71a, n, 72g, o, 73b, g, h, j–l, 124a, g, 125a, 126a). Macronucleus a long, tortuous, more or less nodulated, 4–8 µm thick strand in over 30 populations routinely identified over the years, rarely in two or three long, tortuous pieces dividing individually (see *A. cultriforme scalpriforme*); nucleoli granular, rod-like, or lobate, up to 6 µm in size. Macronucleus pattern very variable in one of the three Hawaiian populations, viz., a highly tortuous strand (occasionally in two pieces) in 58% out of 75 specimens analysed, a mixture of short strands and small nodules in 23%, and up to 100 nodules in 19% of specimens (Fig. 74a–d, 126c–e; Table 1). In a population from Rwanda, all specimens have many nodules (Fig. 74g). Both populations match ordinary populations in all other features, though the Rwandan specimens are longest on average (Table 25). Micronuclei scattered along macronucleus strand, 2–4 µm across in protargol preparations, often difficult to distinguish from similarly sized and impregnated cytoplasmic inclusions; number thus difficult to determine, likely between 10 and 20.

Table 25 Morphometric data on *Arcuospathidium cultriforme cultriforme* populations (Pop) from Hawaii (HA50, HA12; from XU & FOISSNER 2005b), Austria (AUL from Linz, according to XU & FOISSNER 2005b; AUB from Burgenland, according to FOISSNER 1984 and his original notes), Germany (GER, from SCHADE & FOISSNER, unpubl.), and Rwanda (RW, from XU & FOISSNER 2005b)

Characteristics[1]	Pop	$\overline{x}$	M	SD	SE	CV	Min	Max	n
Body, length	HA50	222.5	220.0	28.0	8.4	12.6	193.0	280.0	11
	HA12	270.3	270.0	30.9	9.3	11.4	233.0	348.0	11
	AUL	275.7	282.0	43.5	8.7	15.8	145.0	382.0	25
	AUB	191.6	180.0	33.7	8.7	17.6	145.0	245.0	15
	GER	210.0	215.0	45.8	13.2	21.8	120.0	290.0	12
	RW	305.8	300.0	48.5	16.2	15.9	250.0	400.0	9
Body, width	HA50	36.3	35.0	8.4	2.5	23.2	25.0	56.0	11
	HA12	51.5	53.0	14.5	4.4	28.2	34.0	76.0	11
	AUL	50.3	49.0	12.0	2.4	23.8	26.0	85.0	25
	AUB	27.5	28.0	4.5	1.2	16.6	20.0	34.0	15
	GER	31.2	32.0	7.8	2.1	23.7	22.0	50.0	12
	RW	44.9	47.0	11.0	3.7	24.6	27.0	62.0	9
Body length:width, ratio	HA50	6.4	6.4	1.5	0.4	23.2	4.4	9.2	11
	HA12	5.7	5.1	2.1	0.6	36.7	3.6	10.2	11
	AUL	5.8	5.4	2.1	0.4	35.8	3.2	11.9	25
	AUB	7.2	6.6	1.9	0.5	26.7	4.5	11.4	15
	RW	7.1	6.5	1.5	0.5	21.3	5.2	9.7	9
Oral bulge, length (cord)	HA50	78.7	75.0	14.6	4.4	18.6	58.0	105.0	11
	HA12	84.6	80.0	10.3	3.1	12.2	70.0	107.0	11
	AUL	93.9	95.0	17.4	3.5	18.5	56.0	138.0	25
	AUB	58.1	60.0	8.3	2.1	14.3	37.0	70.0	15
	GER	66.4	62.0	17.8	5.1	26.8	50.0	110.0	12
	RW	102.8	100.0	26.6	8.9	25.9	65.0	145.0	9
Oral bulge (circumoral kinety), width	HA50	4.0	4.0	0.8	0.4	19.8	3.0	5.0	5
	AUL	3.9	4.0	0.2	0.1	5.7	3.5	4.0	5
Oral bulge, height	HA50	2.4	2.5	0.4	0.1	15.6	2.0	3.0	11
	AUL	2.3	2.5	0.3	0.1	11.5	2.0	2.5	11
Oral bulge length:body length, ratio	HA50	0.35	0.35	–	–	10.4	0.29	0.43	11
	HA12	0.32	0.3	–	–	13.9	0.26	0.38	11
	AUL	0.34	0.34	–	–	11.3	0.28	0.43	25
	AUB	0.31	0.3	–	–	15.7	0.22	0.39	15
	GER				about 0.32				
	RW	0.34	0.31	0.1	–	19.3	0.25	0.45	9
Circumoral kinety to last dikinetid of brush row 1, distance	HA50	52.2	52.0	6.5	2.0	12.4	44.0	62.0	11
	RW	66.3	67.0	20.3	6.8	30.6	45.0	100.0	9
Circumoral kinety to last dikinetid of (the longest) brush row 2, distance	HA50	58.6	58.0	5.4	1.6	9.3	50.0	67.0	11
	AUL	68.5	68.0	9.2	2.6	13.4	51.0	86.0	13
	AUB	50.9	49.0	7.9	2.0	15.6	39.0	70.0	15
	GER	63.1	60.0	7.5	2.7	11.9	55.0	75.0	9
	RW	71.4	73.0	20.6	6.9	28.8	45.0	100.0	9
Circumoral kinety to last dikinetid of brush row 3, distance	HA50	47.7	49.0	4.2	1.3	8.8	40.0	54.0	11
	RW	51.8	52.0	13.4	4.5	25.9	35.0	70.0	9
Anterior body end to macronucleus, distance	HA50	72.8	72.0	10.1	3.0	13.9	58.0	90.0	11
Macronucleus figure, length	HA50	113.9	110.0	28.8	8.7	25.3	60.0	155.0	11
	AUL	158.7	173.0	39.9	9.1	25.1	47.0	220.0	19
	AUB	100.6	94.0	27.5	7.1	27.3	69.0	160.0	15
Macronucleus, length (spread, thus approximate)	GER	370.0	400.0	–	–	–	80.0	550.0	10
Macronucleus nodules, length	RW	10.0	9.0	2.7	0.9	26.9	7.0	14.0	9
Macronucleus, width in mid	HA50	5.6	5.0	1.8	0.5	32.0	4.0	9.0	11
	AUB	4.9	5.0	0.6	0.2	12.3	4.2	5.6	15
	GER	4.8	5.0	0.6	0.2	12.9	4.0	6.0	10
Macronucleus nodules, width	RW	4.2	4.0	0.9	0.3	22.2	3.0	6.0	9
Macronucleus, number	HA50				see footnote[2]				
	HA12	1.0	1.0	0.0	0.0	0.0	1.0	1.0	11

continued

Characteristics[1]	Pop	$\overline{x}$	M	SD	SE	CV	Min	Max	n
	AUL	1.0	1.0	0.0	0.0	0.0	1.0	1.0	25
	AUB	1.0	1.0	0.0	0.0	0.0	1.0	1.0	15
	GER[3]	1.2	1.0	–	–	–	1.0	3.0	12
Macronucleus nodules, number	RW[4]	96.0	100.0	15.2	6.8	15.8	70.0	110.0	5
Micronuclei, across	HA50	2.6	2.5	0.2	0.1	6.9	2.5	3.0	8
	AUL	2.8	3.0	0.4	0.1	15.0	2.0	3.5	19
	AUB	2.7	2.7	0.2	0.1	5.6	2.5	2.8	4
	RW	2.9	3.0	0.5	0.2	18.9	2.0	4.0	9
Micronuclei, number	HA50	15.0	13.0	–	–	–	12.0	20.0	3
	AUL	13.3	12.0	3.3	0.7	24.5	9.0	23.0	19
Ciliary rows, number	HA50	31.4	31.0	1.6	0.5	5.0	30.0	35.0	11
	AUL	36.7	36.0	3.6	0.8	9.8	30.0	44.0	23
	AUB	28.1	29.0	1.3	0.3	4.6	26.0	30.0	15
	GER	27.9	28.0	2.7	0.9	9.6	24.0	32.0	10
	RW	28.4	28.0	2.2	0.7	7.7	25.0	31.0	9
Basal bodies in a right side ciliary row, number	HA50	116.8	113.0	14.5	4.4	12.4	100.0	147.0	11
	RW	161.0	165.0	32.5	14.5	20.2	110.0	200.0	5
Basal bodies in 10 µm of a ciliary row, number	AUB	5.4	6.0	1.0	0.3	18.3	3.0	7.0	15
	GER	9.5	9.0	2.5	0.8	26.4	6.0	15.0	10
Dorsal brush rows, number	HA50	3.0	3.0	0.0	0.0	0.0	3.0	3.0	11
	HA12	3.0	3.0	0.0	0.0	0.0	3.0	3.0	11
	AUL	3.0	3.0	0.0	0.0	0.0	3.0	3.0	25
	AUB	3.0	3.0	0.0	0.0	0.0	3.0	3.0	15
	GER[3]	3.6	4.0	0.5	0.2	14.8	3.0	4.0	9
	RW	3.0	3.0	0.0	0.0	0.0	3.0	3.0	9
Dikinetids in brush row 1, number	HA50	48.7	46.0	11.4	3.4	23.4	35.0	76.0	11
	RW	64.6	61.0	15.7	5.9	24.3	44.0	91.0	7
Dikinetids in brush row 2, number	HA50	55.3	54.0	8.8	2.6	15.9	48.0	78.0	11
	AUL	52.4	50.0	7.8	3.0	15.0	47.0	70.0	7
	RW	59.6	58.0	9.9	3.7	16.6	47.0	72.0	7
Dikinetids in brush row 3, number	HA50	40.2	39.0	7.8	2.3	19.3	28.0	51.0	11
	RW	35.6	38.0	6.1	2.3	17.2	24.0	41.0	7

[1] Data based on mounted, protargol-impregnated (FOISSNER's method), and randomly selected specimens from non-flooded Petri dish cultures. Measurements in µm. CV – coefficient of variation in %, M – median, Max – maximum, Min – minimum, n – number of individuals investigated, SD – standard deviation, SE – standard error of arithmetic mean, $\overline{x}$ – arithmetic mean.

[2] Four types are recognizable: Of 75 specimens analysed, 58% have the usual macronucleus pattern, while 19% have many (up to 100) scattered nodules, and 23% have a mixture of nodules and one to several short strands. The fourth type is found in the Rwandan specimens, which have only nodules.

[3] Obviously, some monster specimens with additional one or two macronucleus strands and/or a fourth (supernumerary) row outside of the ordinary brush rows were included, as suggested by SCHADE's drawings.

[4] See text for explanation.

Contractile vacuole in rear body end, forms several adventive vacuoles merging to a large blister during diastole (Fig. 71a, c–e, z, 74c). About 10 excretory pores in pole area (Fig. 71z, 73i, 74f). When defecating, the fecal mass migrates through the contractile vacuole leaving the cell in the pole centre (Fig. 72c, l).

One type of extrusomes in oral bulge. Extrusome shape and size rather similar in the populations investigated, differences thus recognizable only on careful observation; conspicuous both in vivo and in protargol preparations, though small as compared to body size, because forming a row each in right and left half of oral bulge, as already described and illustrated by PENARD (Fig. 71b), producing a conspicuously bright (in vivo) or black (protargol preparations) fringe in ventral anterior third of body; often less narrowly spaced, rarely even lacking in right bulge half, for instance, in 3 out of 43

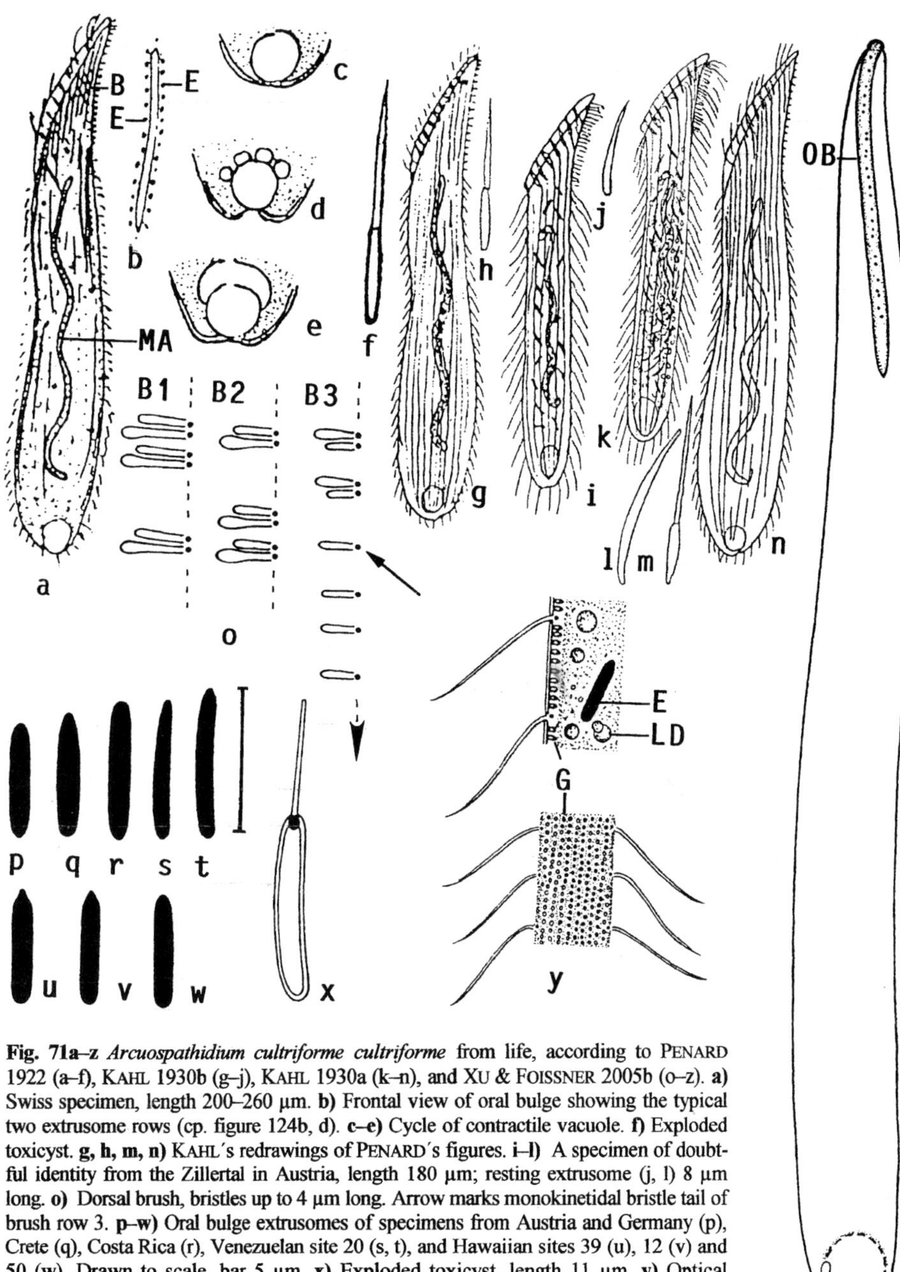

Fig. 71a–z *Arcuospathidium cultriforme cultriforme* from life, according to PENARD 1922 (a–f), KAHL 1930b (g–j), KAHL 1930a (k–n), and XU & FOISSNER 2005b (o–z). **a)** Swiss specimen, length 200–260 µm. **b)** Frontal view of oral bulge showing the typical two extrusome rows (cp. figure 124b, d). **c–e)** Cycle of contractile vacuole. **f)** Exploded toxicyst. **g, h, m, n)** KAHL's redrawings of PENARD's figures. **i–l)** A specimen of doubtful identity from the Zillertal in Austria, length 180 µm; resting extrusome (j, l) 8 µm long. **o)** Dorsal brush, bristles up to 4 µm long. Arrow marks monokinetidal bristle tail of brush row 3. **p–w)** Oral bulge extrusomes of specimens from Austria and Germany (p), Crete (q), Costa Rica (r), Venezuelan site 20 (s, t), and Hawaiian sites 39 (u), 12 (v) and 50 (w). Drawn to scale, bar 5 µm. **x)** Exploded toxicyst, length 11 µm. **y)** Optical section and surface view showing dense cortical granulation. **z)** Ventral view of a specimen from Linz, Austria, length 300 µm. B (1-3) – dorsal brush (rows), E – extrusomes, G – cortical granules, LD – lipid droplets, MA – macronucleus, OB – oral bulge.

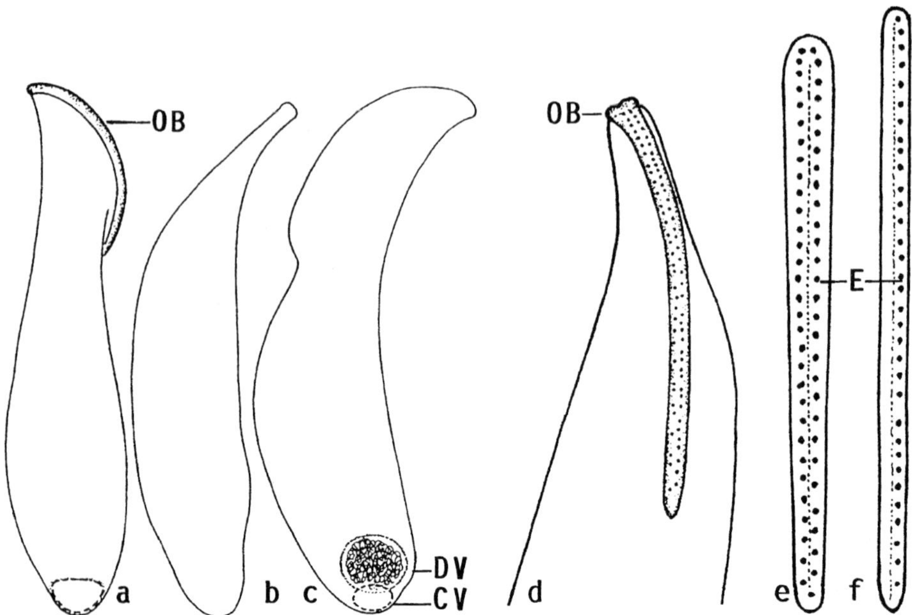

Fig. 72a–f *Arcuospathidium cultriforme cultriforme* from life (a–c, e, f) and after protargol impregnation (d). **a–c)** Right side, dorsal, and left side view of specimens from Hawaiian site 12 (redrawn from video records). **d)** Ventral oral portion of a specimen from Linz, Austria. **e, f)** Variability in shape of oral bulge and arrangement of extrusomes (semi-schematic). Usually, a row of extrusomes each occurs in the right and left bulge half; rarely, extrusomes are lacking in right bulge half. CV – contractile vacuole, DV – defecation vacuole, E – extrusomes, OB – oral bulge.

specimens of the Linz population (Fig. 71a, b, f, h, p–w, 72d–f, i, 73h, q, t–v, 124a–d, h–k, 125b–e, h, 126a, b, f, g, 127c, g). Size very similar in eight populations investigated, that is, 4–6 × 0.7–1.1 µm; shape also highly similar, viz., oblong to very narrowly ellipsoidal with rounded ends, anterior end conical in some populations, making extrusomes looking like blunt pencils (Fig. 71p–r, u–w, 124h–k); thinner, slightly acicular, and inconspicuously curved in the two Venezuelan populations (Fig. 71s, t, 124c, 127d–f). Fully exploded extrusomes 8–12 µm long and of typical toxicyst structure, that is, with a proximal, toxin-containing capsule of similar size and shape as the resting organelle, and a distal tube about as long as capsule and bent by up to 90°; between capsule and tube a refractive granule deeply impregnating with protargol and staining red with methyl green-pyronin. Partially exploded toxicysts, usually found in preserved specimens, about 7–10 µm long and often distinctly knife-shaped, as already mentioned and illustrated by PENARD (Fig. 71a, f–h, x, 72i, o, 73f, 74g, 125d, h, 126b, g). Developing (?) extrusomes studded in cytoplasm, of similar size and shape as mature ones, sometimes form small aggregates, impregnate with protargol (Fig. 71y, 73b, 124g, 125e, 126b).

Cortex colourless and very flexible, contains 5–10 granule rows between each two kineties. Individual granules 0.3–0.6 × 0.2–0.3 µm in size and moderately to strongly refractive, depending on population (Fig. 71y, 72k, p, 124e, f). Cytoplasm colourless,

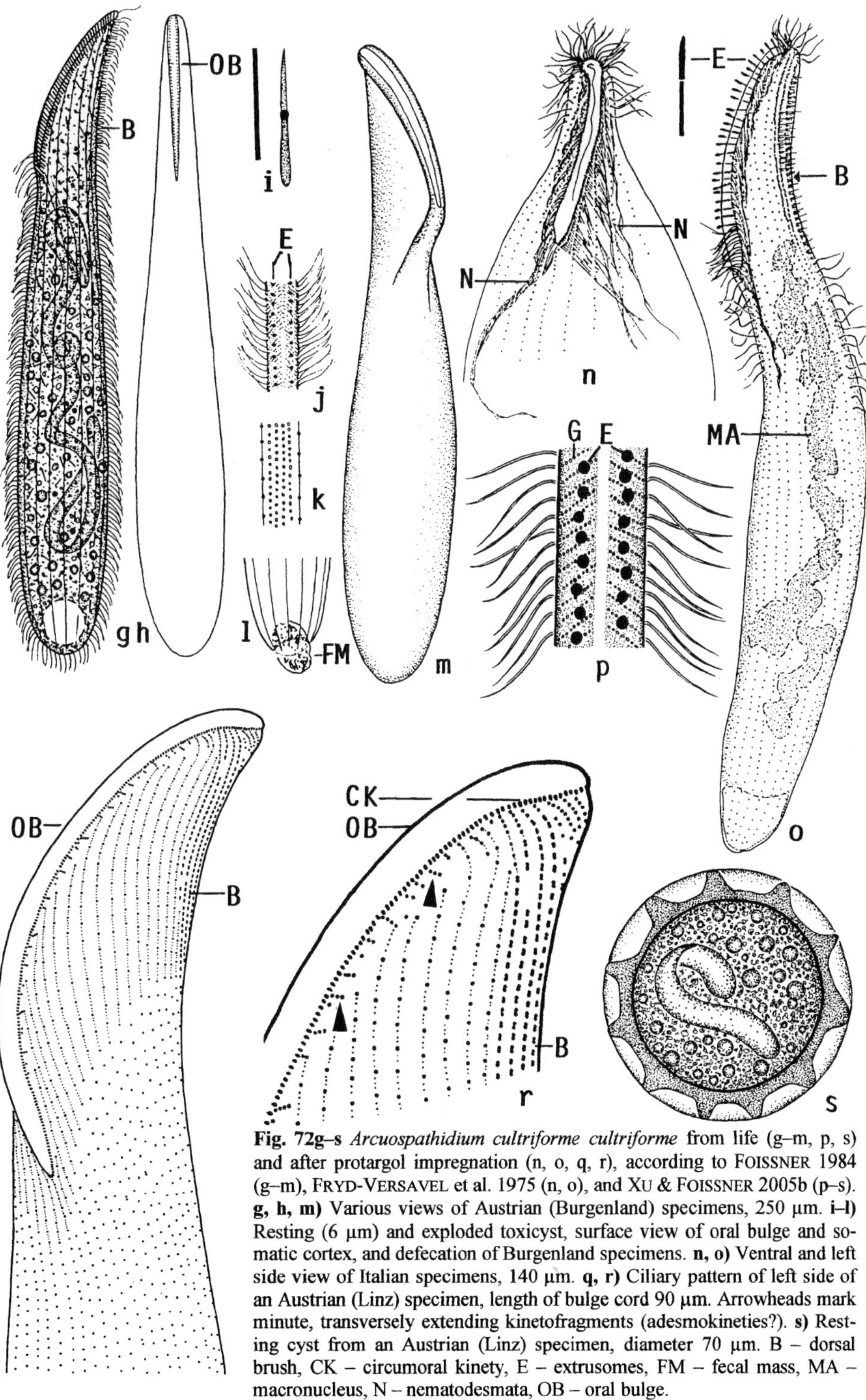

Fig. 72g–s *Arcuospathidium cultriforme cultriforme* from life (g–m, p, s) and after protargol impregnation (n, o, q, r), according to FOISSNER 1984 (g–m), FRYD-VERSAVEL et al. 1975 (n, o), and XU & FOISSNER 2005b (p–s). **g, h, m)** Various views of Austrian (Burgenland) specimens, 250 µm. **i–l)** Resting (6 µm) and exploded toxicyst, surface view of oral bulge and somatic cortex, and defecation of Burgenland specimens. **n, o)** Ventral and left side view of Italian specimens, 140 µm. **q, r)** Ciliary pattern of left side of an Austrian (Linz) specimen, length of bulge cord 90 µm. Arrowheads mark minute, transversely extending kinetofragments (adesmokineties?). **s)** Resting cyst from an Austrian (Linz) specimen, diameter 70 µm. B – dorsal brush, CK – circumoral kinety, E – extrusomes, FM – fecal mass, MA – macronucleus, N – nematodesmata, OB – oral bulge.

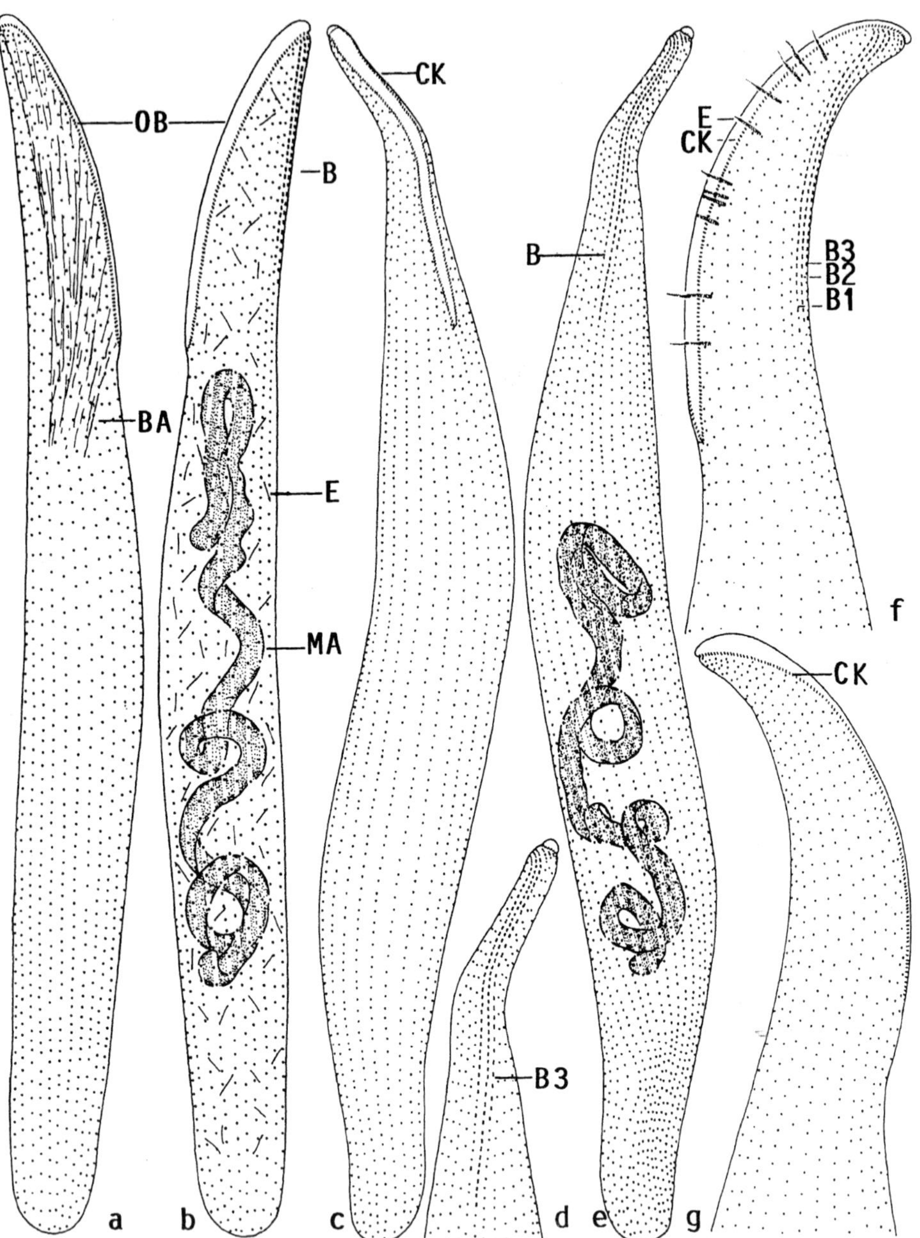

Fig. 73a–g *Arcuospathidium cultriforme cultriforme*, Austrian (Burgenland) specimens after protargol impregnation (from FOISSNER 1984). **a, b)** Right and left side ciliary pattern and macronucleus of a representative specimen, length 190 µm. **c–e)** Somatic and oral ciliary pattern of ventral and dorsal side, length 230 µm. **f, g)** Left and right side ciliary pattern in oral body portion, length 85 µm. Most extrusomes (E) are partially extruded. B (1-3) – dorsal brush (rows), BA – oral basket, CK – circumoral kinety, E – extrusomes, MA – macronucleus, OB – oral bulge.

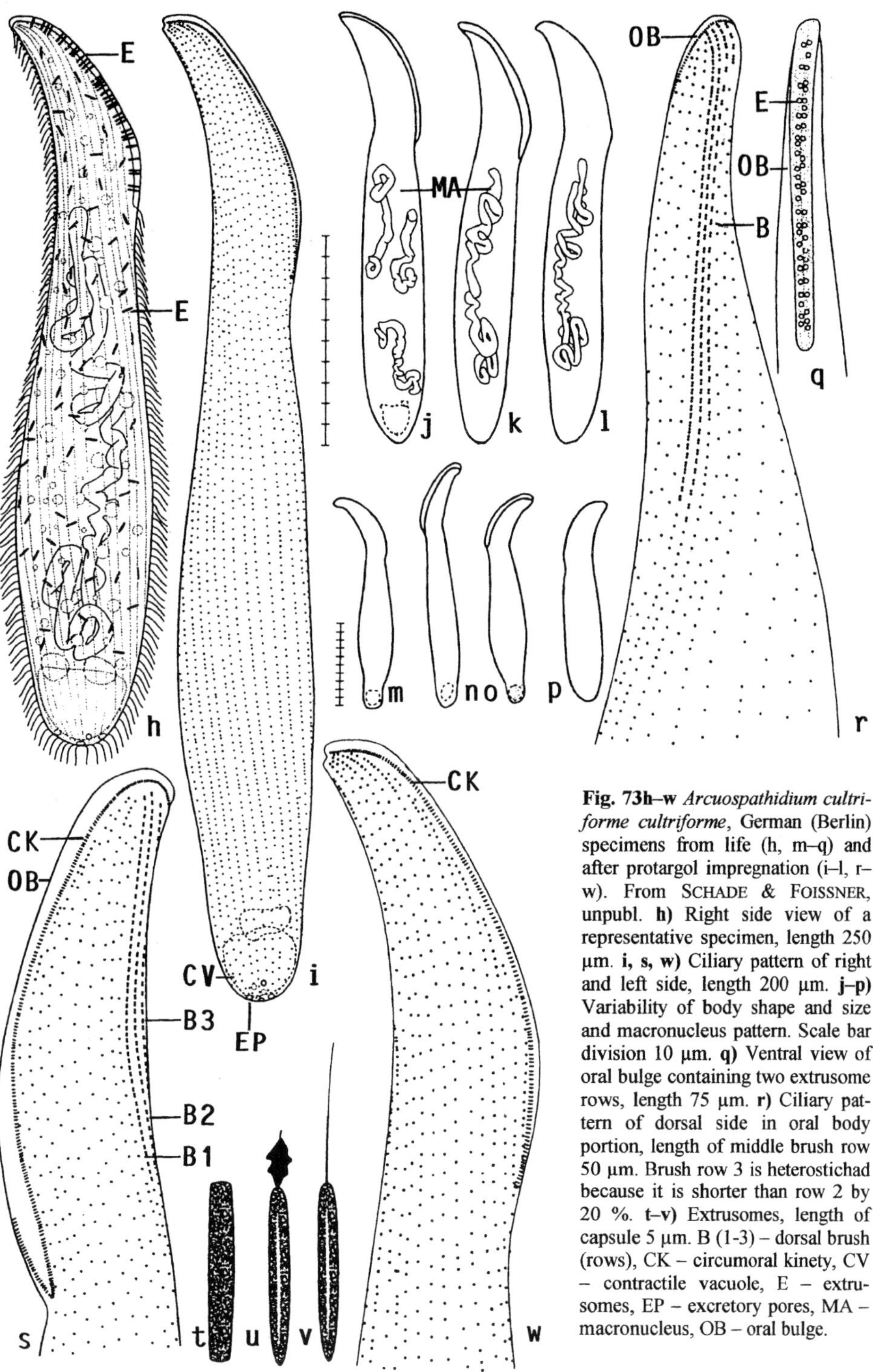

Fig. 73h–w *Arcuospathidium cultriforme cultriforme*, German (Berlin) specimens from life (h, m–q) and after protargol impregnation (i–l, r–w). From SCHADE & FOISSNER, unpubl. **h)** Right side view of a representative specimen, length 250 µm. **i, s, w)** Ciliary pattern of right and left side, length 200 µm. **j–p)** Variability of body shape and size and macronucleus pattern. Scale bar division 10 µm. **q)** Ventral view of oral bulge containing two extrusome rows, length 75 µm. **r)** Ciliary pattern of dorsal side in oral body portion, length of middle brush row 50 µm. Brush row 3 is heterostichad because it is shorter than row 2 by 20 %. **t–v)** Extrusomes, length of capsule 5 µm. B (1-3) – dorsal brush (rows), CK – circumoral kinety, CV – contractile vacuole, E – extrusomes, EP – excretory pores, MA – macronucleus, OB – oral bulge.

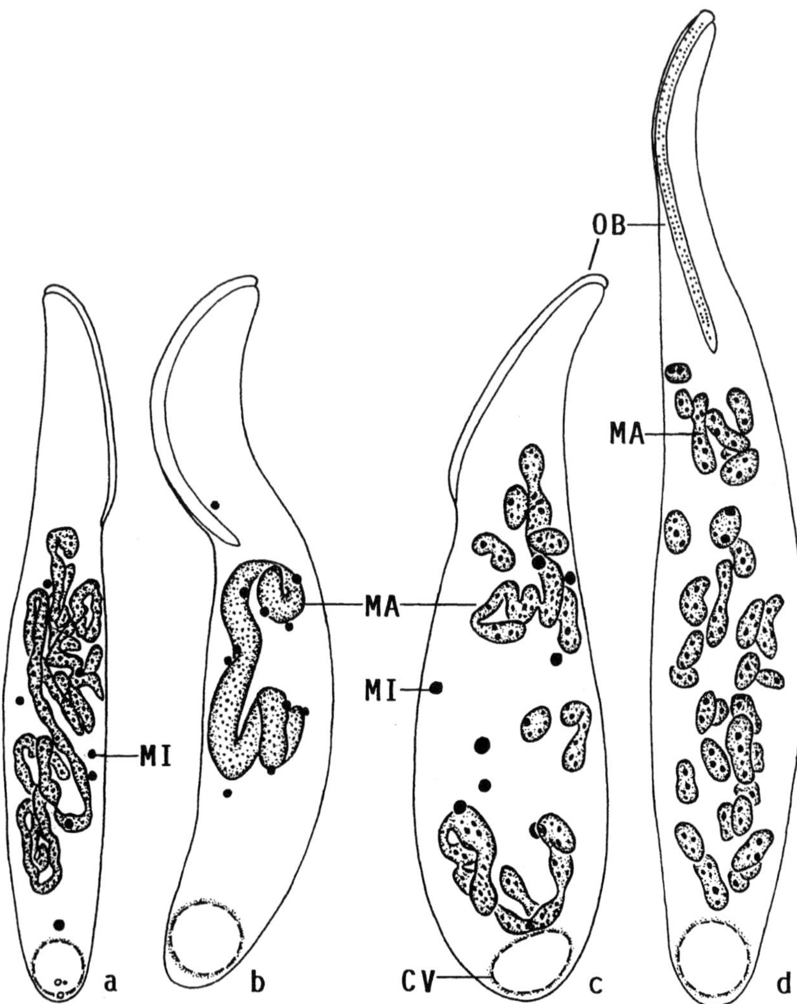

Fig. 74a–d *Arcuospathidium cultriforme cultriforme*, variability of body shape and nuclear apparatus in Hawaiian site (50) specimens after protargol impregnation, length 200 µm, 200 µm, 170 µm, 280 µm. Two thirds of the specimens have an ordinary, more or less tortuous macronucleus strand (a, b), while the others have fragmented the nucleus in many nodules (d); transitions occur (c), showing that they do not represent different species. CV – contractile vacuole, MA – macronucleus (nodules), MI – micronuclei, OB – oral bulge.

postorally packed with lipid droplets 1–6 µm across and food vacuoles containing ciliates and rotifers; specimens thus turbid and brownish at low magnification (≤ ×100). Swims and glides slowly to moderately rapidly on microscope slides and among soil particles, showing great flexibility, especially the oral area, which moves to and fro, as already mentioned by PENARD (1922). Appears somewhat flag-like due to the curved oral area and the cylindroidal postoral portion, when rotating about main body axis.

Somatic cilia about 8–10 µm long in vivo, arranged in 24–44, on average 25 – 37

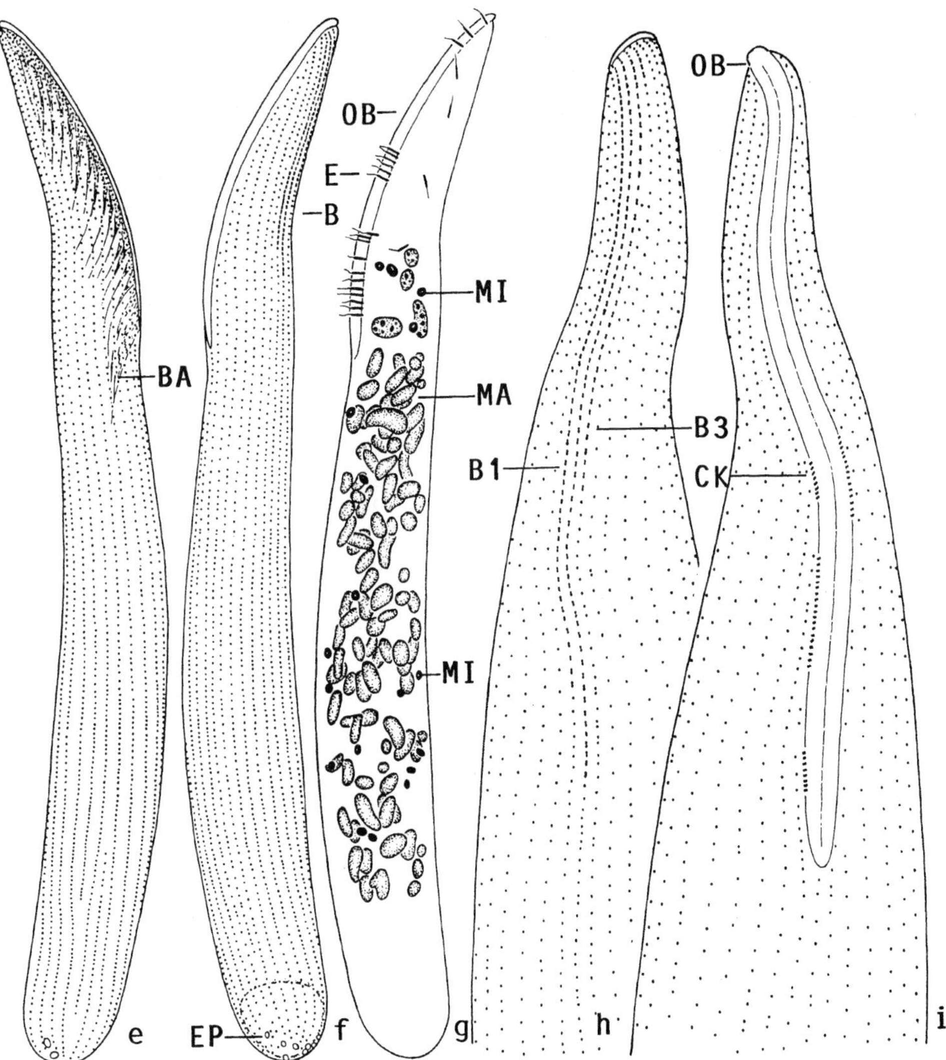

Fig. 74e–i *Arcuospathidium cultriforme cultriforme*, Rwandan specimens after protargol impregnation. The about 20 specimens found in the protargol slides, all have many macronucleus nodules (g), similar to some Hawaiian cells (Fig. 74d). Thus, this population might represent a distinct species. **e–g)** Right and left side ciliary pattern and nuclear apparatus of a representative specimen, length 315 µm. **h, i)** Ciliary pattern of dorsal and ventral side in oral body portion, length of circumoral kinety (oral bulge) 97 µm. B (1, 3) – dorsal brush (rows), BA – oral basket, CK – circumoral kinety, E – extrusomes, EP – excretory pores of contractile vacuole, MA – macronucleus, MI – micronuclei, OB – oral bulge.

equidistant, densely ciliated rows slightly loosened in oral area. Ciliary rows arranged in typical *Arcuospathidium* pattern, that is, curve dorsally on both sides of oral bulge, except for brush kineties which, as usual, slightly curve ventrally (Fig. 71a, g, 72g, o, 73a–i, 74e, f, 125b, c, e–g, 126a, b, 127a, b; Table 25). Short rows, each composed of 1–5 kinetids, extend slightly obliquely from left side circumoral kinety either in line with a certain circumoral dikinetid or separated from circumoral kinety by a minute space; present in all populations of *A. cultriforme cultriforme* and *A. cultriforme scalpriforme*, but not in all specimens (Fig. 72 q, r, 125h, 126b, 129e): for instance, absent in three out of 12 specimens of the Linz population, indistinct (composed of 1–2 kinetids) in four specimens, and distinct (composed of 2–5 kinetids) in five specimens; values similar for Hawaiian site (50) population (n 18): 6/2/10. These short kineties, which suggest a *Dileptus*-like ancestor of the spathidiids (see Vol. II), were neither mentioned by FRYD-VERSAVEL et al. (1975) nor FOISSNER (1984), where they are recognizable on a micrograph (Fig. 129e). Brush exactly on dorsal side, three-rowed, dikinetidal, heterostichad, and slightly shorter than oral bulge; in vivo inconspicuous, though occupying 25–30% of body length, because bristles only 2–3 μm long and pairs ordinarily spaced (1–1.5 μm apart). Bristle rows frequently with small irregularities, such as minute breaks and/or some kinetids out of line, rarely occurs a more or less complete fourth row right of row 1; all rows commence with some ordinary cilia anteriorly and continue as somatic kineties posteriorly. Brush row 1 usually slightly shorter than row 2; row 2 longer by 20–30% (heterostichad) than row 3 which, however, has a monokinetidal tail extending to second third to half of body with about 2 μm long bristles. Details of bristle pairs difficult to analyse, even in the scanning electron microscope, likely as shown in figure 71o, that is, anterior bristle of pairs slightly shorter (1.5–2 μm) than clavate posterior bristle (2–3 μm, rarely up to 5 μm), while vice versa in row 3 (Fig. 71a, 72g, o, q, r, 73b, d–f, r, s, 74f, h, 124d, 125c, f, g, 126a, b, 127a–c, 128a–e; Table 25). Bristle pairs obliquely arranged in upper half of rows in Venezuelan site (20) specimens, a conspicuous feature seemingly doubling row number (129a–d, 131a–c).

Oral bulge conspicuous due to the highly refractive extrusomes and the strong slope (40°–80°, on average 65°, n 15) occupying averages of 31% to 35% of body length in five populations; bulge length thus comparatively constant, though ranges (25%–45%) approach *A. cultriforme scalpriforme*; flat to strongly convex, in frontal view very narrowly cuneate to oblong with proximal end often bluntly pointed, indistinctly apart from body proper because gradually merging into ventral surface and in vivo only about 3 μm high and 5 μm wide. Bulge cortex with arrowhead-like pattern of cortical granules. Circumoral kinety of same shape as oral bulge, continuous, composed of narrowly spaced dikinetids each associated with a cilium and a basket rod (nematodesma). Oral basket hardly recognizable in vivo, but prominent after French protargol impregnation (Fig. 72n, o), while rather inconspicuous with FOISSNER's method (Fig. 73a, 74e, 125e, 129e), composed of cuneate bundles of nematodesmata extending into second third of body. Cytopharyngeal entrance neither recognizable in vivo nor preparations (Fig. 71a, b, n, z, 72a, c–h, j, m–r, 73a–c, f–q, s, w, 74a–i, 124a, b, d, 125a–c, e, h, 126a, b, f, 127a–c, g, 128e; Table 25).

Resting cyst (Fig. 72o, 132a–i): Resting cysts were studied in the populations from Austria (Linz) and the Dominican Republic. They match well, but the ridges of the Dominican specimens are thinner (Fig. 132g, h) and the cysts considerably smaller than those from the Austrian specimens ($\overline{x}$ 68.8 μm, n 22 vs. $\overline{x}$ 45.2 μm, n 6). As concerns the ciliary pattern, see *A. cultriforme scalpriforme*.

Only the Austrian specimens were studied in detail. For encystment, cells were transferred to Eau de Volvic, where they produced a distinct, but not yet faceted wall within six hours; the facets developed slowly over a week. Thus, observations and measurements were performed on two weeks old cysts.

All components of the cyst are colourless at a magnification of $\geq \times 400$, while brown to yellow-brown at low magnifications ($\leq \times 100$) due to the strong light refraction. The cysts are perfectly spherical and have an outer diameter of 55–95 µm ($\overline{x}$ 68.8 µm, M 65, SD 9.6, CV 13.9, n 22), while the inner diameter (cyst proper) is considerably smaller ($\overline{x}$ 48.9 µm, M 48, SD 8.8, CV 18.1, n 22). Thus, the wall has an average thickness of 10 µm (!) and its volume is almost thrice (twice when the facet concavities are detracted) that of the cyst proper (170,430 μm^3 vs. 61,570 μm^3). This is an outstanding feature found, for instance, also in desert strains of *Exocolpoda augustini* (FOISSNER et al., 2002). The most conspicuous part of the cyst is the deeply and roughly faceted middle layer (mesocyst?) which is compact, flexible, and highly refractive. A 0.5–1 µm thick membrane each is found on the facets (ectocyst?) and the cyst proper (endocyst?); both are also recognizable in squashed, split cysts, showing that the inner membrane is not the cortex of the cell, but likely the endocyst. The cyst contents consist of colourless lipid droplets 1–5 µm across and the macronucleus shortened to a semicircular strand, as described in *A. cultriforme scalpriforme*. The contractile vacuole and all extrusomes disappear.

Ontogenesis (130a–c, 131a–c, e, f): The SEM micrographs of dividers from the Venezuelan site (20) population will be included in the detailed description of the ontogenesis of *A. cultriforme scalpriforme*. No differences are recognizable in the two subspecies.

Occurrence and ecology: There are surprisingly few records in the literature, although *A. cultriforme cultriforme* is rather common in terrestrial habitats, especially in Europe (FOISSNER et al. 2005a). PENARD (1922) discovered *A. cultriforme cultriforme* in wall mosses from Switzerland. KAHL (1930a, b) observed a single specimen in moss from the Zillertal in Tyrol, Austria. However, the shape of the extrusomes suggests that it was a different species (Fig. 71i–l). It was only 50 years later, that FRYD-VERSAVEL et al. (1975) reported *A. cultriforme cultriforme* from moss of Italy and provided a brief redescription based on protargol impregnation (Fig. 72n, o). Slightly later, FOISSNER (1984) redescribed *A. cultriforme cultriforme* from soil of a coniferous forest in Austria (Fig. 72g–m, 73a–g, 129e; Table 25). Since then, it was reported from several terrestrial habitats of Austria (FOISSNER 1987a, 2004b, AESCHT & FOISSNER 1993, FOISSNER et al. 2005a); Germany (GORALCZYK & VERHOEVEN 1999, FOISSNER 2000b); Africa, that is, litter and soil (pH 6.5) from the forest surrounding the Sheldrick waterfalls in Kenya (FOISSNER 1999a); and Australia, that is, in the upper litter and soil layer (pH 5.1) of a secondary coniferous forest in the surroundings of the town of Adelaide (BLATTERER & FOISSNER 1988). XU & FOISSNER (2005b) provided 10 additional sites listed in the following paragraphs.

Austria, Linz: mud and soil from a flat, ephemeral pond at the foot of the Pöstlingberg, that is, in the surroundings of the town of Linz, capital of Upper Austria.

Germany, Kassel: litter and soil from the upper 0–5 cm layer of a beech forest on the "Kleinen Gudenberg" about 30 km NW of the town of Kassel. Sample kindly provided by Dr. M. BONKOWSKI (Göttingen University).

Dominican Republic, site 22: litter and soil from the upper 0–3 cm layer of a mangrove forest near the village of Maimon, about 10 km west of the town of Puerto Plata. Soil almost black and spongious, contains many grass and mangrove roots, slightly acidic (pH 6.2 in water) and saline (5‰).

Costa Rica, site 11: tree mosses from the evergreen rain forest in the Braulio-Carillo National Park, Mirador area; pH 4.7 in water.

South America, Venezuelan site 8: soil from the upper 3 cm of a banana plantation at the farm of Mr. EISENBERG, surroundings of the town of Puerto Ayacucho. The field was founded 15 years ago and organically fertilized. The slightly acidic (pH 6 in water) soil was dark and contained many fine roots, but the layer was only about 5 cm thick and followed by yellow sand.

South America, Venezuelan site 20: tree mosses from a gallery forest in the surroundings of the town of Puerto Ayacucho.

Tropical Africa, Rwanda: savannah litter and soil from the surroundings of the village of Gabiro, that is, at the foot of the mountains with the Gorilla National Park. Sample kindly provided by Dr. E. STÜBER (Salzburg).

Hawaii, Big Island site 12: highly saline (~ 20 ‰) mud and soil from dry rockpools in the lower part of a stream between the Kahua Ranch and Kawaihae.

Hawaii, Big Island site 39: tree bark with mosses and lichens along the Bird Trail in the Volcano National Park; pH 7.2 in water.

Hawaii, Oahu Island site 50: litter and soil from the upper 0–3 cm layer of a dry swamp overgrown with ferns, Ihi'lhilanãkea crater area, Koko Head, south of the Hanauma Bay. Sample kindly provided by Dr. HUBERT BLATTERER (Linz).

These data show that *A. cultriforme cultriforme* and *A. cultriforme cultriforme* - like taxa are cosmopolitan, except of Antarctica (FOISSNER 1998). However, it is likely rare in Africa because it was not found in 73 samples from Namibia, where *A. cultriforme megastoma* and *A. lorjeae* were discovered (FOISSNER et al. 2002). In Rwanda and Venezuela, other (sub)species may occur, but more detailed investigations are required (see also biogeography section of *A. cultriforme*!). The ecological range is obviously wide and includes terrestrial, semiterrestrial, and saline habitats. However, true limnetic and extreme habitats, such as alpine soil above the timberline (FOISSNER 1987b), Antarctica (FOISSNER 1998), and hot deserts (FOISSNER et al. 2002) are avoided.

Arcuospathidium cultriforme scalpriforme (KAHL, 1930) FOISSNER, 2003 (Fig. 75a–v, 76a–q, 77a–x, 78a–n, 79a–n, 133–136; Tables 26, 27)

1930 *Spathidium scalpriforme* KAHL, Arch. Protistenk., 70: 381.

1930 *Lionotus scalpriforme* KAHL, 1930 – KAHL, Tierw. Dtl., 18: 165 (The figure explanation and the index indicate that KAHL put it to *Lionotus* per lapsus.).

1984 *Arcuospathidium lionotiforme* (KAHL, 1930) nov. comb. – FOISSNER, Stapfia, 12: 78 (misidentification in this and all subsequent papers!).

2002 *Arcuospathidium cultriforme lionotiforme* (KAHL, 1930) FOISSNER, 1984 nov. stat. – FOISSNER, AGATHA & BERGER, Denisia, 5: 300 (misidentification, see above; ranked as a subspecies).

2003 *Arcuospathidium cultriforme scalpriforme* (KAHL, 1930) comb. n., stat. n. – FOISSNER, Acta Protozool., 42: 54 (rectification of misidentification and rank lowering).

2005 *Arcuospathidium cultriforme scalpriforme* (KAHL, 1930) FOISSNER, 2003 – XU & FOISSNER,

Protistology, 4: 16 (species monograph; voucher slides of the populations mentioned in Table 26 are deposited in the Oberösterreichische Landesmuseum in Linz (LI), Upper Austria).

Improved diagnosis: Mouth (oral bulge) about half of body length. Extrusomes scattered in right and left bulge half.

Type locality: Mosses from the Zillertal in Tyrol, Austria, E11°55' N47°20'. KAHL (1930a) first observed the species in mosses from Mittenwald, a Bavarian village near the northern border of Tyrol. However, the figure he provided seems to be from the single specimen he found in Zillertal moss (reproduced here as figures 75a–c). Thus and because KAHL did not fix a type locality, XU & FOISSNER (2005b) chose the Zillertal.

Synonymy: FOISSNER (2003b) explains his misidentification as follows: "The present study shows that my former redescription of *S. lionotiforme* KAHL (1930a, b) is based on an unfortunate misidentification caused by insufficient experience and the widespread opinion that *Spathidium* is highly variable. The species I investigated in 1984 obviously belongs to the *Spathidium cultriforme* group, specifically to *S. scalpriforme* KAHL 1930a, a species which I never found in soil and moss (FOISSNER 1998), although it lives there (KAHL 1930a, b), simply because I continuously misidentified it as *S. lionotiforme*".

We adopt this explanation, and the senior author emphasizes that all his previous *A. lionotiforme* identifications are wrong (for a review, see FOISSNER 1998) and belong to *A. cultriforme scalpriforme*. This is quite reasonable because the real *Spathidium* (now *Cultellothrix*) *lionotiforme* has a rather different organization making misidentification unlikely; further, *C. lionotiforme* must be a very rare species because it has not been mentioned in the literature since the original description.

Description: The following description contains the data from KAHL (1930a, b), FOISSNER (1984), and XU & FOISSNER (2005b). KAHL (1930a, b) and GELLÉRT (1956) mention small varieties which, in our opinion, belong to other species. KAHL's "local form" from roof moss in northern Germany is 160–180 µm long and has only about half (15) the number of ciliary rows typical for this species (Fig. 75e). GELLÉRT's variety from moss humus in Hungary is only 90 µm long, performs a contractile vacuole cycle within 35–40 sec, and feeds on minute zooflagellates.

Arcuospathidium cultriforme scalpriforme is very similar to *A. cultriforme cultriforme*, except for the characteristics mentioned in the diagnosis and some other minor features, which thus will be emphasized in the following description. Some of these might be more important than presently recognized.

Size and length:width ratio highly similar to those of *A. cultriforme cultriforme* (*Acc*), that is, about 230 × 40 µm in vivo, matching data by KAHL 1930a, b (250–330 µm, length:width ratio 5:1) and WENZEL 1953 (length about 260 µm); highly dependent on culture conditions, as shown by the South African population: on average 207 × 39 µm in protargol-impregnated specimens from a non-flooded Petri dish (raw) culture, while 314 × 66 µm in specimens from a flourishing semipure culture; number of ciliary rows, in contrast, increases only slightly from 33 to 35 (Tables 26, 27).

Body usually more distinctly knife-shaped than in *Acc* due to the longer oral area, widest at proximal end of oral bulge and in mid of trunk (Fig. 75a, c, g, h, r–v, 133a, b,

Table 26 Morphometric data on *Arcuospathidium cultriforme scalpriforme* populations (Pop) from South Africa (SA, from XU and FOISSNER 2005b) and Austria (AUS, from FOISSNER 1984 and his original notes)

Characteristics[1]	Pop	$\overline{x}$	M	SD	SE	CV	Min	Max	n
Body, length	SA	206.8	200.0	25.6	5.6	12.4	167.0	275.0	21
	AUS	251.3	260.0	30.8	8.9	12.3	205.0	290.0	12
Body, width	SA	39.3	42.0	7.7	1.7	19.5	25.0	51.0	21
	AUS	35.3	34.5	3.9	1.1	11.0	30.0	43.0	12
Body length:width, ratio	SA	5.4	5.3	1.0	0.2	18.0	4.0	7.8	21
	AUS	7.2	7.0	1.2	0.4	16.9	5.2	9.7	12
Oral bulge, length (cord)	SA	92.1	94.0	12.8	2.8	13.9	61.0	113.0	21
	AUS	105.8	106.0	7.8	2.2	7.4	92.0	120.0	12
Oral bulge, height	SA	2.6	2.5	0.3	0.1	11.1	2.0	3.0	21
Oral bulge length:body length, ratio	SA	0.45	0.46	0.1	–	11.8	0.35	0.52	21
	AUS	0.42	0.43	–	–	8.7	0.37	0.49	12
Circumoral kinety to last dikinetid of brush row 1, distance	SA	53.6	53.0	7.9	1.7	14.7	41.0	65.0	21
Circumoral kinety to last dikinetid of brush row 2, distance	SA	64.0	63.0	8.1	1.8	12.7	50.0	80.0	21
	AUS	88.4	88.0	8.3	2.4	9.4	70.0	100.0	12
Circumoral kinety to last dikinetid of brush row 3, distance	SA	51.6	49.0	9.1	2.0	17.7	35.0	72.0	21
Anterior body end to macronucleus, distance	SA	78.8	78.0	16.8	3.7	21.3	47.0	113.0	21
Macronucleus figure, length	SA	87.5	90.0	21.5	4.7	24.6	46.0	134.0	21
	AUS	106.3	108.0	25.9	7.5	24.4	42.0	142.0	12
Macronucleus, length (spread, thus approximate)	SA	187.4	200.0	–	–	–	130.0	230.0	21
Macronucleus, width in central third	SA	6.1	6.0	1.4	0.3	22.6	4.0	9.0	21
	AUS	7.2	7.0	0.8	0.2	10.7	6.0	8.5	12
Macronucleus, number	SA	1.0	1.0	0.0	0.0	0.0	1.0	1.0	21
	AUS	1.0	1.0	0.0	0.0	0.0	1.0	1.0	12
Micronuclei, across	SA	3.0	3.0	0.3	0.1	10.6	2.5	4.0	21
	AUS	2.9	2.8	0.1	0.0	4.0	2.8	3.1	12
Micronuclei, number	SA	11.5	12.0	2.6	0.6	22.2	7.0	18.0	21
Ciliary rows, number[2]	SA	32.7	32.0	3.0	0.7	9.1	27.0	40.0	21
	AUS	32.5	32.0	1.2	0.4	3.8	31.0	35.0	12
Basal bodies in a right ciliary row, number	SA	124.4	130.0	25.0	5.5	20.1	85.0	180.0	21
Basal bodies within 10 µm of a somatic kinety in mid-body, number	AUS	6.1	6.0	0.9	0.3	14.8	5.0	8.0	12
Dorsal brush rows, number	SA	3.0	3.0	0.0	0.0	0.0	3.0	3.0	21
	AUS	3.0	3.0	0.0	0.0	0.0	3.0	3.0	12
Dikinetids in brush row 1, number	SA	40.6	39.0	7.3	1.6	18.0	30.0	56.0	21
Dikinetids in brush row 2, number	SA	55.0	55.0	7.0	1.5	12.8	40.0	67.0	21
Dikinetids in brush row 3, number	SA	34.9	35.0	6.1	1.3	17.4	23.0	48.0	21

[1] Data based on mounted, protargol-impregnated (FOISSNER's method), and randomly selected specimens from non-flooded Petri dish cultures. Measurements in µm. CV – coefficient of variation in %, M – median, Max – maximum, Min – minimum, n – number of individuals investigated, SD – standard deviation, SE – standard error of arithmetic mean, $\overline{x}$ – arithmetic mean.

[2] Row number of specimens from semipure culture, which are much larger (Table 27): $\overline{x}$ 35.0, SD 3.6, CV 10.3, Min 30, Max 41, n 21.

d; Table 26); shape similar to that of certain *Dileptus* species (e.g., *Dileptus conspicuus*, redescribed by FOISSNER 1989) and various large, limnetic pleurostomatids, such as *Amphileptus carchesii* and *Litonotus varsaviensis* (for a review, see FOISSNER et al. 1995). Ratio of body and oral bulge length 0.42 in Austrian and 0.45 in South African populations, and thus distinctly higher than in the five populations of *Acc* (0.31–0.35; Tables 26, 27).

Nuclear apparatus as in *Acc*, but specimens/populations with many macronucleus

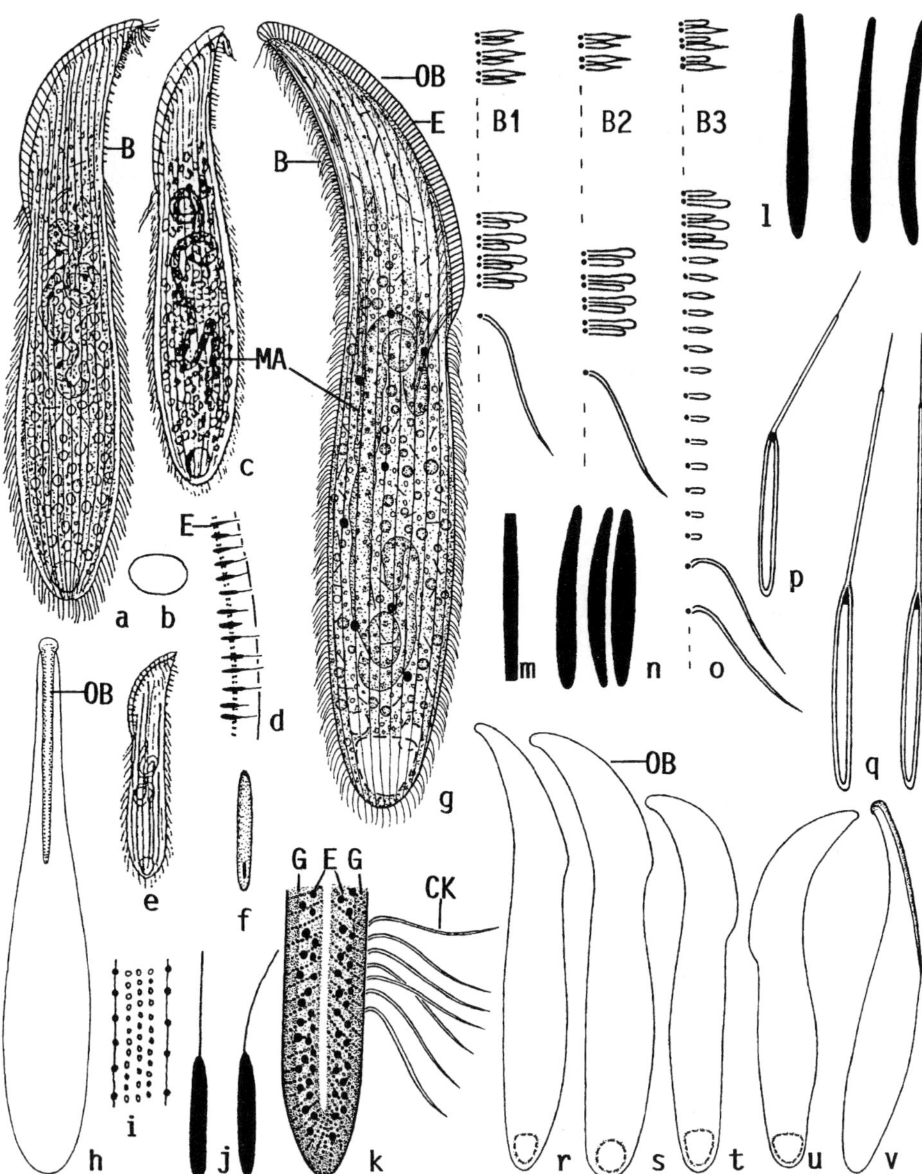

Fig. 75a–v *Arcuospathidium cultriforme scalpriforme* from life (a–c, e, g–i, k–v) and after protargol impregnation (d, f, j), according to KAHL 1930a (a, b), KAHL 1930b (c, e), FOISSNER 1984 (d, f–j, m), and XU & FOISSNER 2005b (k, l, n–v). **a–c)** Left side and transverse (b) view of a specimen likely from the Zillertal in Austria, 260 µm. **d)** Part of oral bulge of specimen shown in figure 76a, e. **e)** Another (?) species from northern Germany, 160 µm. **f, j)** Cytoplasmic extrusomes, capsule 4 µm long. **g, h)** Right side and ventral view of a representative specimen from Austria, 270 µm. **i)** Cortical granulation (schematic). **k)** Proximal portion of the 5 µm wide oral bulge, showing the scattered extrusomes and the arrowhead-like pattern produced by the cortical granules. **l–n)** Resting oral bulge extrusomes from specimens of Brazil (l), Austria (m), and South Africa (n), length 5–8 µm. **o)** Dorsal brush, bristles (up to 5 µm) and somatic cilia (10 µm) drawn to scale. **p, q)** Exploded toxicysts, lenght 15–20 µm. **r–v)** South African specimens, redrawn from video records, oral bulge 37–51 % of body length. B (1-3) – dorsal brush (rows), CK – cilia of circumoral kinety, E – extrusomes, G – cortical granules, MA – macronucleus, OB – oral bulge

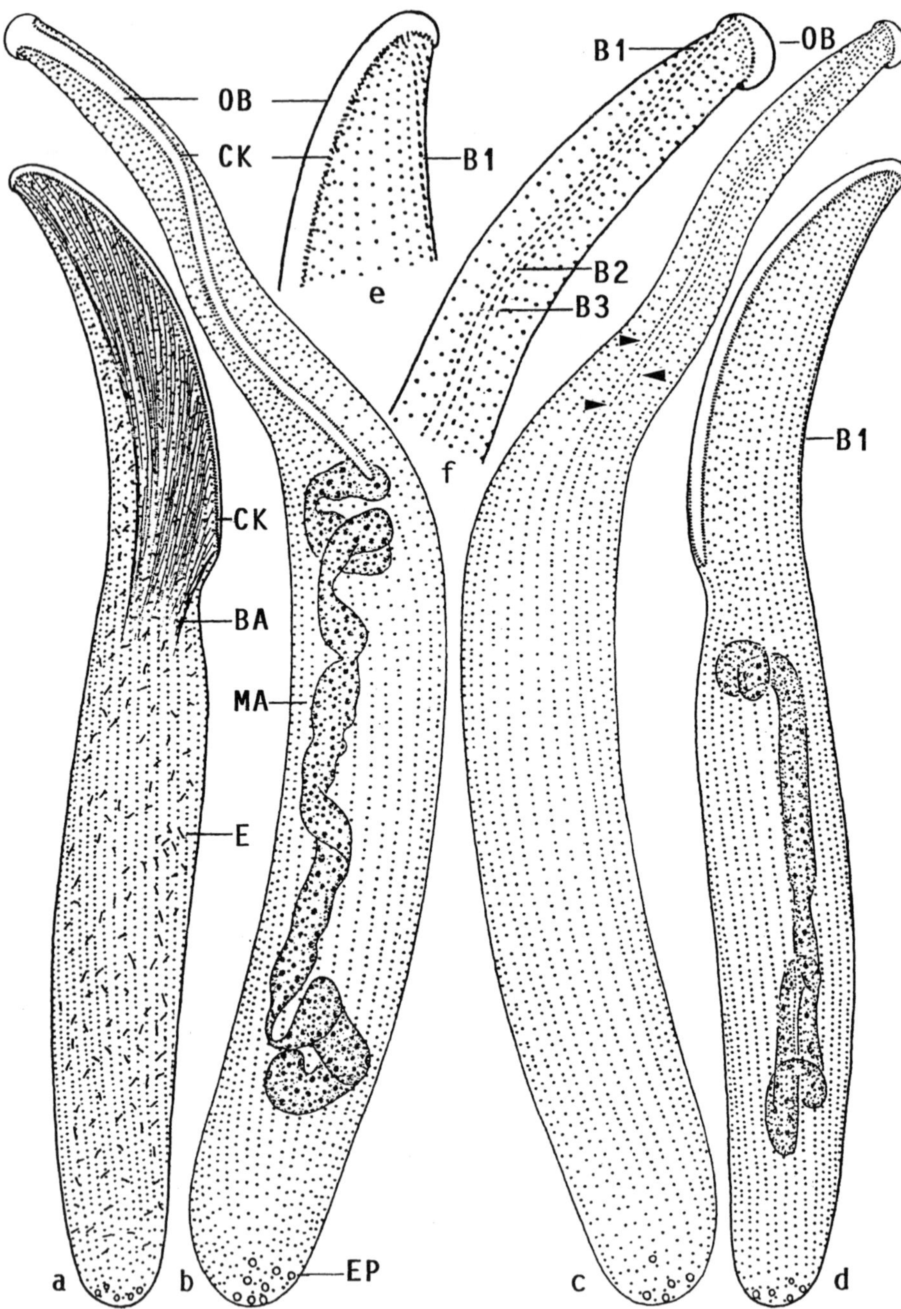

Fig. 76a–f *Arcuospathidium cultriforme scalpriforme*, protargol-impregnated ciliary pattern and macronucleus of the Austrian population studied by FOISSNER (1984). **a, d, e)** Right and left side view, length 295 µm. **b, c, f)** Ventral and dorsal view, length 240 µm. Arrowheads denote proximal end of brush rows. Note dorsal brush exactly on dorsal side of cell (c, f). B 1-3 – dorsal brush rows, BA – oral basket, CK – circumoral kinety, E – extrusomes, EP – excretory pores, MA – macronucleus, OB – oral bulge.

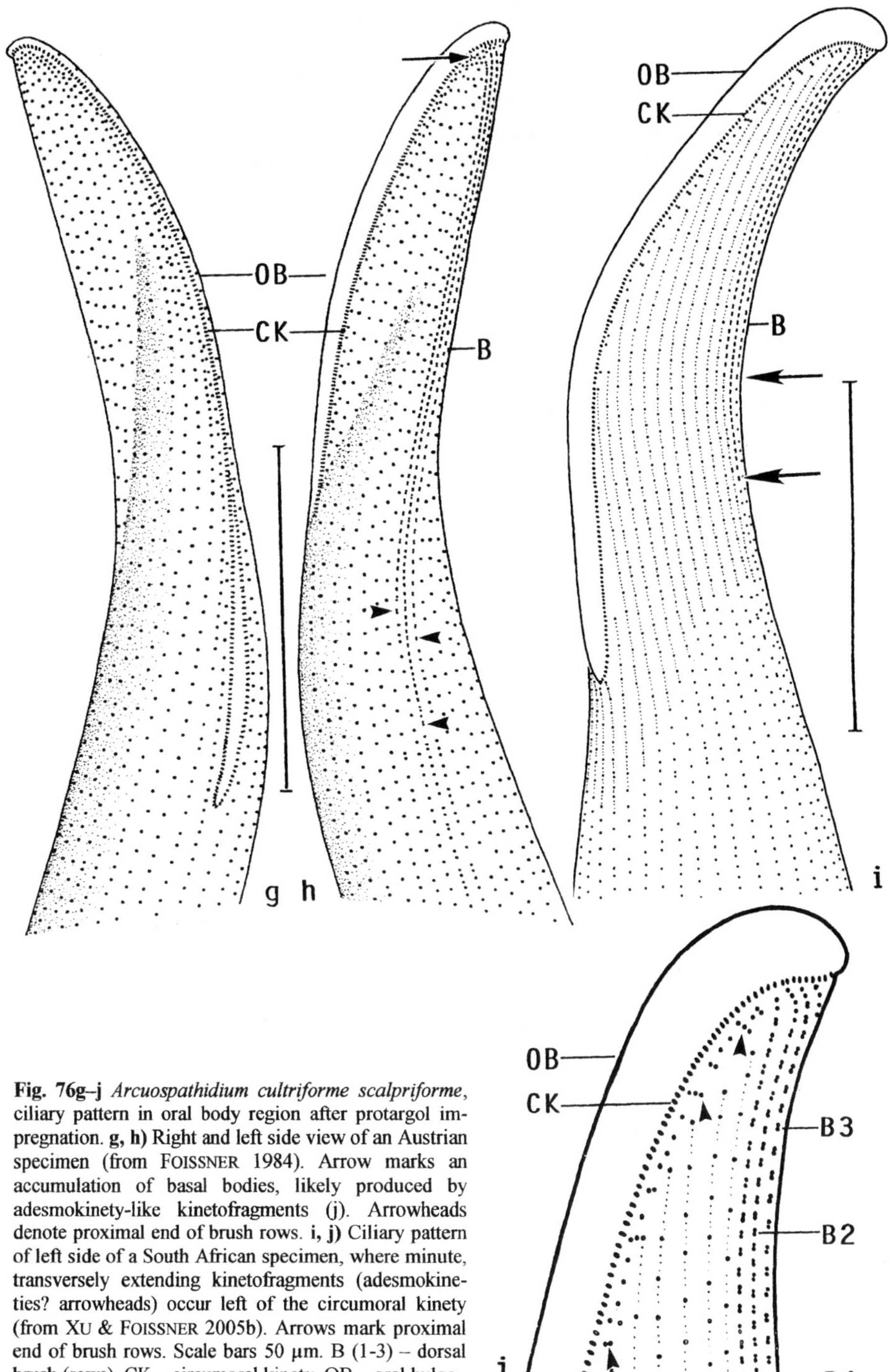

Fig. 76g–j *Arcuospathidium cultriforme scalpriforme*, ciliary pattern in oral body region after protargol impregnation. **g, h)** Right and left side view of an Austrian specimen (from FOISSNER 1984). Arrow marks an accumulation of basal bodies, likely produced by adesmokinety-like kinetofragments (j). Arrowheads denote proximal end of brush rows. **i, j)** Ciliary pattern of left side of a South African specimen, where minute, transversely extending kinetofragments (adesmokineties? arrowheads) occur left of the circumoral kinety (from XU & FOISSNER 2005b). Arrows mark proximal end of brush rows. Scale bars 50 µm. B (1-3) – dorsal brush (rows), CK – circumoral kinety, OB – oral bulge.

nodules were not observed (Fig. 75c, g, 76b, d, 77a, b, 133d). Macronucleus strand in two pieces dividing individually in about 5% of specimens (Fig. 78g–m). Contractile vacuole and cortex also as in *Acc*. Cortical granules very distinct and in about 8 rows between each two kineties in South African specimens (Fig. 135a, b).

Extrusome shape and size very similar in the populations investigated and thus rather dissimilar to the shorter and thus comparatively thicker extrusomes of *Acc*, as already mentioned by KAHL (1930a); data of FOISSNER (1984) do not meet the modern standard, but basically match the observations of XU & FOISSNER (2005b). Extrusomes conspicuous both in vivo and protargol preparations, though small as compared to body size, because forming a bright stripe each in right and left half of oral bulge, especially in broader dorsal half (Fig. 75a, g, h, k, 135a–e). Size about 7 μm in specimens studied by KAHL (1930a), 5 μm in Austrian cells (FOISSNER 1984), 6 × 0.8 μm in Brazilian population, and 7–8 × 0.7–0.9 μm in South African specimens (XU & FOISSNER 2005b). Shape very similar in Brazilian and South African specimens, that is, indistinctly acicular to very narrowly ellipsoidal and somewhat curved and asymmetric (Fig. 75l, n). Fully exploded extrusomes 15–20 × 1 μm in size and of typical toxicyst structure (Fig. 75p, q); partially exploded toxicysts less distinctly knife-shaped than in *Acc*, usually more or less clavate (Fig. 75d, 133f). Developing (?) extrusomes in cytoplasm as shown in figures 75f, j.

Cytoplasm and movement as described in *Acc*. Like that, feeds on rotifers and middle-sized hypotrichs, such as *Gastrostyla steinii*, *Gonostomum affine* and a large *Urosoma* species. Appears elegant and conspicuous when gliding on microscope slide due to the nicely curved dorsal outline and the large size.

Somatic ciliature highly similar to that of *Acc*, except of dorsal brush (Fig. 75a, c, g, 76a–d, 133a, b; Tables 25, 26). Cilia 8–10 μm long, more loosely spaced (~ 3 μm) in oral than postoral area (~ 1.5 μm), arranged in about 30 rows according to KAHL (1930a, b), matching averages (32–33) of Austrian and South African populations (Table 26). Row number conspicuously constant, that is, 33 rows in South African specimens from the non-flooded Petri dish (raw) culture and 35 rows in the 100 μm longer specimens from the semipure culture (Table 26). At right anterior end of dorsal brush of most Austrian specimens an accumulation of scattered basal bodies (Fig. 76h), likely adesmokineties as in the South African specimens (Fig. 76i, j). Dorsal brush basically as in *Acc*, that is, almost heterostichad (row 3 shorter than row 2 by about 20%) occupying 31–35% of body length; rarely occur specimens with one or two more or less distorted bristle rows right of row 1; tail of row 3 extends into second body third (Fig. 75a, g, o, 76c, d, f, h–j, 133d, e, i, 134a–f; Table 26). Details of bristle pairs studied only in South African specimens, likely as shown in figure 75o, that is, slightly clavate anterior bristle longer (3–5 μm) than posterior (2–3 μm) in rows 1 and 2, while vice versa in row 3 (Fig. 75o, 134a–f).

Oral bulge conspicuous, though only about 3 μm high near dorsal end, because almost half as long as body (42–45%; Table 26), strongly slanted (50°–70°, on average 63°, n 10), rather distinctly projecting from body proper at proximal end (Fig. 75g, r–v, 76a, 133d), and studded with refractive extrusomes (Fig. 135a, b); in lateral view slightly to distinctly convex, in frontal view about 5 μm wide and very narrowly cuneate to oblong with proximal end bluntly pointed. Circumoral kinety and oral basket as described in *A. cultriforme cultriforme* (Fig. 75a, g, h, k, r–v, 76a, b, d, e, g–j, 77a, b, 78a,

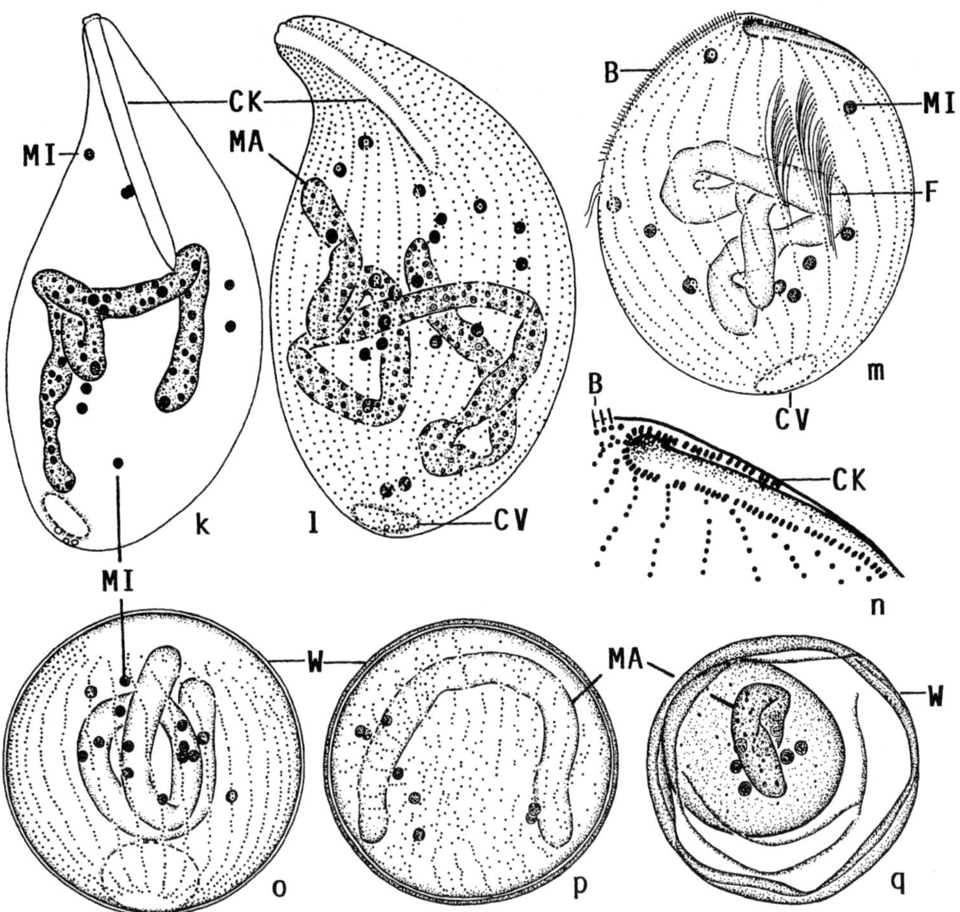

Fig. 76k–q *Arcuospathidium cultriforme scalpriforme*, encystment of South African specimens reconstructed from protargol-impregnated cells (from Xu & Foissner 2005b). **k, l)** Early stages showing reduction of body size and diminution of circumoral kinety, length 120 μm and 108 μm. **m, n)** Next, the cell rounds up and the resorption of the somatic and oral ciliature becomes distinct, length 71 μm. **o, p)** Formation of the cyst wall and shortening of the macronucleus, diameter 67 μm and 65 μm. Resorption of the ciliary structures proceeds. **q)** Young, completed cyst, diameter 58 μm. The macronucleus is now semicircular, most micronuclei and the infraciliature have been resorbed, and a faceted cyst wall developed. Both, wall and cell are distinctly shrunken due to the preparation procedures. B – dorsal brush, CK – circumoral kinety, CV – contractile vacuole, F – (postciliary?) fibres, MA – macronucleus, MI – micronuclei, W – cyst wall.

133a–h, 135a–e; Table 26).

Encystment (Fig. 76k–q): Encystment was investigated by Xu & Foissner (2005b). However, the description is not very detailed because few well-prepared specimens were found in the protargol slides.

When encystment commences, cells become smaller and stouter (Fig. 76a, b). This diminution is associated with the gradual resorption of somatic and oral kinetids, especially those of the dorsal brush and circumoral kinety, both having reduced about half of their length and kinetids. The macronucleus appears more tortuous, but is basically unchanged, while the micronuclei move (?) apart from the macronucleus and distribute throughout the body, as in early dividers. The contractile vacuole is still recognizable.

In the next stage, cells become almost globular and greatly reduce the number of somatic and oral kinetids (Fig. 76m, n). The remaining, still ciliated kinetids are somewhat irregularly arranged within the rows which loose the *Arcuospathidium* pattern and arrange meridionally. The dorsal brush kinetids still have bristles, but lost the dikinetidal appearance. The postciliary microtubule ribbons are very distinct in this stage. Now, a really remarkable process commences, viz., a strong shortening of the macronucleus which is eventually reduced by about 80% of its length and becomes semicircular in the encysted specimen (Fig. 76o–q). This length reduction is not achieved by thickening of the strand or chromatin extrusion, but by resorption of nuclear mass, as evident from simple volume calculations, that is, the nuclear volume decreases from about 5,000 μm^3 in morphostatic cells to 1,000 μm^3 in the mature cyst. The micronuclei are still scattered, and the contractile vacuole is still recognizable.

The next stage is characterized by the appearance of the cyst wall, which is about 1 µm thick and does not impregnate with the protargol method used (Fig. 76o). The specimens are now globular and lost the oral apparatus, while some somatic basal bodies are still recognizable, becoming more and more scattered. The shortening of the macronucleus is now distinct, and the micronuclei arrange around the macronucleus strand. The contractile vacuole is still recognizable.

During the last stages, micronuclei resorption commences and the faceted ectocyst develops, as described in *A. cultriforme cultriforme*. The infraciliature likely disappears completely, though this is difficult to prove because the maturating cysts are rather poorly impregnated (Fig. 76p). The mature cyst wall, although appearing quite stable in vivo, is usually distorted and the cell distinctly shrunken in the preparations, leaving back a more or less wide space between wall and cortex. The macronucleus is shortened to a semicircular strand, and about half of the micronuclei have been resorbed. The nucleoli are very small. The contractile vacuole disappeared (Fig. 76q).

Occurrence and ecology: KAHL (1930a) discovered *A. cultriforme scalpriforme* in mosses from Mittenwald (a village in southern Bavaria near the Tyrolean border) and the Zillertal in Austria (a valley in the Tyrol Alps). The "local form" from roof moss in northern Germany likely belongs to another species (see discussion of synonymy). The same is supposed for GELLÉRT's (1956) variety, which is only 90 µm long. The next reliable record is from the surroundings of Erlangen in Bavaria, where WENZEL (1953) observed *A. cultriforme scalpriforme* in only two out of over 100 moss samples, viz., in dry moss at pH 5.1 and in leaf litter. In 1984, FOISSNER redescribed the species from soil of a mixed forest in the surroundings of the village of Baumgarten, Lower Austria (brown-earth soil, pH 7 in water). More recently, some unsubstantiated records were added from Bulgaria (DETCHEVA 1972b) and Germany (FOISSNER 2000b). XU & FOISSNER (2005b) provided two further, substantiated records, viz., from South America, Brazil site 30 (Litter and soil from the upper 0–5 cm of the floodplain rainforest near the town of Manaus, Janauari region. Soil brown, humic, pH 5.1 in water.) and the Republic of South Africa, site 44 (Litter and soil from the margin of a flat pond, that is, the Sirkelsvlei in the Cape Peninsula National Park. Soil dark, very sandy, with many grass roots, flooded after heavy rain, pH 5.4 in water.).

All other records are from limnetic habitats, viz., the mud and Aufwuchs of various polluted rivers in Bulgaria (DETCHEVA 1972a, c, 1975a, 1979a, b, 1986, 1993, RUSSEV et al. 1976, 1984) and some natural freshwaters and a sewage plant in China (WANG 1977, NING et al. 1993, WANG AND MA 1994). We consider all these unsubstantiated reports as questionable because the senior author did not find this species in more than 1,000 river samples investigated over the years. Likely, the above mentioned authors confused *A. cultriforme scalpriforme* with other ciliates, especially pleurostomatids, some of which have a similar size and shape, as mentioned in the descriptive section. Thus, we do not document the many abiotic environmental data provided by DETCHEVA and her students. On the other hand, our South African habitat, that is, soil from the margin of a flat pond, indicates that this subspecies may occur as a guest in true limnetic habitats.

Reliable records of *A. cultriforme scalpriforme* are available from Germany, Austria, Brazil, and the Republic of South Africa. FOISSNER (1998) lists also a record from Australia. Basically, the reliable records indicate that *A. cultriforme scalpriforme* is a terrestrial cosmopolite rarer than *A. cultriforme cultriforme.*

Ontogenesis: Divisional morphogenesis was described in great detail by XU & FOISSNER (2005b), using raw material from a non-flooded Petri dish culture and specimens from a semipure culture. They emphasized that most stages depicted were seen in at least two, usually three or four specimens. This is also evident from the second series of figures showing division in specimens with two macronucleus pieces (Fig. 78g–n). XU & FOISSNER (2005b) could not follow the origin of the oral basket because the nematodesmata impregnated too faintly. The parental oral apparatus and dorsal brush do not show any changes.

Ordinary specimens (Fig. 77a–x, 78a–f, 130a–d, 131a–f; Table 27).

Stage 1 (very early dividers commencing production of oral kinetofragments and dorsal brush row 2; Fig. 77a–c, i, j, m; Table 27). When ontogenesis commences, the cells are on average slightly larger than morphostatic specimens (Table 27). Basal bodies are produced distinctly underneath mid-body, as evident from middle and late dividers which show that the proter is considerably longer than the opisthe (Table 27). The division plane is not transverse but slightly oblique from dorsal to ventral. Soon, some of the new basal bodies form minute kinetofragments, each likely consisting of monokinetids, except of a dikinetid at anterior end. There is also basal body proliferation in the opisthe's ciliary rows, as recognizable by the narrow and uneven spacing of the kinetids.

When entering stage 2, a remarkable change occurs in body shape, viz., a slight indentation develops in the prospective fission area (Fig. 77b, c), as in *Protospathidium serpens* and *Cultellothrix coemeterii* in which, however, the indentation develops slightly later and is transient. Basal body proliferation continues in all opisthe ciliary rows; the oral kinetofragments become dikinetidal and associated with minute blebs described below; and some somatic dikinetids appear to form a short dorsal brush row 2 (Fig. 77j, m). The macronucleus appears smoothened and the micronuclei distribute throughout the cell.

Stage 2 (early dividers separating kinetofragments from proter kineties and producing brush rows 3 and 1 and the excretory pores; Fig. 77d, k, l, n, o, 130a–d; Table 27).

Early dividers are distinctly longer and narrower than morphostatic specimens, that is, size increases from about 300 × 65 µm to 350 × 80 µm (Table 27). Body indentation and division blebs become more distinct in the prospective fission area (Fig. 77d, k). The division blebs are about 5 µm-sized convexities left and slightly above of the kinetofragments; they are hardly recognizable in vivo but are conspicuous in the scanning electron microscope (Fig. 130a–d, 131e). The blebs disappear in late dividers. The kinetofragment belt is now clearly recognizable and distinctly oblique (30°–45° in three appropriate specimens), likely due to faster growth of the dorsal than ventral side (Fig. 77d, 130a, b, 131a). The newly formed oral kinetofragments, which are longer on the ventral than dorsal side and detach from the proter's ciliary rows in a dorsoventral gradient, commence to curve rightwards on the dorsal surface and later also on the ventral (Fig. 77d, k, l, 130b). Slightly later, when three brush rows are recognizable, the fragments arrange transversely and become concave on the left side, while the right side fragments remain almost straight, but become more distinctly slanted (Fig. 77n, o). In this stage, the newly formed kinetofragments each consists of six to eleven, on average eight dikinetids, that is, they are almost complete; only one dikinetid is added in middle and late dividers (Table 27). This is evident from the following calculation: taking the averages of the fragment dikinetids (9) and ciliary rows (35), 315 dikinetids are available for the new circumoral kinety, which is close to the number found in proter dividers (Tables 26, 27). Thus, there is no second round of dikinetid proliferation, not even during post-divisional growth.

Many new somatic basal bodies have been produced in two regions of the ciliary

Table 27 Morphometric data on *Arcuospathidium cultriforme scalpriforme* from cultivated morphostatic and dividing specimens of South Africa (from XU & FOISSNER 2005b)

Characteristics[1]	Stages	$\overline{x}$	M	SD	SE	CV	Min	Max	n
Body, length	Morphostatic	313.9	300.0	45.4	10.4	14.5	240.0	410.0	19
	Very early	353.9	350.0	50.2	19.0	14.2	270.0	420.0	7
	Early	420.0	440.0	–	–	–	380.0	440.0	3
	Early middle	367.1	370.0	44.8	16.9	12.2	295.0	435.0	7
	Middle–early late	316.4	330.0	42.2	16.0	13.3	255.0	370.0	7
Body, width (maximum)	Morphostatic	65.8	67.0	15.0	3.4	22.7	44.0	95.0	19
	Very early	76.1	80.0	12.7	4.8	16.7	55.0	90.0	7
	Early	79.3	80.0	–	–	–	78.0	80.0	3
	Early middle	79.6	75.0	17.8	6.7	22.4	58.0	110.0	7
	Middle–early late	78.3	78.0	14.5	5.5	18.5	61.0	97.0	7
Body length:width, ratio	Morphostatic	4.9	4.9	0.9	0.2	17.4	3.2	6.4	19
	Very early	4.7	4.7	0.6	0.2	13.4	3.9	5.7	7
	Early	5.3	5.5	–	–	–	4.8	5.6	3
	Early middle	4.8	4.8	1.1	0.4	23.2	3.4	6.5	7
	Middle–early late	4.1	4.1	0.4	0.1	9.4	3.5	4.8	7
Proter, length	Middle	217.6	220.0	41.0	15.5	18.8	145.0	265.0	7
	Late	221.7	205.0	45.9	17.3	20.7	170.0	292.0	7
Proter, width	Middle	80.6	76.0	17.1	6.5	21.2	61.0	110.0	7
	Late	81.6	80.0	13.2	5.0	16.1	62.0	97.0	7

continued

Characteristics[1]	Stages	$\overline{x}$	M	SD	SE	CV	Min	Max	n
Proter length:width, ratio	Middle	2.8	2.5	0.6	0.2	23.1	2.1	4.0	7
	Late	2.7	2.7	0.6	0.2	20.0	2.1	3.5	7
Opisthe, length	Middle	145.7	150.0	20.4	7.7	14.0	108.0	165.0	7
	Late	143.6	150.0	28.5	10.8	19.9	95.0	180.0	7
Opisthe, width	Middle	79.1	75.0	19.3	7.3	24.3	53.0	110.0	7
	Late	78.7	80.0	13.2	5.0	16.7	57.0	92.0	7
Opisthe length:width, ratio	Middle	1.9	1.9	0.4	0.1	19.9	1.4	2.5	7
	Late	1.8	1.8	0.4	0.1	19.3	1.4	2.4	7
Proter:opisthe length, ratio	Middle	1.5	1.5	0.1	–	6.1	1.3	1.6	7
	Late	1.6	1.5	0.2	0.1	11.7	1.4	1.9	7
Macronucleus, length (spread, thus approximate)	Morphostatic	351.4	350.0	–	–	–	230.0	430.0	7
	Very early	368.6	360.0	–	–	–	320.0	430.0	7
	Early	370.0	390.0	–	–	–	320.0	400.0	3
Micronucleus, largest diameter	Morphostatic				about 3 µm				19
	Very early	3.4	3.0	0.8	0.3	22.9	3.0	5.0	7
	Early-middle				up to 6 µm				7
Circumoral dikinetids of proter, number	Middle-late	332.8	330.0	24.9	10.2	7.5	305.0	370.0	6
Dikinetids in corresponding opisthe's oral kinetofragments, number	Middle	275.0	285.0	–	–	–	240.0	300.0	3
	Late	330.0	320.0	–	–	–	320.0	350.0	3
Dikinetids in opisthe's individual oral kinetofragments, number	Early	8.6[2]	8.0	1.3	0.3	15.6	6.0	11.0	6
	Middle	8.8[2]	9.0	1.6	0.3	17.8	6.0	13.0	9
	Late	9.3[2]	9.0	1.0	0.2	10.8	7.0	12.0	9
Dikinetids in proter's brush row 2, number	Early to late stages	62.2	62.0	4.9	1.1	7.9	54.0	75.0	21
Dikinetids in opisthe's brush row 2, number	Early	21.0	20.0	9.9	3.8	47.4	10.0	41.0	7
	Middle	40.8	40.0	3.9	1.3	9.6	36.0	49.0	9
	Late	40.6	41.0	3.0	1.1	7.4	35.0	45.0	7

[1] Data based on mounted and protargol-impregnated (WILBERT's method) specimens from a semipure culture. Measurements in µm. CV – coefficient of variation in %, M – median, Max – maximum, Min – minimum, n – number of individuals investigated, SD – standard deviation, SE – standard error of arithmetic mean, $\overline{x}$ – arithmetic mean.

[2] Based on four kinetofragments each from 6 or 9 specimens.

rows, viz., the posterior area of the proter and the anterior of the opisthe. As usual, the new basal bodies are produced in front of the parental ones. However, some basal body production likely occurs throughout the division process and even in morphostatic specimens, as suggested by barren kinetids found throughout the ciliary rows of morphostatic cells. Possibly, these extra kinetids are a reservoir for cell growth and division. Dorsal brush row 2 is still growing and has now about one third of the dikinetids found in morphostatic specimens (Table 27). Next, brush row 3 is generated, and finally row 1. This curious sequence was observed in several specimens. All brush rows have an anterior tail of ordinary monokinetids with single ordinary cilia (Fig. 77l, q). Now, the

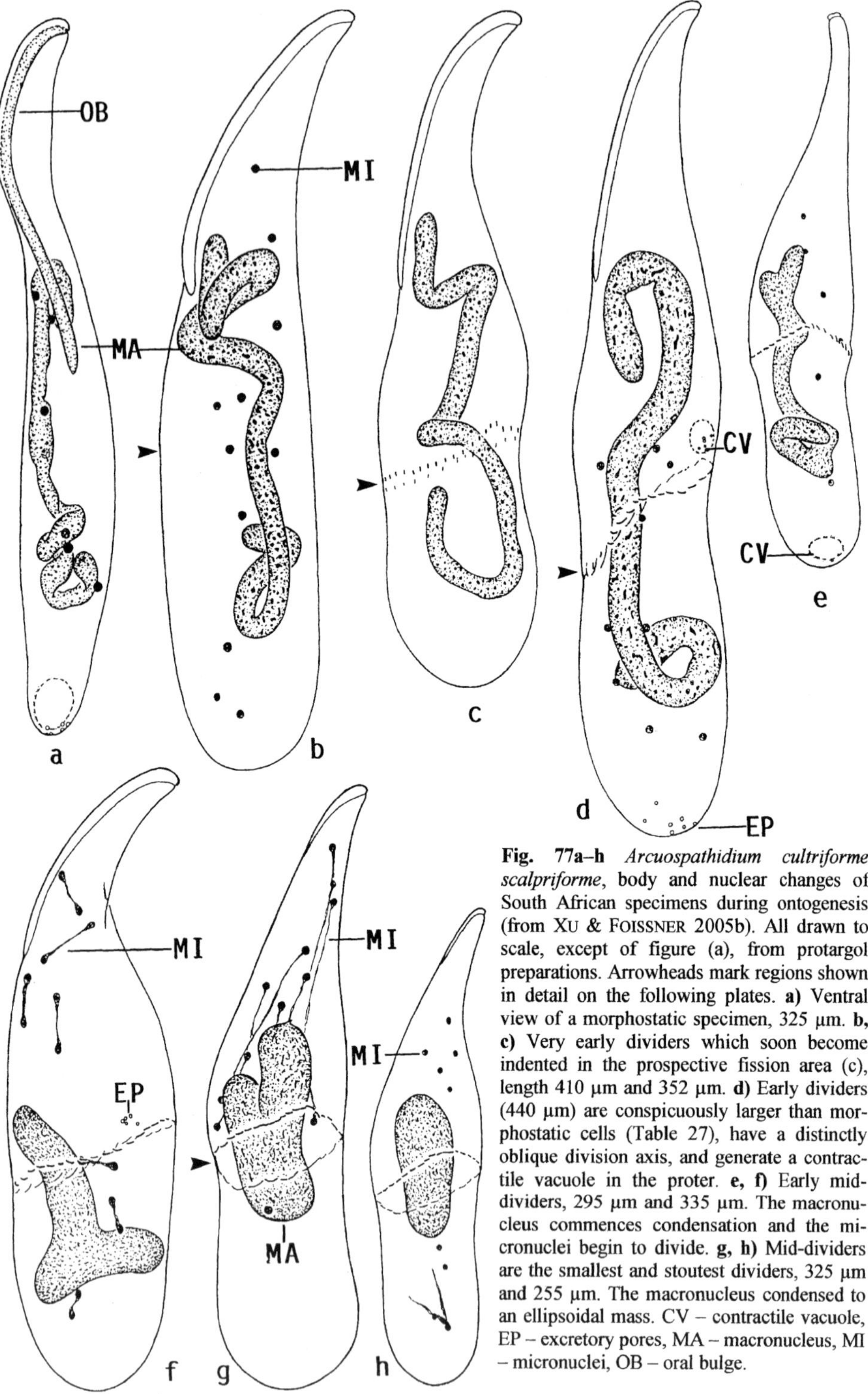

Fig. 77a–h *Arcuospathidium cultriforme scalpriforme*, body and nuclear changes of South African specimens during ontogenesis (from Xu & Foissner 2005b). All drawn to scale, except of figure (a), from protargol preparations. Arrowheads mark regions shown in detail on the following plates. **a)** Ventral view of a morphostatic specimen, 325 µm. **b, c)** Very early dividers which soon become indented in the prospective fission area (c), length 410 µm and 352 µm. **d)** Early dividers (440 µm) are conspicuously larger than morphostatic cells (Table 27), have a distinctly oblique division axis, and generate a contractile vacuole in the proter. **e, f)** Early mid-dividers, 295 µm and 335 µm. The macronucleus commences condensation and the micronuclei begin to divide. **g, h)** Mid-dividers are the smallest and stoutest dividers, 325 µm and 255 µm. The macronucleus condensed to an ellipsoidal mass. CV – contractile vacuole, EP – excretory pores, MA – macronucleus, MI – micronuclei, OB – oral bulge.

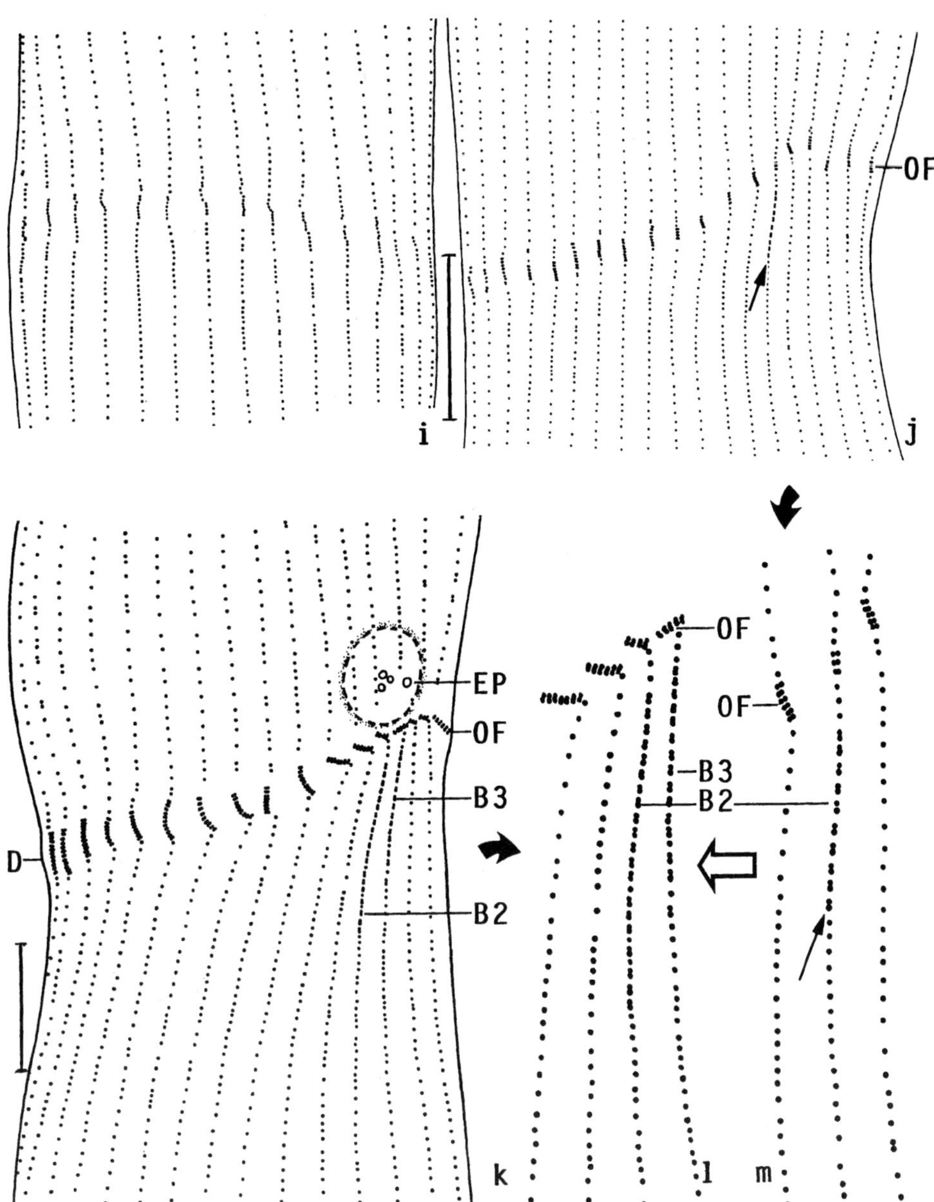

Fig. 77i–m *Arcuospathidium cultriforme scalpriforme*, ciliary pattern of very early and early dividers from South African population after protargol impregnation (from XU & FOISSNER 2005b). **i,** cp. Fig. 77b**)** Very early divider commencing basal body proliferation underneath mid-body. **j, m,** cp. Fig. 77c**)** Very early divider developing a body indentation in the prospective fission area and dorsal brush row 2 (arrows). The opisthe oral kinetofragments become dikinetidal. **k, l)** Early divider with growing dorsal brush rows 2 and 3, while row 1 is generated later. The opisthe oral kinetofragments separate from the parental ciliary rows in a dorsoventral gradient. The kinetofragments orientate ventrally and become curved due to the developing division blebs. A new proter contractile vacuole developed above the opisthe's dorsal brush rows. B2, 3 – dorsal brush rows, D – division blebs, EP – excretory pores, OF – oral kinetofragments. Scale bars 20 μm.

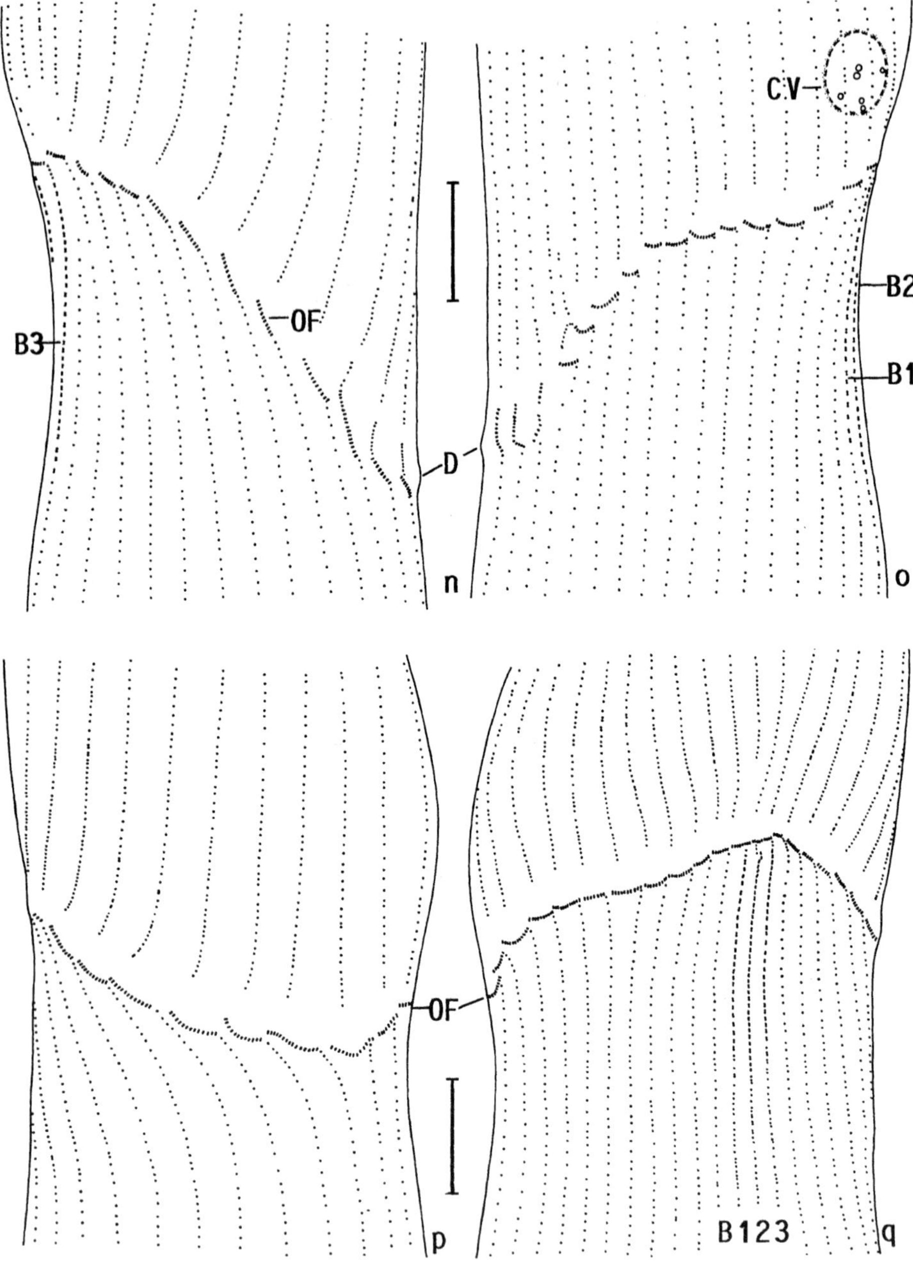

Fig. 77n–q *Arcuospathidium cultriforme scalpriforme*, ciliary pattern of an early divider and an early mid-divider from South African population after protargol impregnation (from Xu & Foissner 2005b). **n, o,** cp. Fig. 77d) Right and left side view of an early divider with fully developed oral kinetofragments and growing brush rows. The kinetofragments are longer on the right than left side, where they arranged horizontally. **p, q,** cp. Fig. 77g) Ventrolateral and dorsolateral view of a mid-divider. The opisthe ciliature is almost complete, that is, the oral kinetofragments are now transversely arranged and the new brush rows have about two thirds of their dikinetids; the final number is obtained in post-dividers or morphostatic cells. B1-3 – dorsal brush rows, CV – contractile vacuole, D – division blebs, OF – oral kinetofragments. Scale bars 20 μm.

nuclear apparatus shows some changes (Fig. 77d): the macronucleus becomes less tortuous, that is, commences to contract, and the micronuclei appear more or less distinctly inflated. Further, a new contractile vacuole and new excretory pores are generated at the proter's posterior end. Usually, the new pores are located dorsally or dorsolaterally, but in some specimens they are near the ventral side.

Stage 3 (early mid-dividers with condensing macronucleus and dividing micronuclei; Fig. 77e, f, 131a–c, e, f; Table 27). Early mid-dividers are smaller and stouter than early dividers, but larger than morphostatic cells (Table 27). The indentation underneath the prospective fission area is still recognizable, but becomes indistinct, especially in cells prepared for scanning electron microscopy (Fig. 77e, f, 131a). The macronucleus commences to shorten towards the centre, as indicated by the swollen ends; the nucleoli gradually disappear. The micronuclei grow up to 6 μm, that is, double size. In the post-brush area, one can observe the origin of dikinetids within the parental ciliary rows: a bristle is generated ahead of each parental cilium which gradually decreases in length eventually becoming a bristle (Fig. 77f).

Stage 4 (mid-dividers with transversely oriented, concave kinetofragments and globular macronucleus; Fig. 77g, h, p, q, 131a–c, e, f; Table 27). These cells are the smallest and stoutest dividers, approaching the length of morphostatic specimens (Table 27). The subequatorial indentation is indistinct, while the division blebs are still distinct (Fig. 77g, h, 131a, b, e). All kinetofragments are now transversely arranged and more or less distinctly concave. The new brush rows have about two thirds of their dikinetids. No further increase of brush dikinetids occurs in late dividers and early post-dividers, suggesting that the final number is obtained only in late post-dividers. The macronucleus condenses to a globular, fibrous mass with lightly impregnated karyoplasm. The dividing micronuclei move apart, still being connected by a bundle of fibres.

Stage 5 (late dividers with forming division furrow and extending macronucleus; Fig. 77t, u; Table 27). Late dividers grow in length, the division blebs disappear, and the division furrow becomes recognizable, dividing the cell in an ellipsoidal opisthe shorter by about one third than the conspicuously conical proter (Table 27). The body constriction causes the onset of kinetofragment alignment, that is, the distances between the individual kinetofragments become smaller. The globular macronuclear mass of the mid-dividers extends and grows to a more or less curved rod with fibrous contents. Most micronuclei completed fission and commence to condense.

Stage 6 (very late dividers with aligning kinetofragments, forming oral bulge, and separating daughter cells; Fig. 78r, s, v, w, x, 79n; Table 27). Very late dividers may grow to a length of 565 μm and are distinctly furrowed. The developing opisthe oral area is conspicuously clavate and the oral bulge, which develops along its margin, becomes recognizable as a bare protuberance. The oral kinetofragments align to the new opisthe circumoral kinety. This process is obviously complex because many fragments overlap more or less and some become strongly curved, as in *A. muscorum* (BERGER et al. 1983). Now, the ciliary rows are distinctly separate from the kinetofragments but still straight, showing that the genus-specific pattern in obtained only in post-dividers. The macronucleus becomes a long, tortuous strand and shows fibrous contents with many minute globules, likely developing nucleoli (Fig. 77w). Finally, the daughters separate. The anterior daughter is longer than the posterior by about one third; the latter is more or less distinctly pointed in the split area. The deeply impregnated micronuclei are still scattered through the cell.

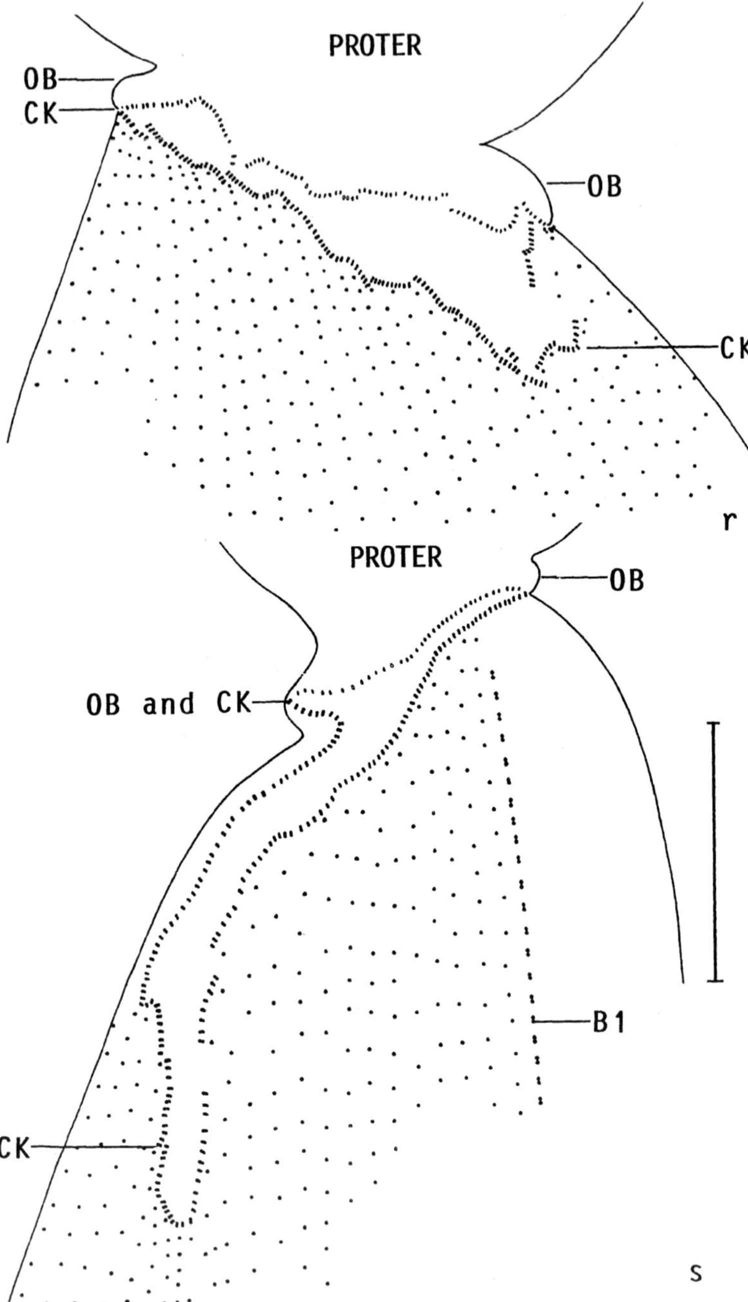

Fig. 77r, s *Arcuospathidium cultriforme scalpriforme*, ciliary pattern of very late dividers from South African population after protargol impregnation (from XU & FOISSNER 2005b). These are details of the specimens shown in figures 77v, w. They demonstrate the complex processes of oral bulge growth and alignment of the kinetofragments to the opisthe's circumoral kinety. B1 – dorsal brush row 1, CK – circumoral kinety, OB – oral bulge. Scale bar 20 µm.

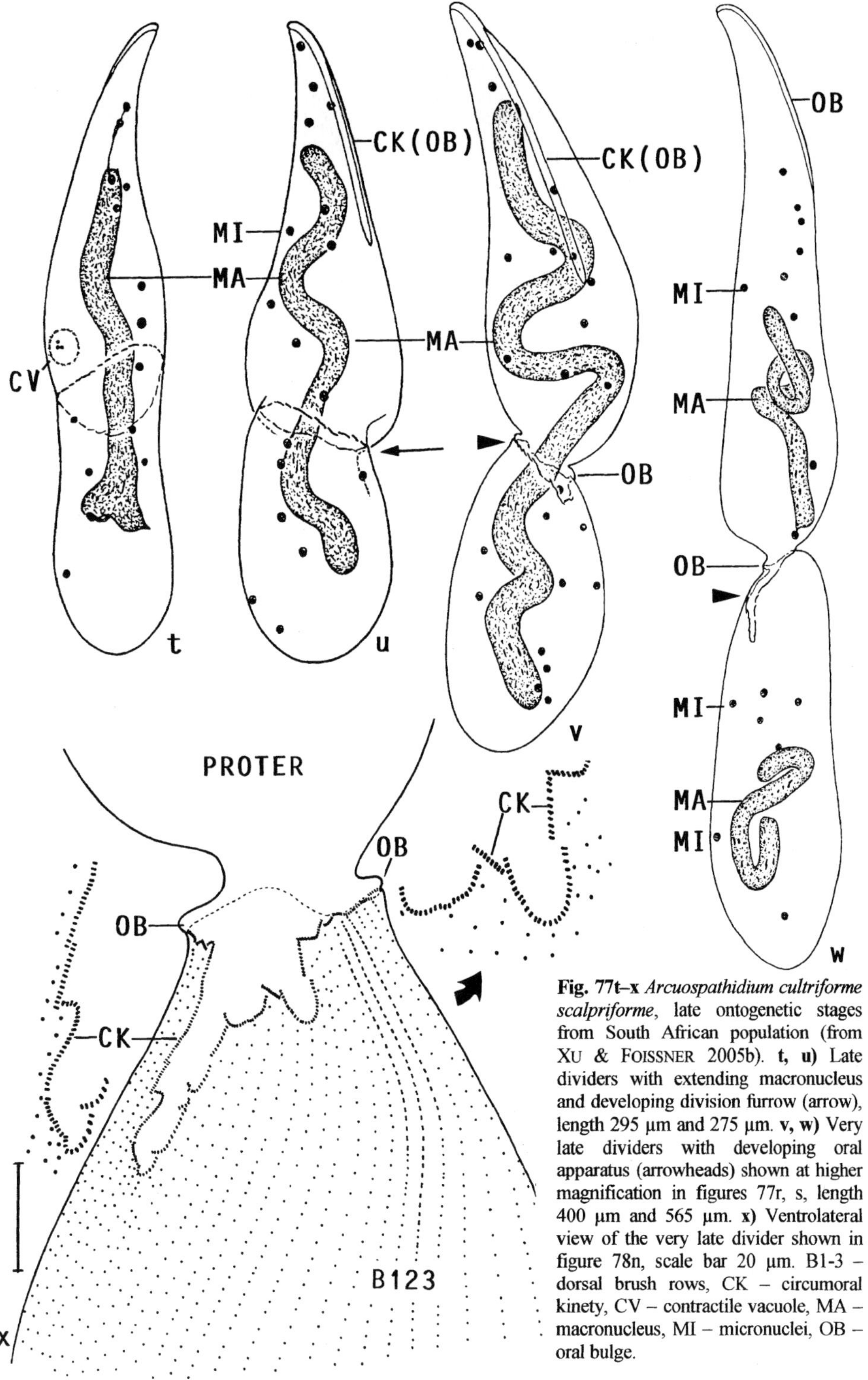

Fig. 77t–x *Arcuospathidium cultriforme scalpriforme*, late ontogenetic stages from South African population (from XU & FOISSNER 2005b). **t, u)** Late dividers with extending macronucleus and developing division furrow (arrow), length 295 µm and 275 µm. **v, w)** Very late dividers with developing oral apparatus (arrowheads) shown at higher magnification in figures 77r, s, length 400 µm and 565 µm. **x)** Ventrolateral view of the very late divider shown in figure 78n, scale bar 20 µm. B1-3 – dorsal brush rows, CK – circumoral kinety, CV – contractile vacuole, MA – macronucleus, MI – micronuclei, OB – oral bulge.

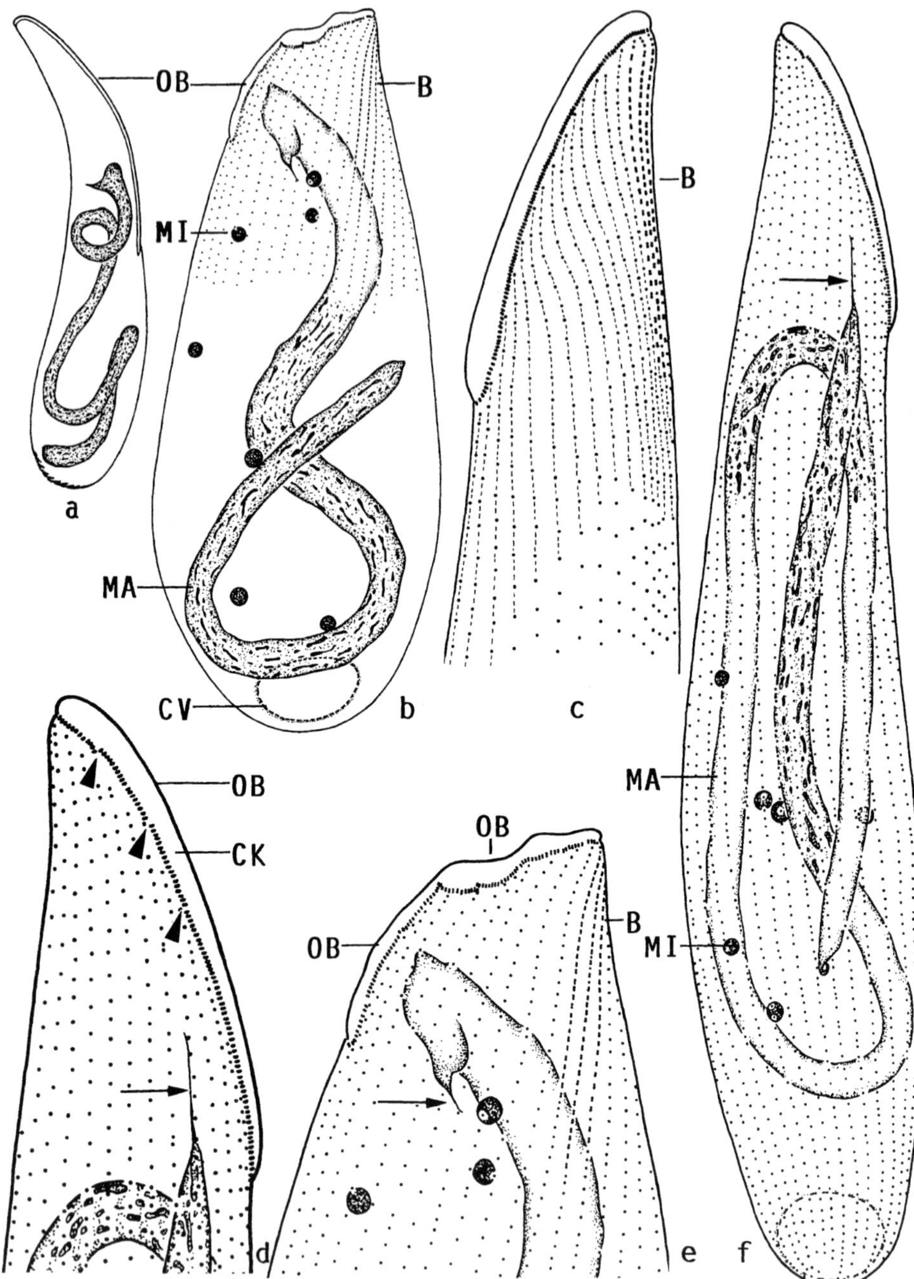

Fig. 78a–f *Arcuospathidium cultriforme scalpriforme*, ciliary pattern of post-dividers from South African population after protargol impregnation (from XU & FOISSNER 2005b). **a)** A proter post-divider recognizable by the long oral bulge, length 290 µm. **b, e)** Very early opisthe post-divider with pointed macronucleus end (arrow), length 142 µm. **c, d, f)** A late opisthe post-divider recognizable by the short oral bulge (26% of body length) and the pointed anterior macronucleus end (arrow), length 215 µm. The circumoral kinety still shows small irregularities (d, arrowheads) and the ciliary rows commence to curve dorsally (c). B – dorsal brush, CK – circumoral kinety, CV – contractile vacuole, MA – macronucleus, MI – micronuclei, OB – oral bulge.

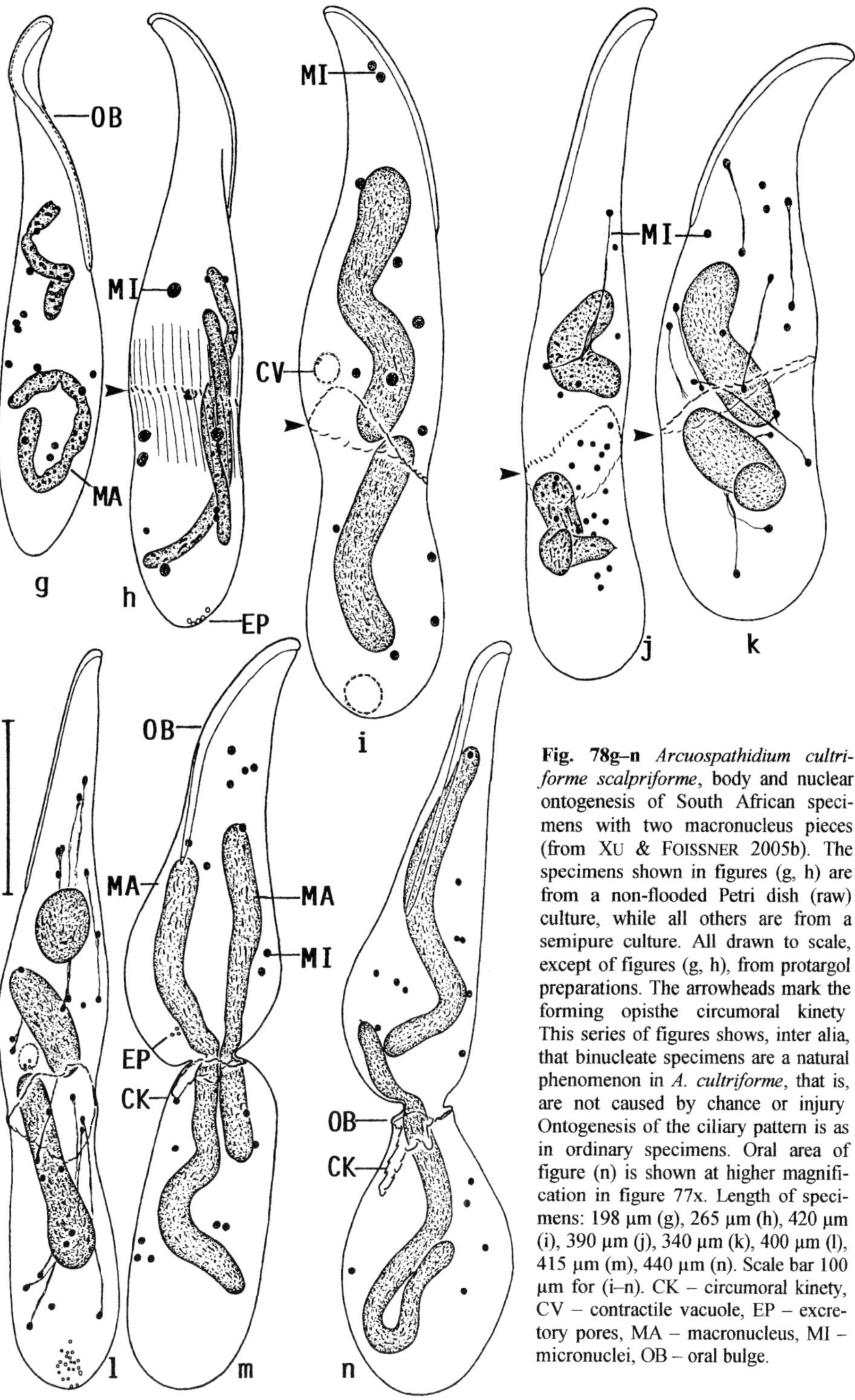

Fig. 78g–n *Arcuospathidium cultriforme scalpriforme*, body and nuclear ontogenesis of South African specimens with two macronucleus pieces (from XU & FOISSNER 2005b). The specimens shown in figures (g, h) are from a non-flooded Petri dish (raw) culture, while all others are from a semipure culture. All drawn to scale, except of figures (g, h), from protargol preparations. The arrowheads mark the forming opisthe circumoral kinety This series of figures shows, inter alia, that binucleate specimens are a natural phenomenon in *A. cultriforme*, that is, are not caused by chance or injury Ontogenesis of the ciliary pattern is as in ordinary specimens. Oral area of figure (n) is shown at higher magnification in figure 77x. Length of specimens: 198 µm (g), 265 µm (h), 420 µm (i), 390 µm (j), 340 µm (k), 400 µm (l), 415 µm (m), 440 µm (n). Scale bar 100 µm for (i–n). CK – circumoral kinety, CV – contractile vacuole, EP – excretory pores, MA – macronucleus, MI – micronuclei, OB – oral bulge.

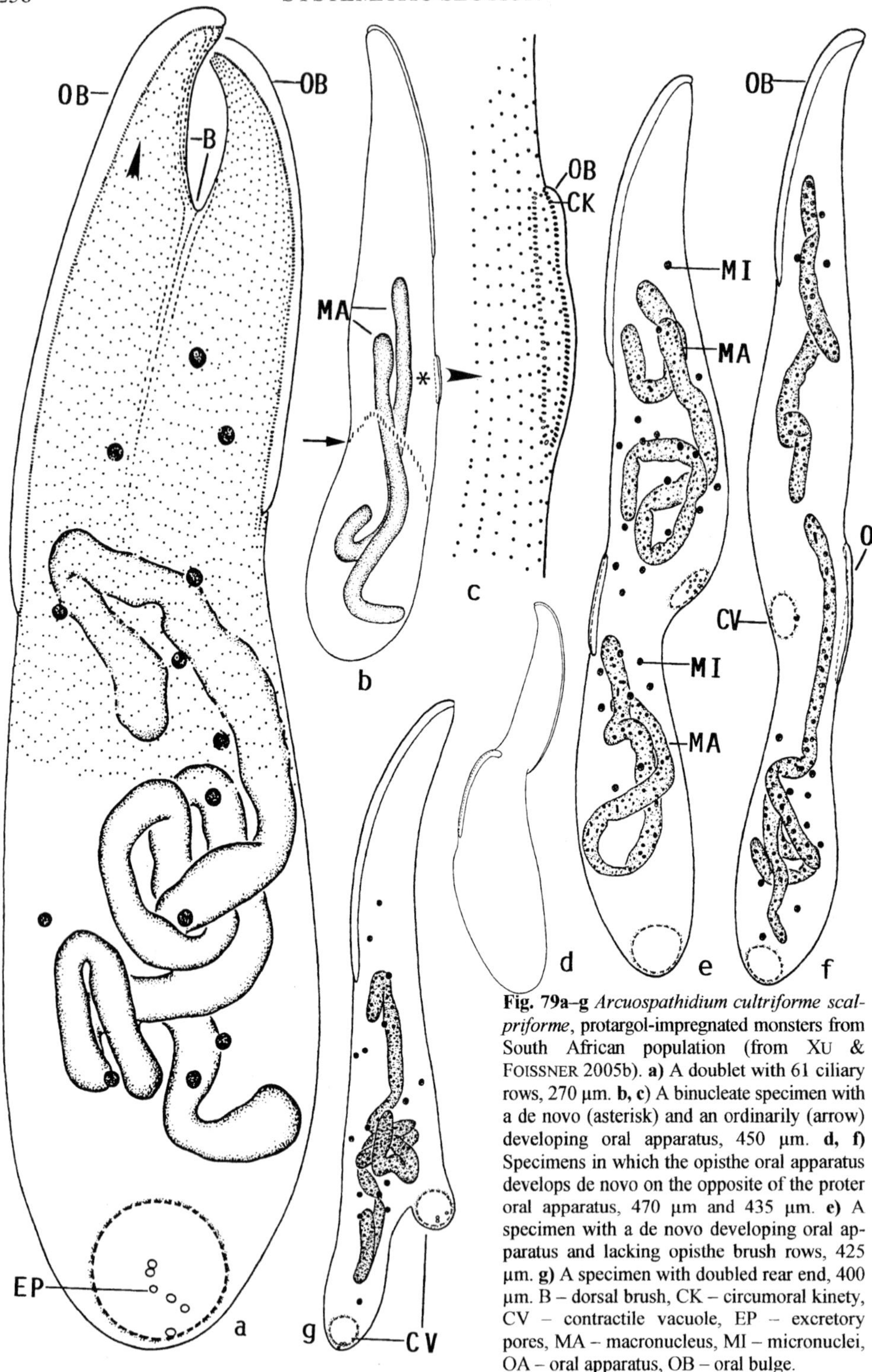

Fig. 79a–g *Arcuospathidium cultriforme scalpriforme*, protargol-impregnated monsters from South African population (from Xu & Foissner 2005b). **a)** A doublet with 61 ciliary rows, 270 µm. **b, c)** A binucleate specimen with a de novo (asterisk) and an ordinarily (arrow) developing oral apparatus, 450 µm. **d, f)** Specimens in which the opisthe oral apparatus develops de novo on the opposite of the proter oral apparatus, 470 µm and 435 µm. **e)** A specimen with a de novo developing oral apparatus and lacking opisthe brush rows, 425 µm. **g)** A specimen with doubled rear end, 400 µm. B – dorsal brush, CK – circumoral kinety, CV – contractile vacuole, EP – excretory pores, MA – macronucleus, MI – micronuclei, OA – oral apparatus, OB – oral bulge.

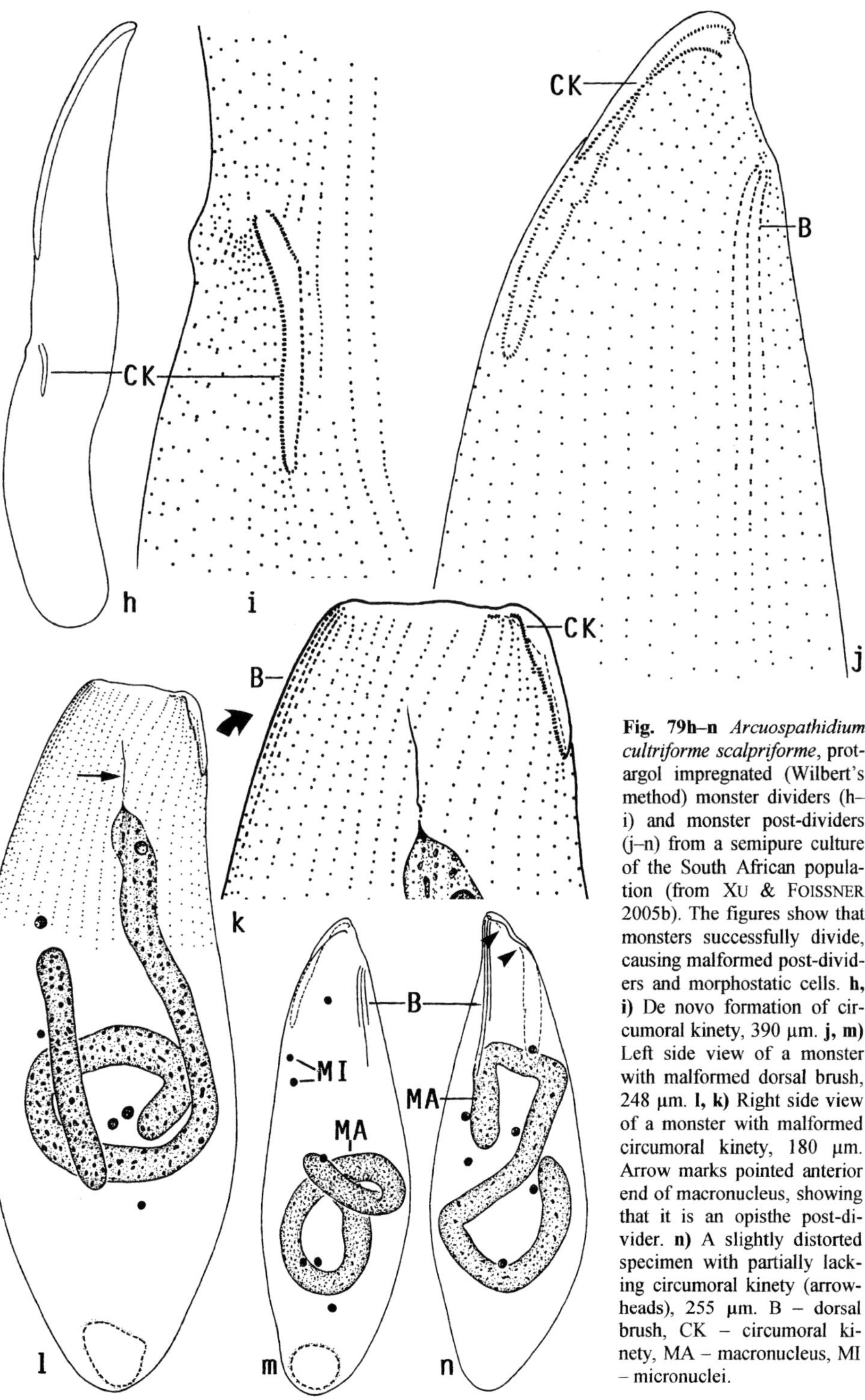

Fig. 79h–n *Arcuospathidium cultriforme scalpriforme*, protargol impregnated (Wilbert's method) monster dividers (h–i) and monster post-dividers (j–n) from a semipure culture of the South African population (from XU & FOISSNER 2005b). The figures show that monsters successfully divide, causing malformed post-dividers and morphostatic cells. **h, i)** De novo formation of circumoral kinety, 390 µm. **j, m)** Left side view of a monster with malformed dorsal brush, 248 µm. **l, k)** Right side view of a monster with malformed circumoral kinety, 180 µm. Arrow marks pointed anterior end of macronucleus, showing that it is an opisthe post-divider. **n)** A slightly distorted specimen with partially lacking circumoral kinety (arrowheads), 255 µm. B – dorsal brush, CK – circumoral kinety, MA – macronucleus, MI – micronuclei.

Stage 7 (early opisthe post-dividers with small size, pointed anterior macronucleus end, and short dorsal brush and circumoral kinety/oral bulge with more or less conspicuous irregularities; Fig. 78b–f). Very early opisthe post-dividers are bursiform and still have a rather distorted circumoral kincty and almost meridionally arranged ciliary rows (Fig. 78b, e). The oral dikinetids are more narrowly spaced than in morphostatic specimens, viz., there are about 17 vs. 11 in an area of 10 µm. This explains why the bulge can grow (become longer) without a second round of basal body production.

Early (to late) post-dividers are already spatulate and have a distinct but short oral bulge, while the circumoral kinety still shows some irregularities. The ciliary rows commence to assume the typical *Arcuospathidium* pattern, that is, curve dorsally in the anterior portion. The dorsal brush is still incomplete, as explained in stage 4. The macronucleus still has a pointed or even filiform anterior end and grows to a long, tortuous strand with distinct nucleoli. The micronuclei commence to arrange near the macronucleus (78c, d, f).

Specimens with two macronucleus strands (Fig. 77x, 78g–n): Specimens with two, comparatively short macronucleus strands occur in both the non-flooded Petri dish culture (5%, Fig. 78g, h) and the semipure culture, from which XU & FOISSNER (2005b) could reconstruct a full division series (Fig. 78g–n) showing the following: (i) the two macronucleus pieces divide individually, that is, do not fuse to a single, globular mass as in ordinary specimens (Fig. 77g, h) and multimacronucleate species (FOISSNER et al. 2002); (ii) the binucleate state is stable and heritable, although figure 78n indicates that the opisthe might become mononucleate; (iii) thus, binucleate specimens are a natural phenomenon in *A. cultriforme*, that is, are not caused by chance or injury; and (iv) ontogenesis of the oral and somatic ciliary pattern as well as the contractile vacuole is as in ordinary specimens (Fig. 77x).

Monsters (Fig. 79a–n): Monster formation was observed by several authors (WENZEL 1955, WILLIAMS 1980; see general section), but only XU & FOISSNER (2005) provided detailed data on their ciliary pattern, as described in the following paragraph.

Monsters occurred in both the non-flooded Petri dish (raw) culture, albeit very rarely (< 1%), and in the semipure culture, where they were frequent (~14%). XU & FOISSNER (2005b) could not clarify the reason(s) for monster formation, but they were most common during a "bloom" of the culture. The following observations were made from protargol slides: (i) monsters occur in specimens with a single macronucleus strand (Fig. 79a, e–g, l) and in cells with two strands (Fig. 79b), as described above; (ii) monsters can divide and grow to ordinary size, as evident, for instance, by post-dividers (Fig. 79l–n) and the mirror-image doublet shown in figure (79a); (iii) the new oral apparatus can develop from underneath the parental oral bulge to near mid-body on the ventral or dorsal, but never lateral surface (Fig. 79b, c–i); (iv) there is indication that specimens can produce an ordinary opisthe and a monster proter by generating a third oral apparatus above the fission area (Fig. 79b); (v) the new circumoral kinety, likely invariably composed of dikinetids, does not develop from somatic kinetofragments but interkinetally, that is, de novo and grows by proliferation of dikinetids within the forming kinety (Fig. 79c, i); and (vi) the new oral apparatus, even if minute, and in an early stage of development has associated extrusomes, while ordinary dividers show extrusomes only in late stages.

Arcuospathidium cultriforme megastoma FOISSNER, AGATHA & BERGER, 2002 (Fig. 80e–m, 136h; Table 28)

2002 *Arcuospathidium cultriforme megastoma* FOISSNER, AGATHA & BERGER, Denisia, 5: 300 (Type slides with protargol-impregnated specimens from type locality are deposited in the Oberösterreichische Landesmuseum in Linz, Upper Austria.).

Improved diagnosis: Mouth (oral bulge) about two thirds of body length. Extrusomes in a row each right and left of oral bulge midline.

Type locality: Mud and soil from road puddles in the Bambatsi Guest Farm, Namibia, E15°25' S20°10'.

Etymology: Apposite noun composed of the Greek prefix *mega* (very large) and the Greek noun *stoma* (mouth), referring to the extraordinarily long mouth.

Description: Size 180–300 × 20–45 μm in vivo, usually near 250 × 30 μm; length:width ratio 6.2–9.6:1, on average 8.8:1 in protargol preparations, and thus slightly more slender than *A. cultriforme cultriforme* and *A. cultriforme scalpriforme* (both about 7:1). Knife-shaped to very narrowly spatulate with anterior body region distinctly narrowed and more or less widely curved dorsally; flattened only in anterior half (Fig. 80e–g; Table 28). Macronucleus in posterior body half, about 90 μm long and tortuous. Micronuclei attached and near macronucleus, about 3 μm across in vivo, that is, small as compared to body and macronucleus size (Fig. 80e, g). Contractile vacuole in rear end, several excretory pores in posterior pole area. Extrusomes in a row each right and left of midline of oral bulge, basically rod-shaped with slightly narrowed and rounded ends, making them indistinctly ellipsoidal or very narrowly cuneate, 5–7 × 0.6–0.8 μm in size, that is, rather thick and highly refractive, impregnate occasionally with protargol (Fig. 80i, l, m, 136h); immature cytoplasmic extrusomes distinctly fusiform, impregnate black. Cytoplasm colourless, in some specimens packed with lipid droplets, food vacuoles up to 10 μm across, and about 8 μm long pharyngeal baskets of prey ciliates (*Drepanomonas*, *Leptopharynx*); none of the 15 specimens seen contained large prey. Creeps vividly on microscope slide curving mouth area to and fro.

Densely ciliated, except in anterior third. On average 30 ciliary rows distinctly curved dorsally at both sides of oral bulge and successively abutting on circumoral kinety, as typical for the genus. Three dorsal rows anteriorly differentiated to a long, heterostichad brush with up to 5 μm long bristles; a minute fourth row, likely a vestige from patterning of rows 1–3, composed of only 3–5 dikinetids right of row 1 in about half of specimens; rows frequently have some irregularities, such as short overlapping or non-overlapping breaks (Fig. 80e, g, h, j, m; Table 28). Brush row 1 slightly shorter than row 2, composed of an average of 80 dikinetids slightly more narrowly spaced than those of row 1; brush row 3 composed of an average of 33 dikinetids more widely spaced than those of rows 1 and 2 (Fig. 80g, j, m; Table 28).

Oral bulge conspicuous because bright due to the thick extrusomes contained and its enormous size occupying almost 60% of body length on average, straight to distinctly curved, depending on state of anterior body third; only 5 μm wide and 2–3 μm high, and thus indistinctly separated from body proper. Circumoral kinety very narrowly oblong,

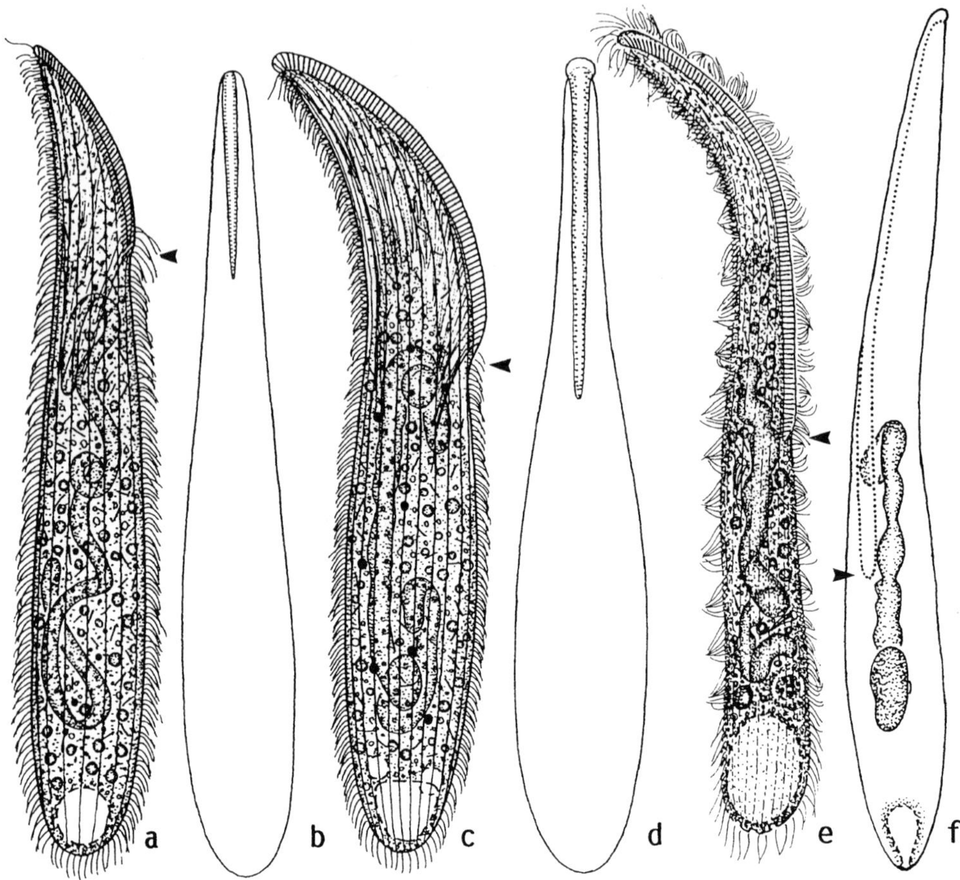

Fig. 80a–f Comparison of the subspecies of *Arcuospathidium cultriforme* from life (a–e) and after protargol impregnation (f). From FOISSNER et al. (2002). Arrowheads mark posterior end of oral bulge. **a, b)** *Arcuospathidium cultriforme cultriforme*, right side and ventral view, length 250 µm. **c, d)** *Arcuospathidium cultriforme scalpriforme*, right side and ventral view, length 270 µm. **e, f)** *Arcuospathidium cultriforme megastoma*, right side and ventrolateral view, length 250 µm.

not widened anteriorly, composed of densely spaced dikinetids associated with very fine nematodesmata producing a narrow, somewhat branched bundle along oral bulge (Fig. 80e–m).

Occurrence and ecology: To date found only at type locality, where it was rare in the non-flooded Petri dish culture. The slender shape indicates that it prefers terrestrial habitats as do the other subspecies.

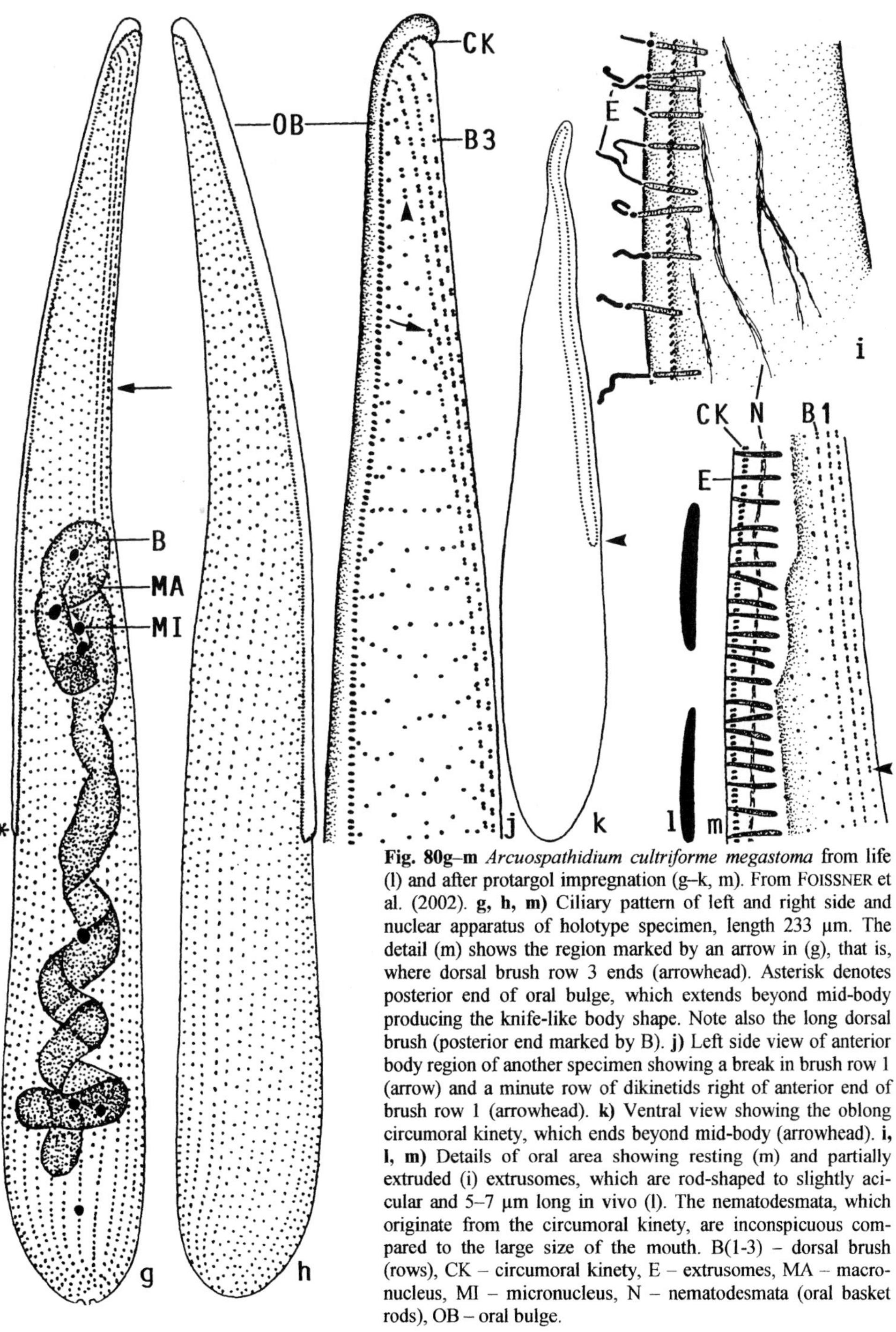

Fig. 80g–m *Arcuospathidium cultriforme megastoma* from life (l) and after protargol impregnation (g–k, m). From FOISSNER et al. (2002). **g, h, m)** Ciliary pattern of left and right side and nuclear apparatus of holotype specimen, length 233 µm. The detail (m) shows the region marked by an arrow in (g), that is, where dorsal brush row 3 ends (arrowhead). Asterisk denotes posterior end of oral bulge, which extends beyond mid-body producing the knife-like body shape. Note also the long dorsal brush (posterior end marked by B). **j)** Left side view of anterior body region of another specimen showing a break in brush row 1 (arrow) and a minute row of dikinetids right of anterior end of brush row 1 (arrowhead). **k)** Ventral view showing the oblong circumoral kinety, which ends beyond mid-body (arrowhead). **i, l, m)** Details of oral area showing resting (m) and partially extruded (i) extrusomes, which are rod-shaped to slightly acicular and 5–7 µm long in vivo (l). The nematodesmata, which originate from the circumoral kinety, are inconspicuous compared to the large size of the mouth. B(1-3) – dorsal brush (rows), CK – circumoral kinety, E – extrusomes, MA – macronucleus, MI – micronucleus, N – nematodesmata (oral basket rods), OB – oral bulge.

Table 28 Morphometric data on *Arcuospathidium lorjeae* (upper line) and *Arcuospathidium cultriforme megastoma* (lower line). From FOISSNER et al. (2002)

Characteristics[1]	$\overline{x}$	M	SD	SE	CV	Min	Max	n
Body, length	173.50	170.0	25.2	5.8	14.5	138.0	215.0	19
	219.80	210.0	39.8	12.6	18.1	165.0	280.0	10
Body, width	15.60	16.0	2.8	0.6	17.8	9.0	20.0	19
	29.30	27.5	7.3	2.3	24.8	20.0	44.0	10
Body length:width, ratio	11.40	11.3	2.1	0.5	18.5	8.4	16.1	19
	7.70	7.6	1.2	0.4	15.7	6.2	9.6	10
Oral bulge, length	59.00	58.0	11.5	2.6	19.5	38.0	80.0	19
	128.70	124.5	27.4	8.7	21.3	80.0	165.0	10
Body length:oral bulge length, ratio	3.00	2.9	0.5	0.1	15.6	2.3	3.9	19
	1.70	1.7	0.2	0.1	13.3	1.4	2.3	10
Oral bulge length:body length, ratio	0.34	0.34	0.1	0.1	15.0	0.26	0.44	19
	0.59	0.59	0.1	0.1	12.1	0.43	0.70	10
Circumoral kinety to last dikinetid of brush row 1, distance	47.40	50.0	9.0	2.1	19.1	25.0	60.0	19
	75.30	78.0	22.5	7.5	29.8	39.0	110.0	9
Dorsal brush row 1, number of dikinetids	33.80	33.0	6.7	1.5	19.7	23.0	44.0	19
	52.30	46.0	15.4	5.1	29.4	35.0	80.0	9
Circumoral kinety to last dikinetid of brush row 2, distance	53.40	53.0	8.0	1.8	15.0	37.0	68.0	19
	80.10	83.0	19.3	6.4	24.1	48.0	110.0	9
Dorsal brush row 2, number of dikinetids	53.20	53.0	11.1	2.6	20.9	36.0	75.0	19
	62.00	60.0	14.5	4.8	23.3	42.0	82.0	9
Circumoral kinety to last dikinetid of brush row 3, distance	42.80	44.0	7.3	1.7	17.1	27.0	52.0	19
	51.70	52.0	12.2	4.1	23.5	30.0	67.0	9
Dorsal brush row 3, number of dikinetids	24.40	25.0	4.4	1.0	17.9	18.0	33.0	19
	32.70	30.0	8.0	2.7	24.5	25.0	50.0	9
Anterior body end to macronucleus, distance	74.20	71.0	13.6	3.1	18.3	47.0	105.0	19
	87.80	89.0	22.3	7.1	25.4	46.0	120.0	10
Macronucleus figure, length	63.20	62.0	11.0	2.5	17.4	46.0	87.0	19
	93.40	79.5	30.2	9.5	32.3	55.0	150.0	10
Macronucleus, length (spread; approximate)	86.10	90.0	–	–	–	60.0	110.0	19
	121.00	100.0	–	–	–	80.0	200.0	10
Macronucleus, width	6.40	6.0	0.8	0.2	12.0	5.0	8.0	19
	7.70	7.5	1.0	0.3	12.3	7.0	10.0	10
Micronuclei, length	2.80	3.0	–	–	–	2.2	3.3	19
	2.50	2.6	–	–	–	2.0	3.0	10
Micronuclei, width	2.60	2.5	–	–	–	2.0	3.2	19
	2.50	2.5	–	–	–	2.0	3.0	10
Micronuclei, number	8.90	8.5	2.8	0.7	31.0	5.0	16.0	18
	13.10	15.0	4.4	1.7	33.6	8.0	19.0	7
Ciliary rows, number	17.10	17.0	1.4	0.3	8.0	14.0	20.0	19
	30.20	30.0	3.2	1.1	10.4	24.0	35.0	0
Dorsal brush rows, number	3.00	3.0	0.0	0.0	0.0	3.0	3.0	19
	3.0[2]	3.0	0.0	0.0	0.0	3.0	3.0	9
Ciliated kinetids in a lateral kinety, number	81.30	82.5	22.0	5.2	27.1	46.0	120.0	18
	99.60	100.0	24.8	9.4	24.9	70.0	150.0	7

[1] Data based on mounted, protargol-impregnated (FOISSNER's method), and randomly selected specimens from non-flooded Petri dish cultures. Measurements in µm. CV – coefficient of variation in %, M – median, Max – maximum, Min – minimum, n – number of individuals investigated, SD – standard deviation, SE – standard error of arithmetic mean, $\overline{x}$ – arithmetic mean.

[2] A fourth, short row right of row 1 occurs in about half of the specimens (see text).

Arcuospathidium lorjeae FOISSNER, AGATHA & BERGER, 2002 (Fig. 80 n–z; 136a–g; Table 28)

2002 *Arcuospathidium lorjeae* FOISSNER, AGATHA & BERGER, Denisia, 5: 295 (Type slides with protargol-impregnated specimens from type locality are deposited in the Oberösterreichische Landesmuseum in Linz, Upper Austria. Further, voucher specimens from the Brazilian population are marked on the type slides of *Cephalospathula brasiliensis*; see occurrence and ecology section.).

Diagnosis: Size about 200 × 18 µm in vivo. Cylindroidal with strongly oblique, very slenderly oblong oral bulge occupying approximately 1/3 of body length. Macronucleus long and tortuous; multimicronucleate. Extrusomes rod-shaped, about 6 × 0.5 µm, arranged in a row each right and left of oral bulge midline. On average 17 ciliary rows, 3 anteriorly modified to conspicuous, heterostichad dorsal brush occupying 31% of body length. Brush bristles up to 10 µm long: row 1 composed of an average of 34 dikinetids, row 2 of 53, and row 3 of 24 dikinetids followed by a long, monokinetidal bristle tail.

Type locality: Mud and soil from road puddles in the Bambatsi Guest Farm, Namibia, E15°25' S20°10'.

Dedication: Dr. SABINE AGATHA dedicated this species to her teacher and friend, Dr. JEANNETTE CORNELIE RIEDEL-LORJÉ, Hamburg.

Description: Size 150–250 × 10–25 µm in vivo, usually near 200 × 18 µm; length:width ratio 8.4–16.1:1, on average about 11:1 in vivo and protargol preparations (Table 28). Cylindroidal with steep, hardly projecting oral bulge; oral area, that is, anterior body third more or less distinctly curved dorsally and laterally flattened up to 2:1 (Fig. 80n, o, u–x; 136a). Macronucleus in central body portion, long and up to 2:1 flattened, more or less distinctly moniliform, coiled and tortuous; nucleoli minute, numerous. On average 13 globular micronuclei attached and near macronucleus. Contractile vacuole in rear end, about six excretory pores in posterior pole area. Extrusomes in an indistinct row each right and left of oral bulge midline and scattered in cytoplasm, do not impregnate with the protargol method used, except for a certain cytoplasmic developmental stage; individual extrusomes rod-shaped with slightly narrowed and rounded ends, about 6 × 0.5 µm in size (Fig. 80o, r; 136b). Extrusomes of Brazilian population very similar, that is, about 5 × 0.5 µm in size and with slightly narrowed ends; a second, minute type (1.5–2 µm long) was probably overlooked in the type population. Cortex flexible, conspicuous because containing narrowly spaced rows of highly refractive granules about 1 × 0.5 µm in size (Fig. 80s, t). Cytoplasm colourless and hyaline, especially in oral area, contains few to many lipid droplets up to 10 µm across. Food not known, likely other ciliates. Swims and glides rather rapidly showing great flexibility.

Cilia about 8 µm long in vivo, widely spaced in oral area, arranged in an average of 17 equidistant rows anteriorly directed dorsally at both sides of circumoral kinety, as typical for the genus. Three dorsal ciliary rows anteriorly differentiated to conspicuous dorsal brush occupying 31% of body length on average; brush bristles about 10 µm (up to 8 µm in Brazilian specimens) long in middle region becoming gradually shorter anteriorly and posteriorly, up to twice as thick as ordinary somatic cilia and thus forming

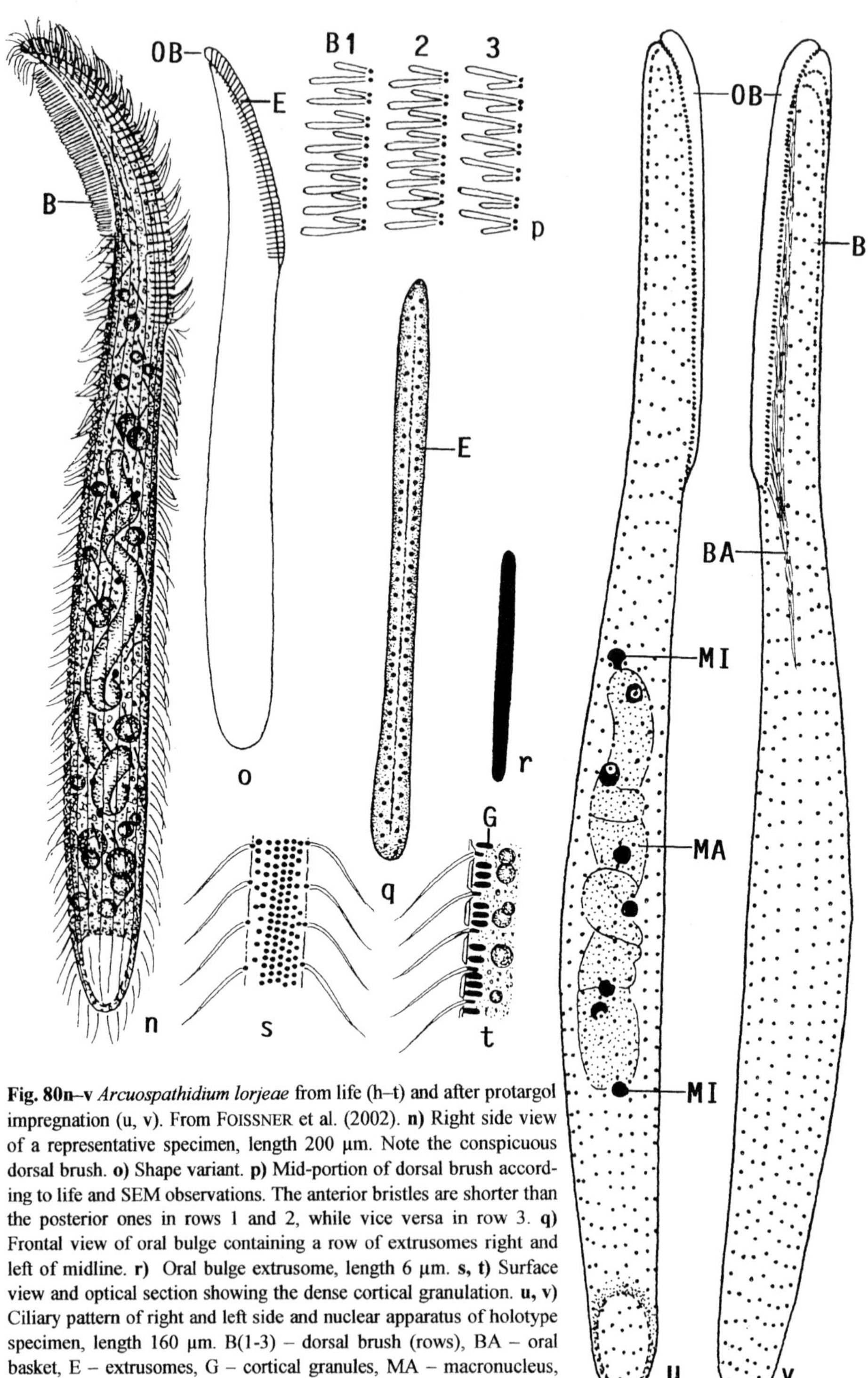

Fig. 80n–v *Arcuospathidium lorjeae* from life (h–t) and after protargol impregnation (u, v). From FOISSNER et al. (2002). **n)** Right side view of a representative specimen, length 200 µm. Note the conspicuous dorsal brush. **o)** Shape variant. **p)** Mid-portion of dorsal brush according to life and SEM observations. The anterior bristles are shorter than the posterior ones in rows 1 and 2, while vice versa in row 3. **q)** Frontal view of oral bulge containing a row of extrusomes right and left of midline. **r)** Oral bulge extrusome, length 6 µm. **s, t)** Surface view and optical section showing the dense cortical granulation. **u, v)** Ciliary pattern of right and left side and nuclear apparatus of holotype specimen, length 160 µm. B(1-3) – dorsal brush (rows), BA – oral basket, E – extrusomes, G – cortical granules, MA – macronucleus, MI – micronuclei, OB – oral bulge.

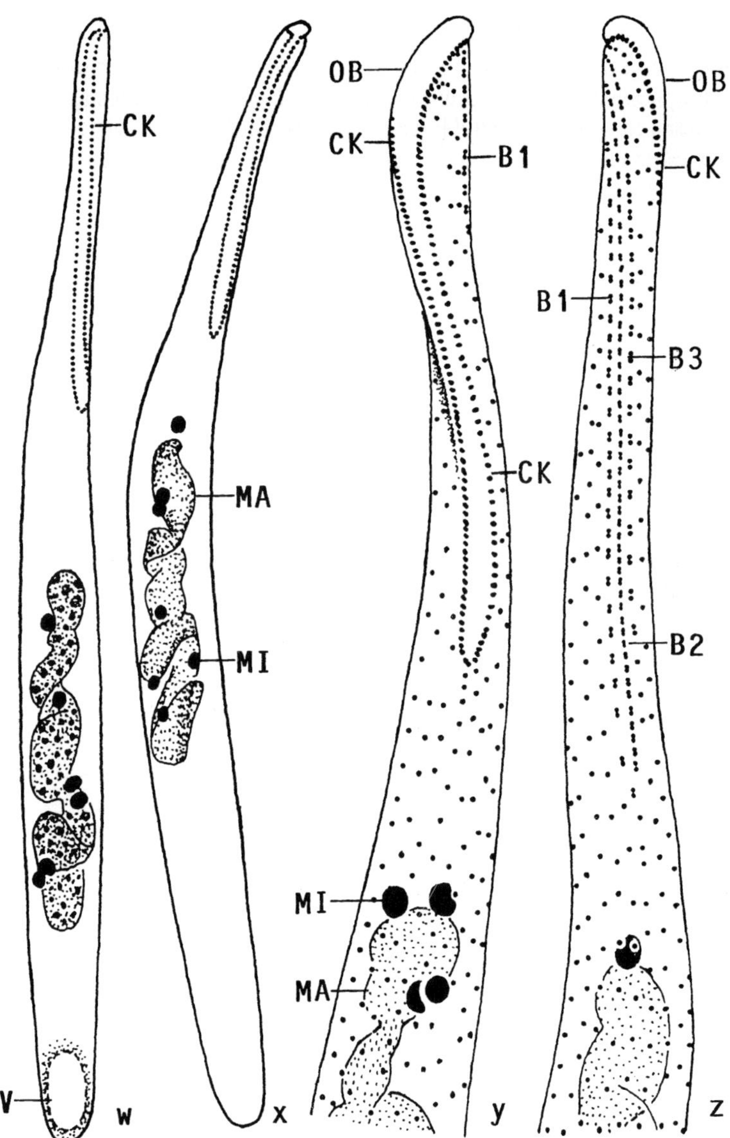

Fig. 80w–z *Arcuospathidium lorjeae* after protargol impregnation (from FOISSNER et al. 2002). **w, x)** Ventral views showing the slenderly oblong circumoral kinety slightly narrowing anteriorly, length 147 µm, 170 µm. **y, z)** Oral and somatic ciliary pattern of ventral and dorsal anterior body region; length of oral bulge 50 µm. Note the slightly dumbbell-shaped oral bulge and the comparatively widely spaced dikinetids of dorsal brush row 3. B1-3 – dorsal brush rows, CK – circumoral kinety, CV – contractile vacuole, MA – macronucleus, MI – micronuclei, OB – oral bulge.

a jelly mass with details difficult to recognize both in vivo and in the scanning electron microscope; distal bristle end becomes inflated on prolonged observation (Fig. 80n, p, 136c–g; Table 28). Brush row 1 slightly shorter than row 2, composed of an average of 34 ordinarily spaced dikinetids with anterior bristles only half as long as posterior. Brush row 2 similar to row 1, but composed of an average of 53 narrowly spaced dikinetids. Dikinetidal portion of brush row 3 distinctly shorter than row 2, composed of an average of 24 comparatively widely spaced dikinetids with posterior bristles about half as long as anterior; has a long, monokinetidal tail made of 1 µm high bristles.

Oral area flattened about 2:1 and usually curved dorsally making oral bulge convex. Oral bulge occupies one third of body length on average, inconspicuous because only about 3 µm wide and high, and thus indistinctly separated from body proper, steep to very steep, that is, extends almost in parallel to main body axis; in frontal view very narrowly oblong or dumbbell-shaped with tapered anterior and narrowly rounded posterior end both in vivo and preparations. Circumoral kinety of same shape as oral bulge, composed of ordinarily spaced dikinetids associated with fine nematodesmata extending posteriorly along oral bulge to form a narrow, straight funnel postorally; cilia of circumoral dikinetids of same length as somatic ones (Fig. 80n, o, q, u–y, 136a, b; Table 28).

Occurrence and ecology: To date found at type locality and in floodplain soil from the Paraná River in Brazil, near the town of Maringá, W53°15' S22°40', where it occurred together with another conspicuous haptorid, viz., *Cephalospathula brasiliensis* FOISSNER, 2003a. *Arcuospathidium lorjeae* was moderately abundant in the non-flooded Petri dish cultures. The slender shape indicates that it is a terricolous species, while the habitats suggest that it might occur also in limnetic environments.

Remarks: *Arcuospathidium lorjeae* belongs to the *Arcuospathidium cultriforme*-group. Within this assemblage, it is most similar to *Arcuospathidium cultriforme cultriforme*, differing distinctly, however, in the length:width ratio (11.4:1 vs. 6.9:1), the number of ciliary rows (17 vs. 28) and, especially, the length of the dorsal bristles (up to 10 µm vs. up to 5 µm; Fig. 136c–g). It is the last feature, which makes this species so conspicuous.

Cultellothrix FOISSNER, 2003

2003 *Cultellothrix* FOISSNER, Acta Protozool., 42: 48 – Type species (by original designation): *Cultellothrix velhoi* FOISSNER, 2003.

Improved diagnosis: More or less distinctly pleurostomatid-shaped Arcuospathidiidae with brush on left side of cell; individual brush rows without anterior tail of ordinary cilia.

Etymology: Composite of the Latin noun *cultellus* (small knife) and the Greek noun *thrix* (hair ~ cilium ~ ciliate), referring to the knife-shape of the type species. Feminine gender.

Remarks: Live specimens of *Cultellothrix* highly resemble pleurostomatid ciliates, such as *Litonotus* and *Acineria*, many of which have a similar body shape and nuclear and ciliary pattern. However, pleurostomatids invariably have the left side ciliature reduced to short bristles (KAHL 1931b, FOISSNER & LEIPE 1995, FOISSNER et al. 1995). *Cultellothrix* has ordinary cilia on both body sides and thus does not belong to the Pleurostomatida. A retrospective analysis of the original notes of the senior author showed that he remarked "body shape pleurostomatid" in the five species he investigated (Fig. 81)! Thus, this feature now has been included in the diagnosis, as already suggested by FOISSNER (2003b).

The body plan and ciliary pattern show that *Cultellothrix* belongs to the Spathidida, as defined by FOISSNER and FOISSNER (1988a). Within this suborder, it resembles members of the family Arcuospathidiidae. *Cultellothrix* and *Latispathidium* are unique among the spathidiids in having the dorsal brush not dorsally or dorsolaterally located, but on the left side. Admittedly, this is a rather inconspicuous difference, but the

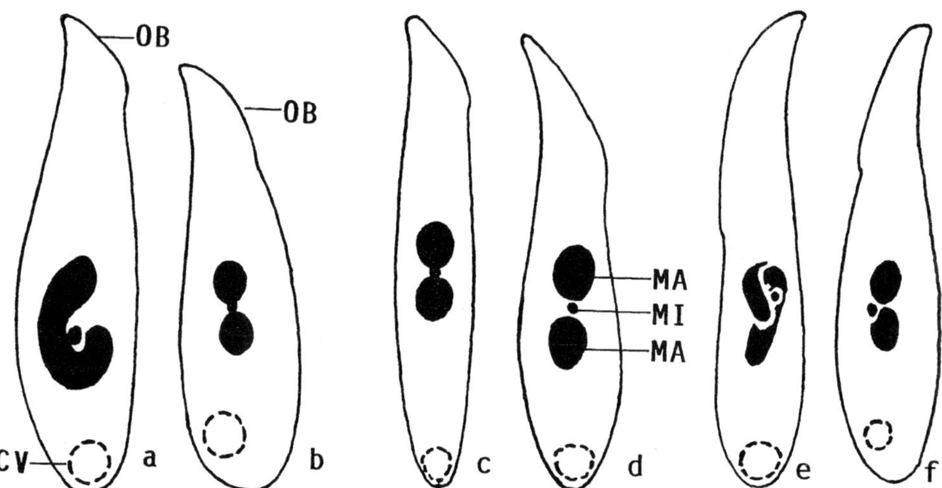

Fig. 81a–f Congruent body shapes and nuclear patterns in *Cultellothrix* and pleurostomatids. **a, b:** *Cultellothrix paucistriata* and *Amphileptus punctatus*. **c, d:** *Cultellothrix tortisticha* and *Acineria incurvata*. **e, f:** *Cultellothrix velhoi* and *Amphileptus fusidens*. CV – contractile vacuole, MA – macronucleus nodules, MI – micronucleus, OB – oral bulge.

occurrence in a considerable number of species and various groups suggests specific evolutionary lineages and thus generic separation. *Latispathidium* has the ciliature in *Spathidium* pattern.

The arrangement of the oral bulge extrusomes in short, transverse rows and the single postoral row in *C. velhoi* resemble several *Bryophyllum* species, whose oral bulge and circumoral kinety, however, usually extend whole body length and curve around the posterior body end to terminate on the dorsal side of the cell (for description of representative species, see FOISSNER et al. 2002). The oral bulge and circumoral kinety of *Cultellothrix*, in contrast, extend only to mid-body. Thus, it is more closely related to the spathidiid than bryophyllid lineage.

Two groups are recognizable within *Cultellothrix*, viz., *C. velhoi* and *C. lionotiformis* on the one hand, and *C. paucistriata*, *C. tortisticha*, *C. coemeterii*, *C. japonica*, and *C. atypica* on the other hand. The first group is knife-shaped and has the brush kinties in parallel with the circumoral kinety. The second group has an oblong shape; an unusually dense cortical granulation forming a plate-like layer in the cortex, which frequently impregnates so heavily with protargol that the ciliary pattern is obscured; and a nuclear apparatus composed of two globular macronucleus nodules and a micronucleus in between, except of *C. coemeterii* and *C. paucistriata*, which have a C-shaped macronucleus. These are considerable differences, suggesting split at subgeneric level, especially if further group 1 species are discovered. Indeed, we know of another, apparently new "typical" *Cultellothrix* species, that is, we found a single specimen in protargol slides of a soil sample from the Dominican Republic. This species has a very similar overall organization as *C. velhoi*, but is more slender (155×25 µm) and has the posterior end tail-like narrowed; the oral bulge is much shorter (30 µm); and the macronucleus is a long, highly tortuous strand.

The five species of the second group resemble *Arcuospathidium*, and we cannot exclude that they belong to this genus. This is why we do not separate them from *Cultellothrix* s. str. at genus level. The arrangement of the dorsal brush shows a transition from a perfect *Cultellothrix* pattern to a rather distinct *Arcuospathidium* pattern: *C. tortisticha* → *C. paucistriata* → *C. japonica* → *C. atypica* → *C. coemeterii* → *Arcuospathidium muscorum*. It is this continuum and the great overall similarity of the five species mentioned above, which suggest to include them in *Cultellothrix*, that is, to consider the similarities with *Arcuospathidium* as resulting from convergent evolution. This is emphasized by a sophistic feature discovered during the preparation of this monograph: only *Cultellothrix*, most *Apertospathula* species, *Longispatha, Armatospathula,* and *Apospathidium* species lack the monokinetidal tail at the anterior end of the dorsal brush rows. Thus, we included this feature in the genus diagnosis because it helps to classify difficult species, such as *C. coemeterii* more properly.

Cultellothrix was established with only two species. However, there are several taxa in the literature that might also belong to this genus, viz., an *Amphileptus* sp. (Fig. 82k), of which KAHL (1931) supposes that it is related to *C. lionotiformis*; *Cranotheridium stilleri* LEPSI, 1959 (Fig. 82n); and *Litonotus dubius* VUXANOVICI 1962c (Fig. 82o). As all are described insufficiently, a definite transfer should await redescription. We add to *Cultellothrix* three species from the more recent *Spathidium* literature and two new taxa from Brazil, South America. This is a considerable gain, showing that *Cultellothrix* contains much more species than originally supposed.

Key to species

1 Distinctly knife-shaped and usually >110 µm long in vivo. Macronucleus moderately long and tortuous. 2
– More or less distinctly spatulate and usually <110 µm long in vivo. Macronucleus reniform or in two globules with a micronucleus in between. 3
2 Length 140–200 µm in vivo. Two ordinary ciliary rows between circumoral kinety and dorsal brush row 1. Postoral extrusome row likely lacking .... *C. lionotiformis*
– Length 80–150 µm in vivo. No ordinary ciliary rows between circumoral kinety and dorsal brush row 1. With postoral extrusome row (difficult to recognize!). Extrusomes about 6 µm long, curved rods .... *C. velhoi*
3 Macronucleus reniform 4
– Macronucleus in two globules with a micronucleus in between 5
4 Size about 95 × 23 µm in vivo. Extrusomes about 6 µm long, curved rods. On average 12 ciliary rows .... *C. coemeterii*
– Size about 65 × 20 µm in vivo. Extrusomes narrowly ovate and about 3 µm long. On average 7 ciliary rows .... *C. paucistriata*
5 Extrusomes rod-shaped or oblong 6
– Extrusomes narrowly ovate. On average about 12 ciliary rows .... *C. japonica*
6 Size about 95 × 25 µm in vivo. On average 12 ciliary rows .... *C. atypica*
– Size about 55 × 10 µm in vivo. On average 7 ciliary rows .... *C. tortisticha*

Cultellothrix lionotiformis (KAHL, 1930) FOISSNER, 2003 (Fig. 82l, m)

1930 *Spathidium lionotiforme* KAHL, Arch. Protistenk., 70: 383.
1930 *Spathidium lionotiforme* spec. n. KAHL, Tierwelt Dtl., 18: 165 (described as species nova a second time!).
1984 *Arcuospathidium lionotiforme* (KAHL, 1930) nov. comb. – FOISSNER, Stapfia, 12: 78 (misidentification! The species identified as *A. lionotiforme* by the FOISSNER group since 1984, is *A. cultriforme scalpriforme*; see that species for details.).

Diagnosis: Not given in original description. Should await reinvestigation with modern methods.

Type locality: *Sphagnum* moss in the Eppendorfer Moor near Hamburg, Germany, E10° N53°30'.

Etymology: Not given in original description. However, *lionotiforme* obviously refers to the overall similarity with the pleurostomatid genus *Litonotus* (formerly *Lionotus*, per lapsus).

Description (according to KAHL 1930a): Length 140–200 µm. Slenderly lanceolate (about 7:1), posterior portion like a rod or slightly flattened laterally: rear half usually oblong or spindle-like narrowed posteriorly; dorsal outline sigmoidally curved anteriorly. Oral bulge steeply extending to mid-body, where it ends in a small prominence or smoothly merges into the ventral margin. A hyaline, very characteristic depression along oral bulge (undotted area in Fig. 82l, m), providing the oral area with a knife-shaped appearance. One rather large micronucleus attached to the moderately long, tortuous macronucleus. Contractile vacuole in rear end. Extrusomes about 8 µm long and rather thick, extend into the hyaline depression described above; closely spaced stippling broad bulge margin in four rows. Cytoplasm colourlessly granulated. Glides or roots metabolically.

Cilia long and loosely spaced, arranged in widely spaced rows, viz., about five on left side and, possibly, slightly more on right. Dorsal brush long, three-rowed, one row extends along dorsal outline, the two others on the left side; bristles long, thick, and loosely spaced.

Oral bulge very steep, extends to mid-body as described above. Bulge broader on left side than right, where the "mane" (circumoral cilia) is near the bulge margin, while about 5 µm off the margin on the left side (transition to *Amphileptus*).

Occurrence and ecology: KAHL (1930a, b) discovered two specimens of *C. lionotiformis* in a road drain with *Hottonia palustris* and, later, in *Sphagnum* moos near Hamburg, where he found 10 specimens and thus assumed that the limnetic record was by chance. However, all latter, unsubstantiated reports are from limnetic or semiterrestrial habitats, viz., the Priest Pot in England (FINLAY & MABERLY 2000); the Tundža River in Bulgaria (DETCHEVA 1986); and infusions of rice-stubble and soil from drained paddy fields in Japan (TAKAHASHI & SUHAMA 1991). This suggests that *C. lionotiformis* is a limnetic and/or semiterrestrial species, similar to *C. velhoi*.

Remarks: Redescription is urgently needed to confirm the differences to *C. velhoi*, viz., the two ordinary ciliary rows between the circumoral kinety and the dorsal brush, the larger size, and the arrangement of the extrusomes.

Cultellothrix velhoi FOISSNER, 2003 (Fig. 82a–j, 137a–g; Table 29)

2003 *Cultellothrix velhoi* FOISSNER, Acta Protozool., 42: 49 (Type slides with protargol-impregnated specimens from type locality are deposited in the Oberösterreichische Landesmuseum in Linz, Upper Austria.).

Diagnosis: Size about 120 × 25 µm in vivo. Knife-shaped with blade (oral portion) approximately half of body length. Macronucleus about 40 µm long and usually tortuous; single micronucleus. Extrusomes rod-shaped and slightly curved, approximately 6 × 0.3 µm in size, form short, oblique rows in oral bulge and a long postoral row extending to dorsal side of cell. On average 12 ciliary rows, those left of circumoral kinety anteriorly modified to strongly heterostichad dorsal brush occupying about 31% of body length. Brush bristles up to 5 µm long: row 1 composed of an average of 29 dikinetids, row 2 of 23, and row 3 of 7 dikinetids followed by a short monokinetidal bristle tail.

Type locality: Floodplain soil of Paraná River in Brazil, near the town of Maringá, W53°15' S22°40'.

Dedication: Dedicated to Dr. Luiz Felipe MACHADO VELHO, Universidade Estadual de Maringá, Brazil, who provided the sample containing this and other new species.

Description: Size 80–150 × 20–40 µm in vivo, laterally flattened up to 2:1, fragile and thus frequently more or less inflated in protargol preparations (Table 29). Slenderly to broadly knife-shaped with blade (oral portion), however, often indistinctly separated from oblong handle; dorsal outline more or less distinctly sigmoidal, ventral slightly convex; anterior end bluntly pointed, posterior more or less broadly rounded (Fig. 82a, d, g, j, 137a). Nuclear apparatus slightly underneath mid-body in postoral portion of cell. Macronucleus about 40 µm long, basically rod-shaped, usually, however, more or less tortuous and spiralized; contains many small and medium-sized nucleoli. Micronucleus attached to middle third of macronucleus, broadly ellipsoidal, 3–4 µm in vivo. Contractile vacuole in posterior body end; excretory pore(s) not impregnated.

Table 29 Morphometric data on *Cultellothrix velhoi* (from FOISSNER 2003b)

Characteristics[1]	$\overline{x}$	M	SD	SE	CV	Min	Max	n
Body, length	105.7	103.0	18.1	4.4	17.1	77.0	140.0	17
Body, width	29.2	25.0	9.7	2.4	33.4	21.0	50.0	17
Anterior body end to end of circumoral kinety (mouth length), distance	53.8	55.0	10.4	2.5	19.3	38.0	75.0	17
Mouth, width in anterior third	3.8	4.0	0.6	0.2	16.6	3.0	5.0	10
Anterior body end to macronucleus, distance	56.3	58.0	12.1	2.9	21.5	41.0	80.0	17
Dorsal brush row 1, length[2]	32.2	32.0	7.6	1.9	23.7	20.0	43.0	17
Dorsal brush row 1, number of dikinetids	29.0	28.0	7.3	1.8	25.1	18.0	40.0	17
Dorsal brush row 2, length[2]	32.7	35.0	8.0	2.0	24.6	20.0	47.0	17
Dorsal brush row 2, number of dikinetids	23.2	24.0	4.4	1.1	18.8	16.0	29.0	17
Dorsal brush row 3, length[2]	9.2	10.0	1.5	0.4	16.0	7.0	12.0	17
Dorsal brush row 3, number of dikinetids	7.1	7.0	1.3	0.3	19.1	5.0	10.0	17
Macronucleus figure, length	26.9	22.0	10.7	2.6	39.7	17.0	56.0	17
Macronucleus, length (spread and thus approximate)	40.7	40.0	–	–	–	25.0	56.0	17
Macronucleus, width	5.8	6.0	1.2	0.3	20.8	4.0	9.0	17
Macronucleus, number	1.0	1.0	0.0	0.0	0.0	1.0	1.0	17
Micronucleus, length	2.9	3.0	0.4	0.1	14.6	2.2	4.0	17
Micronucleus, width	2.6	2.5	0.5	0.1	20.8	2.0	4.0	17
Micronucleus, number	1.0	1.0	0.0	0.0	0.0	1.0	1.0	17
Ciliary rows, total number	11.7	12.0	1.1	0.3	9.4	10.0	14.0	17
Kinetids in a right side ciliary row, number	27.2	28.0	5.0	1.2	18.5	18.0	35.0	17
Dorsal brush rows, number	3.0	3.0	0.0	0.0	0.0	3.0	3.0	17

[1] Data based on mounted, protargol-impregnated (FOISSNER's method), selected (cells with large rotifers ingested strongly inflated and thus excluded) specimens from a non-flooded Petri dish culture. Measurements in µm. CV – coefficient of variation in %, M – median, Max – maximum, Min – minimum, n – number of individuals investigated, SD – standard deviation, SE – standard error of arithmetic mean, $\overline{x}$ – arithmetic mean.

[2] Distance from circumoral kinety to last brush dikinetid.

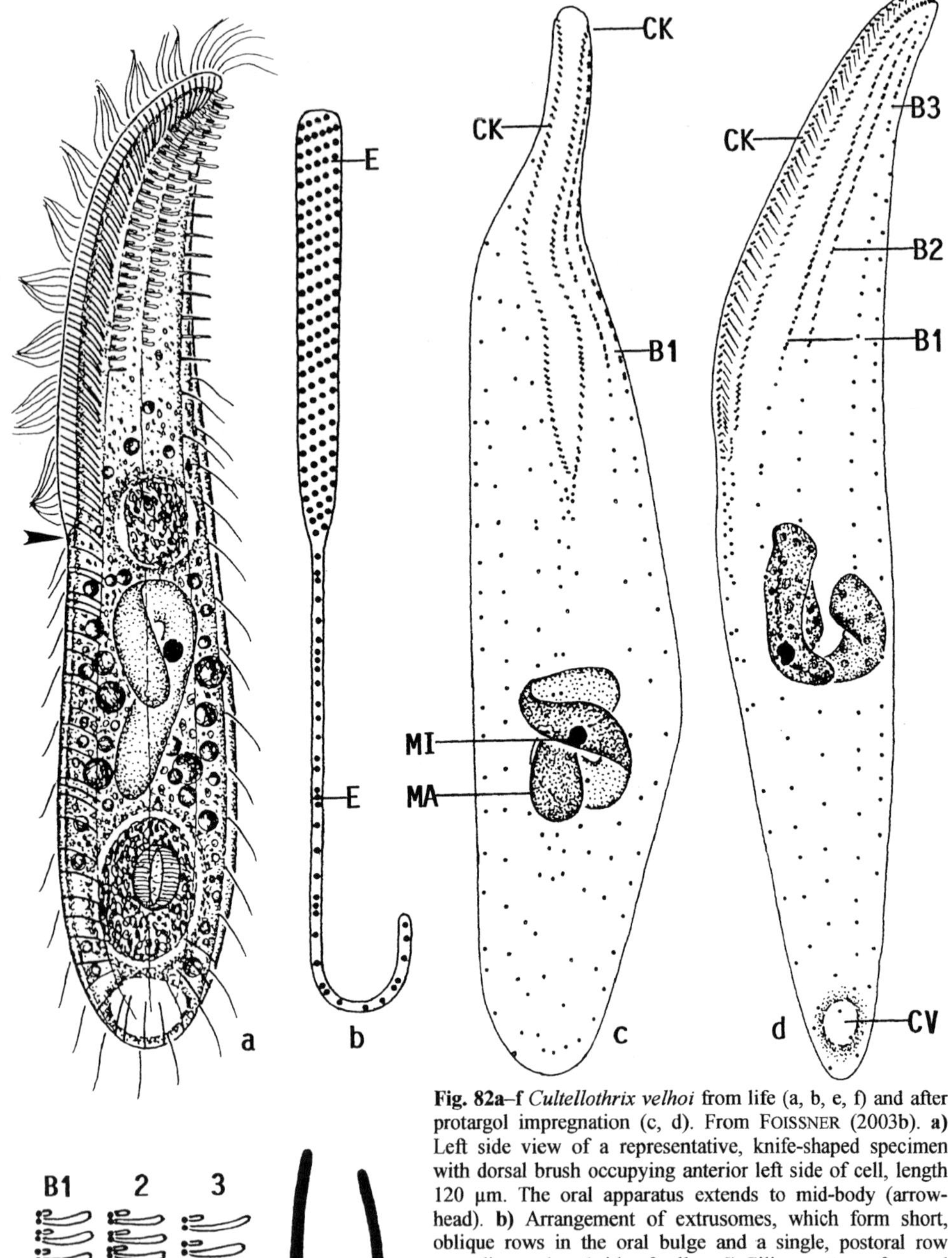

Fig. 82a–f *Cultellothrix velhoi* from life (a, b, e, f) and after protargol impregnation (c, d). From FOISSNER (2003b). **a)** Left side view of a representative, knife-shaped specimen with dorsal brush occupying anterior left side of cell, length 120 µm. The oral apparatus extends to mid-body (arrowhead). **b)** Arrangement of extrusomes, which form short, oblique rows in the oral bulge and a single, postoral row extending to dorsal side of cell. **c, d)** Ciliary pattern of ventral and left side, length 120 µm, 133 µm. Note lack of an ordinary ciliary row between circumoral kinety and dorsal brush, a main difference to *C. lionotiformis*, which has at least two such rows (Fig. 82m). **e)** Middle region of dorsal brush. **f)** Oral bulge extrusomes, length 6–7 µm. B(1-3) – dorsal brush (rows), CK – circumoral kinety, CV – contractile vacuole, E – extrusomes, MA – macronucleus, MI – micronucleus.

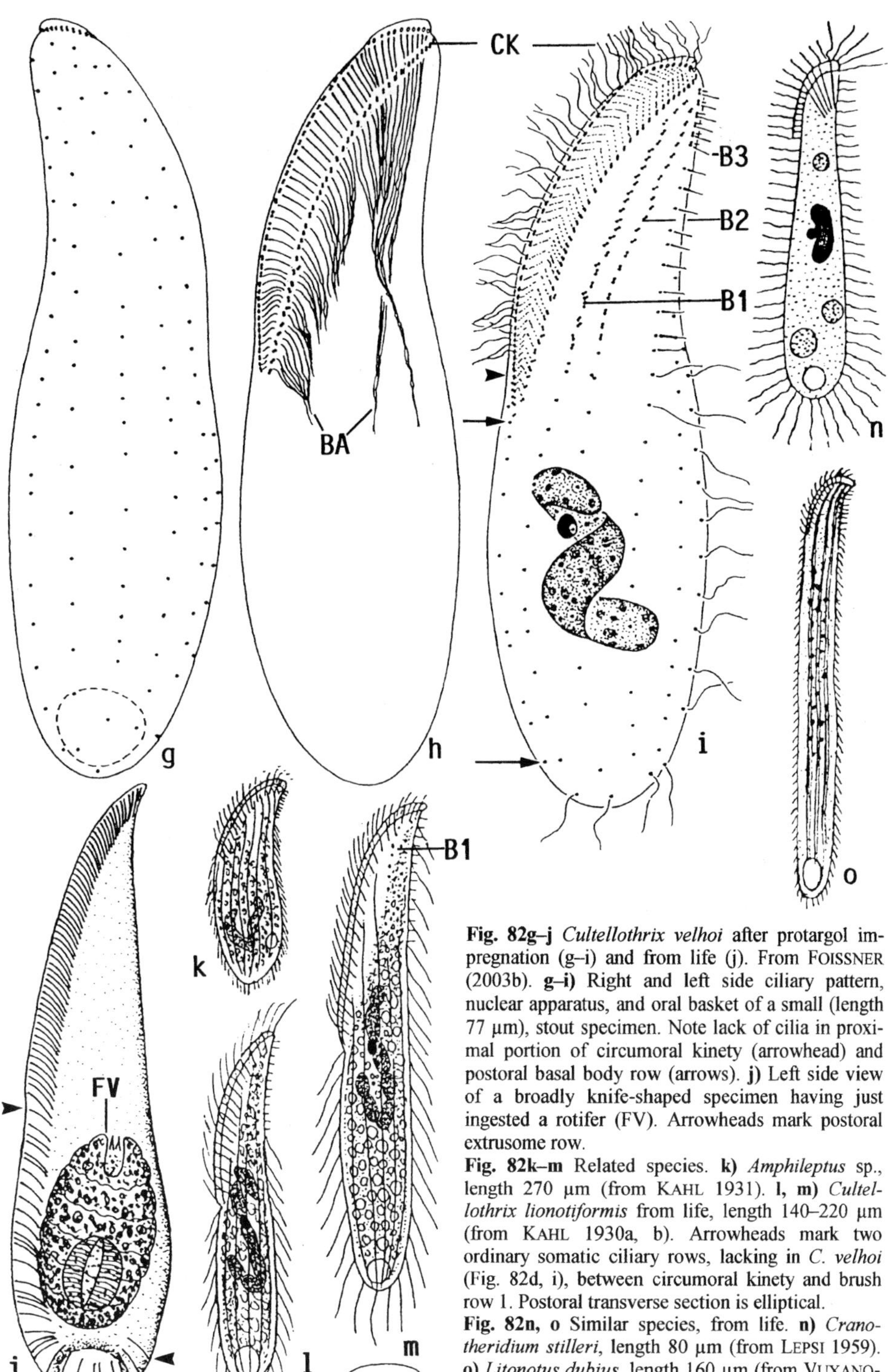

Fig. 82g–j *Cultellothrix velhoi* after protargol impregnation (g–i) and from life (j). From FOISSNER (2003b). **g–i)** Right and left side ciliary pattern, nuclear apparatus, and oral basket of a small (length 77 µm), stout specimen. Note lack of cilia in proximal portion of circumoral kinety (arrowhead) and postoral basal body row (arrows). **j)** Left side view of a broadly knife-shaped specimen having just ingested a rotifer (FV). Arrowheads mark postoral extrusome row.

Fig. 82k–m Related species. **k)** *Amphileptus* sp., length 270 µm (from KAHL 1931). **l, m)** *Cultellothrix lionotiformis* from life, length 140–220 µm (from KAHL 1930a, b). Arrowheads mark two ordinary somatic ciliary rows, lacking in *C. velhoi* (Fig. 82d, i), between circumoral kinety and brush row 1. Postoral transverse section is elliptical.

Fig. 82n, o Similar species, from life. **n)** *Cranotheridium stilleri*, length 80 µm (from LEPSI 1959). **o)** *Litonotus dubius*, length 160 µm (from VUXANOVICI 1962b).

Cortex thin, bright and very flexible; cortical granules and/or extrusomes not recognizable, not even with interference contrast optics. Oral bulge extrusomes rod-shaped and slightly curved in vivo, 6–7 × 0.3 µm in size, do not impregnate with the protargol method used, form many short, oblique rows in oral bulge and a single postoral row extending across posterior body end to dorsal side of cell; some scattered in cytoplasm (Fig. 82a, e, f, 137d–g). Oral area flat and hyaline, postoral portion, depending on state of nutrition, also flattened and hyaline or packed with lipid droplets 1–5 µm across and some large food vacuoles containing rotifers in various stages of digestion (Fig. 82j). Slowly glides among and on soil particles and the microscope slide showing great flexibility.

Somatic cilia 7–8 µm long in vivo, widely spaced, arranged in an average of 12 meridional, ordinarily spaced rows commencing at and near curved anterior end of body and thus more or less distinctly curved dorsally at both sides of circumoral kinety (oral bulge), as in *Arcuospathidium*, except of a single or two, rather irregular postoral rows, which, interestingly, lack cilia; all kineties equidistantly spaced, except of more narrowly arranged post-oral and dorsal brush rows 1 and 2 (Fig. 82a, g, i, 137a, b; Table 29). Dorsal brush occupies left anterior body third, composed of paired bristles at anterior end of first three ciliary rows left of circumoral kinety; posterior bristle of dikinetids 3–5 µm long and slightly clavate, anterior bristle granule-like because only 1–2 µm long; bristle length gradually decreases in posterior third of rows 1 and 2 and is rather variable in individual cells; one or few monokinetids may occur at anterior end of one or all brush rows. Brush row 1 composed of an average of 29 narrowly spaced dikinetids, of similar length as row 2 comprising only 23 dikinetids; row 3 extends along dorsal margin of cell, strongly shortened, composed of only 7 dikinetids followed by a monokinetidal tail extending to second body third with 3–4 µm long bristles (Fig. 82a, c, e, i, 137a, c, d; Table 29).

Oral area more or less distinctly curved dorsally, bulge thus convex; hyaline because flattened about 2:1 and containing only few, small lipid droplets. Mouth occupies anterior body half on average, very steep because extending parallel to main body axis, conspicuous due to the oblique rows of refractive extrusomes contained. Oral bulge inconspicuous because less than 3 µm high and thus hardly separate from body proper, contains short, anteriorly directed fibres, originating from circumoral dikinetids, forming arrowhead-like pattern; in frontal view elongate cuneate to oblong with proximal portion wedge-like narrowed, about 5 µm wide in vivo. Circumoral kinety of same shape as oral bulge, composed of comparatively widely spaced dikinetids each associated with a slightly elongated, about 10 µm long cilium lacking in proximal, wedge-shaped portion of kinety. Nematodesmata originate only from right side dikinetids of circumoral kinety, gradually decrease in length from anterior to posterior, the long anterior rods form bundles extending to mid-body; we could not decide whether the left side basket rods are lacking or not impregnated (Fig. 82a, c, d, g–i, 137a, b, d, e; Table 29).

Occurrence and ecology: *Cultellothrix velhoi* is a rare species because we did not find it in about 1000 other soil and moss samples, including about 100 samples from similar habitats, collected globally. Although looking fragile, *C. velhoi* is a rapacious predator feeding mainly on large rotifers, showing that the oral apparatus can open widely and the body is very flexible (Fig. 82j).

Remarks: *Cultellothrix velhoi* differs from *C. lionotiformis* by body length (80–150 µm vs. 140–200 µm), the postoral extrusome row (lacking in *C. lionotiformis*), and in lacking ordinary ciliary rows (vs. two in *C. lionotiformis*) between circumoral kinety and dorsal brush, that is, all left side kineties of *C. velhoi* are dorsal brush rows. As KAHL's description of *C. lionotiformis* is rather detailed, there is no reason to assume that the differences are based on incomplete observation, although the postoral extrusome row is difficult to recognize. But even if it were present in *C. lionotiformis*, the different left side kinety pattern would be sufficient to distinguish it from *C. velhoi*.

Cultellothrix coemeterii (KAHL, 1943) **nov. comb.** (Fig. 83a–r, 84a–h, 85a–x, 138a–x, 139a–n; Tables 30, 31)

1943 *Spathidium coemeterii* KAHL, Infusorien: 27.

2004 *Arcuospathidium coemeterii* (KAHL 1943) FOISSNER et al. 2004 – FOISSNER & LEI, Linzer biol. Beitr., 36: 187 (description of ontogenesis from protargol preparations deposited at the locality described in the next entry).

2005 *Arcuospathidium coemeterii* (KAHL, 1943) nov. comb. – FOISSNER, BERGER, XU & ZECHMEISTER-BOLTENSTERN, Biodiv. Conserv. 14: 652 (Redescription and neotypification. Slides with protargol-impregnated specimens from neotype locality and a voucher locality are deposited in the Oberösterreichische Landesmuseum in Linz, Upper Austria.).

Diagnosis: Size about 95 × 23 µm in vivo. Narrowly spatulate with oblique, oblong to slightly cuneate oral bulge about as long as widest trunk region. Macronucleus slenderly reniform; single micronucleus. Extrusomes slightly curved, about 6 × 0.5 µm-sized rods. On average 12 ciliary rows, 3 anteriorly differentiated to strongly heterostichad dorsal brush occupying about 23% of body length. Brush bristles up to 3 µm long: row 1 composed of an average of 12 dikinetids, row 2 of 14, and row 3 of 5 dikinetids followed by a monokinetidal bristle tail extending to mid-body.

Type and neotype localities: Type locality unknown, likely the Hamburg area in Germany. The neotype locality is a *Pinus nigra* forest soil in the Stampfltal near Vienna, Austria, N47°53' E16°02'.

Type material: No original type material is available. Well-impregnated specimens are available from the neotype locality, as described above, and from a population of a site rather near to the neotype locality; the latter was cultivated and used to study ontogenesis.

Etymology: The Latin noun *cemeteri* (cementery) obviously refers to the site KAHL discovered this species, viz., in moss from a churchyard.

Description (of Austrian neotype): Size 70–110 × 15–30 µm in vivo, usually near 95 × 23 µm, as calculated from some in vivo measurements and the morphometric data; length:width ratio 3–4:1 in vivo and about 4.3:1 in protargol preparations (Table 30); oral and neck area flattened laterally up to 3:1; well-fed specimens thicker and less flattened than hungry ones, as usual. Shape conspicuously asymmetrical, that is,

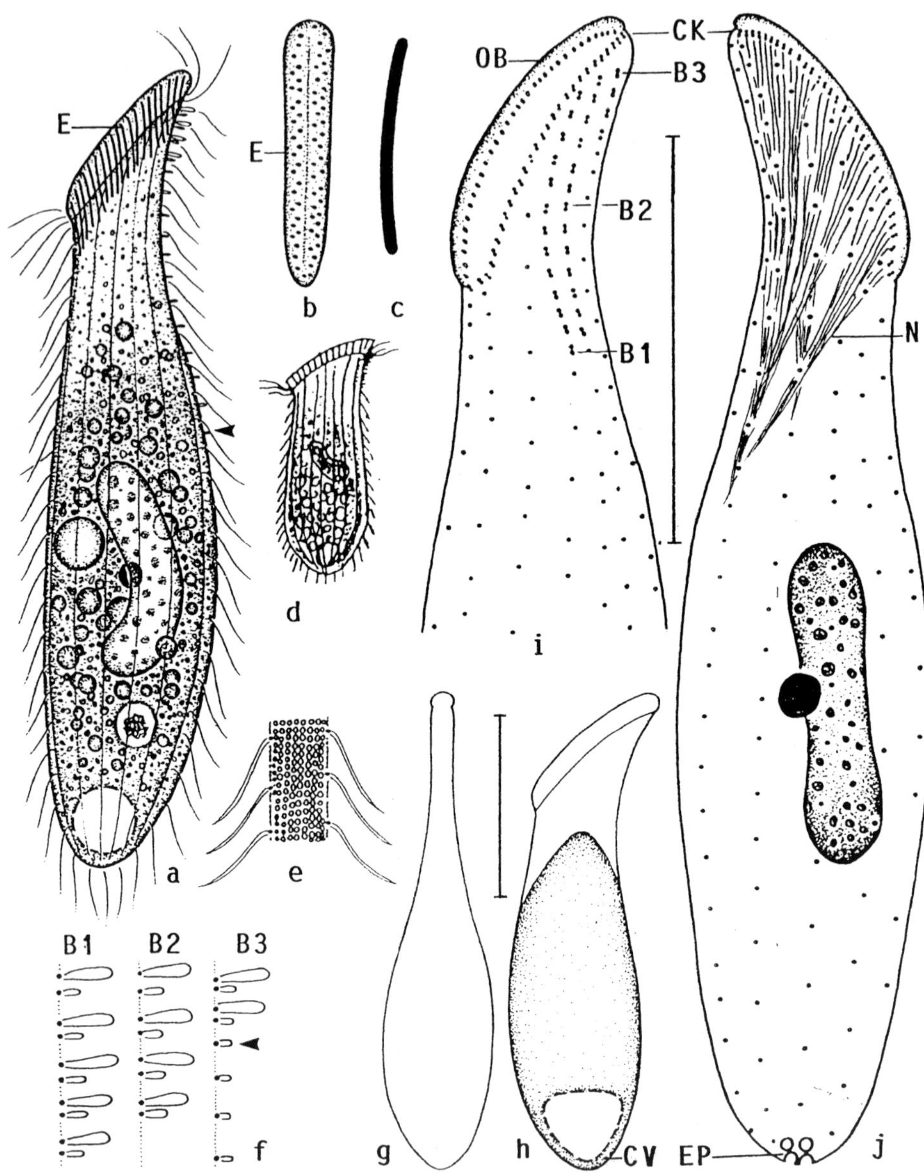

Fig. 83a–j *Cultellothrix coemeterii* from life (a–h) and after protargol impregnation (i, j). **a)** Left lateral view of a representative specimen, length 95 µm. From FOISSNER et al. (2005a). Arrowhead marks bristle tail of brush row 3. **b)** Frontal view of oral bulge studded with extrusomes. **c)** Oral bulge extrusome, length 6 µm. **d)** Original figure by KAHL (1943), length 100 µm. **e)** Surface view showing the very narrowly spaced rows of cortical granules. **f)** Middle region of dorsal brush, bristles up to 3 µm long. Arrowhead marks monokinetidal tail of row 3. **g, h)** Dorsal and lateral view of a well-nourished specimen, length 80 µm. **i, j)** Ciliary pattern of left and right side and nuclear apparatus of main neotype specimen, length 85 µm. B1-3 – dorsal brush rows, CK – circumoral kinety, CV – contractile vacuole, E – extrusomes, EP – excretory pores of contractile vacuole, N – nematodesmata (oral basket rods originating from circumoral dikinetids), OB – oral bulge. Scale bars 30 µm.

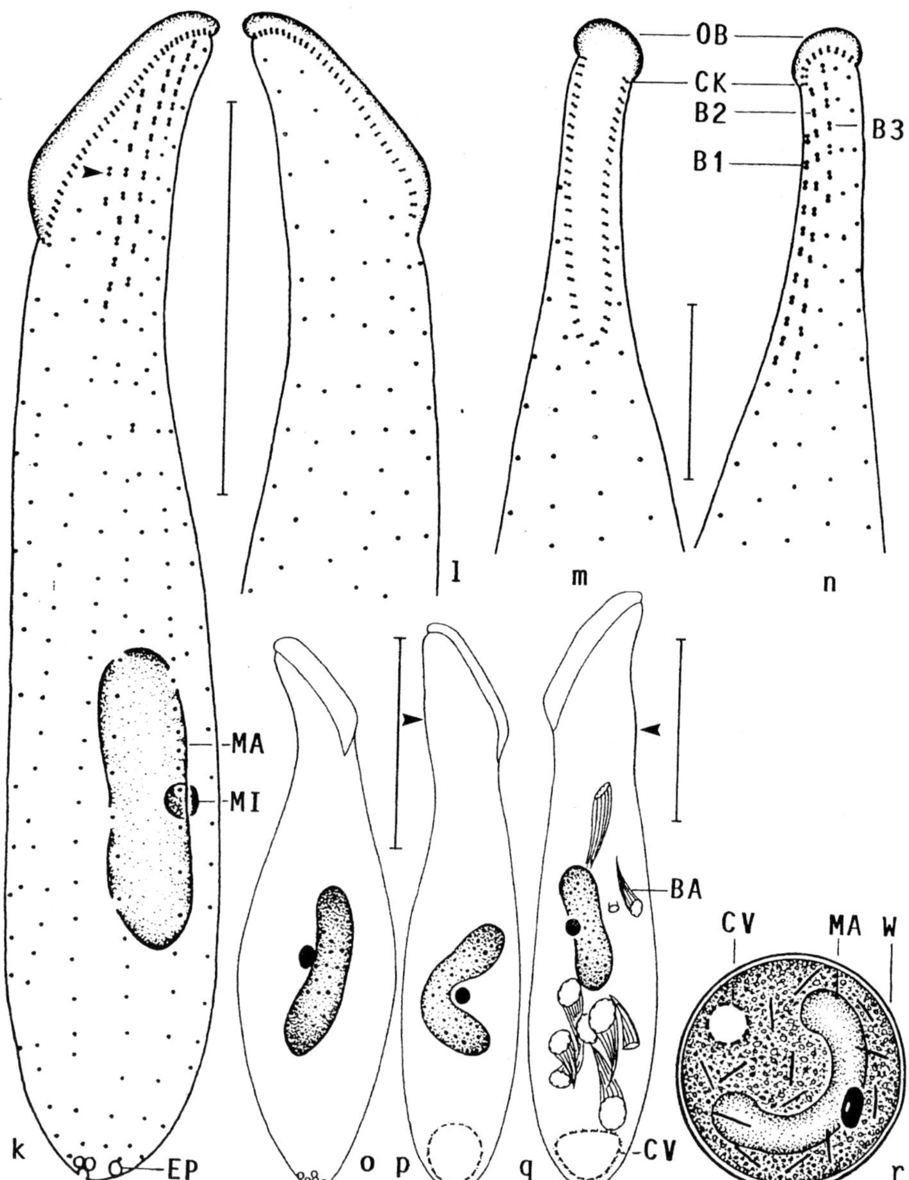

Fig. 83k–r *Cultellothrix coemeterii* after protargol impregnation (from FOISSNER et al. 2005a). **k, l)** Ciliary pattern of left and right side of a specimen with two dikinetids right of brush row 1 (arrowhead), length 90 µm. Note slightly sigmoidal oral bulge. **m, n)** Ventral and dorsal anterior body portion, showing the slightly cuneate circumoral kinety and the dorsal brush with row 3 only about half as long as rows 1 and 2. **o–q)** Variations of body outline and nuclear pattern. One specimen (q) contains oral baskets of microthoracid or euglenid prey. Arrowheads mark neck region slightly inflated due to the preparation procedures. **r)** The resting cysts are about 30 µm across and have a very thin wall. The extrusomes are maintained. B1-3 – dorsal brush rows, BA – oral basket of prey, CK – circumoral kinety, CV – contractile vacuole, EP – excretory pores, MA – macronucleus, MI – micronucleus, OB – oral bulge, W – cyst wall. Scale bars 30 µm (k, l, o–q) and 10 µm (m, n).

Table 30 Morphometric data on *Cultellothrix coemeterii* (from FOISSNER et al. 2005a)

Characteristics [a]	$\overline{x}$	M	SD	SE	CV	Min	Max	n
Body, length	87.2	86.0	11.2	2.4	12.8	67.0	103.0	21
Body, width	20.8	20.0	3.7	0.8	17.8	16.0	28.0	21
Body length:width, ratio	4.3	4.3	0.7	0.2	16.1	3.0	5.4	21
Oral bulge, length (measured as length of cord of circumoral kinety)	21.8	22.0	2.5	0.5	11.5	18.0	28.0	21
Oral bulge, height	4.4	4.3	0.6	0.2	14.4	3.5	5.0	10
Dorsal brush row 1, length (distance circumoral kinety to last dikinetid)	17.1	17.0	2.3	0.5	13.2	14.0	21.0	21
Dorsal brush row 2, length (distance circumoral kinety to last dikinetid)	20.4	19.0	3.8	0.8	18.6	15.0	30.0	21
Dorsal brush row 3, length (distance circumoral kinety to last dikinetid)	6.7	6.0	1.4	0.3	20.6	5.0	10.0	21
Anterior body end to macronucleus, distance	45.0	45.0	8.2	1.8	18.1	33.0	67.0	21
Macronucleus figure, length	21.9	22.0	3.0	0.7	13.8	16.0	26.0	21
Macronucleus, width	6.1	6.0	0.9	0.2	13.9	5.0	7.0	21
Macronucleus, number	1.0	1.0	0.0	0.0	0.0	1.0	1.0	21
Micronucleus, largest diameter	3.0	3.0	0.4	0.1	14.9	2.0	4.0	21
Micronucleus, number	1.0	1.0	0.0	0.0	0.0	1.0	1.0	21
Ciliary rows, number	11.6	12.0	1.2	0.3	10.4	9.0	14.0	21
Ciliated kinetids in a right side row, number	29.3	30.0	4.6	1.0	15.8	22.0	36.0	21
Dorsal brush rows, number	3.0	3.0	0.0	0.0	0.0	3.0	3.0	21
Dikinetids in brush row 1, number	12.0	12.0	1.6	0.4	13.4	9.0	15.0	21
Dikinetids in brush row 2, number	14.6	14.0	2.6	0.6	17.6	11.0	19.0	21
Dikinetids in brush row 3, number	5.6	5.0	1.1	0.2	19.1	4.0	7.0	21

[a] Data based on mounted, protargol-impregnated (FOISSNER 1991, protocol A), and randomly selected specimens from a non-flooded Petri dish culture. Measurements in µm. CV – coefficient of variation in %, M – median, Max – maximum, Min – minimum, n – number of individuals investigated, SD – standard deviation, SE – standard error of arithmetic mean, $\overline{x}$ – arithmetic mean.

dorsal side distinctly longer and more sigmoidal than ventral side due to the narrowed neck and slanted (40–60°) oral bulge about as long as widest trunk region; neck hyaline and rather pronounced dorsally, in protargol preparations often slightly inflated in brush area; widest usually near unflattened mid-region of body, posterior end moderately broadly rounded (Fig. 83a, g, h, j, o–p, 138a, c, d, k, l, n, o, q, t). Macronucleus usually beneath mid-body, basically slenderly reniform, occasionally horseshoe-shaped or oblong; contains numerous small nucleoli. Micronucleus globular, very near or attached to mid-macronucleus (Fig. 83a, j, k, o–q, 138b–d; Table 30). Contractile vacuole in rear body end, four to five excretory pores in pole area. Oral extrusomes rod-shaped, slightly curved, about 5.5–6.5 × 0.5 µm in size and thus longer than oral bulge height, never impregnate with the protargol method used (Fig. 83a–c, 138b, f, m, p). Cortex flexible and often distinctly furrowed in SEM preparations, likely due to postciliary microtubule bundles, as indicated by the oblique arrangement of the furrows; also conspicuous in vivo because about 1 µm thick due to the plate-like packed cortical granules; individual granules < 1 µm across and hardly recognizable because pale, colourless, and very closely spaced (Fig. 83e, 138b, i). Cytoplasm colourless, contains few to many lipid droplets 1–6 µm across and few to many 3–12 µm-sized food vacuoles with heterotrophic flagellates (*Polytomella*) and microthoracid ciliates or euglenoid flagellates, whose oral baskets can be seen in protargol-impregnated specimens (Fig. 83q, 138b, h, k–m). Glides slowly on microscope slide and swims by rotation about main body axis.

Somatic cilia about 7 µm long in vivo, fairly widely spaced, arranged in pronounced *Arcuospathidium*-pattern, that is, in an average of 12 longitudinal rows anteriorly curved dorsally and distinctly separate from circumoral kinety at both sides of cell (Fig. 83a, i–l, 138a, c, d, j, k; Table 30). Dorsal brush on anterior dorsal half of left side, distinctly heterostichad and three-rowed, in some specimens a few bristles right of row 1, all rows continue as ordinary somatic kineties posteriorly and lack ordinary cilia anteriorly; in vivo rather conspicuous, although occupying only 23% of body length, because bristles 2–3 times as thick as ordinary somatic cilia, most bristles, however, collapse in SEM preparations becoming more or less wrinkled (Fig. 138u, f). Anterior bristle of pairs about 3 µm long and bluntly fusiform, posterior granule-like because only 1 µm long. Brush row 1 slightly shorter than row 2, both extend somewhat beyond oral bulge region; row 3 distinctly shorter than rows 1 and 2, composed of an average of only five dikinetids, but followed by a monokinetidal tail extending to mid-body with 1 µm long bristles (Fig. 83a, f, i, k, n, 138a–e, g, j, t–x; Table 30).

Oral bulge occupies anterior body end slanted by 40°–50°, on average as long as widest trunk region, rather indistinct because hardly separated from body proper and only about 3 µm high; outline indistinctly convex or sigmoidal, rarely flat; in frontal view oblong to indistinctly cuneate, rarely rather distinctly cuneate because slightly narrowed ventrally. Bulge surface with arrowhead-like pattern of crests and furrows forming small whirl near dorsal bulge end; whirl not recognizable in vivo and in protargol preparations, but likely corresponds to the temporary cytostome (Fig. 83a, b, i–m, 138a, b, i–l, n, q–u, w, 139a, d, e, f; Table 30). Circumoral kinety not oblong as oral bulge, but oblong with narrowed, acute ventral end; continuous, composed of comparatively widely spaced, viz., about five to six dikinetids between two somatic kineties each; individual dikinetids associated with an about 10 µm long cilium and a basket rod. Nematodesmata (oral basket rods) fairly distinct and bundled, forming rather conspicuous basket in protargol-impregnated specimens (Fig. 83a, i–m, 138c, d, g, r, w).

Type I resting cysts spherical ($\bar{x}$ 30.9 µm, M 31, SD 2.5, SE 0.5, CV 8, Min 26, Max 35 µm; n 23) and colourless; nuclear apparatus, contractile vacuole, and extrusomes maintained; cytoplasm densely granulated by lipid droplets ≤ 2 µm across (Fig. 83r, 139b, c, g–i). Cyst wall smooth, thin (≤ 1 µm).

Occurrence and ecology: KAHL (1943) mentioned that he discovered *C. coemeterii* in moss, but did not indicate the exact site; likely it is the Hamburg area in Germany, where KAHL lived and worked. We found this species in nine out of twelve sites investigated in the surroundings of Vienna, Austria (FOISSNER et al. 2005a), both in deciduous and coniferous forest soils, showing that it is common and has a wide ecological range. Abundances were low to moderate. Likely, we did not identify *C. coemeterii* in our previous studies or misidentified it as a "small variety" of *Spathidium spathula*, which is occasionally rather similar (see below). Indeed, a *Spathidium spathula* from the Shimba Hills of Kenya turned out to be *C. coemeterii* on reinvestigation (FOISSNER 1999a).

Remarks: KAHL's (1943) description is extremely brief: "Size 100 μm, similar to *Spathidium muscicola*, but with shorter toxicysts". However, the figure provided (Fig. 83d) suggests that the population studied by FOISSNER et al. (2005a) belongs to that species; at least, it would be difficult to find a significant difference, except of body shape, which is considerably stouter in KAHL's specimen (2.75 vs. 3–4:1 in vivo), providing the Austrian population with a rather different general appearance (Fig. 83a, d, g–j, k, o–q, 138b–d, k, n); further, the oral bulge, which is slightly convex in KAHL's figure, is frequently indistinctly sigmoidal in the Austrian specimens, that is, slightly depressed in or near the centre (Fig. 83d, k, 138b, t). These differences appear inconspicuous when compared with the matching features, viz., body size (~ 100 μm), the number of ciliary rows (about 10–12, according to KAHL's figure) and, especially, the reniform macronucleus. Furthermore, most of these small differences are not found in another population described below. On the other hand, the identity of *C. coemeterii* is threatened by several species, such as *Spathidium spathula* and *Arcuospathidium muscorum.* This and the geographic neighbourhood justify the neotypification performed by FOISSNER et al. (2005a) with an Austrian population. See volume II for neotypification of ciliates.

KAHL (1943) classified his population in *Spathidium*. However, scanning electron microscopy shows the dorsal brush on the left side of the cell and the ciliary rows arranged in *Arcuospathidium* pattern, that is, the ciliary rows are directed dorsally on both sides of the oral bulge (Fig. 138j, t). Thus, the species belongs to *Cultellothrix*, though the brush is occasionally dorsolaterally located (Fig. 83m, n), as in *Arcuospathidium.* As discussed in the introdution to the genus, *C. coemeterii* is in between *Cultellothrix* and *Arcuospathidium* as concerns the dorsal brush location, while the densely arranged cortical granules indicate a close relationship with typical *Cultellothrix* species, for instance, *C. tortisticha.*

Cultellothrix coemeterii differs from *C. paucistriata*, which has the same type (reniform) of macronucleus, by the extrusomes (rod-shaped vs. narrowly ovate), the number of ciliary rows (12 vs. 7), and body length (95 μm vs. 65 μm). All other congeners have a different macronucleus pattern.

There are two other spathidiids with a close resemblence to *C. coemeterii*, except for the location of the dorsal brush, viz., *Arcuospathidium muscorum* and *Spathidium spathula.* At first glance, *C. coemeterii* is indistinguishable from *A. muscorum*, especially from the Austrian and Venezuelan populations studied by FOISSNER (1981a, 2000a), because most obvious features are rather similar. However, a more detailed comparison reveals significant differences, viz., the shape of the macronucleus (reniform vs. a rather long, tortuous strand), the ratio of oral bulge to body length (25% vs. 41–49%), and the ratio of oral bulge to the longest brush row (about 1:1 vs. about 1:0.3). Thus, the oral bulge is much longer and the brush distinctly shorter in *A. muscorum* than in *C. coemeterii.* Certain populations of *Spathidium spathula* are in vivo also rather similar to *Cultellothrix coemeterii.* However, the ciliary pattern is different (*Spathidium* pattern), the number of kineties distinctly higher (18–30 vs. 9–14), and the oral bulge is more massive and less slanted.

Certainly, the differences between *C. coemeterii* and *Arcuospathidium muscorum* and *Spathidium spathula* are rather inconspicuous in some populations; if in doubt, make protargol preparations.

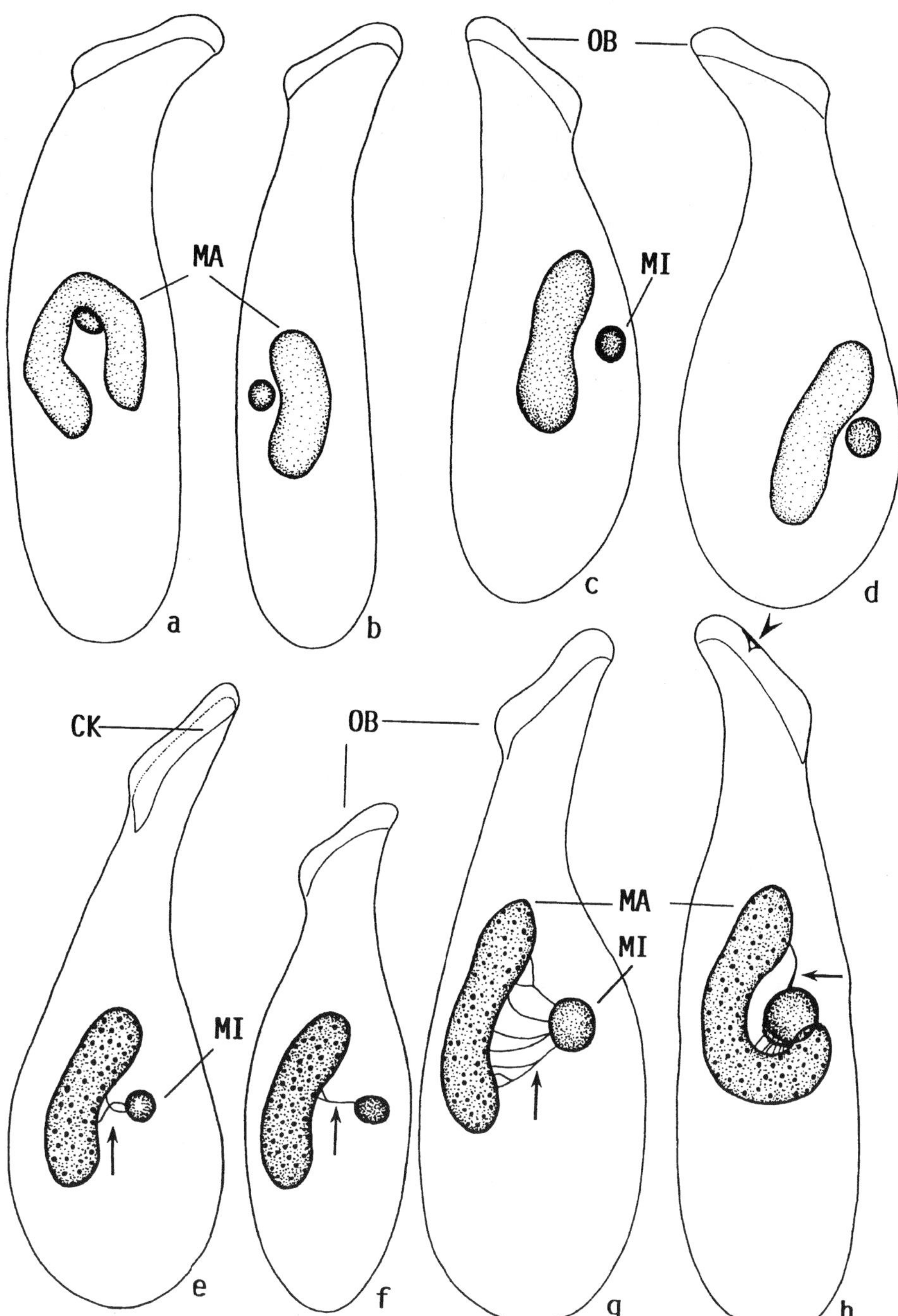

Fig. 84a–h *Cultellothrix coemeterii*, morphostatic specimens (a–f), very early (g), and early (h) dividers after protargol impregnation; drawn to scale, except for figure (g). From FOISSNER & LEI (2004). **a–d)** Variability of body shape and nuclear pattern, length 62–70 µm. **e–h)** The macronucleus and micronucleus are connected by fine fibres (arrows), likely originating from the micronucleus, length 65–95 µm. Arrowhead denotes the temporary cytostome. CK – circumoral kinety, MA– macronucleus, MI – micronucleus, OB – oral bulge.

Table 31 Morphometric data on *Cultellothrix coemeterii* in morphostatic and dividing specimens from the Klausen-Leopoldsdorf population. From FOISSNER & LEI (2004)

Characteristics[1]	Stages	$\overline{x}$	M	SD	SE	CV	Min	Max	n
Body, length	Morphostatic	64.2	64.0	4.0	1.0	6.3	58.0	72.0	15
	Very early	71.4	71.0	5.2	1.3	7.3	61.0	82.0	15
	Early	79.0	80.0	7.3	2.2	9.3	66.0	89.0	11
	Early middle	77.2	80.0	10.6	4.3	13.8	63.0	87.0	6
	Middle	70.3	69.0	–	–	–	68.0	74.0	3
	Late	72.0	75.0	–	–	–	65.0	76.0	3
	Very late	67.5	–	–	–	–	60.0	75.0	2
Body, width	Morphostatic	20.1	21.0	2.9	0.8	14.5	16.0	24.0	15
	Very early	21.6	22.0	3.2	0.8	15.0	17.0	28.0	15
	Early	20.0	19.0	2.8	0.9	14.1	17.0	26.0	11
	Early middle	25.3	24.0	2.4	1.0	9.6	24.0	30.0	6
	Middle	25.3	25.0	–	–	–	25.0	26.0	3
	Late	24.0	23.0	–	–	–	23.0	26.0	3
	Very late	22.5	–	–	–	–	22.0	23.0	2
Body length:width, ratio	Morphostatic	3.2	3.0	0.5	0.1	14.6	2.7	4.2	15
	Very early	3.3	3.2	0.6	0.2	18.1	2.6	4.5	15
	Early	4.0	4.3	0.8	0.2	18.9	2.8	4.9	11
	Early middle	3.1	3.2	0.6	0.2	19.4	2.2	3.6	6
	Middle	2.8	2.8	–	–	–	2.7	2.8	3
Oral bulge, length	Morphostatic	16.3	17.0	2.0	0.5	12.4	12.0	19.0	15
Circumoral kinety to last dikinetid of brush row 1, distance	Morphostatic	12.1	12.0	1.2	0.3	9.8	10.0	14.0	13
Circumoral kinety to last dikinetid of brush row 2, distance	Morphostatic	13.3	13.0	1.5	0.4	11.2	12.0	17.0	13
Circumoral kinety to last dikinetid of brush row 3, distance	Morphostatic	6.4	6.0	1.2	0.3	18.7	5.0	8.0	13
Macronucleus figure, length	Morphostatic	19.7	18.0	4.0	1.0	20.0	16.0	30.0	15
	Very early	29.1	27.5	5.3	1.4	18.3	21.0	40.0	15
	Early	31.3	31.0	4.8	1.4	15.3	24.0	42.0	11
	Early middle	25.8	29.0	6.2	2.5	23.9	17.0	31.0	6
	Middle	15.0	15.0	–	–	–	13.0	17.0	3
Macronucleus, width	Morphostatic	5.5	5.0	0.6	0.2	11.7	5.0	7.0	15
Macronucleus, number	Morphostatic	1.0	1.0	0.0	0.0	0.0	1.0	1.0	15
Micronucleus, largest diameter	Morphostatic	3.7	3.5	0.6	0.2	15.9	3.0	5.0	15
	Very early	6.9	6.8	1.1	0.3	15.4	5.5	9.5	15
	Early	9.9	10.0	1.7	0.5	17.1	7.5	13.0	11
	Early middle	10.4	10.5	1.6	0.7	15.6	8.5	13.0	6
Micronucleus, number	Morphostatic	1.0	1.0	0.0	0.0	0.0	1.0	1.0	15
Ciliary rows, number	Morphostatic	14.4	14.0	0.9	0.2	6.0	13.0	16.0	13
Basal bodies in a right side ciliary row, number	Morphostatic	19.8	19.0	3.6	1.2	18.0	15.0	26.0	9

continued

Characteristics[1]	Stages	$\overline{x}$	M	SD	SE	CV	Min	Max	n
Dorsal brush rows, number	Morphostatic	3.0	3.0	0.0	0.0	0.0	3.0	3.0	15
Dikinetids in brush row 1, number	Morphostatic	10.9	11.0	2.3	0.7	21.5	7.0	15.0	11
Dikinetids in opisthe's brush row 1, number	Early	6.5	7.0	1.0	0.5	15.4	5.0	7.0	4
	Middle	7.3	8.0	–	–	–	6.0	8.0	3
	Late & very late	8.5	8.5	1.3	0.6	15.2	7.0	10.0	4
	Post-divider	8.6	8.0	2.6	1.2	30.3	5.0	12.0	5
Dikinetids in brush row 2, number	Morphostatic	12.1	12.0	1.4	0.4	12.0	10.0	15.0	11
Dikinetids in opisthe's brush row 2, number	Early	7.5	8.0	1.9	1.0	25.5	5.0	9.0	4
	Middle	11.0	11.0	–	–	–	10.0	12.0	3
	Late & very late	10.3	10.0	0.5	0.3	4.9	10.0	11.0	4
	Post-divider	8.8	9.0	2.0	0.9	23.3	7.0	12.0	5
Dikinetids in brush row 3, number	Morphostatic	5.7	5.0	1.3	0.4	23.5	4.0	9.0	11
Dikinetids in opisthe's brush row 3, number	Early	4.3	4.0	0.5	0.3	11.8	4.0	5.0	4
	Middle	4.7	4.0	–	–	–	4.0	6.0	3
	Late & very late	5.5	6.0	1.0	0.5	18.2	4.0	6.0	4
	Post-divider	5.4	5.0	1.5	0.7	28.1	4.0	7.0	5
Dikinetids in circumoral kinety, number	Morphostatic	67.4	65.0	10.6	3.5	15.8	52.0	80.0	9
Dikinetids in opisthe's oral kinetofragments, number	Early	49.0	48.0	–	–	–	44.0	55.0	3
	Middle	51.0	50.0	–	–	–	49.0	54.0	3
	Late & very late	48.5	48.0	8.9	4.4	18.2	40.0	58.0	4
	Post-divider	49.3	48.5	5.4	2.7	10.9	44.0	56.0	4

[1] Data based on mounted, protargol-impregnated (FOISSNER's method), and randomly selected specimens from a non-flooded Petri dish culture. Measurements in µm. CV – coefficient of variation in %, M – median, Max – maximum, Min – minimum, n – number of individuals investigated, SD – standard deviation, SE – standard error of arithmetic mean, $\overline{x}$ – arithmetic mean.

Ontogenesis: Division was studied in a population from soil of a *Hordelymo-Fagetum* (Woodruff-beech) forest in the surroundings of the town of Vienna, Austria, viz., in the suburb Klausen-Leopoldsdorf (FOISSNER & LEI 2004). Cultures were established in Eau de Volvic (French table water) enriched with some drops of percolate from the non-flooded Petri dish culture and a few crushed wheat grains to stimulate growth of bacteria and prey protozoa (*Colpoda inflata*). *Cultellothrix coemeterii* developed readily in such a culture, but the cortical granules impregnated rather strongly and obscured part of the ciliary pattern. A second preparation, performed some days later, worked well although it was performed like the first one.

The Klausen-Leopoldsdorf specimens match the neotype population in most main features (nuclear and ciliary pattern, extrusomes, brush details), but differ considerably in size and shape (64 × 20 µm vs. 87 × 21 µm, that is, 3.2:1 vs. 4.3:1 in protargol preparations; Table 31) and, especially, in the shape of the circumoral kinety, which is distinctly cuneate (Fig. 84e, 85h–j, n, s), while oblong to indistinctly cuneate in the neotype specimens (Fig. 83b, i, m, o, 138g, j, q, r, w, 139a, e). These differences might support ranking the population as a distinct subspecies. However, when considering the rather pronounced overall variability and the agreement in all other main features, such split would appear too progressive at the state of knowledge. As concerns shape, the Klausen-Leopoldsdorf specimens match almost perfectly KAHL's type specimen (length: width ratio 2.75:1), further supporting the identification of FOISSNER et al. (2005a). Other differences and observations (Fig. 84a, 85a–d; Table 31): (i) macronucleus frequently in middle body third (vs. beneath mid-body); (ii) micronucleus (possibly two in some specimens) usually rather distant from (vs. close to) macronucleus; (iii) 14 vs. 12 ciliary rows; (iv) oral bulge shorter than widest trunk region by 20% (vs. as long as) and invariably slightly depressed in centre (vs. slightly concave, convex, or sigmoidal); (v) temporary cytostome occasionally (Fig. 84h) recognizable in protargol preparations (vs. seen only in SEM micrographs).

An outstanding feature of the Klausen-Leopoldsdorf population of *C. coemeterii* are very fine, fibrous connections between the micronucleus and macronucleus, recognizable in five out of forty-four specimens analyzed (Fig. 84e–h). The frequency of these structures, which likely originate from the micronucleus, increases in early dividers, where they are recognizable in eight out of twenty-one appropriately oriented specimens, suggesting some function in nuclear division. Unfortunately, we could not follow the origin of these structures during division. In the neotype specimens, which we checked for such connections, the micronucleus is too close to the macronucleus.

Very early dividers are slightly larger than morphostatic specimens (Fig. 85b–d, 139j; Table 31). Division commences with the production of basal bodies in or slightly underneath mid-body in those kineties which bear the dorsal brush in proter and opisthe. Usually, even an oblique dikinetid, obviously belonging to the prospective oral kinetofragments, is recognizable at the anterior end of each developing brush kinety (Fig. 85c). Soon after, basal body proliferation occurs in all ciliary rows, dikinetids assemble in the opisthe brush kineties, and minute oral kinetofragments, each usually composed of one to five dikinetids, develop at the anterior end of the broken ciliary rows (Fig. 85d). The macronucleus elongates from about 20 µm to 29 µm, but the nucleoli appear unchanged. The micronucleus almost doubles its size from an average of 3.7 µm to 6.9 µm (Fig. 85b–d, 139j, k, m; Table 31).

Early dividers are the largest and slenderest cells within the population because they are significantly longer, but not wider than morphostatic cells. Interestingly, they have a slight indentation in the prospective fission area (Fig. 85e–g, 139k; Table 31). The new oral kinetofragments now consist of three to six dikinetids, that is, are basically completed, but the final, total number of kinetids is possibly obtained only in late post-dividers, as indicated by counts of the circumoral dikinetids (Table 31). The newly formed kinetofragments detach from the somatic kineties and commence to curve rightwards, those on the left side slightly earlier than those on the right (Fig 85e, f). The opisthe's dorsal brush is almost completed, except of one or two dikinetids, which are added in mid-dividers; no monokinetids remain between circumoral kinety and the individual brush rows. An early divider with a fourth brush row right of row 1 in the proter lacks this row in the opisthe (Fig. 85f). The macronucleus now elongates to its maximum length of about 31 µm and assumes an oblong, slightly curved shape; the nucleoli still appear unchanged. The micronucleus, which is now close to the macronucleus in nine out of eleven appropriate stages, increases to about 10 µm in size and becomes somewhat spongious (Fig. 85g, 139k; Table 31).

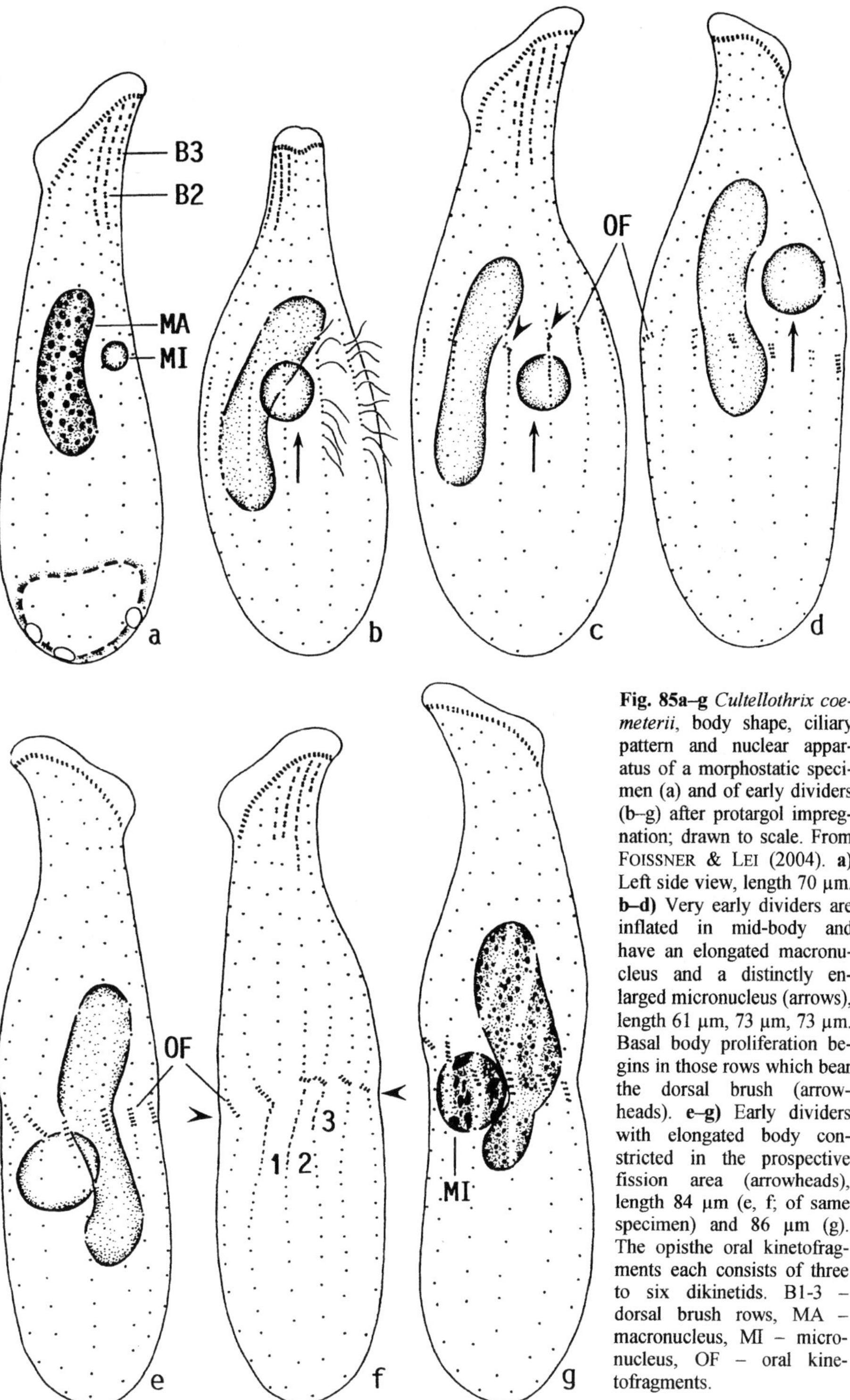

Fig. 85a–g *Cultellothrix coemeterii*, body shape, ciliary pattern and nuclear apparatus of a morphostatic specimen (a) and of early dividers (b–g) after protargol impregnation; drawn to scale. From FOISSNER & LEI (2004). **a)** Left side view, length 70 µm. **b–d)** Very early dividers are inflated in mid-body and have an elongated macronucleus and a distinctly enlarged micronucleus (arrows), length 61 µm, 73 µm, 73 µm. Basal body proliferation begins in those rows which bear the dorsal brush (arrowheads). **e–g)** Early dividers with elongated body constricted in the prospective fission area (arrowheads), length 84 µm (e, f; of same specimen) and 86 µm (g). The opisthe oral kinetofragments each consists of three to six dikinetids. B1-3 – dorsal brush rows, MA – macronucleus, MI – micronucleus, OF – oral kinetofragments.

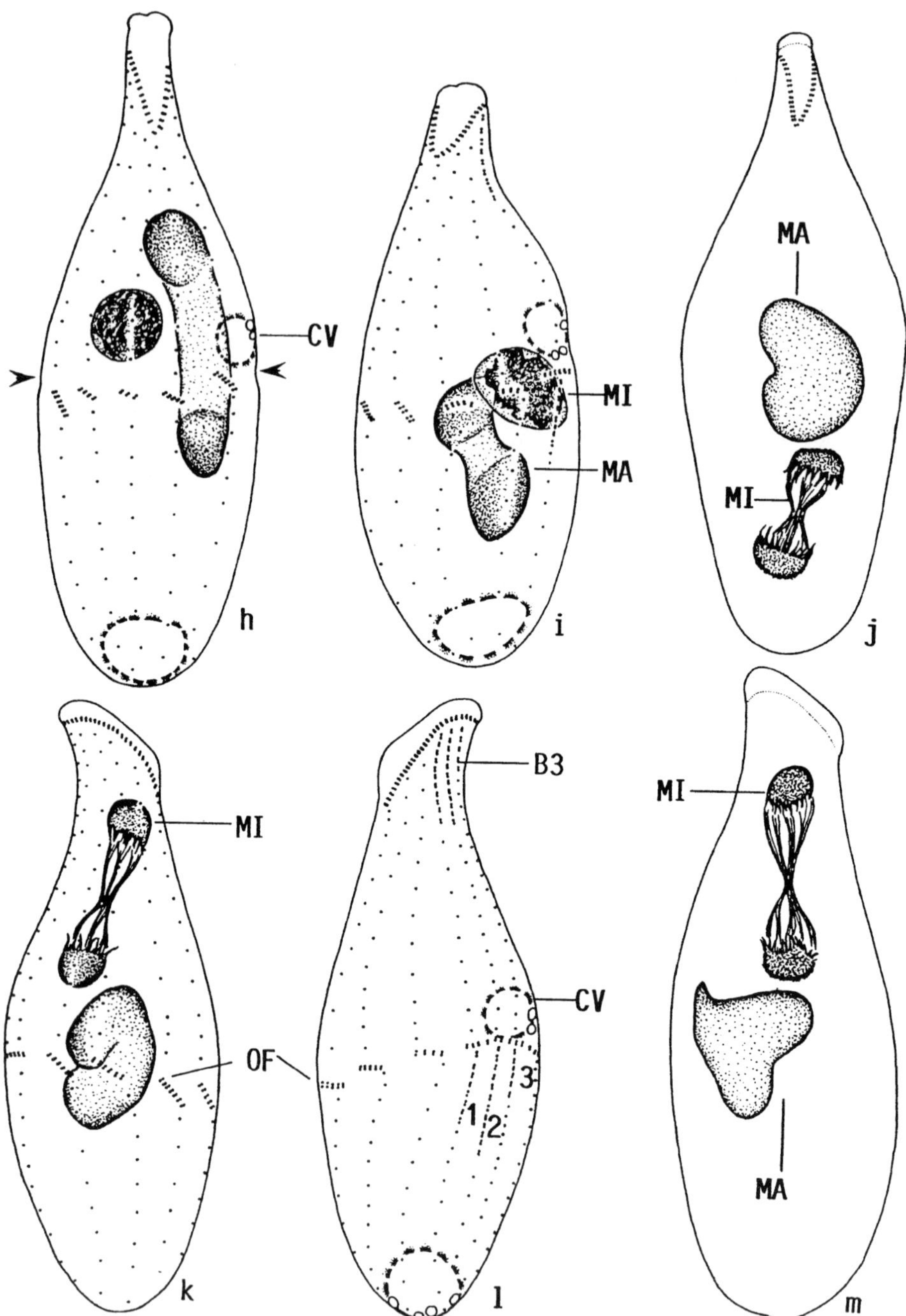

Fig. 85h–m *Cultellothrix coemeterii*, middle dividers after protargol impregnation; drawn to scale. From FOISSNER & LEI (2004). **h)** Early mid-divider with inflated body and newly formed proter's contractile vacuole, length 74 µm. Arrowheads mark the disappearing central constriction. **i)** Early mid-divider with condensing macronucleus and elongating micronucleus, length 63 µm. **j–m)** Middle dividers with condensed macronucleus and dividing micronucleus, length 68 µm (j), 69 µm (k, l; of same specimen), 74 µm (m). B1-3 – dorsal brush rows, CV – contractile vacuole, MA – macronucleus, MI – micronucleus, OF – oral kinetofragments.

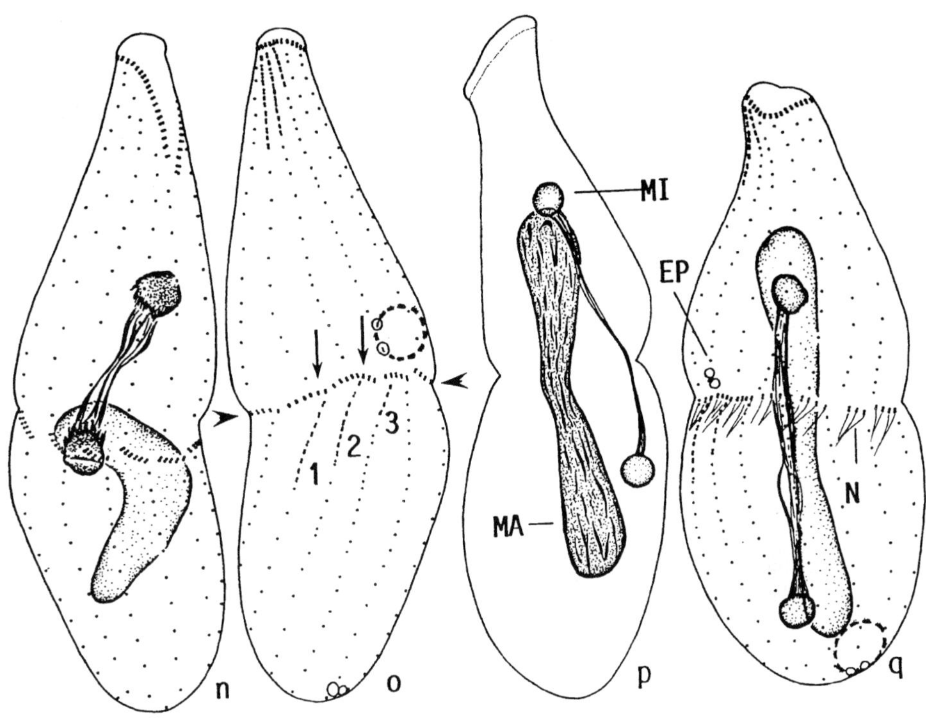

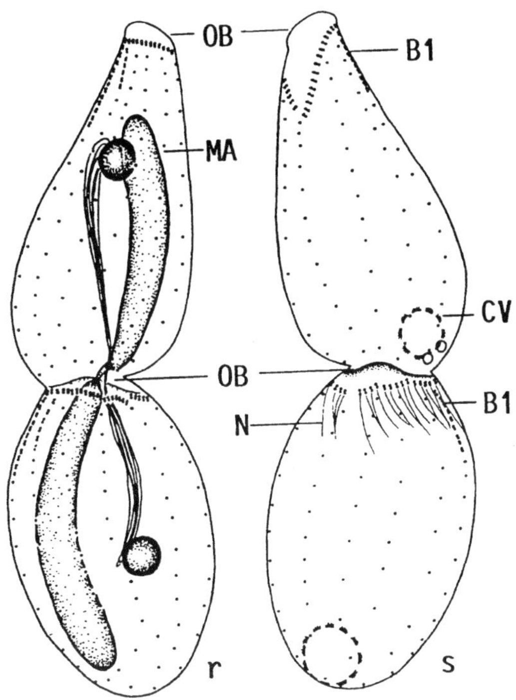

Fig. 85n–s *Cultellothrix coemeterii*, body shape, ciliary pattern and nuclear apparatus of late dividers after protargol impregnation; drawn to scale. From FOISSNER & LEI (2004). **n, o)** Ventral and dorsal view of an early late divider showing the developing division furrow (arrowheads), the elongating macronucleus mass and the dividing micronucleus, length 75 µm. Arrows mark opisthe's oral kinetofragments aligning horizontally in the fission area. **p, q)** Late dividers, length 76 µm and 65 µm. The dividing macronucleus becomes dumb-bell shaped and the divided micronuclei become globular. The opisthe's nematodesmata develop from the oral kinetofragments. **r, s)** Dorsal and ventral view of a very late divider, length 75 µm. The opisthe has an incomplete, convex oral bulge. The rod-shaped daughter's macronuclei and the globular micronuclei are still connected by fine fibres. B1-3 – dorsal brush rows, CV – newly formed contractile vacuole, EP – excretory pores, MA – macronucleus, MI – micronucleus, N – nematodesmata, OB – oral bulge.

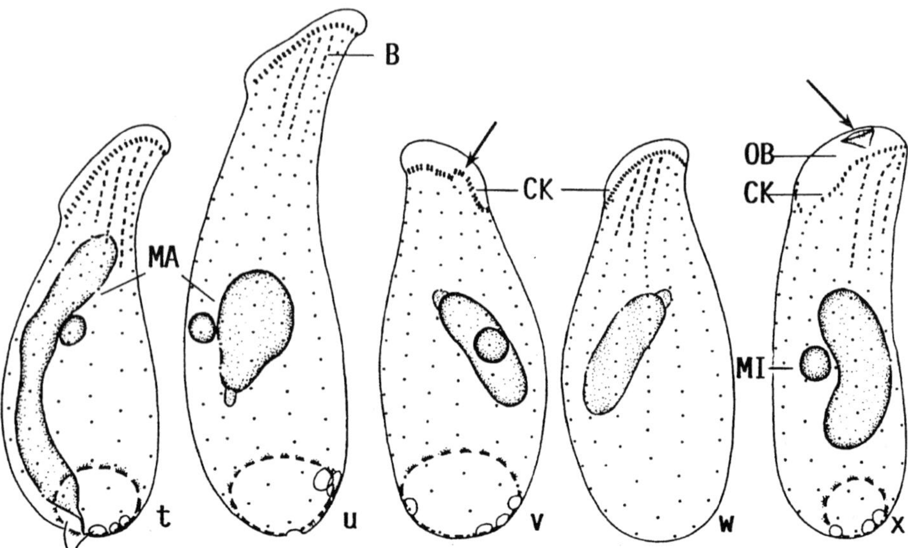

Fig. 85t–x *Cultellothrix coemeterii*, body shape, ciliary pattern and nuclear apparatus of post-dividers after protargol impregnation; drawn to scale. From FOISSNER & LEI (2004). **t)** Left side view of a just separated proter post-divider with irregular posterior body end and long, posteriorly tapered macronucleus, length 45 µm. **u)** A late proter post-divider recognizable by the posteriorly tapered macronucleus, length 58 µm. **v, w)** Right and left side view of same opisthe post-divider with small, slightly oblique oral bulge and anteriorly tapered macronucleus, length 44 µm. Arrow denotes slight disorder in the right side circumoral kinety. **x)** Left side view of another opisthe post-divider (length 45 µm) with well developed macronucleus, but not yet finished oral bulge with a conical indentation, likely the temporary cytostome or the site where the daughters separated. B – dorsal brush, CK – circumoral kinety, MA – macronucleus, MI – micronucleus, OB – oral bulge.

Early mid-dividers have the same length as early dividers, but are inflated in mid-body, where the indentation in the prospective fission area is still recognizable (Fig. 85h, i; Table 31). While the ciliary pattern is very similar to that of early dividers, the nuclear apparatus shows distinct changes: the macronucleus commences to shorten towards the centre, as indicated by the inflated ends; the nucleoli disappear; and the micronucleus, whose spongious content detaches from the nuclear membrane, approaches its maximum size of 10.5 µm on average, that is, shows a really conspicuous, three-fold increase compared to the morphostatic size. A new proter contractile vacuole and excretory pores are generated dorsolaterally above the prospective fission area.

Middle dividers keep the maximum body width of about 25 µm, but shorten to a length of 70 µm becoming bluntly fusiform (Fig. 85j–m, 139l; Table 31). The new oral kinetofragments are horizontally arranged on the left side, while still slanted on the right. The dorsal brush is now complete, that is, no further increase occurs in the number of dikinetids (Table 31). The macronucleus condenses to a globular, homogenously impregnated mass. The micronucleus begins to divide, soon becoming distinctly dumb-bell-shaped because the hemispherical halves are connected by a conspicuous, centrally constricted fibre bundle.

Late dividers are slightly larger than morphostatic cells and reform the division furrow in mid-body (Fig. 85n–q; Table 31). The newly produced somatic basal bodies have fully developed cilia and become evenly distributed in the rows. The opisthe's oral kinetofragments, which are slightly concave ventrally and convex dorsally, loosely align horizontally in the prospective fission area and develop oral basket rods (nematodesmata). The macronuclear mass, which now contains fibrous structures, becomes C-shaped and, somewhat later, extends to a long rod constricting in the mid, that is, in the prospective fission area. The daughter's micronuclei, which are still connected by long, parallel fibres, move apart; gradually smoothen their outline; and condense to compact, spherical masses.

Very late dividers have a similar length as morphostatic specimens and are distinctly furrowed in the prospective fission area (Fig. 85r, s; Table 31). A bare protuberance, that is, the precursor of the oral bulge, develops at the anterior end of the opisthe. The ciliary rows are distinctly separate from the opisthe's kinetofragments and still meridionally arranged, viz., do not show the *Arcuospathidium* pattern typical for the genus. The opisthe's oral kinetofragments now move together, obviously due to the body's constriction in the fission area, forming a slightly irregular, continuous circumoral kinety. The macronucleus has divided into two rod-shaped pieces, each almost as long as the daughter's body and pointed in the fission area. The micronuclei, which are still connected by a long, distinct fibre bundle, are smoothly globular and increase in size.

Very late dividers are only slightly larger than morphostatic cells (Table 31). Thus, early post-dividers are recognizable by the small size and pointed macronucleus (Fig. 85t, u, 139n); opisthe post-dividers, additionally, have a small, somewhat irregular and transverse-truncate oral bulge (Fig. 85v, w). Unfortunately, only two late opisthe post-dividers were found (Fig. 85v–x). Both have an indistinct *Cultelothrix*/*Arcuospathidium* ciliary pattern, which is thus obviously obtained only in fully grown cells, possibly in that the dorsal bulge half grows faster and pulls up the ciliary rows. Further, some circumoral dikinetids are possibly produced post-divisionally, as explained above. Post-divisional development of the nuclear apparatus is the same in proter and opisthe, viz., the macronucleus gradually shortens to obtain the species-specific shape, and the micronucleus moves to mid-macronucleus.

The parental oral apparatus and dorsal brush do not show any changes during division, as in the few other members of the family investigated so far. No bulbs are recognizable in the division area, at least with the light microscope.

Ontogenetic comparison: Ontogenesis of *C. coemeterii* is similar to that of *Arcuospathidium muscorum* (BERGER et al. 1983) and other species of the group (see Table on page 26), but shows the following peculiarities, some of which, however, might have been overlooked in previous studies: (i) early dividers show a transient indentation in the prospective fission area (Fig. 85e–h), a curious feature as yet found only in *Protospathidium serpens* and *Arcuospathidium cultriforme scalpriforme*; (ii) the macronucleus distinctly elongates in early dividers (Table 31), while it remains unchanged in all other species investigated so far; (iii) the micronucleus shows a three-fold size increase and thus becomes very conspicuous in early middle dividers (Fig. 85g–i, 139j, k, n; Table 31), while it merely doubles size in other species; (iv) the left side oral kinetofragments curve rightwards earlier than the right side ones (Fig. 85e, f, k, l), a curious feature described also in some other species; (v) the oral kinetofragments detach from the somatic kineties and curve rightwards earlier in *C. coemeterii* (in early dividers; Fig. 85k, l) than in *A. muscorum* (in middle dividers); (vi) very likely, there is some post-divisional growth of the circumoral kinety (Table 31); (vii) *C. coemeterii* and most other spathidiids form the circumoral kinety in a simple way, viz., by alignment of the newly produced kinetofragments one after the other (Fig. 85n–s), while complex shaping and fusion processes occur in *A. muscorum* and *A. cultriforme scalpriforme*; (viii) shaping of the oral bulge and circumoral ciliature occurs distinctly later in *Arcuospathidium*, *Cultellothrix* (Fig. 85r–x) and *Spathidium* than in *Protospathidium*, viz., in late post-dividers, respectively, very late dividers (see *P. serpens*); (ix) the *Arcuospathidium* ciliary pattern develops via a *Protospathidium* pattern in late post-dividers, suggesting that it is an apomorphic feature.

***Cultellothrix paucistriata* nov. spec.** (Fig. 86a–m, 140a–g; Table 32)

Diagnosis: Size about 65 × 20 μm in vivo. Narrowly spatulate with oblique, slightly cuneate oral bulge shorter than widest trunk region by 40%. Macronucleus usually C-shaped; single micronucleus. Two types of extrusomes: type I narrowly ovate and slightly curved, 3 × 0.7 μm in size; type II oblong and about 1.5 μm long. On average 7 ciliary rows, those on left side anteriorly modified to heterostichad, three-rowed dorsal brush occupying about 20% of body length. Brush bristles up to 5 μm long: row 1 composed of an average of 8 dikinetids, row 2 of 8, and row 3 of 6 dikinetids followed by a monokinetidal bristle tail extending to mid-body..

Type locality: Soil from the surroundings of Rio de Janeiro, Brazil, viz., the shrub zone of the Restingha area about 1 km off the Atlantic sea coast, W43° S23°30'.

Etymology: Composite of the Latin adjectives *pauci* (few) and *striatus* (striated ~ ciliary rows), referring to the small number of ciliary rows, a main feature of the species.

Description: Size moderately variable, viz., 50–80 × 18–25 μm in vivo, usually approximately 65 × 20 μm, as calculated from some in vivo measurements and the morphometric data; length:width ratio 2.6–3.8:1, on average 3.1:1 in protargol preparations, while 3–4:1 in vivo (Fig. 86a; Table 32). Shape very similar to certain pleurostomatids, e.g., *Amphileptus punctatus* (Fig. 81a, b), narrowly spatulate with obliquely truncate anterior (oral) end projecting fairly distinctly dorsally and ventrally producing a rather distinct neck (Fig. 86b). Trunk ellipsoidal with often nearly flat ventral outline and more or less distinctly convex dorsal side, both margins more distinctly convex in protargol preparations than in vivo; laterally flattened about 2:1, in hyaline oral area up to 3:1; posterior end usually rounded, occasionally bluntly pointed, possibly due to the contractile vacuole contained (Fig. 86a–c, j–m, 140 a, e). Macronucleus underneath mid-body, usually C-shaped with ends slightly inflated, occasionally cylindroidal or tortuous; never in two nodules; contains many nucleoli up to 1 μm across. Micronucleus usually embraced by macronucleus, globular to ellipsoidal, about 2 μm across in preparations (Fig. 86a, c, j–m, 140e, f; Table 32). Contractile vacuole in posterior body end, excretory pore(s) not impregnated. Two types of extrusomes studded in oral bulge and scattered in the cytoplasm (Fig. 86a, f–h, 140b–d): type I narrowly ovate and slightly

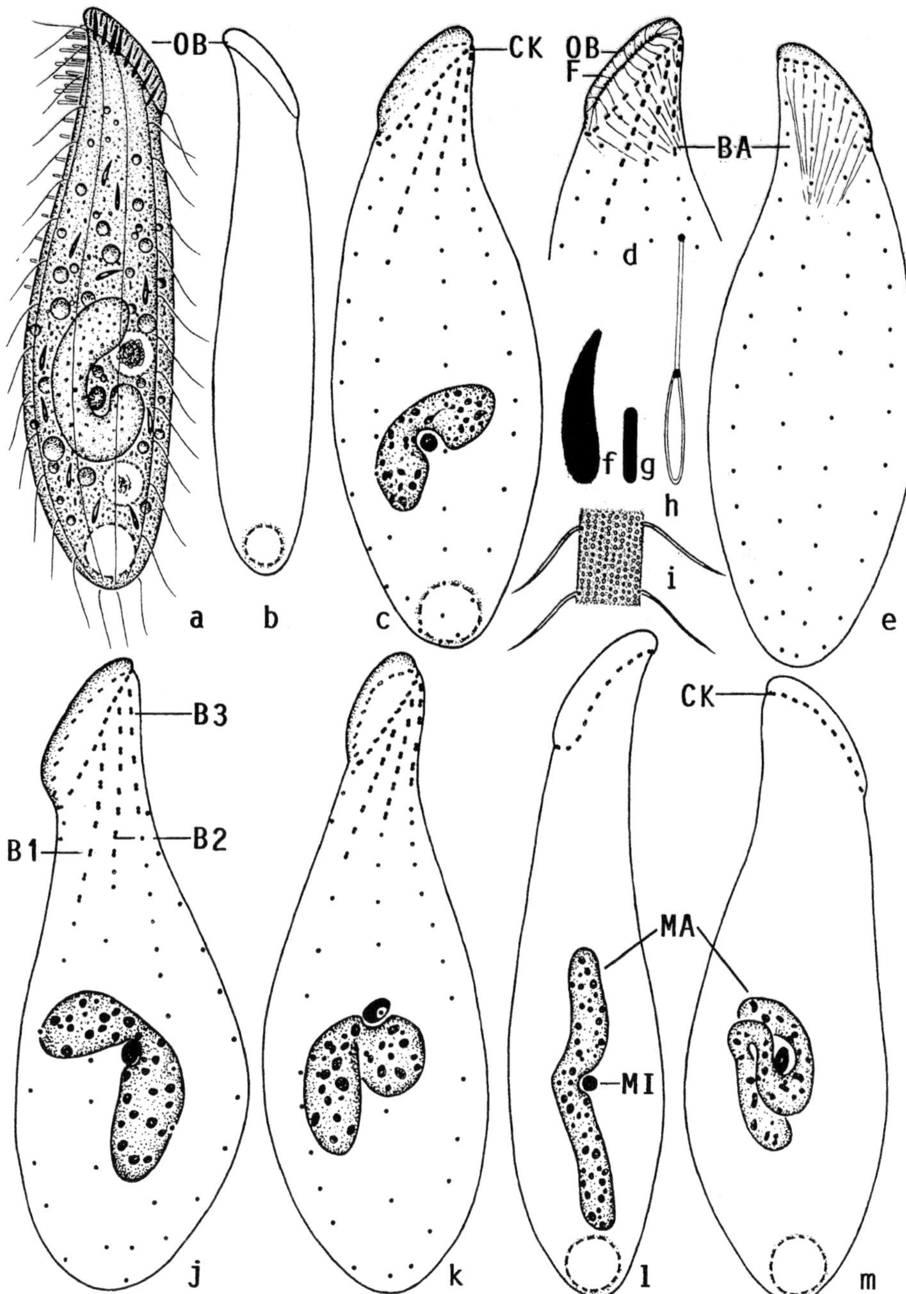

Fig. 86a–m *Cultellothrix paucistriata* nov. spec. from life (a, b f–i) and after protargol impregnation (c–e, j–m). **a, b)** Right side views of a representative and a slender specimen, length 65 µm. **c–e)** Right and left side view of holotype specimen, length 62 µm. **f, h)** Resting and exploded type I extrusome, length 3 µm. **g)** Resting type II extrusome, length 1.5 µm. **i)** Cortical granulation. **j–m)** Ciliary pattern and nuclear apparatus of other specimens, length 46 µm, 68 µm, 64 µm, 66 µm. BA – oral basket, B1-3 – dorsal brush rows, CK – circumoral kinety, F – fibres, MA – macronucleus, MI – micronucleus, OB – oral bulge.

curved, about 3 × 0.7 µm in size, compact and thus highly refractive; type II oblong and inconspicuous because only 1.5 µm long; both do not impregnate with the protargol method used. Exploded type I extrusomes up to 7 µm long, of typical toxicyst structure, contain a refractive granule each in distal end of tube and capsule. Cortex very flexible, contains about nine rows of colourless, narrowly spaced granules forming plate-like layer between each two ciliary rows; individual granules only about 0.2 µm across, usually impregnate with protargol obscuring ciliary pattern (Fig. 140a). Cytoplasm colourless and often very hyaline, contains some lipid droplets up to 3 µm across and food vacuoles with mainly *Protocyclidium muscicola*, which is digested slowly because still identifiable; rarely, diatoms are ingested (Fig. 86a, 140a, b, e, f). Glides and swims conspicuously slowly.

Somatic cilia about 8 µm long in vivo, arranged in an average of seven widely spaced and loosely ciliated, bipolar rows anteriorly curved dorsally, that is, in *Arcuospathidium*, respectively, *Cultellothrix* pattern with dorsal brush occupying left half to left two thirds of left anterior body side; none or only one ordinary ciliary row between dorsal brush and ventral kinety abutting to summit of circumoral kinety (Fig. 86a, c–e, j, k; Table 32). Dorsal brush heterostichad and three-rowed, in vivo rather conspicuous, though extending merely about 20% of body length, because on left body side and bristles up to 5 µm long; all rows without monokinetids anteriorly, rows 1 and 2 continue with ordinary cilia posteriorly. Longest brush row 1 usually abutting on circum-oral kinety in body midline and extending to second body third; row 2 slightly shorter than row 1, both rows composed of an average of eight dikinetids with rod-shaped or slightly clavate bristles; anterior (?) bristle of dikinetids about 3 µm long in middle brush portion, posterior (?) circa 5 µm, bristle length decreases to 2 µm anteriorly and to 1 µm posteriorly. Row 3 near to dorsal margin of cell, slightly shorter than rows 1 and 2, composed of an average of six dikinetids followed by a monokinetidal tail extending to mid-body with comparatively narrowly spaced bristles decreasing in length from about 3 µm anteriorly to 2 µm posteriorly (Fig. 86a, c, d, j, k; Table 32).

Oral bulge slanted by 40–60°, occupies approximately 60% of trunk width and about 21% of body length; distinct in vivo, especially in dorsal half, due to the refractive extrusomes contained and because rather distinctly separated from body proper; about 3 µm high and slightly to distinctly convex (Fig. 86a, b, l, m, 140a, e, g; Table 32). Circumoral kinety cuneate to almost elliptical, continuous, composed of an average of 23 comparatively widely spaced dikinetids each associated with a cilium, a nematodesma, and a faintly impregnated fibre extending to bulge midline. Oral basket funnel-shaped, inconspicuous because extending only in anterior quarter of body and faintly impregnated with protargol (Fig. 86c–e, j–m; Table 32).

Occurrence and ecology: As yet found only at type locality, where it was abundant in the non-flooded Petri dish culture. The sample, which consisted of surface litter, fine roots, and strongly decayed organic material sieved off the upper 10 cm very sandy soil layer, was humic, slightly saline (<10‰), and had pH 5.3 in water. *Cultellothrix paucistriata* is small and thus well adapted to live in sandy soil.

Remarks: The overall appearance of *C. paucistriata* highly resembles that of the other members of *Cultellothrix* group two. It differs from *C. tortisticha*, inter alia, by the macronucleus (C-shaped vs. two nodules) and the shape of the extrusomes

Table 32 Morphometric data on *Cultellothrix paucistriata* (upper line) and *C. tortisticha* (lower line)

Characteristics[1]	$\overline{x}$	M	SD	SE	CV	Min	Max	n
Body, length	59.0	59.0	7.0	1.5	11.9	46.0	75.0	21
	47.8	47.0	5.4	1.2	11.4	37.0	58.0	21
Body, width	19.2	18.0	2.7	0.6	13.8	16.0	26.0	21
	9.4	9.0	2.1	0.5	22.6	6.0	15.0	21
Body length:width, ratio	3.1	3.1	0.4	0.1	12.5	2.6	3.8	21
	5.3	5.3	1.0	0.2	18.7	3.7	7.0	21
Oral bulge, length	12.3	12.0	1.7	0.4	13.4	9.0	15.0	21
	10.7	11.0	1.1	0.2	10.2	9.0	13.0	21
Oral bulge (circumoral kinety), width	3.5	3.5	0.4	0.2	12.8	3.0	4.0	6
	2.8	3.0	0.3	0.1	9.2	2.5	3.0	10
Oral bulge, height	2.1	2.0	–	–	–	2.0	2.5	21
	1.9	2.0	0.4	0.1	21.9	1.0	3.0	21
Oral bulge length:body width, ratio	0.6	0.6	0.1	–	15.0	0.5	0.9	21
	1.2	1.2	0.2	0.1	20.2	0.8	1.7	21
Circumoral kinety to last dikinetid of brush row 1, distance	12.0	13.0	3.3	1.0	27.9	8.0	19.0	11
	5.5	6.0	0.7	0.1	12.3	4.0	7.0	21
Circumoral kinety to last dikinetid of brush row 2, distance	10.7	11.0	1.7	0.5	15.6	8.0	13.0	11
	5.5	5.0	0.6	0.1	11.0	5.0	7.0	21
Circumoral kinety to last dikinetid of brush row 3, distance	8.1	8.0	1.2	0.4	15.1	6.0	10.0	11
	4.4	4.0	1.0	0.2	23.2	3.0	7.0	21
Anterior body end to macronucleus, distance[2]	29.3	29.5	3.6	0.8	12.1	23.0	35.0	20
	21.2	21.5	4.7	1.1	22.1	14.0	32.0	20
Macronucleus/macronucleus nodules, length[2]	17.2	18.0	4.2	0.9	24.6	11.0	27.0	21
	7.2	6.0	2.9	0.6	40.6	3.0	15.0	21
Macronucleus/macronucleus nodules, width[2]	5.2	5.0	1.1	0.2	20.8	3.0	8.0	21
	4.2	4.0	0.8	0.2	18.3	3.0	5.0	21
Macronuclei, number	1.0	1.0	0.0	0.0	0.0	1.0	1.0	21
	1.7	2.0	–	–	–	1.0	2.0	121
Micronucleus, across	2.0	2.0	–	–	–	1.5	2.0	21
	1.3	1.0	0.3	0.1	23.8	1.0	2.0	21
Micronucleus, number	1.0	1.0	0.0	0.0	0.0	1.0	1.0	21
	1.0	1.0	0.0	0.0	0.0	1.0	1.0	21
Ciliary rows, number	6.7	7.0	0.6	0.1	8.7	6.0	8.0	21
	7.0	7.0	0.2	–	3.1	6.0	7.0	21
Basal bodies in a right side ciliary row, number	14.6	15.0	2.8	1.0	18.9	10.0	19.0	7
	14.9	13.0	4.0	0.9	27.2	8.0	21.0	21
Dorsal brush rows, number	3.0	3.0	0.0	0.0	0.0	3.0	3.0	21
	3.0	3.0	0.0	0.0	0.0	3.0	3.0	21
Dikinetids in brush row 1, number	7.8	8.0	1.1	0.3	13.8	6.0	9.0	11
	4.5	5.0	0.5	0.1	11.3	4.0	5.0	21
Dikinetids in brush row 2, number	7.6	8.0	0.8	0.2	10.6	6.0	9.0	11
	4.3	4.0	0.6	0.1	13.3	4.0	6.0	21
Dikinetids in brush row 3, number	5.7	6.0	0.6	0.2	11.3	5.0	7.0	11
	3.3	3.0	0.6	0.1	19.6	3.0	5.0	21
Circumoral dikinetids, number	23.1	23.0	2.2	0.6	9.4	20.0	27.0	13
	26.5	27.0	2.5	0.6	9.5	21.0	31.0	21

[1] Data based on mounted, protargol-impregnated (FOISSNER's method), and randomly selected specimens from non-flooded Petri dish cultures. Measurements in µm. CV – coefficient of variation in %, M – median, Max – maximum, Min – minimum, n – number of individuals investigated, SD – standard deviation, SE – standard error of arithmetic mean, $\overline{x}$ – arithmetic mean.

[2] In *C. tortisticha*, values contain specimens with one or two macronuclear nodules.

(narrowly ovate vs. oblong). *Cultellothrix japonica*, *C. atypica*, and *C. coemeterii* all have about 12 (vs. 7) ciliary rows and rod-shaped or acicular (vs. narrowly ovate) extrusomes.

In vivo, *C. paucistriata* is rather similar to *Apertospathula armata*, differing mainly in the size and shape of the extrusomes (3 µm and narrowly ovate vs. 1.5 µm and ellipsoidal). Small specimens of *Spathidium spathula* also highly resemble *C. paucistriata*, but have rod-shaped extrusomes. *Apertospathula swarezewskyi* differs from *C. paucistriata* by the longer extrusomes (6–8 µm vs. 3 µm), the higher number (about 12 vs. 7) of ciliary rows, and the habitat (saline ponds vs. soil). Generally, spathidiids of this shape and size are frequent and thus difficult to identify, the best features being the size and shape of the extrusomes, the nuclear pattern, and the number of ciliary rows.

***Cultellothrix tortisticha* nov. spec.** (Fig. 87a–s, 140h–o; Table 32)

Diagnosis: Size about 55 × 10 µm in vivo. Narrowly spatulate with oblique, cuneate oral bulge about 1.2 times as long as widest trunk region; bulge and circumoral kinety screwed like a propeller blade. Two macronucleus nodules with one micronucleus in between. Extrusomes oblong and slightly curved, 2.5–3 × 0.5 µm in size. On average 7 ciliary rows, those (three) on left side anteriorly modified to isostichad dorsal brush occupying about 12% of body length. Brush bristles up to 4 µm long: row 1 composed of an average of 5 dikinetids, row 2 of 4, and row 3 of only 3 dikinetids followed by a monokinetidal bristle tail extending to near posterior body end.

Type locality: Terra firma secondary rain forest soil from bank of Rio Negro in the surroundings of Hotel Tropical at Manaus, Brazil, W60° S3°.

Etymology: Composite of the Latin adjective *tortus* (twisted) and the Greek noun *sticha* (row), referring to the propeller blade-like twist of the circumoral kinety and oral bulge, a main feature of this species.

Description: This is the smallest species of the genus, viz., a tiny, inconspicuous ciliate measuring 40–65 × 7–15 µm in vivo, usually about 55 × 10 µm, as calculated from some in vivo measurements and the morphometric data; length:width ratio 3.7–7.1 in preparations, on average near 5.5:1 both in vivo and protargol slides (Fig. 87a; Table 32). Overall appearance *Litonotus*-like because up to 2:1 flattened laterally and narrowly spatulate with obliquely slanted anterior (oral) end smoothly merging into neck usually less pronounced in vivo than in preparations, where cells are slightly contracted or shrunken subapically (Fig. 87a, g, h–m, 140a, b; Table 32). Nuclear apparatus underneath mid-body on average, basically composed of two globular (rarely ellipsoidal) macronucleus nodules with a globular micronucleus in between, as recognizable in early dividers (Fig. 87q); pattern, however, highly variable (Fig. 87e, g–i, k, m–p, 140k–o; Table 32): of 121 specimens analysed, 80 have the ordinary pattern described above, while 40 have only one ellipsoidal (rarely globular) nodule with a micronucleus attached, and 1 (likely an exconjugant) even has four nodules and two micronuclei; in a later preparation without dividers, even half of the specimens have only one macronucleus nodule, showing that the high variability has other than the usual post-divisional

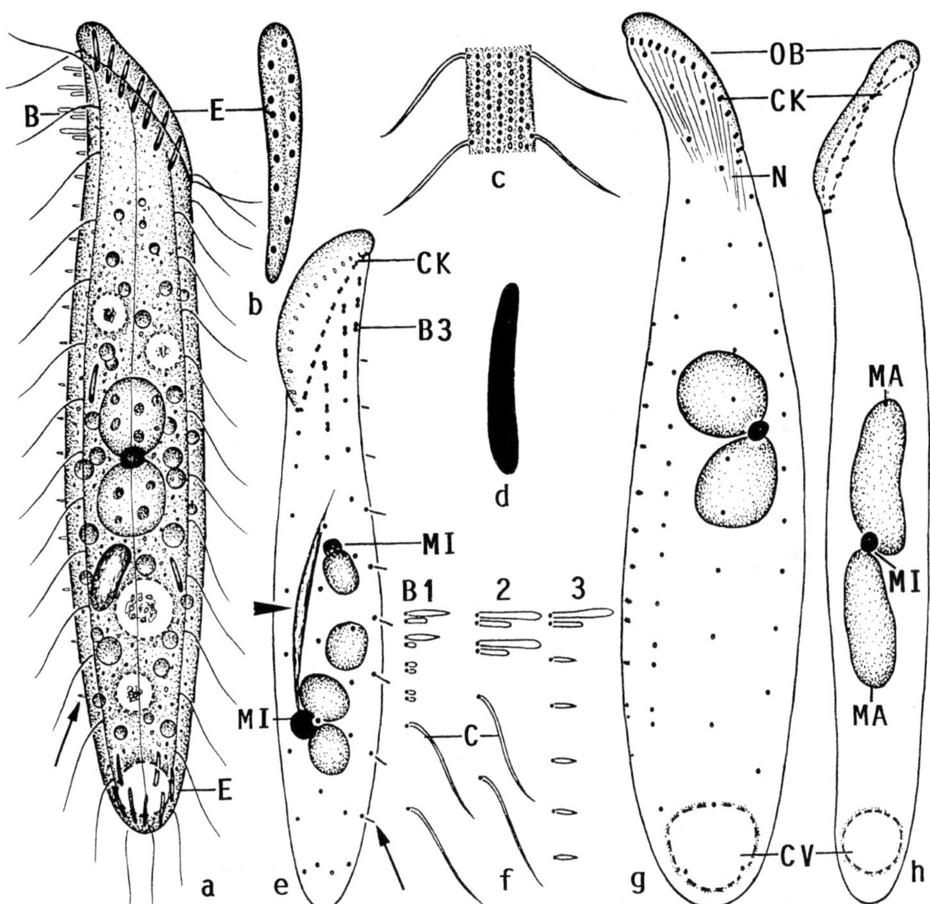

Fig. 87a–h *Cultellothrix tortisticha* nov. spec. from life (a–d, f) and after protargol impregnation (e, g, h). **a)** Right side view of a representative, 55 µm long specimen with two macronucleus nodules and a micronucleus in between. Arrow marks posterior end of monokinetidal bristle tail of brush row 3. Note extrusomes in posterior body end, similar as in some pleurostomatids; thus and because of the body shape, *C. tortisticha* is easily confused with species of this group. **b)** Frontal view of slightly curved and cuneate oral bulge. **c)** Surface view showing dense cortical granulation. **d)** Oral bulge extrusome, length 3 µm. **e)** Ventrolateral view of a 37 µm long post-conjugant (?) with four macronucleus nodules and two micronuclei connected by a fibrous structure (arrowhead). The arrow marks the posterior end of the monokinetidal bristle tail of brush row 3. Note the circumoral kinety whose bow-like pattern is caused by the strongly convex right half of the oral bulge. The dorsal brush occupies the anterior left side of the cell, showing that this species belongs to *Cultellothrix*. **f)** Posterior portion of dorsal brush, longest bristles 3 µm; row 1 is completely shown and consists of two "ordinary" and two minute (length ~0.5 µm) bristle pairs; row 3 has a monokinetidal bristle tail, occasionally recognizable also in protargol-impregnated cells (e, i, m). **g, h)** Right and left side view of specimens with typical nuclear apparatus, length 57 µm and 55 µm. Note the distinctly curved right half of the circumoral kinety (g) causing its bow- or ∞-like appearance in ventrolateral (e) and lateral (i–m) view. B(1-3) – dorsal brush (rows), C – ordinary somatic cilia, CK – circumoral kinety, CV – contractile vacuole, E – extrusomes, MA – macronucleus nodules, MI – micronucleus, N – nematodesmata, OB – oral bulge.

reasons. Contractile vacuole in rear body end and surrounded by extrusomes, some excretory pores in pole area (Fig. 87a, i, k, l). Only six to twelve extrusomes form a rather loose row each in right and left half of oral bulge, and about the same number invariably

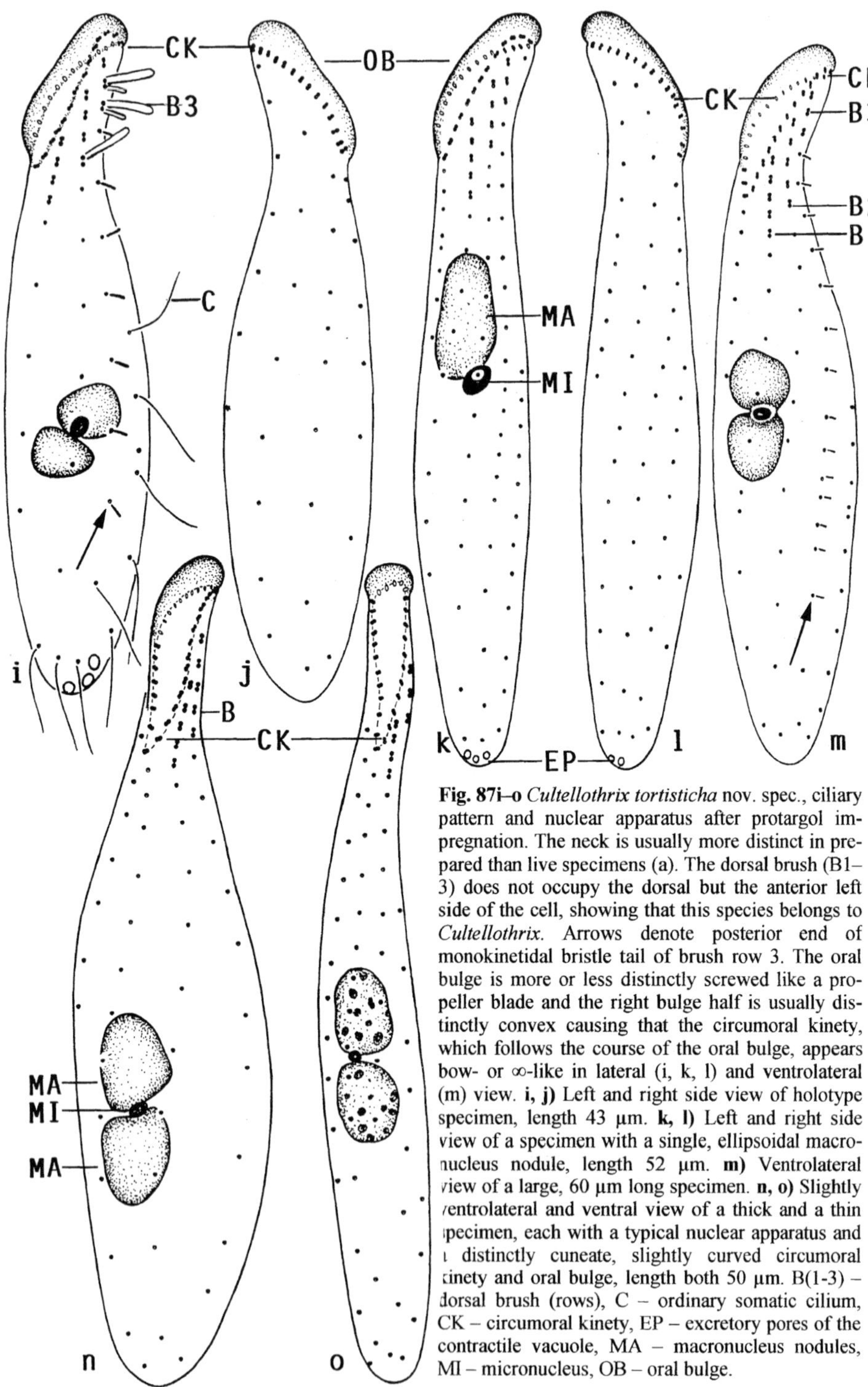

Fig. 87i–o *Cultellothrix tortisticha* nov. spec., ciliary pattern and nuclear apparatus after protargol impregnation. The neck is usually more distinct in prepared than live specimens (a). The dorsal brush (B1–3) does not occupy the dorsal but the anterior left side of the cell, showing that this species belongs to *Cultellothrix*. Arrows denote posterior end of monokinetidal bristle tail of brush row 3. The oral bulge is more or less distinctly screwed like a propeller blade and the right bulge half is usually distinctly convex causing that the circumoral kinety, which follows the course of the oral bulge, appears bow- or ∞-like in lateral (i, k, l) and ventrolateral (m) view. **i, j)** Left and right side view of holotype specimen, length 43 μm. **k, l)** Left and right side view of a specimen with a single, ellipsoidal macronucleus nodule, length 52 μm. **m)** Ventrolateral view of a large, 60 μm long specimen. **n, o)** Slightly ventrolateral and ventral view of a thick and a thin specimen, each with a typical nuclear apparatus and a distinctly cuneate, slightly curved circumoral kinety and oral bulge, length both 50 μm. B(1-3) – dorsal brush (rows), C – ordinary somatic cilium, CK – circumoral kinety, EP – excretory pores of the contractile vacuole, MA – macronucleus nodules, MI – micronucleus, OB – oral bulge.

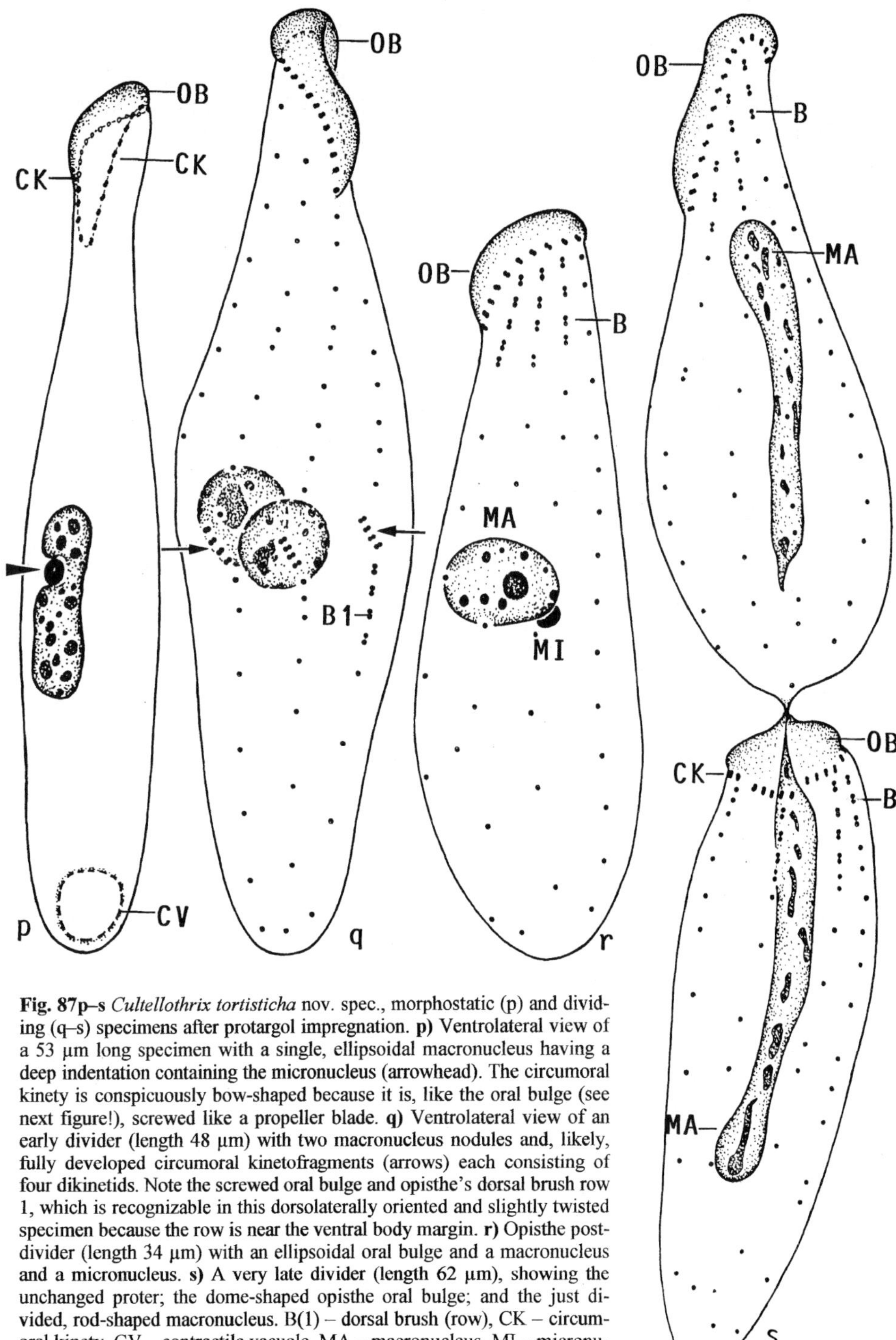

Fig. 87p–s *Cultellothrix tortisticha* nov. spec., morphostatic (p) and dividing (q–s) specimens after protargol impregnation. **p)** Ventrolateral view of a 53 μm long specimen with a single, ellipsoidal macronucleus having a deep indentation containing the micronucleus (arrowhead). The circumoral kinety is conspicuously bow-shaped because it is, like the oral bulge (see next figure!), screwed like a propeller blade. **q)** Ventrolateral view of an early divider (length 48 μm) with two macronucleus nodules and, likely, fully developed circumoral kinetofragments (arrows) each consisting of four dikinetids. Note the screwed oral bulge and opisthe's dorsal brush row 1, which is recognizable in this dorsolaterally oriented and slightly twisted specimen because the row is near the ventral body margin. **r)** Opisthe post-divider (length 34 μm) with an ellipsoidal oral bulge and a macronucleus and a micronucleus. **s)** A very late divider (length 62 μm), showing the unchanged proter; the dome-shaped opisthe oral bulge; and the just divided, rod-shaped macronucleus. B(1) – dorsal brush (row), CK – circumoral kinety, CV – contractile vacuole, MA – macronucleus, MI – micronucleus, OB – oral bulge.

surrounds the contractile vacuole (Fig. 87a, b); individual oral bulge extrusomes oblong to indistinctly ovate and slightly curved, 2.5–3 × 0.5 μm in size (Fig. 87d); do not impregnate with the protargol method used. Cortex very flexible, contains about five rows of colourless, narrowly spaced granules forming plate-like layer between each two ciliary rows; individual granules only 0.1–0.2 μm across and usually not impregnated with the protargol method used, in contrast to those of *C. paucistriata*, *C. atypica*, and *C. japonica* which impregnate so heavily that the ciliary pattern is obscured. Cytoplasm colourless and often very hyaline, contains some lipid droplets and food vacuoles 3–7 μm across with muddy contents. Swims rapidly by rotation about main body axis, but may also slowly glide under the water surface, just as several pleurostomatids.

Somatic cilia about 7 μm long in vivo, arranged in an average of seven equidistant, bipolar, loosely ciliated rows anteriorly arranged in *Arcuospathidium*, respectively, *Cultellothrix* pattern with dorsal brush rows occupying entire left anterior side (Fig. 87a, e, g, i–p, 140k–o; Table 32). Dorsal brush isostichad and three-rowed, inconspicuous, though occupying left anterior body side, because extending only about 12% of body length and bristles merely up to 4 μm long in vivo; all rows without monokinetids anteriorly, rows 1 and 2 continue with ordinary cilia posteriorly; bristles of dikinetids occasionally impregnate with protargol, anterior bristle of dikinetids about twice as thick as posterior one (Fig. 87i). Brush row 1 extends right of body midline, composed of an average of five dikinetids, of which the posterior two to three pairs have bristles reduced to 0.5 μm long stumps, while the anterior two to three pairs have ordinary, bluntly fusiform, up to 3 μm long bristles. Brush row 2 left of body midline, about as long as row 1, composed of four bristles on average, anterior bristle of dikinetids slightly clavate and 3–4 μm long, posterior oblong and 1–2 μm long. Brush row 3 near to dorsal margin of cell, slightly shorter than rows 1 and 2 and thus composed of an average of only three dikinetids having bristles as described for row 2; dikinetidal portion followed by a monokinetidal tail extending to almost posterior body end with about 1 μm long, acicular bristles (Fig. 87a, e, f, I, k, m–p, 140k–n; Table 32).

Oral bulge slanted by 50–60°, approximately 1.2 times as long as widest trunk region, respectively, occupying about 22% of body length (Table 32); inconspicuous in vivo, though somewhat refractive due to the rather thick extrusomes contained, because hardly separated from body proper and merely up to 3 μm high at dorsal end; cuneate and slightly curved with concave margin directed to left side (Fig. 87b, n, o). Bulge and circumoral kinety distinctly twisted about main axis, making both screwed like a propeller blade, a rather unusual pattern in this kind of spathidiids (Fig. 87h–l, p, q). Circumoral kinety of same shape as oral bulge when viewed ventrally (Fig. 87b, n–p), while often conspicuously bow-like in laterally oriented cells due to the strongly curved (screwed) right bulge half (Fig. 87e, m, p); continuous, composed of an average of 27 comparatively widely spaced dikinetids each associated with an about 7 μm long cilium, an approximately 5 μm long nematodesma, and a faintly impregnated transverse microtubule ribbon extending to bulge midline. Oral basket inconspicuous, right side nematodesmata more distinctly impregnated than left side ones (Fig. 87a, e, g, i–m).

Ontogenesis: Only few dividers were found among the many specimens present in the protargol slides. Two very early dividers and one early divider (Fig. 87q) show the "real" macronucleus pattern, viz., two globules; the already completed dorsal brush; and the small circumoral kinetofragments, each consisting of four dikinetids, that is, 28 dikinetids in an average specimen with seven ciliary rows. This figure matches the 27 dikinetids present in morphostatic specimens, suggesting that *C. tortisticha* does not produce oral dikinetids in post-dividers, in contrast to *C. coemeterii*. A very late divider shows a macronucleus rod in each daughter and a roughly circular oral bulge (Fig. 87s), which becomes ellipsoidal in post-dividers (Fig. 87r). This shows that shaping of body, oral bulge and macronucleus occurs only in late post-dividers and, as concerns the nuclear apparatus, even in interphase specimens (see description of nuclear pattern). This matches data from *C. coemeterii*.

Occurrence and ecology: As yet found only at type locality. It was very abundant in the non-flooded Petri dish culture, showing that many cysts were present and specimens reproduced readily. *Cultellothrix tortisticha* is the smallest member of the genus (about 55 × 10 µm), and thus well adapted to inhabitate tiny soil pores.

Remarks: For comparison with congeners and other species, see *C. paucistriata*. *Cultellothrix tortisticha* is conspicuous and unique by the distinctly screwed oral bulge and circumoral kinety, a feature found, if present at all, mainly in larger species, especially in pharyngospathidiids (see Vol. II). Accordingly, it can be considered as a third group (subgenus?) within the genus.

Cultellothrix atypica (WENZEL, 1953) **nov. comb.** (Fig. 88a–u, 141a–l, n, o, 142a–d; Table 33)

1953 *Spathidium atypicum* WENZEL, Arch. Protistenk., 99: 81.
1988 *Arcuospathidium australe* FOISSNER, Stapfia, 17: 99 (Type slides with protargol-impregnated specimens from type locality are deposited in the Oberösterreichische Landesmuseum in Linz, Upper Austria.).
1998 *Arcuospathidium atypicum* (WENZEL, 1953) nov. comb. – FOISSNER, Europ. J. Protistol., 34: 199.

Synonymy and generic assignment: When comparing *A. australe* with similar species, FOISSNER (1988) overlooked *S. atypicum*. Both have the same nuclear pattern and extrusomes as well as a similar size and shape, though WENZEL's specimens are slightly thinner (length:width ratio 5–6:1 vs. 3.2–6.8:1, on average 4.3:1; Table 33). Thus, we agree with the synonymy suggested by FOISSNER (1998).

As concerns generic assignment, *S. atypicum* has an *Arcuospathidium* ciliary pattern with the dorsal brush on the left side. Thus, it belongs to *Cultellothrix*, a genus not yet established in 1998, when FOISSNER transferred *S. atypicum* to *Arcuospathidium*.

Type material: No type material is available from WENZEL's population. FOISSNER (1988) deposited type slides of *A. australe*, which can serve as a neotype because the Australian population is very similar to several Austrian populations studied recently.

Diagnosis: Size usually 80–90 × 18–22 µm in vivo. Narrowly spatulate with oblique oral bulge about as long as widest trunk region. Two macronucleus nodules with one

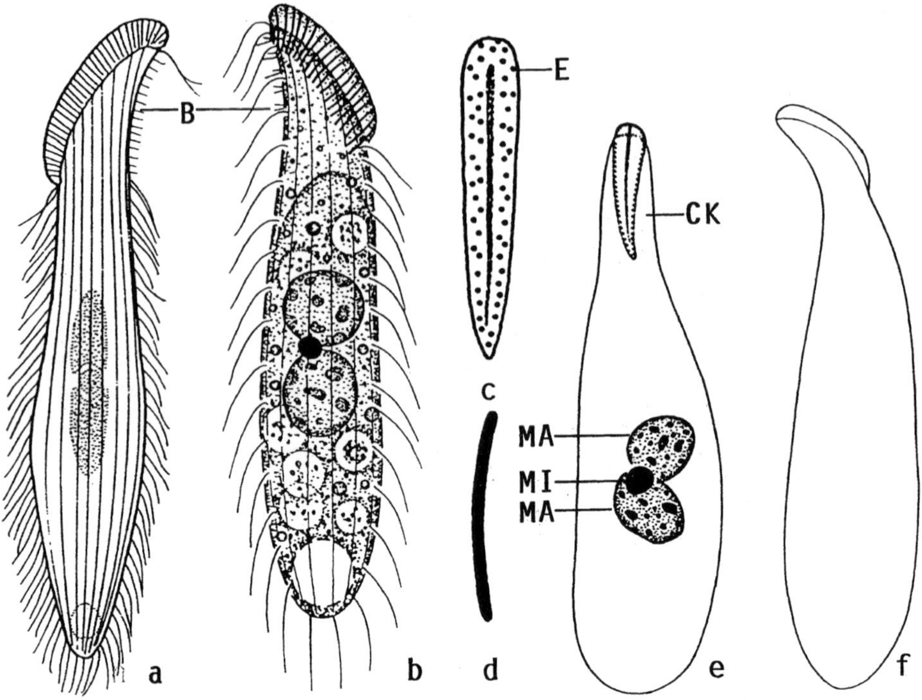

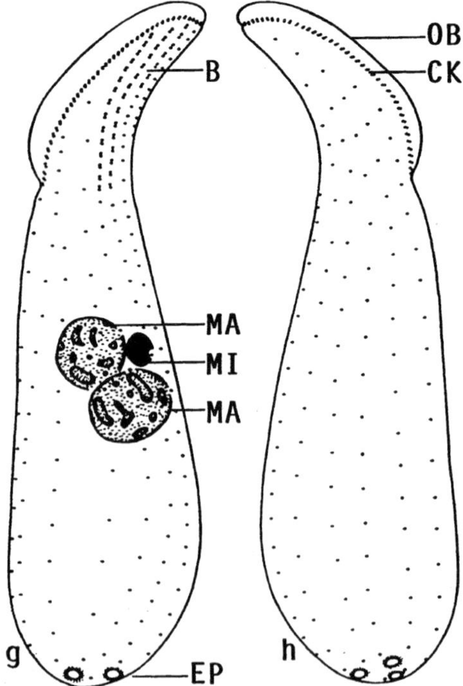

Fig. 88a–h *Cultellothrix atypica* from life (a–d) and after protargol impregnation (e–h). From WENZEL 1953 (a) and FOISSNER 1988 (b–h). **a, b)** Left and right side view of representative specimens, length 120 µm, 85 µm. **c)** Frontal view of oral bulge studded with extrusomes. **d)** Oral bulge extrusome, length 5 µm. **e)** Ventral view showing the cuneate circumoral kinety and the typical nuclear pattern, viz., two macronucleus nodules with a micronucleus in between, length 67 µm. **f)** Slender shape variant resembling WENZEL's specimen (a), length 92 µm. **g, h)** Ciliary pattern of left and right side and nuclear apparatus of main neotype specimen, length 80 µm. B – dorsal brush, CK – circumoral kinety, E – extrusomes, EP – excretory pores of contractile vacuole, MA – macronucleus nodules, MI – micronucleus, OB – oral bulge.

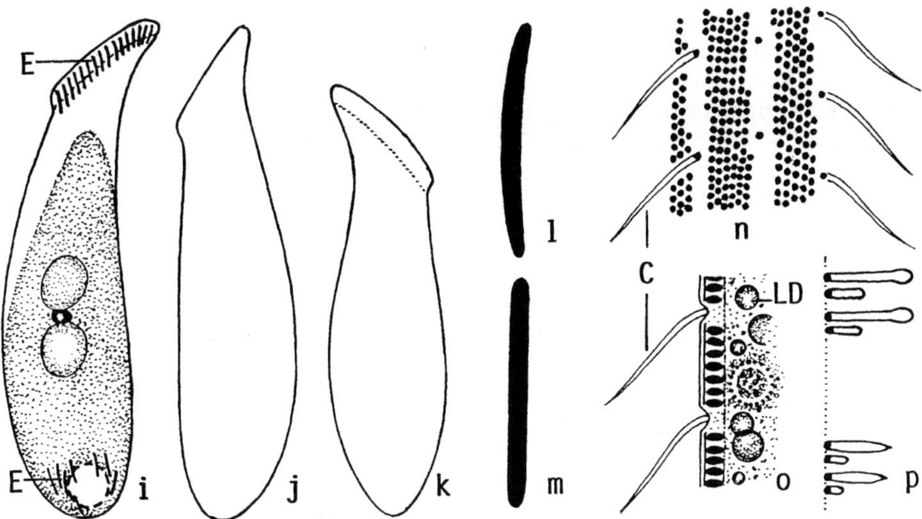

Fig. 88i–p *Cultellothrix atypica* from life. Original figures from Austrian (i, j, l–p) and Brazilian (k) specimens. **i)** Left side view of a representative specimen, length 120 µm. Note extrusomes surrounding the contractile vacuole. **j, k)** Shape variants. **l, m)** Oral bulge extrusome seen from two sides, length 6 µm. **n, o)** Surface view and optical section showing the dense cortical granulation; individual granules about 1 × 0.4 µm in size. **p)** Anterior and posterior portion of brush row 2, longest bristles 3 µm. C – ordinary somatic cilia, E – extrusomes, LD – lipid droplet.

micronucleus in between. Extrusomes rod-shaped, about 5 µm long. Approximately 12 ciliary rows, three left side rows anteriorly modified to heterostichad dorsal brush occupying about 20% of body length. Brush bristles up to 4 µm long; row 1 composed of an average of 7 dikinetids, row 2 of 10, and row 3 of 6 dikinetids followed by a monokinetidal bristle tail extending to mid-body.

Type and neotype localities: Type location not definitely stated by WENZEL (1953). However, most of the mosses he investigated are from the surroundings of Erlangen in Bavaria, Germany. The neotype location is soil from an Eucalyptus forest in the Belair National Park near Adelaide, Australia, E138°30' S35°.

Etymology: Not given in original description. *Atypica* obviously refers to the unusual nuclear pattern which, however, meanwhile has been found in various spathidiids.

Description: Size 65–120 × 15–30 µm in vivo, usually 80–90 × 18–22 µm, as calculated from seven in vivo measurements of specimens from four populations and the morphometric data shown in Table 33; length:width ratio 3–6:1 in vivo, usually near 4:1 both in vivo and preparations. Overall appearance *Litonotus*-like because up to 2:1 flattened laterally and narrowly spatulate with shortly rounded posterior end and usually distinctly slanted anterior (oral) portion with neck often less pronounced in vivo than preparations, where cells are slightly contracted or shrunken subapically (Fig. 88a, b, e–k, 141a–d; Table 33). Nuclear apparatus underneath mid-body, in Australian neotype population and in recently studied Austrian specimens invariably composed of two globular (rarely ellipsoidal) macronucleus nodules with a globular to ellipsoidal micronucleus in between, as in the German population studied by WENZEL (Fig. 88a, e, g, c, h, i, 141c, d, h, i; Table 33); pattern, however, highly variable in specimens from Malaysia (see description below) and another Australian population, which has two to four nodules ($\overline{x}$ 2.6; M 2.0; SD 0.7; CV 27.8; n 30) and occasionally (two out of 30 cells) even two micronuclei (FOISSNER 1988). Contractile vacuole in rear body end and surrounded by extrusomes in an Austrian population, three to six excretory pores in pole area (Fig. 88g–i). Extrusomes scattered in oral bulge and cytoplasm, shape and size very similar in populations from Germany, Austria, Australia, and South America, viz., slightly curved, 4–6 µm long rods indistinctly narrowed at both ends; do not impregnate with the protargol method used, but heavily stain with silver carbonate (Fig. 88a–d, i, l, m, 141f–h, k, l, o), with which a second type impregnates, viz., about 2 µm long rods, possibly cortical mucocysts (Fig. 88n, p, 141n). Cortex very flexible and conspicuous because about 1.5 µm thick, gelatinous, bright due to the granules contained, and having a reticular alveolar pattern (Fig. 88n, o, 141e, i). Cortical granules arranged in about six rows between two kineties each, so densely spaced that a plate-like layer results; individual granules difficult to recognize due to their narrow spacing, moderately refractive,

Table 33 Morphometric data on *Cultellothrix atypica* (upper line) and *Cultellothrix japonica* (lower line). From FOISSNER (1988)

Characteristics[1]	$\overline{x}$	M	SD	SE	CV	Min	Max	n
Body, length	83.6	88.0	12.5	3.9	14.9	61	98	10
	68.9	70.0	6.0	1.7	8.6	56	77	13
Body, width	20.6	21.0	5.0	1.6	24.5	9	27	10
	15.2	14.0	2.3	0.6	15.0	13	21	13
Body length:width, ratio	4.3	4.0	1.0	0.3	23.9	3.2	6.8	10
	4.9	4.5	0.6	0.2	11.7	3.3	5.4	13
Oral bulge, length	21.4	22.5	3.4	1.1	15.8	16	27	10
	12.7	13.0	1.6	0.4	12.6	10	15	13
Oral bulge length:body width, ratio	1.1	1.0	0.3	0.1	24.7	0.8	1.8	10
	0.9	0.9	0.1	0.1	14.7	0.6	1.1	13
Macronucleus nodules, length	9.5	9.5	1.8	0.6	18.7	7	12	10
	8.5	8.0	1.8	0.5	20.6	7	13	13
Macronucleus nodules, width	7.2	7.5	1.3	0.4	18.4	5	9	10
	6.2	6.0	1.1	0.3	17.4	4	7	13
Macronucleus nodules, number	2.0	2.0	0.0	0.0	0.0	2	2	10
	2.2	2.0	0.7	0.1	31.0	1	4	28
Micronucleus, largest diameter	2.7	2.7	0.4	0.1	14.7	2	3	10
	2.9	3.0	0.2	0.1	7.6	2	3	13
Micronuclei, number	1.0	1.0	0.0	0.0	0.0	1	1	10
	1.0	1.0	0.0	0.0	0.0	1	1	13

[1] Data based on mounted, protargol-impregnated (FOISSNER's method), and randomly selected specimens from non-flooded Petri dish cultures. Measurements in µm. CV – coefficient of variation in %, M – median, Max – maximum, Min – minimum, n – number of individuals investigated, SD – standard deviation, SE – standard error of arithmetic mean, $\overline{x}$ – arithmetic mean.

about 1 × 0.4 µm in size, usually impregnate so strongly with protargol that the ciliary pattern is obscured. Cytoplasm colourless, contains few to many lipid droplets 1–5 µm across and some 6µm-sized food vacuoles with fungal spores and unidentifiable contents (Fig. 141a, c, i); *Colpoda maupasi* occurs in the food vacuoles of the Malaysian specimens. Glides slowly on microscope slide and in surface of water drops.

Details of somatic and oral ciliary pattern usually obscured by the strongly impregnated cortical granules (see above). Somatic cilia 6–9 µm long in vivo, in all populations arranged in about 12–14 equidistant, bipolar, loosely ciliated rows anteriorly forming an *Arcuospathidium*, respectively, *Cultellothrix* pattern with dorsal brush occupying left half of left anterior body side (Fig. 88a, b, g, h). Dorsal brush of usual structure, inconspicuous because extending only about 20% of body length and bristles merely up to 4 µm long in vivo. Brush row 3 about half as long as rows 1 and 2, dikinetidal portion followed by a monokinetidal tail extending to mid-body with 1–2 µm long bristles (Fig. 88a, b, g, 141d, j). Individual bristle pairs of Austrian populations composed of a 3–4 µm long anterior cilium slightly inflated distally and a 1–2 µm long, oblong posterior cilium; bristle length decreases posteriorly to 1–2 µm, and bristles become acicular (Fig. 88p, 141j). Posterior bristles longer than anterior bristles in Australian specimens (Fig. 88b).

Oral bulge slanted by 50–60°, about as long as widest trunk region, respectively, occupying approximately 25% of body length, rather conspicuous in vivo because ventrally sharply set off from body proper, 4–5 µm high, bright due to the extrusomes contained, and distinctly cuneate (Fig. 88a–c, e, g, i, 141a, b, n; Table 33). Circumoral kinety cuneate, composed of narrowly spaced dikinetids (Fig. 88e, g, h, 141n).

Observations on a Malaysian population (Fig. 88q–u, 142a–d; Table 34): In Malaysia, *C. atypica* occurred in soil mosses from the fog rainforest on top of Mount G. Brichang, Cameroon Highlands. This population highly resembles the holarctic and Australian populations, except for the macronucleus which consists of 3–11 nodules in most specimens (Table 34). Two early dividers and some well nourished cells have four nodules, strongly suggesting this as the usual pattern, and thus subspecies rank at least; a late divider shows that the nodules fuse and only one micronucleus is present (Fig. 88u). However, specimens with four macronucleus nodules occur also in an Australian population (see above) as well as in *C. japonica* and *C. tortisticha*. Thus, ranking of the Malaysian population would be premature, which is emphasized by the highly similar ciliary pattern and morphometrics (Fig. 88q–t; Table 34). Further, more than four macronucleus nodules, usually of very different size, occur mainly in specimens with few or no food vacuoles, indicating reorganization; however, conjugating cells were not present in the about 100 individuals observed.

The ciliary pattern of the Malaysian specimens could be investigated in detail because the cortical granules did not as strongly impregnate as in the Australian and Austrian cells (Fig. 88q–t, 142a–d; Table 34). This shows that *C. atypica* has a pronounced *Arcuospathidium* pattern and the dorsal brush rows lack an anterior tail, perfectly matching the Australian population investigated by FOISSNER (1988).

Occurrence and ecology: *Cultellothrix atypica* likely feeds on heterotrophic flagellates and small ciliates and occurs in various terrestrial habitats. Likely, it has

Table 34 Morphometric data on a Malaysian population of *Cultellothrix atypica*

Characteristics[1]	$\overline{x}$	M	SD	SE	CV	Min	Max	n
Body, length	67.2	69.5	9.6	2.3	14.3	51.0	91.0	18
Body, width	20.4	19.5	4.2	1.0	20.5	15.0	28.0	18
Body length:width, ratio	3.3	3.3	0.5	0.1	14.9	2.6	4.3	18
Oral bulge, length	17.8	17.5	2.2	0.5	12.4	14.0	23.0	18
Oral bulge, height	3.2	3.0	–	–	–	2.5	4.0	18
Oral bulge length:body width, ratio	0.9	0.9	0.1	0.1	12.5	0.7	1.1	18
Circumoral kinety to last dikinetid of brush row 1, distance	8.7	9.0	1.5	0.4	17.2	7.0	14.0	18
Circumoral kinety to last dikinetid of brush row 2, distance	11.1	11.0	1.5	0.4	13.3	9.0	14.0	18
Circumoral kinety to last dikinetid of brush row 3, distance	6.6	7.0	0.9	0.2	14.0	5.0	8.0	18
Anterior body end to first macronucleus nodule, distance	31.1	29.5	6.5	1.5	20.8	24.0	43.0	18
Macronucleus nodules, length	5.1	5.0	1.7	0.4	33.8	2.5	8.0	18
Macronucleus nodules, width	4.8	5.0	1.4	0.3	29.3	2.5	7.0	18
Macronucleus nodules, number[2]	5.9	6.0	2.2	0.3	36.5	2.0	11.0	40
Micronuclei, number[3]	1.2	1.0	–	–	–	1.0	2.0	16
Ciliary rows, number	12.4	12.5	0.9	0.2	7.4	11.0	14.0	18
Basal bodies in a right side ciliary row, number	19.1	18.0	5.9	1.4	31.0	11.0	33.0	14
Dorsal brush rows, number	3.2	3.0	–	–	–	3.0	4.0	22
Dikinetids in brush row 1, number	6.8	7.0	1.2	0.3	17.1	6.0	11.0	18
Dikinetids in brush row 2, number	9.7	10.0	1.2	0.3	12.7	7.0	11.0	18
Dikinetids in brush row 3, number	5.9	6.0	0.7	0.2	12.3	5.0	7.0	18

[1] Data based on mounted, protargol-impregnated (FOISSNER's method), and randomly selected specimens from a non-flooded Petri dish culture. Measurements in µm. CV – coefficient of variation in %, M – median, Max – maximum, Min – minimum, n – number of individuals investigated, SD – standard deviation, SE – standard error of arithmetic mean, $\overline{x}$ – arithmetic mean.

[2] Only one, very small (length 32 µm) specimen has two nodules.

[3] Uncertain because confusion with similar-sized and impregnated cytoplasmic inclusions cannot be excluded.

cosmopolitan distribution because we have reliable records from Central Europe, Australia, Malaysia (see above) and South America. In Europe, records are available from Germany, where WENZEL (1953) discovered *C. atypica* in two dry, rewetted moss samples; and from Austria, where FOISSNER et al. (2005a) found this species in acidic to circumneutral (pH 4–7.4) litter and soil samples from deciduous and pine forests in Vienna (abundant in litter and soil from the Johannser Kogel), Lower Austria, Styria, and Salzburg. In Australia, *C. atypica* occurred in litter and soil from a secondary pine forest (pH 5.1) and from an *Eucalyptus* forest pH (5.7) in the surroundings of Adelaide (FOISSNER 1988, BLATTERER & FOISSNER 1988). In South America, *C. atypica* was found in litter and soil (pH 5.1) from the rainforest in the surroundings of Manaus, Brazil (FOISSNER 1997a). Obviously, *C. atypica* has a wide ecological range and is well

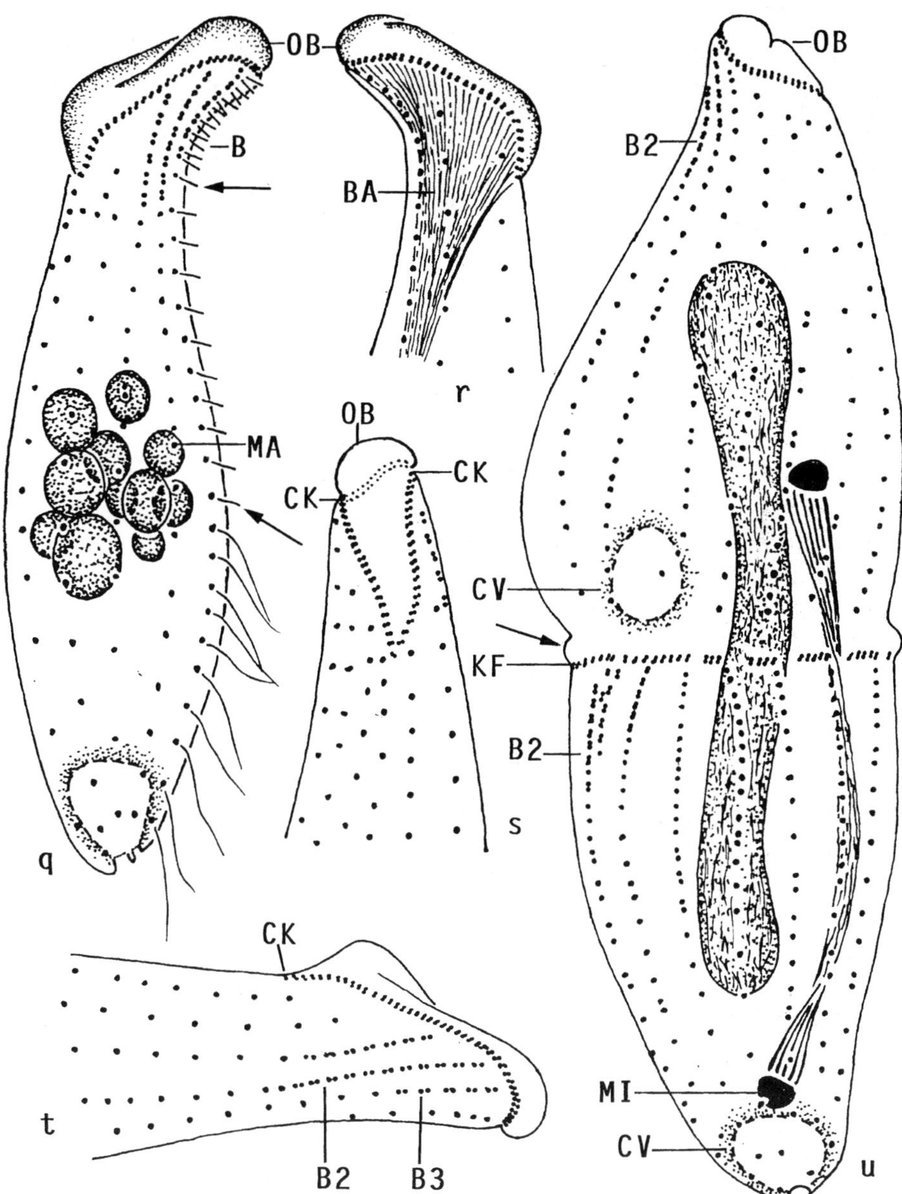

Fig. 88q–u *Cultellothrix atypica*, Malaysian specimens after protargol impregnation. **q, r)** Left and right side view of a representative specimen with many macronucleus nodules and conspicuous oral bulge with propeller-like screwed surface. Arrows mark monokinetidal bristle tail of brush row 3. Length 60 µm. **s)** Ventral view showing the cuneate circumoral kinety. Length of oral bulge plus circumoral kinety 15 µm. **t)** Dorsolateral view of anterior body portion, length of oral bulge 18 µm. **u)** Late divider, length 88 µm. Arrow marks blebs. B(1-3) – dorsal brush rows, BA – oral basket, CK – circumoral kinety, CV – contractile vacuole, KF – oral kinetofragments, MA – macronucleus nodules, MI – micronucleus, OB – oral bulge.

adapted to soil life by its small, slender, flexible body.

Remarks: *Cultellothrix atypica* highly resembles *C. japonica* and *C. tortisticha.* Obviously, these three species are closely related, all having similar size and shape and the same, highly characteristic type of nuclear apparatus. In spite of this, they can be clearly distinguished by the shape of the extrusomes (rod-shaped vs. narrowly ovate) and the number of ciliary rows (12 vs. 7). *Arcuospathidium vermiforme*, which also has two macronucleus nodules with a micronucleus in between, is much more slender and lacks extrusomes.

Cultellothrix japonica (FOISSNER, 1988) **nov. comb.** (Fig. 89a–i, 141m; Table 33)

1988 *Spathidium japonicum* FOISSNER, Stapfia, 17: 102 (Type slides with protargol-impregnated specimens from type locality are deposited in the Oberösterreichische Landesmuseum in Linz, Upper Austria.).

Diagnosis: Size about 75 × 17 μm in vivo. Narrowly spatulate with oblique oral bulge shorter by 16% than widest trunk region. Two macronucleus nodules with one micronucleus in between. Extrusomes acicular to narrowly ovate, about 5 μm long. Approximately 12 ciliary rows, three left side rows anteriorly modified to inconspicuous dorsal brush with bristles up to 3 μm long; brush row 3 indistinctly shortened.

Type locality: Soil of a deciduous forest on Mt. Kado-yama, Amakusa Islands, Japan (Kumamoto Prefecture), E130° N32°10'.

Etymology: Named after the country where it was discovered.

Description and remarks: This species is transferred to *Cultellothrix* for the reasons explained in *C. atypica*. Generally, data are incomplete because the species is difficult to impregnate, viz., usually only the cortical mucocysts are impregnated, as in several species of this group. *Cultellothrix japonica* highly resembles *C. atypica*, except of the extrusomes which are similar to those of *C. paucistriata*. Thus, the user is referred to the detailed description of *C. atypica*, the figures, and the morphometric data, which overlap considerably, except of the length of the oral bulge (Table 33). The nuclear pattern is as variable as in *C. atypica* and *C. tortisticha*. However, most cells have two globular macronucleus nodules with a micronucleus in between (Fig. 89a, f); rarely occur specimens with three to four globular nodules or with only one globular to elongate ellipsoidal nodule (Fig. 89g, i).

FOISSNER (1988) carefully checked the acicular shape of the extrusomes, that is, the main distinguishing feature of *C. japonica*; fortunately, we can confirm the acicular to narrowly ovate extrusome shape and the nuclear pattern from a Brazilian population. The extrusomes of the Brazilian specimens have a usual size of 5–6 × 0.7–0.8 μm (Fig. 89c), in a very small specimen (length 40 μm) they measured only 3 × 0.7 μm and thus highly resembled the extrusomes of *C. paucistriata*. The extrusomes of the Japanese specimens are thinner than those of the Brazilian cells; however, a re-evaluation of the original notes showed that FOISSNER (1988) did not measure extrusome thickness; thus, they might be thicker than shown.

Occurrence and ecology: This is a rare species as yet found only in Japan (type locality; pH 3.5) and Brazil, where it occurred in terra firma soil (pH 5.1) of a primary (?) rain forest on a small island of the Anavilhanas archipelago in the Rio Negro near the town of Manaus (FOISSNER 1997a). Abundances were low in the non-flooded Petri dish cultures. This small and tiny species is well adapted to live in tiny soil pores.

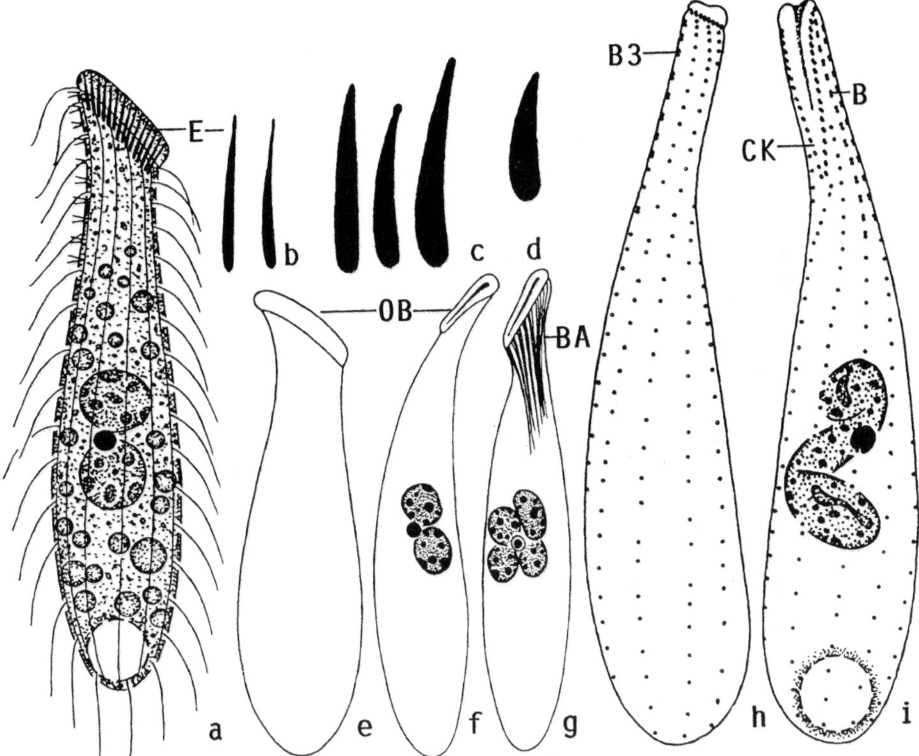

Fig. 89a–i *Cultellothrix japonica* from life (a–e) and after protargol impregnation (f–i). From FOISSNER 1988 (a, b, e–i) and originals (c, d). **a)** Right side view of a representative specimen, length 75 µm. Note the two main features of this species, viz., the acicular extrusomes and the nuclear apparatus which consists of two globular macronucleus nodules with a micronucleus in between. **b)** Oral bulge extrusomes of Japanese specimens, length 6 µm (likely drawn too thin, see text). **c)** Oral bulge extrusomes of Brazilian specimens, length 5–6 µm. **d)** Oral bulge extrusome of a small (length 40 µm) specimen from Brazil, length 3 µm. **e)** Shape variant. **f, g)** Variability of nuclear apparatus, length about 75 µm. **h, i)** Ciliary pattern of dorsal and ventral side, length 64 µm. B(3) – dorsal brush (row), BA – oral basket, CK – circumoral kinety, E – extrusomes, OB – oral bulge.

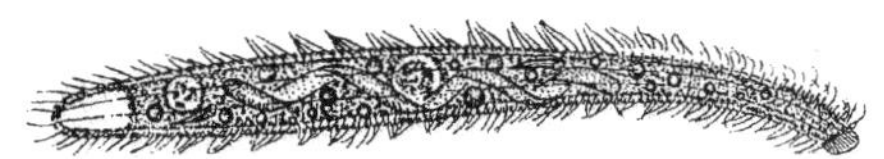

Armatospathula nov. gen.

Diagnosis: Arcuospathidiidae with body (cortical, somatic) extrusomes.

Type species: *Armatospathula periarmata* nov. spec.

Etymology: Composite of the Latin nouns *armatus* (armament) and *spathula* (small spatula), referring to both the increased number of extrusomes and the similarity with the genus *Spathidium*.

Remarks: This new genus unites four *Arcuospathidium* species with body extrusomes, that is, toxicysts attached to the somatic cortex. One such species was described by FOISSNER et al. (2002), while the three others are new to science. Body extrusomes are a very distinct feature found only in one other spathidiid, viz., *Apospathidium*, a genus probably not belonging to the Spathidida because it has oralized somatic kinetids. *Armatospathula*, in contrast, contains typical *Arcuospathidium* species, except for the body extrusomes. A similar differentiation occurs in the family Bryophyllidae, where *Apobryophyllum* has highly differentiated body extrusomes (FOISSNER et al. 2002). Interestingly, *Armatospathula* spp. lack the tail of ordinary cilia at the anterior end of the dorsal brush rows, a feature typically found in *Cultellothrix*. This suggests that *Armatospathula* developed from a *Cultellothrix*-like ancestor. Furthermore, all described *Armatospathula* species have as yet found only in Gondwanan areas.

The four species united in the genus are considerably different, suggesting that many more remain to be discovered and those known might be not congeneric, for instance, the large *A. novaki* and *A. plurinucleata* which lacks the monokinetidal bristle tail usually associated with brush row 3.

Key to species

1 Body length >200 µm in vivo, massive. Oral extrusomes very narrowly ovate, body extrusomes rod-shaped. Macronucleus a long, tortuous strand .... *A. novaki*
– Body length <150 µm in vivo, slender 2
2 Macronucleus a tortuous strand or cylindroidal 3
– Many macronucleus nodules. Oral and body extrusomes very narrowly ovate......... .... *A. plurinucleata*
3 Oral and body extrusomes rod-shaped, about 3 µm long .... *A. periarmata*
– Oral extrusome very narrowly ovate and about 4 µm long, body extrusomes rod-shaped and about 2 µm long.. *A. costaricana*

Armatospathula periarmata nov. spec. (Fig. 90a–p, 143a–i; Table 35)

Diagnosis: Size about 130 × 17 µm in vivo. Very narrowly spatulate with steep to very steep, cuneate oral bulge about as long as widest trunk region. Macronucleus cylindroidal; single micronucleus. Oral and body extrusomes rod-shaped, about 3 µm long.

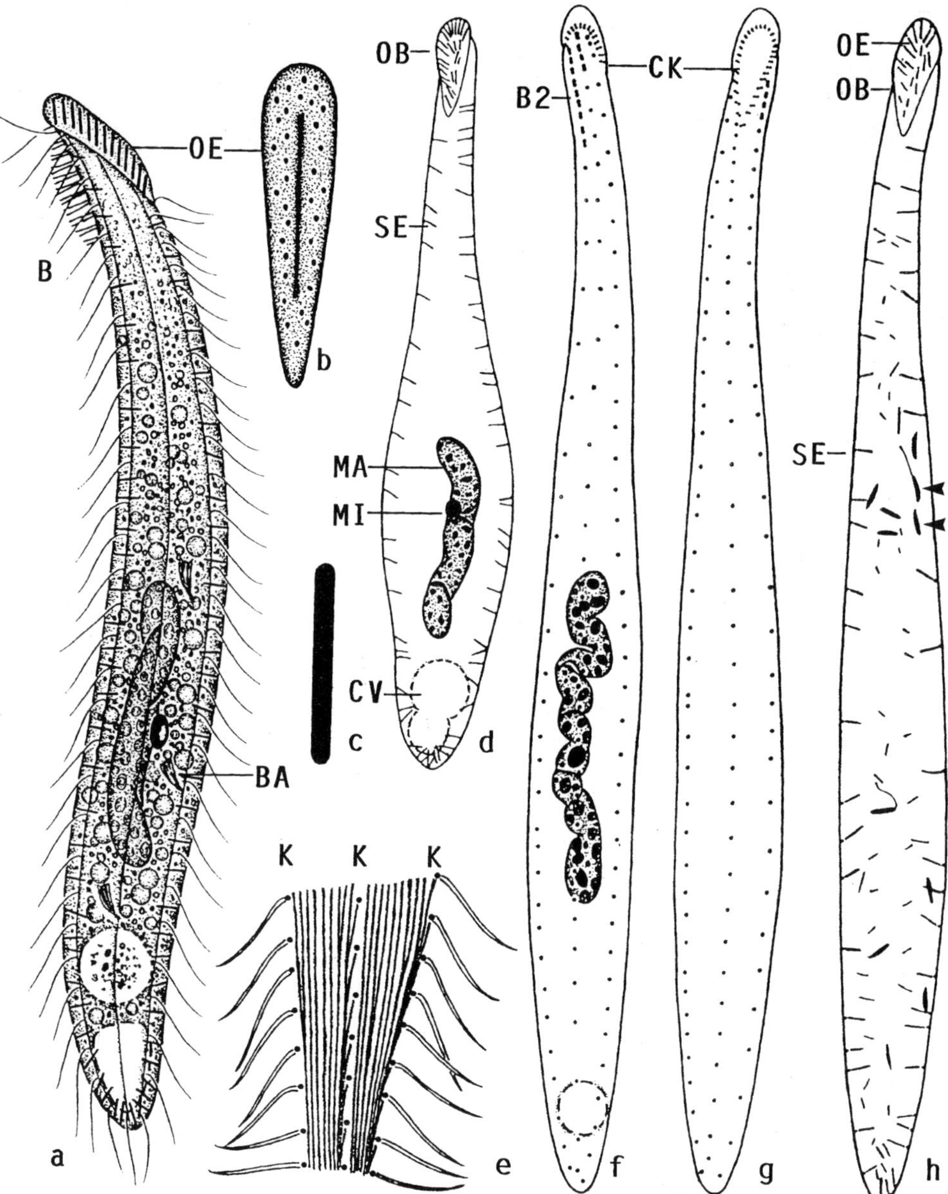

Fig. 90a–h *Armatospathula periarmata* nov. spec. from life (a–c) and after protargol impregnation (d–h). **a)** Right side view of a representative specimen, length 120 µm. **b)** Frontal view of oral bulge. **c)** Oral bulge and somatic extrusomes have the same shape and size (length ~3 µm). **d)** Ventral view of a broad specimen showing the faintly impregnated somatic extrusome rows, length 100 µm. **e)** Cortical fibre system. **f–h)** Ciliary pattern of dorsal and ventral side, nuclear apparatus, and extrusome pattern of holotype specimen, length 110 µm. Details see figures 90m, n. Arrowheads mark developing and exploded cytoplasmic extrusomes. B(2) – dorsal brush (row), BA – oral basket of microthoracid ciliate prey, CK – circumoral kinety, CV – contractile vacuole, K – somatic kineties, MA – macronucleus, MI – micronucleus, OB – oral bulge, OE, SE – oral and somatic extrusomes.

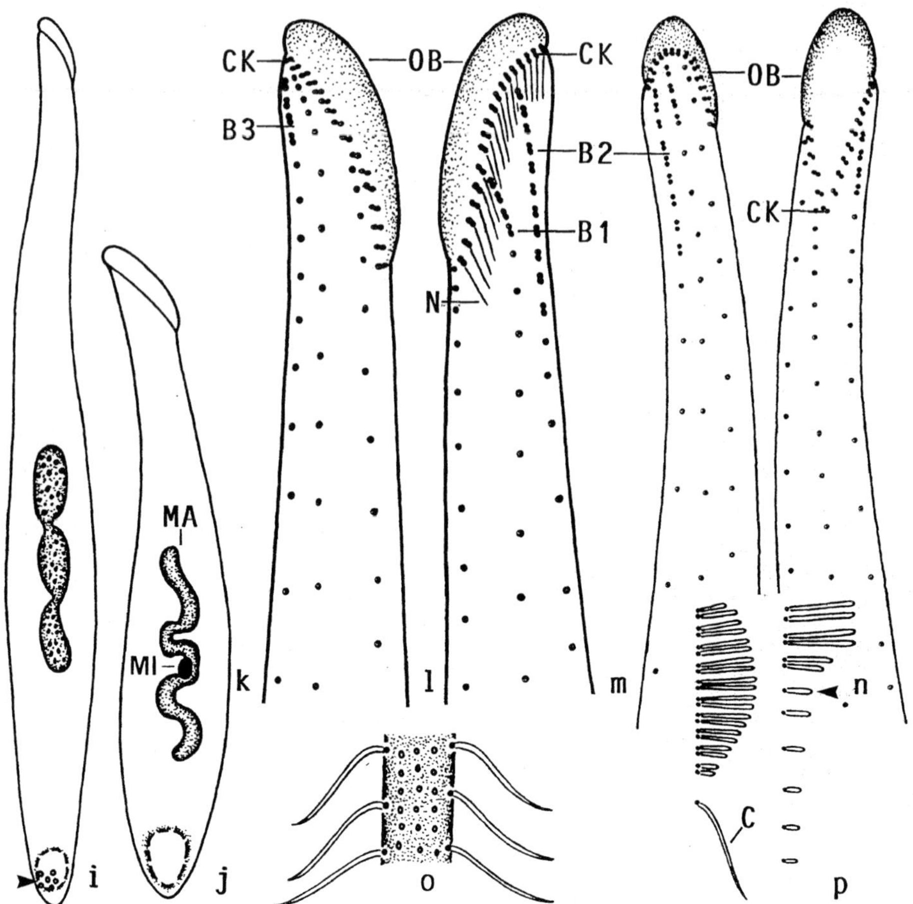

Fig. 90i–p *Armatospathula periarmata* nov. spec. from life (j, o, p) and after protargol impregnation (i, k–n). **i, j)** Variability of body shape and size and macronucleus pattern, length 143 µm, 90 µm. Arrowhead marks excretory pores of the contractile vacuole. **k, l)** Ciliary pattern of right and left side in anterior body region, length of oral bulge 14 µm. **m, n)** Ciliary pattern of dorsal and ventral side of holotype specimen, length of oral bulge/circumoral kinety 12 µm (overviews, see figures 90f, g). **o)** Surface view showing loose cortical granulation. **p)** Dorsal brush rows 2 and 3, longest bristles 5 µm. The monokinetidal bristle tail of row 3 (arrowhead) extends to second body third. B(1-3) – dorsal brush (rows), C – ordinary somatic cilium, CK – circumoral kinety, MA – macronucleus, MI – micronucleus, N – nematodesmata, OB – oral bulge.

On average 6 ciliary rows, those on left side anteriorly modified to distinctly heterostichad dorsal brush occupying about 10% of body length. Brush bristles up to 5 µm long: row 1 composed of an average of 4 dikinetids, row 2 of 10, and row 3 of 4 dikinetids followed by a short, monokinetidal bristle tail.

Type locality: Soil from a *Tachypogon* savannah in the surroundings of the town of Puerto Ayacucho, Venezuela, W68° N5°.

Etymology: Composite of the Latin prefix *peri* (surrounding) and the adjective *armatus* (armed), referring to a main feature of the species, viz., the peripheral body extrusomes.

Description: Size 100–160 × 10–20 µm in vivo, usually about 130 × 17 µm, as calculated from some in vivo measurements and the morphometric data; length:width ratio highly variable, that is, 5–12.8:1 in protargol preparations because specimens are sometimes distinctly inflated in mid-body, usually, however, near 8:1 both in vivo and in preparations (Table 35). Narrowly spatulate to cylindroidal, occasionally obclavate, frequently slightly curved, especially in flattened anterior third. Anterior body end (oral bulge) slanted by about 60° and about as long as widest trunk region; neck indistinct; trunk usually widest underneath mid-body and more or less densely packed with lipid droplets; posterior end narrowly rounded, rarely bluntly pointed (Fig. 90a, d, f, i, j, 143a, f–i). Macronucleus in third body quarter on average, oblong and more or less curved, moniliform in one out of 80 specimens analysed; nucleoli numerous, globular to lobate. Micronucleus attached to mid-region of macronucleus, slightly ellipsoidal, in vivo inconspicuous because only 4 × 3 µm in size (Fig. 90a, d, f, i, j, 143a, f–i; Table 35). Contractile vacuole in rear body end, excretory pores slightly subterminal (Fig. 90i, 143a, f). Oral bulge and body extrusomes of same shape and size, that is, 2.5–3 × 0.3–0.4 µm-sized rods in vivo; body extrusomes in a rough row between two kineties each; impregnate more or less intensely with the protargol method used; exploded extrusomes of typical toxicyst structure. Developing cytoplasmic extrusomes bluntly fusiform and comparatively thick, viz., 2.5–3.5 × 0.8 µm in protargol-prepared cells (Fig. 90a, c, d, h, 143a–i). Cortex very flexible, contains the extrusomes described above; loosely spaced rows of colourless granules about 0.2 µm across; and more or less distinctly impregnated fibres extending from basal bodies posteriorly, forming about ten bundles between each two ciliary rows (Fig. 90e, o). Cytoplasm colourless, usually packed with mature and developing extrusomes, lipid droplets up to 3 µm across, and oral baskets of ciliate or euglenid prey; in rear end frequently a vacuole with indigestible debris (Fig. 90a, 143a–e). Swims and glides rather rapidly showing worm-like flexibility.

Cilia about 8 µm long in vivo and rather widely spaced (~4 µm), arranged in an average of six equidistant, straight rows distinctly separate from circumoral kinety. *Arcuospathidium* ciliary pattern rather indistinct due to the low number of ciliary rows and the slightly ventrally curved brush rows 1 and 2 (Fig. 90a, k–n; Table 35). Dorsal brush strongly heterostichad and three-rowed, inconspicuous because occupying only about 10% of body length and bristles merely 4–5 µm long and rod-shaped; all rows continue as somatic kineties posteriorly and lack ordinary cilia anteriorly, a feature typically found only in *Cultellothrix*. Brush rows 1 and 3 very short and of similar length, each consisting of an average of only four dikinetids; row 1 so distinctly curved ventrally that a V-shaped pattern is formed with row 2; middle row 2 distinctly longer than rows 1 and 3, consisting of an average of ten dikinetids with bristles up to 4 µm long in middle third, gradually decreasing to 1–2 µm at ends of row; bristles of row 3 decreasing in length from 4–5 µm anteriorly to 2–3 µm posteriorly, continues with 1–1.5 µm long, monokinetidal bristles to second body third, where it transforms to an ordinary somatic ciliary row (Fig. 90a, f, g, k–n, p, 143b, c, g, h; Table 35).

Oral bulge of ordinary appearance, about as long as widest trunk region, oblique to strongly oblique, that is, slanted by 40°–60°; surface flat to slightly convex, cuneate in frontal view. Circumoral kinety of same shape as oral bulge, definitely continuous and separate from ciliary rows, composed of dikinetids comparatively widely spaced in ventral half of kinety; individual dikinetids associated with a single cilium and a fine nematodesma contributing to the short, inconspicuous oral basket recognizable only in over-impregnated specimens (Fig. 90a, b, g, k–n, 143a–d, h, i; Table 35). Oralized somatic monokinetids definitely lacking.

Occurrence and ecology: As yet found only at type locality, that is, in very dry and hard savannah soil, which had 52°C on the surface when the sample was taken; the site is in the surroundings of the farm of Mr. Aranguren in an area called Fundo el Desoro. The savannah is burnt, and the soil consists mainly of yellowish sand because the site is in the former floodplain of the Orinoco river. *Armatospathula periarmata* became moderately abundant in the non-flooded Petri dish culture and is well adapted to the soil environment by its slender body.

Remarks: The family classification of this ciliate is problematic because the low number of ciliary rows makes the ciliary pattern difficult to recognize. The slightly ventrally curved left side rows argue for a classification in the Spathidiidae. However, all these kineties bear a dorsal brush and are slightly curved ventrally also in typical *Arcuospathidium* species. Thus, we emphasize the other arcuospathidiid features, viz., the steep, cuneate oral bulge and the continuous circumoral kinety, which is distinctly separate from the ciliary rows.

Armatospathula periarmata differs from the congeners by body size and the uniform extrusomes (see key to species). However, at first glance, *A. periarmata* resembles a tailless *Apospathidium atypicum*, an amphoriform species with 9–12 (vs. 6) ciliary rows and a widely subterminal (vs. terminal) contractile vacuole (see Vol. II). Species of the *Spathidium claviforme* group may also resemble *A. periarmata.* They have more ciliary rows (>10 vs. 6) and dikinetids in dorsal brush row 1 (≥ 10 vs. 4).

Table 35 Morphometric data on *Armatospathula periarmata* (upper line) and *A. costaricana* (lower line)

Characteristics[1]	$\bar{x}$	M	SD	SE	CV	Min	Max	n
Body, length	111.9	109.0	14.9	3.3	13.3	94.0	154.0	21
	111.4	130.0	30.5	13.6	27.4	70.0	139.0	5
Body, width	15.2	15.0	3.3	0.7	22.0	10.0	21.0	21
	14.4	15.0	3.5	1.6	24.4	10.0	19.0	5
Body length:width, ratio	7.8	7.2	2.2	0.5	28.3	5.0	12.8	21
	8.0	8.7	2.3	1.0	29.0	4.6	10.8	5
Oral bulge (circumoral kinety), length	14.1	14.0	2.0	0.4	13.8	11.0	18.0	21
	29.4	30.0	7.5	3.3	25.4	20.0	38.0	5
Oral bulge length:body width, ratio	1.0	0.9	0.2	0.1	23.8	0.6	1.5	21
	2.1	2.0	0.6	0.3	28.6	1.3	2.9	5
Oral bulge, (circumoral kinety), width	3.7	3.5	–	–	–	3.5	4.0	7
	–	–	–	–	–	–	–	–
Oral bulge, height	2.5	2.5	–	–	–	2.0	3.0	21
	1.5	1.5	–	–	–	1.0	2.0	5
Circumoral kinety to last dikinetid of brush row 1, distance	4.7	5.0	0.7	0.2	15.5	3.5	6.0	21
	18.2	22.0	8.7	3.9	47.8	9.0	28.0	5
Circumoral kinety to last dikinetid of brush row 2, distance	11.4	12.0	1.4	0.3	12.6	8.0	14.0	21
	18.6	22.0	8.2	3.7	43.9	10.0	28.0	5

continued

Characteristics[1]	$\overline{x}$	M	SD	SE	CV	Min	Max	n
Circumoral kinety to last dikinetid of brush row 3, distance	4.4	4.0	0.6	0.1	14.1	3.5	6.0	21
	11.0	13.0	4.2	1.9	38.0	6.0	15.0	5
Anterior body end to macronucleus, distance	54.0	54.0	7.3	1.6	13.4	42.0	66.0	21
	46.8	51.0	17.6	7.9	37.5	27.0	66.0	5
Macronucleus figure, length	26.8	26.0	5.4	1.2	20.1	20.0	39.0	21
	42.4	47.0	14.7	6.6	34.6	23.0	61.0	5
Macronucleus, length (spread and thus approximate)	34.3	33.0	–	–	–	24.0	45.0	21
	–	–	–	–	–	–	–	–
Macronucleus, width	4.4	4.0	0.8	0.2	18.0	3.0	6.0	21
	3.9	4.0	0.7	0.3	19.0	3.0	5.0	5
Macronucleus, number	1.0	1.0	0.0	0.0	0.0	1.0	1.0	21
	1.0	1.0	0.0	0.0	0.0	1.0	1.0	5
Micronuclei, length	3.3	3.0	0.7	0.1	20.5	2.5	5.0	21
	3.3	3.0	0.4	0.2	13.6	3.0	4.0	5
Micronucleus, width	2.5	2.5	0.5	0.1	21.3	2.0	4.0	21
	2.6	3.0	0.7	0.3	25.1	1.5	3.0	5
Micronuclei, number	1.0	1.0	0.0	0.0	0.0	1.0	1.0	21
	3.2	3.0	0.8	0.4	26.1	2.0	4.0	5
Ciliary rows, number	6.1	6.0	–	–	–	6.0	7.0	21
	10.6	10.0	0.9	0.4	8.4	10.0	12.0	5
Basal bodies in a right side ciliary row, number	34.0	34.0	5.1	1.1	15.0	24.0	42.0	21
	43.2	48.0	9.9	4.4	23.0	32.0	53.0	5
Dorsal brush rows, number	3.0	3.0	0.0	0.0	0.0	3.0	3.0	21
	3.0	3.0	0.0	0.0	0.0	3.0	3.0	5
Dikinetids in brush row 1, number	4.2	4.0	0.5	0.1	12.7	3.0	5.0	21
	19.4	21.0	7.1	3.2	36.6	11.0	26.0	5
Dikinetids in brush row 2, number	10.5	10.0	1.4	0.3	13.0	8.0	15.0	21
	21.6	23.0	8.3	3.7	38.5	12.0	30.0	5
Dikinetids in brush row 3, number	3.8	4.0	0.6	0.1	15.8	3.0	5.0	21
	11.4	12.0	3.5	1.6	30.8	7.0	16.0	5
Circumoral kinetids, number				not counted				
	56.0	49.0	–	–	–	40.0	79.0	3

[1] Data based on mounted, protargol-impregnated (FOISSNER's method), and randomly selected specimens from non-flooded Petri dish cultures. Measurements in µm. CV – coefficient of variation in %, M – median, Max – maximum, Min – minimum, n – number of individuals investigated, SD – standard deviation, SE – standard error of arithmetic mean, $\overline{x}$ – arithmetic mean.

***Armatospathula costaricana* nov. spec.** (Fig. 91a–n, 144a–d; Table 35)

Diagnosis: Size about 130 × 16 µm in vivo. Very narrowly spatulate with strongly oblique oral bulge about twice as long as widest trunk region. Macronucleus a moniliform, tortuous strand with several micronuclei. Oral bulge extrusomes narrowly ovate and about 4 × 1 µm in size; body (somatic) extrusomes rod-shaped and thin, about 2 µm long. On average 10 ciliary rows, 3 anteriorly modified to heterostichad dorsal brush occupying about 17% of body length. Brush bristles up to 4 µm long: row 1 composed of an average of 19 dikinetids, row 2 of 22, and row 3 of 11 dikinetids followed by a monokinetidal bristle tail extending to mid-body.

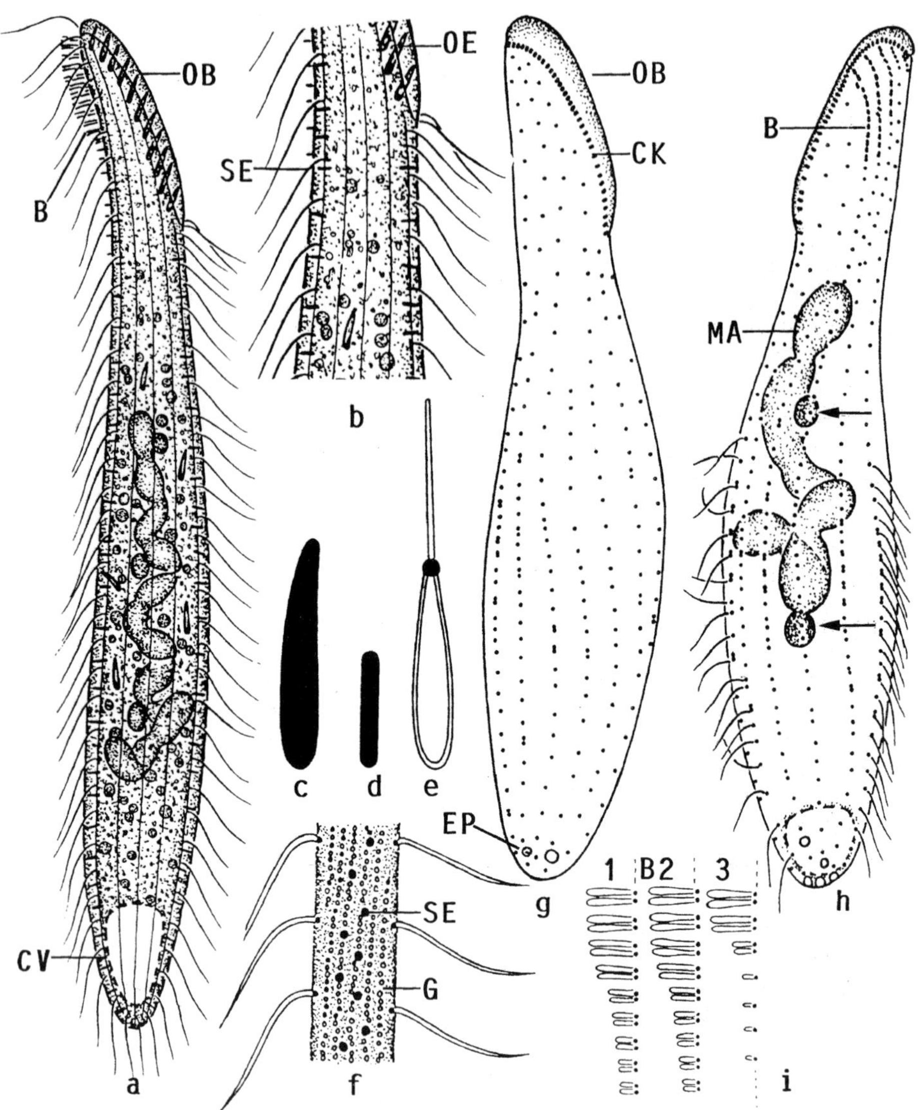

Fig. 91a–i *Armatospathula costaricana* nov. spec. from life (a–f, i) and after protargol impregnation (g, h). **a, b)** At first glance, this species is highly similar to *A. periarmata*. However, the oral bulge is twice as long, the oral extrusomes are different (vs. the same) from the somatic ones, the number of ciliary rows is considerably higher (10 vs. 6), and dorsal brush row 1 is much longer. Thus, these two species are easily distinguished both in vivo and preparations. Length of the specimen shown: 130 µm. **c)** Oral bulge extrusome, about 4 × 1 µm. **d)** Body extrusome, about 2 µm long. **e)** Exploded oral bulge extrusome, length 8 µm. **f)** Surface view showing arrangement of cortical granules (G) and body extrusomes (SE), which are conspicuously bright (refractive) in vivo. **g, h)** Ciliary pattern of right and left side of a blunt specimen with rather many dikinetids in mid-body, indicating that it is an early divider; length 88 µm. Arrows mark micronuclei. **i)** Posterior portion of dorsal brush. B(1-3) – dorsal brush (rows), CK – circumoral kinety, CV – contractile vacuole, EP – excretory pores, G – cortical granules, MA – macronucleus, OB – oral bulge, OE – oral extrusomes, SE – somatic extrusomes.

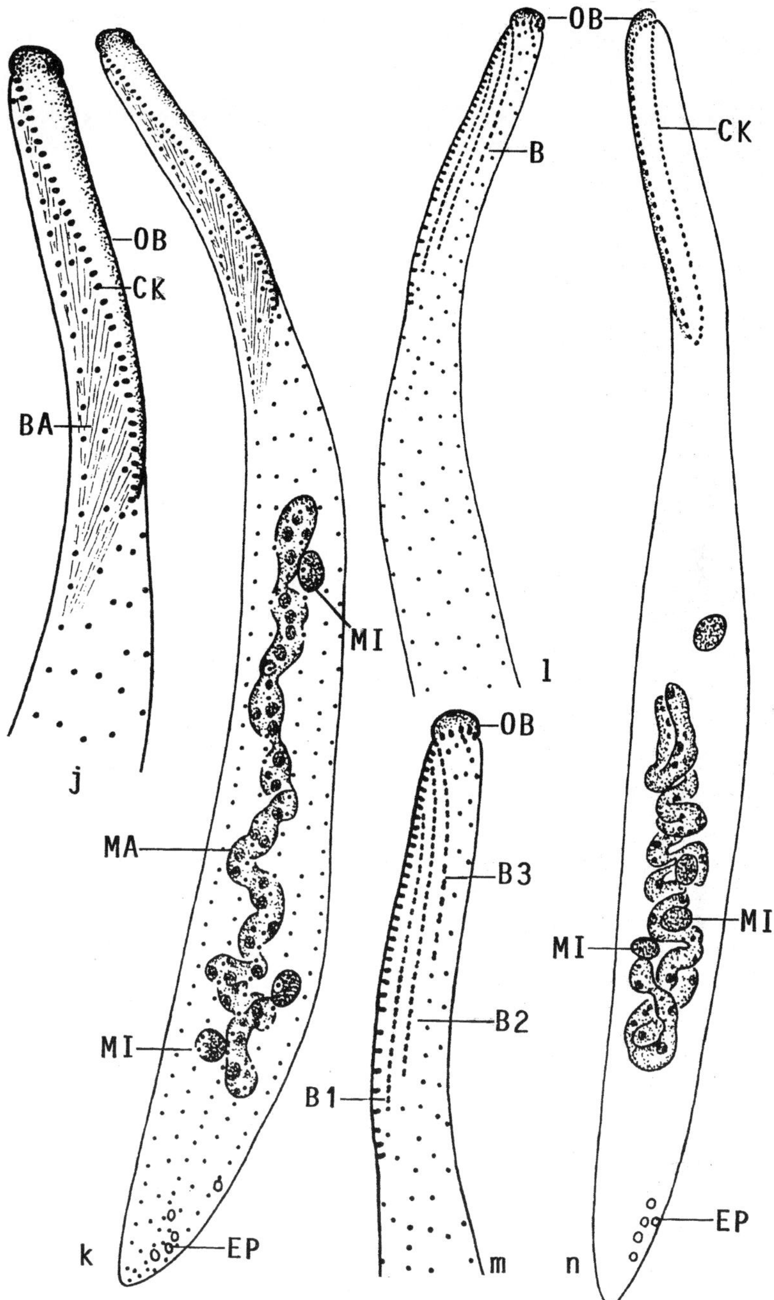

Fig. 91j–n *Armatospathula costaricana* nov. spec., oral and somatic ciliary pattern and nuclear apparatus after protargol impregnation. **j–m)** Ventrolateral and dorsolateral view of holotype specimen, length 130 µm. **n)** Ventral view of another specimen, length 130 µm. B(1-3) – dorsal brush (rows), BA – oral basket, CK – circumoral kinety, EP – excretory pores of contractile vacuole, MA – macronucleus, MI – micronuclei, OB – oral bulge.

Type locality: Soil from a horse pasture in the surroundings of the Selva Verde Lodge, Pto. Viejo, Costa Rica, W84° N10°27'.

Etymology: Composite of Costa Rica and the Latin adjective *anus* (belonging to), referring to the country the species was discovered.

Description: This ciliate was very rare, and only five mediocre impregnated specimens were found in the protargol slides. Thus, morphometry is incomplete. Nevertheless, we describe and name the population because it has several highly distinct features facilitating identification and contributing to the general morphology of the spathidiids.

Size 80–160 × 11–22 µm in vivo, usually about 130 × 16 µm, as calculated from in vivo measurements and the morphometric data (Table 35). Narrowly to very narrowly spatulate with strongly oblique oral bulge (anterior end) occupying first quarter of body and hardly separate from body proper; widest underneath mid-body; posterior end usually bluntly pointed, never tail-like; distinctly flattened only in hyaline oral region (Fig. 91a, g, k, 144a). Macronucleus in middle third of body, usually a moniliform, twisted strand composed of about 14 nodules; individual nodules circa 7 × 4 µm in size and with few nucleoli up to 2 µm across. On average three ellipsoidal, 3 µm large micronuclei near or attached to macronucleus (Fig. 91a, h, k, n; Table 35). Contractile vacuole in posterior body end, excretory pores terminal in one out of five specimens analysed, in the others rather distinctly subterminal (Fig. 91a, g, h, k, n). Extrusomes in two types, do not impregnate with the protargol method used (Fig. 91a–f, 144a–d). Oral extrusomes attached with narrow end to bulge and scattered in cytoplasm, very narrowly to narrowly ovate, slightly asymmetrical, compact and thus highly refractive in vivo, about 4 × 0.8 µm in size. Body extrusomes attached to somatic cortex, forming two rough rows between each two kineties; distinct in vivo, although fine and only about 2 µm long, because highly refractive, especially in surface view, where they appear as bright dots among the pale cortical granules. Exploded oral bulge extrusomes of typical toxicyst structure and up to 8 µm long. Cortex flexible, contains about five rows of colourless, minute (about 0.2 µm) granules between each two ciliary rows and extrusomes, as described above (Fig. 91f); postciliary fibres more faintly impregnated than in *A. periarmata*. Cytoplasm colourless, usually contains some small lipid droplets. Food not known. Glides and swims slowly on microscope slide.

Somatic cilia about 8 µm long in vivo, arranged in an average of ten equidistant, bipolar, ordinarily ciliated rows distinctly separate from circumoral kinety; *Arcuospathidium* ciliary pattern as indistinct as in *A. periarmata* (Fig. 91a, g, h, j–m; Table 35). Dorsal brush heterostichad and three-rowed, inconspicuous because occupying only about 17% of body length and bristles merely up to 4 µm long in vivo; all rows lack monokinetids anteriorly and continue as somatic kineties posteriorly. Brush bristles similar in all rows, that is, slightly clavate and about 4 µm long anteriorly, gradually decreasing to about 1.5 µm long rods posteriorly. Brush rows 1 and 2 of similar length, composed of an average of 21, respectively, 23 rather narrowly spaced dikinetids; row 3 distinctly shorter than rows 1 and 2, composed of an average of 12, comparatively loosely spaced dikinetids followed by about 1 µm long bristles forming a monokinetidal tail extending to mid-body (Fig. 91a, h, i, l, m, 144b; Table 35).

Oral bulge steeply slanted by more than 60° and approximately twice as long as widest trunk region; very indistinct in vivo because less than 3 µm high and hardly separate from body proper (Fig. 91a, 144a, b, d). Circumoral kinety very narrowly oblong to cuneate, continuous, composed of about 56 dikinetids more widely spaced ventrally than dorsally; each dikinetid associated with a cilium, a nematodesma, and a faintly impregnated fibre extending to midline of oral bulge. Oral basket rather voluminous, but faintly impregnated (Fig. 91a, g, h, j, k, n; Table 35).

Occurrence and ecology: As yet found only at type locality, where it was very rare in the non-flooded Petri dish culture. The sample consisted of surface soil (0–5 cm) with many grass roots, had pH 5.5 in water, and contained several new species. *Armatospathula costaricana* is well adapted to soil life by the slender body.

Remarks: As concerns the generic allocation, see *A. periarmata*. *Armatospathula costaricana* is easily distinguished from *A. novaki* by the smaller size (130 × 16 μm vs. 270 × 55 μm) and thus lower number of ciliary rows (10 vs. 20); and from *A. periarmata* by the two types of extrusomes (vs. one type). It differs from *A. periarmata* also in other main features, such as the doubled length of the oral bulge, the number of micronuclei (about three vs. invariably one) and ciliary rows (10–12 vs. 6–7), and details of the dorsal brush.

***Armatospathula plurinucleata* nov. spec.** (Fig. 92a–k, 145a–e; Table 36)

Diagnosis: Size about 110 × 18 μm in vivo. Narrowly spatulate to almost knife-shaped with steep to very steep, narrowly cuneate oral bulge about twice as long as widest trunk region. On average 43 macronucleus nodules; multimicronucleate. Oral and body extrusomes very narrowly ovate and slightly curved, 3–4 μm long. On average 8 ciliary rows, those on left side anteriorly modified to isostichad dorsal brush occupying about 18% of body length. Brush bristles up to 3 μm long; row 1 composed of an average of 10 dikinetids, row 2 of 15, and row 3 of 11 dikinetids not followed by a monokinetidal bristle tail.

Type locality: Soil from a secondary rain forest in the surroundings of Tropical Hotel Manaus, Brazil, W60° S3°.

Etymology: Composite of the Latin adjectives *pluri* (many) and *nucleatus* (with a nucleus), referring to the many macronucleus nodules, a main feature of this species.

Description: Size 90–130 × 14–22 μm in vivo, usually about 110 × 18 μm, as calculated from some in vivo measurements and the morphometric data (Table 36); length:width ratio moderately variable, that is 4.5–7.7:1, on average near 6:1 both in vivo and prepared cells. Narrowly spatulate to almost knife-shaped because anterior (oral) end distinctly slanted and occupying about 30% of body length, posterior end rounded, widest usually in or underneath mid-body; in preparations, frequently slightly inflated in mid-body and shrunken in neck area, providing specimens with a bluntly

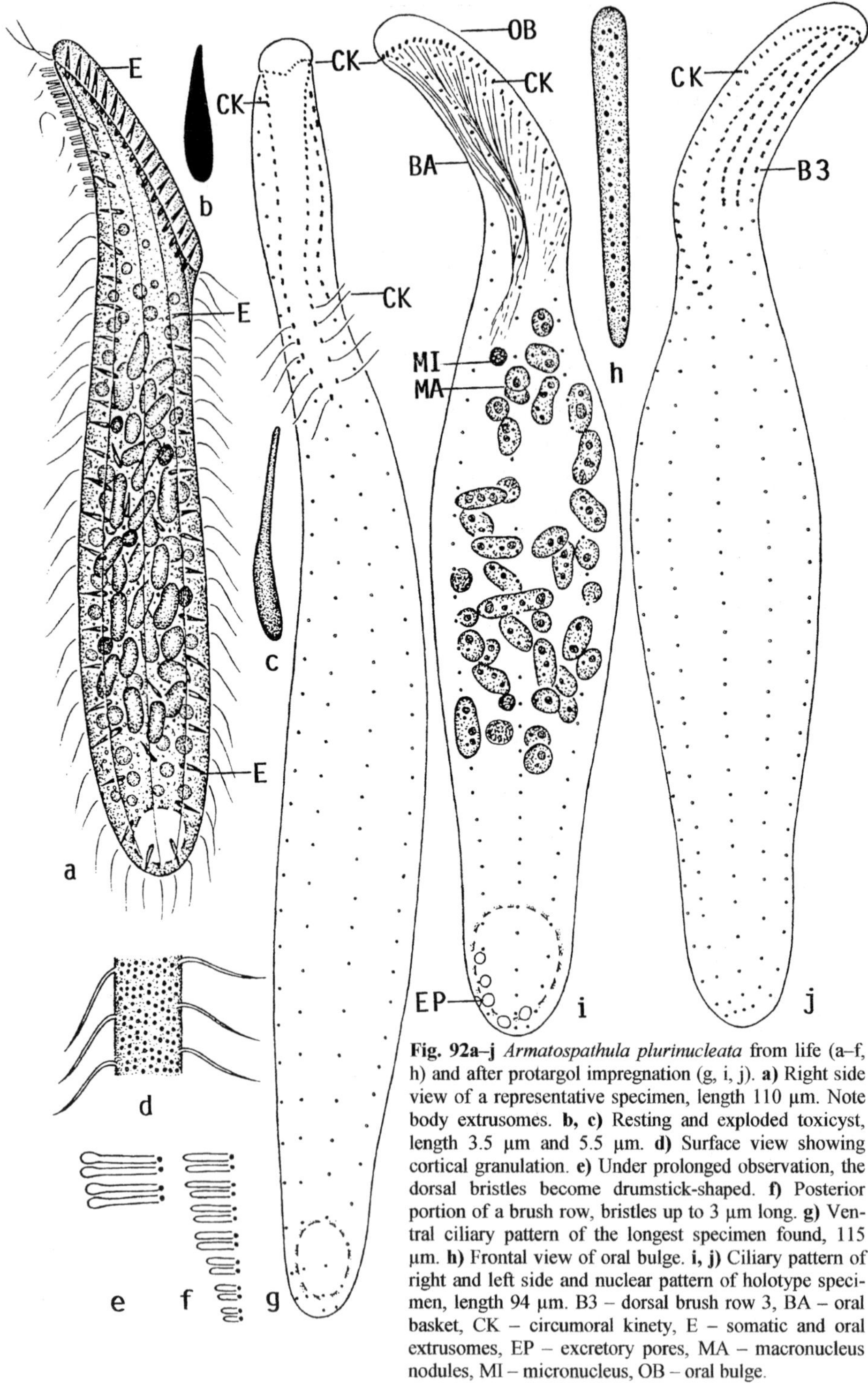

Fig. 92a–j *Armatospathula plurinucleata* from life (a–f, h) and after protargol impregnation (g, i, j). **a)** Right side view of a representative specimen, length 110 µm. Note body extrusomes. **b, c)** Resting and exploded toxicyst, length 3.5 µm and 5.5 µm. **d)** Surface view showing cortical granulation. **e)** Under prolonged observation, the dorsal bristles become drumstick-shaped. **f)** Posterior portion of a brush row, bristles up to 3 µm long. **g)** Ventral ciliary pattern of the longest specimen found, 115 µm. **h)** Frontal view of oral bulge. **i, j)** Ciliary pattern of right and left side and nuclear pattern of holotype specimen, length 94 µm. B3 – dorsal brush row 3, BA – oral basket, CK – circumoral kinety, E – somatic and oral extrusomes, EP – excretory pores, MA – macronucleus nodules, MI – micronucleus, OB – oral bulge.

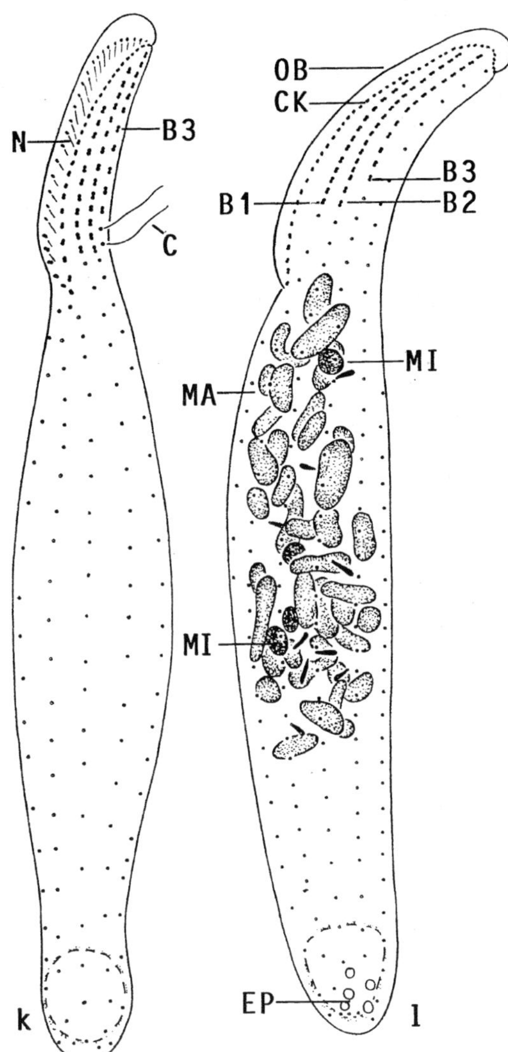

Fig. 92k, l *Armatospathula plurinucleata*, left side ciliary pattern and nuclear apparatus after protargol impregnation. The left specimen is bluntly fusiform due to the shrunken neck and contractile vacuole areas. Note that the dorsal brush is not on the dorsal side but on the right side. Further, brush row 3 lacks a posterior bristle tail and continues with ordinary cilia (C). B1-3 – dorsal brush rows, C – cilia, CK – circumoral kinety, EP – excretory pores, MA – macronucleus nodules, MI – micronuclei, N – nematodesmata, OB – oral bulge.

fusiform appearance; laterally flattened about 2:1 (Fig. 92a, j, k, l, 145a, b; Table 36). Nuclear apparatus in middle quarters of body, composed of an average of 43 scattered macronucleus nodules and five micronuclei. Macronucleus nodules ellipsoidal, spherical, dumbbell-shaped or ampulliform, usually about 5 × 2 µm in silver preparations; contain few nucleoli about 1 µm across; occasionally, some nodules connected by a fine thread producing short strands. Micronuclei ellipsoidal to spherical, up to 3 µm across, scattered among macronucleus nodules, often difficult to identify because faintly impregnated and of similar size as macronucleus nodules (Fig. 92a, i, l). Contractile vacuole in rear body end, some subterminal excretory pores. Oral and body extrusomes of same shape and size, that is, narrowly ovate and sligthly curved, 3–4 µm long, compact and thus distinct in vivo; body extrusomes in a rough row between two kineties each; do not impregnate with the protargol method used, except of a cytoplasmic developmental stage 2–3 × 0.8 µm in size. Exploded extrusomes of typical toxicyst structure and usually slightly curved, 5–6 µm long (Fig. 92a–c, 146d, e). Cortex very flexible, contains extrusomes as described above and many densely spaced rows of pale, colourless granules about 0.3 µm across and often faintly impregnated with protargol (Fig. 92d). Cytoplasm colourless, rather hyaline because containing only few and small lipid droplets. Food not known. Movement without peculiarities.

Somatic cilia about 8 µm long in vivo and ordinarily spaced (~ 3 µm), arranged in an

Table 36 Morphometric data on *Armatospathula plurinucleata*

Characteristics[1]	$\overline{x}$	M	SD	SE	CV	Min	Max	n
Body, length	96.6	94.0	8.5	2.1	8.8	86.0	115.0	17
Body, width	16.2	16.0	2.2	0.5	13.5	12.0	19.0	17
Body length:width, ratio	6.0	6.0	0.9	0.2	15.0	4.5	7.7	17
Oral bulge, length	28.6	29.0	2.4	0.6	8.3	24.0	34.0	17
Oral bulge, height	2.6	2.5	0.3	0.1	12.3	2.0	3.0	17
Oral bulge length:body width, ratio	1.8	1.7	0.3	0.1	17.8	1.3	2.4	17
Circumoral kinety to last dikinetid of brush row 1, distance	13.4	13.5	1.9	0.5	14.0	11.0	16.0	14
Circumoral kinety to last dikinetid of brush row 2, distance	17.5	17.0	1.6	0.4	8.9	15.0	21.0	14
Circumoral kinety to last dikinetid of brush row 3, distance	14.2	14.0	1.6	0.4	11.1	12.0	17.0	14
Anterior body end to first macronucleus nodule, distance	27.0	27.0	3.3	0.8	12.4	18.0	32.0	17
Macronucleus figure, length	52.5	54.0	10.2	2.5	19.4	40.0	72.0	17
Macronucleus nodules, length	5.0	5.0	1.9	0.5	38.1	2.0	8.0	17
Macronucleus nodules, width	2.3	2.0	0.6	0.1	25.2	2.0	4.0	17
Macronucleus nodules, number	43.4	38.0	12.2	3.0	28.1	30.0	78.0	17
Micronuclei, across	2.5	2.5	0.5	0.1	21.1	1.5	3.0	13
Micronuclei, number	5.6	5.0	1.9	0.6	33.8	3.0	9.0	9
Ciliary rows, number	8.2	8.0	0.4	0.1	5.3	8.0	9.0	17
Basal bodies in a right side ciliary row, number	34.8	35.0	4.3	1.1	12.4	28.0	44.0	15
Dorsal brush rows, number	3.0	3.0	0.0	0.0	0.0	3.0	3.0	17
Dikinetids in brush row 1, number	10.5	10.0	1.9	0.5	18.2	7.0	14.0	14
Dikinetids in brush row 2, number	15.4	15.5	2.3	0.6	15.0	12.0	19.0	14
Dikinetids in brush row 3, number	11.1	11.0	1.7	0.4	14.9	8.0	14.0	14
Circumoral kinetids, number	65.2	64.5	6.2	2.5	9.6	60.0	77.0	6

[1] Data based on mounted and protargol-impregnated (FOISSNER's method) specimens from a non-flooded Petri dish culture. All specimens found in eight slides were used. Measurements in µm. CV – coefficient of variation in %, M – median, Max – maximum, Min – minimum, n – number of individuals investigated, SD – standard deviation, SE – standard error of arithmetic mean, $\overline{x}$ – arithmetic mean.

average of eight equidistant, bipolar rows anteriorly distinctly curved dorsally and separate from circumoral kinety, forming very pronounced *Arcuospathidium* pattern, except of a ventral row invariably commencing underneath proximal end of circumoral kinety (Fig. 92a, i–l, 145a, b; Table 36). Dorsal brush isostichad and three-rowed, conspicuous, though occupying only about 18% of body length and bristles merely 3 µm long in vivo, because occupying entire left oral surface and extending in parallel with circumoral kinety; thus, no ordinary ciliary rows between ventral kinety and first brush row, as in *Cultellothrix velhoi*; furthermore, all brush rows lack anterior and posterior tails. Brush bristles of similar configuration in all rows, that is, anterior and posterior bristle of dikinetids rod-shaped decreasing in length from about 3 µm anteriorly to 1.5 µm posteriorly; become drumstick-shaped and up to 5 µm long under prolonged observation. Brush rows 1 and 2 end at same level, though row 1 is slightly shorter anteriorly than row 2, composed of an average of 10 and 15 ordinarily spaced dikinetids, respectively; row 3 of similar length as row 1, composed of an average of 11 dikinetids (Fig. 92a, e, f, j–l, 145a–c; Table 36).

Oral area more or less distinctly curved dorsally, bulge thus convex, especially in prepared cells, and steeply to very steeply (>60°) slanted and about twice as long as widest trunk region; fairly distinct in vivo due to the refractive extrusomes contained; narrowly cuneate in frontal view (Fig. 92a, g–j, 145a–c; Table 36). Circumoral kinety of same shape as oral bulge, continuous, composed of about 65 dikinetids more widely spaced ventrally (distance 1.5–2 µm) than dorsally (< 1 µm). Nematodesmata recognizable only in right side dikinetids, as in *Cultellothrix velhoi,* those of dorsal half form distinct bundles extending over anterior third of body, while length of basket rods sharply decreases in ventral half; we could not determine whether the left side basket rods are absent or present but not impregnated (Fig. 92g, i–l).

Occurrence and ecology: As yet found only at type locality. *Armatospathula plurinucleata* was very rare in the non-flooded Petri dish culture, where it occurred 20 days after rewetting. The type locality is a secondary rain forest in the original floodplain of the Rio Negro; however, floods now occur rarely because the bank is protected. Soil brownish, acidic (pH 5.1 in water), and with root carpet in the upper 5 cm.

Remarks: *Armatospathula plurinucleata* has many distinct features, for instance, the many macronucleus nodules and, in particular, the very narrowly ovate oral and body extrusomes. These features distinguish *A. plurinucleata* from the congeners and all other spathidiids. In contrast, the generic allocation is difficult because it has features used to define two arcuospathidiid genera, viz., the left laterally located dorsal brush (*Cultellothrix*) and the body extrusomes (*Armatospathula*). Thus, the Brazilian population could be a *Cultellothrix* with body extrusomes or an *Armatospathula* with left laterally located brush. Furthermore, *A. plurinucleata* strongly resembles *Cultellothrix velhoi*, especially in the location of the dorsal brush and the lack (?) of oral basket rods in the left half of the circumoral kinety. Possibly, these two species constitute a distinct genus different from the other *Cultellothrix* and *Armatospathula* species.

Armatospathula novaki (Foissner, Agatha & Berger, 2002) **nov. comb.** (Fig. 93a–k; Table 37)

2002 *Arcuospathidium novaki* Foissner, Agatha & Berger, 2002, Denisia, 5: 303 (Type slides with protargol-impregnated specimens from type locality are deposited in the Oberösterreichische Landesmuseum in Linz, Austria.).

Diagnosis: Size about 270 × 55 µm in vivo. Narrowly spatulate with distinctly oblique, cuneate oral bulge about as long as widest trunk region. Macronucleus long and tortuous. Oral bulge extrusomes very narrowly ovate and about 7 × 0.8 µm in size;

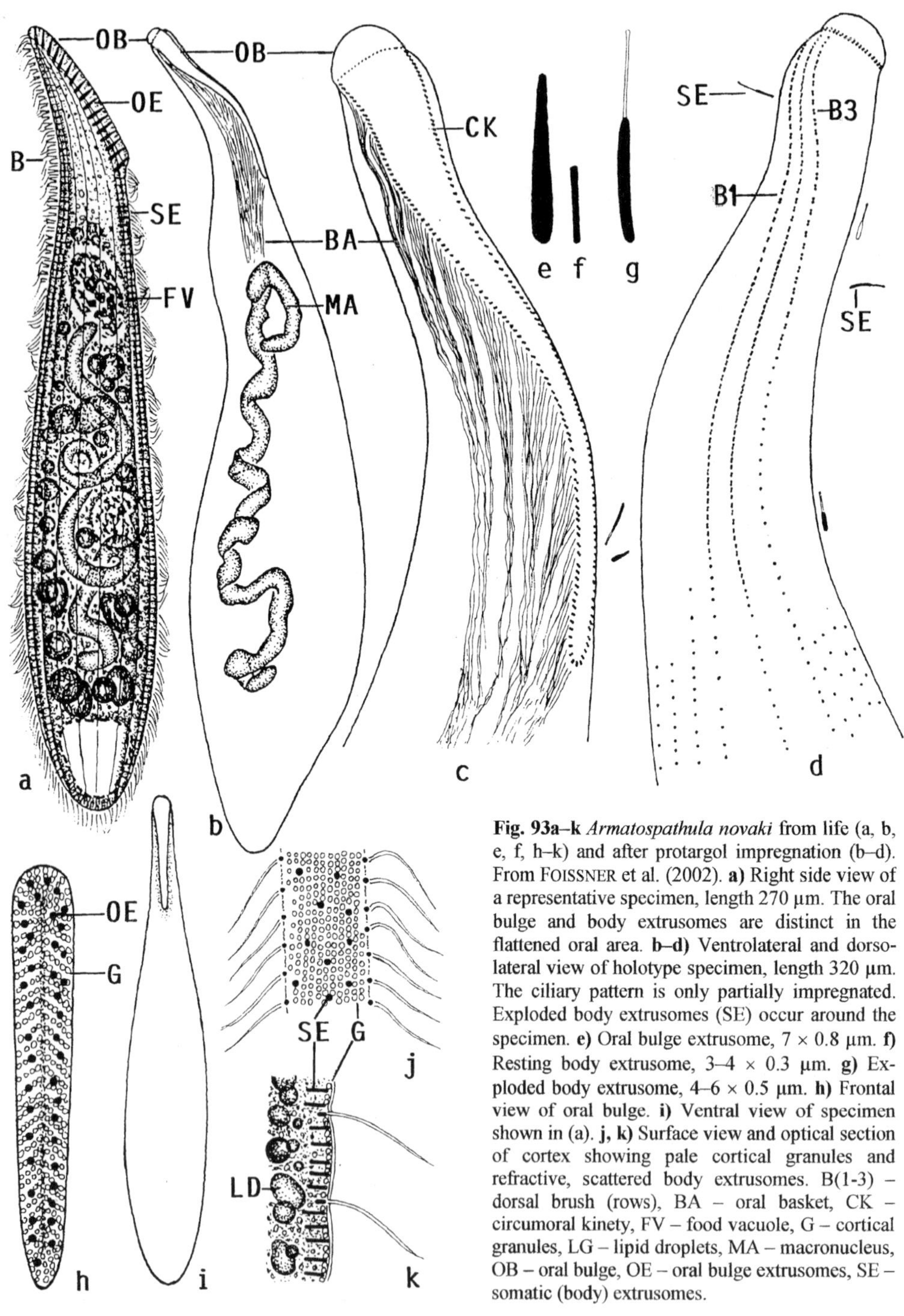

Fig. 93a–k *Armatospathula novaki* from life (a, b, e, f, h–k) and after protargol impregnation (b–d). From FOISSNER et al. (2002). **a)** Right side view of a representative specimen, length 270 μm. The oral bulge and body extrusomes are distinct in the flattened oral area. **b–d)** Ventrolateral and dorsolateral view of holotype specimen, length 320 μm. The ciliary pattern is only partially impregnated. Exploded body extrusomes (SE) occur around the specimen. **e)** Oral bulge extrusome, 7 × 0.8 μm. **f)** Resting body extrusome, 3–4 × 0.3 μm. **g)** Exploded body extrusome, 4–6 × 0.5 μm. **h)** Frontal view of oral bulge. **i)** Ventral view of specimen shown in (a). **j, k)** Surface view and optical section of cortex showing pale cortical granules and refractive, scattered body extrusomes. B(1-3) – dorsal brush (rows), BA – oral basket, CK – circumoral kinety, FV – food vacuole, G – cortical granules, LG – lipid droplets, MA – macronucleus, OB – oral bulge, OE – oral bulge extrusomes, SE – somatic (body) extrusomes.

body extrusomes rod-shaped, approximately 4 × 0.3 µm. About 20 ciliary rows, 3 anteriorly modified to heterostichad dorsal brush.

Type locality: Mud and soil from road puddles in the Bambatsi Guest Farm, Namibia, E15°25' S20°10'.

Dedication: Dedicated Dr. RUDOLF NOVAK, scientific manager of the Austrian Science Foundation.

Description: This large ciliate was rare and the few cells found in the protargol slides were poorly impregnated. Thus, data are incomplete. In spite of this, FOISSNER et al. (2002) described and named the population because it has several outstanding features facilitating identification and contributing to the general morphology of the spathidiids.

Size conspicuous, that is, 200–350 × 40–70 µm in vivo, usually near 270 × 55 µm; length:width ratio about 5:1 in vivo, while 2.3–5.9:1, on average 3.8:1 in protargol slides, indicating some inflation during preparation (Table 37). Shape inconspicuous, that is, spatulate to narrowly spatulate with distinctly oblique oral bulge occupying 1/4–1/5 of body length; acontractile. Oral area hyaline because flattened approximately 3:1, oral and body extrusomes thus distinct; postoral region ellipsoidal and dark at low magnification (≤×100) due to numerous cytoplasmic inclusions (Fig. 93a, b, i). Macronucleus in central quarters of cell, highly tortuous, when spread about as long as body. Many micronuclei difficult to distinguish from similar-sized and impregnated cytoplasmic inclusions. Contractile vacuole in rear end; no second contractile vacuole above mid-body. Extrusomes highly remarkable because a special type each in oral bulge and somatic cortex (Fig. 93a, d–g, h, j, k). Oral bulge extrusomes form a rough row each in right and left half of oral bulge, very narrowly cuneate, about 7 × 0.8 µm in vivo, compact and thus highly refractive, do not impregnate with protargol. Body extrusomes randomly attached to somatic cortex, lacking in oral bulge, rod-shaped, 3–4 × 0.3 µm in vivo, difficult to recognize, except in hyaline oral area; when extruded, occasionally impregnate with protargol showing a dark, thick posterior half and a brownish, thin anterior half with a minute globule on distal end. Cortex very flexible, contains about 15 rows of minute (~ 0.2 µm) granules between two kineties each and extrusomes, as described above, appearing as bright dots among the pale granules (Fig. 93k). Cytoplasm colourless, usually packed with globular and irregular, highly refractive lipid inclusions up to 30 µm across. Feeds on ciliates, for instance, *Colpoda minima*, recognizable in a food vacuole 90 × 55 µm in size. Glides and swims slowly showing great flexibility.

Cilia 10 µm long and ordinarily spaced in vivo, arranged in about 20 widely spaced rows three anteriorly modified to an about 70 µm long, ordinary dorsal brush with up to 4 µm long, rod-shaped bristles. Brush dikinetids more closely spaced in rows 1 and 2 than in row 3, which is also distinctly shorter, but has a monokinetidal tail extending to second third of cell with 1 µm long bristles (Fig. 93a, d).

Oral bulge about 60 µm long in vivo and 5 µm high at dorsal end, distinctly oblique and slightly convex, narrowly to very narrowly cuneate in frontal view, covered by oblique rows of cortical granules. Circumoral kinety of similar shape as oral bulge, composed of narrowly spaced dikinetids, distinctly separate from somatic ciliary rows. Oral basket voluminous (Fig. 93a–c, h, i; Table 37).

Table 37 Morphometric data on *Armatospathula novaki* (from FOISSNER et al. 2002)

Characteristics [1]	$\overline{x}$	M	SD	SE	CV	Min	Max	n
Body, length	242.5	245.0	60.6	24.7	25.0	175.0	325.0	6
Body, width	66.1	67.0	13.5	5.5	20.4	48.0	80.0	6
Body length:width, ratio	3.8	3.5	1.2	0.5	32.7	2.3	5.9	6
Oral bulge, length	52.0	49.0	8.9	4.5	17.1	45.0	65.0	4
Macronucleus figure, length	129.2	130.0	29.7	12.1	23.0	90.0	165.0	6
Macronucleus, length (spread; approximate)	248.0	250.0	–	–	–	150.0	320.0	6
Macronucleus, width	8.0	8.0	1.1	0.4	13.7	7.0	10.0	6

[1] Data based on mounted and protargol-impregnated (FOISSNER's method) specimens from a non-flooded Petri dish culture. All species found were used. Measurements in µm. CV – coefficient of variation in %, M – median, Max – maximum, Min – minimum, n – number of individuals investigated, SD – standard deviation, SE – standard error of arithmetic mean, $\overline{x}$ – arithmetic mean.

Occurrence and ecology: To date found only at type locality, where it was very rare in the non-flooded Petri dish culture. The massive body indicates that it is a limnetic species.

Remarks: *Armatospathula novaki* has several outstanding features, viz., body (somatic) extrusomes (toxicysts) and, in spite of the huge size, only about 20 ciliary rows. Thus, it is easily distinguished from all other *Spathidium* s.l. species, except of the congeners and *Apospathidium* spp., which also have body extrusomes and few ciliary rows. However all these species are smaller (< 200 µm) and have a much shorter macronucleus.

The generic classification of this population is somewhat uncertain because the ciliary pattern is poorly impregnated. However, the narrowly cuneate oral bulge and the continuous circumoral kinety exclude *Apospathidium* and suggest that it belongs to the arcuospathidiids.

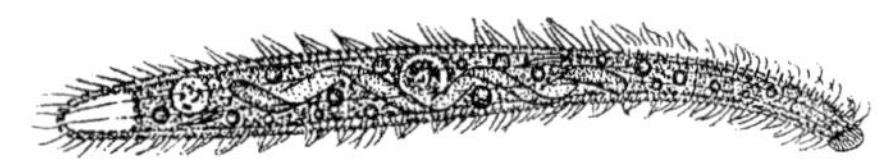

Spathidiodes KAHL, 1926

1926 *Spathidiodes* KAHL, Arch. Protistenk., 55: 276 – Type species (by monotypy): *Spathidiodes hyalina* KAHL, 1926.
1930 *Spathidiella* KAHL, Arch. Protistenk., 70: 392 – Type species (by original designation): *Spathidiella hyalina* (KAHL, 1926).
1930 *Spathidiella* KAHL, 1930 – KAHL, Tierwelt Dtl., 18: 169 – Type species (by designation): *Spathidiella hyalina* (KAHL, 1926).

Diagnosis: Arcuospathidiidae (?) with hyaline, opalescent, somewhat rigid cortex and verruciform oral bulge.

Etymology: Not given in original description. *Spathidiodes* is a composite of the generic name *Spathidium* (small spathula) and *oïdes* (similis), meaning a *Spathidium*-like ciliate. The gender is feminine because KAHL (1926) combined the name with the feminine adjective hyalina (ICZN 30.1.4.4.).

Nomenclature: KAHL (1930b) emphasized that the original spelling is caused by a typing error, that is, the lack of an "i". Further, he noticed that the corrected name would be a junior homonym of *Spathidioides* BRODSKY, 1925, another spathidiid (?) genus. Thus, he renamed the genus *Spathidiella* (KAHL 1930b). However, this was not accepted by CORLISS (1979) and AESCHT (2001) because a one-letter-difference is sufficient to prevent homonymy (ICZN, article 56.2.).

Remarks: KAHL (1930b) already noted that this genus is insufficiently separated from *Spathidium*. We agree. The size and shape of the body and oral bulge of the type species resemble *Apertospathula verruculifera*, and the two species added by KAHL (1930b) also resemble *Apertospathula, Arcuospathidium,* and *Cultellothrix*. Unfortunately, a proper classification needs details of the circumoral kinety, which are not available because none of the three species has been reinvestigated with modern methods. Further, KAHL (1926, 1930a, b) provided rather different descriptions of the type species (see below). Thus, the classification of *Spathidiodes* remains obscure, and it is not known whether the three species are congeneric and need a distinct genus at all. None the less, we keep the genus at the present state of knowledge because of the bright (opalescent), comparatively rigid cortex, an unusual feature in spathidiids.

Key to species

1 Length 40–70 µm. Posterior end narrowly to broadly rounded 2
– Length 90–130 µm. Posterior end sharply pointed .... *S. euglenivora*
2 Body outline elliptical, narrowly to broadly rounded posteriorly. Oral bulge triangular in lateral view and distinctly shorter than widest trunk region .... *S. hyalina*
– Body outline pyriform, posterior end narrowly rounded. Oral bulge cuneate in lateral view and about as long as widest trunk region .... *S. rigida*

Spathidiodes hyalina KAHL, 1926 (Fig. 94a–c, h)

1926 *Spathidiodes hyalina* KAHL, Arch. Protistenk., 55: 276 (no type material available).
1930 *Spathidiella hyalina* n. comb. – KAHL, Arch. Prostistenk., 70: 392 (brief redescription).
1930 *Spathidiella (Spathidiodes) hyalina* KAHL, 1926 – KAHL, Tierwelt Dtl., 18: 169 (revision).

Diagnosis: Should await redescription, which is urgently needed.

Type locality: Not definitely mentioned by KAHL (1926). However, it is obvious that it is the Hamburg area, Germany, E10° N53°30'.

Etymology: Not given in original description. The Latin adjective *hyalina* (hyaline) obviously refers to the "very transparent" cytoplasm.

Description (after KAHL 1926; Fig. 94a, b): Length 40–50 µm (likely 50–60 µm due to a calibration error; FOISSNER & WENZEL 2004). Macronucleus roundish, accompanied by a small micronucleus. Contractile vacuole in posterior end. Oral bulge trichocysts short. Ectoplasm (cortex) very bright (opalescent). Cytoplasm transparent, contains food vacuoles with small rhodobacteria. Swims slowly rotating about main body axis.

Cilia long, loosely arranged. Ciliary rows widely spaced and in distinct furrows the margins of which appear as additional rows on superficial observation. Dorsal brush likely present. Oral bulge distinctly shorter than widest trunk region, according to figure 94a, bulge height decreases from dorsal to ventral side of cell, lateral view thus triangular, contains short extrusomes; appears as indistinct beak when cell is viewed dorsally.

Differs from other *Spathidium* species by the strongly opalescent ectoplasm and the more rigid body.

Description (after KAHL 1930a; Fig. 94c): Length 40–60 µm. Shape oval (2:1 – 3:2), more rarely narrowed posteriorly as shown in 1926. Nucleus roundish or short-oval. Contractile vacuole in posterior end. Oral bulge trichocysts about 4–5 µm long. Ectoplasm (cortex) bright, widely striated. Cytoplasm with bright, colourless granules and usually with some single rhodobacteria likely serving as food.

Cilia long and widely spaced. Dorsal brush, not clearly seen in 1926, with 3–4 µm long bristles and thus comparatively distinct. Oral bulge beak-like projecting dorsally, contains about 4–5 µm long trichocysts along whole length.

Description (after KAHL 1930b; Fig. 94h): Length 40–60 µm. Shape oval, obovoid, or ellipsoidal, sometimes slightly narrowed around contractile vacuole. Macronucleus round. Ectoplasm (cortex) rigid, bright, and widely striated, with slight furrows ("Zwischenstreifen") extending between ciliary rows. Cytoplasm usually with some vacuoles containing rhodobacteria.

Cilia long and loosely spaced. Dorsal brush distinct. Oral bulge very short-triangular, appears beak-like.

Occurrence and ecology: Frequent, but never abundant in the saprobic mud

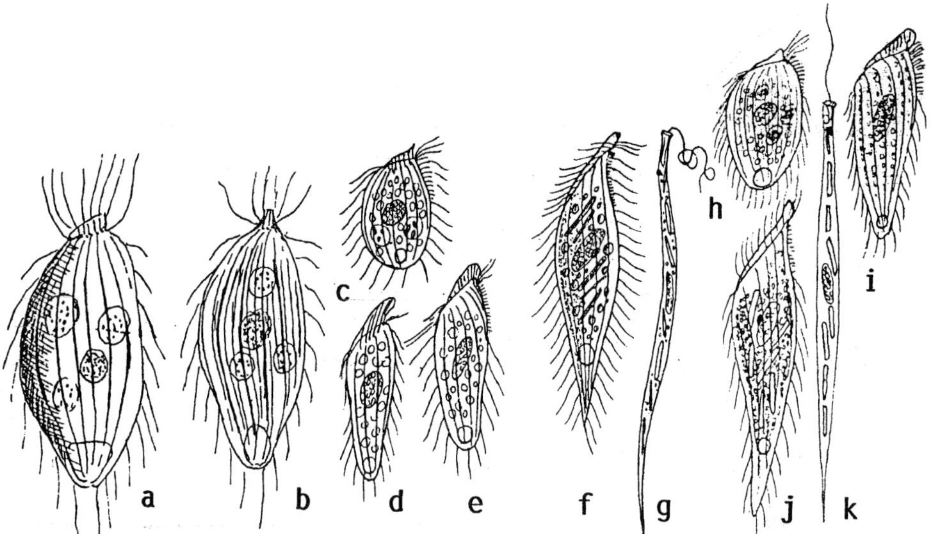

Fig. 94a–i *Spathidiodes* species from life. **a, b, c, h)** *S. hyalina*, left side and dorsal view, length 50–60 µm (a, b, from KAHL 1926; c, from KAHL 1930a; h, from KAHL 1930b). **e, i)** *S. rigida*, left side view, length 60–70 µm (e, from KAHL 1930a; i, from KAHL 1930b). **f, j)** *S. euglenivora*, left side view, length 90–130 µm (f, from KAHL 1930a; j, from KAHL 1930b). **d)** A species similar or identical with *S. rigida* (from KAHL 1930a). **g, k)** *Euglena* sp., length 150–240 µm, the preferred food of *S. euglenivora* (from KAHL 1930a, b).

of small ponds, during summer also between aquatic plants. No further records known.

Remarks: The most important difference in the descriptions concerns the trichocysts which are not mentioned in KAHL (1930b). Further, the shape of the body and oral bulge are markedly different in KAHL (1926) and KAHL (1930a, b). Thus, it cannot be excluded that KAHL (1926, 1930a, b) mixed two species. On the other hand, all populations obviously possessed a bright (opalescent) cortex, a curious feature emphasized by KAHL. Further, the "very transparent" cytoplasm is unusual for this kind of sapropelic spathidiids.

The rigidity of the cortex, an important feature, is mentioned in KAHL (1926) and KAHL (1930b), but not in KAHL (1930a), where he somewhat relinquished the feature in the comment to the genus: "The ectoplasm is hyaline, slightly armoured and thus slightly rigid, but is not ametabolic"; further, KAHL (1930b) remarks: "species with hyaline ectoplasm occur also in *Spathidium*".

Spathidiodes rigida (KAHL, 1930) **nov. comb.** (Fig. 94e, i)

1930 *Spathidiella rigida* KAHL, Arch. Protistenk., 70: 393 (no type material available).
1930 *Spathidiella rigida* KAHL, 1930 – KAHL , Tierwelt Dtl., 18: 169 (revision).

Diagnosis: Should await redescription, which is urgently needed.

Type locality: Sapropel of the "Eppendorfer Schießstand" in the surroundings of Hamburg, Germany, E10° N53°30'.

Etymology: Not given in original description. The Latin adjective *rigida* (rigid) likely refers to the rigid cortex.

Description: Length 60–70 µm. Body very slenderly pyriform (3.5:1) to ovoid, rigid and bright. Macronucleus elongate ellipsoidal. Extrusomes rather thick, 3–4 µm long. Ectoplasm (cortex) bright and distinctly furrowed by ciliary rows. Cytoplasm contains colourless reserve bodies; food not observed, but does not contain rhodobacteria.

Cilia loosely arranged, ciliary rows widely spaced. Dorsal brush with 4–5 µm long bristles and thus distinct. Oral bulge (anterior body end) slanted by about 45°, about as long as widest trunk region, distinctly higher dorsally than ventrally where it smoothly merges into body proper, rarely projects beak-like dorsally.

Occurrence and ecology: Known only from type locality, where it was rare in the sapropelic mud. May become rather abundant in blades of grass or rotting pieces of *Glyceria* put into the sampling jar.

Remarks: KAHL (1930b) emphasized the stability of the characteristics separating *S. rigida* and *S. hyalina*, which he discovered in the same pond. Further, he found this or a similar species with longer trichocysts in a pond of the botanical garden of Hamburg (Fig. 94d).

There are several species resembling *S. rigida*, for instance, *Apertospathula armata, A. cuneata,* and *Arcuospathidium pelobium*; however, all have a very flexible body and thus cannot be identified as *S. rigida*. Further, these species differ also in other features, such as the length of the brush bristles (4–5 µm vs. 1 µm; *Apertospathula armata*), body shape (pyriform vs. clavate, that is, 3.5:1 vs. ~5:1; *Arcuospathidium pelobium*), and the size and shape of the extrusomes (4–5 long rods vs. 2.5 µm and narrowly cuneate; *Apertospathula cuneata*). Generally, there are many of these small spathidiids. At first glance, all look very similar, but on more detailed observation they show many distinct features. Many, perhaps most of these species wait to be discovered.

Spathidiodes euglenivora (KAHL, 1930) **nov. comb.** (Fig. 94f, g, j, k)

1930 *Spathidiella euglenivora* KAHL, Arch. Protistenk., 70: 393 (no type material available).
1930 *Spathidiella euglenivora* KAHL, 1930 – KAHL, Tierwelt Dtl. 18: 170 (revision).

Diagnosis: Should await redescription, which is urgently needed.

Type locality: Mesosaprobic to polysaprobic streamlet in the surroundings of Hamburg, Germany, E10° N53°30'.

Etymology: Not given in original description. The name is a composite of the generic name *Euglena* and the verb *vorare* (feeding), referring to the preferred food of this ciliate, viz., *Euglena* sp.

Description: Length 90–130 µm. Body lanceolate or fusiform, circular in transverse view, widest near second fifth gradually tapering to acute posterior end. Narrowed anterior portion shorter than posterior, tapers asymmetrically forming a distinctly sigmoidal dorsal outline. Macronucleus elongate ellipsoidal, micronucleus oval and attached to macronucleus. Contractile vacuole distinctly subterminal, according to the figures (Fig. 94f, j). Trichocysts not recognizable, but oral bulge transversely striated by narrowly spaced rows of granular protrichocysts ("cortical granules", authors). Ectoplasm (cortex) glass-like shining, but not or only slightly rigid. Cytoplasm often rather hyaline, well-fed specimens dark in middle quarters by numerous granules, large lipid droplets, and yellowish paramylum rods from a colourless euglenid on which the species feed. When swimming almost serpentinely curved and rotating about main body axis.

Cilia long and arranged in about 10 widely spaced, rather loosely ciliated rows. Dorsal brush distinct, bristles short. Oral bulge of usual spathidiid structure, slanted by about 45° gradually merging into ventral surface, dorsal end beak-like projecting.

Occurrence and ecology: *Spathidiodes euglenivora* was discovered in the mud of the outlet of a small pond heavily polluted by food remnants of an inn (KAHL 1927, 1930a). This pond was in the so-called Eppendorfer Moor in the surroundings of Hamburg, Germany. Later, *S. euglenivora* was reported from the Rio Bella in Peru (CAIRNS 1965).

Spathidiodes euglenivora was rather abundant at type locality and fed exclusively on the paramylum rods of a colourless, very slender euglenid with a length of 150–240 µm (Fig. 94g, k). KAHL (1930a) supposed that this *Euglena* could be an undescribed species.

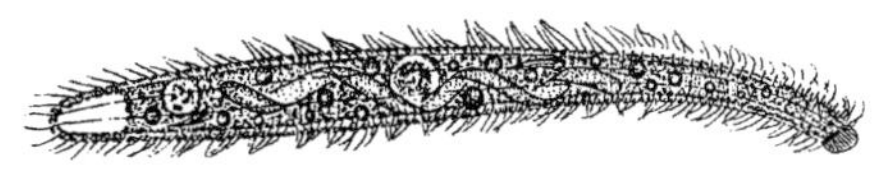

Family Apertospathulidae FOISSNER, XU & KREUTZ, 2005

2005 Apertospathulidae FOISSNER, XU & KREUTZ, J. Eukaryot. Microbiol. 52: 12 – Type genus (by original designation): *Apertospathula* FOISSNER, AGATHA & BERGER, 2002.

Diagnosis: Spathidiina with lasso-shaped oral bulge and circumoral kinety open ventrally.

Remarks: FOISSNER et al. (2005b) classified the genera *Apertospathula, Longispatha,* and *Rhinothrix* into a new family, the Apertospathulidae, characterized by a lasso-shaped oral bulge and circumoral kinety, a unique feature not found in any of the described haptorid and spathidiid families. Thus, this family is more distinct than several other spathidiid families, for instance, the Arcuospathidiidae and Pharyngospathidiidae.

Apertospathula highly resembles small *Spathidium* species, except of the ventrally opened circumoral kinety. In contrast, *Longispatha* and *Rhinothrix* resemble the spathidiid families Bryophyllidae and Perispiridae because the oral bulge and the circumoral kinety basically extend whole body length (FOISSNER & LEI 2004, WIRNSBERGER et al. 1984). However, the Bryophyllidae and Perispiridae have a complete oral bulge and circumoral kinety, while the left half of the bulge and circumoral kinety is distinctly shortened in the Apertospathulidae, which thus possibly evolved from a *Bryophyllum*-like ancestor. This appears comprehensible in *Longispatha* and *Rhinothrix*, while *Apertospathula* spp. look highly similar to *Spathidium* spp. Thus, we cannot exclude that *Apertospathula* evolved from a different ancestor and the family Apertospathulidae is biphyletic.

The sigmoidal oral apparatus of *Rhinothrix* resembles that of *Perispira*, another aberrant spathidiid genus (WIRNSBERGER et al. 1984). However, *Perispira* has not only a complete oral bulge and circumoral kinety but also spiralized somatic ciliary rows, while those of *Rhinothrix* extend meridionally. This is achieved by the insertion of some shortened kineties in the curves of the bulge (Fig. 105q, s) and suggests a different evolutionary history of the two genera.

The family consists of 3 genera with a total of 14 species, most of which have been described recently or will be described here for the first time. Fourteen species are a considerable gain, confirming FOISSNER et al. (2002) that, especially, *Apertospathula* contains many species most of which are still undescribed.

Key to genera

1 Oral bulge distinctly projecting dorsally, forming the palpus oralis. Oral bulge and circumoral kinety extend spirally .... *Rhinothrix*
– Without palpus oralis. Oral bulge and circumoral kinety extend meridionally 2
2 Mouth and circumoral kinety (oral bulge) distinctly shorter than postoral area. Right half of circumoral kinety only slightly longer than left. Appears *Spathidium*-like .... *Apertospathula*
– Right half of circumoral kinety and oral bulge extend to posterior body end, left half confined to anterior body half. Appears *Arcuospathidium*-like ..... *Longispatha*

Apertospathula FOISSNER, AGATHA & BERGER, 2002

2002 Apertospathula FOISSNER, AGATHA & BERGER, Denisia, 5: 318 – Type species (by original designation): *Apertospathula inermis* FOISSNER, AGATHA & BERGER, 2002.

Improved diagnosis: Indistinctly spatulate, flexible Apertospathulidae with brush located dorsally or slightly dorsolaterally; individual brush rows without or with very short anterior tail of ordinary cilia. Mouth (oral bulge) and circumoral kinety distinctly shorter than body proper, extend meridionally; right half of circumoral kinety slightly longer than left.

Etymology: Composite of the Latin nouns *apertum* (open field) and *spatha* (spatula), and the diminutive suffix *ula*, referring to the ventrally opened circumoral kinety and the similarity to small species of the genus *Spathidium*. Feminine gender.

Remarks: The main feature of this genus is the circumoral kinety, which is shortened at left and thus open ventrally. The ciliary pattern of *Apertospathula* resembles that of *Arcuospathidium* (somatic and oral ciliature distinctly separate) and *Spathidiodes*, a still poorly known genus differing from other spathidiids mainly by the bright, rigid cortex (Fig. 94).

Three well-defined *Apertospathula* species have been described (FOISSNER et al. 2002). One of these, *A. dioplites*, has been referred to the new genus *Longispatha* by FOISSNER et al. (2005b). Five new species are added, and the reinvestigation of *Spathidium swarezewskyi* showed that it belongs to *Apertospathula* too.

Key to species (careful live observation indispensable)

1 Body ≥ 100 µm long in vivo. Extrusomes present. Six or more ciliary rows 2
– Length about 55 µm. Extrusomes lacking (observe carefully!). About six ciliary rows .... *A. inermis*
2 Body ≥ 100 µm long in vivo. Extrusomes rod-shaped and longer than 5 µm 3
– Body < 80 µm long in vivo. Extrusomes of various shapes and inconspicuous because ≤ 3 µm long 5
3 Two size types (12 µm and 7 µm) of rod-shaped extrusomes 4
– Extrusomes 5–8 µm long, rod-shaped, and slightly curved .... *A. swarezewskyi*
4 Many minute palps between dorsal brush rows .... *A. lajacola*
– Oral bulge with distinct concavity .... *A. pelobia*
5 Body < 80 µm long in vivo. 5–7, usually 6 ciliary rows 6
– Body < 80 µm long in vivo. 8–16 ciliary rows 7
6 Extrusomes about 1.5 µm long. Brush bristles about 1 µm long .... *A. armata*
– Extrusomes about 0.7 µm long. Brush bristles up to 5 µm long .... *A. similis*
7 Body clavate. About 15 ciliary rows. Brush bristles very conspicuous because up to 15 µm long (do not mix with somatic cilia!) .... *A. longiseta*
– Body usually oblong or obovate, rarely clavate. 8–12 ciliary rows. Brush bristles up to 3 µm long 8

8 Usually 9 ciliary rows. Brush bristles up to 3 μm long. Extrusomes narrowly cuneate .... *A. cuneata*
– Usually 11 ciliary rows. Brush bristles about 1 μm long. Extrusomes rod-shaped and about 2 μm long. With 2–3 μm long palpus dorsalis .... *A. verruculifera*

Apertospathula inermis FOISSNER, AGATHA & BERGER, 2002 (Fig. 95a–i, 146a, d–f; Table 38)

2002 *Apertospathula inermis* FOISSNER, AGATHA & BERGER, Denisia, 5: 318 (Type slides with protargol-impregnated specimens from type locality are deposited in the Oberösterreichische Landesmuseum in Linz, Upper Austria.).

Diagnosis: Size about 55 × 12 μm in vivo. Indistinctly clavate to very narrowly spatulate with oblique, obovate oral bulge shorter than widest trunk region by about 30%. Macronucleus oblong; single micronucleus. No extrusomes recognizable. On average 6 ciliary rows and 17 circumoral dikinetids. Dorsal brush isostichad and three-rowed, occupies 8% of body length, each row composed of 3–4 dikinetids with 1 μm long bristles.

Type locality: Sand dune near the town of St. Anthony, Idaho, USA, N 43° W112°.

Etymology: The Latin adjective *inermis* (unarmed) refers to the lacking extrusomes.

Description: Size 40–75 × 8–15 μm, usually near 55 × 12 μm in vivo; length:width ratio highly variable, that is, 2.7–7:1, on average near 5:1 in protargol preparations (Table 38). Shape usually slightly clavate because more or less distinctly inflated subapically and narrowing posteriorly; more rarely very narrowly spatulate or bluntly fusiform (Fig. 95a, d, f, h); laterally flattened indistinctly. Macronucleus in widened anterior body half, usually oblong, in five (likely post-conjugants) out of 30 specimens composed of two to five small globules, and in one cell even a tortuous strand; nucleoli globular and of ordinary size. Micronucleus attached to macronucleus, slightly ellipsoidal (Fig. 95a, c, e, g, h). Contractile vacuole in posterior body end. No extrusomes recognizable, not even with interference contrast optics, in the three populations investigated. Cortex thin and flexible. Cytoplasm colourless, contains some lipid droplets. Food not known. Swims rather rapidly by rotation about main body axis.

Cilia 7–8 μm long in vivo, loosely spaced, arranged in six to seven meridional, equidistant rows anteriorly distinctly separated from circumoral kinety. Three dorsolateral rows anteriorly differentiated to short, very inconspicuous, isostichad dorsal brush composed of paired basal bodies bearing only 1 μm long bristles appearing as bright dots in vivo: rows 1 and 3 each composed of three dikinetids, row 2 with four bristle pairs on average; we did not check whether row 3 has a monokinetidal bristle tail; one or two ordinary cilia between anterior end of individual brush rows and circumoral kinety (Fig. 95a, d–i).

Oral bulge moderately distinct, surface slightly convex to concave and obliquely truncate with dorsal portion higher than ventral; about 8 μm long and up to 3 μm high in vivo; cuneate in lateral and spatulate in frontal view because gradually merging into

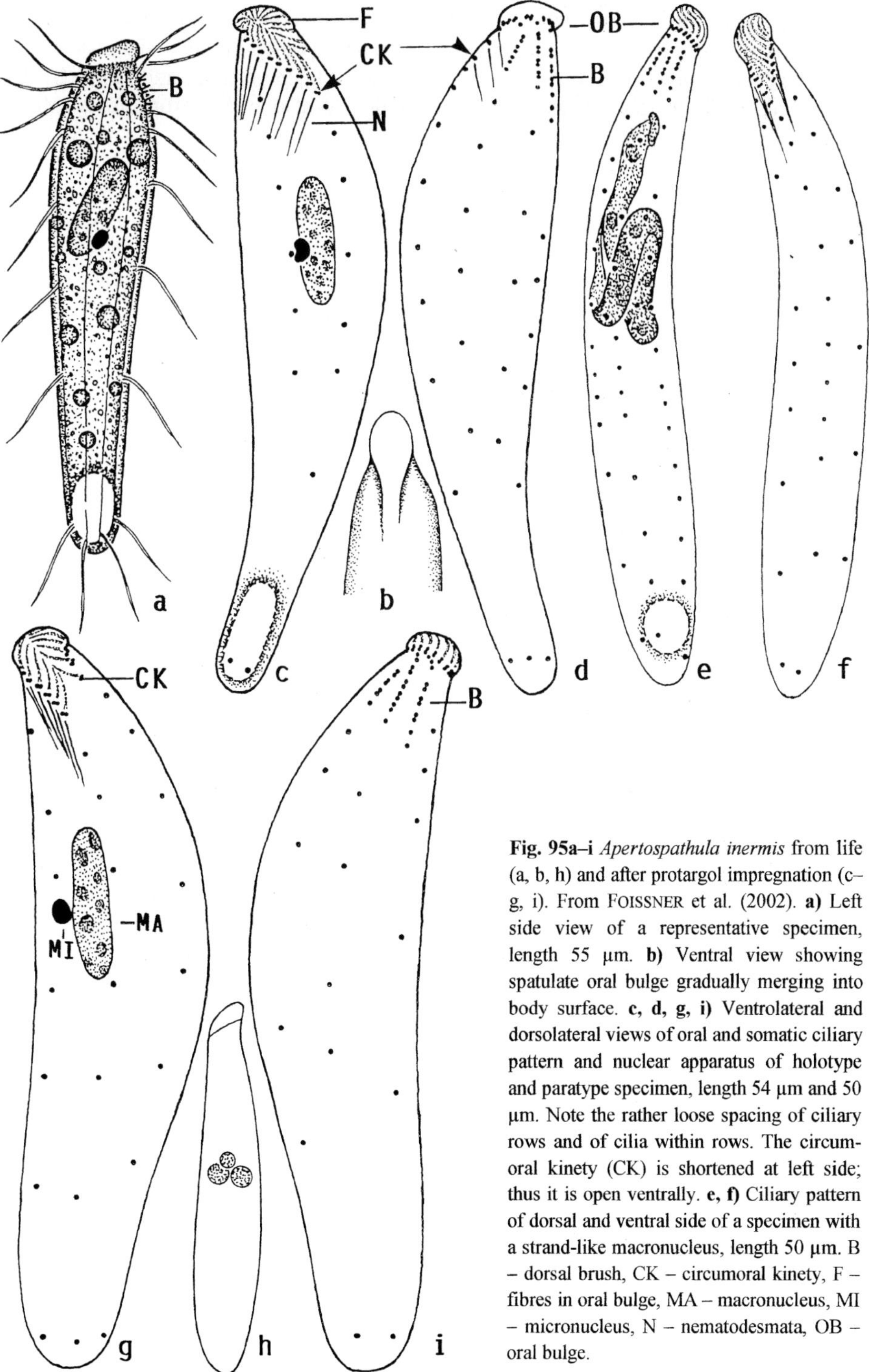

Fig. 95a–i *Apertospathula inermis* from life (a, b, h) and after protargol impregnation (c–g, i). From FOISSNER et al. (2002). **a)** Left side view of a representative specimen, length 55 µm. **b)** Ventral view showing spatulate oral bulge gradually merging into body surface. **c, d, g, i)** Ventrolateral and dorsolateral views of oral and somatic ciliary pattern and nuclear apparatus of holotype and paratype specimen, length 54 µm and 50 µm. Note the rather loose spacing of ciliary rows and of cilia within rows. The circumoral kinety (CK) is shortened at left side; thus it is open ventrally. **e, f)** Ciliary pattern of dorsal and ventral side of a specimen with a strand-like macronucleus, length 50 µm. B – dorsal brush, CK – circumoral kinety, F – fibres in oral bulge, MA – macronucleus, MI – micronucleus, N – nematodesmata, OB – oral bulge.

ventral surface. Circumoral kinety at base of oral bulge, composed of an average of 17 comparatively widely spaced dikinetids each associated with a cilium and a short nematodesma contributing to the inconspicuous oral basket; open ventrally and longer by two to four dikinetids at right than left side of oral bulge, as also evident from fibres extending into the bulge (Fig. 95a–i).

Occurrence and ecology: The type population was discovered in sieved litter (pH 7.5) from a (likely volcanic) sand dune in Idaho, USA. The Namibian population occurred in a different habitat, viz., in soil from the swampy margin of a pond. In Australia, *A. inermis* was found in a mud sample from a rock-pool at the entrance to the Sydney Harbour National Park. Recently, we found *A. inermis* in floodplain soil of Austria (FOISSNER et al. 2005a) and in a small bog pond near Constance (sample kindly provided by Dr. MARTIN KREUTZ). In the latter habitat, the cytoplasm contained many crystalline (?) inclusions with a curious shape explained in figures 146d–f. *Apertospathula inermis* was rare at all sites, at least in the non-flooded Petri dish cultures. The data indicate that it is a cosmopolitan species occurring in both terrestrial and limnetic habitats.

Table 38 Morphometric data on *Apertospathula inermis* (upper line; from FOISSNER et al. 2002), *A. armata* from USA type locality (middle line; from FOISSNER et al. 2002), and *A. armata* from Greece (lower line; original)

Characteristics[1]	$\overline{x}$	M	SD	SE	CV	Min	Max	n
Body, length	50.2	50.0	8.5	2.0	16.9	32.0	65.0	19
	51.0	52.0	7.6	2.2	14.9	38.0	66.0	12
	41.8	39.0	7.9	2.4	18.9	28.0	57.0	11
Body, width	11.0	11.0	2.2	0.5	19.8	7.0	14.0	19
	15.4	15.0	2.4	0.7	15.8	12.0	20.0	12
	8.3	8.0	1.2	0.4	14.4	7.0	11.0	11
Body length:width, ratio	4.8	4.6	1.2	0.3	24.4	2.7	7.0	19
	3.4	3.5	0.6	0.2	16.7	2.3	4.1	12
	5.1	5.1	0.9	0.3	18.1	3.5	6.9	11
Anterior body end to macronucleus, distance	13.4	12.0	4.5	1.0	33.9	7.0	24.0	19
	13.6	13.0	4.4	1.3	32.3	8.0	21.0	12
	12.2	10.0	5.0	1.5	40.9	7.0	22.0	11
Circumoral kinety to last dikinetid of brush row 1, distance	3.2	3.0	0.5	0.1	16.7	2.5	4.0	19
	4.7	5.0	0.7	0.2	14.0	4.0	6.0	12
	3.8	4.0	0.6	0.2	15.8	3.0	5.0	11
Circumoral kinety to last dikinetid of brush row 2, distance	4.0	4.0	0.6	0.1	14.6	3.0	5.0	19
	5.0	5.0	0.7	0.2	14.8	4.0	6.0	12
	4.7	5.0	0.6	0.2	13.7	4.0	6.0	11
Circumoral kinety to last dikinetid of brush row 3, distance	3.6	3.5	0.7	0.2	19.4	2.0	5.0	19
	4.8	5.0	0.8	0.2	15.9	4.0	6.0	12
	4.1	4.0	0.5	0.2	13.2	3.0	5.0	11
Oral bulge (circumoral kinety), maximum length	8.0	8.0	0.9	0.2	11.8	6.0	10.0	19
	7.5	7.0	1.0	0.3	13.3	7.0	10.0	12
	7.9	8.0	0.8	0.3	10.5	7.0	9.0	11
Macronucleus, length	11.8	11.0	2.9	0.7	24.9	7.0	18.0	16
	17.8	17.5	3.0	0.9	16.7	13.0	22.0	12
	11.1	10.0	2.0	0.6	18.2	9.0	15.0	11
Macronucleus, width	3.2	3.0	0.6	0.1	17.9	2.0	4.0	16

continued

Characteristics[1]	$\overline{x}$	M	SD	SE	CV	Min	Max	n
	3.7	4.0	0.7	0.2	17.8	3.0	5.0	12
	3.4	3.0	0.6	0.2	16.4	3.0	4.5	11
Micronucleus, length	2.4	2.0	–	–	–	2.0	3.0	13
	3.2	3.0	–	–	–	2.8	4.0	12
	2.8	3.0	0.5	0.2	18.2	2.0	4.0	11
Micronucleus, width	1.8	2.0	–	–	–	1.5	2.0	13
	2.9	3.0	–	–	–	2.5	3.0	12
	2.8	3.0	0.5	0.2	18.2	2.0	4.0	11
Ciliary rows, number	6.2	6.0	–	–	–	6.0	7.0	19
	6.3	6.0	–	–	–	6.0	7.0	12
	5.9	6.0	–	–	–	5.0	6.0	11
Ciliated kinetids in a right side ciliary row, number	8.6	8.0	1.4	0.3	16.2	6.0	12.0	19
	11.0	11.0	2.5	0.7	22.9	7.0	15.0	12
	9.2	9.0	1.8	0.6	20.0	6.0	13.0	11
Dorsal brush rows, number	3.0	3.0	0.0	0.0	0.0	3.0	3.0	19
	3.0	3.0	0.0	0.0	0.0	3.0	3.0	12
	3.0	3.0	0.0	0.0	0.0	3.0	3.0	11
Dikinetids in brush row 1, number	2.9	3.0	–	–	–	2.0	3.0	19
	3.8	4.0	0.6	0.2	15.1	3.0	5.0	12
	3.3	3.0	0.5	0.1	14.3	3.0	4.0	11
Dikinetids in brush row 2, number	3.8	4.0	–	–	–	2.0	4.0	19
	4.8	5.0	0.9	0.3	18.2	4.0	7.0	12
	4.2	4.0	0.4	0.1	9.7	4.0	5.0	11
Dikinetids in brush row 3, number	3.0	3.0	–	–	–	2.0	4.0	19
	3.9	4.0	0.7	0.2	17.1	3.0	5.0	12
	3.4	3.0	0.5	0.2	15.0	3.0	4.0	11
Circumoral kinetids, number	17.0	17.0	1.2	0.3	7.1	15.0	20.0	19
	21.3	22.0	2.8	0.8	13.1	17.0	26.0	12
	18.0	18.0	1.4	0.4	7.9	15.0	20.0	11

[1] Data based on mounted, protargol-impregnated (FOISSNER's method), and randomly selected specimens from non-flooded Petri dish cultures. Measurements in µm. CV – coefficient of variation in %, M – median, Max – maximum, Min – minimum, n – number of individuals investigated, SD – standard deviation, SE – standard error of arithmetic mean, $\overline{x}$ – arithmetic mean.

Remarks: *Apertospathula inermis* is a really inconspicuous ciliate because it is small and lacks distinct features. The lack of light-microscopically recognizable extrusomes was carefully checked in three populations and is thus a main feature of the species. Several *Spathidium* species also lack extrusomes, and even some rather large *Arcuospathidium* species, viz., *A. vermiforme* and *A. cooperi*. Thus, the absence in *A. inermis* is not too surprising.

Apertospathula inermis resembles several small *Spathidium* and *Spathidiodes* species, all rather superficially described. The most similar species is probably *Spathidium microstomum* VUXANOVICI, 1962c, which, however, has an only 3 µm wide, button-shaped oral bulge.

Apertospathula armata FOISSNER, AGATHA & BERGER, 2002 (Fig. 96a–i, 146b, c, g; Table 38)

2002 *Apertospathula armata* FOISSNER, AGATHA & BERGER, Denisia, 5: 318 (Type and voucher slides with protargol-impregnated specimens from Israel and Greece are deposited in the Oberösterreichische Landesmuseum in Linz, Upper Austria.).

Diagnosis: Size about 60 × 15 µm in vivo. Narrowly spatulate with oblique, obovate oral bulge slightly (~ 10%) to distinctly (~ 30%) shorter than widest trunk region. Macronucleus oblong; single micronucleus. Extrusomes ellipsoidal, 1.5 µm long. On average 6 ciliary rows and 22 circumoral dikinetids. Dorsal brush isostichad and three-rowed, occupies 10% of body length, each row composed of 4–5 dikinetids with 1 µm long bristles.

Type locality: Loamy wheat field soil about 10 km south of Nazareth, Israel, N32°30' E 35°.

Etymology: The Latin adjective *armatus* (armed) refers to the extrusomes.

Description: Size 45–75 × 10–20 µm, usually near 60 × 15 µm in vivo; length:width ratio moderately variable, on average 3.4:1 in protargol preparations (Table 38). Shape narrowly spatulate or fusiform, rather similar to that of *A. inermis*, posterior third frequently wrinkled; oral area slightly flattened laterally (Fig. 96a, d–f, h, i). Macronucleus in anterior body half, oblong (3:1) to cylindroidal (up to 7:1); nucleoli globular to lobate. Micronucleus attached to macronucleus, 3–4 µm across and thus rather conspicuous, even in vivo (Fig. 96a, e, g). Contractile vacuole in posterior end. Oral bulge extrusomes ellipsoidal to indistinctly fusiform, about 1.5 µm long in vivo. Cytoplasm colourless, packed with bright globules 1 µm across and many apparently empty vacuoles. Swims rather rapidly showing great flexibility when squeezed by soil particles.

Somatic and oral ciliary pattern as described in *A. inermis*, with slight morphometric differences shown in Table 38.

New observations on a population from Greece (Fig. 96j–t, 146b, c, g; Table 38): This population differs from the type in two important features, viz., the extrusomes (rod-shaped vs. ellipsoidal) and the habitat (arable soil vs. highly saline coastal soil). Thus, it could be a different, new species. Unfortunately, the exact shape of the extrusomes is difficult to recognize because they are only about 1 µm long. Thus, we assign the Greek population to *A. armata* at the present state of knowledge.

Size 30–80 × 8–12 µm in vivo, usually about 45 × 10 µm, that is, considerably smaller than specimens from type population. Shape as described in Israel specimens. Macronucleus in various patterns due to many post-conjugants: a single globular to oblong nodule in about 45% of specimens (out of 31 cells); two globular nodules in about 45% of specimens; and three nodules, of which at least one shows distinct signs of disintegration in 10% of specimens (Fig. 96l, o–r, 146b, c; Table 38). Extrusomes form a row in oral bulge, attached to frontal bulge cortex, about 1 µm long and likely rod-shaped. Cortical granules not recognizable. Cytoplasm packed with minute granules and lipid droplets 2–5 µm across. Swims slowly by rotation about main body axis.

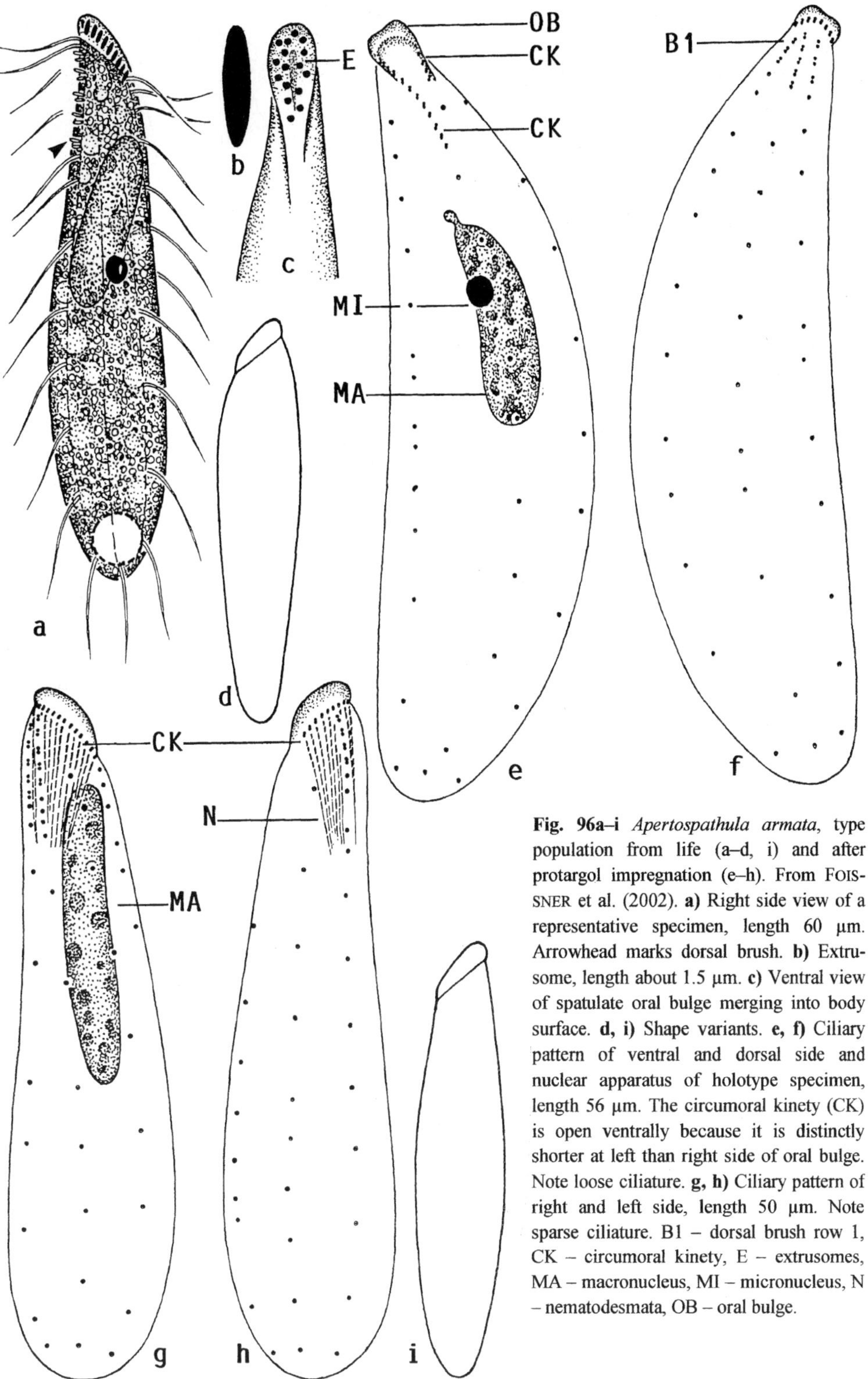

Fig. 96a–i *Apertospathula armata*, type population from life (a–d, i) and after protargol impregnation (e–h). From FOISSNER et al. (2002). **a)** Right side view of a representative specimen, length 60 µm. Arrowhead marks dorsal brush. **b)** Extrusome, length about 1.5 µm. **c)** Ventral view of spatulate oral bulge merging into body surface. **d, i)** Shape variants. **e, f)** Ciliary pattern of ventral and dorsal side and nuclear apparatus of holotype specimen, length 56 µm. The circumoral kinety (CK) is open ventrally because it is distinctly shorter at left than right side of oral bulge. Note loose ciliature. **g, h)** Ciliary pattern of right and left side, length 50 µm. Note sparse ciliature. B1 – dorsal brush row 1, CK – circumoral kinety, E – extrusomes, MA – macronucleus, MI – micronucleus, N – nematodesmata, OB – oral bulge.

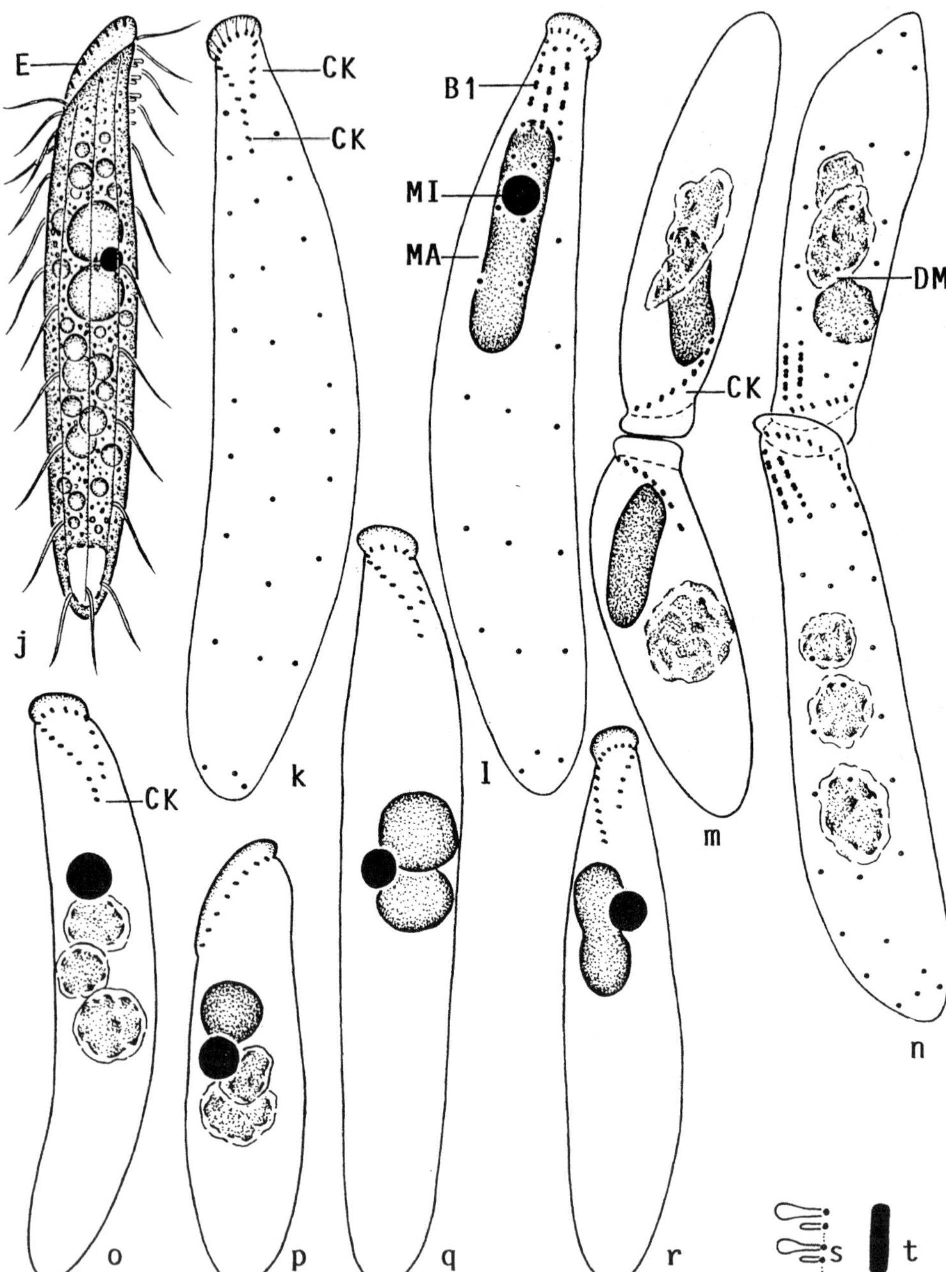

Fig. 96j–t *Apertospathula armata*, original figures of Greek specimens from life (j, s, t) and after protargol impregnation (k–r). **j)** Left side view of a postconjugant with two macronucleus nodules, length 60 µm. **k, l)** Ciliary pattern of ventral and dorsal side of an ordinary specimen, 50 µm. **m, n)** Conjugating specimens, 80 µm, 65 µm. **o–r)** Nuclear figures in postconjugants, length 38 µm, 28 µm, 48 µm, 35 µm. **s)** Dorsal brush, bristles up to 1 µm long. **t)** Extrusome, length about 1 µm, B1 – dorsal brush row 1, CK – circumoral kinety, DM – disintegrating macronucleus, E – extrusomes, MA – macronucleus, MI – micronucleus.

Somatic and oral ciliary pattern as in type and *A. inermis*. Dorsal brush inconspicuous, row 3 has a monokinetidal tail composed of only two rod-shaped bristles; anterior bristle of pairs clavate and about 1 µm long, posterior rod-shaped and approximately 0.5 µm long. Oral bulge about 3 µm high dorsally and thus fairly conspicuous (Fig. 96j–l, 146b, c; Table 38).

Conjugating specimens attach with the oral bulge and form a long, slightly curved rod. The ciliary pattern does not change (Fig. 96m, n, 146g).

Occurrence and ecology: To date found at type locality and in Greece, where it occurred in highly saline, circumneutral (pH 6.9) coastal soil near the town of Nauplia.

Remarks (see also *A. inermis*): *A. armata* is highly similar to *A. inermis*, differing mainly by the possession of extrusomes which are, however, minute and thus easily overlooked. Shape and morphometrics are almost identical or at least overlapping (Table 38).

Apertospathula similis nov. spec. (Fig. 97a–h; Table 39)

Diagnosis: Size about 60 × 10 µm in vivo. Narrowly spatulate with oblique, obovate oral bulge shorter than widest trunk region by about 40%. Macronucleus ellipsoidal; single micronucleus. Extrusomes oblong and only about 0.7 µm long. On average 6 ciliary rows and 16 circumoral dikinetids. Dorsal brush isostichad and three-rowed, occupies 11% of body length, each row composed of an average of 4–5 dikinetids with up to 5 µm long bristles; monokinetidal tail of row 3 composed of two bristles.

Type locality: Soil from a green region ("green river bed") of the Chobe River near the Muchenje Safari Lodge, Botswana, E 24°40' S18°.

Etymology: The Latin adjective *similis* (similar) refers to the similarity with *Apertospathula armata* and *A. inermis*.

Description: Size 45–75 × 7–15 µm in vivo, usually about 60 × 10 µm, as calculated from some in vivo measurements and the morphometric data; length:width ratio highly variable, that is, 4.3–9.6 : 1 in prepared specimens, on average near 6 : 1 both in vivo and protargol preparations (Table 39). Usually narrowly spatulate or cylindroidal, widest in mid-body, slightly flattened laterally; anterior end (oral bulge) oblique, posterior broadly rounded and frequently wrinkled or inflated due to the contractile vacuole contained (Fig. 97a, d, h). Macronucleus in middle body third, ellipsoidal; nucleoli about 1 µm across. Micronucleus in indentation of macronucleus, about 2 µm across in vivo. Contractile vacuole in rear body end, 1–2 excretory pores in pole area. Oral bulge extrusomes difficult to recognize because only about 0.7 µm long in vivo, likely oblong or broadly fusiform, do not impregnate with the protargol method used. Cortex flexible, contains 3–5 rows of minute (~ 0.2 µm), loosely spaced, colourless granules between each two ciliary rows. Cytoplasm colourless, oral body region with conspicuous accumulation of minute granules, like in *Homalozoon vermiculare* (LEIPE et al. 1992);

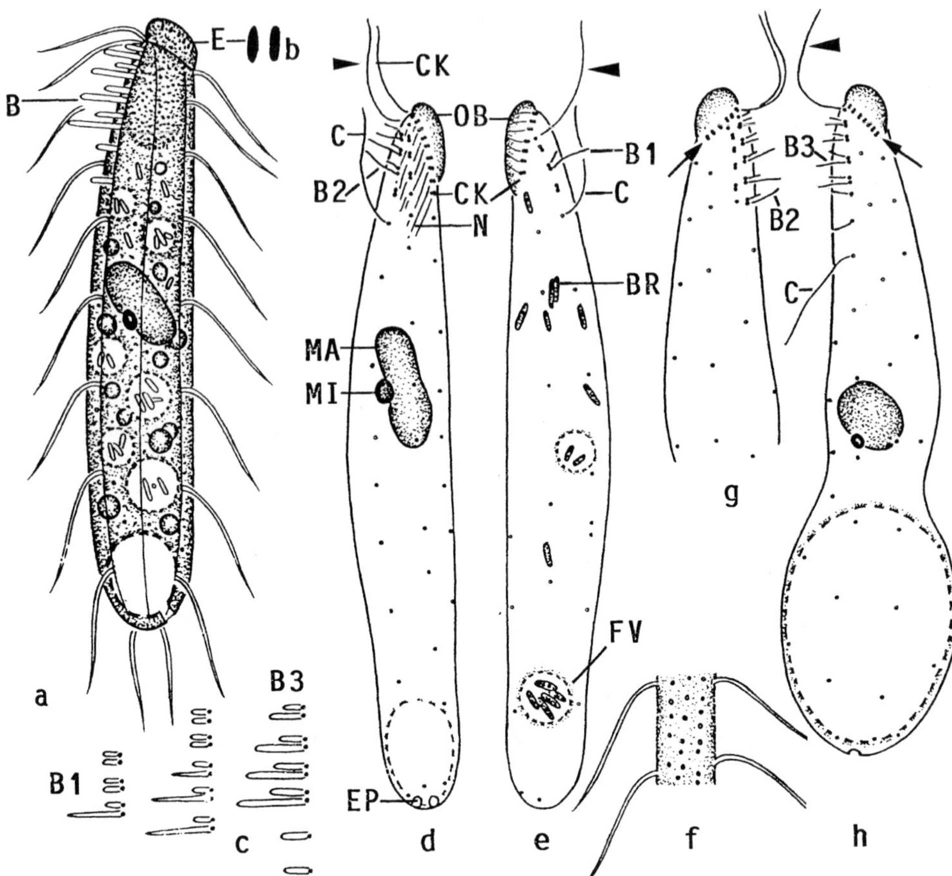

Fig. 97a–h *Apertospathula similis* nov. spec. from life (a–c, f) and after protargol impregnation (d, e, g, h). **a)** Right side view of a representative specimen, length 60 μm. Note the accumulation of minute granules in the oral area. **b)** The extrusomes are minute, that is, about 0.7 μm long and thus very difficult to recognize. **c)** Dorsal brush, whose up to 5 μm long bristles define this species. **d, e)** Dorsolateral and ventrolateral view of ciliary pattern and nuclear apparatus of holotype specimen, length 57 μm. Arrowheads mark the single cilium between circumoral kinety and dorsal brush rows. The dorsal bristles can be seen in the preparation, and the pattern matches that observed in vivo (c). **f)** Surface view showing cortical granulation. **g, h)** Left and right side view of a specimen with filled contractile vacuole deforming the posterior body region, length 57 μm. Arrows mark left and right end of the circumoral kinety which is open ventrally. Arrowhead denotes the single cilium between circumoral kinety and dorsal brush rows. The specific pattern of the brush bristles can be seen in the preparation. B(1-3) – dorsal brush (rows), C – ordinary somatic cilia, CK – circumoral kinety, CV – contractile vacuole, E – extrusomes, EP – excretory pores of the contractile vacuole, MA – macronucleus, MI – micronucleus, N – nematodesmata, OB – oral bulge.

contains some lipid droplets up to 3 μm across; many food vacuoles with bacteria and, occasionally, heterotrophic flagellates; and numerous darkly-impregnated bacterial rods similar to those found in the food vacuoles (Fig. 97a, e). Movement without peculiarities.

Cilia about 10 μm long in vivo, arranged as described in *A. inermis* and *A. armata*, with inconspicuous morphometric differences shown in Table 39. Dorsal brush also similar to that of the species mentioned before, viz., three-rowed, isostichad and short with an ordinary cilium each between brush rows and circumoral kinety. Length of brush bristles, however different (Fig. 97a, c, d, e, g, h): about 1 µm in anterior two pairs of rows 1 and 2, while posterior bristle becomes acicular and gradually increases in length to 3–4 µm posteriorly; both bristles of row 3 cylindroidal and increasing in length from about 1 µm anteriorly to 4–5 µm posteriorly, anterior bristles about half as long as posterior ones; monokinetidal tail of row 3 invariably composed of two circa 2 µm long bristles. Bristle pattern constant and seen both in vivo (Fig. 97c) and protargol-impregnated specimens (Fig. 97d, e, g, h; Table 39).

Oral bulge distinct because knob-like projecting, in vivo about 6 µm long and 3 µm high, slanted by 30°–50°, slightly convex and obovate. Circumoral kinety composed of an average of 16 rather widely spaced dikinetids each associated with a faintly impregnated basket rod and a fibre extending to bulge centre. Ventral opening of circumoral kinety minute because right branch longer than left by only 1–2 dikinetids (Fig. 97a, d, e, g, h; Table 39).

Occurrence and ecology: As yet found only at type locality, where it was very rare in the non-flooded Petri dish culture, that is, only nine specimens were found in eight protargol slides. This indicates that *A. similis* is a limnetic species. Green river beds occur in the tropics during the dry season. The Chobe River was dry since two months when the sample was taken from the upper 5 cm of the bed, which was green by grasses and herbs. The soil and water are highly fertile because both, the wet and the dry river, are densely populated by hippos and elephants whose fecal masses are scattered over the bed. There was few litter, but the dark, slightly acidic (pH 5.1 in water) and very humic soil contained many grass roots. When rewetted, the soil becomes a swampy, fine-grained mass difficult to investigate.

Remarks: Few specimens were found (see above). They match *A. inermis* and *A. armata* both in morphology and ciliary pattern. However, the long brush bristles (up to 5 µm vs. 1 µm) clearly distinguish *A. similis* from these species. Furthermore, *A. inermis* lacks extrusomes.

Table 39 Morphometric data on *Apertospathula similis* (upper line) and *A. longiseta* (lower line)

Characteristics[1]	$\overline{x}$	M	SD	SE	CV	Min	Max	n
Body, length	54.7	55.0	7.8	2.6	14.2	43.0	67.0	9
	60.6	62.0	6.1	1.7	10.1	50.0	69.0	13
Body, width	9.6	10.0	2.1	0.7	22.3	6.0	13.0	9
	16.4	16.0	2.5	0.7	15.5	13.0	22.0	13
Body length:width, ratio	6.0	6.1	1.7	0.6	28.7	4.3	9.6	9
	3.8	3.7	0.5	0.2	14.6	3.0	4.8	13
Oral bulge (circumoral kinety), length	5.8	6.0	0.8	0.3	14.4	5.0	7.0	9
	5.8	6.0	0.7	0.3	12.3	5.0	7.0	8
Oral bulge, height	2.3	2.0	–	–	–	2.0	3.0	9
	1.6	1.5	–	–	–	1.5	2.0	13
Oral bulge length:body width, ratio	0.6	0.6	–	–	–	0.5	1.0	9
	0.4	0.4	–	–	–	0.3	0.4	8

continued

Characteristics[1]	$\overline{x}$	M	SD	SE	CV	Min	Max	n
Circumoral kinety to last dikinetid of brush row 1, distance	5.2	5.0	0.7	0.2	12.8	4.0	6.0	9
	11.1	11.0	1.3	0.4	11.9	9.0	13.0	13
Circumoral kinety to last dikinetid of brush row 2, distance	6.1	6.0	0.9	0.3	15.2	5.0	7.0	9
	12.0	12.0	1.4	0.4	11.3	10.0	14.0	13
Circumoral kinety to last dikinetid of brush row 3, distance	5.0	5.0	0.7	0.2	14.1	4.0	6.0	9
	6.2	6.0	0.9	0.2	14.6	5.0	8.0	13
Anterior body end to macronucleus, distance	18.0	18.0	4.5	1.5	25.0	10.0	24.0	9
	15.7	15.0	4.0	1.1	25.4	9.0	22.0	13
Macronucleus, length	8.4	8.0	1.6	0.5	18.8	6.0	11.0	9
	16.8	16.0	2.7	0.8	16.4	13.0	23.0	13
Macronucleus, width	4.0	4.0	0.6	0.2	14.0	3.0	5.0	9
	5.8	6.0	0.6	0.2	10.4	5.0	7.0	13
Macronucleus, number	1.0	1.0	0.0	0.0	0.0	1.0	1.0	9
	1.0	1.0	0.0	0.0	0.0	1.0	1.0	13
Micronucleus, across	1.7	1.5	–	–	–	1.5	2.0	8
	4.4	4.0	–	–	–	3.5	5.0	12
Micronucleus, number	1.0	1.0	0.0	0.0	0.0	1.0	1.0	9
	1.0	1.0	0.0	0.0	0.0	1.0	1.0	13
Circumoral dikinetids, number	15.8	16.0	2.0	0.7	12.6	13.0	19.0	9
	19.1	19.0	2.3	0.9	11.8	16.0	23.0	7
Ciliary rows, number	6.0	6.0	0.7	0.2	11.8	5.0	7.0	9
	14.5	15.0	1.5	0.4	10.4	12.0	16.0	13
Basal bodies in a right side ciliary row, number	7.9	8.0	1.3	0.4	16.1	5.0	9.0	9
	18.2	18.0	2.8	0.8	15.4	12.0	22.0	13
Dorsal brush rows, number	3.0	3.0	0.0	0.0	0.0	3.0	3.0	9
	3.0	3.0	0.0	0.0	0.0	3.0	3.0	13
Dikinetids in brush row 1, number	4.0	4.0	0.5	0.2	12.5	3.0	5.0	9
	10.3	10.0	0.9	0.2	8.3	9.0	12.0	13
Dikinetids in brush row 2, number	4.6	5.0	–	–	–	4.0	5.0	9
	11.6	12.0	0.9	0.2	7.5	10.0	13.0	13
Dikinetids in brush row 3, number	3.7	4.0	–	–	–	3.0	4.0	9
	5.7	6.0	–	–	–	5.0	6.0	13

[1] Data based on mounted and protargol-impregnated (FOISSNER's method) specimens from non-flooded Petri dish cultures. Measurements in µm. CV – coefficient of variation in %, M – median, Max – maximum, Min – minimum, n – number of individuals investigated, SD – standard deviation, SE – standard error of arithmetic mean, $\overline{x}$ – arithmetic mean.

Apertospathula longiseta **nov. spec.** (Fig. 98a–k; Table 39)

Diagnosis: Size about 70 × 20 µm in vivo. Clavate with oblique, obovate oral bulge about half as long as widest trunk region. Macronucleus oblong; single micronucleus. Extrusomes ovate, about 1.2 × 0.8 µm. On average 15 ciliary rows and 19 circumoral dikinetids. Dorsal brush three-rowed and heterostichad, occupies 20% of body length, conspicuous due to the up to 15 µm long bristles. Brush row 1 composed of an average of 10 dikinetids with up to 10 µm long bristles; row 2 composed of 12 dikinetids with up to 15 µm long bristles; row 3 composed of 6 dikinetids with up to 8 µm long bristles and 4 monokinetidal, 3 µm long bristles.

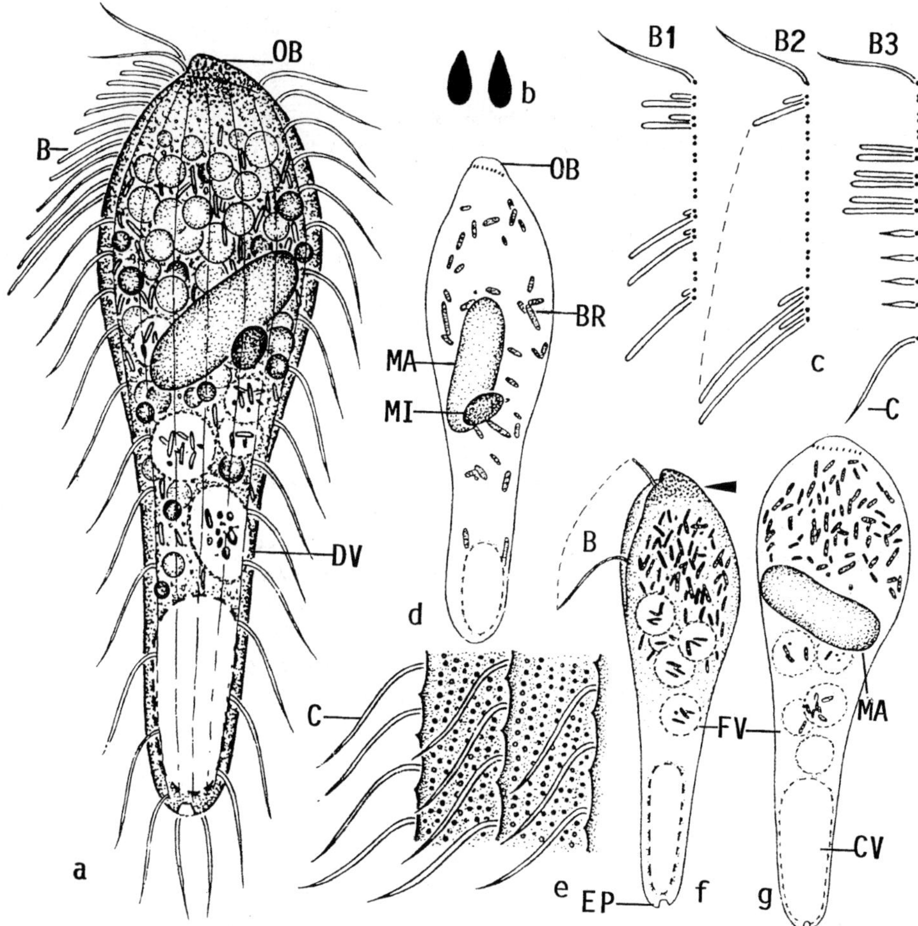

Fig. 98a–g *Apertospathula longiseta* nov. spec. from life (a–c, f) and after protargol impregnation (d, g). **a)** Right side view of a representative specimen, length 70 µm. **b)** Oral bulge extrusomes, 1.2 × 0.4 µm. **c)** Dorsal brush (see text for explanation). **d, f, g)** Shape variability and symbiotic (?) bacteria. Arrowhead in (f) marks an accumulation of minute granules, which faintly impregnate with protargol, in the oral area; length about 60 µm. **e)** Surface view showing cortical granulation and crenellation along the ciliary rows. B(1-3) – dorsal brush rows, BR – bacterial rods, C – somatic cilium, CV – contractile vacuole, DV – defecation vacuole, EP – excretory pore of contractile vacuole, FV – food vacuoles, MA – macronucleus, MI – micronucleus, OB – oral bulge.

Type locality: Soil from Zambezi floodplain about 1.5 km upstream the Victoria Falls, Botswana, E 25°50' S 18°4'.

Etymology: Composed of the Latin adjective *longa* (long) and the Latin noun *seta* (bristle), referring to the conspicuous brush bristles, a main feature of the species.

Description: Size 55–85 × 15–25 µm in vivo, usually about 70 × 20 µm, as calculated from some in vivo measurements and the morphometric data; length:width ratio

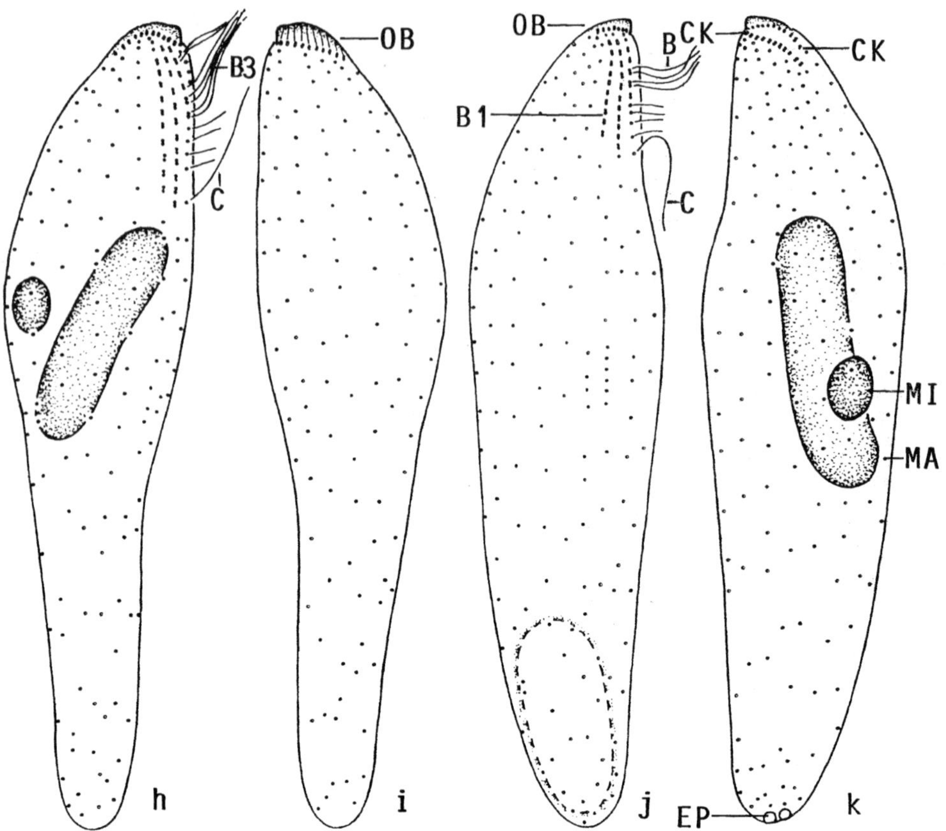

Fig. 98h–k *Apertospathula longiseta* nov. spec., ciliary pattern and nuclear apparatus after protargol impregnation. **h, i)** Left and right side view of holotype specimen, length 66 µm. The long bristles of the dorsal brush impregnated faintly and are thus recognizable in the preparation. **j, k)** Dorsolateral and ventrolateral view of paratype specimen, length 67 µm. The 8 µm long bristles of brush row 3 are faintly impregnated. The ciliature is distinctly loosened in posterior body half. B (1-3) – dorsal brush (rows), C – ordinary somatic cilia, CK – circumoral kinety, EP – excretory pores of the contractile vacuole, MA – macronucleus, MI – micronucleus, OB – oral bulge.

moderately variable, viz., 3.0–4.8:1 in impregnated specimens, on average near 3.5:1 both in vivo and protargol preparations (Table 39). Shape conspicuous because usually rather distinctly clavate, that is, widest in anterior body half; anterior end (oral bulge) slightly slanted, posterior rounded and frequently wrinkled due to the contractile vacuole contained; laterally slightly flattened (Fig. 98a, d, f–h). Macronucleus usually in anterior body half, oblong, contains nucleoli about 1 µm across. Micronucleus ellipsoidal, rather conspicuous because up to 5 µm across. Contractile vacuole large, occupies posterior quarter, with one, rarely two excretory pores in pole area. Oral bulge extrusomes ovate, about 1.2 × 0.8 µm in vivo, do not impregnate with the protargol method used. Cortex flexible, in vivo distinctly furrowed along ciliary rows and crenellated where cilia emerge, contains many rows of colourless granules about 0.3 µm across between two kineties each. Cytoplasm colourless and studded with the following inclusions, mainly in widened anterior half (Fig. 98a, d, f, g): (i) many pale globules 3–6 µm across; (ii) some highly refractive lipid droplets up to 4 µm across; (iii) some 3–5 µm-sized food vacuoles with bacteria and heterotrophic flagellates; (iv) one or two large defecation vacuoles with granular contents; (v) numerous 3–4 µm long rods, likely symbiotic bacteria, impregnating black with protargol; and (vi) an accumulation of minute, refractive granules underneath the oral bulge, similar to those found in *A. similis* and *Homalozoon vermiculare* (LEIPE et al. 1992). Swims rather rapidly rotating about main body axis.

Cilia 10–12 µm long in vivo, distances of cilia within rows gradually increase from about 1 µm anteriorly to 10 µm posteriorly, arranged in an average of 15 meridional, ordinarily and equidistantly spaced rows anteriorly separate from circumoral kinety; occasionally shortened or broken rows (Fig. 98a, h–k). Three dorsolateral rows anteriorly differentiated to very conspicuous, heterostichad dorsal brush with soft, rod-shaped bristles up to 15 µm long in vivo; all rows commence with one to two ordinary cilia anteriorly and continue as somatic kineties posteriorly (Fig. 98a, c, f, h, j; Table 39). Brush row 1 composed of an average of 10 dikinetids with anterior bristle of pairs decreasing in length from about 3 µm anteriorly to 1 µm posteriorly, posterior bristle of pairs increases in length from about 6 µm anteriorly to 10 µm posteriorly. Brush row 2 composed of an average of 12 dikinetids with anterior bristle of pairs about 3 µm long, while length of posterior bristle increases from 7 µm anteriorly to 15 µm posteriorly. Brush row 3 composed of an average of 6 dikinetids with bristles increasing in length from about 5 µm anteriorly to 8 µm posteriorly; posterior tail usually composed of four, 3 µm long, acicular bristles.

Oral bulge difficult to recognize both in vivo and preparations because only 2–3 µm high and gradually merging into ventral surface, in lateral view slightly convex with dorsal end forming right angles with body proper. Circumoral kinety composed of an average of 19 ordinarily spaced dikinetids each associated with a cilium and a fibre extending to bulge centre; ventral opening inconspicuous because right branch longer than the left by only 1–2 dikinetids. Nematodesmata not recognizable (Fig. 98a, h–k; Table 39).

Occurrence and ecology: As yet found only at type locality, where it was very rare in the non-flooded Petri dish culture, that is, only 14 specimens and an early divider were found in six protargol slides. The loamy, sandy soil, taken in the Mopane zone about 30 m off the river bank, was very humic, contained many plant roots, and had pH 6 in water.

Remarks: The generic classification of this population is difficult because the oral area is small and thus difficult to analyse. However, the ventral opening in the circumoral kinety was clearly seen in two specimens (Fig. 98j, k).

The up to 15 µm long brush bristles distinguish *A. longiseta* from all described congeners, whose bristles are not longer than 5 µm. Such long brush bristles are uncommon, but occur, for instance, also in *Arcuospathidium lorjeae* and *A. namibiense.*

***Apertospathula cuneata* nov. spec.** (Fig. 99a–m; Table 40)

Diagnosis: Size about 50 × 15 μm in vivo. Narrowly spatulate to clavate with oblique, obovate oral bulge slightly shorter than widest trunk region. Macronucleus broadly ellipsoidal; single micronucleus. Extrusomes narrowly ovate or cuneate, about 2.5 μm long. On average 9 loosely ciliated kineties and 30 circumoral dikinetids. Dorsal brush heterostichad and three-rowed, occupies 18% of body length, bristles up to 3 μm long: on average seven dikinetids in row 1, eight in row 2, and four in row 3 followed by five monokinetidal bristles.

Type locality: Upper soil layer of a small swamp in the surroundings of the town of Eubenangee, south of the town of Cairns, Australia, E145° S17°.

Etymology: The Latin adjective *cuneata* refers to the shape of the extrusomes, which makes the species easily recognizable.

Description: Size 40–70 × 10–20 μm in vivo, usually about 50 × 15 μm; length:width ratio near 3.5:1 both in vivo and protargol preparations (Table 40). Narrowly spatulate to clavate with oblique anterior (oral) end, rarely slenderly bursiform; clavate shape caused by subapical, dorsal hump recognizable also in some protargol-impregnated specimens; flattened only in oral region (Fig. 99a, c, e, f, j). Macronucleus in or slightly above mid-body, globular to ellipsoidal (2:1), on average 7.7 × 5.8 μm in protargol preparations, rarely C-shaped or in two small nodules; nucleoli distinct. Micronucleus attached to macronucleus, about 2 μm across in vivo, rarely impregnates with the protargol method used (Fig. 99a, c, e, g, h; Table 40). Contractile vacuole in rear end, invariably with two excretory pores in pole area (Fig. 99a, c, d, h). Extrusomes scattered in oral bulge, narrowly ovate or cuneate, inconspicuous because only 2.5 μm long (Fig. 99a, b, i). Cortex flexible and studded with about 1 × 0.5 μm-sized granules forming a circa 1 μm thick, viscous layer and distinct rows between two kineties each (Fig. 99k, l). Cytoplasm colourless, contains few to many highly refractive, crystal-like inclusions making anterior body half dark at low magnification; individual inclusions recognizable also in protargol-impregnated specimens, cubic to oblong, sometimes ring-shaped, 3–4 × 2–3 μm in size (Fig. 99a, k); impregnated cells contain some darkly stained, 2–3 μm long, methanogenic (?) bacteria (Fig. 99e). Food not known. Swims rapidly by rotation about main body axis with anterior end swinging wider than posterior, producing conspicuous, conical movement figures (Fig. 99m).

Cilia about 12 μm long in vivo, circa 5 μm apart and thus loosely spaced, arranged in an average of nine meridional, ordinarily (~ 5 μm) and equidistantly spaced rows distinctly separate from circumoral kinety and hardly curved dorsally at anterior end. Anterior sixth of three dorsal ciliary rows differentiated to inconspicuous, heterostichad brush with bristles up to 3 μm long in vivo. Monokinetidal tail of row 3 short, that is, composed of about five bristles usually impregnating with the protargol method used; one out of 20 specimens analysed has some ordinary cilia at anterior end of brush rows (Fig. 99a, c–e, g, h; Table 40).

Oral bulge rather distinct because up to 4 μm high and 10 μm wide in vivo; length highly variable, on average slightly shorter than widest trunk region; oblique with slightly convex surface; conspicuously spatulate with right handle margin longer and

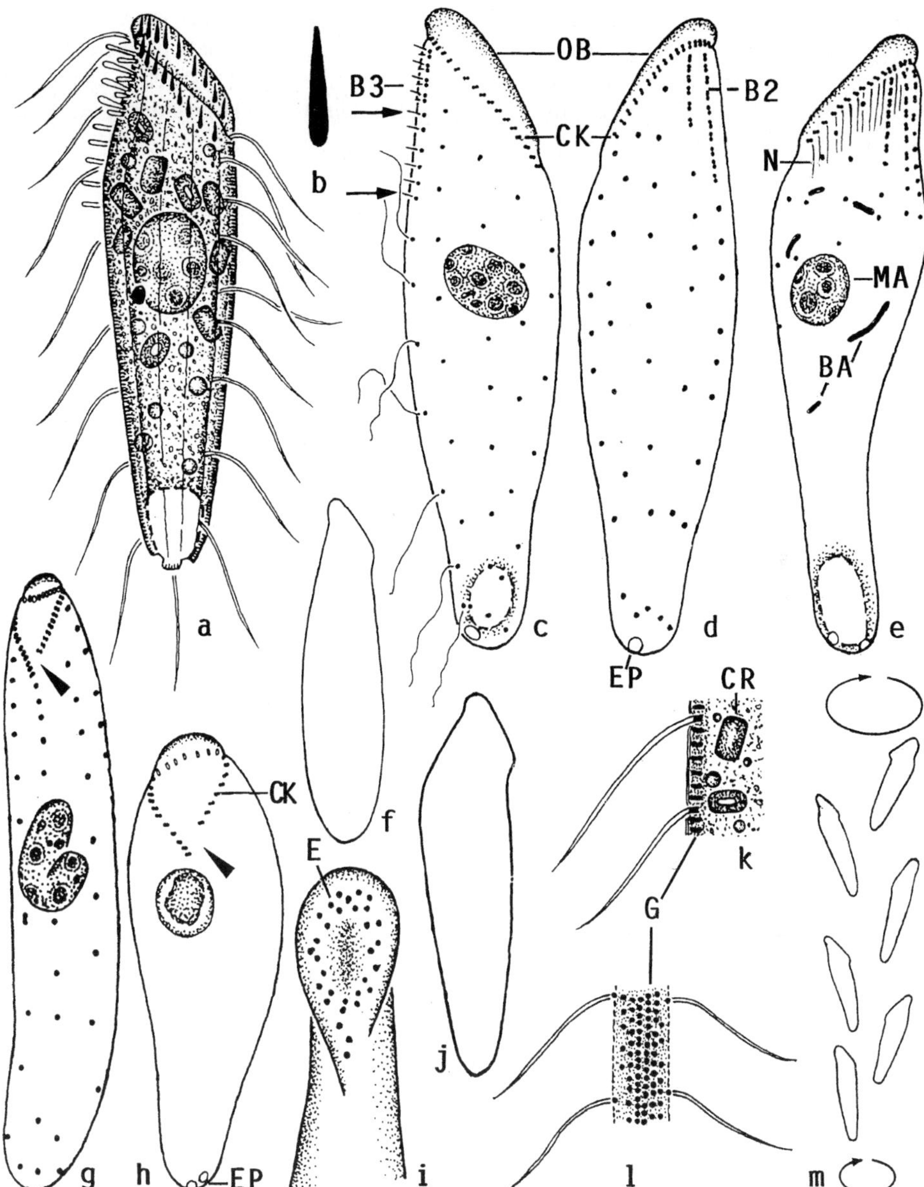

Fig. 99a–m *Apertospathula cuneata* nov. spec. from life (a, b, f, i–m) and after protargol impregnation (c–e, g, h). **a)** Right side view of a representative specimen with crystal-like inclusions, length 50 µm. **b)** Oral bulge extrusome, length 2.5 µm. **c, d)** Right and left side view of ciliary pattern and macronucleus of holotype specimen, length 57 µm. Arrows mark tail of brush row 3. **e)** Left side ciliary pattern of a clavate specimen, length 54 µm. **f, j)** Shape variants. **g, h)** Ventral views showing the ventrally opened circumoral kinety (arrowheads). **i)** Frontal view of oral bulge in vivo. **k, l)** Optical section and surface view of cortex. **m)** When swimming, the anterior end swings wider than the posterior, producing conical swimming figures. B2, 3 – dorsal brush rows, BR – bacteria, CK – circumoral kinety, CR – crystals (?), E – extrusomes, EP – excretory pores, G – cortical granules, MA – macronucleus, N – nematodesmata, OB – oral bulge.

more distinct than left, spoon rather distinctly concave making extrusome-containing bulge margin collar-like (Fig. 99a, c, d, i; Table 40). Circumoral kinety obovate, ventral opening distinct, composed of an average of 30 ordinarily spaced dikinetids each associated with a cilium and a fine, faintly impregnated nematodesma producing an inconspicuous oral basket (Fig. 99c–e, g, h; Table 40).

Occurrence and ecology: As yet found only at type locality. Rare in the non-flooded Petri dish culture, suggesting that it is a limnetic, probably microaerobic species with methanogenic (?) bacteria (Fig. 99e). In Africa occurs a similar species (*A. similis*) in the same type of habitat.

Remarks: *Apertospathula cuneata* differs from most similar-sized and shaped congeners by the minute, narrowly cuneate or ovate extrusomes; the African *A. similis* has similar extrusomes, but is oblong and has a distinctly shorter oral bulge and circumoral kinety comprising only 16 (vs. 30) dikinetids. There are several *Spathidium* s.l. species which look very similar at first glance (Vol. II): *S. deforme* (longer, rod-shaped extrusomes; more ciliary rows), *S. striatum* (oral bulge minute, about 16 ciliary rows), and *Spathidiodes rigida* (cortex rigid; see family Arcuospathidiidae).

Table 40 Morphometric data on *Apertospathula verruculifera* (upper line; from FOISSNER et al. 2005b) and *Apertospathula cuneata* (lower line)

Characteristics[1]	$\overline{x}$	M	SD	SE	CV	Min	Max	n
Body, length	40.4	40.0	4.3	0.9	10.6	32.0	48.0	21
	47.4	46.0	6.4	1.5	13.5	38.0	62.0	19
Body, width	19.8	20.0	2.6	0.6	13.3	15.0	24.0	21
	14.1	14.0	2.4	0.5	16.7	10.0	18.0	19
Body length:width, ratio	2.1	2.0	0.2	0.1	10.7	1.7	2.5	21
	3.5	3.5	0.7	0.2	20.2	2.3	4.8	19
Oral bulge (circumoral kinety), maximum length	7.1	7.0	1.0	0.3	14.4	6.0	9.0	16
	11.9	12.0	2.1	0.5	17.3	7.0	15.0	19
Oral bulge, dorsal height	2.1	2.0	–	–	–	1.5	3.0	19
	2.5	2.5	–	–	–	2.0	3.0	19
Oral bulge length:body width, ratio	0.4	0.3	0.1	0.1	17.6	0.3	0.5	16
	0.9	0.9	0.2	0.1	22.6	0.4	1.2	19
Circumoral kinety to last dikinetid of brush row 1, distance	5.2	5.0	0.8	0.2	14.5	4.0	6.0	11
	6.5	6.0	0.8	0.2	13.0	5.0	8.0	19
Circumoral kinety to last dikinetid of brush row 2, distance	6.0	6.0	1.2	0.4	19.7	4.0	8.0	11
	8.1	8.0	1.2	0.3	14.2	6.0	10.0	19
Circumoral kinety to last dikinetid of brush row 3, distance	3.6	4.0	–	–	–	3.0	4.0	10
	4.8	5.0	0.8	0.2	15.8	3.0	6.0	19
Anterior body end to macronucleus, distance	11.9	12.0	4.1	0.9	34.8	6.0	23.0	21
	17.2	17.0	2.9	0.7	17.1	12.0	23.0	19
Macronucleus nodules, length	7.6	8.0	2.1	0.5	27.9	4.0	11.0	21
	7.7	8.0	1.5	0.4	19.7	5.0	12.0	19
Macronucleus nodules, width	4.8	5.0	0.7	0.1	14.1	4.0	6.0	21

continued

Characteristics[1]	$\overline{x}$	M	SD	SE	CV	Min	Max	n
	5.8	6.0	0.6	0.1	10.9	5.0	7.0	19
Macronucleus nodules, number	1.0	1.0	0.0	0.0	0.0	1.0	1.0	21
	1.0	1.0	0.0	0.0	0.0	1.0	1.0	19
Ciliary rows, number	10.7	11.0	0.7	0.2	7.0	10.0	12.0	19
	8.9	9.0	0.7	0.2	8.3	8.0	10.0	19
Kinetids in a right side ciliary row, number	12.5	12.0	3.4	0.8	27.1	8.0	21.0	17
	10.8	10.0	3.3	0.8	31.0	7.0	20.0	18
Dorsal brush rows, number	3.0	3.0	0.0	0.0	0.0	3.0	3.0	15
	3.0	3.0	0.0	0.0	0.0	3.0	3.0	19
Dikinetids in brush row 1, number[2]	5.8	6.0	0.7	0.2	12.6	5.0	7.0	13
	6.6	7.0	0.9	0.2	13.5	4.0	8.0	19
Dikinetids in brush row 2, number[2]	6.9	7.0	1.1	0.3	16.1	5.0	8.0	13
	7.6	8.0	1.1	0.2	14.0	6.0	9.0	19
Dikinetids in brush row 3, number[2]	3.6	4.0	0.7	0.2	18.0	3.0	5.0	13
	4.1	4.0	0.5	0.1	11.2	3.0	5.0	19
Circumoral dikinetids, number	–	–	–	–	–	–	–	–
	30.4	30.0	4.3	1.0	14.2	24.0	40.0	19

[1] Data based on mounted, protargol-impregnated (FOISSNER's method), and randomly selected specimens from non-flooded Petri dish cultures. Measurements in µm. CV – coefficient of variation in %, M – median, Max – maximum, Min – minimum, n – number of individuals investigated, SD – standard deviation, SE – standard error of arithmetic mean, $\overline{x}$ – arithmetic mean.

[2] Partially based on specimens from SEM preparations.

Apertospathula verruculifera FOISSNER, XU & KREUTZ, 2005 (Fig. 100 a–n, 147j–u; Table 40)

2005 *Apertospathula verruculifera* FOISSNER, XU & KREUTZ, J. Eukaryot. Microbiol., 52: 360, 371 (Type slides with protargol-impregnated specimens from type locality are deposited in the Oberösterreichische Landesmuseum in Linz, Upper Austria.).

Diagnosis: Size about 50 × 25 µm in vivo. Spatulate to clavate with oblique, obovate oral bulge occupying about one third of trunk wide; palpus dorsalis between anterior ends of brush rows 1 and 2, 2–3 µm high. Macronucleus broadly ellipsoidal; single micronucleus. Extrusomes oblong, about 2 µm long. On average 11 ciliary rows and about 25 circumoral dikinetids. Dorsal brush heterostichad and three-rowed, occupies 15% of body length, inconspicuous because bristles only 1 µm long: on average six dikinetids in row 1, seven in row 2, and four in row 3 followed by a short, monokinetidal bristle tail.

Type locality: Saline mud and soil from flooded grassland in the Maracay National Park, north coast of Venezuela, W68° N10°.

Etymology: The Latin adjective *verruculifera* (with tubercular process) refers to a main feature of the species, viz., the palpus between dorsal brush rows 1 and 2.

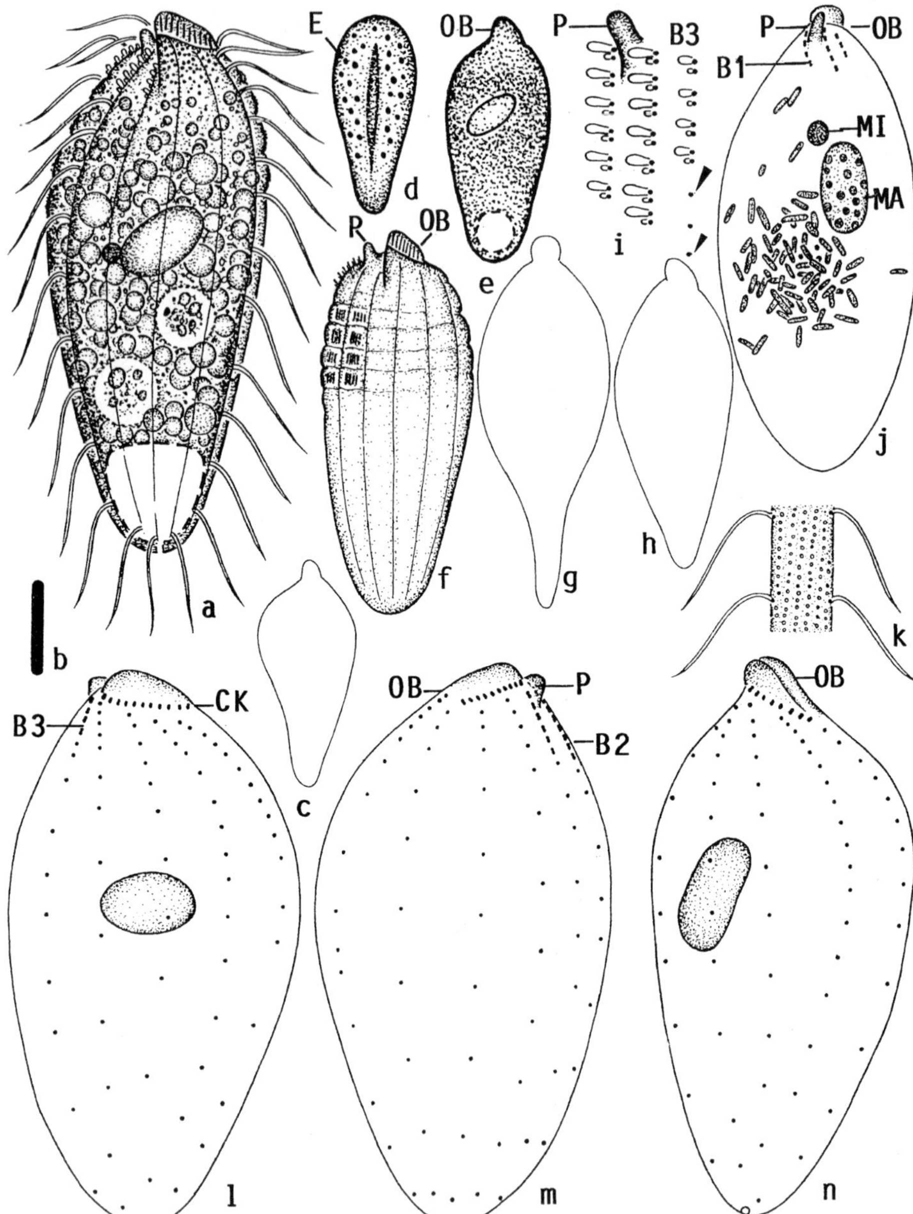

Fig. 100a–n *Apertospathula verruculifera* from life (a–i, k) and after protargol impregnation (j, l–n). From FOISSNER et al. (2005b). **a, f)** Dorsolateral views of representative specimens, length 50 µm. **b)** Extrusome, length 2 µm. **c, g, h)** Right side (c, h) and dorsal (g) views showing shape variability. **d)** Frontal view of oral bulge. **e)** Cells are dark at low magnification due to countless lipid droplets. **i)** Dorsal brush and palpus dorsalis, bristles up to 1 µm long. Arrowheads mark tail of row 3. **j)** Dorsal view showing body centre packed with bacterial rods, length 36 µm. **k)** Cortical granulation. **l, m)** Right and left side view of holotype specimen, length 40 µm. **n)** Ventrolateral view, length 43 µm. B (1, 3) – dorsal brush (rows), CK – circumoral kinety, E – extrusomes, MA – macronucleus, MI – micronucleus, OB – oral bulge, P – palpus dorsalis.

Description: Size 40–60 × 18–30 μm in vivo, usually about 50 × 25 μm, as calculated from some in vivo measurements and the protargol preparations; length: width ratio near 2:1 both in vivo and prepared specimens (147j, m, n, p, t; Table 40). Obovate, bluntly fusiform, or clavate after systole of contractile vacuole, dorsal side usually more distinctly convex than ventral; anterior (oral) body end broadly rounded with oblique oral bulge appearing as a minute triangular process; gradually narrowed posteriorly with rear end occasionally bluntly pointed and/or wrinkled after systole of contractile vacuole. Palpus dorsalis between anterior ends of brush rows 1 and 2, about 2–3 × 1–2 μm in size, well recognizable in vivo and some prepared cells, but inconspicuous in scanning preparations, where it tends to collapse (Fig. 100a, c, f–j, 147m, q, r, t, u). Macronucleus usually above mid-body, globular to ellipsoidal, on average 7.6 × 4.8 μm in protargol preparations, composed of two nodules in 2 out of 52 specimens analysed; many nucleoli about 1 μm across. Micronucleus not unequivocally identified due to many cytoplasmic inclusions, likely globular and near to macronucleus (Fig. 100a, j, l, n, 147j, n; Table 40). Contractile vacuole in rear end, likely with one excretory pore in pole centre. Extrusomes studded in oral bulge, oblong, indistinct in vivo because pale and only about 2 × 0.3 μm in size (Fig. 100a, b, d, f); do not impregnate with the method used. Cortex thin and flexible, conspicuously furrowed in anterior body third, as shown in figures 100f and 147k, n, o; cortex bolsters filled with ordinary cytoplasmic inclusions, do not contain extrusomes and are not preserved in SEM preparations. Cortical granules minute (< 0.5 μm) and pale, difficult to recognize. Cytoplasm colourless, middle body third usually dark at low magnification because packed with lipid droplets 1–5 μm across and (methanogenic?) 2–3 μm long bacteria intensely impregnating with protargol (Fig. 100a, j). Underneath oral bulge an accumulation of 0.5 μm-sized granules resembling those found in *Homalozoon* (147q). Food not known, likely small ciliates considering the numerous lipid droplets (147k, o, t). Swims moderately rapid rotating about main body axis.

Cilia about 10 μm long in vivo, distances of cilia within rows gradually increase from about 1 μm anteriorly to 5 μm posteriorly, arranged in an average of 11 meridional, widely (~ 7 μm) and equidistantly spaced rows clearly separate from circumoral kinety and only indistinctly curved dorsally at anterior end. Dorsal brush heterostichad and three-rowed, very inconspicuous because extending only 15% of body length and bristles merely up to 1 μm long in vivo. Bristles clavate and comparatively thick, posterior bristle of pairs < 0.4 μm long and thus appearing dot-like. Brush rows 1 and 2 of similar length each composed of six to seven dikinetids on average; row 3 about half as long as row 2, composed of an average of only four dikinetids with minute bristles, followed by a short, monokinetidal tail (Fig. 100a, i, j, l–n, 147k–m, p; Table 40).

Oral bulge indistinct because occupying only 40% of body width, gradually merging into body proper ventrally, and merely 2–3 μm high in dorsal region appearing as a wart-like eminence in vivo; obovate in frontal view with bulge halves separated by a rather distinct furrow (Fig. 100a, d–h, m, n, 147m, n, p; Table 40). Circumoral kinety of similar shape as oral bulge, ventral opening distinct in protargol and SEM preparations, composed of about 25 dikinetids (up to 70 in Fig. 147s, and thus likely a different species) each associated with a cilium and a nematodesma. Oral basket rods fine and rarely impregnated, oral basket thus inconspicuous (Fig. 100a, l–n, 147s; Table 40).

Occurrence and ecology: *Apertospathula verruculifera* was moderately abundant in the non-flooded Petri dish culture set up with mud and soil from the surface of a flat, dry puddle covered with halophytes and crusts of cyanobacteria. The sample, which was collected in 1996 and studied in April 1997, had pH 7.3 (in water) and was highly saline (> 30‰). Later, we found *A. verruculifera* in a similar habitat of the Dominican Republic, viz., in soil from the margin of a Mangrove forest. Thus, *A. verruculifera* possibly prefers ephemerally flooded coastal puddles and/or soils.

Remarks: *Apertospathula verruculifera* is a distinct species easily recognizable by the palpus dorsalis, the wart-like oral bulge, and the stout body (Fig 100a). The palpus dorsalis is a palpus oralis and distinctly longer in *Rhinotrix* spp. which are, additionally, considerably larger (≥ 80 μm vs. ~ 50 μm) and have long extrusomes. *Apertospathula lajacola*, which has many minute palps, is much more slender and has up to 12 μm long toxicysts. There is also a considerable similarity with *Spathidiodes hyalina* which, however, has a rigid, bright cortex and lacks the palpus dorsalis. Further details, see genus *Spathidiodes.*

Apertospathula swarezewskyi (TUCOLESCO, 1962) **nov. comb.** (Fig. 101 a–w; 147a–i; Table 41)

1962 *Spathidium swarezewskyi* TUCOLESCO, Arch. Protistenk., 106: 9 (No type material mentioned. We deposit four slides with protargol-impregnated specimens from an African population in the Oberösterreichische Landesmuseum in Linz, Upper Austria.).

Improved diagnosis (includes original data and the redescription provided below): Size about 100 × 30 μm in vivo. Spatulate to narrowly spatulate with oblique, cuneate oral bulge about half as long as widest trunk region. Macronucleus oblong; one or two micronuclei. Extrusomes rod-shaped with narrowed ends, slightly curved and 5–8 μm long. On average 13 ciliary rows and about 80 circumoral dikinetids. Dorsal brush three-rowed and heterostichad, occupies 15% of body length, bristles up to 2 μm long: rows 1 and 2 each composed of an average of 15 dikinetids, row 3 of nine dikinetids followed by a short monokinetidal bristle tail.

Type locality: Brackwater lake Agigea near the town of Constanta on the Black Sea coast, Rumania, N44° E28°20'.

Etymology: Dedicated to B. SWAREZEWSKY, who investigated the epizoic ciliates of Lake Baikal gammarids.

Description: Our redescription is based on a population from Botswana, Africa. TUCOLESCO's description is provided in the "remarks" section.

Size 70–150 × 25–55 μm in vivo, usually near 110 × 35 μm (n 13); length:width ratio 1.4–5:1, on average about 3.2:1 both in vivo and protargol preparations (Table 41). Shape also highly variable, especially in flourishing cultures containing many overfed, stout specimens (Fig. 101g, s); ordinarily nourished cells oblong to bluntly fusiform with more or less distinctly narrowed, oblique anterior end gradually merging into trunk; anterior dorsal end usually narrowly rounded, rarely slightly rostrate; posterior

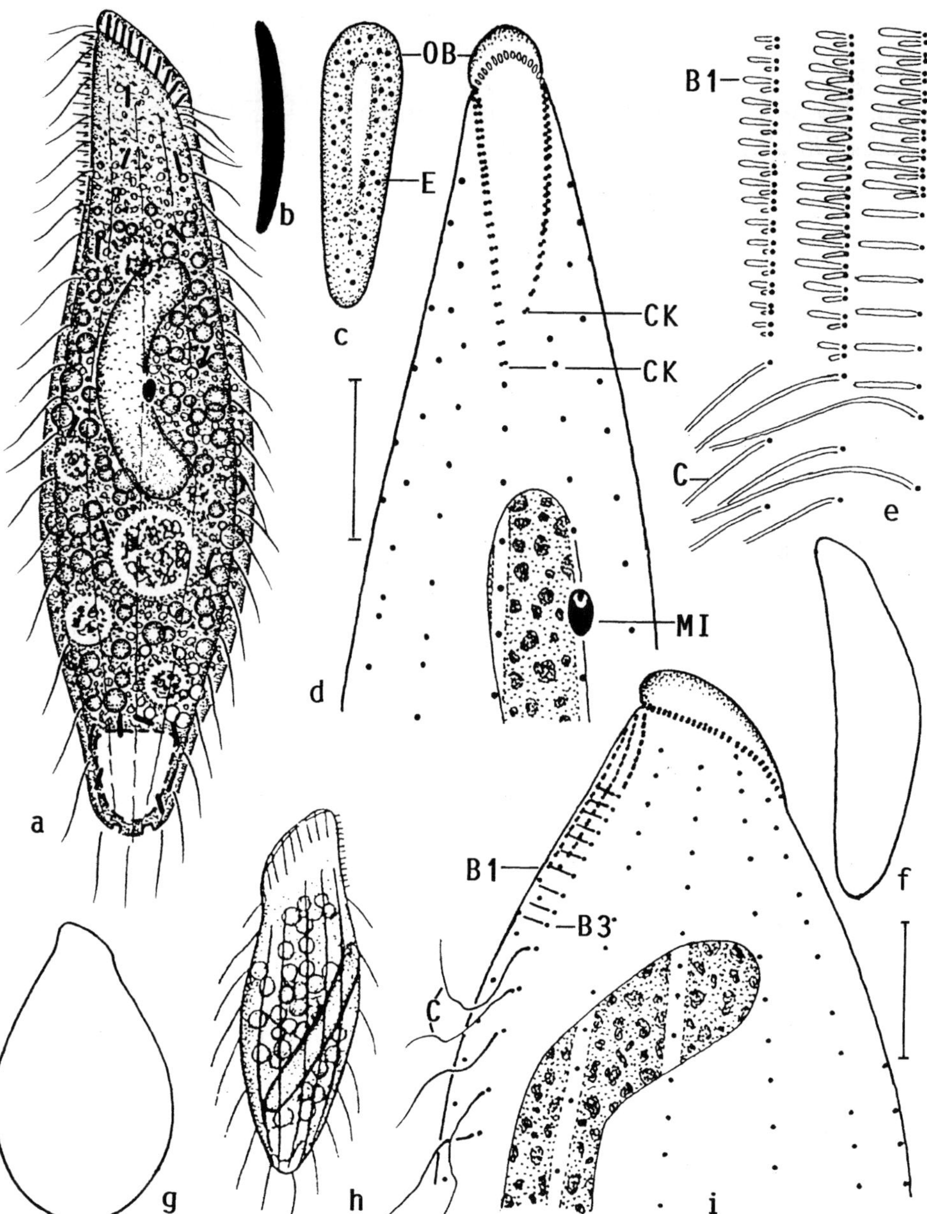

Fig. 101a–i *Apertospathula swarezewskyi*, African specimens (a–g, i) and Rumanian type (h) from life (a–c, e–h) and after protargol impregnation (d, i). **a, h)** Right and left side view of representative specimens, length 110 µm and 80 µm. **b)** Extrusome, length 6 µm. **c)** Frontal view of oral bulge. **d)** Ventral view of anterior body portion, showing the open circumoral kinety composed of clearly discernible dikinetids. **e)** Dorsal brush; bristles drawn to scale, those of the short, monokinetidal tail of row 3 are 2 µm long. Ordinary somatic cilia 10 µm long and thus only partially shown. **f, g)** A slender and a stout, overfed specimen. **i)** Right side view showing the impregnated bristles of the monokinetidal tail of brush row 3. B1, 3 – dorsal brush rows, C – ordinary somatic cilia, CK – circumoral kinety, E – extrusomes, MA – macronucleus, MI – micronucleus, OB – oral bulge. Scale bars 10 µm.

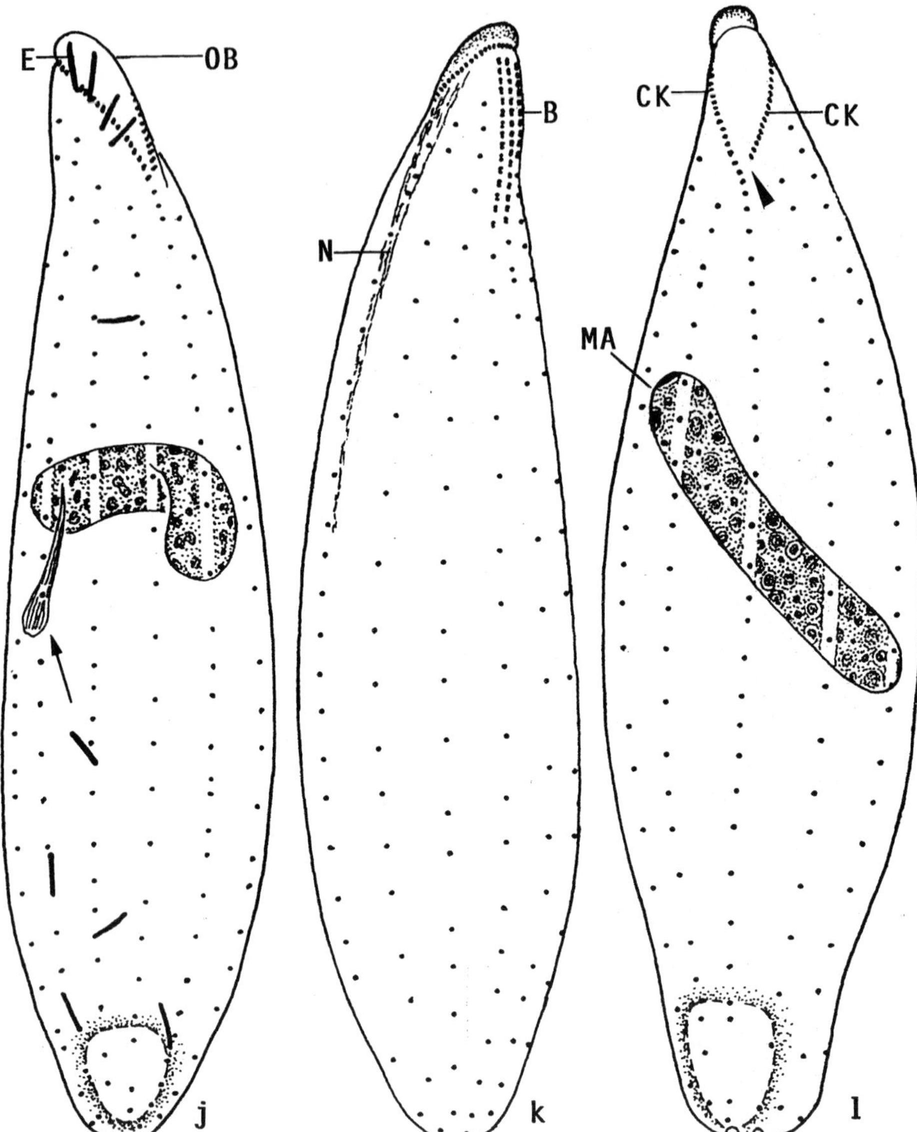

Fig. 101j–l *Apertospathula swarezewskyi*, somatic and oral ciliary pattern and nuclear apparatus after protargol impregnation. **j, k)** Right and left side view of a slender specimen, length 112 µm. Note the short oral bulge. Arrow marks oral basket of a prey ciliate. **l)** Ventral view of main voucher specimen showing the open circumoral kinety (arrowhead), length 104 µm. Dorsal view, see Fig. 101m. B – dorsal brush, CK – circumoral kinety, E – extrusomes, MA – macronucleus, N – nematodesma bundle.

end narrowly to broadly rounded; laterally flattened only in anterior and posterior quarter (Fig. 101a, f, g, p–s, v, w). Nuclear apparatus in or slightly above mid-body, stands out as a bright structure from strongly granulated cytoplasm. Macronucleus oblong to cylindroidal with ends often slightly inflated, length:width ratio 3–7:1, on average 4.5:1

in protargol preparations, rod-shaped to C-like curved; nucleoli irregularly globular, numerous and small. One to two micronuclei attached to macronucleus, globular to broadly ellipsoidal, often not distinguishable from similar-sized and impregnated cell inclusions (Fig. 101a, j, l, 147a, d; Table 41). Contractile vacuole in posterior end, usually three excretory pores, often surrounded by extrusomes very similar to those found in oral bulge. Extrusomes studded in oral bulge and scattered in cytoplasm and around contractile vacuole, rod-shaped with narrowed ends, 5–7 × 0.6–0.8 µm in size, occasionally impregnate with the protargol method used (Fig. 101a, b, j, u, 147c, g, h; Table 41). Cortex thin and flexible, cortical granules not recognizable, not even with interference contrast optics. Cytoplasm colourless, in flourishing cultures packed with refractive, 0.5 to 7 µm-sized lipid droplets making cells dark at low magnification (≤ x100). Feeds on heterotrophic flagellates and middle-sized ciliates digested in up to 60 µm large vacuoles. Swims rather slowly by rotation about main body axis, rarely glides on soil particles and microscope slides, never rests.

Cilia 10 µm long in vivo, about 5 µm apart and thus loosely spaced, arranged in an average of 13 meridional, widely (8µm) and equidistantly spaced rows distinctly separate from circumoral kinety and not curved dorsally at anterior end. Anterior sixth of three (rarely four) dorsal ciliary rows differentiated to short, heterostichad dorsal brush with bristles only up to 2 µm long in vivo; details shown in figure 101e and Table 41; anterior tail lacking in all rows. Monokinetidal bristle tail of row 3 short, not extending beyond first body third, occasionally impregnates with protargol. Anterior thirds of brush frequently on a flat convexity hardly recognizable in vivo, but rather distinct in some protargol-impregnated specimens (Fig. 101a, i–m, µ, 147a–c, e, f; Table 41).

Oral bulge indistinct in vivo and preparations because only 3 µm high at dorsal end, merely half as long as widest trunk region, and hardly separate from body proper with ventral half gradually merging into body surface; slightly (30°) to moderately (55°) oblique and convex; cuneate in frontal view and slightly depressed in midline (Fig. 101a, c, f, g, p–s, u–w; Table 41). Circumoral kinety composed of about 80 dikinetids more loosely spaced ventrally than dorsally, ventral opening usually distinct, right branch longer than left by one to four dikinetids; individual dikinetids associated with a 10 µm long cilium, a distinct fibre lining the oral bulge, and a fine nematodesma. Oral basket rarely impregnated, basically inconspicuous, but remarkable in that only the left side rods form bundles one even extending to mid-body (Fig. 101d, i–l, n, o, 147a–c, e, g, I; Table 41).

Occurrence and ecology: TUCOLESCO (1962a, b) discovered *A. swarezewskyi* in a brackish lake on the west coast of the Black Sea. Unfortunately, he did not mention whether it occurred in the plankton or benthos; we assume the later because most of the associated ciliates are benthic species (see "remarks" section).

Our population is from mud and surface soil of a dry puddle near the Khwai River Lodge, that is, from an area covered by *Sphaeranthus incisus* and flooded two to three months each year. The lodge is in the Moremi Wildlife Reserve, Okovango Basin, Botswana (E20° S20°). The black, very fine-grained soil becomes a muddy, viscous mass when flooded. Thus, it was enveloped in gauze and the water outside the soil bag was enriched with some wheat grains. A rich ciliate community with at least five undescribed species developed. *Apertospathula swarezewskyi* became abundant in sapropelic, microaerobic patches of the culture. Several encystment trials failed. The isolated

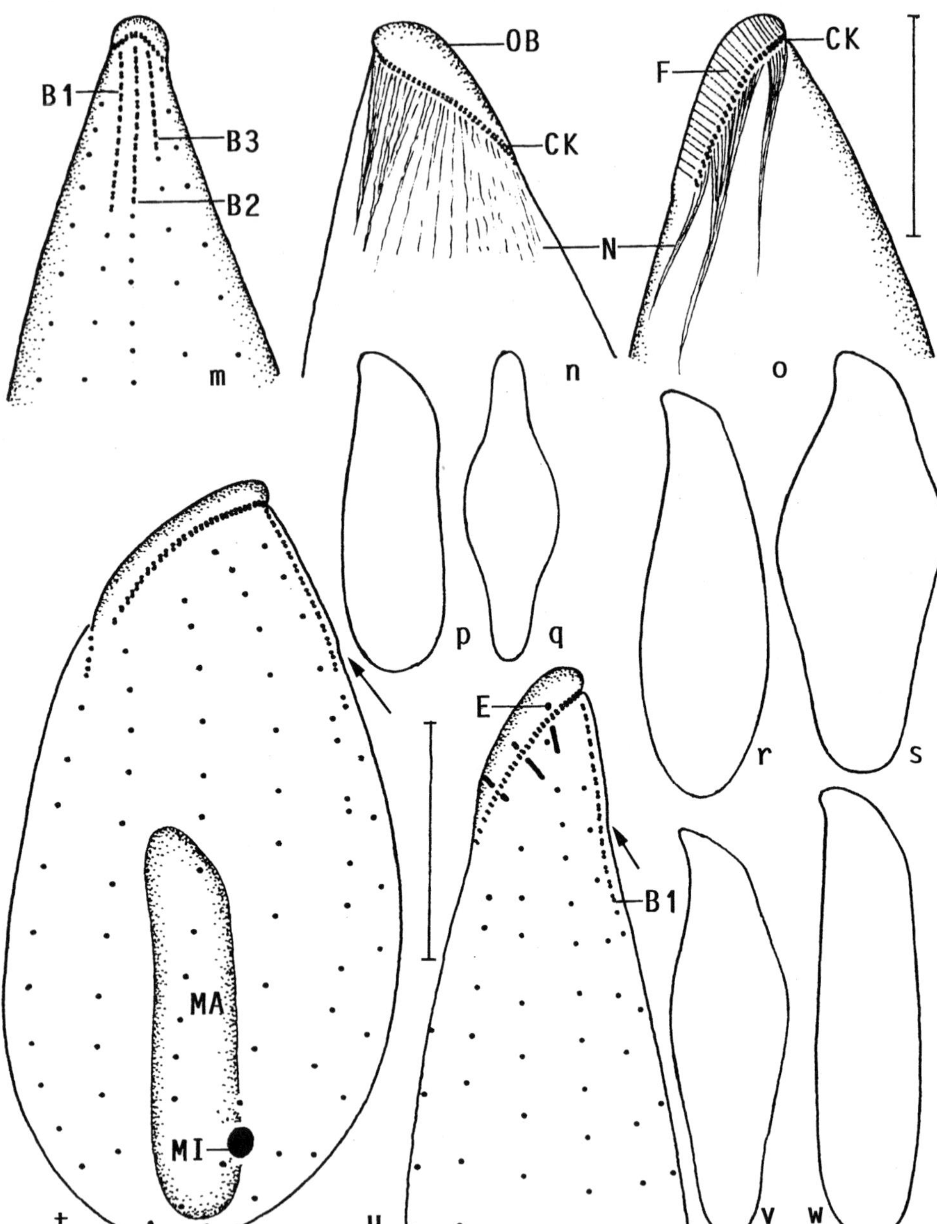

Fig. 101m–w *Apertospathula swarezewskyi* from life (p–s, v, w) and after protargol impregnation (m–o, t, u). **m)** Dorsal view of main voucher specimen, length 35 µm (ventral view, see Fig. 101l). The brush is exactly on the dorsal side of the cell. **n, o)** Right and left side view of anterior body portion showing the oral basket made of single (right) and bundled (left) nematodesmata. **p–s, v, w)** Shape variants. **t)** Proter post-divider. Note the stout shape and the inconspicuous convexity (arrow) along the dorsal brush. **u)** Left side view showing ciliary pattern and the inconspicuous convexity (arrow) in upper half of dorsal brush. B1-3 – dorsal brush rows, CK – circumoral kinety, E – extrusome, F – fibres, MA – macronucleus, MI – micronucleus, N – nematodesmata, OB – oral bulge. Scale bars 20 µm.

specimens died after one to two days, regardless they were kept in filtered culture water or pure Eau de Volvic.

Remarks: Our data show that *Spathidium swarezewskyi* belongs to *Apertospathula*, that is, has an open circumoral kinety (Fig. 101d, 147e, g, h). In vivo, this feature is difficult to recognize because the oral bulge looks like in *Spathidium* (Fig. 101c). Thus, the feature is useless for separating the Rumanian and African populations.

TUCOLESCO (1962) described *S. swarezewskyi* as follows (Fig. 101h): "Taille 75–80 µ. Pôle postérieur acuminé. Troncature antérieure, avec, ses deux angles régulièrement arrondis; les lèvres de la fente buccale portent des trichocystes, de longueur modérée (8 µ). Un peigne d'épines très courtes. Noyau en forme de boudin, allongé et légèrement recourbé, situé obliquement, dont les deux bouts atteignent latéralement la face interne de la pellicule. Vacuole contractile terminale. Striations assez espaceées. Lac saumâtre d'Agigea, dans l'ensemble biocénotique suivant: *Frontonia marina, Metopus contortus, Cristigera cirrifera, Metopus* sp., *Spathidium* sp., *Metacystis tesselata, Euplotes trisulcatus, Perispira ovum*".

Several main features are obviously very similar in the Rumanian and African specimens, viz., the number of ciliary rows (about 12, as estimated from TUCOLESCO's figure), the shape of the macronucleus and extrusomes, and the inconspicuous dorsal brush. Other characteristics match also rather well, viz., body shape and size (80 × 27 µm vs. 115 × 20 µm), body length:width ratio (2.9 vs. 3.2:1), extrusomes (slightly shorter and thicker in the African specimens), and the habitat (brackish water vs. ephemeral river, where the water becomes highly "concentrated" during desiccation). Two features, however, are rather different, viz., the oral area is less conspicuously separated from body proper and the oral bulge is distinctly shorter (ratio body length:oral bulge length 6:1 vs. 3:1) in the African specimens. Unfortunately, both characteristics are related to body shape which is highly variable in the African population (Table 41). Thus, it appears reasonable to emphasize the matching features and to identify the African population as *S. swarezewskyi.* On the other hand, neotypification would be too premature.

Table 41 Morphometric data on *Apertospathula swarezewskyi* (upper line) and *Apertospathula lajacola* (lower line)

Characteristics[1]	$\overline{x}$	M	SD	SE	CV	Min	Max	n
Body, length	101.9	104.0	19.1	4.2	18.8	70.0	145.0	21
	104.7	102.0	16.5	4.6	15.7	79.0	127.0	13
Body, width	35.1	32.0	7.8	1.7	22.4	24.0	54.0	21
	19.0	19.0	3.7	1.0	19.6	14.0	26.0	13
Body length:width, ratio	3.2	3.3	1.0	0.2	32.3	1.3	6.0	21
	5.7	5.6	1.4	0.4	24.8	3.5	7.9	13
Anterior body end to macronucleus, distance	34.3	33.0	11.2	2.5	32.7	19.0	55.0	21
	36.0	32.0	8.9	2.5	24.7	27.0	52.0	13
Circumoral kinety to last dikinetid of brush row 1, distance	14.9	15.0	1.7	0.4	11.5	10.0	18.0	21
	11.1	11.0	1.5	0.4	13.5	9.0	14.0	13
Circumoral kinety to last dikinetid of brush row 2, distance	14.9	15.0	1.9	0.4	12.7	10.0	18.0	21
	10.9	10.0	1.5	0.4	13.5	9.0	13.0	13

continued

Characteristics[1]	$\overline{x}$	M	SD	SE	CV	Min	Max	n
Circumoral kinety to last dikinetid of brush row 3,	9.2	9.0	1.4	0.3	14.9	7.0	12.0	21
distance	5.7	5.0	1.1	0.3	19.5	4.0	8.0	13
Oral bulge, length	16.8	17.0	1.5	0.3	9.1	13.0	19.0	21
	16.0	16.0	1.8	0.5	11.4	13.0	20.0	13
Body width: oral bulge length, ratio	2.1	1.9	0.5	0.1	25.0	1.3	3.1	21
	1.2	1.2	0.3	0.1	25.7	0.8	1.6	13
Oral bulge, height	3.3	3.0	–	–	–	3.0	4.0	15
	–	–	–	–	–	–	–	–
Oral bulge, width in ventral view	6.0	6.0	1.3	0.4	22.4	4.0	8.0	11
	–	–	–	–	–	–	–	–
Anterior body end to end of right branch of circum-	–	–	–	–	–	–	–	–
oral kinety, distance	26.8	27.0	3.8	1.0	14.1	20.0	33.0	13
Anterior body end to end of left branch of circum-	–	–	–	–	–	–	–	–
oral kinety, distance	16.0	16.0	1.8	0.5	11.4	13.0	20.0	13
Macronucleus figure, length[2]	31.8	32.0	4.6	1.0	14.5	21.0	43.0	21
	41.5	45.0	–	–	–	25.0	60.0	13
Macronucleus, width in mid	7.0	7.0	0.8	0.2	12.0	6.0	9.0	21
	6.1	6.0	1.4	0.4	22.7	4.0	8.0	13
Macronucleus, number	1.0	1.0	0.0	0.0	0.0	1.0	1.0	21
	1.0	1.0	0.0	0.0	0.0	1.0	1.0	21
Micronucleus, length	3.3	3.3	–	–	–	3.0	4.0	10
	–	–	–	–	–	–	–	–
Micronucleus, width	3.0	3.0	–	–	–	2.7	3.5	10
	–	–	–	–	–	–	–	–
Micronuclei, number	1.5	1.5	–	–	–	1.0	2.0	10
	–	–	–	–	–	–	–	–
Ciliary rows, number	12.9	13.0	0.7	0.2	5.7	12.0	15.0	21
	12.5	12.0	1.3	0.4	10.2	11.0	15.0	13
Kinetids in a right side ciliary row, number	22.2	22.0	3.2	0.7	14.2	16.0	26.0	21
	17.5	18.0	4.3	1.2	24.7	10.0	26.0	13
Dorsal brush rows, number	3.0	3.0	0.0	0.0	0.0	3.0	3.0	21
	3.0	3.0	0.0	0.0	0.0	3.0	3.0	21
Dikinetids in brush row 1, number	14.6	15.0	1.6	0.4	11.1	12.0	17.0	21
	10.9	11.0	1.7	0.5	15.5	8.0	14.0	13
Dikinetids in brush row 2, number	15.3	16.0	1.9	0.4	12.4	11.0	19.0	21
	10.3	10.0	1.8	0.5	17.0	8.0	14.0	13
Dikinetids in brush row 3, number	9.2	9.0	1.0	0.2	11.3	8.0	11.0	21
	5.7	5.0	1.0	0.3	16.6	5.0	8.0	13
Excretory pores, number	3.1	3.0	0.9	0.3	28.2	2.0	5.0	10
	–	–	–	–	–	–	–	–
Extrusomes, length	5.1	5.0	0.8	0.2	15.4	4.0	7.0	21
	–	–	–	–	–	–	–	–
Extrusomes, width	0.7	0.6	–	–	–	0.5	1.0	21
	–	–	–	–	–	–	–	–
Circumoral dikinetids, number	–	–	–	–	–	–	–	–
	76.4	74.0	12.7	3.5	16.7	60.0	100.0	13

[1] Data based on mounted, protargol-impregnated (FOISSNER's method), and randomly selected specimens from flooded (*A. swarezewskyi*) and non-flooded (*A. lajacola*) Petri dish cultures. Measurements in µm. CV – coefficient of variation in %, M – median, Max – maximum, Min – minimum, n – number of individuals investigated, SD – standard deviation, SE – standard error of arithmetic mean, $\overline{x}$ – arithmetic mean.

[2] Macronucleus "uncoiled" in *A. lajacola*; values thus approximate.

There are several species which resemble *A. swarezewskyi*, viz., *Spathidium fresenburgi, S. modestum, S. obliquum*, and *S. paucistriatum* (Vol. II). They differ, inter alia, by the ellipsoidal (~2:1) macronucleus which is oblong to cylindroidal (4–7:1) in *A. swarezewskyi* (Fig. 101a, h, l, 147a, d).

Apertospathula lajacola **nov. spec.** (Fig. 102a–o, 148a–c; Table 41)

Diagnosis: Size about 115 × 20 µm in vivo. Narrowly to very narrowly spatulate to clavate with oblique, cuneate oral bulge shorter than widest trunk region by about 15%. Macronucleus oblong to cylindroidal, usually U-shaped. Two size types (12 µm and 7 µm) of filiform extrusomes. On average 12 loosely ciliated kineties and 76 circumoral dikinetids. Dorsal brush three-rowed and heterostichad, occupies 11% of body length, bristles up to 3 µm long, individual rows accompanied by minute palps: rows 1 and 2 each composed of an average of 11 dikinetids, row 3 consists of six dikinetids followed by about five monokinetidal bristles. Right branch of circumoral kinety distinctly longer than left.

Type locality: Ephemeral puddles (Lajas) on granitic outgrowths between the Agricultural Research Institute and the airport of Pto. Ayachuco, Venezuela, W75° N6°.

Etymology: Latin noun in apposition composed of *Laja* (Venezuelan name for puddles on granitic outgrowths in the savannah) and *cola* (dweller).

Description: Size 80–140 × 15–30 µm in vivo, usually near 115 × 20 µm; length:width ratio 3.5–7.9:1, on average about 5.5:1 both in vivo and protargol preparations (Table 41). Shape also rather variable, that is, narrowly to very narrowly spatulate to clavate with oblique to strongly oblique anterior (oral) end; laterally flattened only in anterior and posterior quarter (Fig. 102a, d, k, o). Macronucleus in or slightly above mid-body, oblong to cylindroidal, usually U-shaped or semicircular, rarely circular or spiralized, in most specimens ribbon-like flattened (2:1); nucleoli globular, rather large (Fig. 102a, k, o, 148a; Table 41). Micronucleus(i) not recognizable, not even in perfectly impregnated specimens. Contractile vacuole in rear end, with two to three excretory pores in pole area. Extrusomes studded in oral bulge, though fine rather conspicuous because up to 13 µm long; filiform, straight, occur in two size classes: 10–13 × 0.3 µm and 7 × 0.2 µm; do not impregnate with the protargol method used (Fig. 102a–c). Cortex thin and flexible, contains rows of inconspicuous granules ≤ 0.5 µm across. Cytoplasm colourless, contains about 20 curious, crystal-like inclusions, forming conspicuous stacks in some species (Fig 102a, e, 148a). Food not known. Movement conspicuously slow wriggling between soil particles.

Cilia about 10 µm long in vivo, circa 6 µm apart and thus loosely spaced, arranged in an average of 12 meridional, ordinarily (~ 5 µm) and equidistantly spaced rows distinctly separate from circumoral kinety and not curved dorsally at anterior end. Anterior tenth of three dorsal ciliary rows differentiated to inconspicuous, very short brush with bristles up to 3 µm long in vivo; details similar as in *A. swarezewskyi.* Monokinetidal tail of row 3 very short, that is, composed of about five bristles usually impregnating

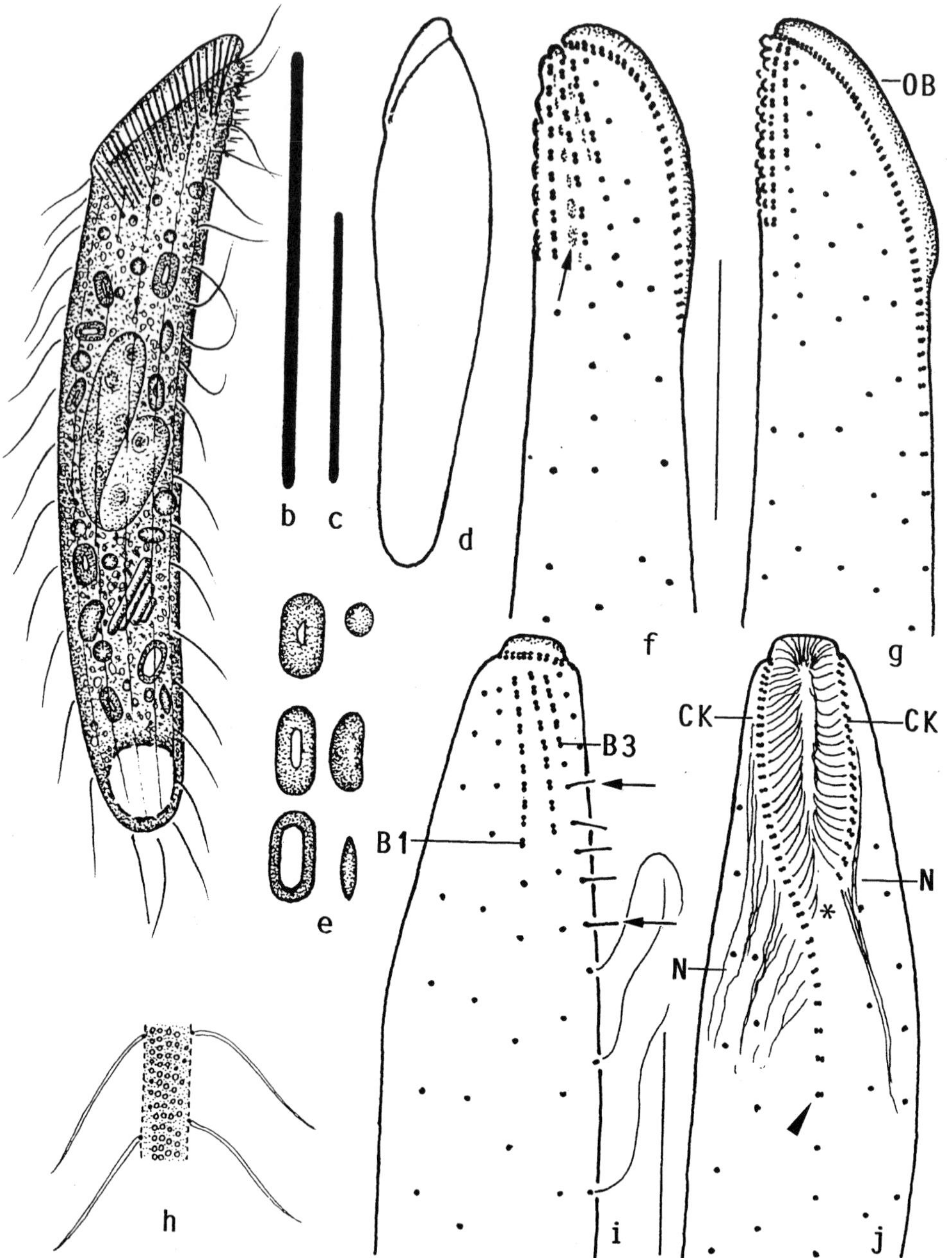

Fig. 102a–j *Apertospathula lajacola* nov. spec. from life (a–e, h) and after protargol impregnation (f, g, i, j). **a)** Left side view of a slightly cuneate specimen, length 115 µm. **b, c)** Long (12 µm) and short (7 µm) oral bulge extrusomes. **d)** Shape variant. **e)** Crystalline (?) inclusions, up to 5 µm long. **f, g)** Dorsolateral and right side view showing the minute tubercles right of the brush rows (arrow). **h)** Cortical granulation. **i, j)** Dorsal and ventral view of oral area showing the widely open circumoral kinety (asterisk), the long right branch of the circumoral kinety (arrowhead), and the monokinetidal bristle tail of brush row 3 (arrows). B1,3 – dorsal brush rows, CK – circumoral kinety, N – nematodesmata, OB – oral bulge. Scale bars 15 µm.

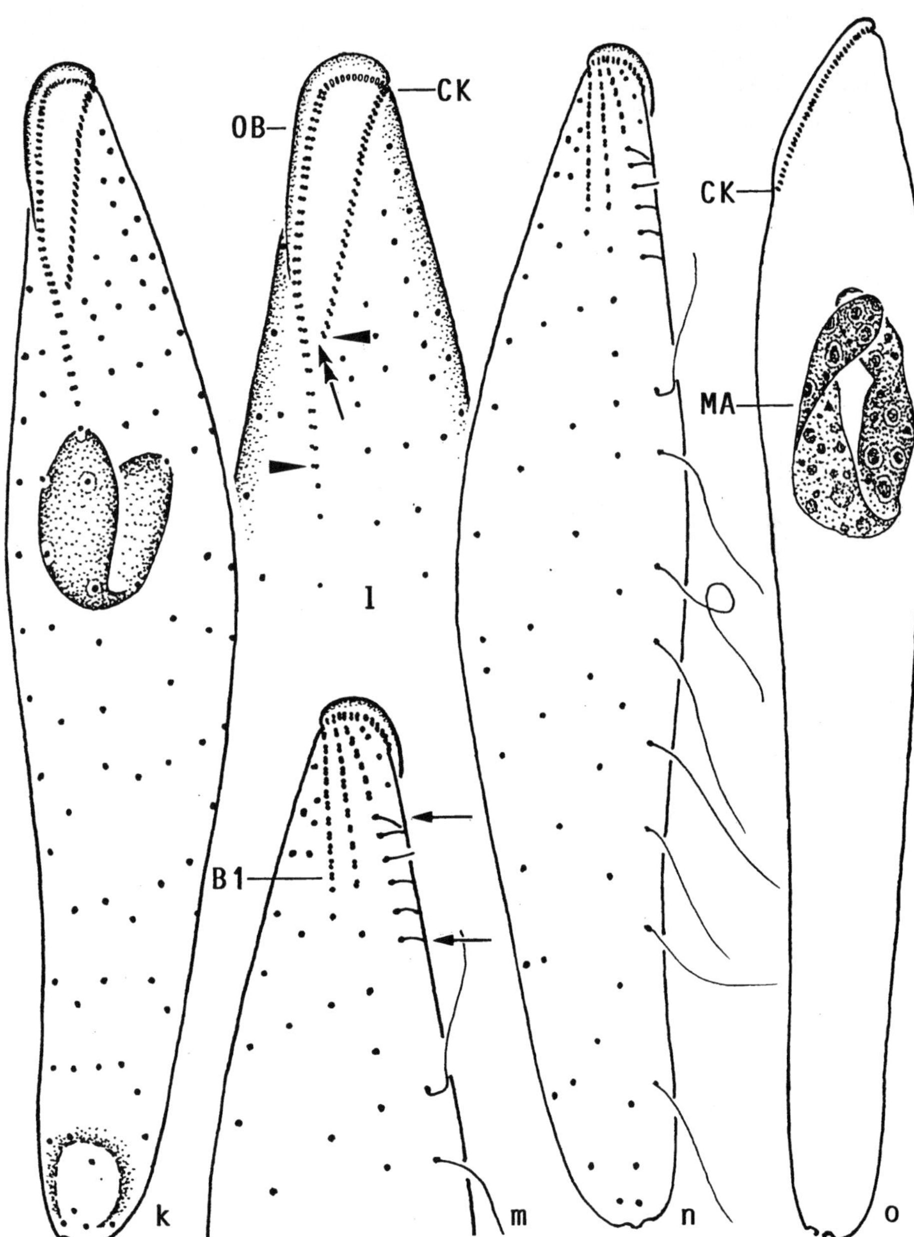

Fig. 102k–o *Apertospathula lajacola* nov. spec., ciliary pattern and macronucleus after protargol impregnation. **k–n)** Ventral and dorsal view of holotype specimen, length 84 μm. Arrows denote monokinetidal bristle tail of dorsal brush row 3. The right branch of the circumoral kinety is considerably longer than the left (arrowheads). Double arrow marks the opening in the circumoral kinety, that is, the main generic feature. **o)** A slender specimen with circular macronucleus, length 127 μm. B1 – brush row 1, CK – circumoral kinety, MA – macronucleus, OB – oral bulge.

with the protargol method used. Individual brush rows at right accompanied by minute palps difficult to recognize and decreasing in height from 2 µm anteriorly to 0.5 µm posteriorly; upper third of tubercles impregnates faintly with protargol (Fig. 102a, f, g, i–n; 148b, c; Table 41).

Oral bulge moderately distinct, in vivo about 4 µm high at dorsal end and gradually merging into body proper ventrally, especially the elongated right bulge half; length rather variable, on average slightly shorter than widest trunk region; oblique (~ 50°) to strongly oblique (~ 70°), cuneate, surface slightly to moderately convex (Fig. 102a, d, f, g, o; Table 41). Circumoral kinety cuneate to oblong in frontal view, composed of an average of 76 dikinetids more loosely spaced ventrally than dorsally, ventral opening distinct, right branch more conspicuous than in congeners because longer than left by 67% on average (Table 41); individual dikinetids associated with a 10 µm long cilium, a distinct fibre lining the oral bulge, and a fine nematodesma. Oral basket rarely impregnated, composed of indistinct nematodesma bundles (Fig. 102a, f, g, j, k, l, o, 148b).

Occurrence and ecology: As yet found only at type locality. The habitat is a tropical speciality, that is, hemispherical granitic outgrowths called "Lajas" in Venezuela. The black surface of the Lajas is caused by a dense cyanobacterial cover. The top is often studded with a variety of rainwater puddles containing a rather rich cyanobacterial and ciliate community (FOISSNER, unpubl.). The dry puddles contain dark cyanobacterial mud mixed with aeolian soil deposits. This material was used for the non-flooded Petri-dish culture which contained very low numbers of *A. lajacola.*

Remarks: *Apertospathula lajacola* is rather similar to *A. swarezewskyi*, differing from that species by the narrower body, the filiform extrusomes, and the very short dorsal brush occupying only 11% (vs. 15%) of body length. *Apertospathula lajacola* also resembles several *Spathidium* species described by KAHL (Vol. II): *S. modestum* (length 50–70 µm, long dorsal bristles, ellipsoidal macronucleus vs. length 115 µm, short dorsal bristles, oblong to cylindroidal macronucleus); *S. paucistriatum* (extrusomes short and stout, macronucleus ellipsoidal vs. filiform and oblong to cylindroidal); and *S. implicatum* (extrusomes 8 µm long and rather thick, macronucleus long and tortuous vs. filiform and oblong to curved cylindroidal).

Apertospathula pelobia nov. spec. (Fig. 102p–w; Table 42)

Diagnosis: Size about 100 × 25 µm in vivo. Narrowly spatulate with distinct, oblique oral bulge shorter than widest trunk region by about 25%. Macronucleus oblong; single micronucleus. Extrusomes rod-shaped and slightly curved, about 12 µm long. On average 15 ciliary rows, 3 anteriorly differentiated to heterostichad dorsal brush occupying 20% of body length. Brush bristles up to 3 µm long; row 1 composed of an average of 15 dikinetids, row 2 of 18, and row 3 of 11 dikinetids followed by some monokinetidal bristles. Oral bulge conspicuous because screwed like a propeller blade and with distinct central concavity.

Type locality: Soil from a green area ("green river bed") of the Chobe River near the Muchenje Safari Lodge, Botswana, E24°40' S18°.

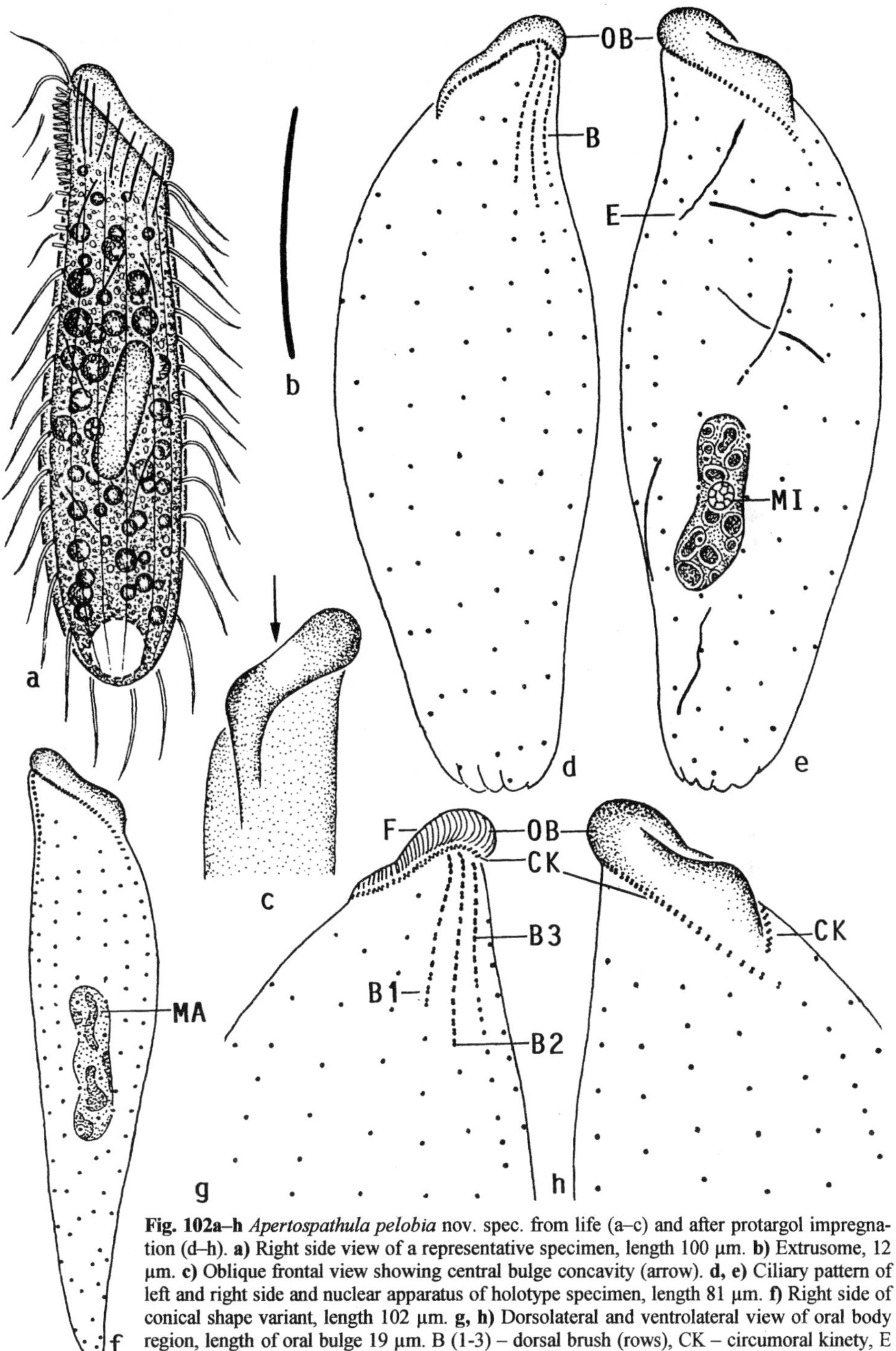

Fig. 102a–h *Apertospathula pelobia* nov. spec. from life (a–c) and after protargol impregnation (d–h). **a)** Right side view of a representative specimen, length 100 µm. **b)** Extrusome, 12 µm. **c)** Oblique frontal view showing central bulge concavity (arrow). **d, e)** Ciliary pattern of left and right side and nuclear apparatus of holotype specimen, length 81 µm. **f)** Right side of conical shape variant, length 102 µm. **g, h)** Dorsolateral and ventrolateral view of oral body region, length of oral bulge 19 µm. B (1-3) – dorsal brush (rows), CK – circumoral kinety, E – extrusome, F – bulge fibres, MA – macronucleus, MI – micronucleus, OB – oral bulge.

Table 42 Morphometric data on *Apertospathula pelobia*

Characteristics	x̄	M	SD	SE	CV	Min	Max	n
Body, length	87.8	89.0	13.0	5.3	14.8	67.0	102.0	6
Body, width	27.2	29.0	4.7	1.9	17.3	21.0	33.0	6
Body length: width, ratio	3.4	3.0	1.1	0.4	31.5	2.0	5.0	6
Oral bulge (circumoral kinety), right side length	20.0	21.0	3.4	1.4	17.0	14.0	24.0	6
Oral bulge (circumoral kinety), left side length	16.5	17.0	2.5	1.0	15.2	12.0	19.0	6
Oral bulge, height	4.2	4.0	–	–	–	4.0	5.0	6
Oral bulge right side length: body width, ratio	0.8	0.7	0.2	0.1	24.6	0.5	1.0	6
Circumoral kinety to last dikinetid of brush row 1, distance	14.8	15.0	1.7	0.7	11.6	12.0	17.0	6
Circumoral kinety to last dikinetid of brush row 2, distance	16.7	17.0	1.9	0.8	11.2	14.9	19.0	6
Circumoral kinety to last dikinetid of brush row 3, distance	11.0	11.0	1.1	0.5	10.0	10.0	12.0	6
Anterior body end to macronucleus, distance	39.7	40.0	2.2	0.9	5.5	37.0	43.0	6
Macronucleus, length	21.0	21.0	4.4	1.8	20.9	16.6	26.0	6
Macronucleus, width	6.4	7.0	0.8	0.3	12.9	5.0	7.0	6
Macronucleus, number	1.0	1.0	0.0	0.0	0.0	1.0	1.0	6
Micronucleus, length	3.1	3.0	–	–	–	3.0	4.0	5
Micronucleus, width	3.0	3.0	–	–	–	3.0	4.0	5
Micronucleus, number	1.0	1.0	0.0	0.0	0.0	1.0	1.0	5
Ciliary rows, number	15.3	16.0	1.2	0.5	7.9	14.0	17.0	6
Basal bodies in a right side row, number	14.7	14.0	4.1	1.7	28.2	10.0	20.0	6
Dorsal brush rows, number	3.0	3.0	0.0	0.0	0.0	3.0	3.0	6
Dikinetids in brush row 1, number	14.8	15.0	1.9	0.8	13.1	13.0	18.0	6
Dikinetids in brush row 2, number	17.8	18.0	1.5	0.6	8.3	15.0	19.0	6
Dikinetids in brush row 3, number	11.0	12.0	1.3	0.5	11.5	9.0	12.0	6

[1] Data based on mounted and protargol-impregnated (FOISSNER's method) specimens from a non-flooded Petri dish culture; all specimens found in eight slides were used. Measurements in µm. CV – coefficient of variation in %, M – median, Max – maximum, Min – minimum, n – number of individuals investigated, SD – standard deviation, SE – standard error of mean, x̄ – arithmetic mean.

Etymology: The Greek *pelobia* (living in mud) refers to the habitat the species was discovered, viz., the mud of a green river bed.

Description: Size 70–120 × 15–30 µm in vivo, usually about 100 × 25 µm, as calculated from some in vivo measurements and the morphometric data; inflated in the preparations, length: width ratio thus about 4:1 in vivo and 3:1 protargol slides (Table 42). Usually narrowly spatulate to oblong, rarely conically narrowing posteriorly; anterior (oral) end oblique, posterior rounded or, in conical specimens, acute (Fig. 102p, s, u). Macronucleus in middle third of body, oblong to elongate reniform; nucleoli large, globular to oblong. Micronucleus attached to macronucleus, of foamy structure, about 3 µm across in vivo (Fig. 102p, t, u; Table 42). Contractile vacuole in rear body end. Oral bulge extrusomes rod-shaped and slightly curved, conspicuous in vivo because 12 ×

0.3 µm in size, do not impregnate with the method used, while a certain cytoplasmic developmental stage impregnates brownish (Fig. 102p, q, t). Cortex flexible, cortical granules inconspicuous, not studied in detail. Cytoplasm colourless, usually packed with lipid droplets up to 7 µm across. Glides and swims moderately rapidly.

Cilia about 14 µm (!) long in vivo, loosely spaced, arranged in an average of 15 meridional, ordinarily and equidistantly spaced rows anteriorly distinctly shortened and thus widely separate from circumoral kinety (Fig. 102p, s–w; Table 42). Three dorsal rows differentiated to an inconspicuous, heterostichad brush occupying about 20% of body length: row 1 composed of an average of 15 dikinetids with up to 1 µm long bristles; row 2 composed of an average of 18 dikinetids with slightly inflated, fusiform bristles up to 3 µm long; row 3 composed of an average of 11 dikinetids with rod-shaped bristles up to 3 µm long, followed by a few monokinetidal bristles. Rarely one ordinary cilium between anterior end of individual brush rows and circumoral kinety (Fig. 102p, s, u, w; Table 42).

Oral bulge shorter than widest trunk region by 25% on average, ordinarily oblique and narrowly obovate in frontal view; conspicuous because about 5 µm high, surface screwed like a propeller blade, and central area distinctly concave, but not open as in the pharyngospathidiids (Fig.102a, c–h; Table 42; Vol. II). Circumoral kinety not propeller-like as oral bulge, ventral opening distinct because right branch 4 µm longer than left and continuing as ordinary somatic ciliary row (Fig. 102s–w; Table 42).

Occurrence and ecology: See *A. similis*, which was discovered in the same sample.

Remarks: Although only six specimens are contained in the protargol slides, the population is described as a new species because it has distinct features and the specimens are well impregnated. Only *A. lajacola* is similar in size, shape, and the extrusomes, but differs by the U-shaped macronucleus, the palps between the dorsal brush rows, and the two types of straight (vs. one type and curved) extrusomes. At first glance, *A. pelobia* also resembles *Cultellothrix coemeterii* and certain populations of *Spathidium spathula*. Fortunately, they can be easily distinguished in vivo by the length of the extrusomes: 12 µm vs. 4–7 µm. In protargol preparations, the genus specific features separate, inter alia, these species.

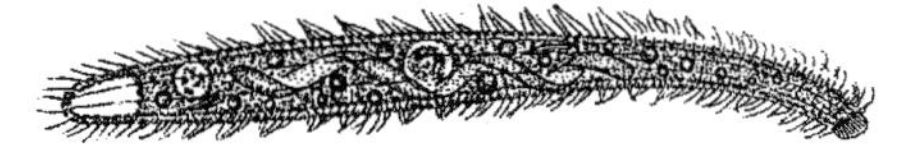

Longispatha FOISSNER, XU & KREUTZ, 2005

2005 *Longispatha* FOISSNER, XU & KREUTZ, J. Eukaryot. Microbiol., 52: 12 – Type species (by original designation): *Longispatha elegans* FOISSNER, XU & KREUTZ, 2005.

Diagnosis: Indistinctly spatulate, flexible Apertospathulidae with brush located dorsally; individual brush rows without anterior tail of ordinary cilia. Mouth (oral bulge) meridionally extending to or near to posterior body end. Right branch of circumoral kinety about as long as body, left branch distinctly shorter extending in anterior body half.

Etymology: Composite of the Latin adjective *longa* (long) and the Latin noun *spatha* (spatula), referring to the long oral bulge and the spathidiid relationship. Feminine gender.

Remarks: FOISSNER et al. (2002) described the first species of this genus, but assigned it to *Apertospathula*. However, they suggested that *A. dioplites* could be the representative of a new genus. This was confirmed by FOISSNER et al. (2005b), who discovered a second species with the same curious organization. Future studies might show that the peculiar impregnation property of brush row 1 should be added to the diagnosis.

Longispatha resembles members of the family Bryophyllidae FOISSNER, 2004 in FOISSNER & LEI (2004) because the oral bulge extends to rear body end. However, only the right half does so, while the left is as in *Apertospathula*, suggesting that *Longispatha* is more closely related to the Spathidiidae s.l. than to the Bryophyllidae. Some specific features, viz., the inconspicuous dorsal brush and the open circumoral kinety suggest *Apertospathula* as the nearest relative of *Longispatha*.

Key to species (careful live observation sufficient)

1 Body narrowly to very narrowly spatulate (~ 6.5:1). One type of oral bulge extrusomes: very narrowly ovate/cuneate and about 4 µm long .... *L. elegans*

– Spatulate to narrowly spatulate (~ 3.7:1). Two types of oral bulge extrusomes: type I asymmetrically obclavate and about 6 µm long; type II oblong and about 2 µm long .... *L. dioplites*

Longispatha dioplites (Foissner, Agatha & Berger, 2002) FOISSNER, XU & KREUTZ, 2005 (Fig. 103a–j; Table 43)

2002 *Apertospathula dioplites* FOISSNER, AGATHA & BERGER, Denisia, 5: 322 (Type slides with protargol-impregnated specimens from type locality are deposited in the Oberösterreichische Landesmuseum in Linz, Upper Austria.).

2005 *Longispatha dioplites* (FOISSNER, AGATHA & BERGER, 2002) n. comb. – FOISSNER, XU & BERGER, J. Eukaryot. Microbiol., 52: 12.

Diagnosis: Size about 95 × 20 μm in vivo. Narrowly spatulate with narrowly cuneate oral bulge curved laterally along anterior body end, left half of bulge shorter than widest trunk region by 60%. Macronucleus elongate reniform; single micronucleus. Two types of extrusomes: type I asymmetrically obclavate and 5–6 × 0.8–1 μm in size; type II oblong and about 2 μm long. On average 9 ciliary rows and 50 circumoral dikinetids. Dorsal brush heterostichad and three-rowed, occupies 13% of body length, rows 1 and 2 each composed of eight dikinetids, row 3 composed of four dikinetids followed by seven monokinetidal bristles on average.

Type locality: Soil from margin of a small pond in the Aubschlucht, Namibia, S23°55' E16°15'.

Etymology: Composite apposition of *di* (two) and *hoplites* (soldier ~ extrusome), referring to a main feature of the species, viz., the two types of extrusomes.

Description: Size 65–135 × 15–30 μm in vivo, usually about 95 × 20 μm; length:width ratio near 4–5:1; laterally flattened up to 2:1, especially in oral region. Shape rather variable because posteriorly narrowed or widened, basically narrowly spatulate with ventral anterior third obliquely truncate (Fig. 103a, f, h; Table 43); very flexible but not contractile; posterior half frequently wrinkled in protargol preparations. Macronucleus usually in anterior body half, elongate reniform with ends often slightly inflated; nucleoli small, globular and numerous. Micronucleus usually attached to concave side of macronucleus, broadly ellipsoidal, rather large and thus easy to recognize in vivo (Fig. 103a, f). Contractile vacuole in rear end, excretory pore(s) not impregnated. Cortex very flexible, thin, contains loosely spaced, pale granules about 0.2 μm across. Two shape and size types of extrusomes in oral bulge and cytoplasm (Fig. 103a–c, i): type I obclavate and 5–6 × 0.7–1 μm in size, less numerous than type II, a 4–5 μm long developmental stage impregnates in the cytoplasm; type II rod-shaped and about 2 μm long. Cytoplasm colourless, usually crammed with lipid droplets 0.2–10 μm across and large food vacuoles containing remnants of ciliates, especially *Metopus hasei*; in protargol preparations studded with minute, black granules obscuring the ciliary pattern. Glides rather rapidly on soil particles and microscope slide.

Cilia about 12 μm long in vivo and protargol preparations, circa 6 μm apart and thus loosely spaced, arranged in an average of nine widely (~ 10 μm) and equidistantly spaced rows distinctly separate from circumoral kinety and hardly curved dorsally at anterior end. Three dorsal rows anteriorly differentiated to short but rather conspicuous brush composed of paired basal bodies with up to 6 μm long bristles, some showing an unusual impregnation capability (Fig. 103a, e, f, g): row 1 composed of an average of eight dikinetids, of which the anterior bristle of the four rear pairs impregnates rather intensely with protargol (Fig. 103d, e); row 2 also composed of an average of eight dikinetids with posterior bristles up to 6 μm long; row 3 consists of an average of four dikinetids with up to 5 μm long bristles and has a monokinetidal tail with 3 μm long bristles impregnating intensely with the protargol method used (Fig. 103f, g, i; Table 43).

Anterior (dorsal) third of oral bulge oblique, slightly convex, and rather conspicuous because bright due to the extrusomes contained, 3–4 μm high, and curved laterally exposing bulge front to the observer; ventral region flattened distinctly, bulge thus

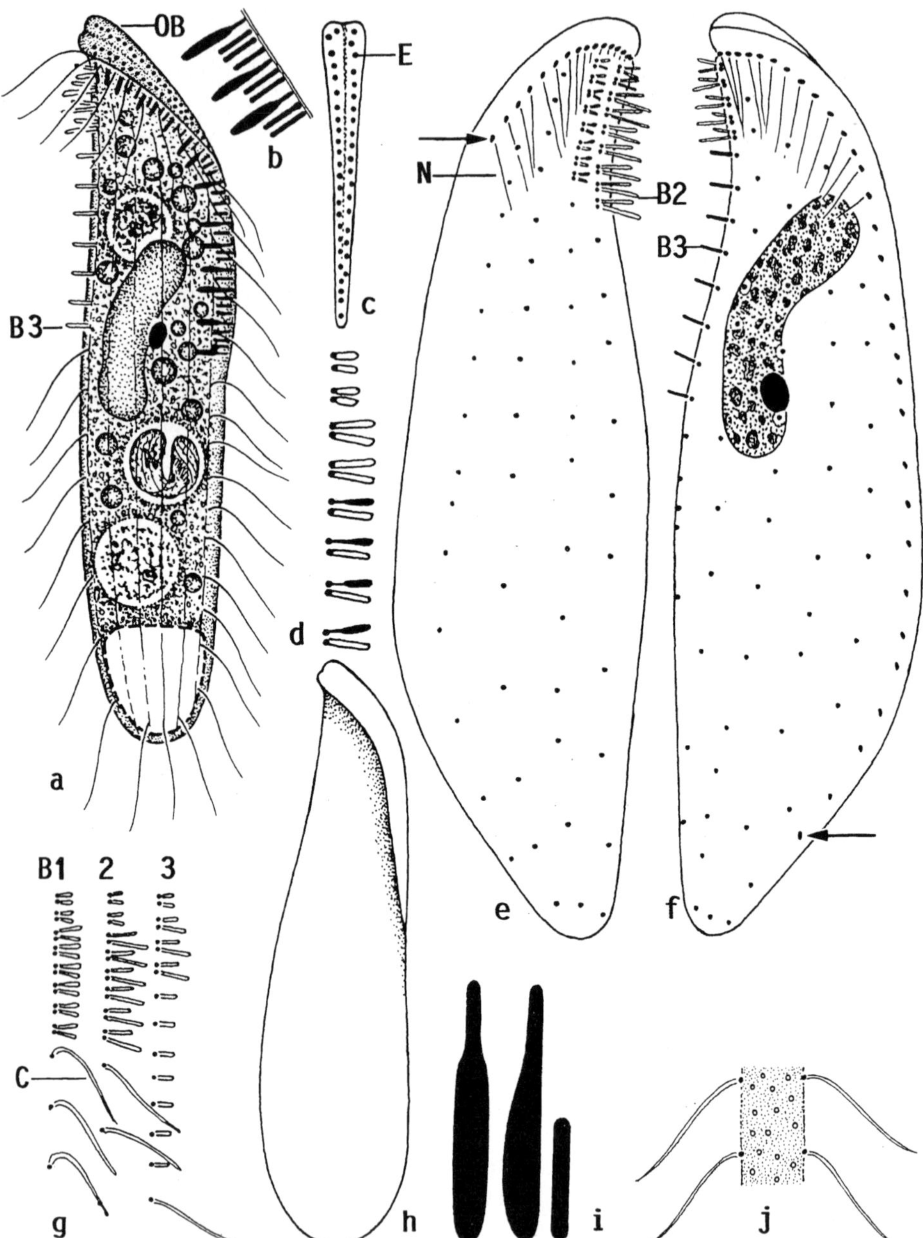

Fig. 103a–j *Longispatha dioplites* from life (a–c, g–j) and after protargol impregnation (d–f). From FOISSNER et al. (2002). **a)** Right side view of a representative specimen, length 100 µm. **b, i)** Arrangement and structure of type I (length 5–6 µm) and type II (2–2.5 µm) extrusomes in the oral bulge. **c)** Frontal view of oral bulge. **d)** Details of brush row 1 (see text). **e, f)** Ciliary pattern and nuclear apparatus of holotype specimen, length 77 µm. Arrows mark ends of circumoral kinety. **g)** Dorsal brush, longest bristles 6 µm. **h)** Shape variant. **j)** Surface view showing cortical granulation. B1-3 – dorsal brush rows, C – ordinary somatic cilium, E – extrusomes, N – nematodesmata, OB – oral bulge.

slenderly cuneate in frontal view (Fig. 103a, c). Circumoral kinety open ventrally, as is typical for the genus, left branch ends subapically, while the right extends to near body end, as indicated by the kinetids of the first kinety, which are slightly enlarged and more intensely impregnated; however, nematodesmata and extrusomes are recognizable only in the anterior body third (Fig. 103e, f). Circumoral kinetids widely spaced, except in dorsal region of kinety, each associated with an about 12 μm long cilium and a short, fine nematodesma contributing to the indistinct oral basket (Fig. 103a, e, f).

Occurrence and ecology: To date found only at type locality, where it was very rare (about 30 specimens in 15 slides), but present for three weeks. Likely, we fixed the culture too early because some dividers were found, indicating onset of exponential growth phase. Several of the specimens contained up to three only partially digested *Metopus hasei*, indicating that *A. dioplites* is a effective predator that rapidly engulfs prey.

Remarks: See *L. elegans*!

Table 43 Morphometric data on *Longispatha dioplites* (upper line, from FOISSNER et al. 2002) and *Longispatha elegans* (lower line, from FOISSNER et al. 2005b)

Characteristics [1]	$\overline{x}$	M	SD	SE	CV	Min	Max	n
Body, length	89.0	90.0	20.0	6.0	22.5	60.0	125.0	11
	99.3	96.0	7.8	2.0	7.9	87.0	113.0	15
Body, width	24.5	25.0	3.5	1.0	14.1	20.0	30.0	11
	15.8	16.0	3.1	0.8	19.8	12.0	23.0	15
Body length:width, ratio	3.7	3.8	0.8	0.2	21.5	2.3	4.8	11
	6.5	6.4	1.4	0.4	21.5	4.1	8.9	15
Body, thickness	19.6	21.0	3.2	1.2	16.1	14.0	23.0	7
	–	–	–	–	–	–	–	–
Anterior body end to macronucleus, distance	27.7	24.0	15.8	4.8	56.9	11.0	58.0	11
	47.5	49.0	9.7	2.5	20.3	27.0	65.0	15
Anterior body end to last left side circumoral kinetid, distance	9.5	10.0	3.1	0.9	32.2	5.0	16.0	11
	12.1	13.0	2.7	0.7	22.4	7.0	17.0	15
Circumoral kinety to last dikinetid of brush row 1, distance	10.4	10.0	1.9	0.6	17.9	7.0	14.0	11
	11.7	12.0	1.3	0.3	10.9	10.0	15.0	15
Circumoral kinety to last dikinetid of brush row 2, distance	11.7	11.0	2.1	0.7	17.7	10.0	16.0	9
	13.1	13.0	1.7	0.4	12.8	10.0	16.0	15
Circumoral kinety to last dikinetid of brush row 3, distance	6.6	7.0	0.9	0.3	13.9	5.0	8.0	11
	6.9	7.0	1.2	0.3	16.8	5.0	10.0	15
Circumoral kinety to last kinetid of monokinetidal tail of brush row 3, distance	30.3	30.0	7.9	2.4	26.1	18.0	42.0	11
	–	–	–	–	–	–	–	–
Macronucleus, length	22.5	23.0	4.1	1.2	18.3	17.0	30.0	11
	17.6	18.0	3.2	0.8	17.9	13.0	26.0	15
Macronucleus, width in mid	6.1	6.0	0.7	0.2	11.5	5.0	7.0	11
	5.3	5.0	0.7	0.2	13.6	4.0	6.0	15
Micronucleus, length	3.5	3.5	–	–	–	2.7	4.0	11

continued

Characteristics [1]	$\overline{x}$	M	SD	SE	CV	Min	Max	n
	2.9	3.0	–	–	–	2.5	3.5	14
Micronucleus, width	3.1	3.0	–	–	–	2.0	3.0	11
	2.6	2.5	–	–	–	2.0	3.0	14
Ciliary rows, number	8.8	9.0	1.0	0.3	11.1	7.0	10.0	11
	7.3	7.5	0.3	0.2	10.6	7.0	9.0	12
Kinetids in a right side ciliary row, number	15.7	16.0	2.2	0.7	14.2	11.0	18.0	11
	12.2	12.0	2.1	0.6	17.2	10.0	15.0	11
Dorsal brush rows, number	3.0	3.0	0.0	0.0	0.0	3.0	3.0	11
	3.0	3.0	0.0	0.0	0.0	3.0	3.0	15
Dikinetids in brush row 1, number	7.6	8.0	–	–	–	7.0	8.0	11
	7.7	8.0	–	–	–	7.0	8.0	15
Dikinetids in brush row 2, number	8.1	8.0	0.8	0.3	9.6	7.0	9.0	9
	8.3	9.0	0.8	0.2	9.8	7.0	9.0	15
Dikinetids in brush row 3, number	4.5	4.0	–	–	–	4.0	5.0	11
	4.4	4.0	–	–	–	4.0	5.0	15
Bristles in monokinetidal tail of brush row 3, number	7.2	7.0	0.9	0.3	12.2	6.0	8.0	11
	–	–	–	–	–	–	–	–
Circumoral kinetids, number	50.1	50.0	5.6	1.9	11.3	42.0	60.0	9
	42.8	42.0	4.8	1.3	11.3	38.0	57.0	15

[a] Data based on mounted and protargol-impregnated (FOISSNER's method) specimens from non-flooded Petri dish cultures. All specimens found were used. Measurements in µm. CV – coefficient of variation in %, M – median, Max – maximum, Min – minimum, n – number of individuals investigated, SD – standard deviation, SE – standard error of arithmetic mean, $\overline{x}$ – arithmetic mean.

Longispatha elegans FOISSNER, XU & KREUTZ, 2005 (Fig. 104a–m, 148d–g; Table 43)

2005 *Longispatha elegans* FOISSNER, XU & KREUTZ, J. Eukaryot. Microbiol., 52: 362, 371 (Type slides with protargol-impregnated specimens are deposited in the Oberösterreichische Landesmuseum in Linz, Upper Austria.).

Diagnosis: Size about 100 × 18 µm in vivo. Narrowly to very narrowly spatulate, oral bulge narrowly cuneate, right half projects sail-like in anterior third, left half shorter than widest trunk region by 25%. Macronucleus oblong; single micronucleus. Extrusomes narrowly ovate, about 4 × 0.8 µm in size. On average 7 ciliary rows and 43 circumoral dikinetids. Dorsal brush heterostichad and three-rowed, occupies 13% of body length, rows 1 and 2 each composed of eight dikinetids on average, row 3 composed of four dikinetids followed by a short monokinetidal bristle tail.

Type locality: Field (Mahada) soil in the surroundings of the village of El Sapo and the Orinoco River, that is, about 50 km north of Pto. Ayachuco, Venezuela, W75° N6°.

Etymology: The Latin adjective *elegans* refers to the elegant shape of the species.

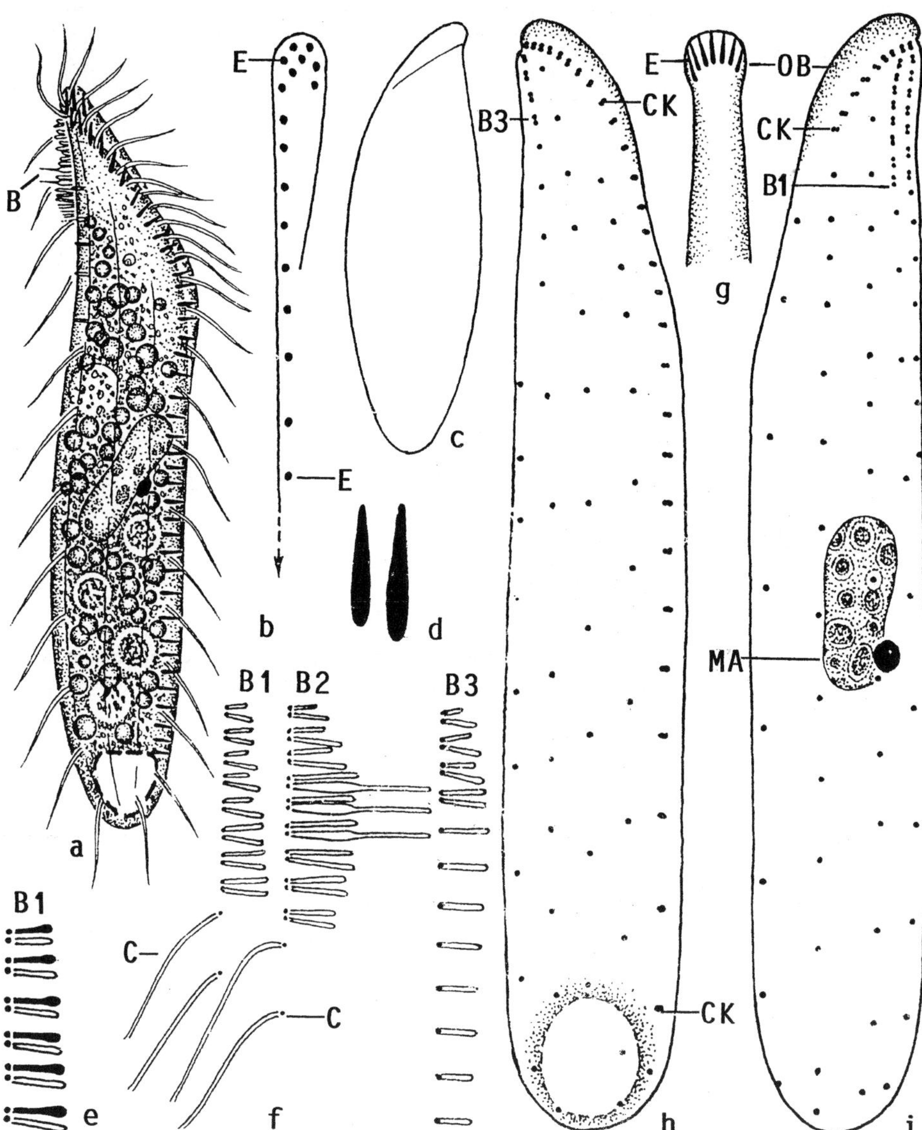

Fig. 104a–i *Longispatha elegans* from life (a–d, f, g) and after protargol impregnation (e, h, i). From FOISSNER et al. (2005b). **a)** Right side view of a representative specimen packed with lipid droplets, length 110 µm. **b)** Frontal view of oral bulge. The right branch of the bulge contains a row of extrusomes and extends to near posterior body end. **c)** Blunt shape variant. **d)** Extrusomes, length 3–4 µm. **e)** Dorsal brush row 1. The anterior bristle of the pairs impregnates darkly. **f)** Dorsal brush, bristles drawn to scale; note the 6 µm long bristles in row 2 and the monokinetidal bristle tail of row 3. Somatic cilia not shown in full length. **g)** Dorsal view of anterior body region. **h, i)** Right and left side view of ciliary pattern and nuclear apparatus of holotype specimen, length 93 µm. B(1-3) – dorsal brush (rows), C – ordinary somatic cilia, CK – circumoral kinety, E – extrusomes, MA – macronucleus, OB – oral bulge.

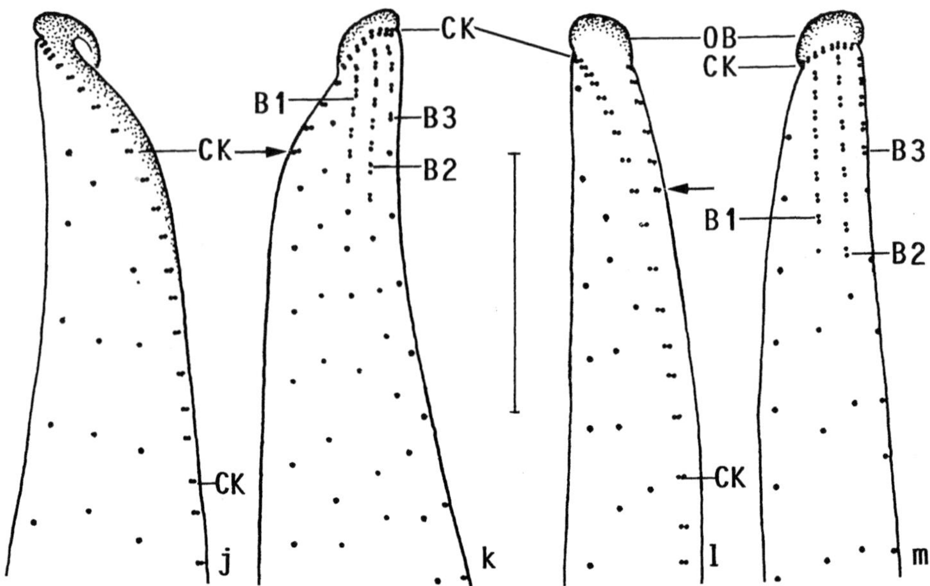

Fig. 104j–m *Longispatha elegans*, somatic and oral ciliary pattern in anterior body region after protargol impregnation (from FOISSNER et al. 2005b). The figures show the main generic feature of *Longispatha*, that is, the right branch of the circumoral kinety extends to near body end (figures 104a, h), while the left ends subapically (arrows). **j, k)** Ventrolateral and dorsolateral view. **l, m)** Ventral and dorsal view. B(1-3) – dorsal brush rows, CK – circumoral kinety, OB – oral bulge. Scale bar 15 µm.

Description: Size 80–120 × 15–25 µm in vivo, usually about 100 × 18 µm; length:width ratio near 6:1 both in vivo and protargol preparations; laterally flattened up to 2:1, especially in hyaline anterior third. Narrowly to very narrowly spatulate to ellipsoidal, anterior third strongly oblique with right oral bulge half sail-like projecting subapically; very flexible but not contractile (Fig. 104a, c, h; Table 43). Nuclear apparatus in mid-body, stands out as a bright structure from strongly granulated cytoplasm. Macronucleus oblong with ends often slightly inflated; nucleoli globular, moderately large. Micronucleus attached to macronucleus, globular to broadly ellipsoidal, about 3 × 2.6 µm in protargol preparations (Fig. 104a, i; Table 43). Contractile vacuole in rear end, excretory pore(s) not impregnated. Cortex flexible and bright, thin, cortical granules not recognizable. Extrusomes studded in widened dorsal end of oral bulge and in right bulge half to form a row extending to rear body end, highly refractive and thus appearing as bright dots when bulge is viewed frontally; lacking in left bulge half. Individual extrusomes narrowly ovate and about 3–4 × 0.8 µm in size, do not impregnate with the protargol method used (Fig. 104a, b, d, g). However, protargol-impregnated cells contain 2 µm long rods and 3 µm long, narrowly ovate to broadly fusiform structures, likely developing extrusomes. Cytoplasm colourless, usually crammed with lipid droplets 1–5 µm across and globular food inclusions, likely heterotrophic flagellates. Movement conspicuous, that is, very slowly gliding on microscope slide.

Cilia about 12 µm long in vivo, circa 10 µm apart and thus very loosely spaced, arranged in an average of seven rather widely (~ 7 µm) and equidistantly spaced rows distinctly separate from circumoral kinety and hardly curved dorsally at anterior end (Fig. 104a, h, i; Table 43). Three dorsal rows anteriorly differentiated to short (about 17% of body length) brush composed of paired, up to 6 µm long bristles showing some specialities (Fig. 104a, e, f, i, k, m, 148d–g; Table 43): row 1 exactly as in *L. dioplites*, that is, composed of an average of eight dikinetids, of which the anterior bristle intensely impregnates with protargol (Fig. 104e); row 2 also composed of an average of eight dikinetids, posterior bristle of two to three middle dikinetids clavate and elongated to 6 µm; row 3 consists of an average of four dikinetids with up to 2 µm long bristles and has a short monokinetidal tail with bristles slightly decreasing in length from about 2 µm anteriorly to 1.5 µm posteriorly, do not impregnate with protargol.

Oral bulge hyaline, right bulge half sail-like protruding subapically, left half very short occupying only 10% of body length, dorsal end head-like inflated, ventral end narrowed, frontal view thus narrowly cuneate. Circumoral kinety composed of an average of 43 dikinetids, open ventrally with short left branch and long right one extending to near rear body end; kinetids widely spaced, except in dorsal region of kinety, each associated with an about 12 µm long cilium and a short, fine nematodesma contributing to the inconspicuous oral basket faintly impregnated in only one specimen (Fig. 104a–c, g–m, 148d–g; Table 43).

Occurrence and ecology: As yet found only at type locality, which is probably sometimes flooded by the Orinoco River, as indicated by the very fine, grey soil. The Mahada *L. elegans* was discovered was in the tenth year of use and planted with bananas. Very low numbers of *L. elegans* occurred in the non-flooded Petri dish culture set up with a mixture of banana and grass litter and surface soil.

Remarks: *Longispatha elegans* differs from *L. dioplites* mainly by the shape and arrangement of the extrusomes. Most other features are highly similar in the African and Venezuelan populations, even the curious impregnation capacity of brush row 1.

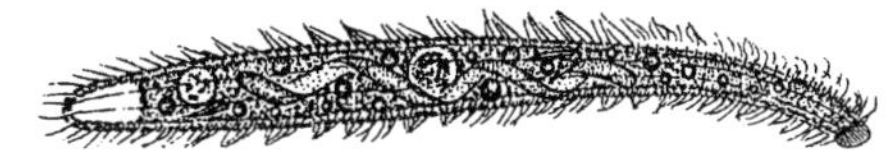

Rhinothrix FOISSNER, XU & KREUTZ, 2005

2005 *Rhinothrix* FOISSNER, XU & KREUTZ, J. Eukaryot. Microbiol., 52: 12 – Type species (by original designation): *Rhinothrix porculus* (PENARD, 1922) FOISSNER, XU & KREUTZ.

Diagnosis: Indistinctly spatulate, flexible Apertospathulidae with dorsally located brush; individual brush rows with short anterior tail of ordinary cilia. Some shortened ciliary rows along right and left side of right branch of circumoral kinety to compensate for the bare, triangular fields caused by the spiral course of the oral bulge. Mouth (oral bulge) spirally extending to or near to posterior body end, dorsal bulge end elongated to a short process (palpus oralis). Right branch of circumoral kinety about as long as body, left branch distinctly shorter extending in anterior body half.

Etymology: Composite of the Greek nouns *rhino* (nose) and *thrix* (hair ~ ciliate s. l.), meaning a ciliate with nose-like process. Feminine gender.

Remarks: FOISSNER et al. (2005b) transfered to this genus also *Spathidium antennatum* and *Holophrya barbatula*. However, they could not exclude that the processes of these species are dorsal palps, as in *Apertospathula verruculifera* or, in the case of *H. barbatula*, the species represents a distinct genus. Likewise, it is uncertain whether the ciliate described by FOISSNER at al. (2005b) is *Spathidium porculus* at all (see *R. barbatula*). All these problems can be solved only by a reinvestigation of various populations of the species involved.

Key to species

1 One palp 2
– Two long (> 10 µm) palps .... *R. barbatula*
2 Palp cylindroidal.. *R. porculus*
– Palp with acute distal end .... *R. antennata*

Rhinothrix porculus (PENARD, 1922) FOISSNER, XU & KREUTZ, 2005 (Fig. 105a–s, 106a–g, n, o, 146a–v; Table 44)

1922 *Legendrea porculus* PENARD, Infusoires: 29.

1926 *Legendrea porculus* PENARD – KAHL, Arch. Protistenk., 55: 269 (brief redescription; name mistakingly written *Sp. porrulus* in figure explanation).

1930 *Spathidium (Legendrea) porculus* PENARD, 1922 – KAHL, Arch. Protistenk., 70: 374 (brief redescription with improved figure).

1930 *Spathidium (Legendrea) porculus* PENARD, 1922 – KAHL, Tierw. Dtl., 18: 154 (revision).

2005 *Rhinothrix porculus* (PENARD, 1922) n. comb. – FOISSNER, XU & KREUTZ, J. Eukaryot. Microbiol. 52: 12 (Detailed redescription of specimens from life and after protargol impregnation. However, the authors mentioned identification problems and thus did not neotypify the species with the slides deposited in the Oberösterreichische Landesmuseum in Linz, Upper Austria. See remarks section for details.).

Improved diagnosis (based on the redescription by FOISSNER et al. 2005b): Size about 100 × 30 µm in vivo. Obovate to oblong with about 10 µm long, cylindroidal palpus oralis. Macronucleus oblong; micronucleus broadly ellipsoidal. Three types of basically rod-shaped extrusomes: type I along ciliary rows, curved, about 7 × 0.5 µm; type II associated with anterior loop of circumoral kinety, circa 50 × 0.5 µm; type III forms oblique rows in oral bulge and palp, oblong, about 1.5 × 0.3 µm. On average 21 meridionally arranged ciliary rows. Dorsal brush with up to 5 µm long bristles, row 1 composed of 15 dikinetids, row 2 of 19, and row 3 of 12 dikinetids followed by an average of 14 monokinetidal bristles. Oral bulge distinct only along oblique anterior body end; left branch of circumoral kinety occupies 27% of body length.

Type locality: Not given; likely a moorland puddle in the surroundings of Geneva, Switzerland, E6° N46°.

Etymology: Not given in the original description. The Latin noun *porculus* (pig) probably refers to both, the pig snout-like palpus oralis and the sapropelic habitat.

Description (by PENARD 1922; Fig. 106d, e): "Corps cylindroïde, trapu, renflé, comprimé seulement à sa partie antérieure, arrondi en arrière, terminé en avant par un bourrelet buccal descendant en diagonale vers le côté ventral. Sous la pointe dorsale, une ligne de soies courtes en brosse, et tout près un appendice ou bâtonnet mobile, mou, plus ou moins extensible. Fente buccale allongée, à parois bordées de trichocystes en petit nombre, en alène, épaissis en arrière et acérés en avant. Lignes ciliaires bien marquées, plus serrées sur l'une des faces que sur l'autre, longitudinales; cils longs, fins, flexueux, plus développés en avant tout autour du bourrelet buccal. Noyau en boudin large, plus ou moins recourbé. Vésicule contractile terminale, susceptible d'une énorme dilatation. Longueur 75 à 125 µ; largeur 25 à 40 µ.

Encore une *Legendrea* qui n'en est peut-être par une! mais elle se rapproche de si près de l'espèce précédente qu'en tout cas le genre est le même. Le caractère spécifique le plus nettement distinctif réside dans cet appendice mobile que l'on rencontre à la commissure supérieure des lèvres; un prolongement étroit, cylindrique, mou, pâle, dirigé tantôt en avant tantôt sur le côté, et dont le rôle est probablement tactile. Peut-être est-il composé de cils soudés fortement entre eux; mais telle n'en est pas l'apparence, et sous une forte compression l'organe ne se résout pas en ses différents éléments. Le corps est lourd et trapu, et l'on y peut distinguer une face ventrale presque plane et une face dorsale fortement convexe; les stries ciliaires sont peu marquées, plus serrées sur l'une des faces que sur l'autre, et couvertes de petites perles sur toute leur longueur. La vésicule contractile se dessine tout d'abord comme une vaste lacune postérieure, traversée par des brides de cytoplasme, puis finit par s'arrondir en une immense vacuole.

Dans la *figure 1*, l'animal est vu par sa face la plus large; et la *fig. 2* le montre par le côte".

Redescription (by FOISSNER et al. 2005b; Fig. 105a–s, 106a–h, n, o, 149a–v; Table 44): "Size 80–130 × 25–40 µm, usually about 100 × 30 µm in vivo, according to some in vivo measurements and the morphometric data. Length:width ratio moderately variable, about 2.5–4.5:1 in vivo, usually 3–4:1 both in vivo and protargol preparations, where specimens tend to become stouter (Table 44); very flexible but not contractile and only slightly flattened laterally, mainly in narrowed posterior half. Shape dominated by the dorsally projecting oral bulge, forming a distinct process (nose or snout), for which we suggest the term palpus oralis; usually more or less narrowly obovate (3–4:1), especially when seen ventrally and dorsally, while narrowly ellipsoidal or oblong in lateral view (Fig. 105a, h–l, 106c, 149a–c); cylindroidal (4–5:1) in starved specimens. Macronucleus in or near mid-body, oblong (~ 3:1), rarely globular or reniform; nucleoli in vivo up to 3 μm across and distinct. Micronucleus in minute depression of mid-macronucleus, conspicuous in vivo because about 6 × 4 μm in size, broadly ellipsoidal to almost hemispherical (Fig. 105a, l, r, 106c, d, h, 149a, c, e, g, s–u; Table 44). Contractile vacuole in rear body end, large, forms from large adventive vacuoles, some excretory pores in pole area (Fig. 105a, h, l, r, 106b, c, 149a–c, g). Cytopyge in posterior pole area, fecal mass globular and compact, migrates through the contractile vacuole (Fig. 149b). Cortex very flexible, of ordinary appearance, contains about eight rows of narrowly spaced, minute (~ 0.5 μm), colourless granules between two kineties each (Fig. 105p); cortex and granules impregnate heavily in ordinarily fixed specimens obscuring the ciliary pattern, but remain unstained in alcohol-fixed cells and silver carbonate preparations.

Extrusome apparatus highly developed, composed of three types of specifically arranged and located toxicysts; inconspicuous in vivo because individual extrusomes thin and/or short and/or obscured by cytoplasmic inclusions; very conspicuous but more or less distorted in silver carbonate preparations; impregnate faintly with protargol, except of some deeply stained cytoplasmic developmental stages. Type I (body) toxicysts within or slightly left of ciliary rows, about as widely spaced as cilia, extend transversely into body proper, basically rod-shaped with narrowed ends, distinctly curved when attached to cortex becoming straight when detached, in vivo about 7 × 0.5 μm in size, circa 15 μm long when exploded; often acicular in silver carbonate preparations (Fig. 105a, c, p, 106c, 149e, g, h, j–o, v). Type II and III toxicysts associated with oral bulge. Type II extrusomes about 50 × 0.5 μm in size, rod-shaped and slightly curved, attached to inner margin of anterior loop of circumoral kinety, extend obliquely backwards; in vivo distinct in flattened cells and very conspicuous, but often wrinkled in silver carbonate preparations; developing type II extrusomes form some small bundles with about five rods each in cytoplasm (Fig. 105a, b, m, o, 106c, 149g, o, r). Type III oral bulge extrusomes oblong and about 1.5 × 0.3 μm in size, arranged in many oblique rows each consisting of 5–10 toxicysts; extrusome rows in both halves of oral bulge, including palpus oralis, form ladder-like pattern extending to rear body end underneath left branch of circumoral kinety; conspicuous in silver carbonate preparations imaging course of oral bulge, fairly distinct also in vivo because individual extrusomes rather refractive; scattered and numerous in cytoplasm, where they are about 2 μm long and often flask-shaped (Fig. 105a, d–f, n, 149f, j–q, v).

Cytoplasm colourless, contains some lipid droplets up to 10 μm across and numerous highly refractive inclusions up to 6 × 4 μm in size in anterior, rarely also in posterior body half (Fig. 105a, l, 149a–e, g, i). Specimens thus dark under low (≤ ×100) bright field magnification and details of oral apparatus and dorsal brush difficult to analyse. Inclusions globular to ellipsoidal, rarely ring-shaped, the larger ones often with a central cleft, do not dissolve when specimens are squashed but disappear in starving

Table **44** Morphometric data on *Rhinothrix porculus* (from FOISSNER et al. 2005b)

Characteristics[a]	$\bar{x}$	M	SD	SE	CV	Min	Max	n
Body (trunk), length	93.1	93.0	9.3	2.0	10.0	77.0	114.0	21
Body, lateral width (ventral to dorsal)	29.6	29.0	2.3	0.5	7.9	25.0	33.0	21
Body, width in ventral or dorsal view	26.7	27.0	3.0	0.7	11.1	22.0	33.0	17
Body length:width, ratio	3.2	3.1	0.3	0.1	9.7	2.5	3.7	21
Palpus oralis, length	8.0	8.0	1.0	0.2	12.1	7.0	10.0	21
Palpus oralis, width	3.6	3.5	0.5	0.1	12.7	3.0	4.0	21
Oral bulge (circumoral kinety), maximum width	7.6	8.0	0.9	0.3	12.0	6.0	9.0	8
Oral bulge (excluding palpus), height	2.7	3.0	0.4	0.1	15.0	2.0	3.0	21
Circumoral kinety to last dikinetid of brush row 1, distance	17.4	18.0	1.6	0.4	9.0	15.0	21.0	19
Circumoral kinety to last dikinetid of brush row 2, distance	19.2	20.0	1.6	0.4	8.4	16.0	22.0	19
Circumoral kinety to last dikinetid of brush row 3, distance	12.2	12.0	0.7	0.2	5.8	11.0	13.0	19
Anterior body end to last left side circumoral kinetid, distance	25.4	25.0	3.6	0.8	14.2	18.0	33.0	21
Anterior body end to macronucleus, distance	38.1	36.0	6.7	1.5	17.6	27.0	51.0	21
Macronucleus figure, length	25.8	25.0	4.1	0.9	16.1	20.0	35.0	21
Macronucleus, maximum width	8.4	9.0	1.3	0.3	15.8	6.0	11.0	21
Macronucleus, number	1.0	1.0	0.0	0.0	0.0	1.0	1.0	21
Micronucleus, across	4.7	5.0	0.5	0.1	9.8	4.0	5.0	21
Micronucleus, number	1.0	1.0	0.0	0.0	0.0	1.0	1.0	21
Ciliary rows, number	20.9	21.0	1.0	0.2	4.8	19.0	23.0	21
Kinetids in a right side complete kinety, number	28.7	29.0	6.8	1.5	23.6	21.0	45.0	21
Kinetids in the left-most short kinety, number	9.6	10.0	2.7	0.7	28.6	5.0	14.0	15
Kinetids in the right-most short kinety, number	12.6	13.0	3.5	1.0	27.5	7.0	18.0	11
Dorsal brush rows, number	3.0	3.0	0.0	0.0	0.0	3.0	3.0	21
Dikinetids in brush row 1, number	14.9	15.0	1.5	0.4	10.3	13.0	17.0	19
Dikinetids in brush row 2, number	18.6	19.0	1.7	0.4	9.0	16.0	21.0	19
Dikinetids in brush row 3, number	12.2	12.0	1.2	0.3	10.1	10.0	14.0	19
Bristles in monokinetidal tail of brush row 3, number	14.4	14.0	2.4	0.7	16.8	12.0	19.0	11

[a] Data based on alcohol-fixed, mounted, protargol-impregnated (FOISSNER's method), and randomly selected specimens from *Sphagnum* mud. Measurements in µm. CV – coefficient of variation in %, M – median, Max – maximum, Min – minimum, n – number of individuals investigated, SD – standard deviation, SE – standard error of arithmetic mean, $\bar{x}$ – arithmetic mean.

cells. Food vacuoles up to 15 µm across, in some specimens contain oral baskets of microthoracid ciliates or euglenids, suggesting that the refractive inclusions described above are paramylon grains; one specimen contained a small desmid. Swims slowly to moderately rapid by rotation about main body axis; highly flexible under slight coverslip pressure.

Unfortunately, no dividers were contained in the protargol slides, likely because the population was disturbed by four days of transport from Constance to Salzburg. However, some conjugants were later seen in the rest of the sample.

Cilia about 10 µm long in vivo, slightly condensed in anterior portion of rows, arranged in an average of 21 meridional, ordinarily (~ 5 µm) and equidistantly spaced rows forming an indistinct *Spathidium* pattern, that is, anterior portion of right side kineties slightly curved dorsally, while left side rows straight or slightly curved ventrally. Some shortened ciliary rows in the curves formed by the sigmoidal oral bulge, that is, above and underneath mid-body (Fig. 105a, q–s, 106a, b, 149a, d, g, t–v; Table 44). Dorsal brush exactly on dorsal side of cell, three-rowed and isostichad, composed of about 1-µm-spaced dikinetids and one or two, rarely three ordinary cilia at anterior end of each row; of ordinary distinctness, longest row 2 extends 21% of body length on average, bristles about as thick as ordinary cilia and decreasing in length from 4–5 µm anteriorly to 1–2 µm posteriorly. Brush row 1 slightly shorter than row 2, composed of an average of 15 dikinetids; longest row 2 composed of 19 dikinetids on average; row 3 shorter by about 30% than rows 1 and 2, composed of an average of 12 dikinetids followed by an average of 14 monokinetidal, 2–3 µm long bristles forming a short tail extending to mid-body (Fig. 105l, q, 106b, h, 149s–v; Table 44).

Oral bulge inconspicuous in vivo, except of palpus oralis, when viewed laterally because merely 3–4 µm high and clearly recognizable only in anterior quarter of cell; conspicuous when viewed ventrally because extending spirally whole body length and clavate with width decreasing from 7–10 µm anteriorly to about 4 µm posteriorly; inflated anterior portion with distinct central cleft dividing bulge in a right half extending to rear body end and a left half lacking in posterior two thirds. Palpus oralis conspicuous because 5–12 µm long in vivo and distinctly projecting from dorsal anterior end of cell; cylindroidal and slightly contractile, belongs to the oral bulge because studded with type III extrusomes (Fig. 105a, h, m–o, q–s, 106a, d–h, 149a–d, k, l, n, p, s–v; Table 44). Circumoral kinety of same shape as oral bulge, that is, lasso-like and widely open ventrally (~ 4 µm), right branch extends full body length, left restricted to anterior 37% of body on average and not continuing posteriorly as ordinary somatic ciliary row; composed of dikinetids very narrowly spaced in anterior third, where faintly impregnated nematodesmata originate from the right and left branch of the kinety, forming a rather conspicuous oral basket (Fig. 105a, m, n, q–s, 106a, 149k, l, n, p, s–v; Table 44)".

Occurrence and ecology: FOISSNER et al. (2005b) found *R. porculus* in the acidic (pH ~ 5) mud of the effluent brook of the Simmelried mire near the village of Hegne, that is, in the surroundings of the town of Constance, Germany, about 250 km NE of Geneva, the supposed type locality. One of us (KREUTZ) investigates this small (diameter < 1 km), eutrophic mire since years, but found *R. porculus* only at two sites, viz., in the effluent mud described above and in an about 70 cm wide puddle filled with *Sphagnum* mud to a deep of over 2 m. When collected in early October 2004, no *R.*

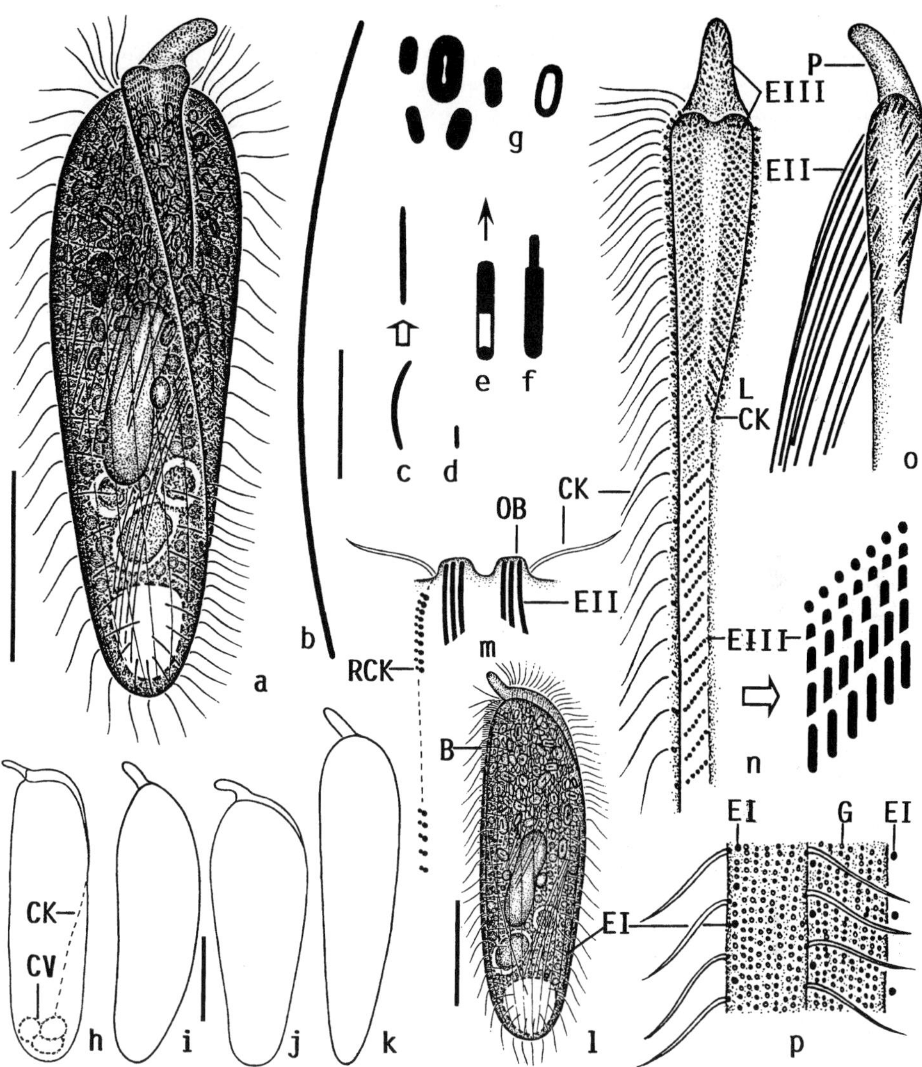

Fig. 105a–p *Rhinothrix porculus* from life (a–e, g–p) and after silver carbonate impregnation (f). From FOISSNER et al. (2005b). **a, l)** Ventral and right side view of representative specimens, length 100 µm. Usually, cells are dark due to refractive inclusions. **b)** Type II oral bulge extrusome, length about 50 µm. **c)** Type I (body) extrusomes attached (curved) and detached (straight) from cortex, length 7 µm. **d)** Type III oral bulge extrusome, length 1.5 µm. **e)** Exploding type II extrusome. **f)** Type III extrusome from cytoplasm, length 2 µm. **g)** Refractive cytoplasmic inclusions, length 2–6 µm. **h–k)** Dorsal (i, k) and right side (h, j) views showing shape and size variability. **m)** Optical section of oral bulge near dorsal end. Circumoral kinety (RCK) from silver carbonate-impregnated specimens. **n)** Ventral view of anterior half of oral bulge; at right part of bulge in oblique view to show the oblong type III extrusomes arranged in many short, oblique rows. **o)** Ventral view of anterior third of oral bulge showing the location of the 50 µm long type II extrusomes not illustrated in full length. **p)** Surface view showing cortical granulation and refractive "pearls" caused by the type I (body) extrusomes. B – dorsal brush, CK – circumoral kinety, CV – contractile vacuole, EI–III – extrusome types, G – cortical granules, LCK – left branch of circumoral kinety, OB – oral bulge, P – palpus oralis, RCK – right branch of circumoral kinety. Scale bars 30 µm (a, h–l) and 10 µm (b–d, drawn to scale).

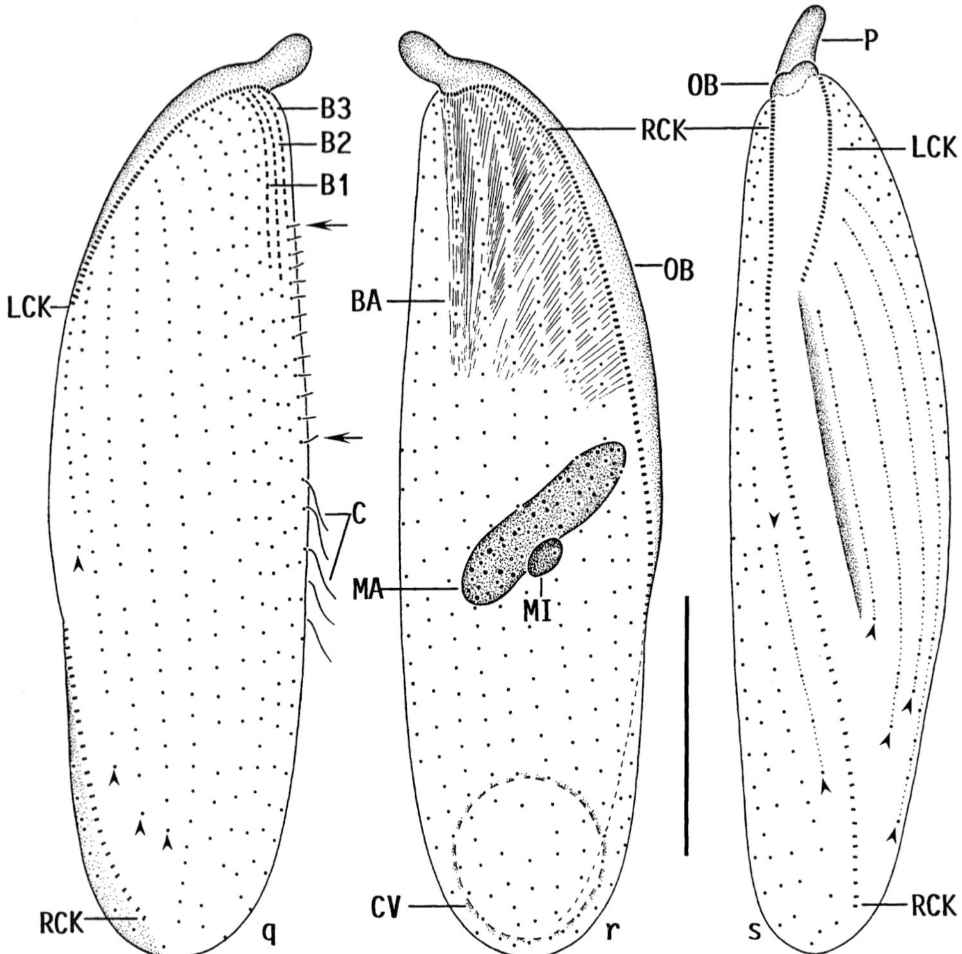

Fig. 105q–s *Rhinothrix porculus*, somatic and oral ciliary pattern after protargol impregnation (from FOISSNER et al. 2005b). Left side, right side, and ventral view of representative specimens, length 100 µm and 103 µm (bar 30 µm). The circumoral kinety (RCK, LCK) is sigmoidal and extends spirally to posterior body end (see also Fig. 106a). The ciliary rows, in contrast, extend meridionally. This is achieved by the insertion of some shortened rows (arrowheads) in the curves of the circumoral kinety. Basically, the circumoral kinety (RCK, LCK) extends whole body length, but the left branch (LCK) is distinctly shortened causing the lasso-like shape of the circumoral kinety. Arrows mark monokinetidal bristle tail of brush row 3. B1-3 – dorsal brush rows, BA – oral basket, C – ordinary somatic cilia, LCK – left branch of circumoral kinety, MA – macronucleus, MI – micronucleus, OB – oral bulge, P – palpus oralis, RCK – right branch of circumoral kinety.

porculus was recognized in the sample. However, after one week standing at room temperature (~ 21 °C), the species became rather abundant, that is, there were 30–50 specimens ml^{-1}. In the sampling jar, *R. porculus* inhabited the about 5 mm thick mud layer, together with various microaerobic (*Spirostomum*, *Pseudoblepharisma*) and anaerobic (*Metopus* spp.) ciliates and a small *Oscillatoria* mat. When isolated on a slide, *R. porculus* survived for some hours but then died, suggesting that it is microaerobic.

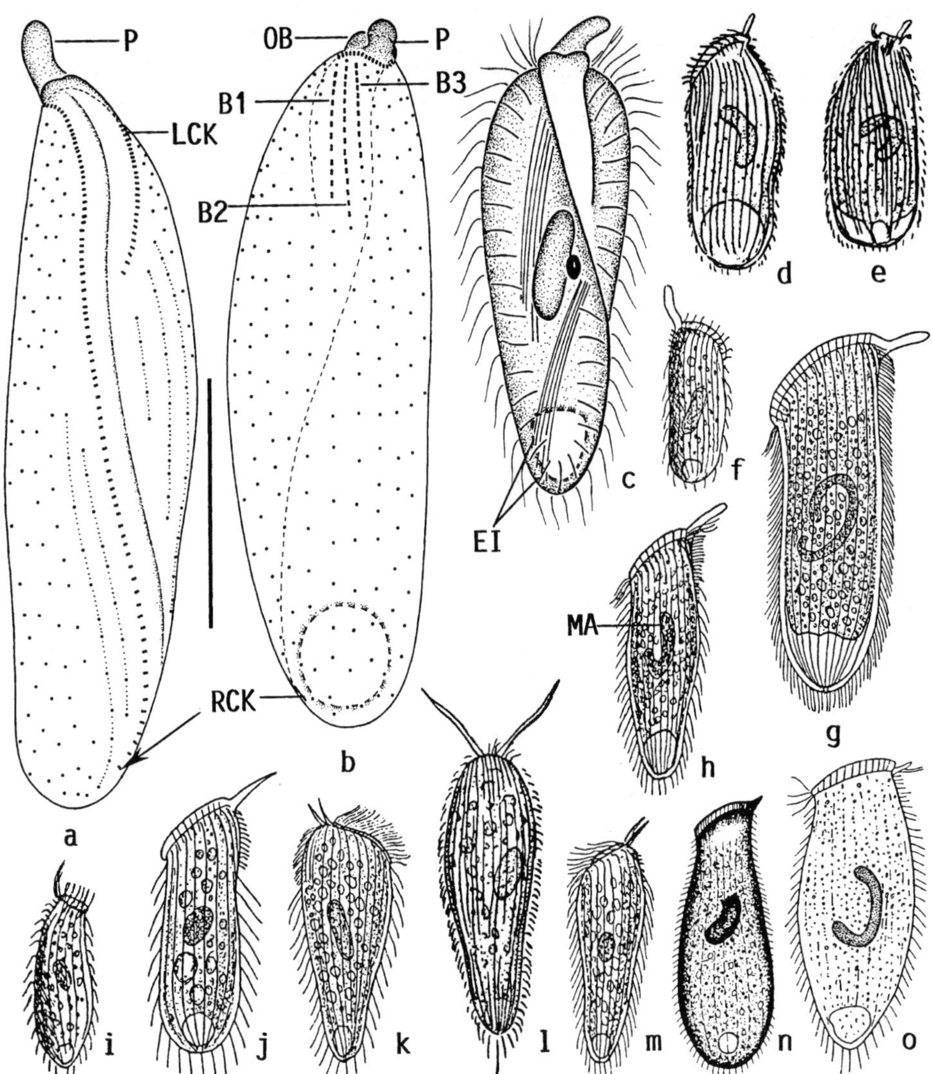

Fig. 106a–g, n, o *Rhinothrix porculus* after protargol impregnation (a, b) and from life (c–g), according to FOISSNER et al. 2005b (a–c), PENARD (d, e), KAHL 1926 (f), 1930a (g) and 1930b (h), MAJEJKAJTKE 1977 (n), and LOKOT 1987 (o). **a–c)** Ciliary pattern of ventral and dorsal side (a, b) and basal organization (c), length 88 µm, 83 µm, 100 µm; scale bar 30 µm for (a, b). The circumoral kinety is sigmoidal and extends spirally to posterior end. The ciliary rows, in contrast, extend meridionally. This is achieved by the insertion of some shortened rows (stippled) in the curves of the circumoral kinety. The dorsal brush is three-rowed and short. B1-3 – dorsal brush rows, EI – type 1 (body) extrusomes, LCK – left branch of circumoral kinety, OB – oral bulge, P – palpus oralis, RCK – right branch of circumoral kinety.

Fig. 106i, j *Rhinothrix antennata* from KAHL (1926) and KAHL (1930a), length 40–90 µm. The acute oral palp distinguishes this species from *R. porculus*.

Fig. 106k, m *Spathidium* ? *(Holophrya) barbatula* after KAHL 1930a, b, length 120 µm.

Fig. 106l *Rhinothrix barbatula* from life, length 95 µm (from PENARD 1922). This species has two palps.

There are few records of *R. porculus*, and all are from Europe and Asia; only three are substantiated by figures. Basically, misidentification of *R. porculus* is unlikely due to the highly characteristic palpus oralis (Fig. 105a, 106c–h, 149a–c). However, there are probably similar, not yet described species/genera, such as *Apertospathula verruculifera*, where the palp is associated with the dorsal brush (Fig. 100i). Unfortunately, PENARD (1922) did not provide information where he discovered *R. porculus*; likely it was a moorland puddle in the surroundings of Geneva, Switzerland. KAHL (1926, 1930a, b) found it several times in sapropelic mud in the surroundings of Hamburg, Germany. He mentioned that the species is common but never abundant and often contains rhodobacteria which, however, could come from prey ciliates. MAJEJKAJTE (1977) reported *R. porculus* from the European part of Russia and illustrated the species (Fig. 106n), but did not give the source; possibly, it is a poor redrawing of PENARD's or KAHL's illustrations. The same is true for the report of LOKOT (1987) from freshwater of the Baikal area, Russia (Fig. 106o).

Further reports (all without illustrations): bottom mud of the Elbe River in Hamburg, Germany (ROY 1938), especially in organically enriched mud above the bottom and in organically enriched bottom sand (BARTSCH & HARTWIG 1984). GRABACKA (1971) found *R. porculus* in the mud of seven out of 10 rearing fish ponds in Poland, both in fertilized and unfertilized ponds. CHORIK & VIKOL (1973) observed *R. porculus* in the bottom mud of cooling plants in Moldavia and classified it as an alpha-mesosaprobic indicator species (VIKOL & CIORIC 1994). OLEKSIV (1985) and BABKO & KOVALCHUK (1992) reported *R. porculus* from some ponds in the Ukraine. ALEKPEROV (1988) found *R. porculus* in nine out of 12 freshwater sites from Azerbaidjan, Russia and calculated a (much too low!) biomass of 0.0005 g for 10^6 individuals (ALEKPEROV et al. 1996).

Remarks: *Rhinothrix porculus*, as redescribed by FOISSNER et al. (2005b), is unequivocally defined by the oral palp, the spiral oral bulge extending to posterior body end, and the highly evolved extrusome apparatus. However, the two last mentioned features were not described by PENARD (1922) and KAHL (1930a, b). Thus FOISSNER et al. (2005b) could not exclude that the species they investigated is not that described by PENARD and KAHL: "Our live observations (Fig. 105a–p, 149a–j) differ markedly from those of PENARD (Fig. 106d, e) and KAHL (Fig. 106f–h), suggesting that the Simmelried population could be a different species or even genus. Neither PENARD (1922) nor KAHL (1926, 1930a, b) mention that the rather distinct oral bulge extends spirally and to posterior body end, although this is easily seen when the specimens slowly rotate about the main body axis. Likewise, they overlooked (?) the highly differentiated extrusome apparatus, though KAHL (1930a) describes 6–8 µm long, rather thick oral bulge extrusomes and PENARD (1922) mentions minute pearls in the ciliary rows, possibly marking the somatic extrusomes present in our population (Fig. 105p). Further, neither PENARD nor KAHL mention the conspicuous accumulation of bright inclusions in the anterior body half. It is difficult to believe that both, PENARD (1922) and KAHL (1926, 1930a, b) overlooked all these distinct features. Our identification is thus based on body size and shape, the conspicuous palpus oralis, and the assumption that PENARD and KAHL studied the species superficially because of the highly characteristic oral palp. Further studies are necessary to validate or falsify our identification. Considering the high diversity of *Spathidium*-like ciliates, we would be not surprised to find a species matching the descriptions of PENARD and KAHL (see also the recent data from *R. barbatula*). Thus, we do not neotypify *Spathidium porculus* with our population".

Based on these problems, we decided to provide both, PENARD's original description and the redescription by FOISSNER et al. (2005b). KAHL (1926, 1930a, b) added few details (see remarks above), but provided excellent figures (Fig. 106f–h).

Rhinothrix antennata (KAHL, 1926) FOISSNER, XU & KREUTZ, 2005 (Fig. 106i, j)

1926 *Spathidium antennatum* KAHL, Arch. Protistenk., 55: 268.
1930 *Spathidium antennatum* KAHL, 1926 – KAHL, Arch. Protistenk., 70: 374.
1930 *Spathidium antennatum* KAHL, 1926 – KAHL, Tierw. Dtl., 18: 154.
2005 *Rhinothrix antennata* (KAHL, 1926) n. comb. – FOISSNER, XU & KREUTZ, J. Eukaryot. Microbiol., 52 : 12.

Diagnosis: Length 40–90 µm. Narrowly spatulate with oblique oral bulge and acicular palpus oralis. Macronucleus ellipsoidal. Oral bulge extrusomes thin, about 8 µm long. Dorsal brush distinct.

Type locality: Road ditch in the city park of Alsterdorf, Hamburg area, Germany, E 10° N 53°30′.

Etymology: Not given. The adjectival noun *antennata* obviously refers to the antenna (feeler)-like dorsal process.

Description: Length 40–90 µm (KAHL 1930b), only "about 30 µm" in KAHL (1926) due to a calibration error of the microscope. Shape elongate bursiform in 1926 (Fig. 106i), while narrowly spatulate in 1930 (Fig. 106j). Anterior end (oral bulge) oblique and with acicular, rather rigid and thin palpus oralis. Macronucleus ellipsoidal to reniform. Contractile vacuole in posterior end. Oral bulge extrusomes thin and about 8 µm long. Cortex conspicuously thick and alveolate. Cytoplasm granulated, specimens inflated anteriorly if granules are numerous. Number of ciliary rows not given; dorsal brush "high", that is, with comparatively long bristles. Very metabolic rooting in organic debris.

Occurrence and ecology: KAHL (1926, 1930a, b) found *R. antennata* several times in flat, astatic rood ditches of the Hamburg area, but numbers where always low. There are two unsubstantiated and thus doubtful records: Black Sea coast (TUCOLESCO 1962) and an unfertilized fish pond in Poland (GRABACKA 1971, SIEMIÑSKA & SIEMIÑSKA 1967).

Remarks: No modern data are available. Thus, the diagnosis is based on the description of KAHL (1930b), where he summarized the observations from KAHL (1926) and (1930a). A remark of KAHL (1930a) shows that the studied *R. antennata* and *R. porculus* several times: "Because of spatial constraints, I provide only figures which are, due to recent observations more detailed than those given in 1926." Redescritpion with modern methods is required (see also remarks to genus *Rhinothrix*!)

Rhinothrix barbatula (PENARD, 1922) FOISSNER, XU & KREUTZ, 2005 (Fig. 106k–m, 150a–e)

1922 *Holophrya barbatula* PENARD, Infusoires: 15.
1930 *Spathidium* sp. – KAHL, Arch. Protistenk., 70: 374.
1930 *Spathidium ? (Holophrya) barbatula* PENARD, 1922 – KAHL, Tierw. Dtl., 18: 154.
1930 *Holophrya barbatula* PENARD, 1922 – KAHL, Tierw. Dtl., 18: 53.
2005 *Rhinothrix barbatula* (PENARD, 1922) n. comb. – FOISSNER, XU & KREUTZ, J. Eukary. Microbiol. 52: 12.

Diagnosis: Not available. Should await redescription.

Type locality: Moorland marsh in the surroundings of Geneva ("marais de Rouelbeau"), Switzerland, E6° N46°.

Description (after PENARD, Fig. 106l): Size about 95 × 30 µm. Elongate obovid and slightly narrowed beyond mid-body. Mouth apical, small, surrounded by two long, supple and flexible palps with acute distal end. Macronucleus in mid-body, large, ellipsoidal. Contractile vacuole in posterior end, with conspicuous excretory pore. Trichocysts lacking, but mucocysts are secreted when carmin is added. Ciliary rows narrowly spaced, extend meridionally, cilia slightly elongated in posterior body region. See type locality for occurrence and ecology.

Remarks: The generic home of this curious species is uncertain because PENARD (1922) observed only a single specimen, no recent data are available, and palps occur also in ciliates with circular, apical mouth opening, viz., in *Holophrya sulcata* PENARD, 1922 and *H. pogonias* SMITH, 1897. Furthermore, KAHL (1930a, b) observed a ciliate with spathidiid oral bulge and two short palps (Fig. 106k, m); he doubts that these specimens are conspecific with PENARDS's population. Redescription urgently needed.

Very recently, M. KREUTZ (Constance; pers. infor.) found a single specimen in the same area as *R. porculus* (Fig. 150a–d). This specimen lacks the highly distinct oral bulge of *R. porculus*, suggesting that *R. barbatula* belongs to another genus. Further, this emphasizes the doubts by FOISSNER et al. (2005b) about the identification of the population they investigated.

References

With few exceptions, titles of journals are given in accordance with the abbreviations found in the 4th edition of the "World List of Scientific Periodicals", published by Butterworths, London, 1963–1965. Also practically without exception, all works cited here have been examined first-hand in order that dates, titles, names of journals or books, and complete pagination could be given with accuracy. No attempts were made to correct title mistakes, even if obvious. Several papers are from journals or books difficult to obtain and to reference. We did or best to cite them in a way that experienced librarians can locate them.

Some authors use (unfortunately) several spellings of their name. For the sake of simplicity, we arbitrarily have choosen one of the spellings and used it throughout the text, while names are written as they stand on the papers in the reference section. This mainly concerns the following authors: DETCHEVA, R.B. (DECHEVA, R., DETSCHEVA, R.B., DETSCHEWA), DRAGESCO-KERNÉIS, A. (DRAGESCO-KERNEIS, A.), GELEI, J. (GELEI, J.v.), GROLIÈRE, C.-A. (GROLIERE, C.-A.), LEPSI, J. (LEPSI, I.), NJINE, T. (NJINÉ, T.).

AESCHT, E. (2001): Catalogue of the generic names of ciliates (Protozoa, Ciliophora). – Denisia, **1:** 1–350.

AESCHT, E., FOISSNER, W. (1993): Effects of organically enriched magnesite fertilizers on the soil ciliates of a spruce forest. – Pedobiologia, **37:** 321–335.

ALEKPEROV, I.K. (1984a): Free-living infusoria of Khachinchay reservoir. – Hydrobiol. J., **20:** 17–22.

ALEKPEROV, I.K. (1984b): New species of ciliates (Gymnostomata) from water bodies of Azerbaijan. – Zool. Zh., **63:** 1417–1420 (in Russian with English summary).

ALEKPEROV, I.K. (1988): Relationships of trophic groups of freshwater infusoria of Azerbaidjan in different biotops and their biocenotic relations with other hydrobionts. – Izv. Akad. Nauka SSR, year 1988: 57–62 (in Russian).

ALEKPEROV, I.K. (1989): Type and rate of sterile substrates colonization by limnetic infusoria. – Zool. Zh., **68:** 15–20 (in Russian with English summary).

ALEKPEROV, I. K., ASADULLAYEVA, E. S., ZAIDOV, T. F. (1996): Methods of Collection and Investigation of Free-living Ciliates and Testate Amoebae. Private edition, St. Petersburg, 51 pp. (in Russian).

AUGUSTIN, H., FOISSNER, W. (1992): Morphologie und Ökologie einiger Ciliaten (Protozoa: Ciliophora) aus dem Belebtschlamm. – Arch. Protistenk., **141:** 243–283.

BABKO, R. V., KOVALCHUK, A. A. (1992): Free-living infusoria of the Sumsoi region. In: MINISTRY OF EDUCATION and PEDAGOGICAL INSTITUTE MAKARENKO OF THE TOWN SUMSOI (eds.): Problems of Protection and Conservation of the Natural Resources of the Sumsoi Region, p. 113–125. Sumsoi (in Ukrainian; ISBN 5-7707-2535-4).

BALTES, W., WENZEL, F. (1966): Nachweis eines chemotaktisch wirksamen Stoffes bei *Spathidium stammeri* WENZEL (Protozoa, Ciliata). – Naturwiss., **53:** 478.

BARTSCH, I., HARTWIG, E. (1984): Die bodenlebende Mikrofauna im Hamburger Hafen. – Arch. Hydrobiol., Suppl., **61:** 543–586.

BAUMEISTER, W. (1962): Planktonkunde für Jedermann. Kosmos, Franckh'sche Verlagshandlung Stuttgart, 121 pp.

BELOKRYS, L.S. (1997): The polytypism of the species *Pseudarcella rhumbleri* Spandel (Tintinnida, Infusoria) in Paleogene of Europe. – Paleont. Zh., **5**: 18–21 (in Russian with English summary).

BERGER, H. (1999): Monograph of the Oxytrichidae (Ciliophora, Hypotrichia). – Monographiae biol., **78:** XII + 1080 pp.

BERGER, H. (2005): Monograph of the Urostylidae (Ciliophora, Hypotrichia). – Monographiae boil. (in press).

BERGER, H., FOISSNER, W., ADAM, H. (1983): Morphology and morphogenesis of *Fuscheria terricola* n. sp. and *Spathidium muscorum* (Ciliophora: Kinetofragminophora). – J. Protozool., **30:** 529–535.

BERGER, H., FOISSNER, W., ADAM, H. (1984): Taxonomie, Biometrie und Morphogenese einiger terricoler Ciliaten (Protozoa: Ciliophora). – Zool. Jb. Syst., **111:** 339–367.

BLATTERER, H., FOISSNER, W. (1988): Beitrag zur terricolen Ciliatenfauna (Protozoa: Ciliophora) Australiens. – Stapfia, Linz, **17:** 1–84.

BOHATIER, J., IFTODE, F., DIDIER, P., FRYD-VERSAVEL, G. (1978): Sur l'ultrastructure des genres *Spathidium* et *Bryophyllum*, ciliés Kinetophragmophora (de Puytorac et all, 1974). – Protistologica, **14:** 189–200.

BÖHME, W. (1978): Das Kühnelt'sche Prinzip der regionalen Stenözie und seine Bedeutung für das Subspezies-Problem: ein theoretischer Ansatz. – Z. zool. Syst. Evolut.-forsch., **16**: 256--266.

BOLTOVSKOY, E. (1954): The species and subspecies concepts in the classification of the Foraminifera. – Micropaleontologist, **8**: 52–56.

BRODSKY, A. (1925): Deux infusoires holotriches nouvelles du Turkestan. – Bull. Univ. Asiae cent., **8:** 40–44 (in Russian with French summary).

BUITKAMP, U. (1977a): Die Ciliatenfauna der Savanne von Lamto (Elfenbeinküste). – Acta Protozool., **16:** 249–276.

BUITKAMP, U. (1977b): Über die Ciliatenfauna zweier mitteleuropäischer Bodenstandorte (Protozoa; Ciliata). – Decheniana (Bonn), **130:** 114–126.

CAIRNS, J., JR. (1965): The protozoa of the Conestoga basin. – Notul. Nat., **375:** 1–14.

CALKINS, G.N. (1919): *Uroleptus mobilis* ENGELM. I. History of the nuclei during division and conjugation. – J. exp. Zool., **27:** 293–357.

CALKINS, G.N., SUMMERS, F.M. (1941; eds.): Protozoa in Biological Research. Columbia Univ. Press, New York, XLI + 1148 pp.

CHORIK, F.P., VIKOL, M.M. (1973): Bottom-dwelling, free-living infusoria of the Kuchurgansk cooling plant of the Moldavian F.R.E.S. – Biol. Resursy Vodonoy Moldavii, **11:** 56–72 (in Russian).

COMMITTEE ON CULTURES, SOCIETY OF PROTOZOOLOGISTS (1958): A catalogue of laboratory strains of free-living and parasitic protozoa. – J. Protozool., **5:** 1–38.

CORLISS, J.O. (1979): The Ciliated Protozoa. Characterization, Classification and Guide to the Literature. 2nd ed. Pergamon Press, Oxford, New York, Toronto, Sydney, Paris, Frankfurt, XIV + 455 pp.

DETCHEVA, R.B. (1972a): Distribution des especes de cilies dans certains affluents Bulgares du Danube, aux eaux polluees. – Annls Stn biol. Besse, **6-7** (years 1971/1972): 261–269.

DETCHEVA, R.B. (1972b): Recherches en Bulgarie sur les protozoaires du sol. – Annls Stn limnol. Besse, **6–7** (years 1971/1972): 273–284.

DETSCHEWA, R.B. (1972c): Beitrag zur Kenntnis der Infusorienfauna (Protozoa-Ciliata) in den Binnengewässern Bulgariens. – Isv. zool. Inst., Sof., **36:** 61–79.

DETCHEVA, R. (1979a). Übersicht über die Ciliatenfauna der bulgarischen Donauzuflüsse. – XXI. Arbeitstagung der Societas Internationalis Limnologiae, Novi Sad, SFR Jugoslawien, pp. 357–364.

DETCHEVA, R. (1979b): Paramètres saprobiologiques et hydrochimiques pour les ciliés de certains affluents Bulgares de la Mer Noire. – Hydrobiology, **9:** 57–73.

DETCHEVA, R.B. (1981): Etude de la composition et de la répartition saisonnière des ciliés de la rivière Maritza. – Hydrobiology, **14:** 16–30.

DETCHEVA, R.B. (1983a): Caracteristiques ecologiques des cilies de la riviere Maritza. – Annls Stn limnol. Besse, **16** (year 1982): 200–219.

DETCHEVA, R.B. (1983b): Distribution des ciliés (Protozoa - Ciliata) par rapport à la pollution hydrochimique de la rivière d'Ossâm - affluent Bulgare du Danube. – Hidrobiologia, **17:** 362–380.

DETCHEVA, R.B. (1986): Composition and distribution of the ciliates (Protozoa, Ciliata) from the Tundža river. – Hydrobiology, **28:** 61–65 (in Russian with English summary).

DETCHEVA, R.B. (1993): Les données écologiques pour les ciliés (Protozoa-Ciliophora) du bassin de la rivière Iskar. – Hydrobiology, **38:** 33–44 (in Russian with French summary).

DINGFELDER, J.H. (1962): Die Ciliaten vorübergehender Gewässer. – Arch. Protistenk., **105:** 509–658.

DRAGESCO, J., DRAGESCO-KERNEIS, A. (1979): Cilies muscicoles nouveaux ou peu connus. – Acta Protozool., **18:** 401–416.

DRAGESCO, J., DRAGESCO-KERNÉIS, A. (1986): Ciliés libres de l'Afrique intertropicale. Introduction à la connaissance et à l'étude des ciliés. – Faune tropicale (Éditions de L'ORSTOM, Paris), **26:** 1–559.

DUDICH, E. (1967): Systematisches Verzeichnis der Tierwelt der Donau mit einer zusammenfassenden Erläuterung. – Limnologie der Donau, **3:** 4–69.

EHRENDORFER, F. (1984): Artbegriff und Artbildung in botanischer Sicht. – Z. zool. Syst. Evolut.-forsch., **22:** 234–263.

FINLAY, B.J., MABERLY, S.C. (2000): Microbial Diversity in Priest Pot: A Productive Pond in the English Lake District. Wilson & Son, Kendal, Cumbria, 73 pp.

FINLAY, B.J., ROGERSON, A., COWLING, A.J. (1988): A Beginner's Guide to the Collection, Isolation, Cultivation and Identification of Freshwater Protozoa. Wilson & Son, Kendal, Cumbria, 78 pp.

FINLAY, B.J., CORLISS, J.O., ESTEBAN, G., FENCHEL, T. (1996): Biodiversity at the microbial level: the number of free-living ciliates in the biosphere. – Q. Rev. Biol., **71:** 221–237.

FOISSNER, W. (1980a): Artenbestand und Struktur der Ciliatenzönose in alpinen Kleingewässern (Hohe Tauern, Österreich). – Arch. Protistenk., **123:** 99–126.

FOISSNER, W. (1980b): Taxonomische Studien über die Ciliaten des Großglocknergebietes (Hohe Tauern, Österreich) IV. Familien Spathidiidae, Podophryidae und Urnulidae. – Verh. zool.-bot. Ges. Wien, **118/119:** 97–112.

FOISSNER, W. (1981a): Morphologie und Taxonomie einiger neuer und wenig bekannter kinetofragminophorer Ciliaten (Protozoa: Ciliophora) aus alpinen Böden. – Zool. Jb. Syst., **108:** 264–297.

FOISSNER, W. (1981b): Die Gemeinschaftsstruktur der Ciliatenzönose in alpinen Böden (Hohe Tauern, Österreich) und Grundlagen für eine Synökologie der terricolen Ciliaten (Protozoa, Ciliophora). – Veröff. Österr. MaB-Programms, **4:** 7–52.

FOISSNER, W. (1984): Infraciliatur, Silberliniensystem und Biometrie einiger neuer und wenig bekannter terrestrischer, limnischer und mariner Ciliaten (Protozoa: Ciliophora) aus den Klassen Kinetofragminophora, Colpodea und Polyhymenophora. – Stapfia, Linz, **12:** 1–165.

FOISSNER, W. (1987a): Soil protozoa: fundamental problems, ecological significance, adaptations in ciliates and testaceans, bioindicators, and guide to the literature. – Progr. Protistol., **2:** 69–212.

FOISSNER, W. (1987b): Faunistische und taxonomische Notizen über die Protozoen des Fuscher Tales (Salzburg, Österreich). – Jber. Haus Nat. Salzburg, **10:** 56–68.

FOISSNER, W. (1987c): Neue terrestrische und limnische Ciliaten (Protozoa, Ciliophora) aus Österreich und Deutschland. – Sber. Akad. Wiss. Wien, **195:** 217–268.

FOISSNER, W. (1988): Gemeinsame Arten in der terricolen Ciliatenfauna (Protozoa: Ciliophora) von Australien und Afrika. – Stapfia, Linz, **17:** 85–133.

FOISSNER, W. (1991): Basic light and scanning electron microscopic methods for taxonomic studies of ciliated protozoa. – Europ. J. Protistol., **27:** 313–330.

FOISSNER, W. (1993): Colpodea (Ciliophora). – Protozoenfauna, **4/1**, 1–798.

FOISSNER, W. (1996a): Faunistics, taxonomy and ecology of moss and soil ciliates (Protozoa, Ciliophora) from Antarctica, with description of new species, including *Pleuroplitoides smithi* gen. n., sp. n. – Acta Protozool., **35:** 95–123.

FOISSNER, W. (1996b): Ontogenesis in ciliated protozoa, with emphasis on stomatogenesis. In: HAUSMANN K., BRADBURY P.C. (eds.), Ciliates: Cells as Organisms, p. 95–177. Fischer, Stuttgart, Jena, Lübeck, Ulm.

FOISSNER, W. (1996c): Terrestrial ciliates (Protozoa, Ciliophora) from two islands (Gough, Marion) in the southern oceans, with description of two new species, *Arcuospathidium cooperi* and *Oxytricha ottowi*. – Biol. Fertil. Soils, **23:** 282–291.

FOISSNER, W. (1997a): Soil ciliates (Protozoa: Ciliophora) from evergreen rain forests of Australia, South America and Costa Rica: diversity and description of new species. – Biol. Fertil. Soils, **25:** 317–339.

FOISSNER, W. (1997b): Protozoa as bioindicators in agroecosystems, with emphasis on farming practices, biocides, and biodiversity. – Agric. Ecosyst. Environ., **62:** 93–103.

FOISSNER, W. (1997c) Faunistic and taxonomic studies on ciliates (Protozoa, Ciliophora) from clean rivers in Bavaria (Germany), with descriptions of new species and ecological notes. – Limnologica, **27:** 179–238.

FOISSNER, W. (1997d): Global soil ciliate (Protozoa, Ciliophora) diversity: a probability-based approach using large sample collections from Africa, Australia and Antarctica. – Biodiv. Concerv., **6:** 1627-1638.

FOISSNER, W. (1998): An updated compilation of world soil ciliates (Protozoa, Ciliophora), with ecological notes, new records, and descriptions of new species. – Europ. J. Protistol., **34:** 195–235.

FOISSNER, W. (1999a): Notes on the soil ciliate biota (Protozoa, Ciliophora) from the Shimba Hills in Kenya (Africa): diversity and description of three new genera and ten new species. – Biodiv. Conserv., **8**: 319–389.

FOISSNER, W. (1999b): Protist diversity: estimates of the near-imponderable. – Protist, **150:** 363–368.

FOISSNER, W. (1999c): Floodplain soils – untouched protozoan biotopes. In: HANSEN, P. J., FENCHEL, T. (eds.): Book of Abstracts. 3rd Europ. Congr. Protistol. and 9th Europ. Conf. Ciliate Biol. Helsingor: Abstract 30.

FOISSNER, W. (2000a): Notes on ciliates (Protozoa, Ciliophora) from *Espeletia* trees and *Espeletia* soils of the Andean Páramo, with descriptions of *Sikorops espeletiae* nov. spec. and *Fragmocirrus espeletiae* nov. gen., nov. spec. – Stud. Neotrop. Fauna, Environm., **35**, 52–79.

FOISSNER, W. (2000b): A compilation of soil and moss ciliates (Protozoa, Ciliophora) from Germany, with new records and descriptions of new and insufficiently known species. – Europ. J. Protistol., **36**, 253–283.

FOISSNER, W. (2000c): Two new terricolous spathidiids (Protozoa, Ciliophora) from tropical Africa: *Arcuospathidium vlassaki* and *Arcuospathidium bulli*. – Biol. Fertil. Soils, **30**, 469–477.

FOISSNER, W. (2002): Neotypification of protists, especially ciliates (Protozoa, Ciliophora). – Bull. Zool. Nom., **59**: 165–169.

FOISSNER, W. (2003a): The Myriokaryonidae fam. n., a new family of spathidiid ciliates (Ciliophora: Gymnostomatea). – Acta Protozool., **42:** 113–143.

FOISSNER, W. (2003b): *Cultellothrix velhoi* gen. n., sp. n., a new spathidiid ciliate (Ciliophora: Haptorida) from a Brazilian floodplain soil. – Acta Protozool., **42:** 47–54.

FOISSNER, W. (2003c): Two remarkable soil spathidiids (Ciliophora: Haptorida), *Arcuospathidium pachyoplites* sp. n. and *Spathidium faurefremieti* nom. n. – Acta Protozool., **42:** 145–159.

FOISSNER, W. (2004a): Ubiquity and cosmopolitanism of protists questioned. – SILnews, **43:** 6–7

FOISSNER, W. (2004b): Some new ciliates (Protozoa, Ciliophora) from an Austrian floodplain soil, including a giant, red "flagship", *Cyrtohymena* (*Cyrtohymenides*) *aspoecki* nov. subgen., nov. spec. – Denisia, **13:** 369–382.

FOISSNER, W., FOISSNER, I. (1988a): The fine structure of *Fuscheria terricola* BERGER et al., 1983 and a proposed new classification of the subclass Haptoria CORLISS, 1974 (Ciliophora, Litostomatea). – Arch. Protistenk., **135:** 213–235.

FOISSNER, W., FOISSNER, I. (1988b): Stamm: Ciliophora. – Catalogus Faunae Austriae, **Ic:** 1–147.

FOISSNER, W., KORGANOVA, G.A.(2000): The *Centropyxis aerophila* complex (Protozoa: Testacea). – Acta Protozool., **39**: 257–273.

FOISSNER, W., LEI, Y.-L. (2004): Morphology and ontogenesis of some soil spathidiids (Ciliophora, Haptoria). – Linzer biol. Beitr., **36:** 159–199.

FOISSNER, W., LEIPE, D. (1995): Morphology and ecology of *Siroloxophyllum utriculariae* (PENARD, 1922) n. g., n. comb. (Ciliophora, Pleurostomatida) and an improved classification of pleurostomatid ciliates. – J. Eukaryot. Microbiol., **42:** 476–490.

FOISSNER, W., PEER, T. (1985): Protozoologische Untersuchungen an Almböden im Gasteiner Tal (Zentralalpen, Österreich). I. Charakteristik der Taxotope, Faunistik und Autökologie der Testacea und Ciliophora. – Veröff. Österr. MaB-Programms, **9:** 27–50.

FOISSNER, W., WENZEL, F. (2004): Life and legacy of an outstanding ciliate taxonomist, Alfred Kahl (1877-1946), including a facsimile of his forgotten monograph from 1943. – Acta Protozool., **43:** 3–69.

FOISSNER, W., PEER, T., ADAM, H. (1985): Pedologische und protozoologische Untersuchung einiger Böden des Tullnerfeldes (Niederösterreich). – Mitt. öst. bodenk. Ges., **30:** 77–117.

FOISSNER, W., BUCHGRABER, K., BERGER, H. (1990): Bodenfauna, Vegetation und Ertrag bei ökologisch und konventionell bewirtschaftetem Grünland: Eine Feldstudie mit randomisierten Blöcken. – Mitt. öst. bodenk. Ges., **41:** 5–33.

FOISSNER, W., BLATTERER, H., BERGER, H. , KOHMANN, F. (1991): Taxonomische und ökologische Revision der Ciliaten des Saprobiensystems – Band I: Cyrtophorida, Oligotrichida, Hypotrichia, Colpodea. – Informa-tionsberichte Bayer. Landesamt für Wasserwirtschaft, **1/91:** 1–478.

FOISSNER, W., BERGER, H. , KOHMANN, F. (1992a): Taxonomische und ökologische Revision der Ciliaten des Saprobiensystems – Band II: Peritrichia, Heterotrichida, Odontostomatida. – Informationsberichte Bayer. Landesamt für Wasserwirtschaft, **5/92:** 1–502.

FOISSNER, W., BERGER, H. , KOHMANN, F. (1994): Taxonomische und ökologische Revision der Ciliaten des Saprobiensystems – Band III: Hymenostomata, Prostomatida, Nassulida. – Informationsberichte Bayer. Landesamtes für Wasserwirtschaft, **1/94:** 1–548.

FOISSNER, W., BERGER, H., BLATTERER, H., KOHMANN, F. (1995): Taxonomische und ökologische Revision der Ciliaten des Saprobiensystems – Band IV: Gymnostomatea, Loxodes, Suctoria. – Informationsberichte Bayer. Landesamtes für Wasserwirtschaft, **1/95:** 1–540.

FOISSNER W., BERGER H., SCHAUMBURG J. (1999): Identification and ecology of limnetic plankton ciliates. – Informationsberichte Bayer. Landesamtes für Wasserwirtschaft, **3/99:** 1–793.

FOISSNER, W., AGATHA, S., BERGER, H. (2002): Soil ciliates (Protozoa, Ciliophora) from Namibia (Southwest Africa), with emphasis on two contrasting environments, the Etosha region and the Namib Desert. – Denisia, **5**: 1–1459.

FOISSNER, W., STRÜDER-KYPKE, M.C., VAN DER STAAY, G.W.M., MOON-VAN DER STAAY, S.-Y., HACKSTEIN, J.H.P. (2003): Endemic ciliates (Protozoa, Ciliophora) from tank bromeliads (Bromeliaceae): a combined morphological, molecular, and ecological study. – Europ. J. Protistol., **39:** 365–372.

FOISSNER, W., BERGER, H., XU, K. & ZECHMEISTER-BOLTENSTERN, S. (2005a): A huge, undescribed soil ciliate (Protozoa: Ciliophora) diversity in natural forest stands of Central Europe. – Biodiv. Conserv., **14:** 617–701.

FOISSNER, W., XU, K., KREUTZ, M. (2005b): The Apertospathulidae, a new family of haptorid ciliates (Protozoa, Ciliophora). – J. Eukaryot. Microbiol., **52:** 360–373.

FRYD-VERSAVEL, G., IFTODE, F., DRAGESCO, J. (1975): Contribution a la connaissance de quelques ciliés gymnostomes II. Prostomiens, pleurostomiens: morphologie, stomatogenèse. – Protistologica, **11:** 509–530.

GELLÉRT, J. (1956): Ciliaten des sich unter dem Moosrasen auf Felsen gebildeten Humus. – Acta biol. hung., **6:** 337–359.

GOLIŃSKA, K. (1995): Formation and orientation of skeletal elements during development of oral territory in a ciliate, *Dileptus*. – Acta Protozool., **34:** 101–113.

GONG, X. (1986): The development of eutrophication in Lake Donghu, Wuhan, during the last two decades based on the investigation of protozoan changes. – Acta hydrobiol. sin., **10:** 340–352 (in Chinese with English summary).

GORALCZYK, K., VERHOEVEN, R. (1999): Bodengenese als Standortfaktor für Mikrofauna – Ciliaten und Nematoden in Dünenböden. In: KOEHLER, H., MATHES, K. , BRECKLING, B. (eds.): Bodenökologie interdisziplinär, p. 105–118. Springer, Berlin, Heidelberg.

GRABACKA, E. (1971): Ciliata of the bottom of rearing fishponds in the Golysz Complex. – Acta hydrobiol., Kraków, **13:** 5–28.

HAUSMANN, K., HÜLSMANN, N., RADEK, R. (2003): Protistology. E. Schweizerbart'sche Verlagsbuchhandlung, Berlin, Stuttgart, XI + 379 pp.

INTERNATIONAL COMMISSION ON ZOOLOGICAL NOMENCLATURE (1999): International Code of Zoological Nomenclature. 4th ed. –Tipografia La Garangola, Padova, XXIX + 306 pp.

KAHL, A. (1926): Neue und wenig bekannte Formen der holotrichen und heterotrichen Ciliaten. Arch. Protistenk., **55:** 197–438.

KAHL, A. (1927): Neue und ergänzende Beobachtungen holotricher Ciliaten. I. – Arch. Protistenk., **60:** 34–129.

KAHL, A. (1930a): Neue und ergänzende Beobachtungen holotricher Infusorien. II. – Arch. Protistenk.,**70:** 313–416.

KAHL, A. (1930b): Urtiere oder Protozoa I: Wimpertiere oder Ciliata (Infusoria) 1. Allgemeiner Teil und Prostomata. – Tierwelt Dtl., **18:** 1–180.

KAHL, A. (1932): Urtiere oder Protozoa I: Wimpertiere oder Ciliata (Infusoria) 3. Spirotricha. – Tierwelt Dtl., **25**: 399–650.

KAHL, A. (1943): Infusorien (1. Teil). Handbücher für die praktische wissenschaftliche Arbeit 31/32, 52 pp. Franckh'sche Verlagshandlung, Stuttgart.

KIRBY, H. (1950): Materials and Methods in the Study of Protozoa. Univ. Calif. Press, Berkeley, X + 72pp.

KUHLMANN, S., PATTERSON, D.J., HAUSMANN, K. (1980): Untersuchungen zu Nahrungserwerb und Nahrungsaufnahme bei *Homalozoon vermiculare*, Stokes 1887 1. Nahrungserwerb und Feinstruktur der Oralregion. – Protistologica, **16:** 39–55.

LEE, J.J., SOLDO, A.T. (1992; eds.): Protocols in Protozoology. Society of Protozoologists, Allen Press, Lawrence.

LEHLE, E. (1989): Beiträge zur Fauna der Ulmer Region II. Ciliaten (Protozoa:Ciliophora) Bioindikatoren in Waldböden. – Mitt. Ver. Naturwiss. Math. Ulm (Donau), **35:** 131–156.

LEIPE, D.D., OPPELT, A., HAUSMANN, K., FOISSNER, W. (1992): Stomatogenesis in the ditransversal ciliate *Homalozoon vermiculare* (Ciliophora, Rhabdophora). – Europ. J. Protistol., **28:** 198–213.

LEPŞI, J. (1959): Über einige neue holotriche Süßwasser-Ciliaten. – Arch. Protistenk., **104:** 254–260.

LOKOT, L.I. (1987): Ökologie der Wimpertiere im Süßwasser des zentralen Baikalgebietes. Akademia Nauka, Novosibirsk, 152 pp. (in Russian).

LUCKOW, M. (1995): Species concepts: assumptions, methods, and applications. – System. Bot., **20:** 589–605.

MADONI, P., GHETTI, P.F. (1981a): Ciliated protozoa and water quality in the torrente Stirone (northern Italy). – Acta hydrobiol., Kraków, **23:** 143–154.

MADONI, P., GHETTI, P.F. (1981b): The structure of ciliated protozoa communities in biological sewage-treatment plants. – Hydrobiologia, **83:** 207–215.

MAJEJKAJTE, S.I. (1977): Ciliata. In: KUTIKOVA, L.A., STAROBOGATOV, J.J. (eds.): Bestimmungsbuch für Süßwasser-Wirbellose im europäischen Teil von Russland (Plankton & Benthos), pp. 46–97. Hydrometer, Leningrad, 510 pp. (in Russian).

MATIS, D., STRAKOVÁ-STRIEŠKOVÁ, M. (1991): Nálevníky (Ciliophora) teplého potoka a termálnych jazierok Bojnických Kúpel'ov. [Ciliates (Ciliophora) of the Teplý brook and thermal lakes in Bojnice spa (Czecho-Slovakia).] – Biológia, Bratisl., **46:** 113–118 (in Czech with English and Russian summaries).

MARGULIS, L., MCKHANN, H.I., OLENDZENSKI, L., HIEBERT, S. (1993): Illustrated Glossary of Protoctista. Jones and Bartlett Publ., Boston, London, XLVIII + 278pp.

MAYR, E. (1963): Animal Species and Evolution. Havard University Press, Cambridge, 797 pp.

MAYR, E. (1975): Grundlagen der Zoologischen Systematik. Pary, Hamburg & Berlin, 370 pp.

MCDADE, K.C. (1995): Species concepts and problems in practice: insight from botanical monographs. – System. Bot., **20**, 606-622.

MOODY, J.E. (1912): Observations on the life-history of two rare ciliates, *Spathidium spathula* and *Actinobolus radians*. – J. Morph., **23:** 349–407.

MOORE, E.L. (1924a): A further study of the effects of conjugation and encystment in *Spathidium spathula*. – J. exp. Zool., **40:** 217–230.

MOORE, E.L. (1924b): Endomixis and encystment in *Spathidium spathula*. – J. exp. Zool., **39:** 317–337.

MOORE, E.L. (1924c): Regeneration at various phases in the life-history of *Spathidium spathula* and *Blepharisma undulans*. – J. exp. Zool., **39:** 249–316.

NING, Y., WANG, S., MA, Z., GAN, Y., YU, Z., GENG, W., GU, X., CAI, G., AN, Q. (1993): Investigations on the freshwater protozoa in Lanzhou. – Chin. J. Zool., **28**: 1–3 (in Chinese).

NIXON, K.C. & WHEELER, Q.D. (1990): An amplification on the phylogenetic species concept. — Cladistics, **6**: 211–223.

O'NEILL, J.P. (1982): The subspecies concept. – Auk, **99**: 609–612.

OLEKSIV, I.T. (1985): Species composition and abundance of planktonic infusoria in ponds. – Gidrobiol J., **21:** 89–93 (in Russian with English summary).

PENARD, E. (1922): Études sur les Infusoires d'Eau Douce. Georg & Cie, Genève, 331 pp.

PETZ, W., FOISSNER, W. (1997): Morphology and infraciliature of some soil ciliates (Protozoa, Ciliophora) from continental Antarctica, with notes on the morphogenesis of *Sterkiella histriomuscorum*. – Polar Rec., **33:** 307–326.

POHLA, H., KASPEROWSKI, E., KANDELER, E., BERTHOLD, A., LÜFTENEGGER, G. (1994): Bodenbiologische, –chemische und –physikalische Erhebungen im Raum Brixlegg: Mikrobiologie und Mikrofauna. – UBA Reports, **94–99b:** 1–115.

PUYTORAC, P. DE (1994; ed.): Infusoires Ciliés. Traité de Zoologie **2(2):** 1–880.

RAIKOV, I.B. (1972): Nuclear phenomena during conjugation and autogamy in ciliates. In: CHEN, T.-T. (ed.): Research in Protozoology, **4**: 146–289. Pergamon Press, Oxford.

RAIKOV, I.B. (1982): The protozoan nucleus. Morphology and evolution. – Cell Biol. Monogr., **9:** XV + 474 pp.

ROLÁN-ALVAREZ, E., ROLÁN, E. (1995): The subspecies concept, its applicability in taxonomy and relationship to speciation. – Argonauta, 9: 1–4.

ROY, H. (1938): Untersuchungen der Detritusfauna im Abwassergebiet bei Hamburg. – Arch. Hydrobiol., **32:** 115–161.

RUSSEV, B., KOVATSCHEV, S., JANEWA, I., KARAPETKOWA, M., UZUNOV, J., DETSCHEWA, R. (1976): Vertreter der bulgarischen Flussfauna als limnosaprobe Bioindikatoren. – Hydrobiology, **4:** 60–66.

RUSSEV, B.K., JANEVA, I.J., DETCHEVA, R.B. (1984): Einige Besonderheiten in der Selbstreinigung des Donauzuflusses Ossam. – Hydrobiology, **21:** 14–27.

SIEMIŃSKA, A., SIEMIŃSKA, J. (1967): Flora i fauna w rejonie zespołu gospodarstw dóswiadczalnych pan i zbiornika Goczałkowickiego na Śląsku. [Flora and fauna in the region of the experimental farms of the Polish Academy of Sciences and of Goczałkowice Reservoir, Silesia.] – Acta hydrobiol., Kraków, **9:** 1–109 (in Polish with English summary).

SMITH, J.C. (1897): Notes on some new, or presumably new, infusoria. – I. – Am. mon. microsc. J., **18:** 141–148.

SONG, W., WEI, J. (1998): Morphological studies on three marine pathogenetic ciliates. – Acta Hydrobiol. Sin., **22**: 361–366 (in Chinese with English Summary).

STEARN, W.T. (1992): Botanical Latin. History, Grammar, Syntax, Terminology and Vocabulary. 4th ed. David & Charles, Newton, Abbot, Devon. XIV + 546 pp.

SUDZUKI, M. (1979): On the microfauna of the Antarctic region III. Microbiota of the terrestrial interstices. – Mem. nat. Inst. Polar Res., Tokyo, Special Issue, **11:** 104–126.

TAKAHASHI, T., SUHAMA, M. (1991): Ciliated protozoan fauna in paddy fields. – Bull. Jap. Soc. Microb. Ecol., **6:** 103–115 (in Japanese with English summary).

TIRJAKOVÁ, E. (1988): Structures and dynamics of communities of ciliated protozoa (Ciliophora) in field communities. I. Species composition, group dominance, communities. – Biológia, Bratisl., **43:** 497–503.

TUCOLESCO, J. (1962a): Écodynamique des infusoires du littoral Roumain de la Mer Noire et des bassins salés para-marins. – Annls Sci. nat. (Zool.), **3:** 785–845.

TUCOLESCO, J. (1962b): Études protozoologiques sur les eaux Roumaines I. Espèces nouvelles d'infusoires de la mer Noire et des bassins salés paramarins. – Arch. Protistenk., **106:** 1–36.

TURNER G. F. (1999): What is a fish species? – Reviews in Fish Biology and Fisheries, **9**: 281–297.

VERNI, F., ERRA, F. (1994): Soil ciliates in soil pollution monitoring. – Boll. Mus. Ist. biol. Univ. Genova, **58–59** (years 1992-1993)**:** 57–67.

VERNI, F., GUALTIERI, P. (1997): Feeding behaviour in ciliated protists. – Micron, **28:** 487–504.

VICOL, M., CIORIC, T. (1994): Valenta saprobică a infuzorilor (Protozoa, Ciliata) in bazinele acvatica din republica Moldova. [Saprobic valency of infusoria (Protozoa, Ciliata) in reservoirs of republic Moldova.] – Buletinul Academiei de Stiinte a Republicii Moldova Stiinte Biologice si Chimice, **2:** 68–72, 77 (in Moldavian).

VUXANOVICI, A. (1962a): Contribuţii la sistematica ciliatelor (Nota I). – Studii Cerc. Biol. (Biol. Anim.), **14:** 197–215 (in Rumanian with Russian and French summaries).

VUXANOVICI, A. (1962b): Contribuţii la sistematica ciliatelor (Nota II). – Studii Cerc. Biol. (Biol. Anim.), **14:** 331–349 (in Rumanian with Russian and French summaries).

VUXANOVICI, A. (1962c): Contribuţii la sistematica ciliatelor (Nota III). – Studii Cerc. Biol. (Biol. Anim.), **14:** 549–573 (in Rumanian with Russian and French summaries).

WANG, J. (1977): Protozoa from some districes of Tibetan Plateau. – Acta zool. sin., **23:** 131–160 (in Chinese with English summary).

WANG, S., MA, Z. (1994): The ciliates from industrial wastewater of Xining and three indicative species of Rhizopoda on high mountains and subhigh mountains. – J. Northwest Normal Univ. (Nat. Sci.), **30:** 60–65 (in Chinese with English summary).

WENZEL, F. (1953): Die Ciliaten der Moosrasen trockner Standorte. – Arch. Protistenk., **99:** 70–141.

WENZEL, F. (1955): Über eine Artentstehung innerhalb der Gattung *Spathidium* (Holotricha, Ciliata). [*S. ascendens* n. sp. und *S. polymorphum* n. sp.]. – Arch. Protistenk., **100:** 515–540.

WENZEL, F. (1959): Ein Beitrag zur Kenntnis der Ciliatengattung *Spathidium* (*S. stammeri* n. sp.). – Zool. Anz., **163:** 209–216.

WENZEL, F., BALTES, W. (1967): Untersuchungen über eine chemotaktische Reaktion von *Spathidium stammeri* (Ciliata, Holotricha) beim Beuteerwerb. – Zool. Anz., **178:** 151–154.

WILLIAMS, D.B. (1958): Growth of *Spathidium spathula*, a predaceous holotrich, in bacteria-free medium. – J. Protozool., Suppl., **5:** 24–25 (abstract).

WILLIAMS, D.B. (1980): Clonal aging in two species of *Spathidium* (Ciliophora: Gymnostomatida). – J. Protozool., **27:** 212–215.

WILLIAMS, D.B. (1989): Does encystment of *Spathidium spathula* have any effect on sexual immaturity? – J. Protozool., **36:** p. 2A, abstract 9.

WILLIAMS, D.B., WILLIAMS, B.D., HOGAN, B.K. (1981): Ultrastructure of the somatic cortex of the gymnostome ciliate *Spathidium spathula* (O.F.M.). – J. Protozool., **28:** 90–99.

WIRNSBERGER, E., FOISSNER, W., ADAM, H. (1984): Morphologie und Infraciliatur von *Perispira pyriformis* nov. spec., *Cranotheridium foliosus* (FOISSNER, 1983) nov. comb. und *Dileptus anser* (O. F. MÜLLER, 1786) (Protozoa, Ciliophora). – Arch. Protistenk., **128:** 305–317.

WOODRUFF, L.L., MOORE, E.L. (1924): On the longevity of *Spathidium spathula* without endomixis or conjugation. – Proc. natn. Acad. Sci., Wash., **10:** 183–186.

WOODRUFF, L.L., SPENCER, H. (1921a): The early effects of conjugation on the division rate of *Spathidium spathula*. – Proc. Soc. exp. Biol. Med., **18:** 240–241.

WOODRUFF, L.L., SPENCER, H. (1921b): The food reactions of the infusorian *Spathidium spathula*. – Proc. Soc. exp. Biol. Med., **18:** 183–184.

WOODRUFF, L.L., SPENCER, H. (1921c): The survival value of conjugation in the life history of *Spathidium spathula*. – Proc. Soc. exp. Biol. Med., **18:** 303–304.

WOODRUFF, L.L., SPENCER, H. (1922): Studies on *Spathidium spathula* I. The structure and behavior of *Spathidium*, with special reference to the capture and ingestion of its prey. – J. exp. Zool., **35:** 189–205.

WOODRUFF, L.L., SPENCER, H. (1924): Studies on *Spathidium spathula* II. The significance of conjugation. – J. exp. Zool., **39:** 133–196.

XU, K., FOISSNER, W. (2003): From the temporary cytostome towards a permanent cytopharynx: a new evolutionary line in spathidiid gymnostomes (Ciliophora). 4th Europ. Cong. Protistol. and 10th Europ. Confer. Ciliate Biol. San Benedetto del Tronto (AP), Italy. Abstract book, p. 20.

XU, K., FOISSNER, W. (2004): Body, nuclear and ciliary changes during conjugation of *Protospathidium serpens* (Ciliophora, Haptoria). – J. Eukaryot. Microbiol., **51**: 605–617.

XU, K., FOISSNER, W. (2005a): Descriptions of *Protospathidium serpens* (Kahl, 1930) and *P. fraterculum* n. sp. (Ciliophora, Haptoria), two species based on different resting cyst morphology. – J. Eukaryot. Microbiol., **52:** 298–309.

XU, K., FOISSNER, W. (2005b): Morphology, ontogenesis, and encystment of a soil ciliate (Ciliophora, Haptorida), *Arcuospathidium cultriforme* (Penard, 1922), with models for the formation of the oral bulge, the ciliary patterns, and the evolution of the spathidiids. – Protistology, **4:** 5–55.

ZUSI, R.L. (1982): Infraspecific geographic variation and the subspecies concept. – Auk, **99**: 606–608.

Systematic Index

The index contains (i) all names of spathidiid ciliates, (ii) food organisms, and (iii) other ciliates mentioned in the book. The index is two-sided, that is, species appear both with the generic name ahead (for example, *Arcuospathidium bulli*) and with the species name first (*bulli, Arcuospathidium*). Valid (in our judgment) spathidiid taxa (species, subspecies, genera, families) are in **boldface print**, all other names are in ordinary font (for example, the synonym *Arcuospathidium australe*). **Boldface page number** indicates the location of the main description of a taxon, while "T" locates the Table(s) with the morphometric data.

Micrographs

Halftone micrographs were collected in plates put at the end of the book to guarantee good reproduction. With few exceptions, all are originals or are from our previous papers. Usually, scale bars are omitted for the reasons explained on pages 55, 56.

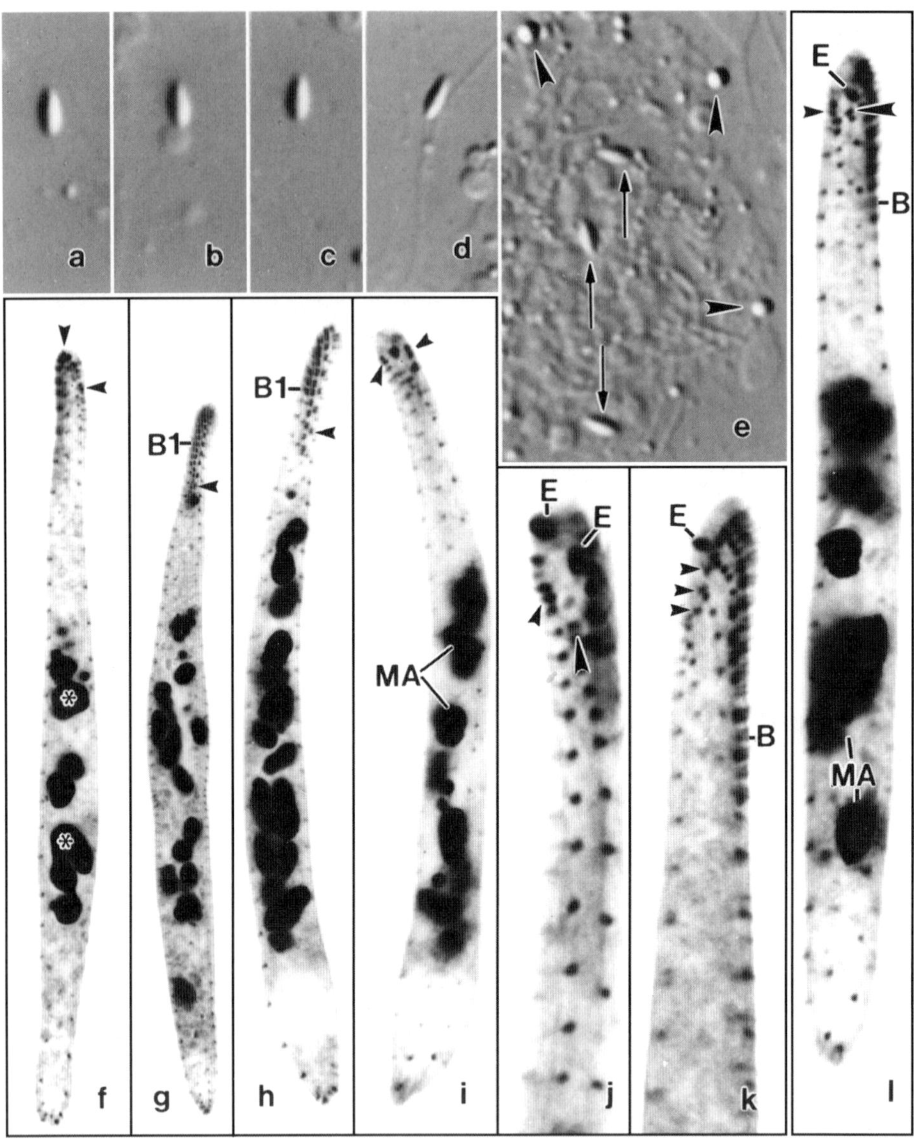

Fig. 107a–l *Edaphospathula fusioplites* from life (a–e) and after protargol impregnation (f–l). From FOISSNER et al. (2005a). **a–d)** Oral bulge extrusomes are fusiform to indistinctly ovate and 1.5–2 μm long. **e:** Oral bulge extrusomes in lateral (arrows) and transverse (arrowheads) view where they appear as conspicuous, highly refractive globules. **f)** Right side overview showing the minute oral bulge (arrowheads) and eight macronuclear nodules (asterisks). **g, h)** Dorsal views of specimens with many scattered macronuclear nodules. Dorsal brush 1 consists of only three dikinetids, while dikinetids of row 3 are very widely spaced (arrowheads). **i–l)** Ventral (i, j) and left side (k, l) views showing the circumoral kinetofragments (arrowheads) composed of only one to two dikinetids. Note dorsal location of brush and the darkly impregnated extrusomes in the oral bulge. B(1) – dorsal brush (row), E – extrusomes, MA – macronuclear nodules.

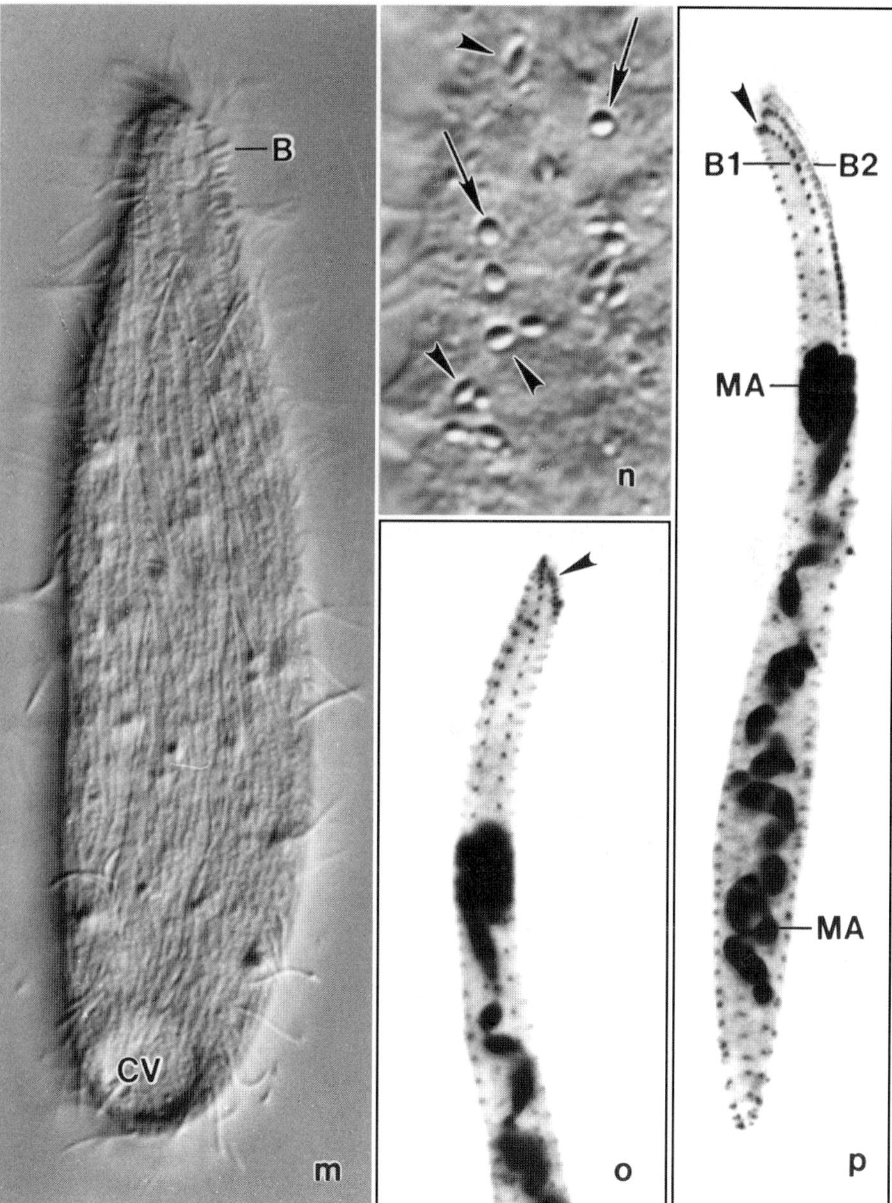

Fig. 107m–p *Edaphospathula paradoxa* nov. spec. from life (m, n) and after protargol impregnation (o, p). **m)** Surface view of a heavily squashed and flattened specimen showing the longitudinal rows of minute cortical granules and part of the dorsal brush. **n)** Cytoplasmic extrusomes in transverse (arrows) and lateral (arrowhead) view, size about 1.5 × 1 µm. **o, p)** Right and left side view of a typical specimen with nodulated and spiralized macronucleus. Arrowhead in (p) marks circumoral kinetofragment at end of kinety bearing brush row 1, which consists of only two dikinetids. Note very dense spacing of brush row 2 dikinetids. B (1, 2) – dorsal brush (rows), CV – contractile vacuole, MA – macronucleus.

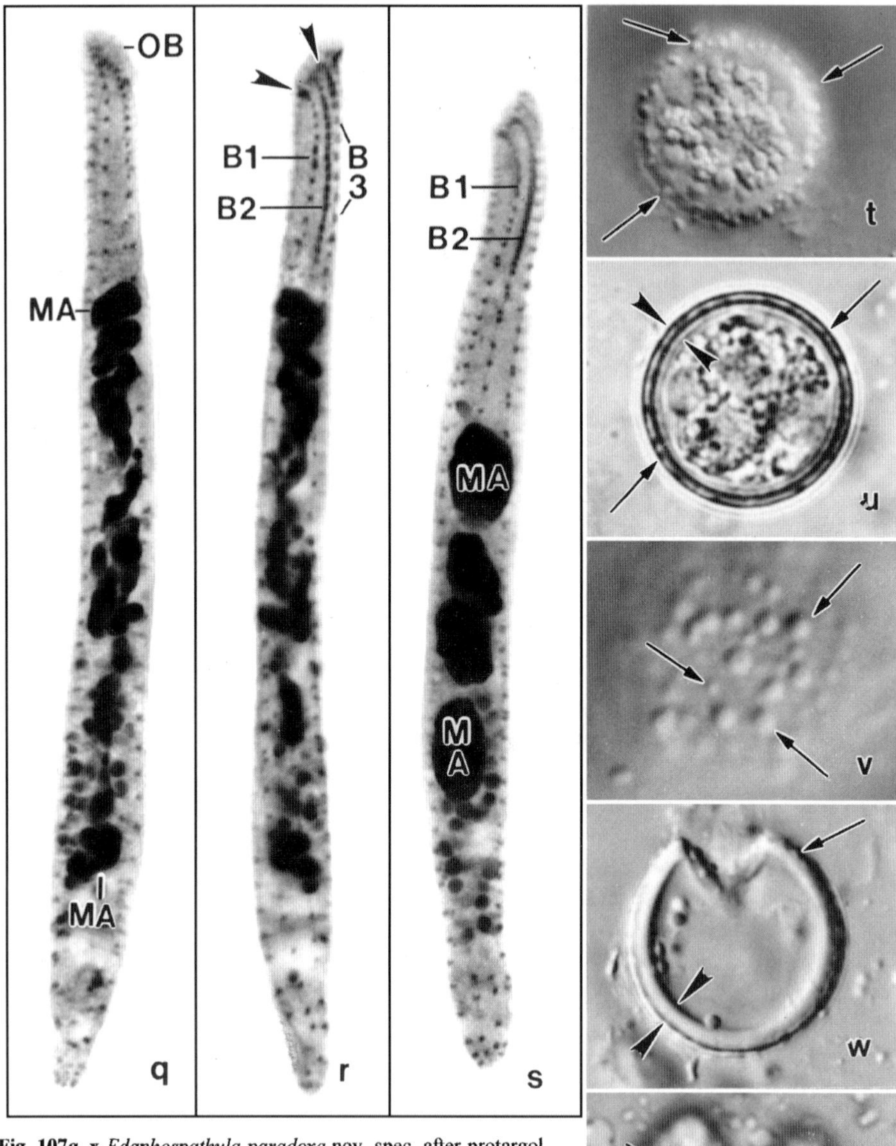

Fig. 107q–x *Edaphospathula paradoxa* nov. spec. after protargol impregnation (q–s) and from life (t–x). **q, r)** Right and left side of a specimen with strongly nodulated macronucleus. Note the circumoral kinetofragments (arrowheads) and the characteristic dorsal brush. **s)** A specimen with four large macronuclear nodules. **t–x)** Intact (t–v) and squashed (w, x) resting cyst showing the thick wall (opposed arrowheads) and the minute hemispheres (arrows) on the surface of the endocyst wall. All figures show the same cyst because only one out of 10 specimens encysted. B1-3 – dorsal brush rows, MA – macronucleus, OB – oral bulge.

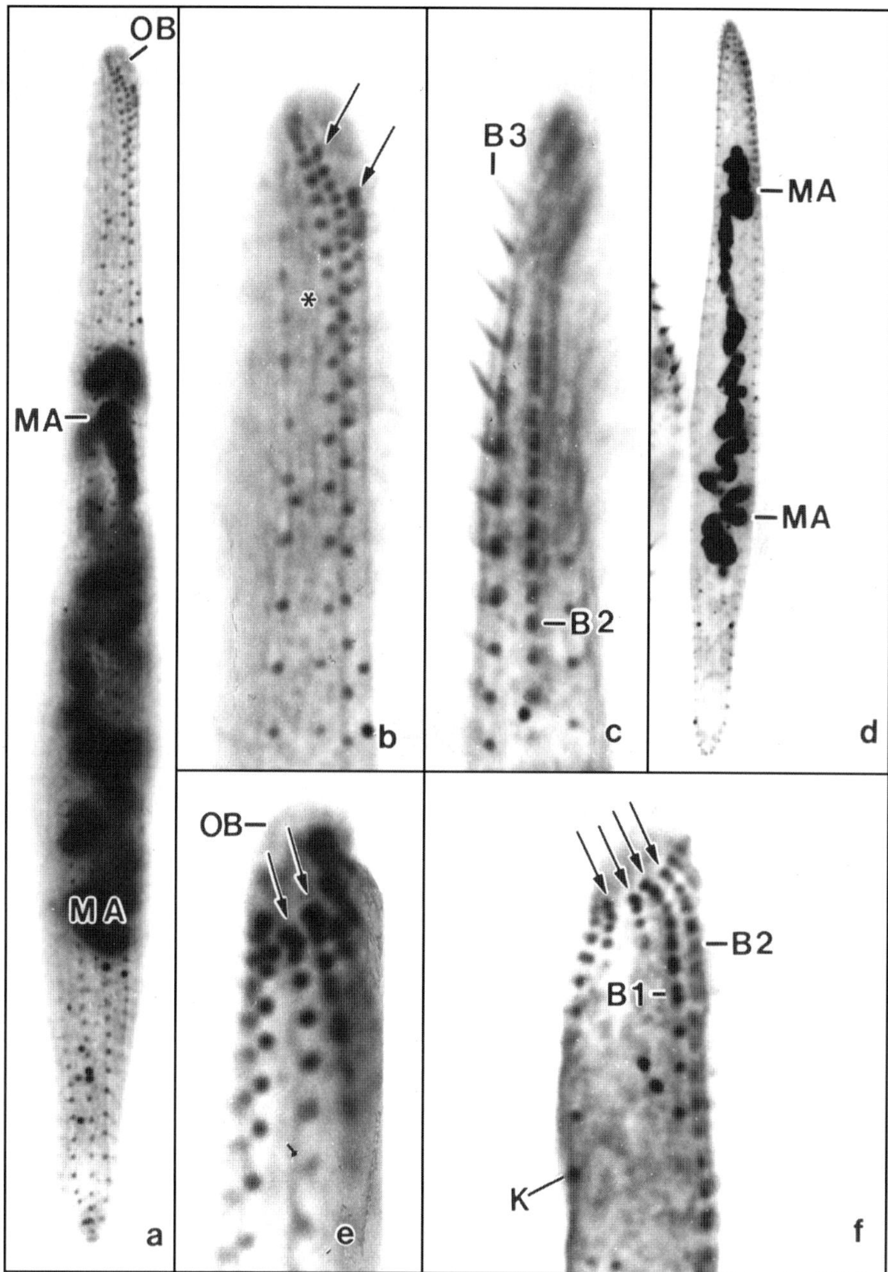

Fig. 108a–f *Edaphospathula espeletiae,* ciliary pattern and macronucleus after protargol impregnation (from FOISSNER 2000a). Figures (a–c) show the holotype specimen. Arrows mark circumoral dikinetid at anterior end of ciliary rows, a highly characteristic feature found also in *E. fusioplites*. Asterisk denotes slightly increased distance between ventral and lateral ciliary rows. B(1-3) – dorsal brush (rows), K – somatic ciliary row, MA – macronucleus, OB – oral bulge.

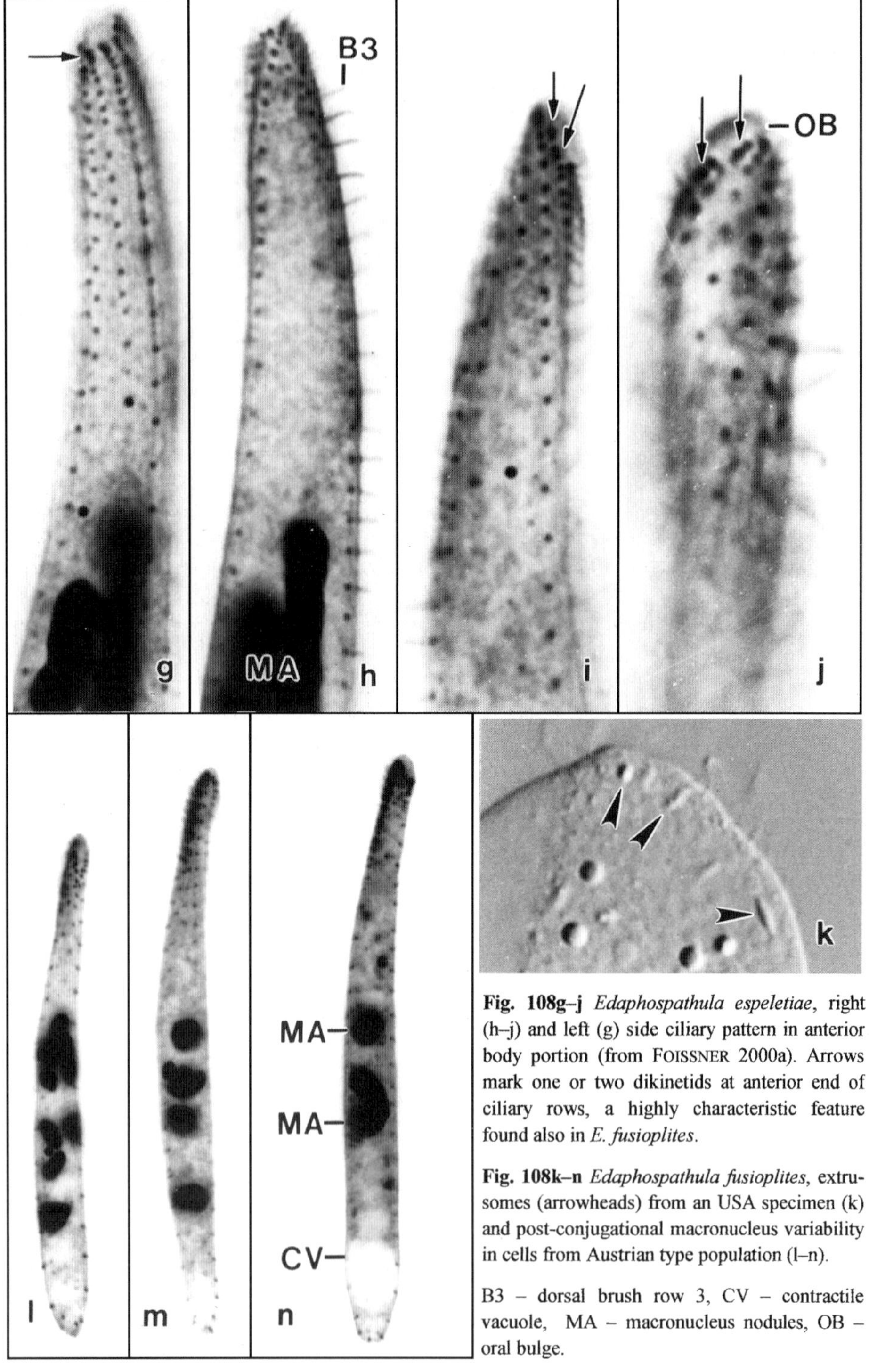

Fig. 108g–j *Edaphospathula espeletiae*, right (h–j) and left (g) side ciliary pattern in anterior body portion (from FOISSNER 2000a). Arrows mark one or two dikinetids at anterior end of ciliary rows, a highly characteristic feature found also in *E. fusioplites*.

Fig. 108k–n *Edaphospathula fusioplites*, extrusomes (arrowheads) from an USA specimen (k) and post-conjugational macronucleus variability in cells from Austrian type population (l–n).

B3 – dorsal brush row 3, CV – contractile vacuole, MA – macronucleus nodules, OB – oral bulge.

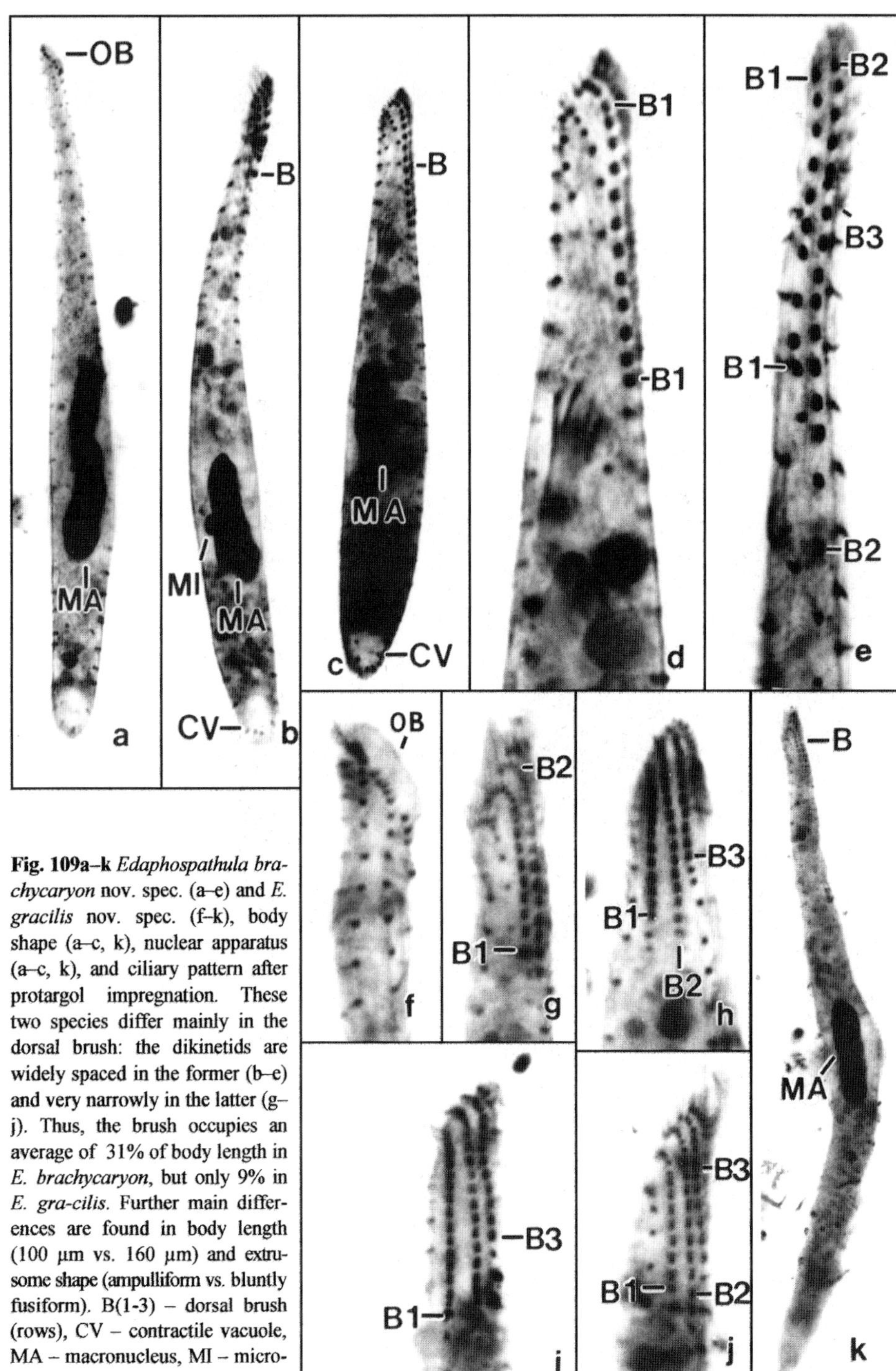

Fig. 109a–k *Edaphospathula brachycaryon* nov. spec. (a–e) and *E. gracilis* nov. spec. (f–k), body shape (a–c, k), nuclear apparatus (a–c, k), and ciliary pattern after protargol impregnation. These two species differ mainly in the dorsal brush: the dikinetids are widely spaced in the former (b–e) and very narrowly in the latter (g–j). Thus, the brush occupies an average of 31% of body length in *E. brachycaryon*, but only 9% in *E. gra-cilis*. Further main differences are found in body length (100 µm vs. 160 µm) and extrusome shape (ampulliform vs. bluntly fusiform). B(1-3) – dorsal brush (rows), CV – contractile vacuole, MA – macronucleus, MI – micronucleus, OB – oral bulge.

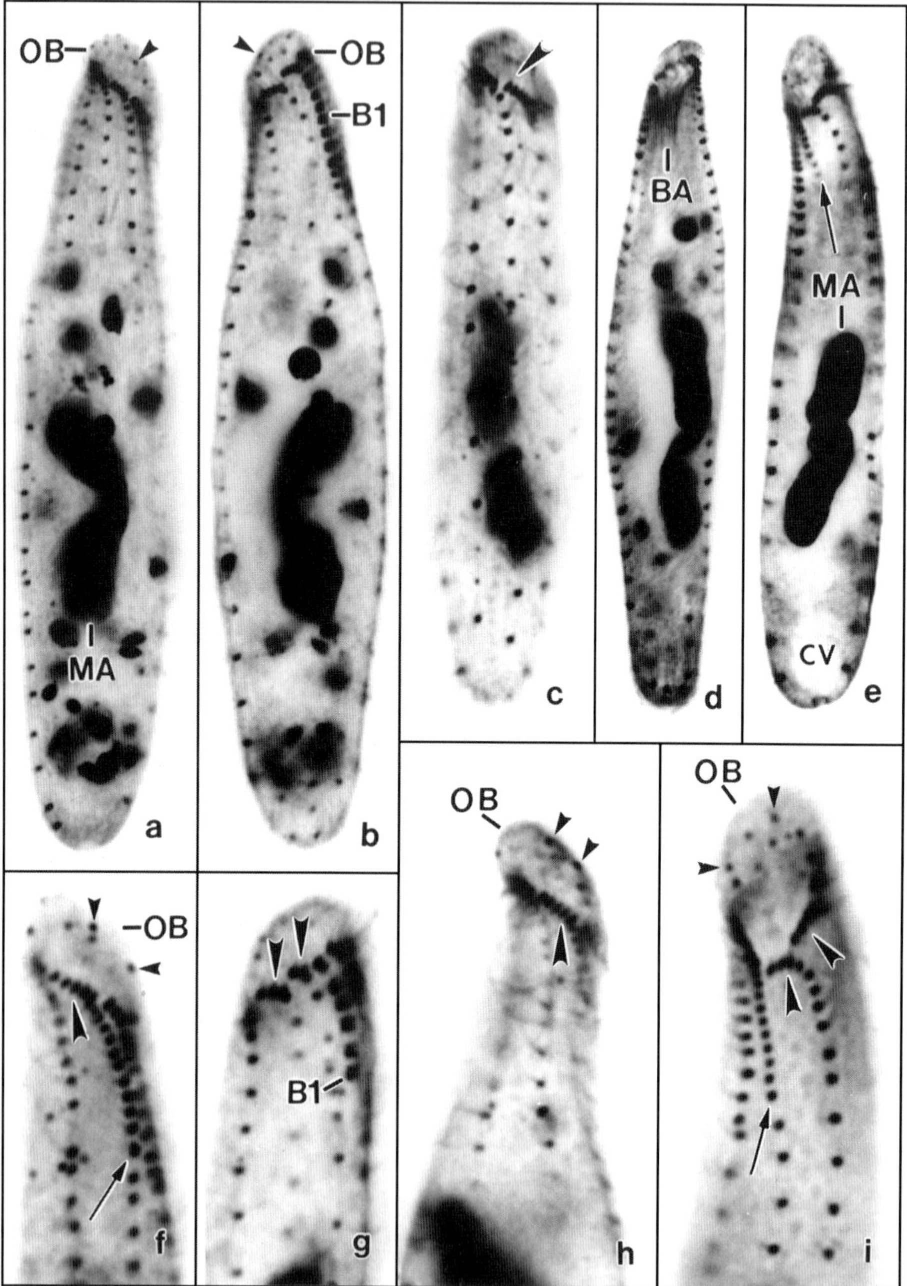

Fig. 110a–i *Protospathidium vermiculus* after protargol impregnation. Arrows mark anterior region of first and/or second right side kinety with very narrowly spaced basal bodies. Large arrowheads denote individual oral kinetofragments. Small arrowheads mark anterior end of oral bulge extrusomes. **a, b)** Right and left side view of same specimen. **c, d, f–h)** Right (c, f, h) and left (d, g) side views of other specimens. **e, i)** Ventral views. B1 – brush row 1, BA – oral basket, CV – contractile vacuole, MA – macronucleus, OB – oral bulge.

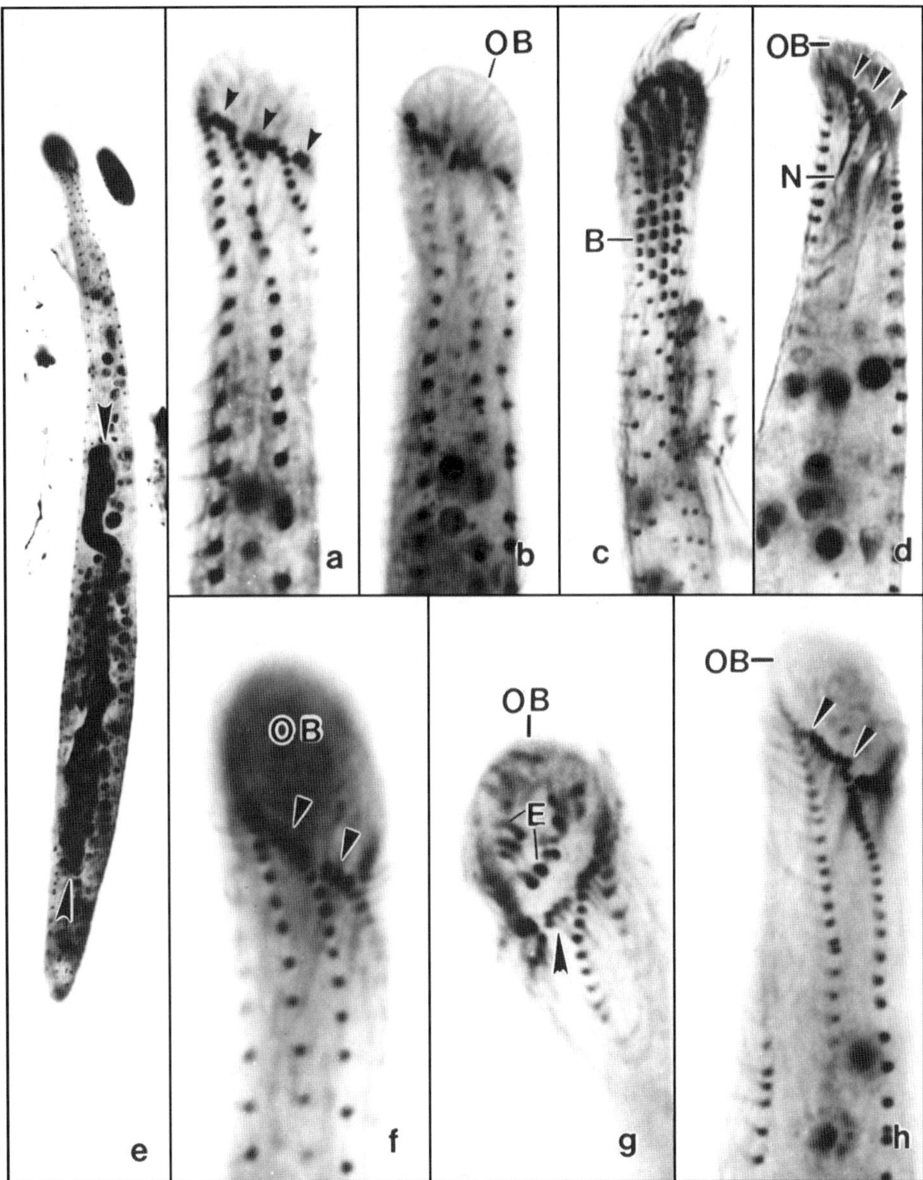

Fig. 111a–c, e, f *Protospathidium namibicola*, body shape and ciliary pattern after protargol impregnation (from FOISSNER et al.. 2002). This is the largest species of the genus. **a, b)** Right side view and optical section of a specimen showing the circumoral kinetofragments (arrowheads) and the hemispherical oral bulge. **c, e, f)** Same specimen in overview (e) and detail to show the long, tortuous macronucleus (arrowheads in e), the circumoral kinetofragments (arrowheads), and the dorsal brush (B), which consists of four rows in this specimen. B – dorsal brush, OB – oral bulge.

Fig. 111d, g, h *Protospathidium arenicola* nov. spec., USA specimens showing the comparatively inconspicuous (cp. figures b, f) and obovate oral bulge (g) surrounded by conspicuous kinetofragments (arrowheads) giving rise to distinct nematodesma bundles (d). E – extrusomes, N – nematodesmata, OB – oral bulge.

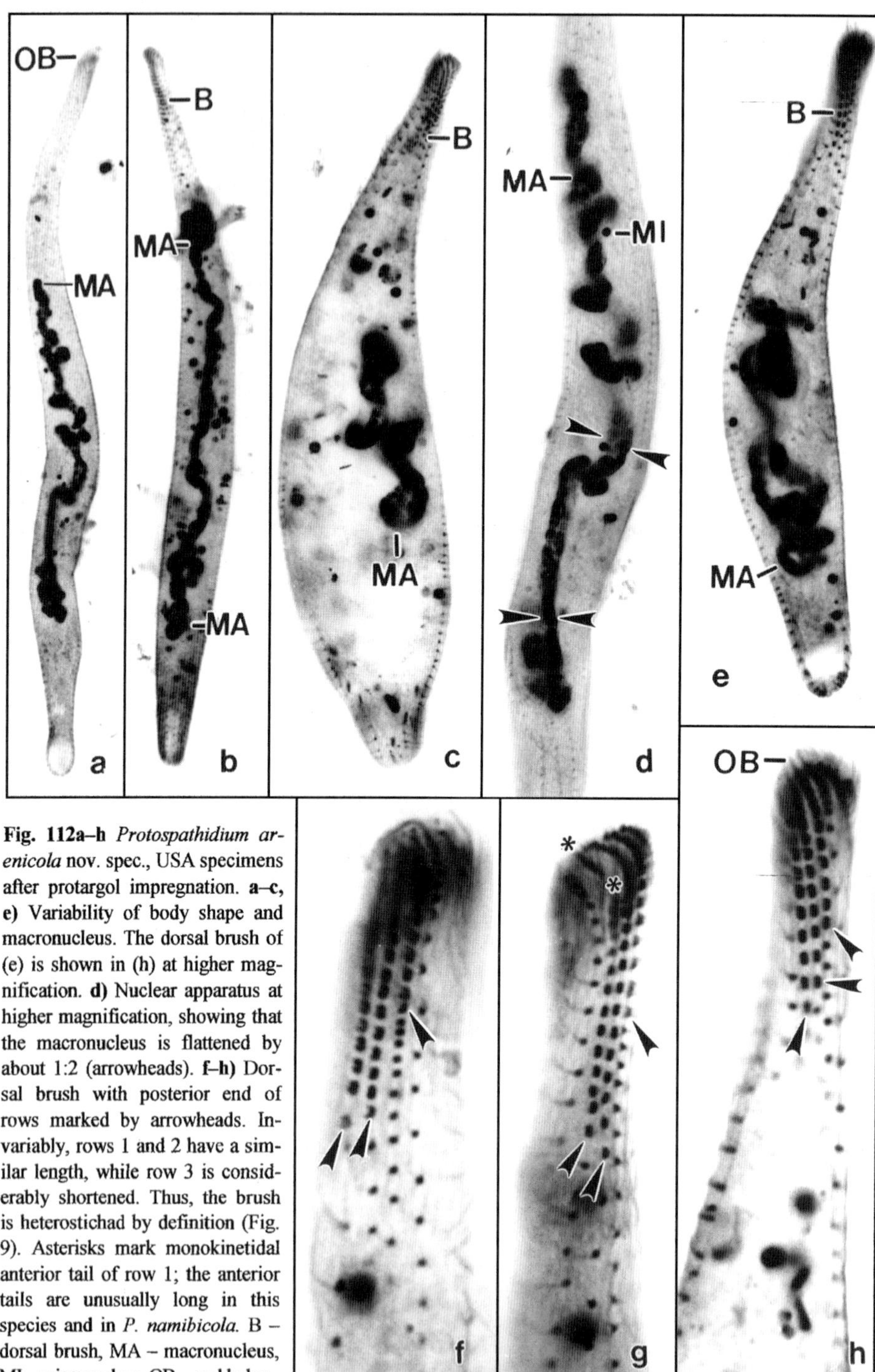

Fig. 112a–h *Protospathidium arenicola* nov. spec., USA specimens after protargol impregnation. **a–c, e)** Variability of body shape and macronucleus. The dorsal brush of (e) is shown in (h) at higher magnification. **d)** Nuclear apparatus at higher magnification, showing that the macronucleus is flattened by about 1:2 (arrowheads). **f–h)** Dorsal brush with posterior end of rows marked by arrowheads. Invariably, rows 1 and 2 have a similar length, while row 3 is considerably shortened. Thus, the brush is heterostichad by definition (Fig. 9). Asterisks mark monokinetidal anterior tail of row 1; the anterior tails are unusually long in this species and in *P. namibicola*. B – dorsal brush, MA – macronucleus, MI – micronucleus, OB – oral bulge.

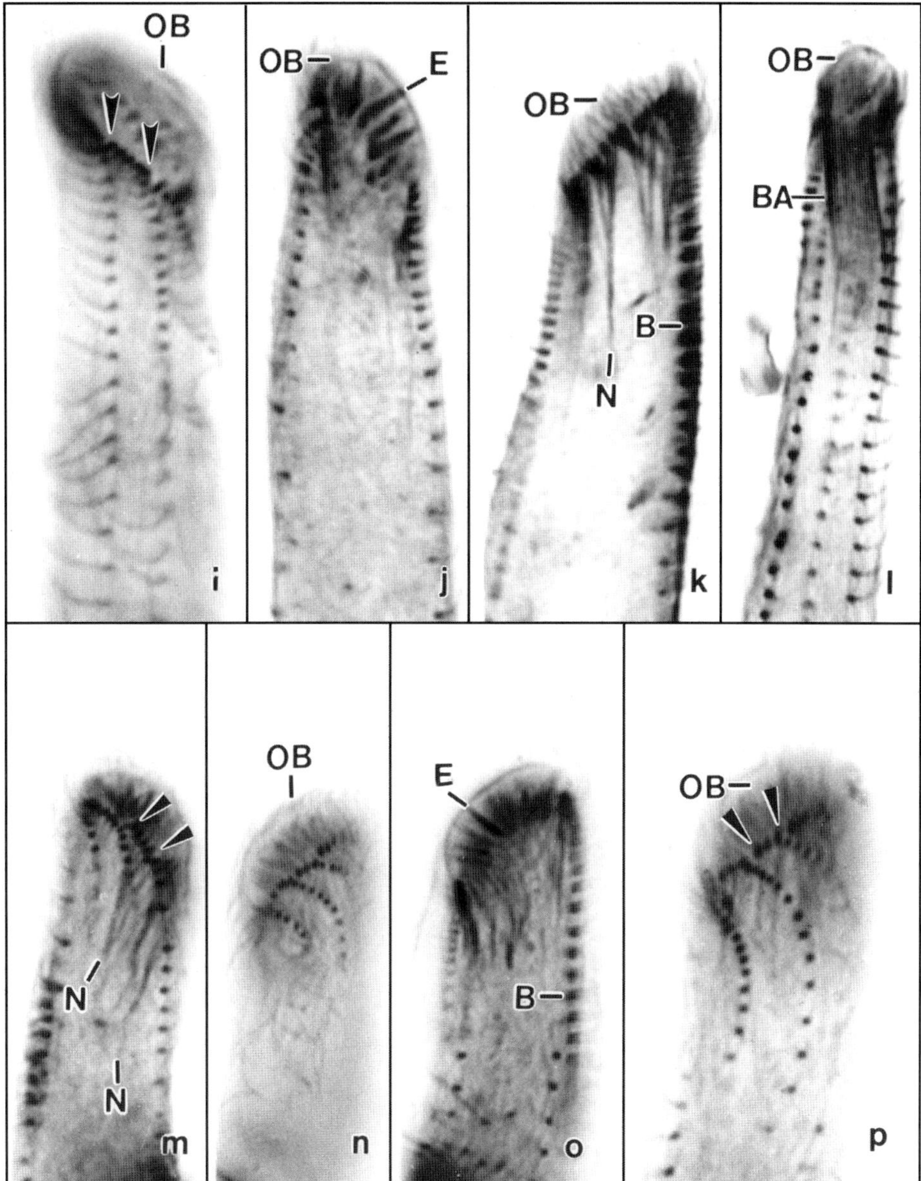

Fig. 112i–p *Protospathidium arenicola* nov. spec., somatic and oral infraciliature of USA specimens after protargol impregnation. **i, j)** Right side view and optical section of same specimen. Arrowheads mark a right side oral kinetofragment. Note the fusiform extrusomes, an important difference to *P. namibicola*, where they are rod-shaped. **k, l)** Left side view and ventral optical section showing body flattening and the nematodesma bundles forming a distinct oral basket. **m–o)** Right and left side view and optical section of same specimen, showing the oral kinetofragments (one marked by arrowheads) and the rather conspicuous oral bulge. **p)** Left side view showing oral kinetofragments (arrowheads) each composed of three dikinetids. B – dorsal brush, BA – oral basket, E – extrusomes, N – nematodesma bundles, OB – oral bulge.

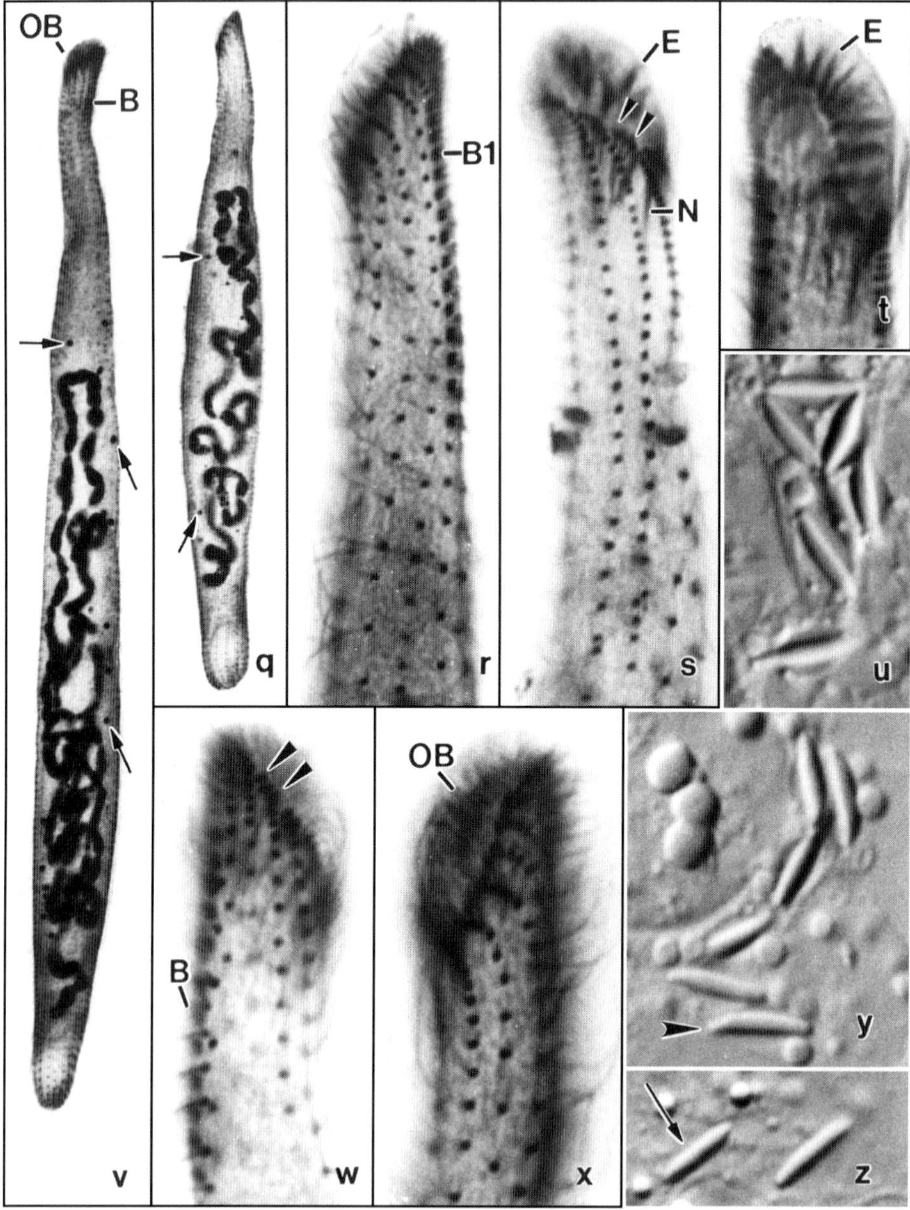

Fig. 112q–z *Protospathidium arenicola* nov. spec., Australian specimens from life (u, y, z) and after protargol impregnation (q–t, v–x). **q, v)** Overviews showing shape variability and the long, highly tortuous macronucleus. Arrows mark some of the many micronuclei. **r, s)** Left and right side view showing the protospathidiid ciliary pattern. Arrowheads mark an oral kinetofragment. **t)** Optical section showing the thick, rather distinctly impregnated extrusomes. **u, y, z)** The narrowly ellipsoidal extrusomes are asymmetrical (arrow and arrowhead show the two aspects/sides) and 4–5 × 1 µm in size. **w, x)** Right and left side view of same specimen. Arrowheads mark a kinetofragment composed of three dikinetids. B(1) – dorsal brush (row), E – extrusomes, N – nematodesma bundle, OB – oral bulge.

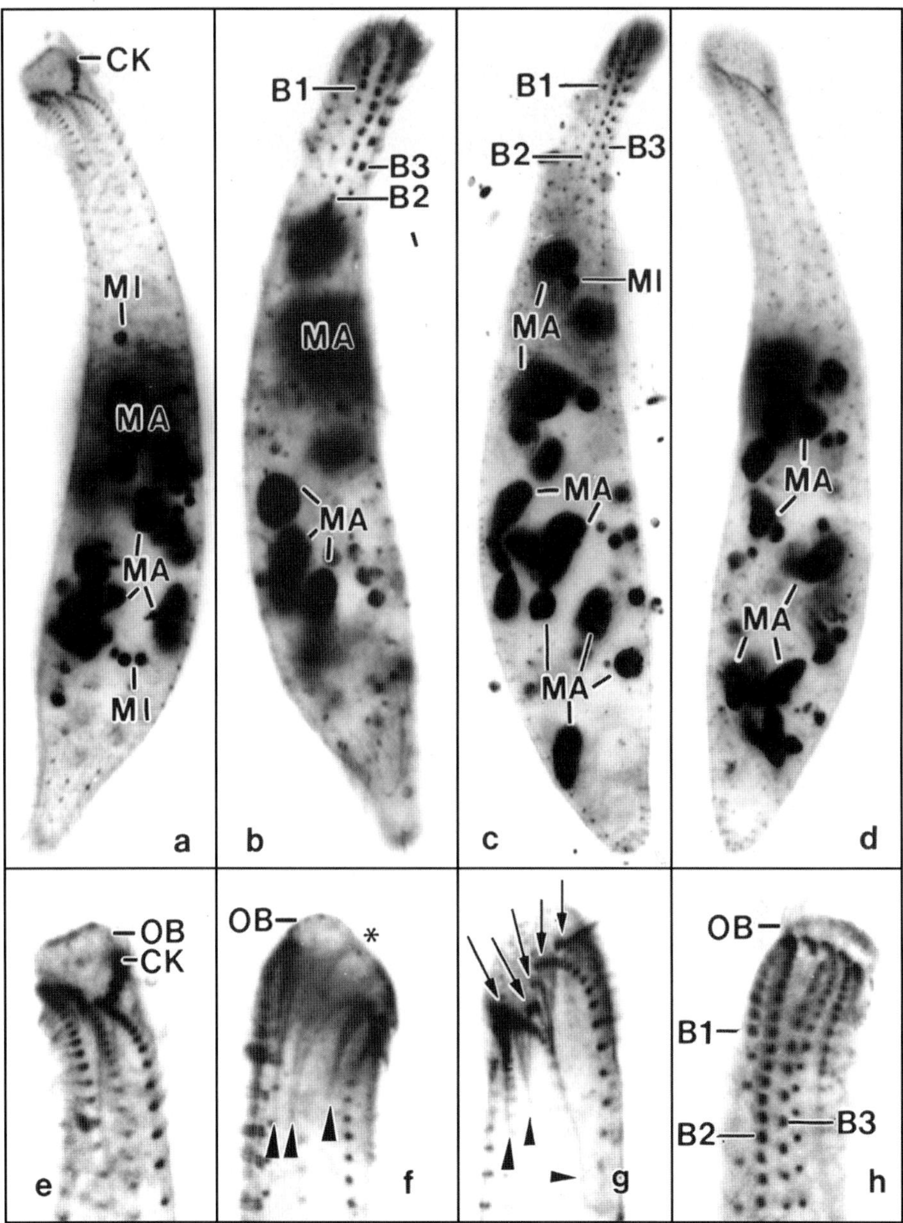

Fig. 113a–h *Protospathidium muscicola*, Venezuelan specimens after protargol impregnation. **a, b, e)** Ventral and dorsal view showing the scattered macronucleus nodules and the typical dorsal brush: row 1 minute consisting of only three dikinetids; row 2 long and composed of many narrowly spaced dikinetids; row 3 slightly shorter than row 2 consisting of eight widely spaced dikinetids. **c, d, h)** Dorsal (c, h) and ventrolateral views of other specimens, showing the same details as explained in figures (a, b, e). **f, g)** Optical section and left side view showing the central bulge cavity (asterisk) and the circumoral kinetofragments (arrows) associated with cuneate nematodesma bundles (arrowheads). B1-3 – dorsal brush rows, CK – circumoral kinetofragments, MA – macronucleus nodules, MI – micronuclei, OB – oral bulge.

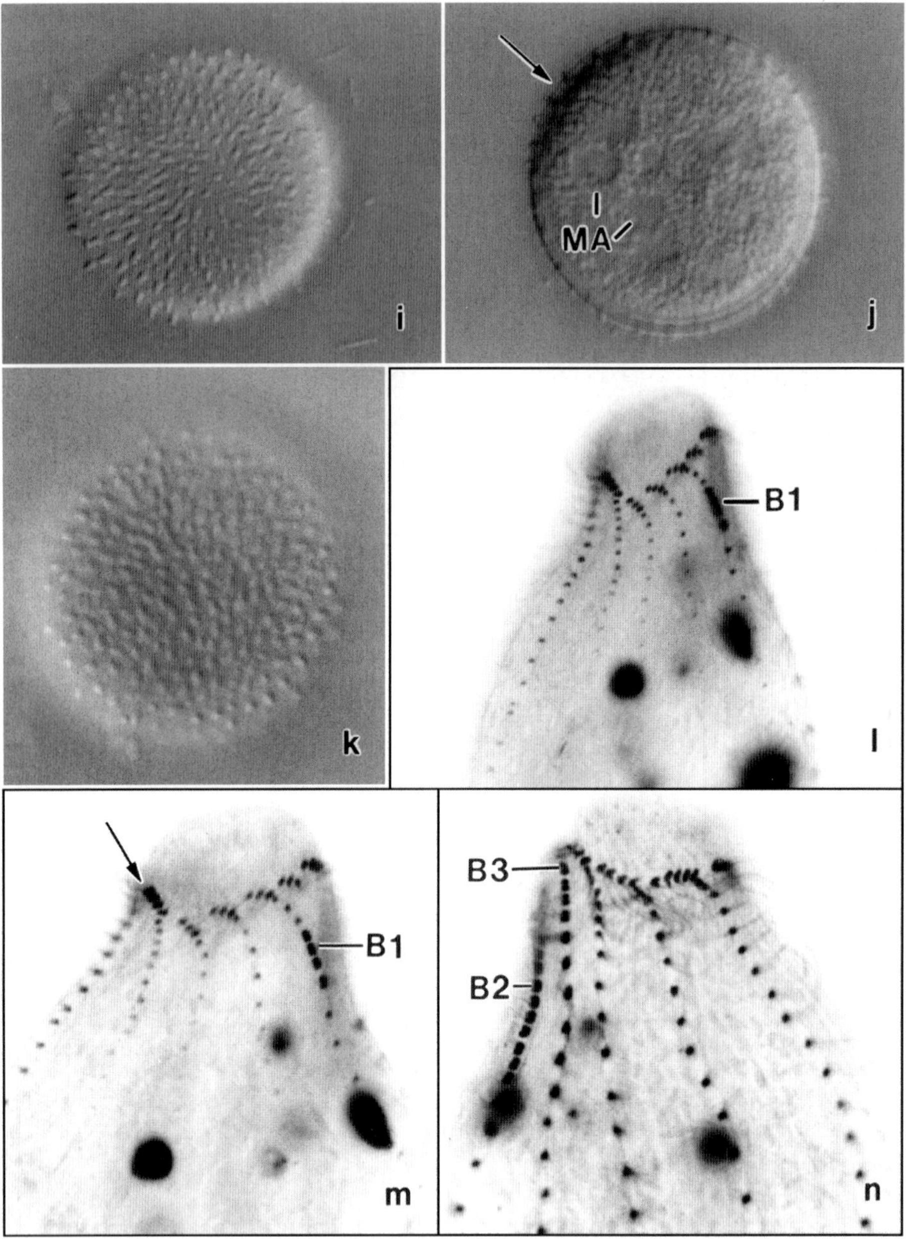

Fig. 113i–k *Protospathidium muscicola*, resting cyst of an African (Botswana) specimen in surface view (i), semi-surface view (k), and optical section (j). Arrow marks a site where the thin membrane covering (?) the spines is recognizable. MA – macronucleus nodules.

Fig. 113l–n *Protospathidium serpens*, ciliary pattern of an Austrian neotype specimen after protargol impregnation. Arrow marks an oral kinetofragment where the dikinetidal organization is well recognizable. B1-3 – dorsal brush rows.

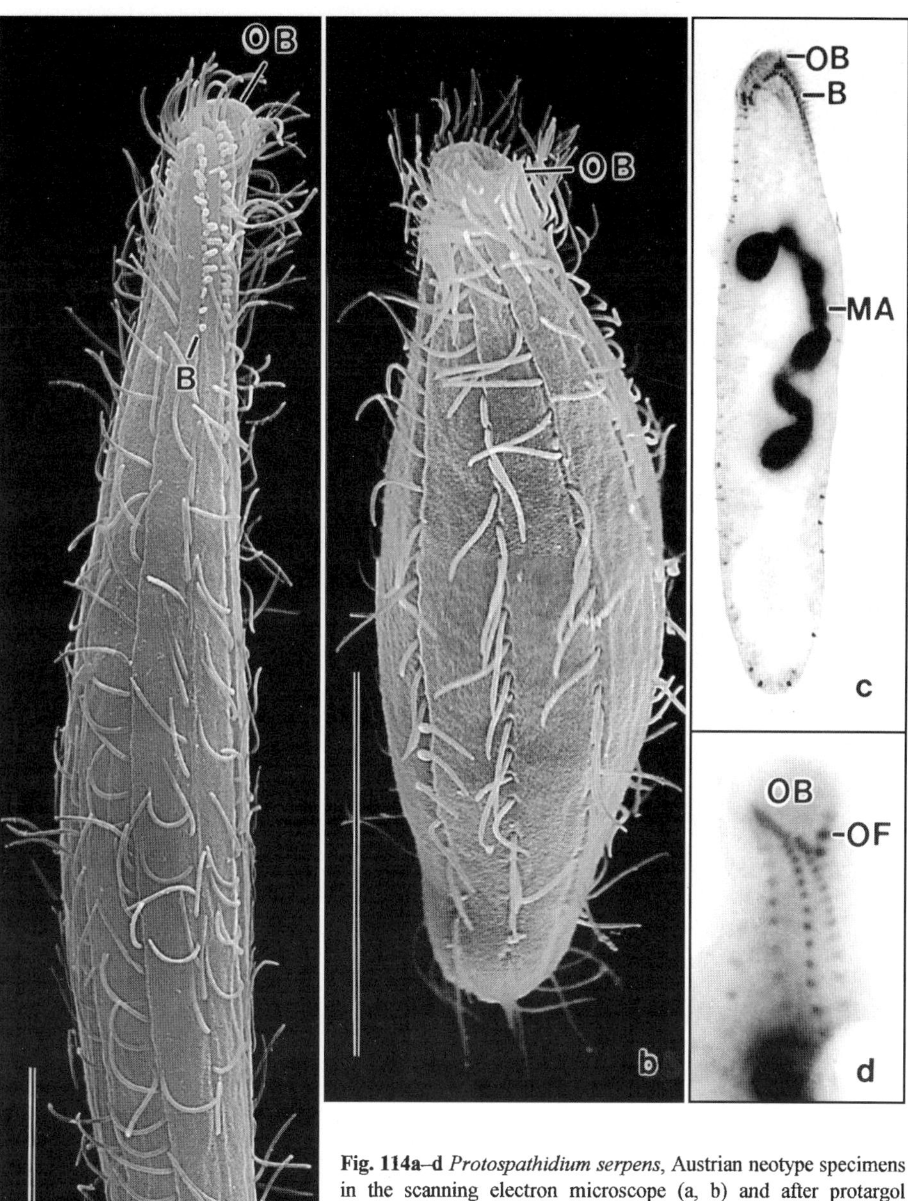

Fig. 114a–d *Protospathidium serpens*, Austrian neotype specimens in the scanning electron microscope (a, b) and after protargol impregnation (c, d). From XU & FOISSNER (2005a). **a)** Dorsal view of a typical, slender specimen. Note the short dorsal brush (B). Scale bar 20 µm. **b)** Ventrolateral view of a small, stout specimen. Scale bar 20 µm. **c)** Left side overview showing the irregularly nodulated macronucleus, a main feature of the species. **d)** Ventral view of oral portion of a specimen from a permanent slide. Note the obovate oral bulge and the circumoral kinetofragments separated from each other by minute gaps. B – dorsal brush, MA – macronucleus, OB – oral bulge, OF – oral kinetofragments.

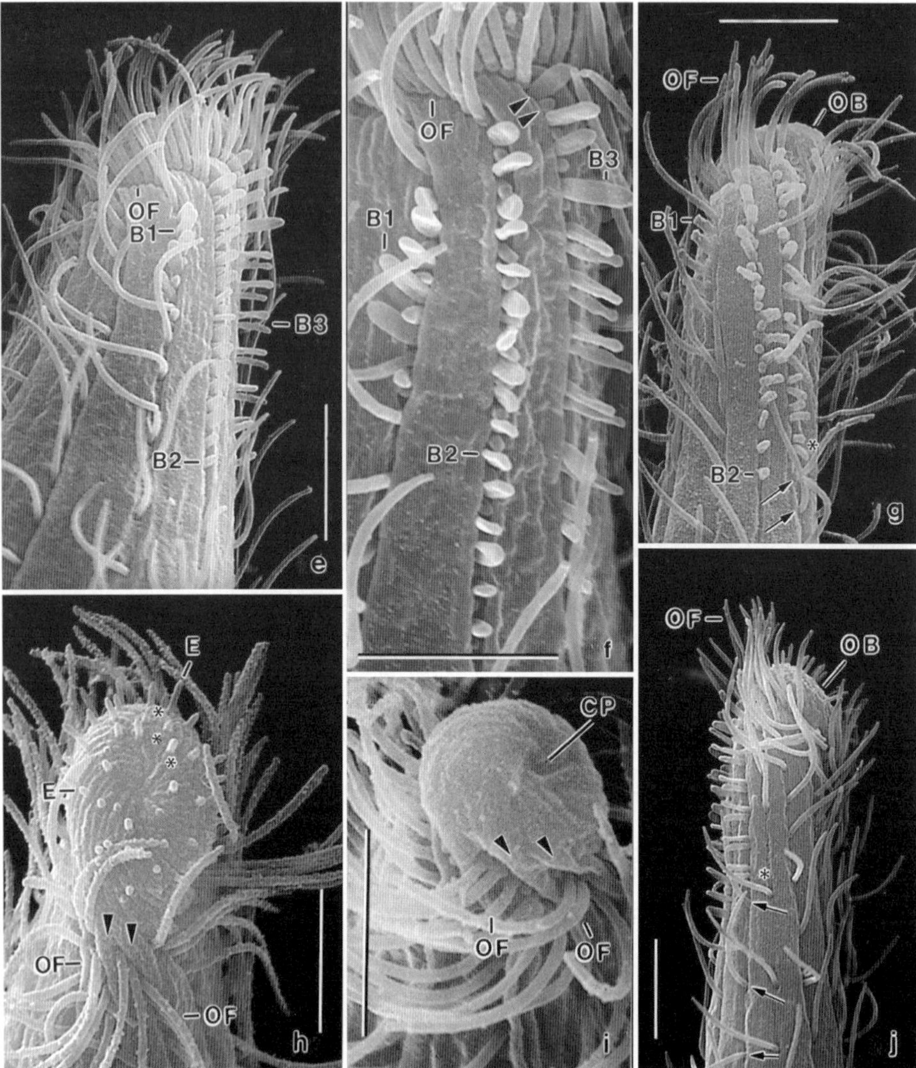

Fig. 114e–j *Protospathidium serpens*, Austrian neotype specimens in the scanning electron microscope (from Xu & Foissner 2005a). **e, f, g, j)** Dorsal and dorsolateral views showing the dorsal brush area. Most brush bristles are inflated, but shrunken as indicated by their wrinkled appearance. Asterisks and arrows in figures (g) and (j) mark the last bristle pair of brush row 3 and the first post-brush cilium, showing that brush row 3 of *P. serpens* lacks a monokinetidal bristle tail, in contrast to the Antarctic population of *P. fraterculum* Xu & Foissner (2005a). The posterior bristle of the apical pairs of brush row 3 is strongly shortened (f, arrowheads). **h, i)** Ventral and ventrolateral views of oral area, showing the obovate oral bulge containing three rough circles of extrusomes (asterisks) and the deepened cytopharyngeal entrance (temporary cytostome) in bulge center. The circumoral kinetofragments are separated from each other by conical ridges (arrowheads) produced by the oblique clefts from which the fragments originate. B1-3 – dorsal brush rows, CP – cytopharyngeal opening, E – extrusomes, OB – oral bulge, OF – oral kinetofragments. Scale bars = 5 µm.

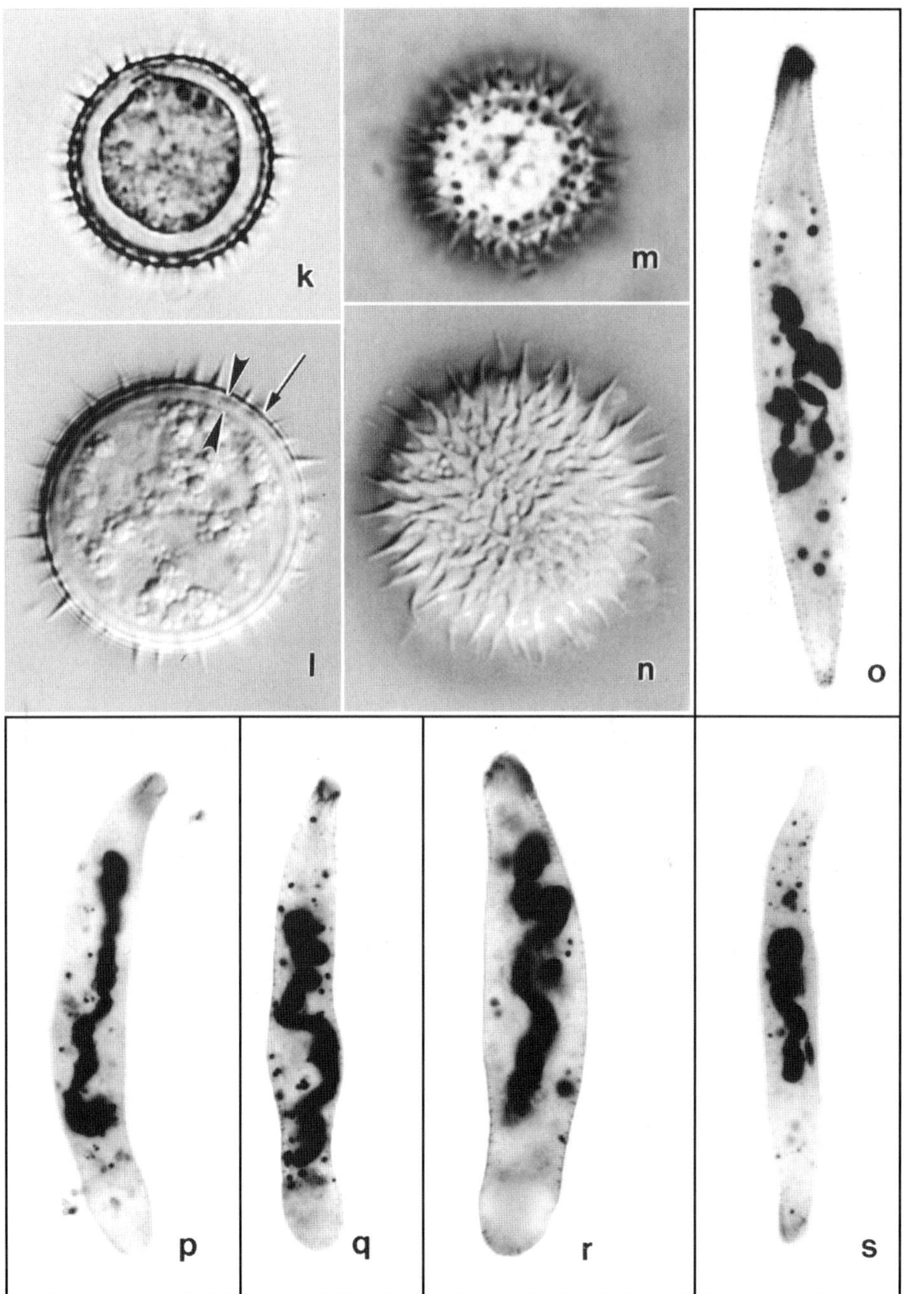

Fig. 114k–s *Protospathidium serpens*, Salzburg neotype specimens from life (k–n) and cultivated Namibian site 16 cells (FOISSNER et al. 2002) after protargol impregnation (o–s). From XU & FOISSNER (2005a) and originals (m, o–s). **k–n)** Resting cysts in optical section (k, l) and surface view (m, n). Arrow marks thin outer wall, opposed arrowheads denote thick inner wall. **o–s)** Variability of body and macronucleus shape. Further examples, see figures 115g–i.

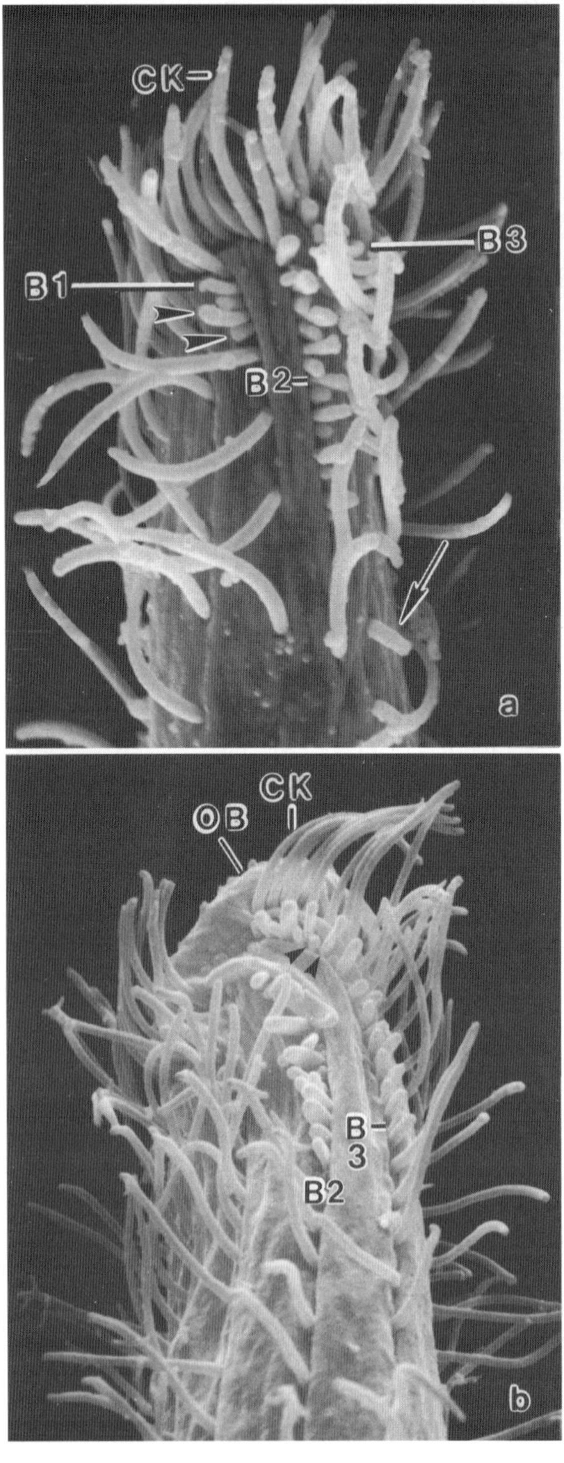

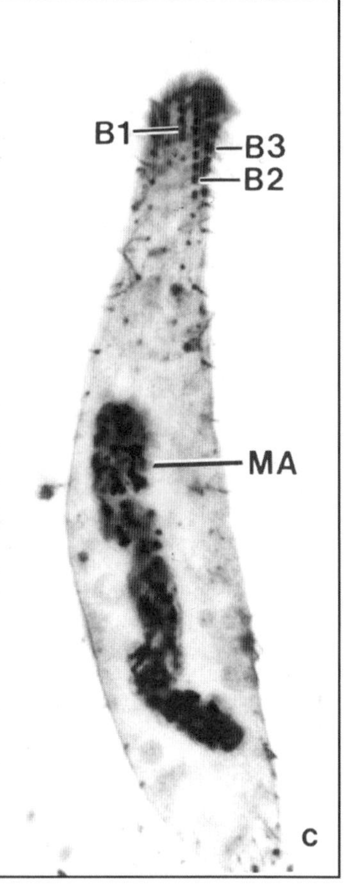

Fig. 115a–c *Protospathidium serpens* in the SEM (a, b) and after protargol impregnation (c). The micrographs show the dorsal brush of Antarctic (a, c) and Namibian site 70 specimens (b). The brush consists of three rows of dikinetids with slightly inflated bristles: rows 1 and 2 are composed of few, respectively, many dikinetids with the anterior bristle longer than the posterior (a, arrowheads); row 3 has a monokinetidal tail in the Antarctic specimen (a, arrow), and the anterior bristle of the dikinetids is shorter than the posterior (b, arrowhead). B1-3 – dorsal brush rows, CK – cilia of circumoral kinetofragments, MA – macronucleus, OB – oral bulge.

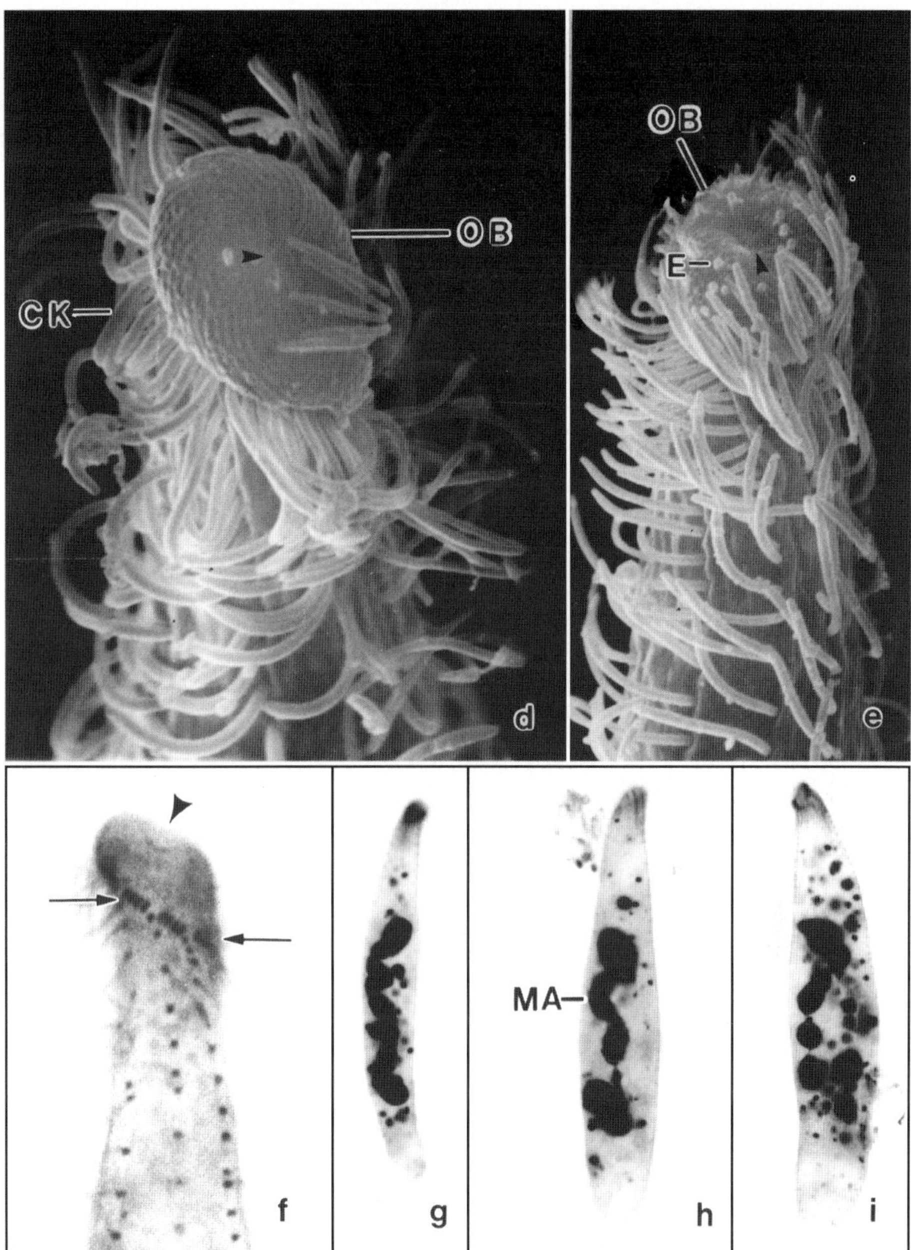

Fig. 115 d–i *Protospathidium serpens* in the scanning electron microscope (d, e) and after protargol impregnation (f–i). **d–f)** Marion (d, e) and Signey (f) Island specimens showing the typical, obovate oral bulge with a central concavity (arrowheads), that is, the "temporary mouth" where the bulge opens during feeding. Arrows denote the distinct circumoral kinetofragments attached to the somatic kineties. These populations are now assigned to *P. fraterculum* (see Xu & Foissner 2005a). **g–i)** Macronucleus variability in Namibian site 16 specimens (see also figures 114o–s). CK – cilia of circumoral kinetofragments, E – extrusomes leaving the oral bulge, MA – macronucleus, OB – oral bulge.

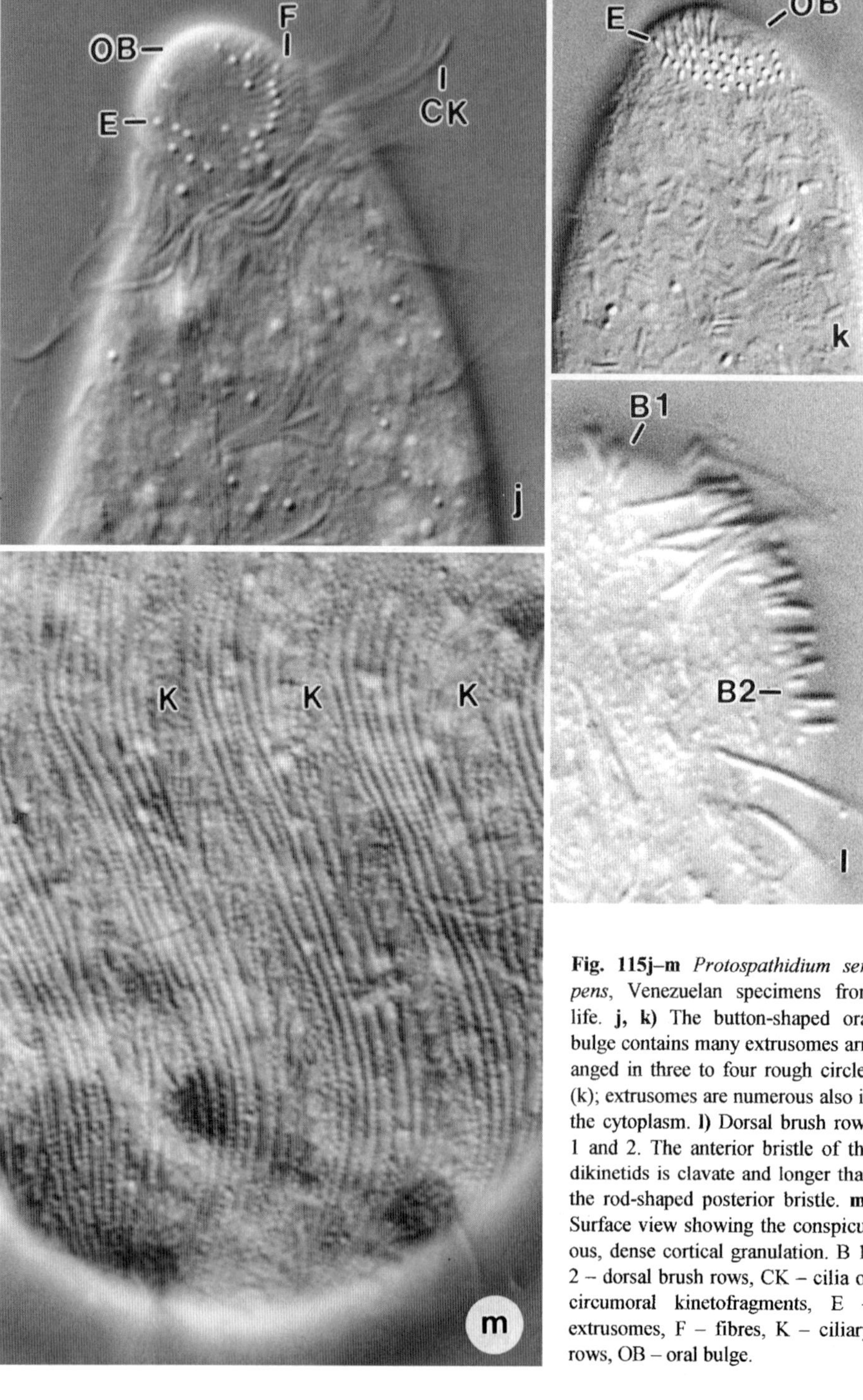

Fig. 115j–m *Protospathidium serpens*, Venezuelan specimens from life. **j, k)** The button-shaped oral bulge contains many extrusomes arranged in three to four rough circles (k); extrusomes are numerous also in the cytoplasm. **l)** Dorsal brush rows 1 and 2. The anterior bristle of the dikinetids is clavate and longer than the rod-shaped posterior bristle. **m)** Surface view showing the conspicuous, dense cortical granulation. B 1, 2 – dorsal brush rows, CK – cilia of circumoral kinetofragments, E – extrusomes, F – fibres, K – ciliary rows, OB – oral bulge.

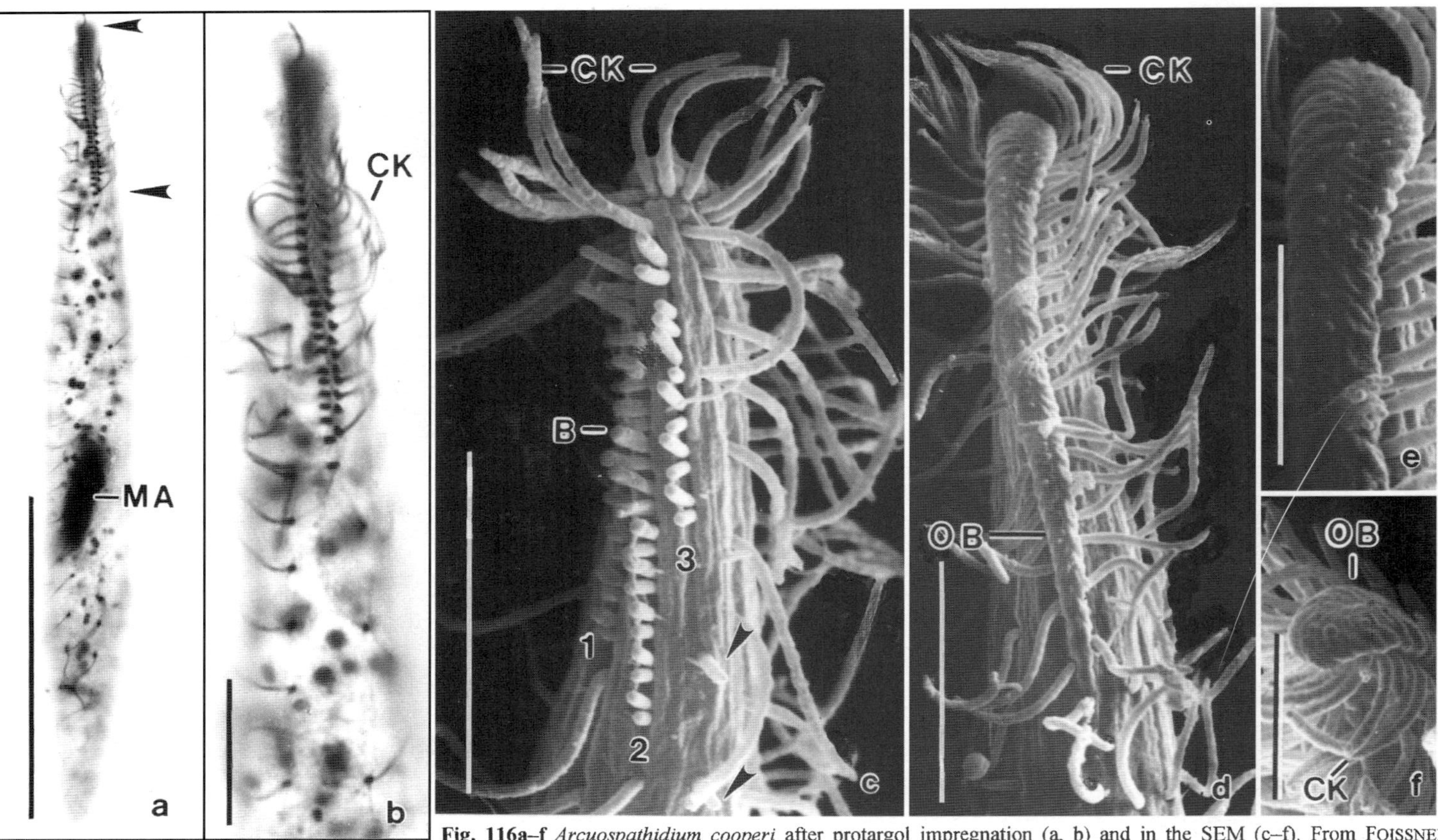

Fig. 116a–f *Arcuospathidium cooperi* after protargol impregnation (a, b) and in the SEM (c–f). From FOISSNER (1996c) and originals. **a, b)** Ventral views showing the short and narrow circumoral kinety (arrowheads). **c, d)** Dorsal and ventral view showing the narrow, cuneate oral bulge and the dorsal brush with row 3 continuing with single, short bristles (arrowheads). **e, f)** Anterior portion of oral bulge in ventral and lateral view. B(1-3) – dorsal brush (rows), CK – cilia of circumoral kinety, MA – macronucleus, OB – oral bulge. Scale bars 50 µm (a), 10 µm (b–d), and 5 µm (e, f).

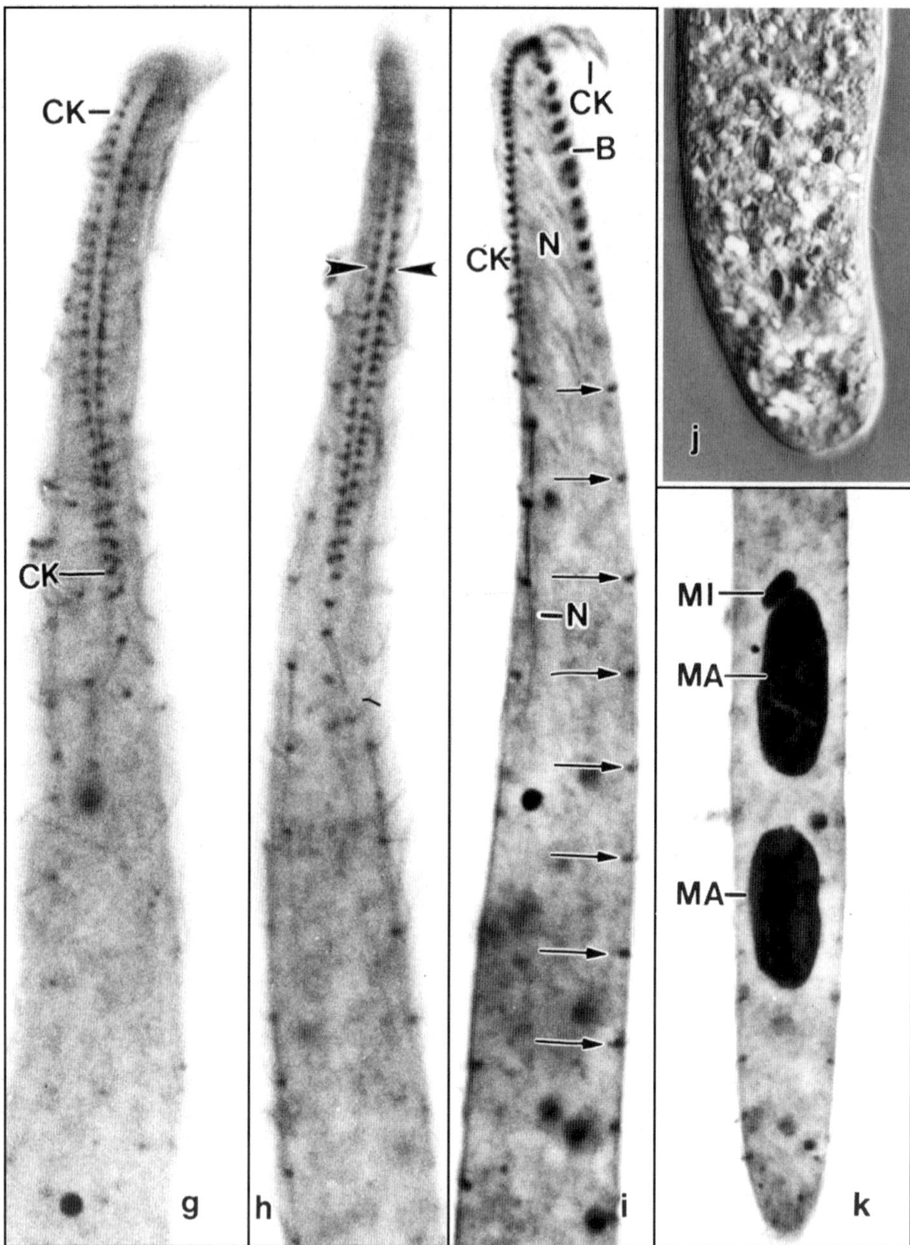

Fig. 116g–k *Arcuospathidium vermiforme*, specimens from Salzburg type population (g–I, k) and Carinthia (j) after protargol impregnation (g–i, k) and from life (j). From FOISSNER (1984) and originals. **g, h)** Ventral views showing the very narrow (arrowheads), but distinctly cuneate circumoral kinety. **i)** Left side view showing the inconspicuous oral basket and the widely spaced somatic cilia (arrows). **j)** The Carinthian specimens are packed with highly refractive inclusions. **k)** Posterior body half showing the typical nuclear pattern. B – dorsal brush, CK – (cilia of) circumoral kinety, MA – macronucleus nodules, MI – micronucleus.

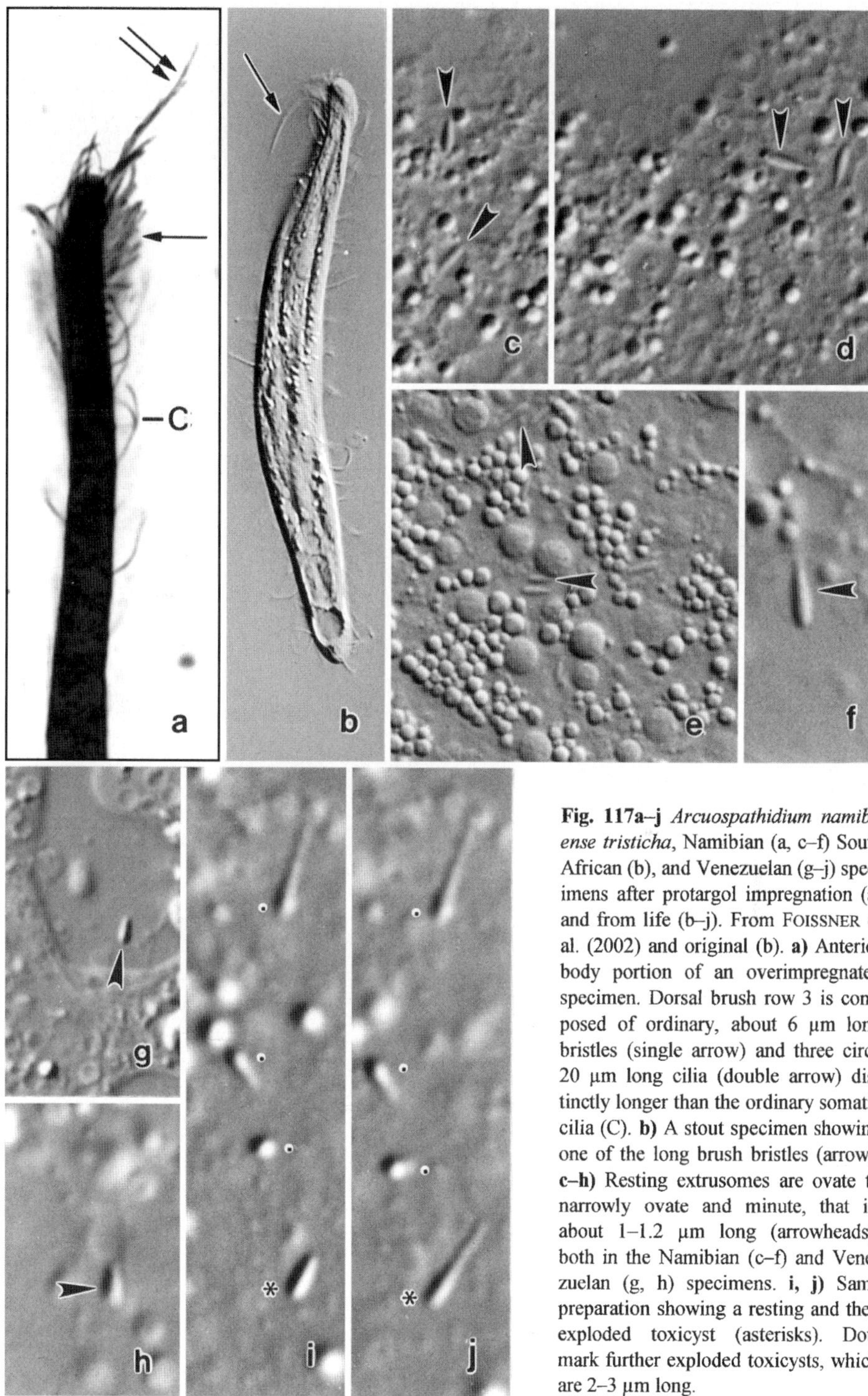

Fig. 117a–j *Arcuospathidium namibiense tristicha*, Namibian (a, c–f) South African (b), and Venezuelan (g–j) specimens after protargol impregnation (a) and from life (b–j). From FOISSNER et al. (2002) and original (b). **a)** Anterior body portion of an overimpregnated specimen. Dorsal brush row 3 is composed of ordinary, about 6 µm long bristles (single arrow) and three circa 20 µm long cilia (double arrow) distinctly longer than the ordinary somatic cilia (C). **b)** A stout specimen showing one of the long brush bristles (arrow). **c–h)** Resting extrusomes are ovate to narrowly ovate and minute, that is, about 1–1.2 µm long (arrowheads), both in the Namibian (c–f) and Venezuelan (g, h) specimens. **i, j)** Same preparation showing a resting and then exploded toxicyst (asterisks). Dots mark further exploded toxicysts, which are 2–3 µm long.

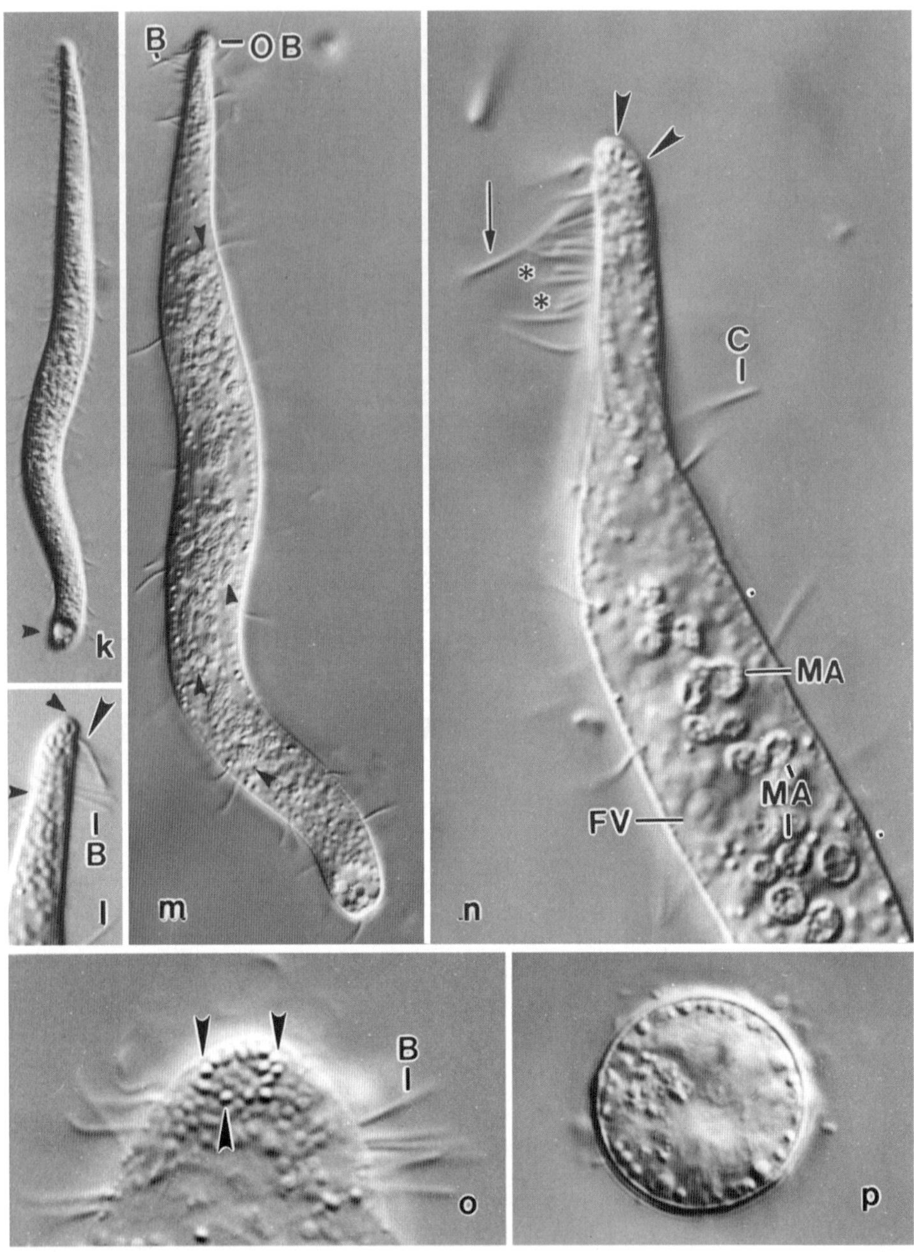

Fig. 117k–p *Arcuospathidium namibiense tristicha*, Austrian specimens from life. **k)** A freely motile specimen with contractile vacuole marked by arrowhead. **l)** Lateral view with oral bulge marked by arrowheads. The brush bristle marked by the large arrowhead is about 15 µm long, but its distal half is out of focal plane. **m)** Slightly squashed specimen with some macronucleus nodules marked by arrowheads. **n)** Brush bristle pairs (asterisks) and one (arrow) out of three about 15 µm long bristles at anterior end of row 3. Arrowheads mark the 1 µm long extrusomes. Dots mark regions where cortical granules are recognizable. **o)** Squashed specimen showing the obovate extrusome accumulation in the oral bulge. **p)** Mature (?) resting cyst. B – brush bristles, C – somatic cilium, FV – food vacuole, MA – macronucleus nodules, OB – oral bulge.

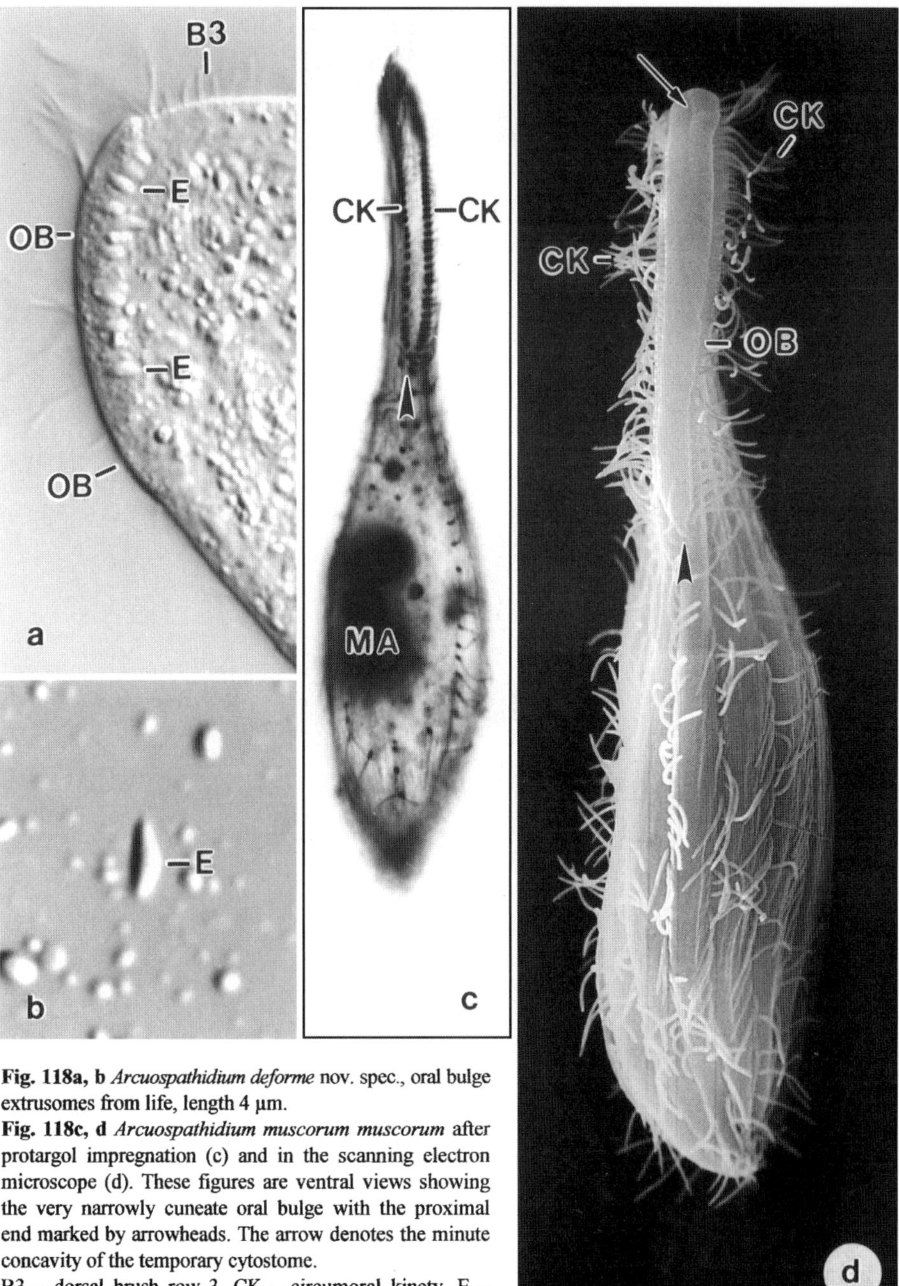

Fig. 118a, b *Arcuospathidium deforme* nov. spec., oral bulge extrusomes from life, length 4 μm.

Fig. 118c, d *Arcuospathidium muscorum muscorum* after protargol impregnation (c) and in the scanning electron microscope (d). These figures are ventral views showing the very narrowly cuneate oral bulge with the proximal end marked by arrowheads. The arrow denotes the minute concavity of the temporary cytostome.

B3 – dorsal brush row 3, CK – circumoral kinety, E – extrusomes, MA – macronucleus, OB – oral bulge.

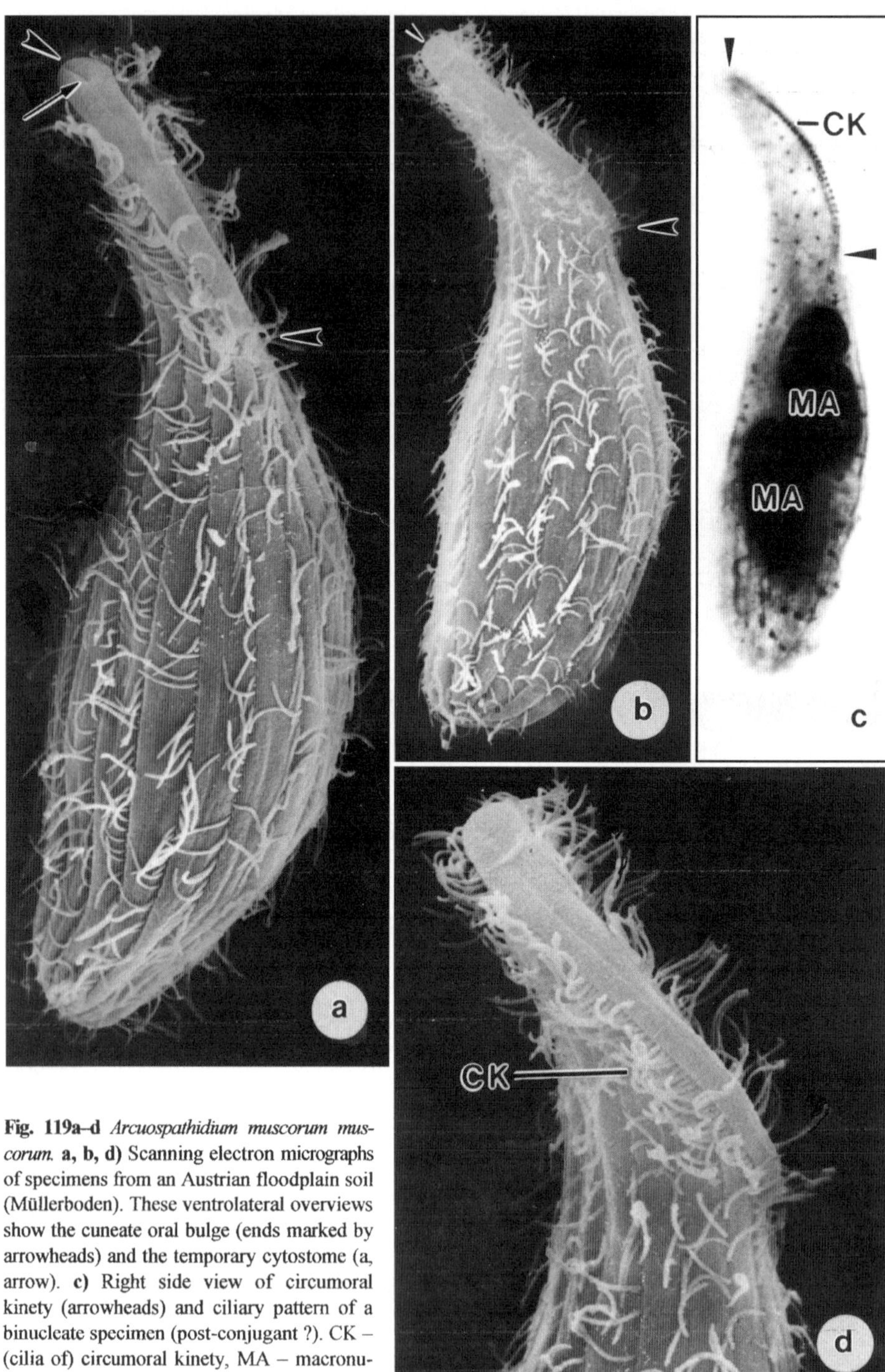

Fig. 119a–d *Arcuospathidium muscorum muscorum.* **a, b, d)** Scanning electron micrographs of specimens from an Austrian floodplain soil (Müllerboden). These ventrolateral overviews show the cuneate oral bulge (ends marked by arrowheads) and the temporary cytostome (a, arrow). **c)** Right side view of circumoral kinety (arrowheads) and ciliary pattern of a binucleate specimen (post-conjugant ?). CK – (cilia of) circumoral kinety, MA – macronucleus nodules.

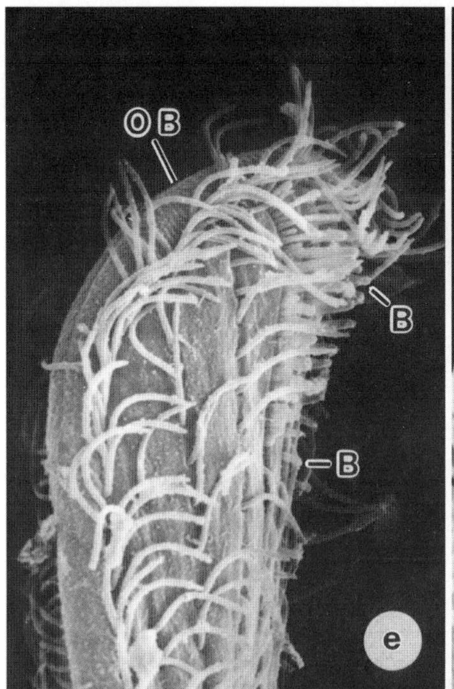

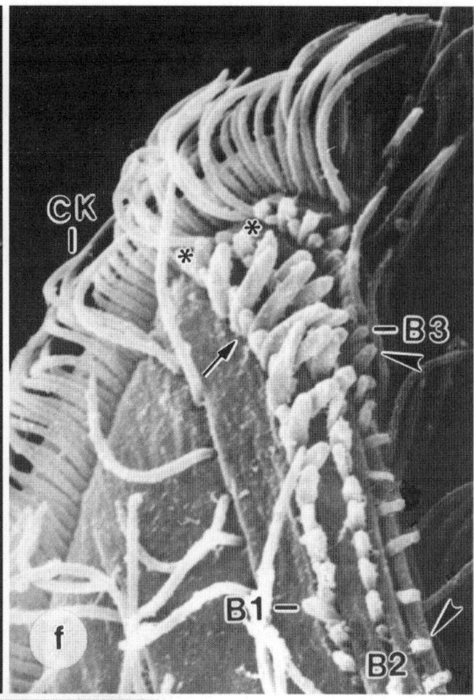

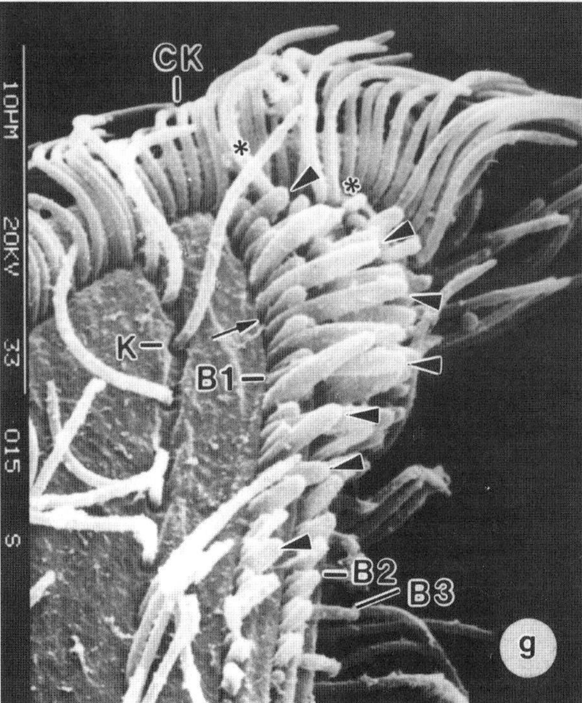

Fig. 119e–g *Arcuospathidium muscorum muscorum*, SEM micrographs of dorsal brush of specimens from Austrian floodplain soil (Müllerboden). The dorsal brush consists of three rows, each composed of dikinetids with slightly clavate bristles. The posterior bristle (arrows) of each pair is distinctly shorter than the anterior. One to two ordinary cilia (asterisks) are at the anterior end of the rows, where they form the anterior tail. **e, g)** Lateral and dorsolateral views showing the typical arch (arrowheads) formed by the anterior bristles, which gradually decrease in length anteriorly and posteriorly. **f)** Dorsal view showing part of the monokinetidal bristle tail (arrowheads) of brush row 3, whose bristles are distinctly shorter than those of rows 1 and 2. B(1-3) – dorsal brush (rows), CK – cilia of circumoral kinety, K – somatic kinety, OB – oral bulge.

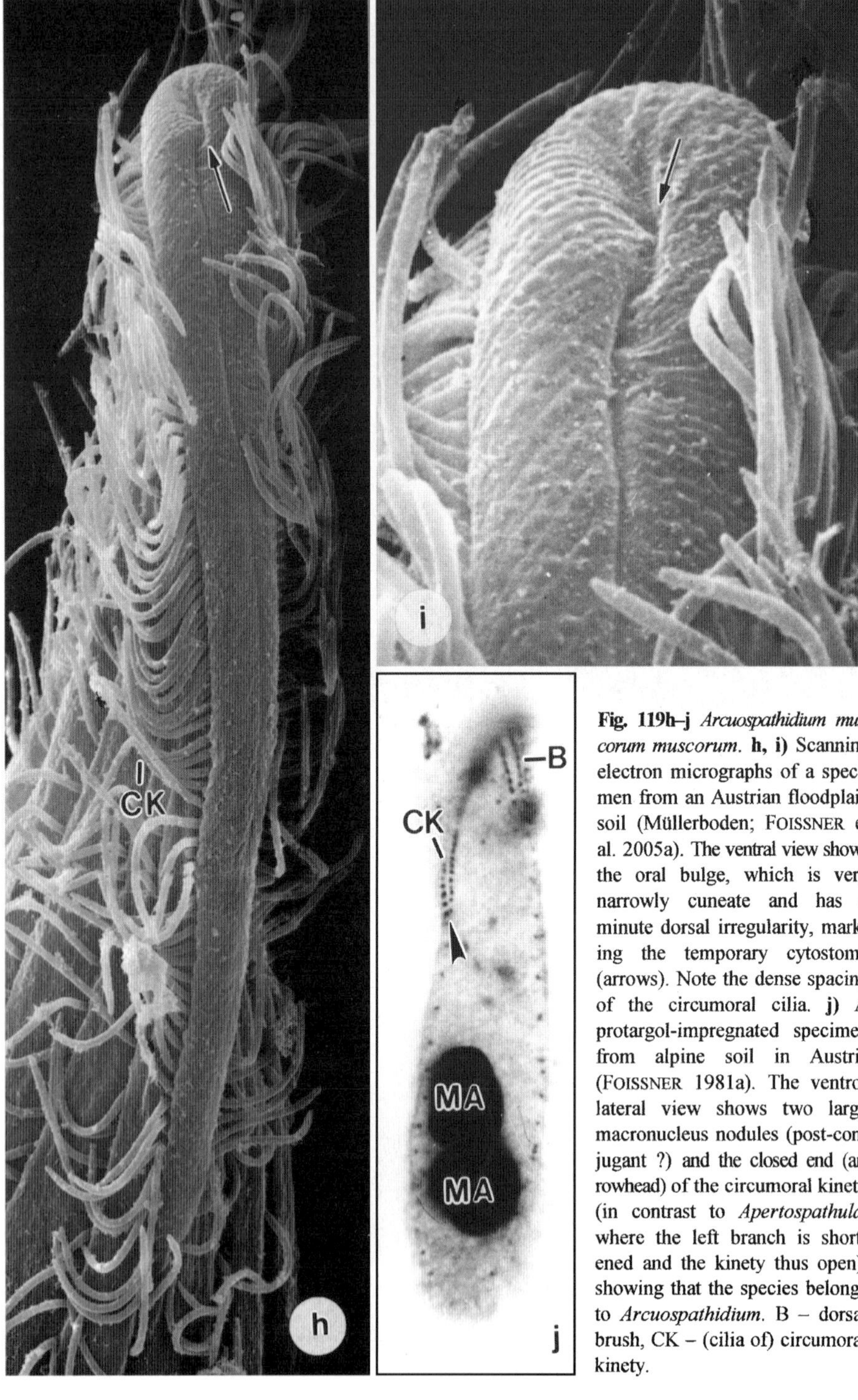

Fig. 119h–j *Arcuospathidium muscorum muscorum*. **h, i)** Scanning electron micrographs of a specimen from an Austrian floodplain soil (Müllerboden; FOISSNER et al. 2005a). The ventral view shows the oral bulge, which is very narrowly cuneate and has a minute dorsal irregularity, marking the temporary cytostome (arrows). Note the dense spacing of the circumoral cilia. **j)** A protargol-impregnated specimen from alpine soil in Austria (FOISSNER 1981a). The ventrolateral view shows two large macronucleus nodules (post-conjugant ?) and the closed end (arrowhead) of the circumoral kinety (in contrast to *Apertospathula*, where the left branch is shortened and the kinety thus open), showing that the species belongs to *Arcuospathidium*. B – dorsal brush, CK – (cilia of) circumoral kinety.

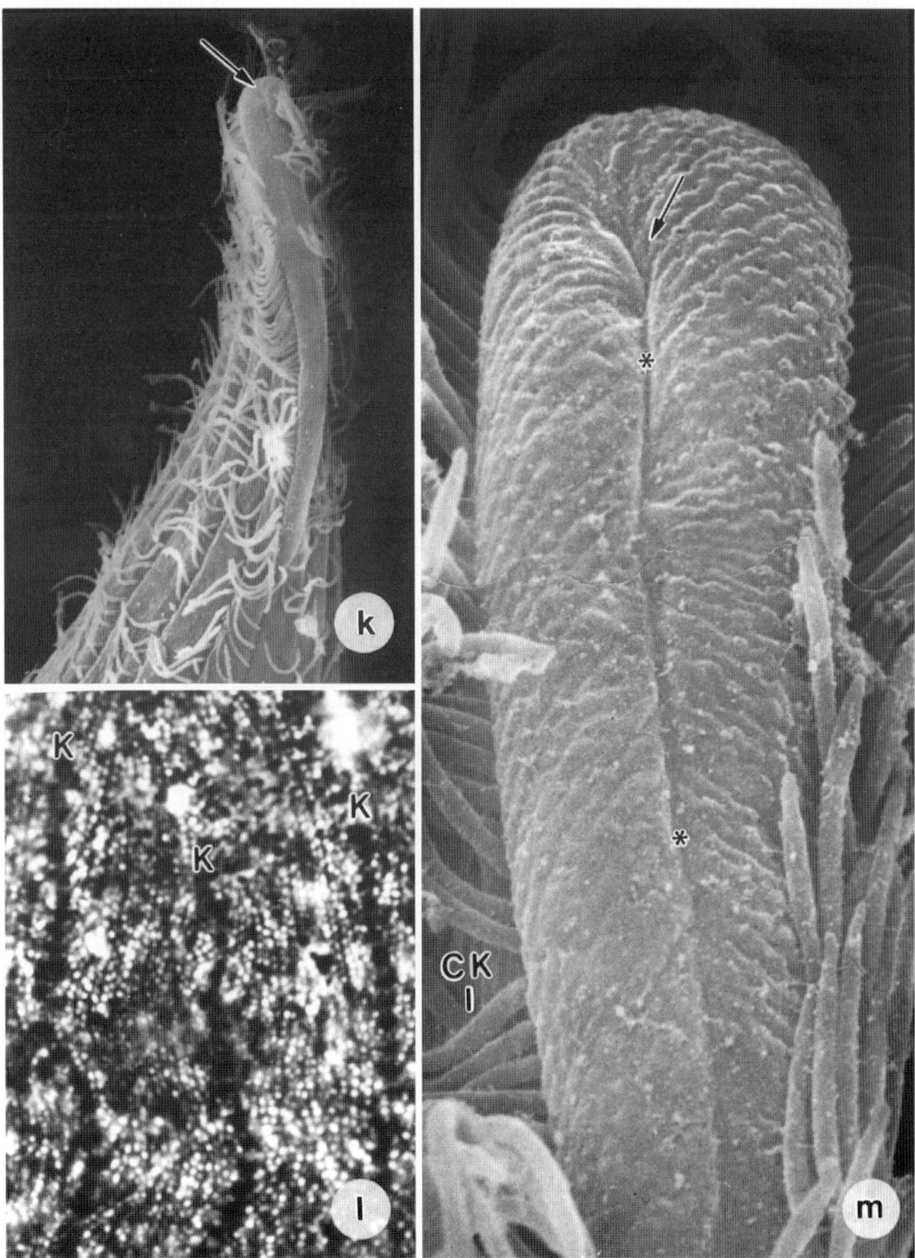

Fig. 119k–m *Arcuospathidium muscorum muscorum*, scanning electron micrographs (k, m) of specimens from an Austrian floodplain soil (Müllerboden), and a silver nitrate-impregnated (l; KLEIN-FOISSNER method) cell from alpine soil (FOISSNER 1981a). **k, m)** Ventral views of oral bulge with temporary cytostome marked by arrows (see also figure (h)). The bulge, which opens in midline (asterisks), shows an oblique striation caused by microtubule bundles originating from the circumoral dikinetids. **l)** Narrowly meshed silverline pattern. CK – cilia of circumoral kinety, K – somatic ciliary rows.

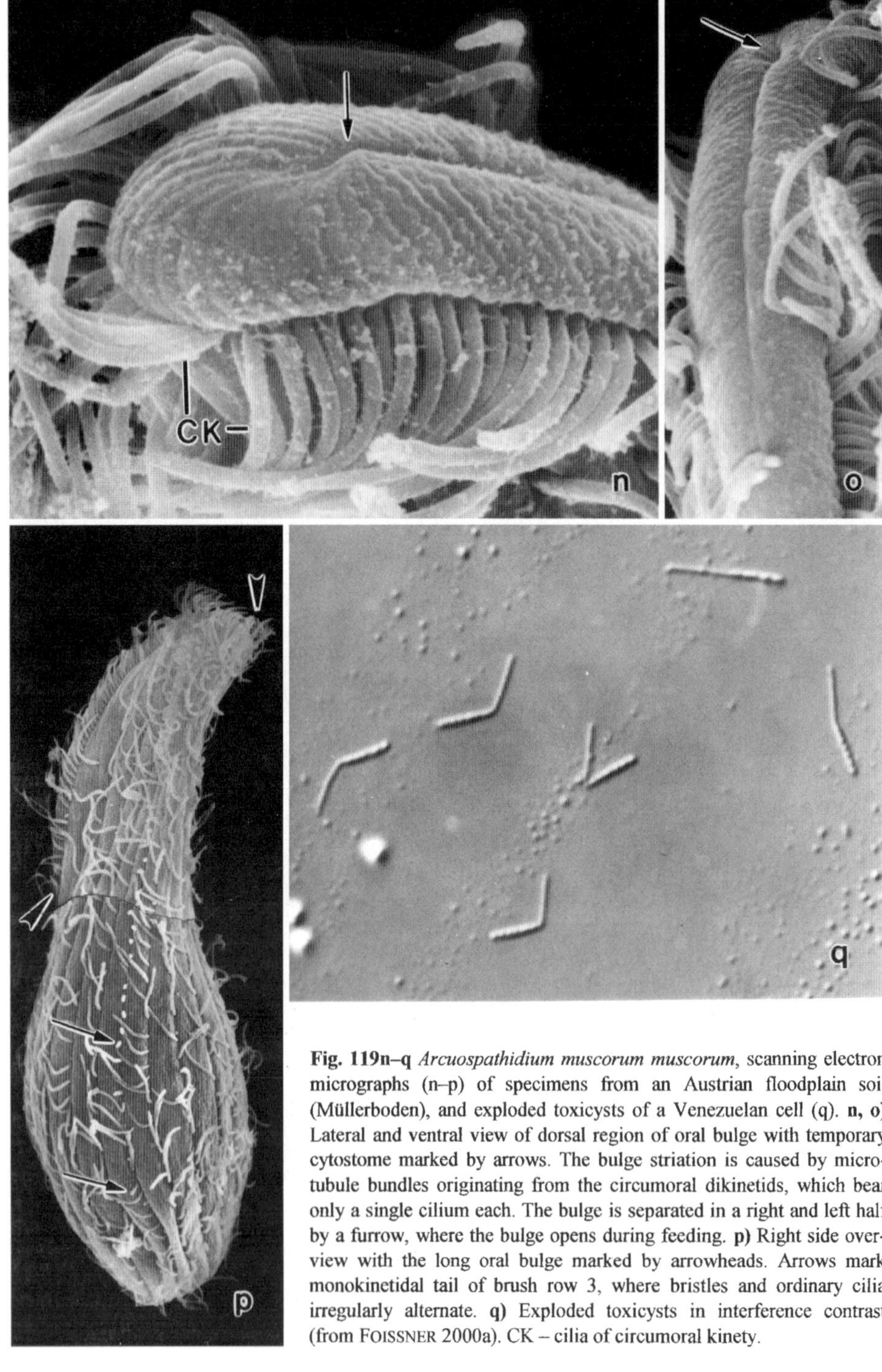

Fig. 119n–q *Arcuospathidium muscorum muscorum*, scanning electron micrographs (n–p) of specimens from an Austrian floodplain soil (Müllerboden), and exploded toxicysts of a Venezuelan cell (q). **n, o)** Lateral and ventral view of dorsal region of oral bulge with temporary cytostome marked by arrows. The bulge striation is caused by microtubule bundles originating from the circumoral dikinetids, which bear only a single cilium each. The bulge is separated in a right and left half by a furrow, where the bulge opens during feeding. **p)** Right side overview with the long oral bulge marked by arrowheads. Arrows mark monokinetidal tail of brush row 3, where bristles and ordinary cilia irregularly alternate. **q)** Exploded toxicysts in interference contrast (from FOISSNER 2000a). CK – cilia of circumoral kinety.

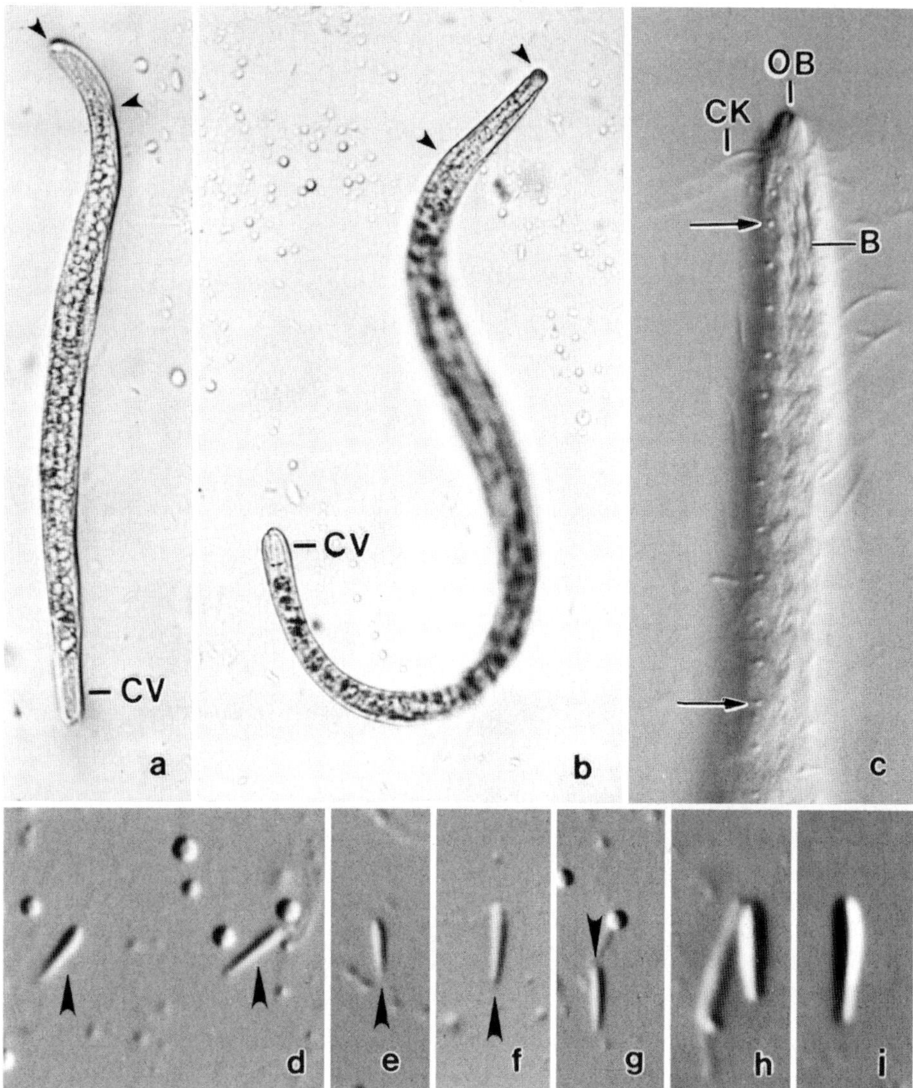

Fig. 120a–i *Arcuospathidium vlassaki*, Namibian type population (a–g) and Saudi Arabian specimens (h, i) from life (bright field and interference contrast). From FOISSNER (2000c). **a, b)** Overviews showing the vermiform shape and movement (cp. figures j–l, t, u). Arrowheads mark the oral bulge which is inconspicuous both in vivo (a, b) and the scanning electron microscope (q–s). **c)** Dorsolateral view of anterior body portion. Arrows mark a row of dots, that is, the proximal end of the cilia of the first kinety right of the dorsal brush (cp. figures 120v–y). **d–g)** The oral bulge extrusomes (arrowheads) are very narrowly obovate and about 5 × 1 µm in size. They are attached with the broad end (!) to only the left half of the oral bulge **h, i)** The extrusomes of the Saudi Arabian specimens are slightly thicker than those of the Namibian type population (d–g) and are also attached with the broad end to the oral bulge. B – dorsal brush, CK – cilia of circumoral kinety, CV – contractile vacuole, OB – oral bulge.

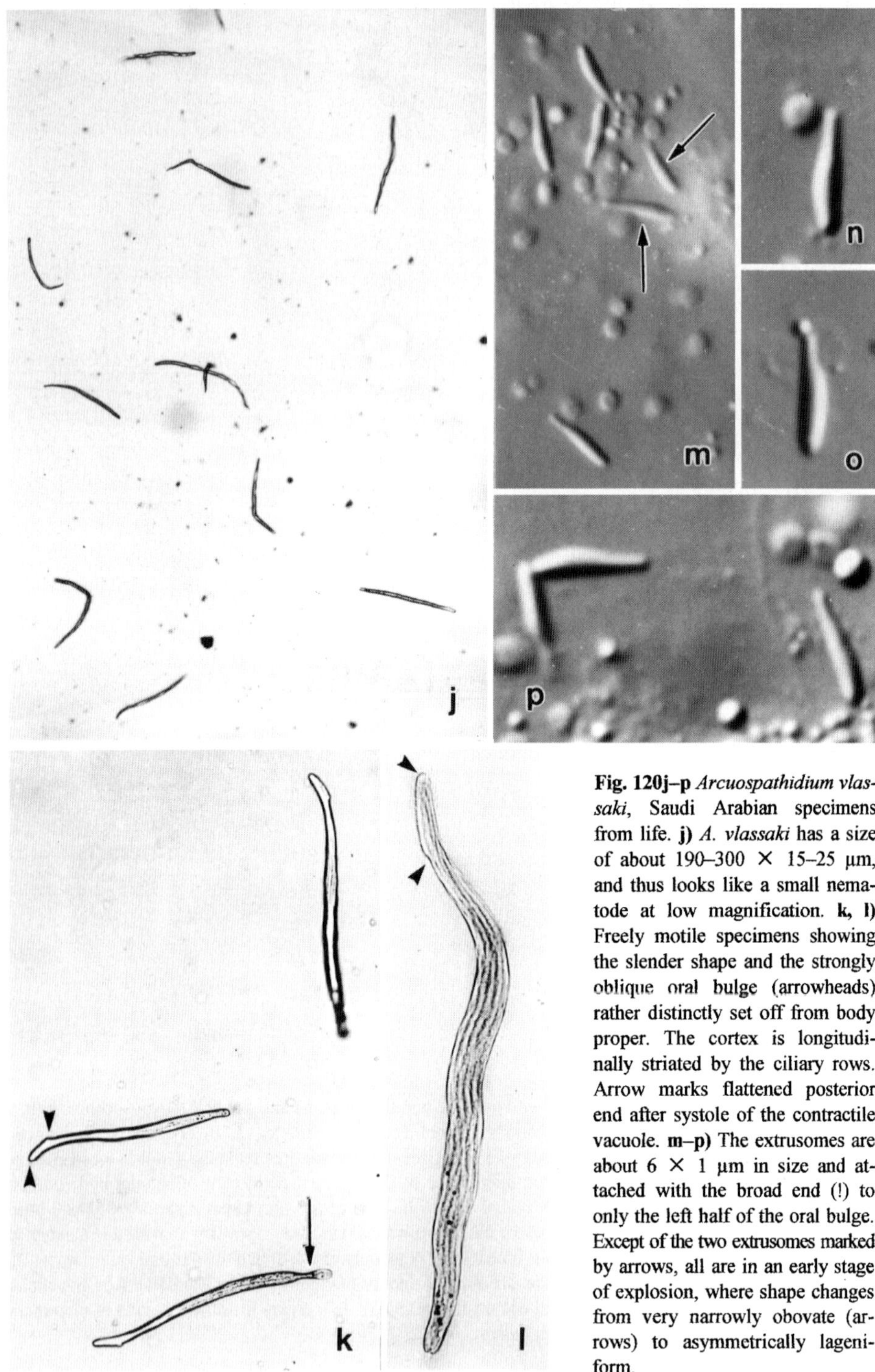

Fig. 120j–p *Arcuospathidium vlassaki*, Saudi Arabian specimens from life. **j)** *A. vlassaki* has a size of about 190–300 × 15–25 µm, and thus looks like a small nematode at low magnification. **k, l)** Freely motile specimens showing the slender shape and the strongly oblique oral bulge (arrowheads) rather distinctly set off from body proper. The cortex is longitudinally striated by the ciliary rows. Arrow marks flattened posterior end after systole of the contractile vacuole. **m–p)** The extrusomes are about 6 × 1 µm in size and attached with the broad end (!) to only the left half of the oral bulge. Except of the two extrusomes marked by arrows, all are in an early stage of explosion, where shape changes from very narrowly obovate (arrows) to asymmetrically lageniform.

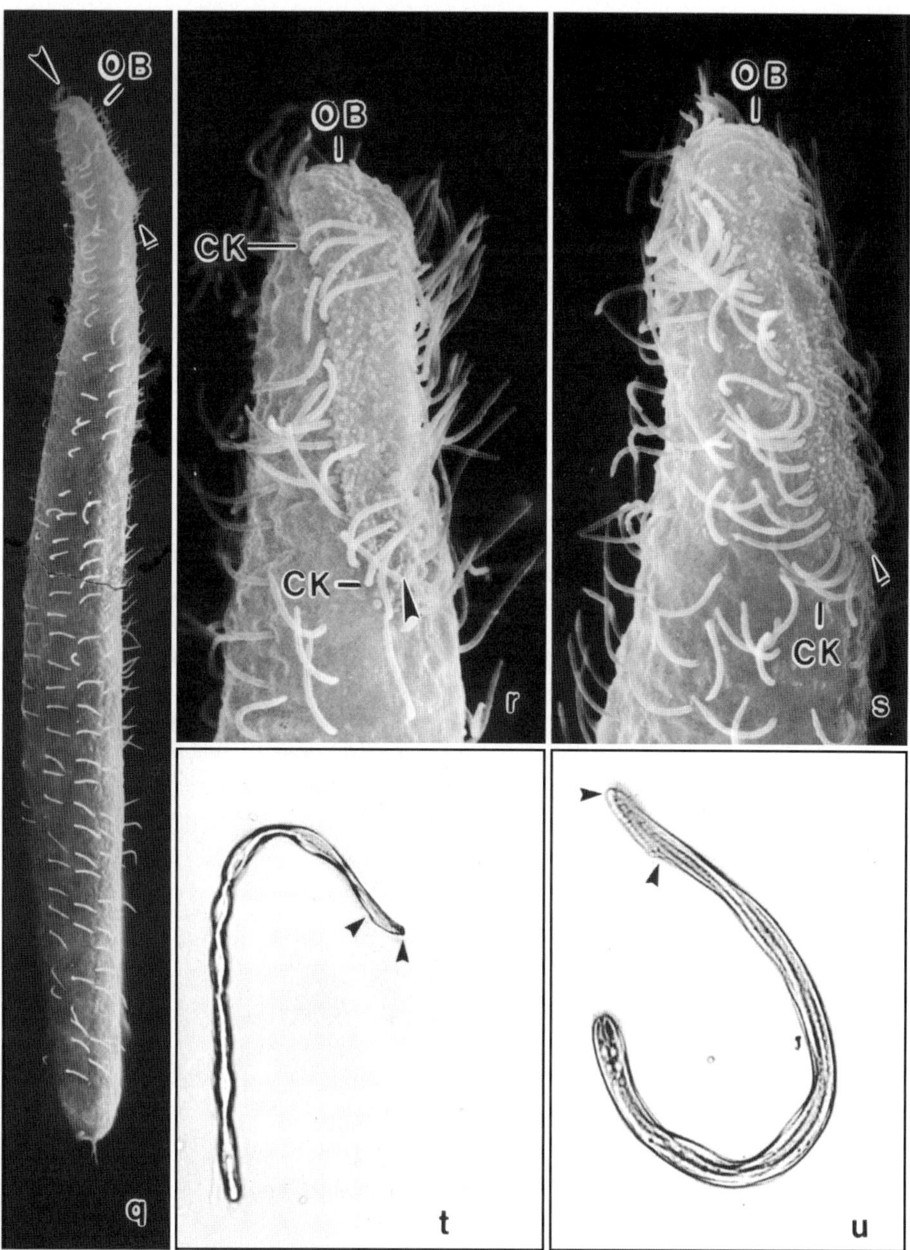

Fig. 120q–u *Arcuospathidium vlassaki*, Saudi Arabian specimens in the scanning electron microscope (q–s) and from life (t, u). **q)** Right side overview with ends of oral bulge marked by arrowheads. Note the widely spaced somatic cilia. **r, s)** Ventral and ventrolateral view showing the flat, slightly cuneate, finely granulated oral bulge (posterior end marked by arrowhead) and the comparatively widely spaced cilia of the circumoral kinety. A temporary cytostome is not recognizable. **t, u)** A highly and a slightly twisted specimen with oral bulge marked by arrowheads. CK – circumoral kinety, OB – oral bulge.

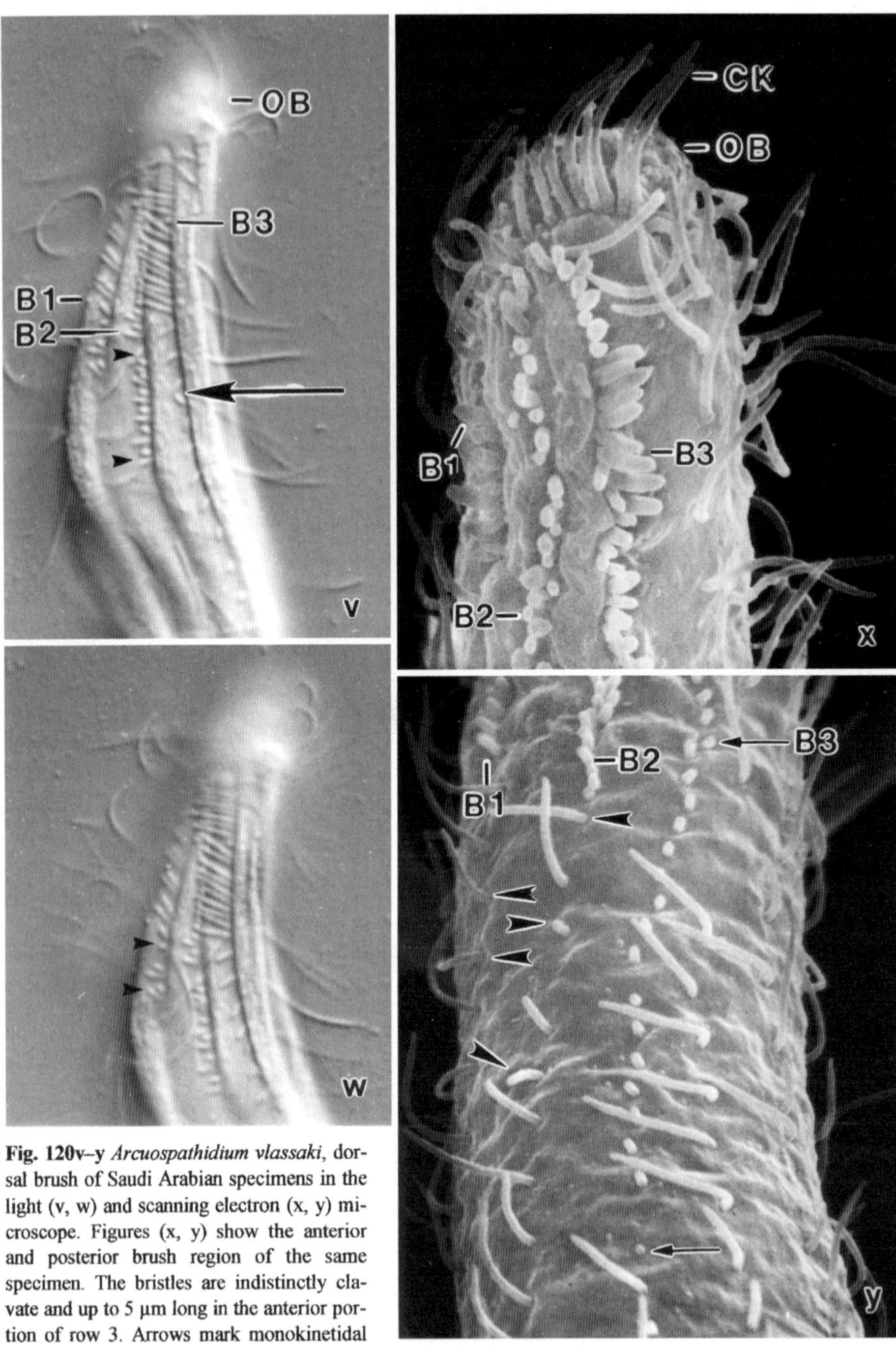

Fig. 120v–y *Arcuospathidium vlassaki*, dorsal brush of Saudi Arabian specimens in the light (v, w) and scanning electron (x, y) microscope. Figures (x, y) show the anterior and posterior brush region of the same specimen. The bristles are indistinctly clavate and up to 5 µm long in the anterior portion of row 3. Arrows mark monokinetidal bristle tail of brush row 3, which continues posteriorly as ordinary ciliary row (y). Arrowheads denote tails of brush rows 1 and 2. B(1-3) – dorsal brush rows, CK – cilia of circumoral kinety, OB – oral bulge.

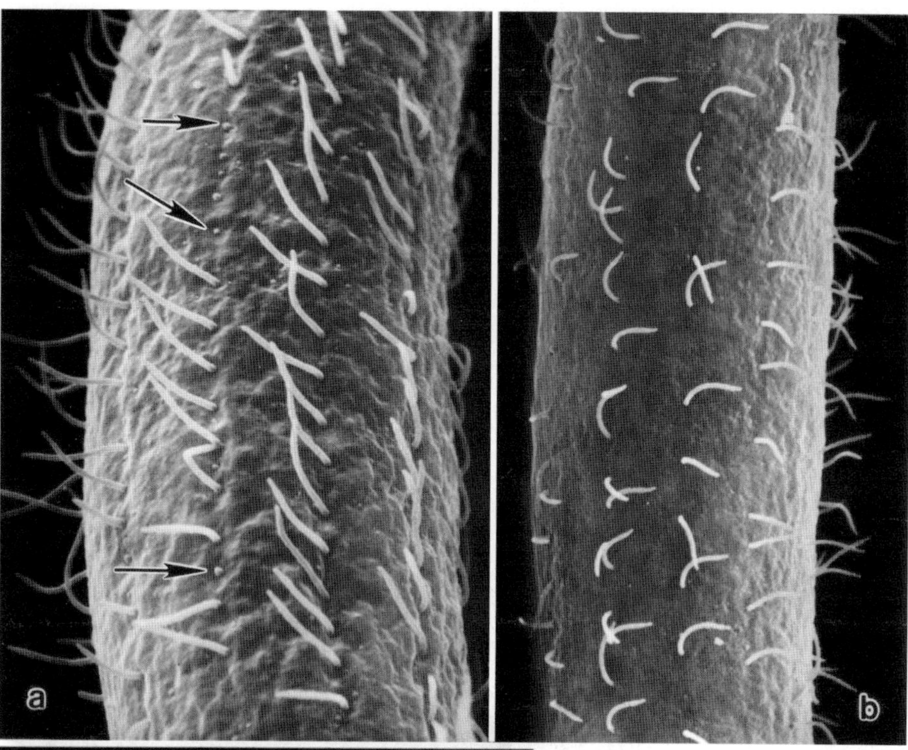

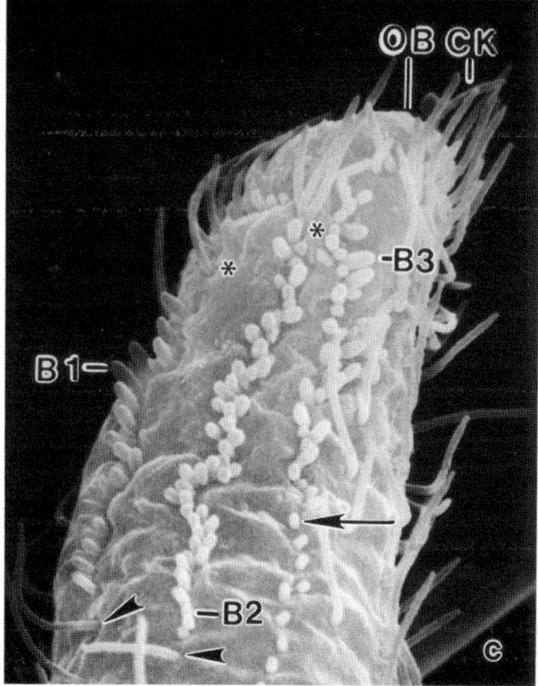

Fig. 121a–c *Arcuospathidium vlassaki*, Saudi Arabian specimens in the scanning electron microscope. **a, b)** Somatic ciliature in mid-body. The cilia are rather widely spaced and some kinetids are unciliated (a; arrows), likely serving as a reservoir during division. **c)** Dorsal view of anterior body region showing the dorsal brush, which consists of three rows of indistinctly clavate, up to 5 μm long bristles. The complex structure of the brush is only partially preserved in the scanning electron microsope (cp. figures v, w), where the oral region appears contracted. Arrow marks begin of monokinetidal bristle tail of row 3. Arrowheads denote first ordinary cilium underneath brush rows 1 and 2, showing that they extend posteriorly as ordinary kineties. Asterisks mark some ordinary cilia (anterior tail) at anterior end of brush rows 1 and 2. B(1-3) – dorsal brush (rows).

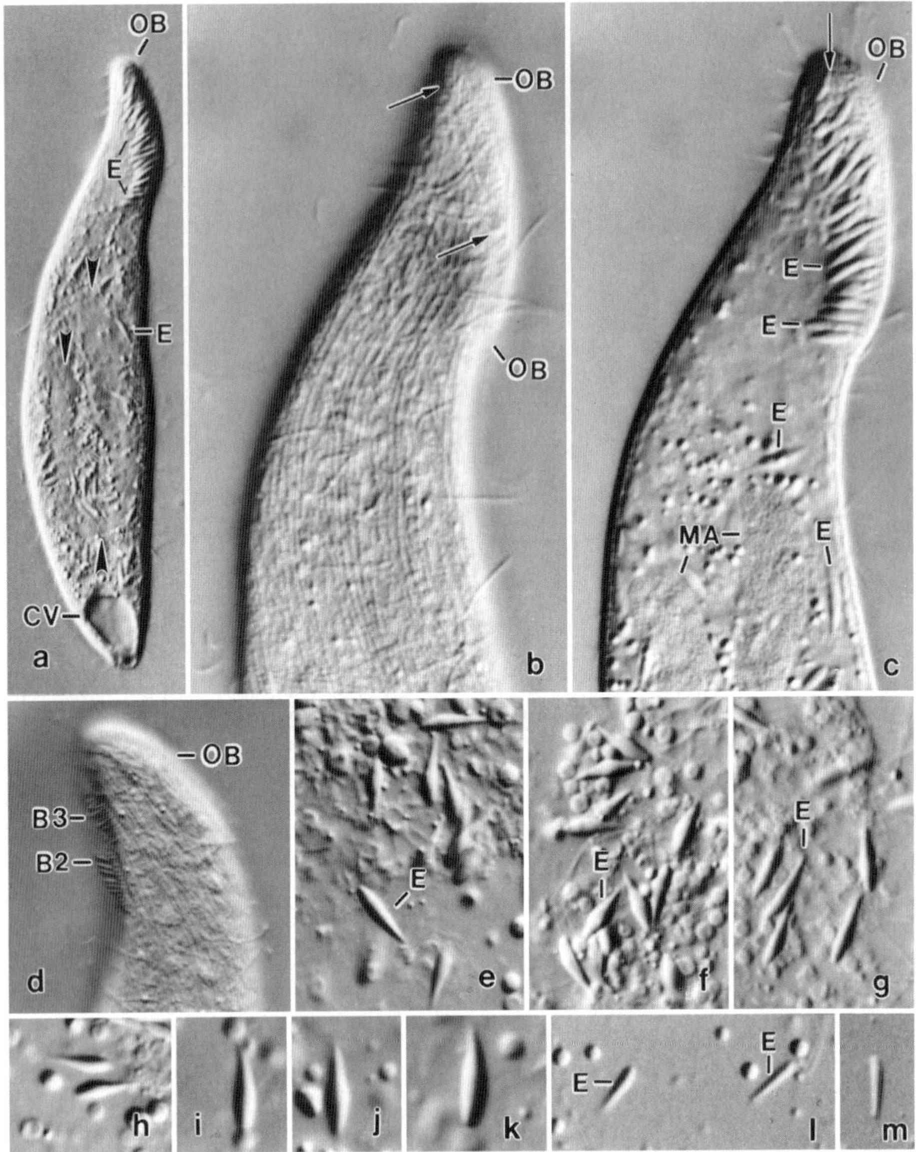

Fig. 122a–m *Arcuospathidium pachyoplites* (a–k) and *A. vlassaki* (l, m) from life. From FOISSNER (2000c, 2003a). **a–c)** Overview and oral details of a slightly squashed and thus broadened specimen. *Arcuospathidium pachyoplites* is conspicuous due to the thick and numerous extrusomes contained in the oral bulge (a, c). Figure (**b**) is a surface view showing the closely spaced rows of cortical granules and the right margin of the oral bulge (arrows). Arrowheads in figure (a) mark the long, tortuous macronucleus. **d)** Right side view showing dorsal brush rows 2 and 3 with bristles up to 4 μm long. **e–h)** Oral bulge extrusomes are 6–8 × 1–1.7 μm in size and rather variable in shape, usually, however, they are narrowly ovate. **i–m)** Same scale comparison of extrusomes of *A. pachyoplites* (i–k) and *A. vlassaki* (l, m). The extrusomes of *A. pachyoplites* are considerably larger (7 × 1.4 μm) than those of *A. vlassaki* (5 × 1 μm). B 2, 3– dorsal brush rows, CV – contractile vacuole, E – extrusomes, MA – macronucleus, OB – oral bulge.

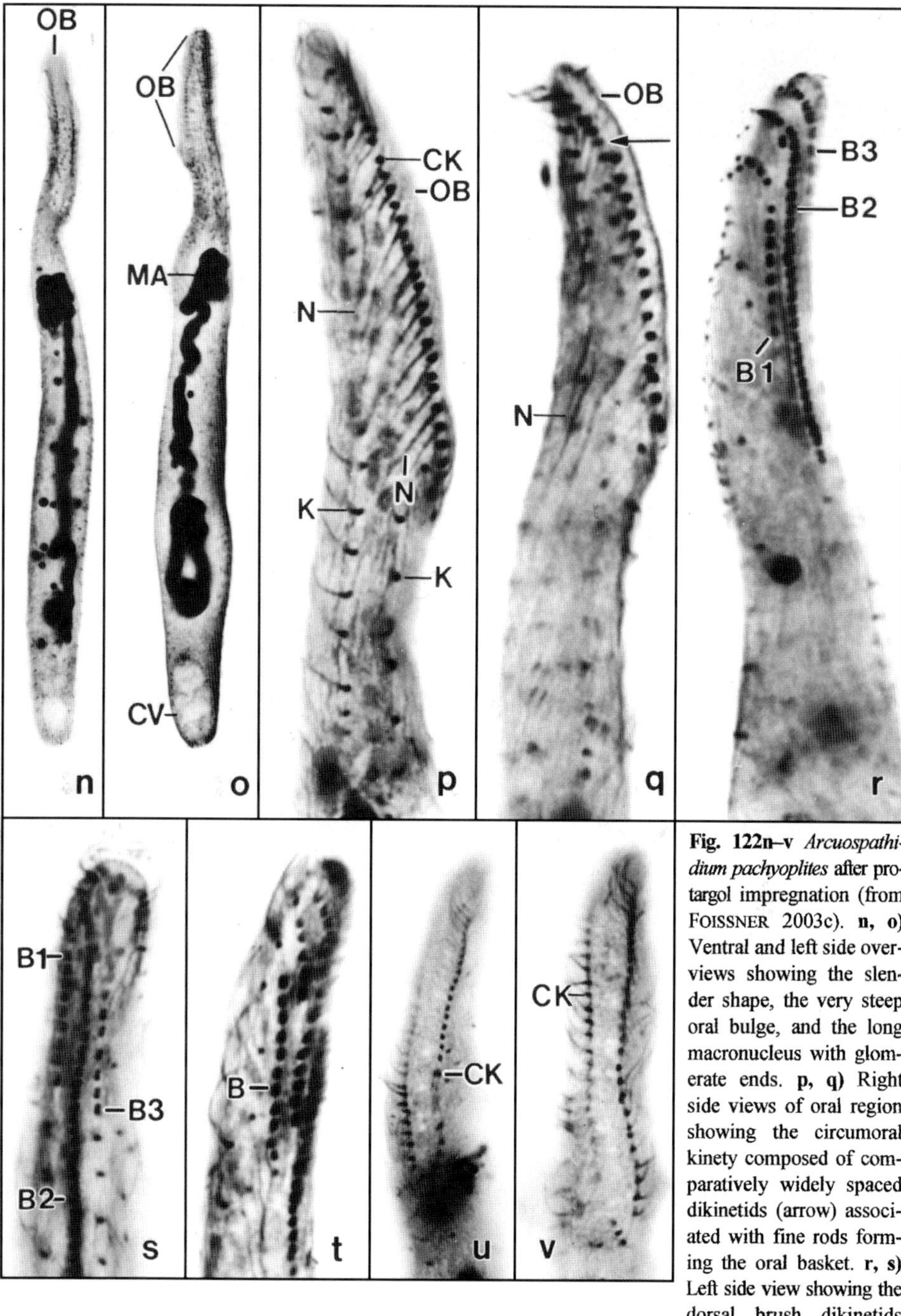

Fig. 122n–v *Arcuospathidium pachyoplites* after protargol impregnation (from FOISSNER 2003c). **n, o)** Ventral and left side overviews showing the slender shape, the very steep oral bulge, and the long macronucleus with glomerate ends. **p, q)** Right side views of oral region showing the circumoral kinety composed of comparatively widely spaced dikinetids (arrow) associated with fine rods forming the oral basket. **r, s)** Left side view showing the dorsal brush dikinetids much more narrowly spaced in row 2 than in rows 1 and 3. Dorsal brush row (2) is shorter than the oral bulge, a main difference to *A. vlassaki*, where it is longer. **t)** A specimen with irregular dorsal brush. **u, v)** Frontal views of circumoral kinety. The white dots within the kinety (oral bulge) are optical transverse sections of the large, unstained extrusomes. B(1-3) – dorsal brush (rows), CK – circumoral kinety, CV – contractile vacuole, K – ciliary rows, MA – macronucleus, N – nematodesmata (oral basket rods), OB – oral bulge.

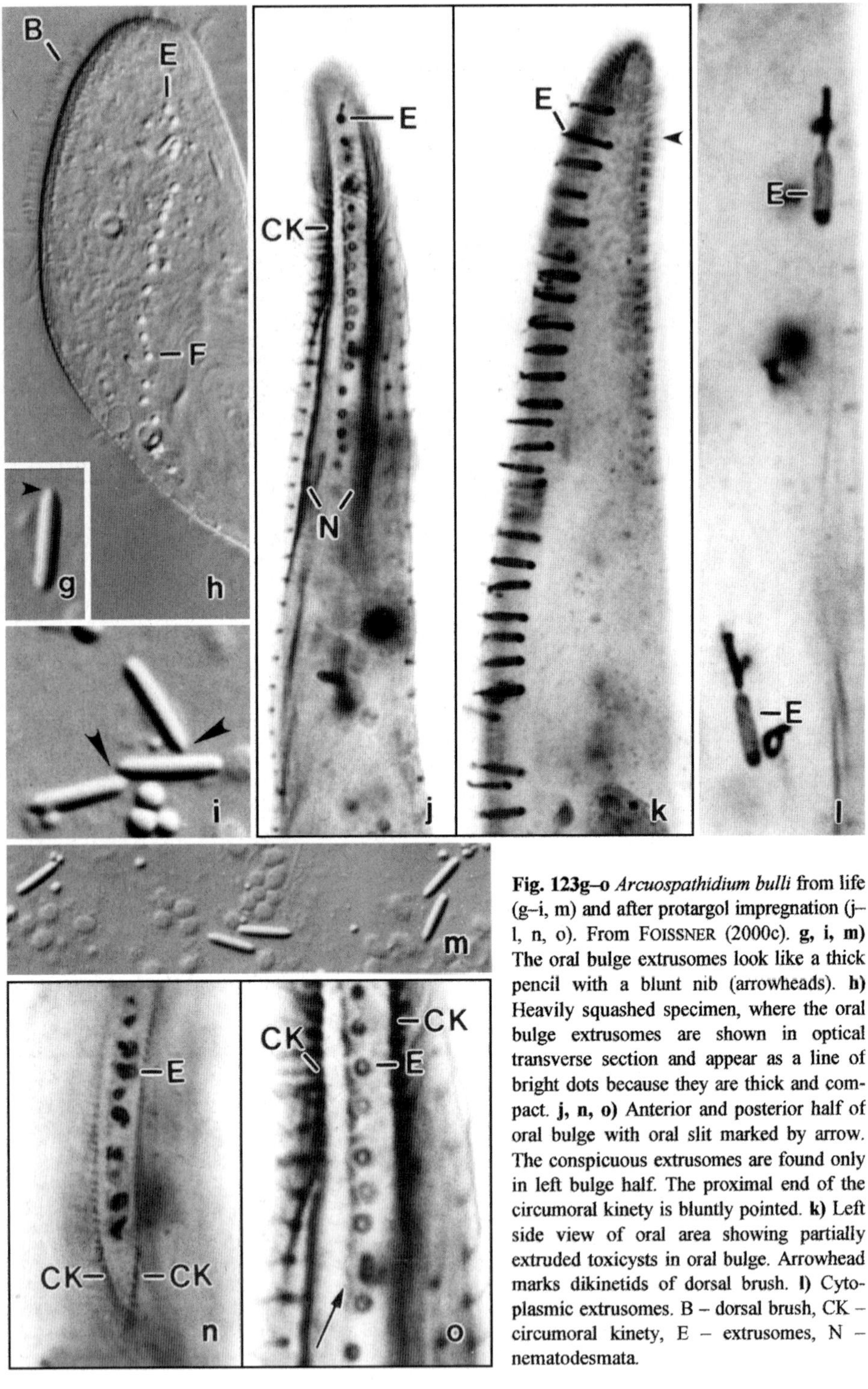

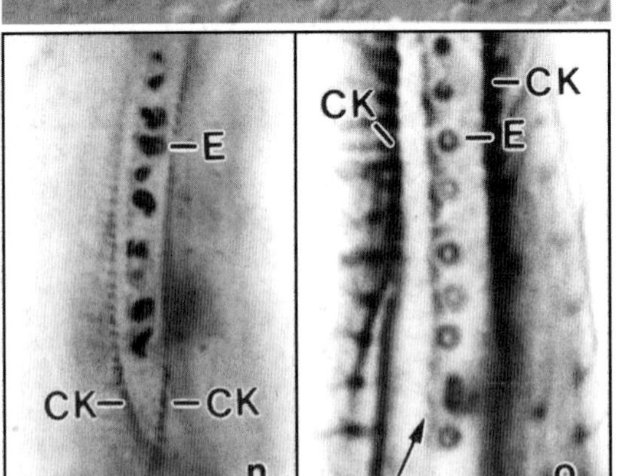

Fig. 123g–o *Arcuospathidium bulli* from life (g–i, m) and after protargol impregnation (j–l, n, o). From FOISSNER (2000c). **g, i, m)** The oral bulge extrusomes look like a thick pencil with a blunt nib (arrowheads). **h)** Heavily squashed specimen, where the oral bulge extrusomes are shown in optical transverse section and appear as a line of bright dots because they are thick and compact. **j, n, o)** Anterior and posterior half of oral bulge with oral slit marked by arrow. The conspicuous extrusomes are found only in left bulge half. The proximal end of the circumoral kinety is bluntly pointed. **k)** Left side view of oral area showing partially extruded toxicysts in oral bulge. Arrowhead marks dikinetids of dorsal brush. **l)** Cytoplasmic extrusomes. B – dorsal brush, CK – circumoral kinety, E – extrusomes, N – nematodesmata.

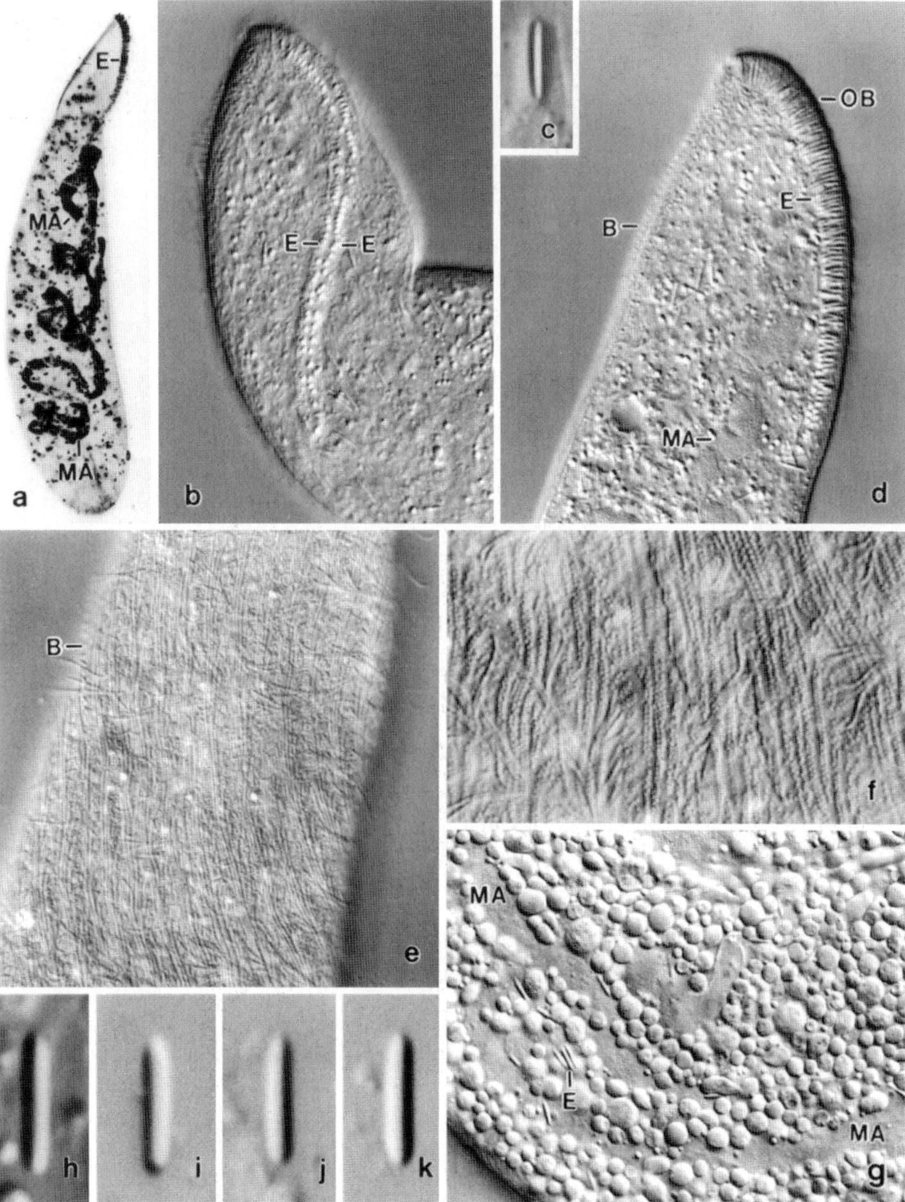

Fig. 124a–k *Arcuospathidium cultriforme cultriforme* from life (b–k) and after protargol impregnation (a). From Xu & Foissner (2005b). **a)** Overview of a specimen from the Linz (Austrian) population, showing the tortuous macronucleus and oral extrusome fringe. **b–d)** Ventral and lateral view of the oral area of a specimen from Venezuelan site (8). The extrusomes, which are 6 µm long and comparatively thin in this population (c), form a row each in right and left bulge half (b, d). **e, f)** Surface view showing cortical granulation in specimens from Venezuelan site 8 (e) and Costa Rica (f). The cortical granules are 0.2–0.4 µm across and form slightly oblique rows paired in the Costa Rican specimen (f). **g)** The Hawaiian site (39) specimens are studded with lipid droplets. **h–k)** Oral bulge extrusomes of specimens from Hawaiian sites 50 (h) and 39 (i–k) are blunt, 4–5 µm long rods. B – dorsal brush, E – extrusomes, MA – macronucleus, OB – oral bulge.

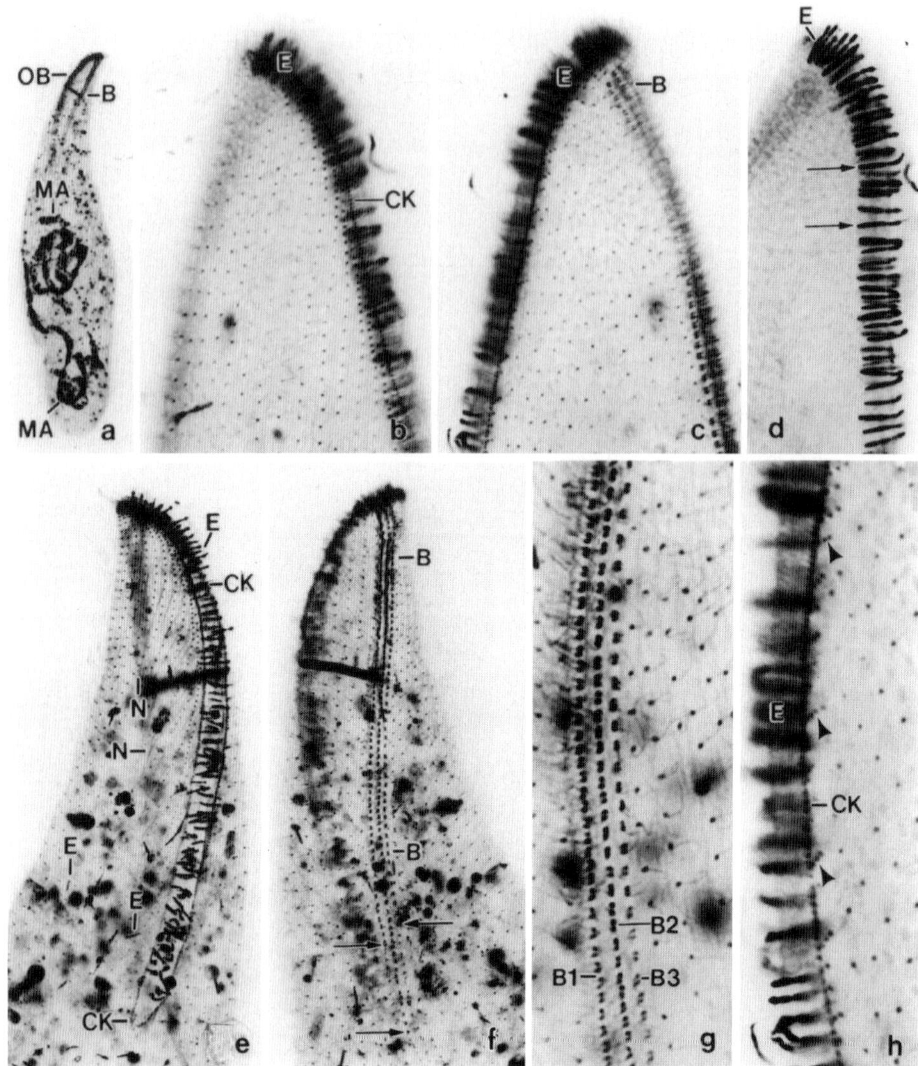

Fig. 125a–h *Arcuospathidium cultriforme cultriforme*, Linz (Austrian) specimens after protargol impregnation. From Xu & Foissner (2005b). All flattened by coverslip pressure to reveal as many details as possible. Compare with figure 126a which shows the ciliary pattern of an ordinarily prepared specimen. **a, e, f)** Overview and details of specimen 1, showing the long, tortuous macronucleus (a); the conspicuous circumoral kinety (e); and the long, three-rowed dorsal brush (f) with posterior row ends marked by arrows. The right side ciliary rows are distinctly curved dorsally along the circumoral kinety (e). **b–d, h)** Details of right and left oral area of specimen 2. On the left side, there are curious, short kinetofragments (c, h; arrowheads) attached to the circumoral kinety, which is composed of very narrowly spaced dikinetids (h). The oral bulge extrusomes, which impregnate intensely and thus form a conspicuous fringe, are knife-shaped when partially extruded (d, arrows). **g)** Details of dorsal brush of specimen 3. The brush is composed of dikinetids wider spaced in row 3 than in rows 1 and 2. B (1-3) – dorsal brush (rows), CK – circumoral kinety, E – extrusomes, MA – macronucleus, N – nematodesmata forming the oral basket, OB – oral bulge.

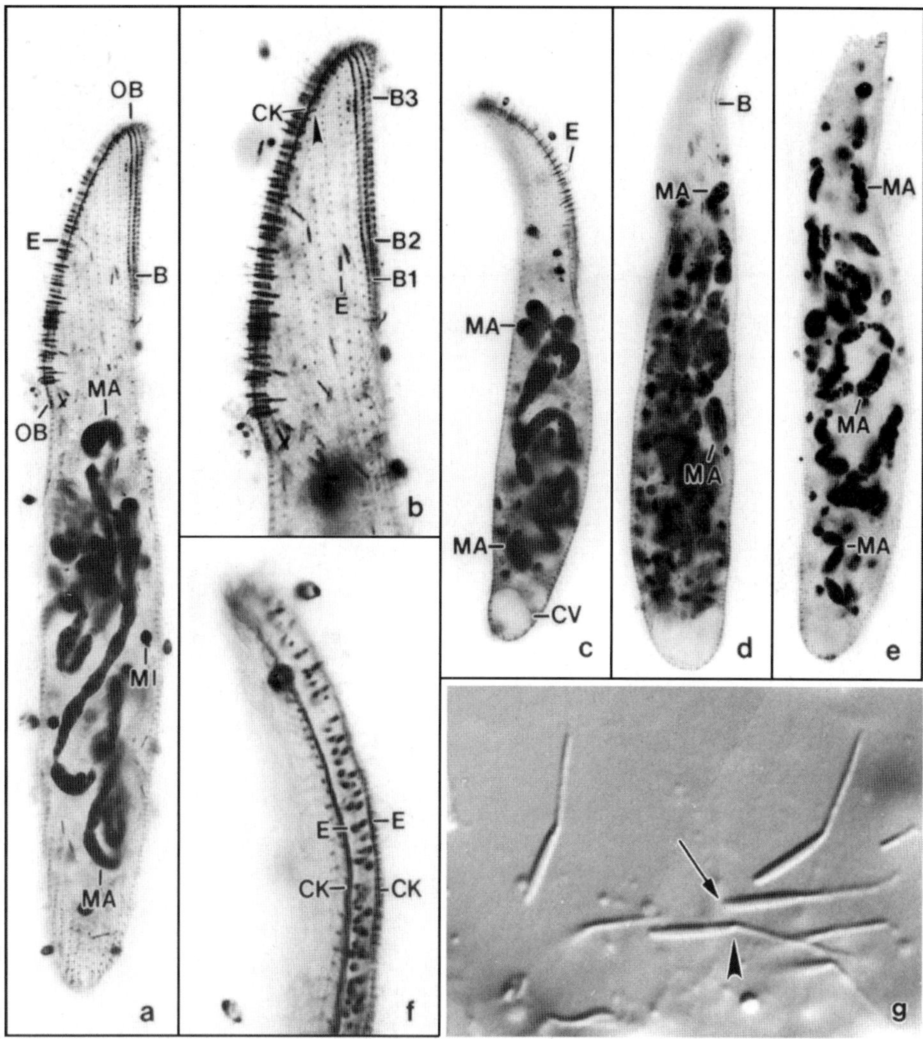

Fig. 126a–g *Arcuospathidium cultriforme cultriforme*, Hawaiian sites 50 (a–f) specimens and exploded toxicysts of a Venezuelan site (20) cell from life (g) and after protargol impregnation (a–f). From XU & FOISSNER (2005b). **a–e)** Two thirds of the Hawaiian site (50) specimens have an ordinary macronucleus, that is, a long, tortuous strand (a, c), while one third have many macronucleus nodules (d). Transitions occur (e), suggesting that both belong to the same species. The oral bulge occupies 30% (a) to 37% (c) of body length, as typical for this subspecies. Short kinetofragments (b, arrowhead) are attached to the left side circumoral kinety, as in the European populations (Fig. 125h). The anterior end of the dorsal brush kineties is slightly curved ventrally (a, b). **f)** The extrusomes form a row each in the right and left half of the oral bulge, as typical for this subspecies. **g)** Exploded toxicysts are 10–15 µm long and knife-shaped (arrow), as mentioned and illustrated by PENARD (Fig. 71f). The arrowhead marks a minute globule between shaft and capsule. B (1-3) – dorsal brush (rows), CK – circumoral kinety, CV – contractile vacuole, E – extrusomes, MA – macronucleus (nodules), MI – micronuclei, OB – oral bulge.

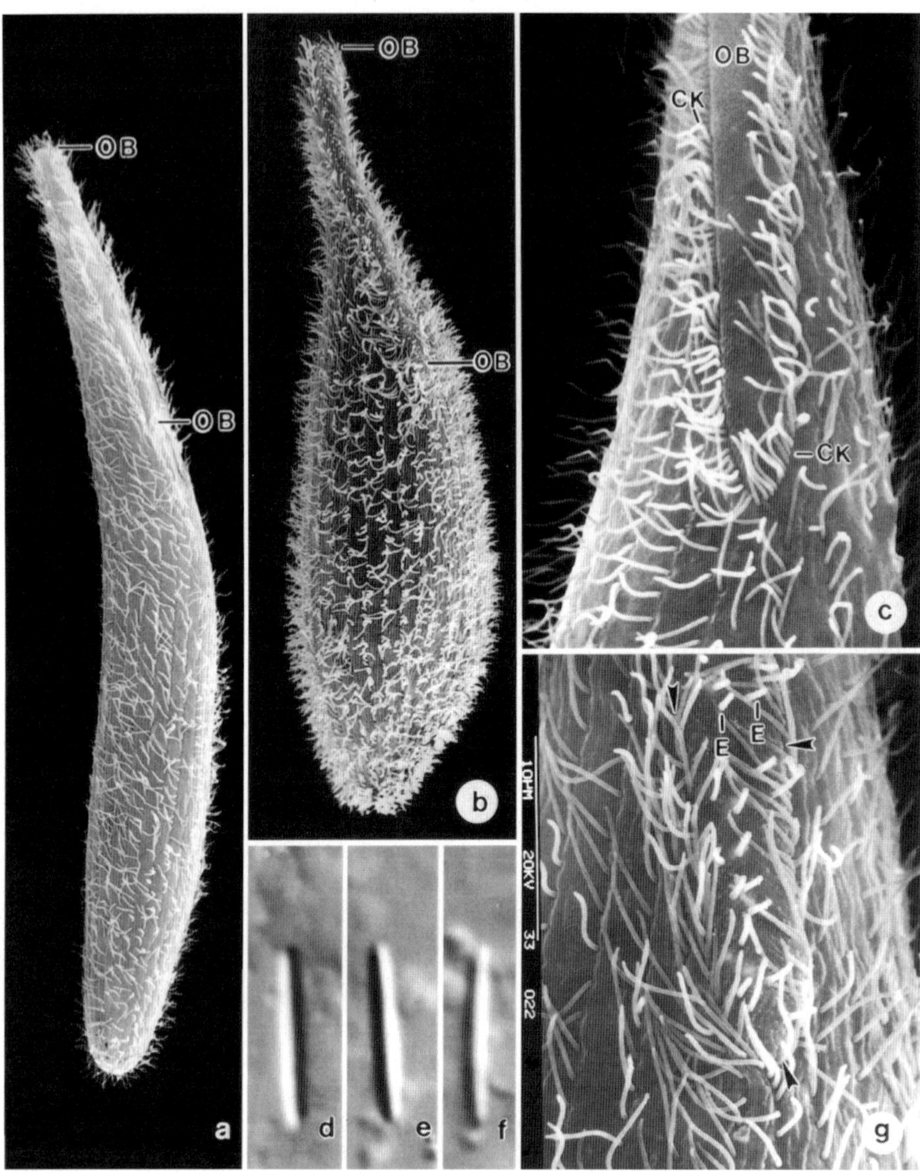

Fig. 127a–g *Arcuospathidium cultriforme cultriforme*, Venezuelan site (20) specimens in the scanning electron microscope (a–c, g) and from life (d–f). From Xu & Foissner (2005b). **a, b)** Ventrolateral views of a slender and a stout specimen with the oral bulge occupying 33% and 42% of body length; the latter value approaches those found in *A. cultriforme scalpriforme*. **c, g)** Ventral views showing posterior portion of oral bulge, which is surrounded by the circumoral kinety (CK and arrowheads), whose cilia have the same length as those on the body. The oral bulge is low, has a flat surface, and contains a row of extrusomes each in right and left half (g). **d–f)** The extrusomes are about 6 µm long and more slender than those of the European and Hawaiian populations (Fig. 124h–k); they are slender also in another Venezuelan population (Fig 124c). Thus, the Venezuelan populations might represent a distinct subspecies. CK – circumoral kinety, E – extrusomes, OB – oral bulge.

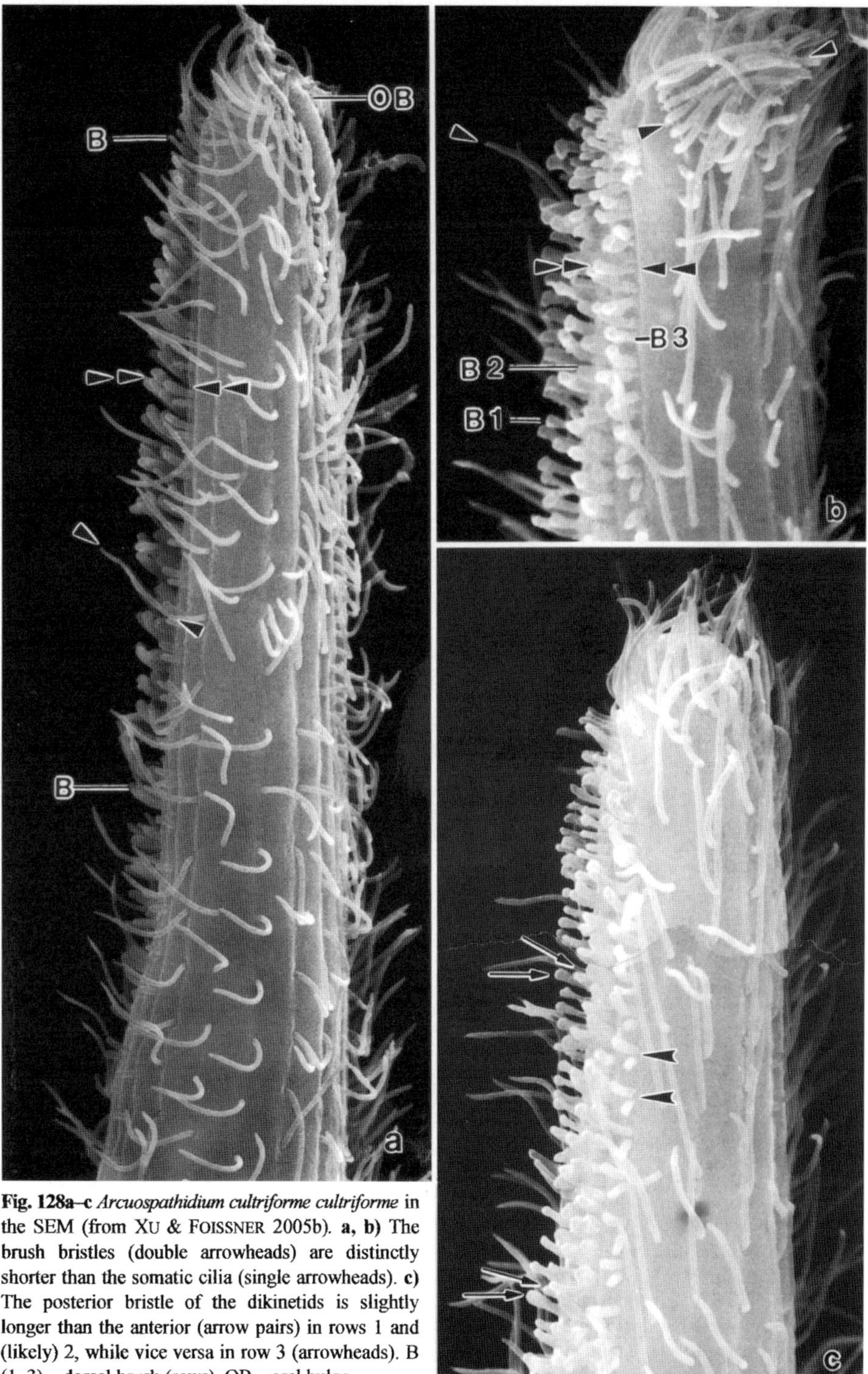

Fig. 128a–c *Arcuospathidium cultriforme cultriforme* in the SEM (from XU & FOISSNER 2005b). **a, b)** The brush bristles (double arrowheads) are distinctly shorter than the somatic cilia (single arrowheads). **c)** The posterior bristle of the dikinetids is slightly longer than the anterior (arrow pairs) in rows 1 and (likely) 2, while vice versa in row 3 (arrowheads). B (1–3) – dorsal brush (rows), OB – oral bulge.

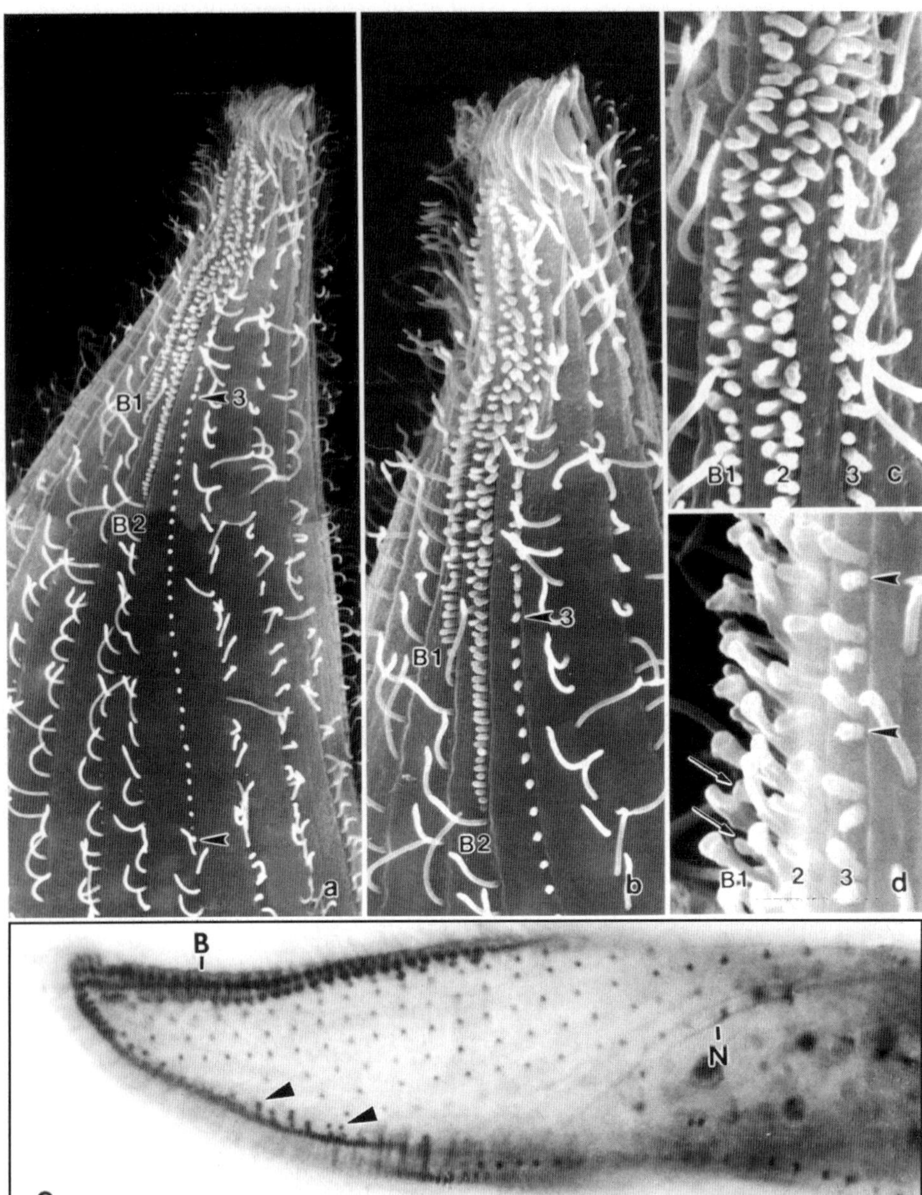

Fig. 129a–e *Arcuospathidium cultriforme cultriforme*, Venezuelan site (20) specimens (a–d; from XU & FOISSNER 2005b) and an Austrian cell (e; from FOISSNER 1984). **a–d)** Dorsal brush in overview (a) and at higher magnifications (b–d). The brush is three-rowed (numerals) and composed of dikinetids with up to 3 µm long, clavate bristles. The posterior bristle of the dikinetids is slightly longer than the anterior (d, arrows) in rows 1 and 2, while vice versa in row 3 (d, arrowheads). The dikinetids are obliquely arranged in the anterior two thirds of the rows, producing a distinct zigzag pattern of the bristles (b, c). Brush row 3 has a monokinetidal bristle tail marked by arrowheads (a, b). **e)** Left side view showing adesmokineties (arrowheads) along the circumoral kintey. B (1-3) – dorsal brush (rows), CK – circumoral kinety, N – nematodesmata.

Fig. 130a–d *Arcuospathidium cultriforme cultriforme* (a–c) from Venezuelan site (20) and *Spathidium stammeri* (**d**) in the scanning electron microscope. From XU & FOISSNER (2005**b**). **a–c)** Early dividers showing the strongly oblique division axis (arrowheads) and the developing division blebs (D) and excretory pores (EP). Asterisks in (a) mark the oral bulge. Note growing cilia in the forming oral kinetofragments (arrows). **d)** Division blebs in a early mid-divider of *S. stammeri*. The forming oral kinetofragments (OF) curve around the right margin of the division blebs (D) and thus assume a concave shape. D – division blebs, EP – excretory pores for the proter, OF – forming oral kinetofragments.

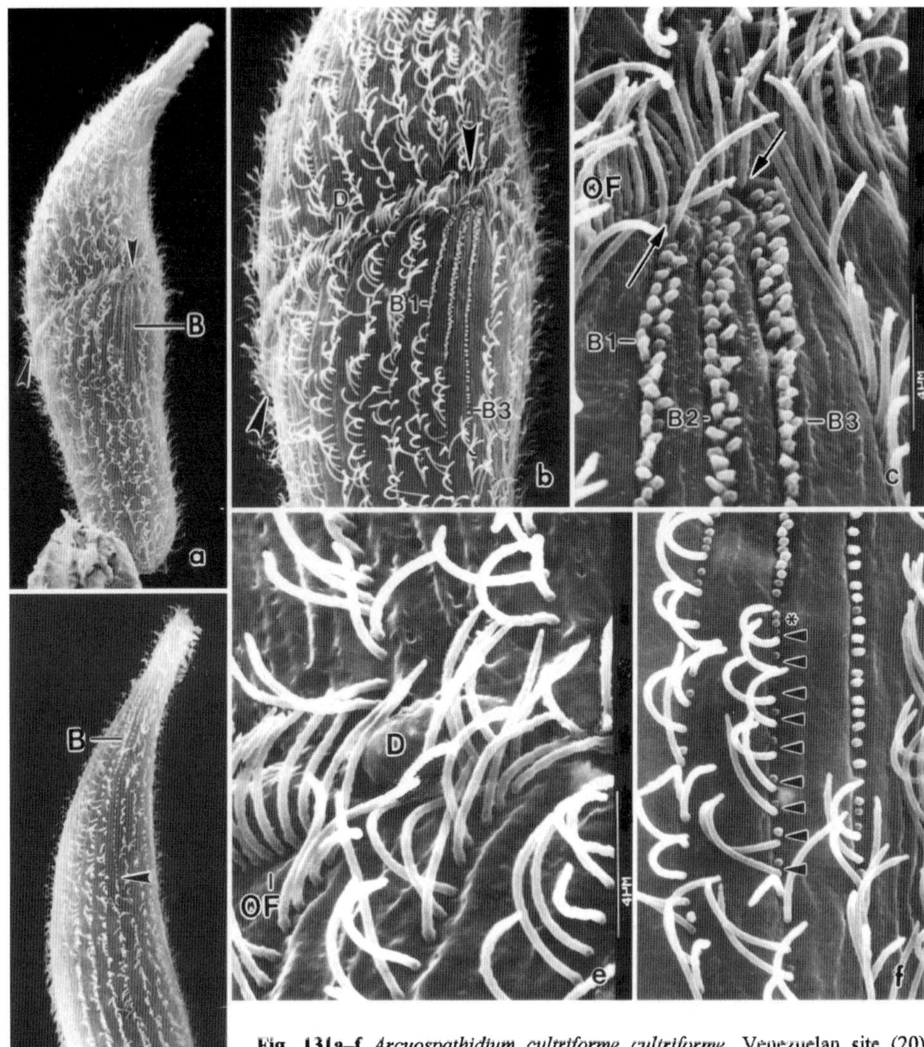

Fig. 131a–f *Arcuospathidium cultriforme cultriforme*, Venezuelan site (20) specimens in the scanning electron microscope. From Xu & Foissner (2005b). **a–c, e, f)** Dorsal overview and details of an early mid-divider showing the strongly oblique division axis (a, b; arrowheads), the division blebs (D), and the growing dorsal brush. There are three brush rows (B1-3) with zigzaging (= obliquely arranged dikinetids) bristles in the anterior portion, a unique feature of the Venezuelan specimens, suggesting that it is a distinct subspecies, like the thin extrusomes (Fig. 124c, 127d–f). Some ordinary cilia (c; arrows) are at the anterior end of the brush rows. Figure (f) shows the formation of the brush dikinetids (asterisk): a new bristle each if formed anteriorly of the parental somatic cilia which become gradually shortened to posterior brush bristles (triangle series). The division blebs (D) are now conspicuous (b, e), and the cilia of the oral kinetofragments reached full length (b, c, e; OF). **d)** Dorsal view of a morphostatic specimen showing the monokinetidal bristle tail of brush row 3 extending to mid-body (arrowhead); then, the row continues as ordinary somatic kinety to rear end. The length of the tail of brush row 3 is an important feature for distinguishing *Spathidium* s. l. species. B (1-3) – dorsal brush (rows), D – division blebs, OF – kinetofragments.

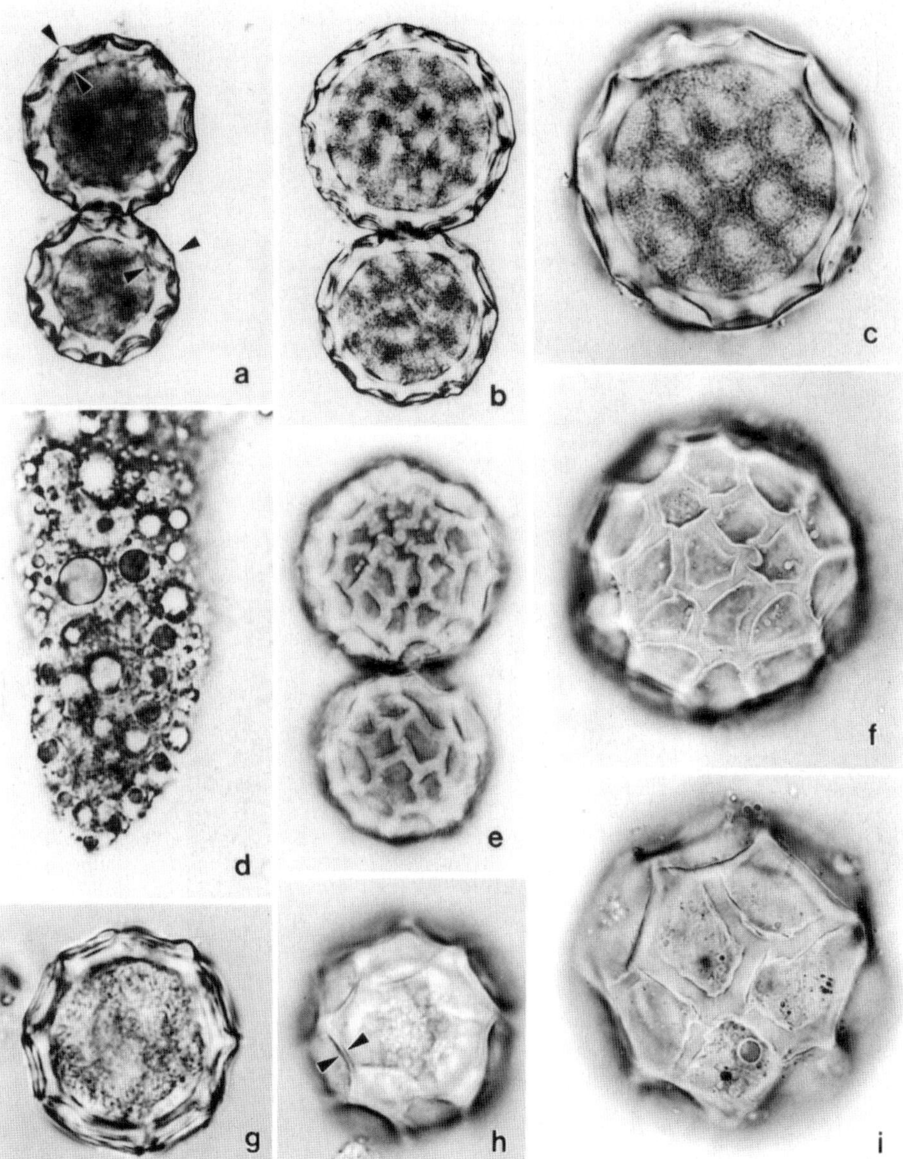

Fig. 132a–i *Arcuospathidium cultriforme cultriforme*, resting cysts (diameter 42–95 µm) of specimens from Linz, Austria (a–f, i) and the Dominican Republic (g, h). From XU & FOISSNER (2005b). **a)** Optical section showing the thick wall delimited by arrowheads. **b, e)** Optical section and surface view showing the thick wall and the polygonally faceted surface. **c, f)** Optical section and surface view of same specimen, showing the broad ridges (f) causing the spotted appearance of the cyst content in optical section (c). **d)** Cyst contents mainly consisting of highly refractive lipid droplets 1–5 µm across. **g, h)** Optical section and surface view of a cyst from the Dominican specimen. Obviously, it is very similar to those of the Austrian specimens, except for the thinner ridges (arrowheads) and the smaller diameter (45.2 vs. 68.8 µm on average). **i)** Surface view of a cyst with facets distinctly larger than those recognizable in (e, f).

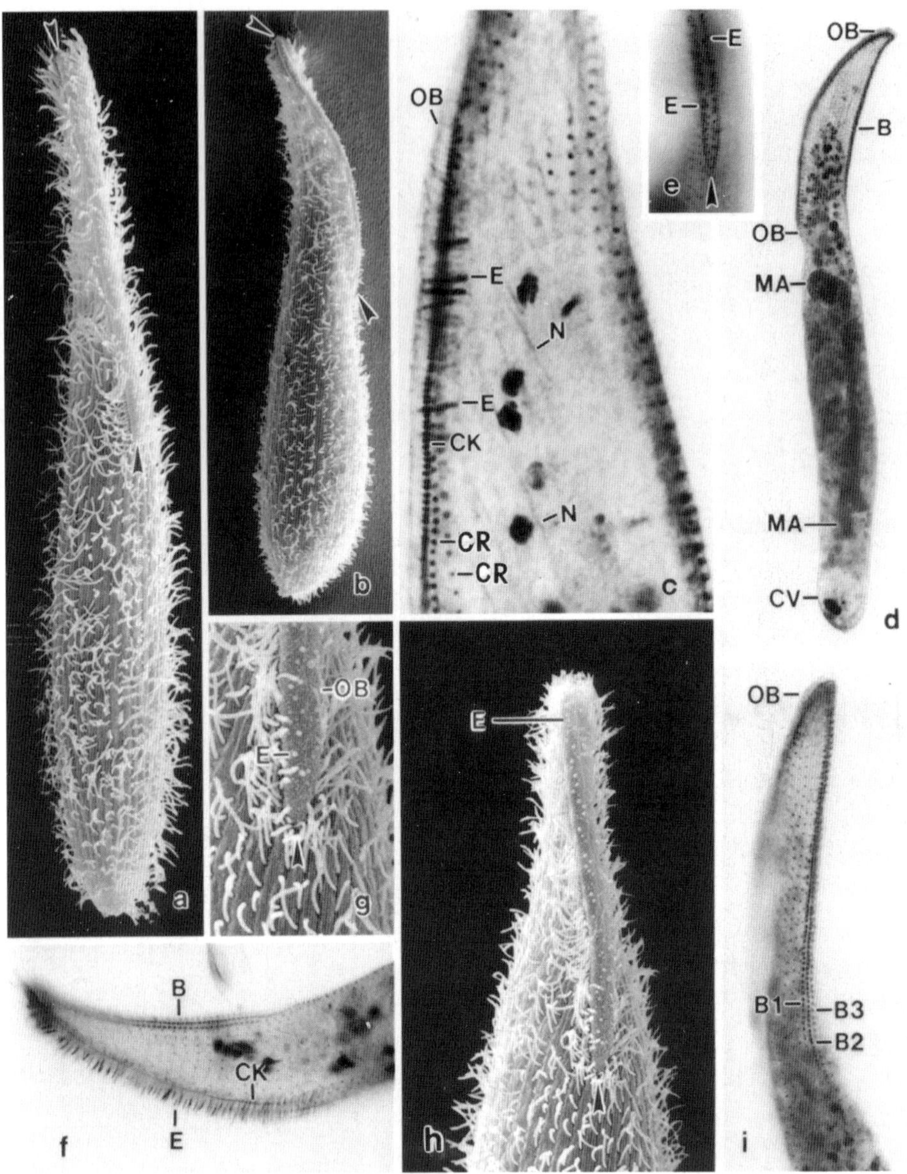

Fig. 133a–i *Arcuospathidium cultriforme scalpriforme*, South African (a, b, e–h; from XU & FOISSNER 2005b) and Austrian (c, d, i; from FOISSNER 1984) specimens in the SEM (a, b, g, h) and after protargol impregnation (c–f, i). **a, b)** Ventrolateral overviews showing the oral bulge extending to almost mid-body (ends marked by arrowheads). **c)** Left side oral area showing the comparatively thin, resting extrusomes and the circumoral kinety composed of very narrowly spaced dikinetids. **d)** Left side overview. **e, g, h)** Ventral views showing the bluntly pointed proximal end of the oral bulge and circumoral kinety (arrowheads). **f)** Left side view of oral area showing the dorsal brush and the partially extruded, thin extrusomes. **i)** Dorsolateral view showing the long dorsal brush. B (1-3) – dorsal brush (rows), CK – circumoral kinety, CR – ciliary rows, CV – contractile vacuole, E – extrusomes, MA – macronucleus, N – nematodesmata, OB – oral bulge.

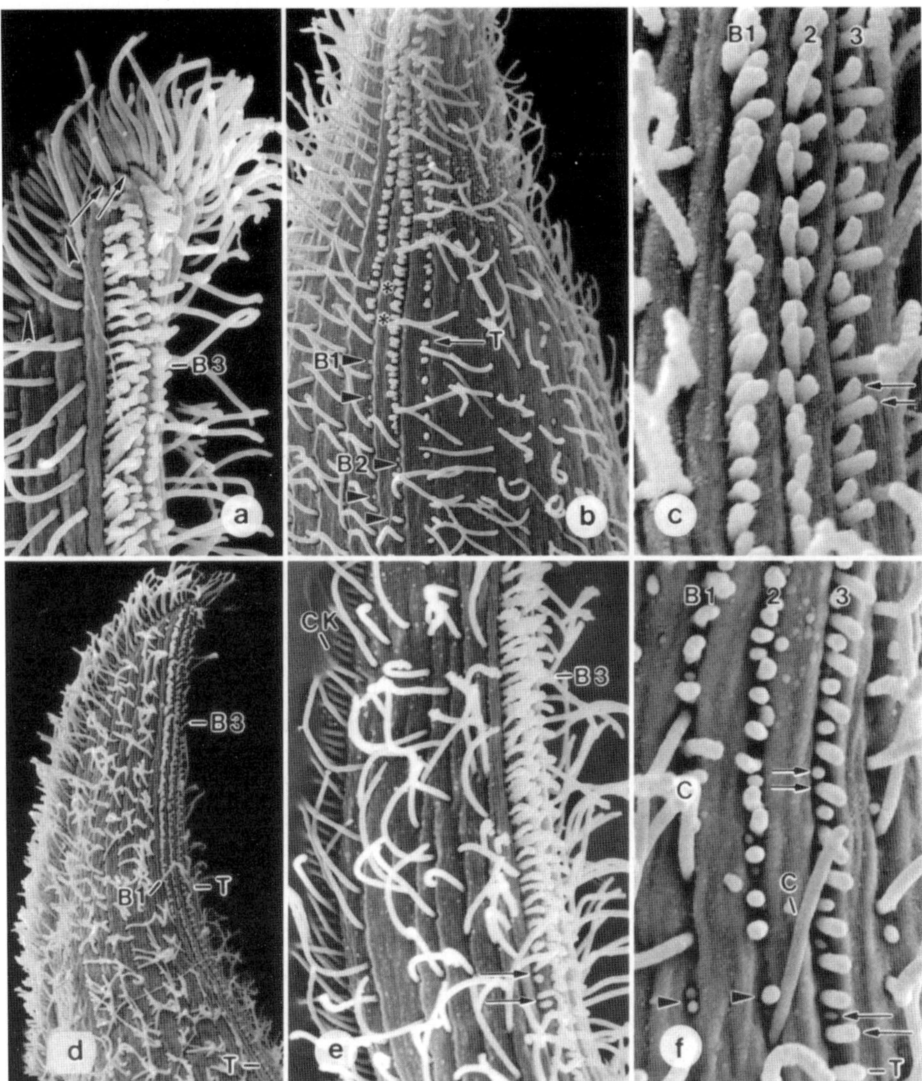

Fig. 134a–f *Arcuospathidium cultriforme scalpriforme*, dorsal brush area of South African specimens in the scanning electron microscope. From Xu & Foissner (2005b). **a)** Bluntly pointed anterior end, where (brush) ciliary rows curve ventrally. The brush rows have some ordinary cilia at anterior end (arrows). Arrowheads mark short kinetofragments (adesmokineties?) left of the circumoral kinety. The brush bristles are three times shorter than the somatic cilia, that is, about 3 µm long in vivo. **b)** Posterior half of brush. The rows end at similar level, but row 3 has a monokinetidal bristle tail (arrow) extending into second third of body. Arrowheads mark heteromorphic tail of brush rows 1 and 2, where minute bristles irregularly alternate with ordinary cilia. Asterisks denote row 1 dikinetids with a very short anterior bristle and an ordinary posterior cilium. **c, f)** Mid-portion and end of dorsal brush showing that, in row 3, the anterior bristle of the dikinetids is shorter than the posterior (arrows). Arrowheads mark strongly shortened bristles. **d)** Oral region showing that brush bristle length decreases at both ends of rows. **e:** Posterior third of brush showing that, in row 1, the posterior bristle of the dikinetids is shorter than the anterior (arrows). B (1-3) – dorsal brush (rows), C – somatic cilia, CK – circumoral kinety, T – tail of brush row 3.

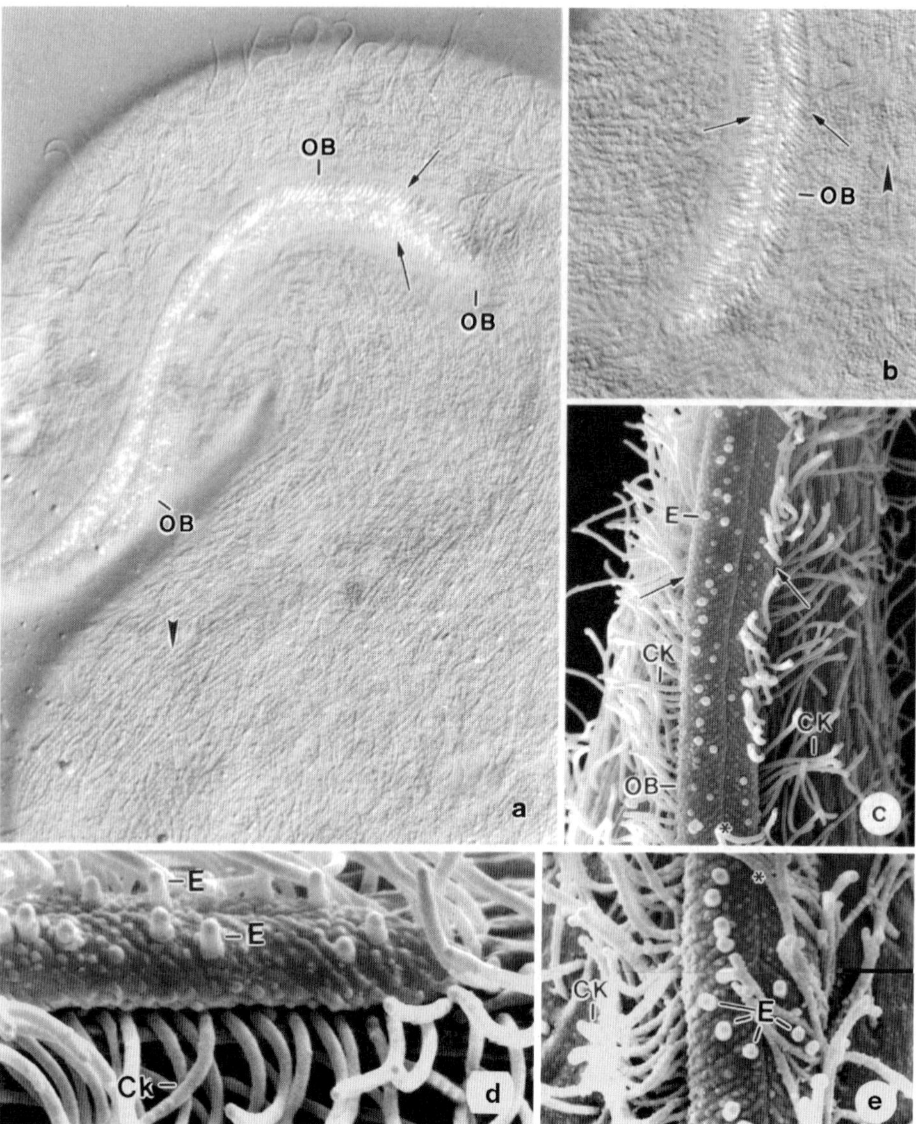

Fig. 135a–e *Arcuospathidium cultriforme scalpriforme*, South African specimens from life (a, b) and in the scanning electron microscope (c–e). From XU & FOISSNER (2005b). **a, b)** Surface views of oral area. The oral bulge appears as a bright band due to the refractive extrusomes which are scattered in both bulge halves, while they form a row each in the right and left bulge half of *A. cultriforme cultriforme* (Fig. 124b). The bulge cortex contains oblique rows of cortical granules arranged in an arrowhead-like pattern (arrows). The same granules occur in the somatic cortex, where they form long rows (arrowheads). **c–e)** In the scanning electron microscope, the oral bulge is more distinct than in vivo. The scattered bulge extrusomes are just exploding and appear as distinct granules in surface view (c, e) and as small domes in side view (d). Note the bulge midline (asterisks), the oblique striation (arrows) caused by the cortical granules, and the narrowly spaced cilia of the circumoral kinety (CK). E – extrusomes, OB – oral bulge.

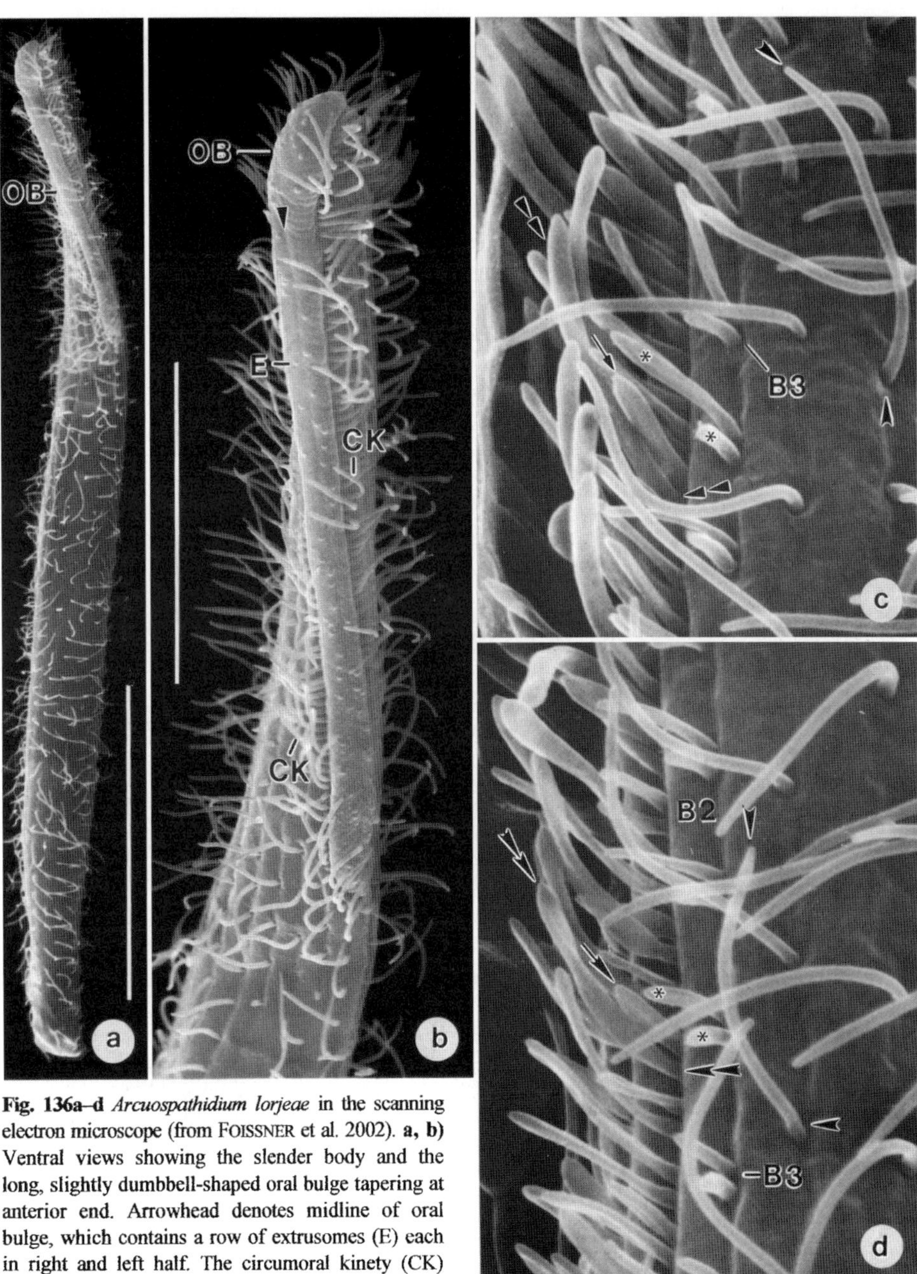

Fig. 136a–d *Arcuospathidium lorjeae* in the scanning electron microscope (from FOISSNER et al. 2002). **a, b)** Ventral views showing the slender body and the long, slightly dumbbell-shaped oral bulge tapering at anterior end. Arrowhead denotes midline of oral bulge, which contains a row of extrusomes (E) each in right and left half. The circumoral kinety (CK) consists of single, narrowly spaced cilia. Scale bars 50 µm and 20 µm. **c, d)** Dorsal brush in middle portion and near end of row 3, whose bristles are much shorter and wider spaced than those of rows 1 and 2. Double arrowheads mark two of many row 2 bristles having almost the same length as ordinary somatic cilia (two marked by single arrowheads). Arrows denote anterior row 2 bristles, which are distinctly shorter than the posterior (double arrowheads) in rows 1 and 2, while the anterior bristle is longer than the posterior in row 3 (asterisks). B2,3 – dorsal brush rows, CK – circumoral kinety, E – extrusome tip, OB – oral bulge.

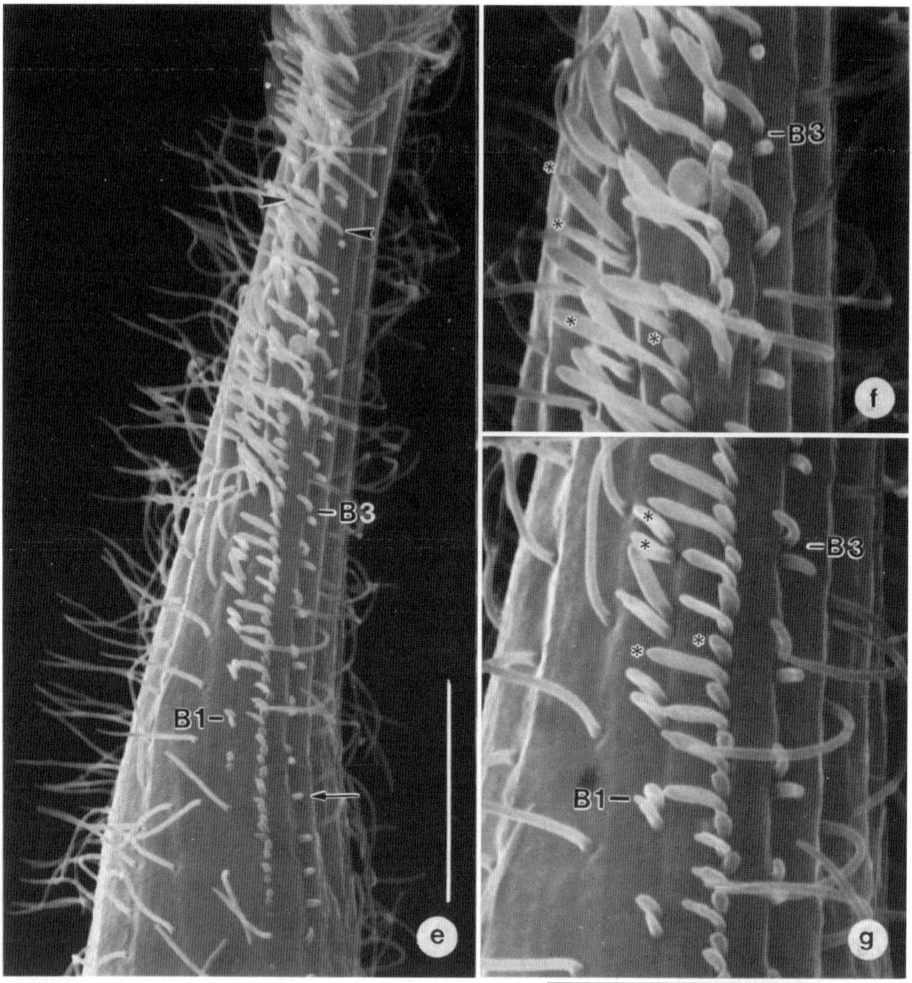

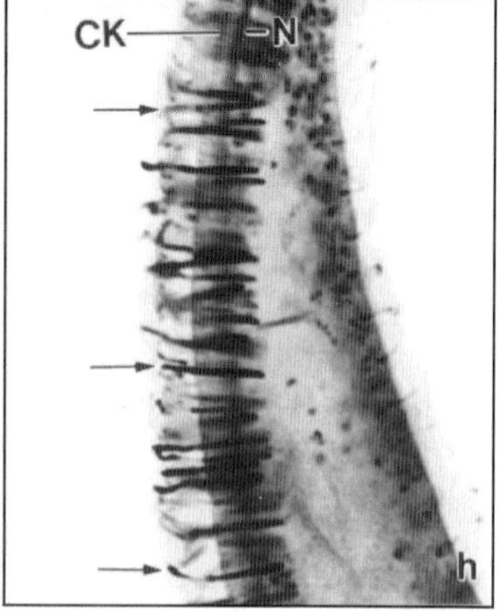

Fig. 136e–h *Arcuospathidium lorjeae* and *A. cultriforme megastoma* (h) in the scanning electron microscope (e–g) and after protargol impregnation (h). From FOISSNER et al. (2002). **e–g)** Dorsal views of anterior body region showing the highly differentiated dorsal brush, scale bar 20 µm. The brush bristles of this specimen are not as long as those of the cell shown in figures 136c, d, but are still conspicuous, for instance, the anterior bristles of row 3 marked by arrowheads. Arrow marks begin of monokinetidal bristle tail of row 3. Asterisks mark bristles of a pair each in brush rows 1 and 2: in row 1, the bristles have similar length; in row 2, the posterior bristle is much longer than the anterior; and in row 3, the anterior bristle is much longer than the posterior (e, arrowheads). **h)** The oral bulge is packed with partially extruded and thus knife-shaped toxicysts (arrows). B1, 3 – dorsal brush rows, CK – circumoral kinety, N – nematodesmata bundle.

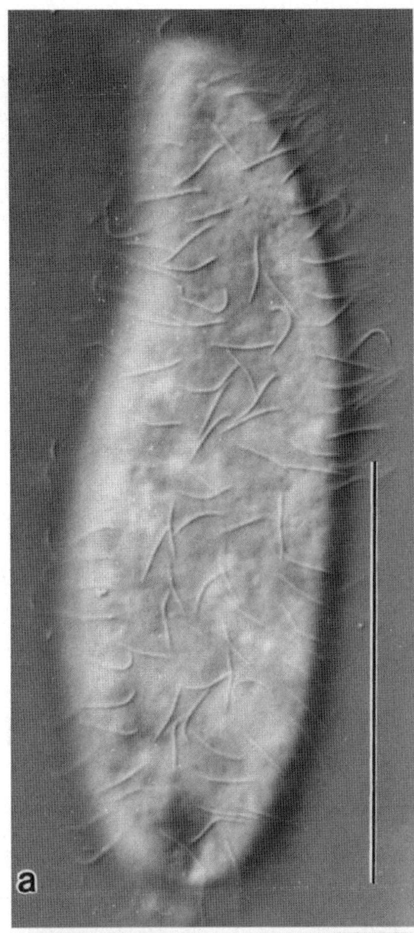

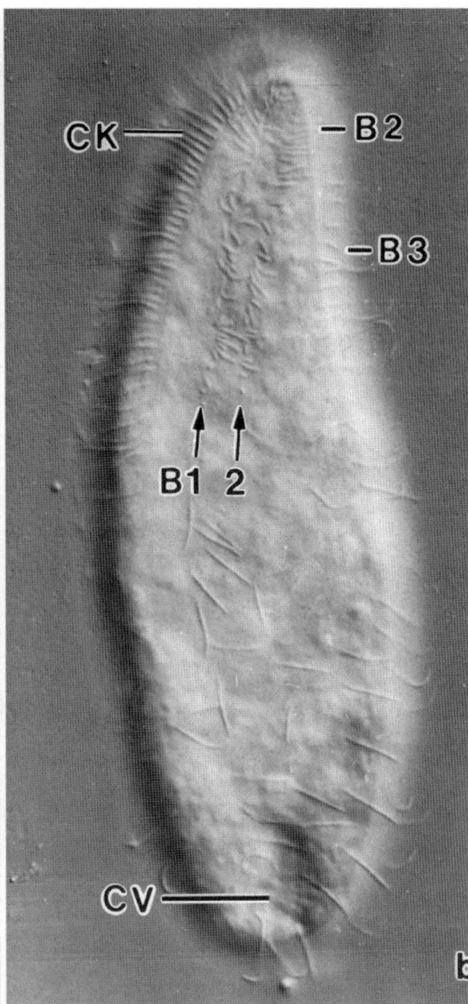

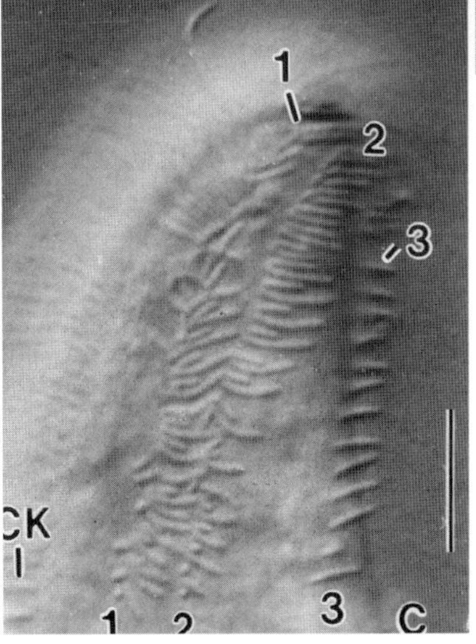

Fig. 137a–c *Cultellothrix velhoi* from life (interference contrast), right side (a) and left side (b, c) views of a slightly squeezed and flattened (by coverslip pressure) specimen (from FOISSNER 2003c). Note the loose ciliation and the three dorsal brush rows (B1-3), which are on the left side of the cell (main generic feature). The brush bristles are up to 5 μm long and the anterior bristle of a pair is distinctly shorter than the posterior; the posterior portion of row 3 consists of monokinetidal, 4 μm long bristles, showing that the species does not belong to the pleurostomatids; the brush rows continue with ordinary somatic cilia to the posterior end of the cell. B(1-3) – dorsal brush rows, C – ordinary somatic cilia, CK – circumoral kinety, CV – contractile vacuole. Scale bars 50 μm (a, b) and 10 μm (c).

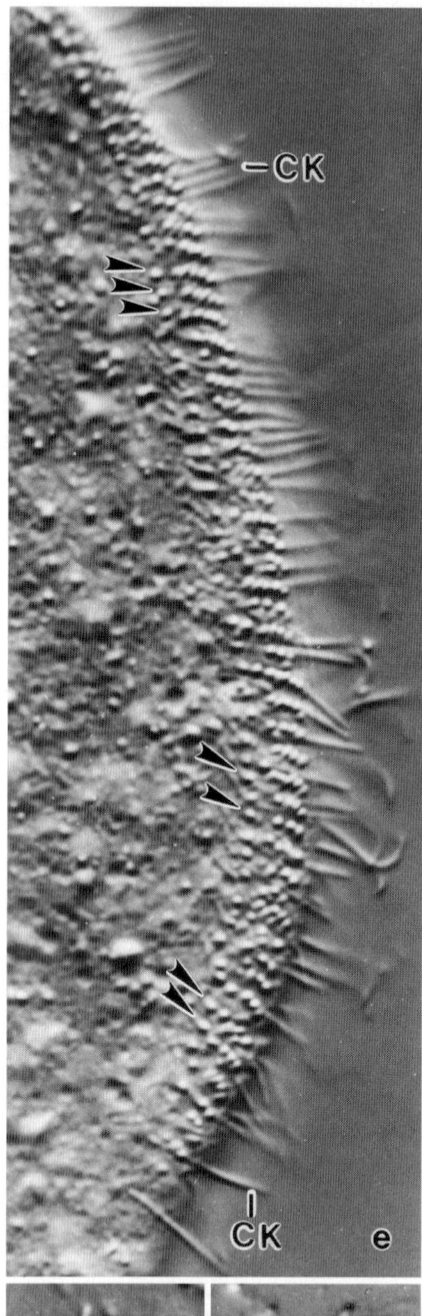

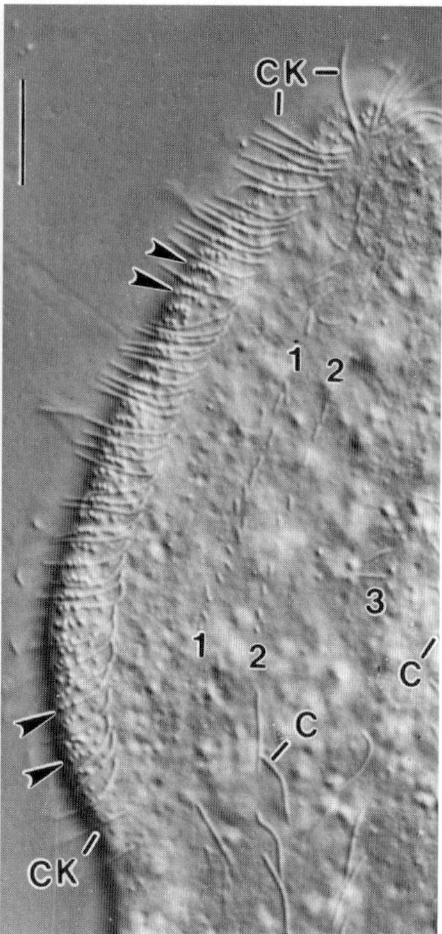

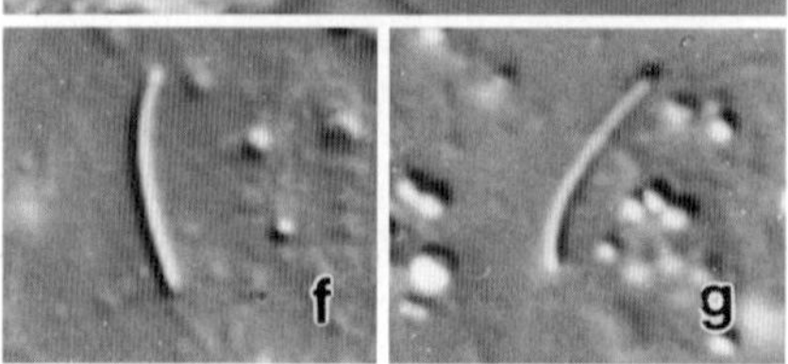

Fig. 137d–g *Cultellothrix velhoi* from life (interference contrast). From FOISSNER (2003c). **d)** A heavily flattened (by coverslip pressure) specimen showing the up to 3 µm long dorsal bristles forming three rows (numbers 1–3) in the anterior portion of the cell's left side; brush row 3 mainly consists of 3 µm long, monokinetidal bristles. Arrowheads mark oblique extrusome rows in oral bulge, shown at higher magnification in figure (e). The cilia of the circumoral kinety (CK) are slightly longer (10 µm) and stiffer than the ordinary somatic cilia (7 µm; C). **e)** Same specimen as depicted in figure (d), showing the oblique oral bulge extrusome rows, some marked by arrowheads, at higher magnification. **f, g)** Oral bulge extrusomes, length 6–7 µm. C – ordinary somatic cilia, CK – circumoral kinety, 1, 2, 3 – dorsal brush rows. Scale bar 10 µm.

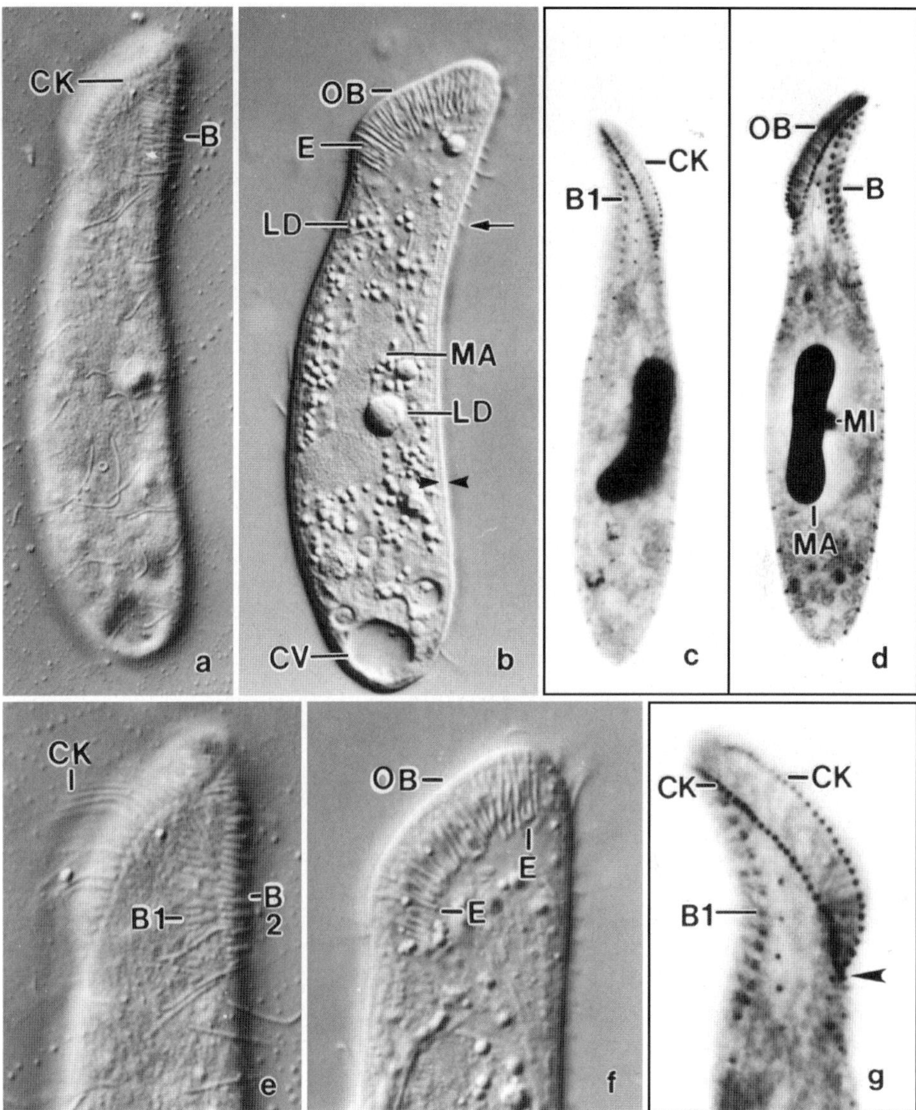

Fig. 138a–g *Cultellothrix coemeterii,* neotype population from life (a, b, e, f) and after protargol impregnation (c, d, g). From Foissner et al. (2005a). **a, b, e)** Left side views of same specimen in several focal planes to show, inter alia, body shape (b), loose ciliation (a, e), and the comparatively widely spaced cilia of the circumoral kinety (e). Arrow marks bristles of tail of brush row 3. Opposed arrowheads denote the cortex which is comparatively thick due to the narrowly spaced cortical granules shown at high magnification in figure 138i. **f)** The oral bulge is studded with rod-shaped, about 6 µm long extrusomes. **c, d, g)** Left side views (c, g, focused through and thus side-inverted) of ciliary pattern and nuclear apparatus, which consists, typically, of a short, reniform macronucleus and a single micronucleus attached to the centre of the macronucleus. Arrowhead marks narrowed and acute ventral end of the circumoral kinety, while the oral bulge is oblong to slightly cuneate (Fig. 138r, w, 139e). The dorsal brush consists of three short rows of paired bristles. B(1, 2) – dorsal brush (rows), CK – circumoral kinety, CV – contractile vacuole, E – extrusomes, LD – lipid droplets, MA – macronucleus, MI – micronucleus, OB – oral bulge.

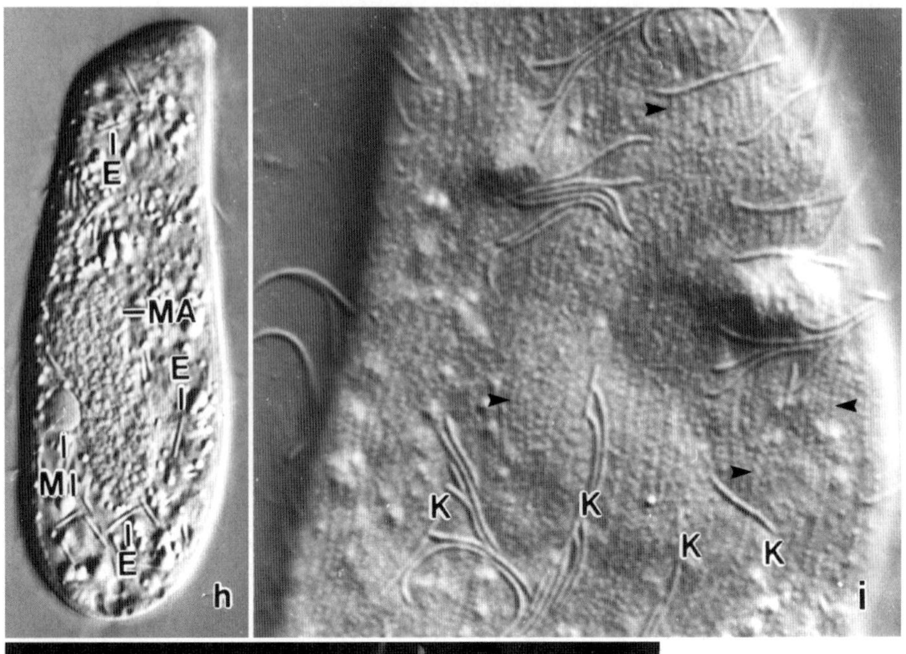

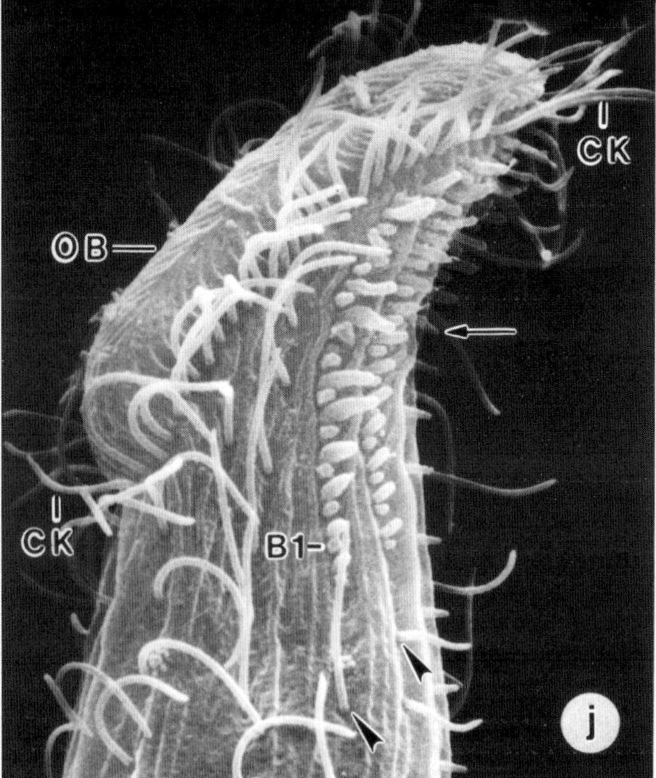

Fig. 138h–j *Cultellothrix coemeterii*, Klausen-Leopoldsdorf specimens from life (a, b) and neotype population in the scanning electron microscope (from FOISSNER et al. 2005a). **h)** Squeezed specimen showing some main organelles. **i)** Surface view showing the narrowly spaced rows, some marked by arrowheads, of minute cortical granules. **j)** Left side view of anterior body portion showing the monokinetidal tail (first bristle marked by arrow) of brush row 3 and the first ordinary cilium (arrowheads) underneath brush rows 1 and 2. B1 – dorsal brush row 1, CK – circumoral kinety, E – extrusomes, K – kineties (somatic ciliary rows), MA – macronucleus, MI – micronucleus, OB – oral bulge.

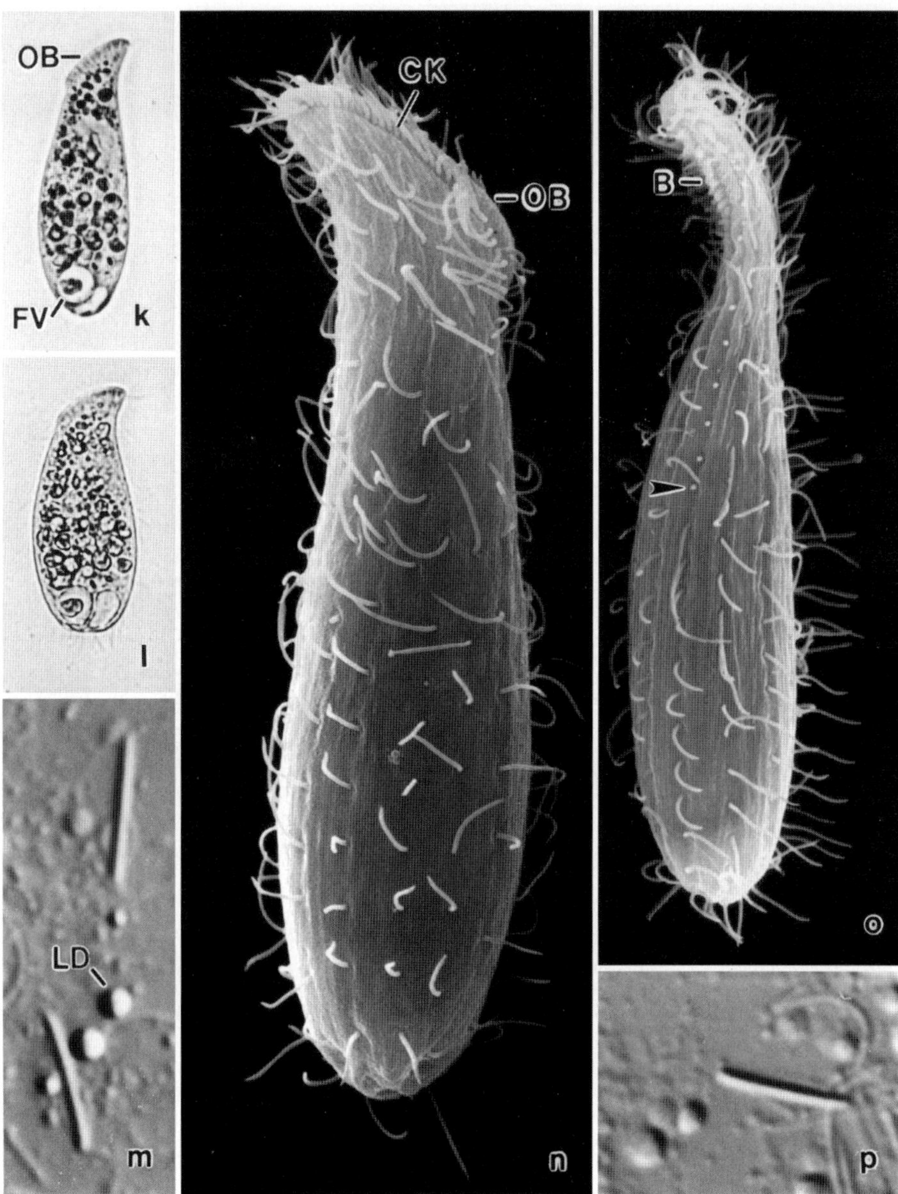

Fig. 138k–p *Cultellothrix coemeterii*, neotype population from life (h–j, m) and in the scanning electron microscope (k, l). From FOISSNER et al. (2005a). **k, l)** Left side overviews of a well-nourished specimen with many lipid (dark) droplets in freely motile (k) and slightly squeezed (l) condition, where shape changes considerably and highly resembles KAHL's figure (Fig. 83d). **m, p)** Extrusomes are slightly curved, about 6 µm long rods. **n, o)** Right side and dorsal view showing body flattening, loose ciliation, slightly convex oral bulge, and last bristle (arrowhead) of monokinetidal tail of brush row 3, which extends to mid-body (see also Fig. 138t, v). B – dorsal brush, CK – circumoral kinety, FV – food vacuole, LD – lipid droplet, OB – oral bulge.

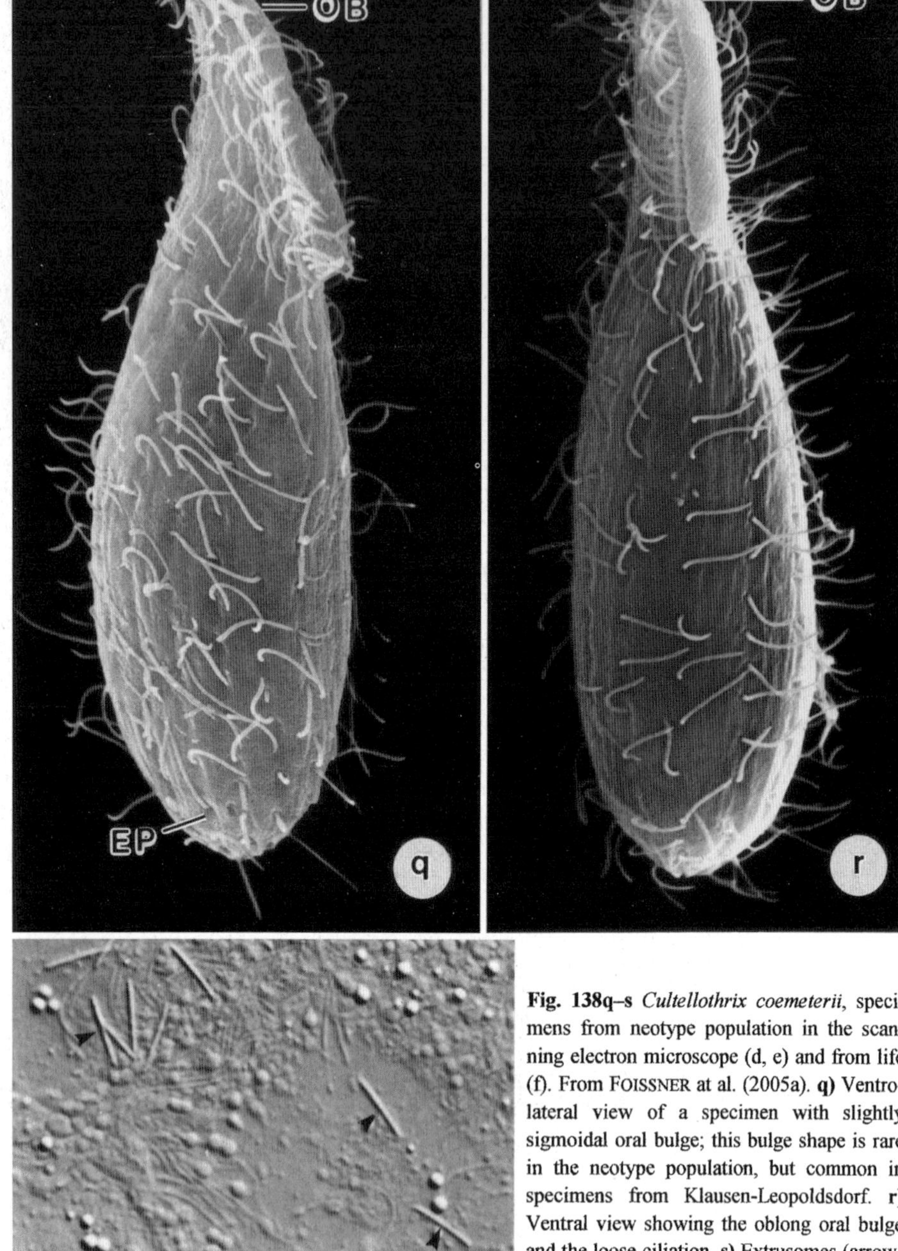

Fig. 138q–s *Cultellothrix coemeterii*, specimens from neotype population in the scanning electron microscope (d, e) and from life (f). From FOISSNER at al. (2005a). **q)** Ventrolateral view of a specimen with slightly sigmoidal oral bulge; this bulge shape is rare in the neotype population, but common in specimens from Klausen-Leopoldsdorf. **r)** Ventral view showing the oblong oral bulge and the loose ciliation. **s)** Extrusomes (arrowheads) are slightly curved, about 6 μm long rods. EP – excretory pores of contractile vacuole, OB – oral bulge.

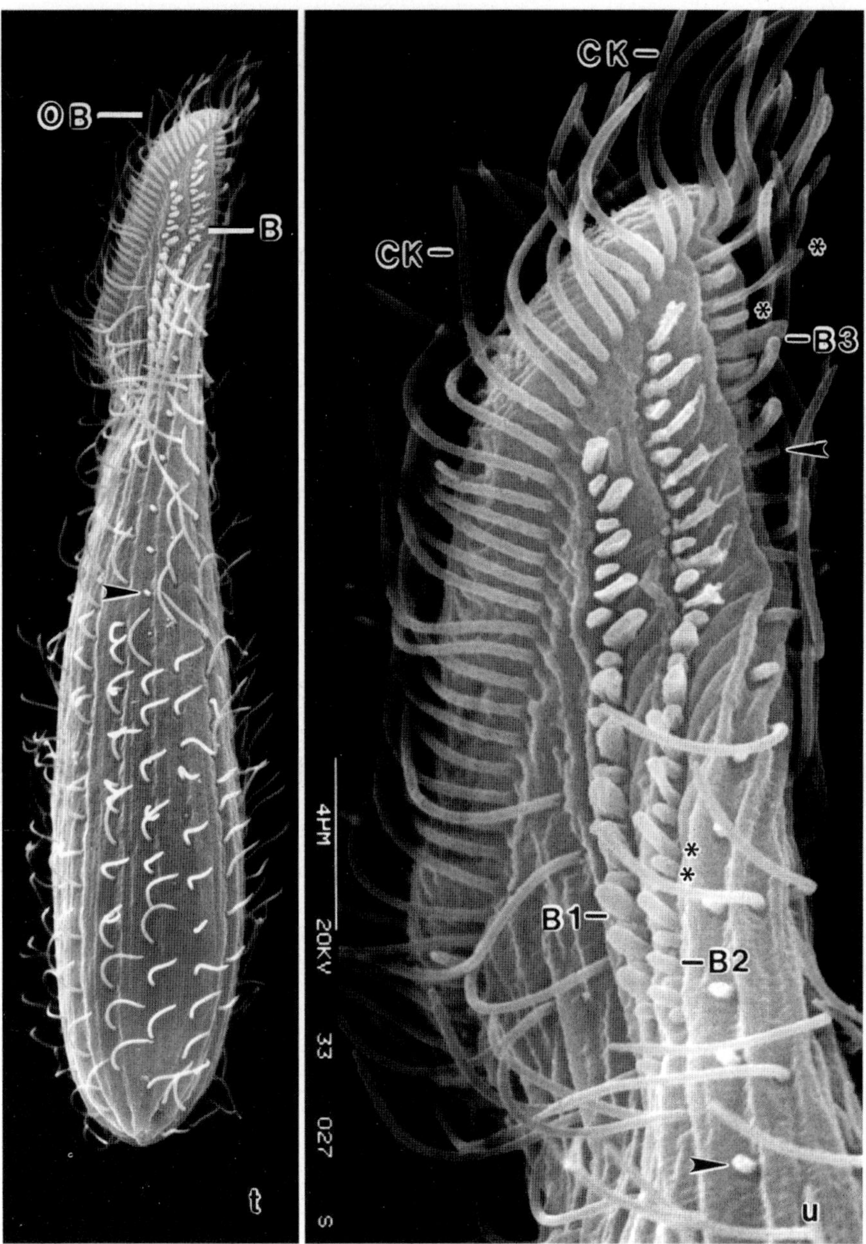

Fig. 138t, u *Cultellothrix coemeterii*, left side views of a specimen from the neotype population in the scanning electron microscope (from FOISSNER et al. 2005a). This well-preserved specimen shows the spatulate body with the steep, slightly sigmoidal oral bulge; the loose ciliature; and the comparatively widely spaced cilia of the circumoral kinety, which is composed of dikinetids each bearing a single cilium. The dorsal brush consists of three rows of paired, distally inflated bristles with the anterior bristle longer than the posterior one (asterisk pairs), especially in anterior portion of row 3, which has a monokinetidal bristle tail (arrowheads). Note that most bristles, which are clavate in vivo, are distinctly flattened (shrunken) due to the preparation procedures. B(1-3) – dorsal brush (rows), CK – circumoral kinety, OB – oral bulge.

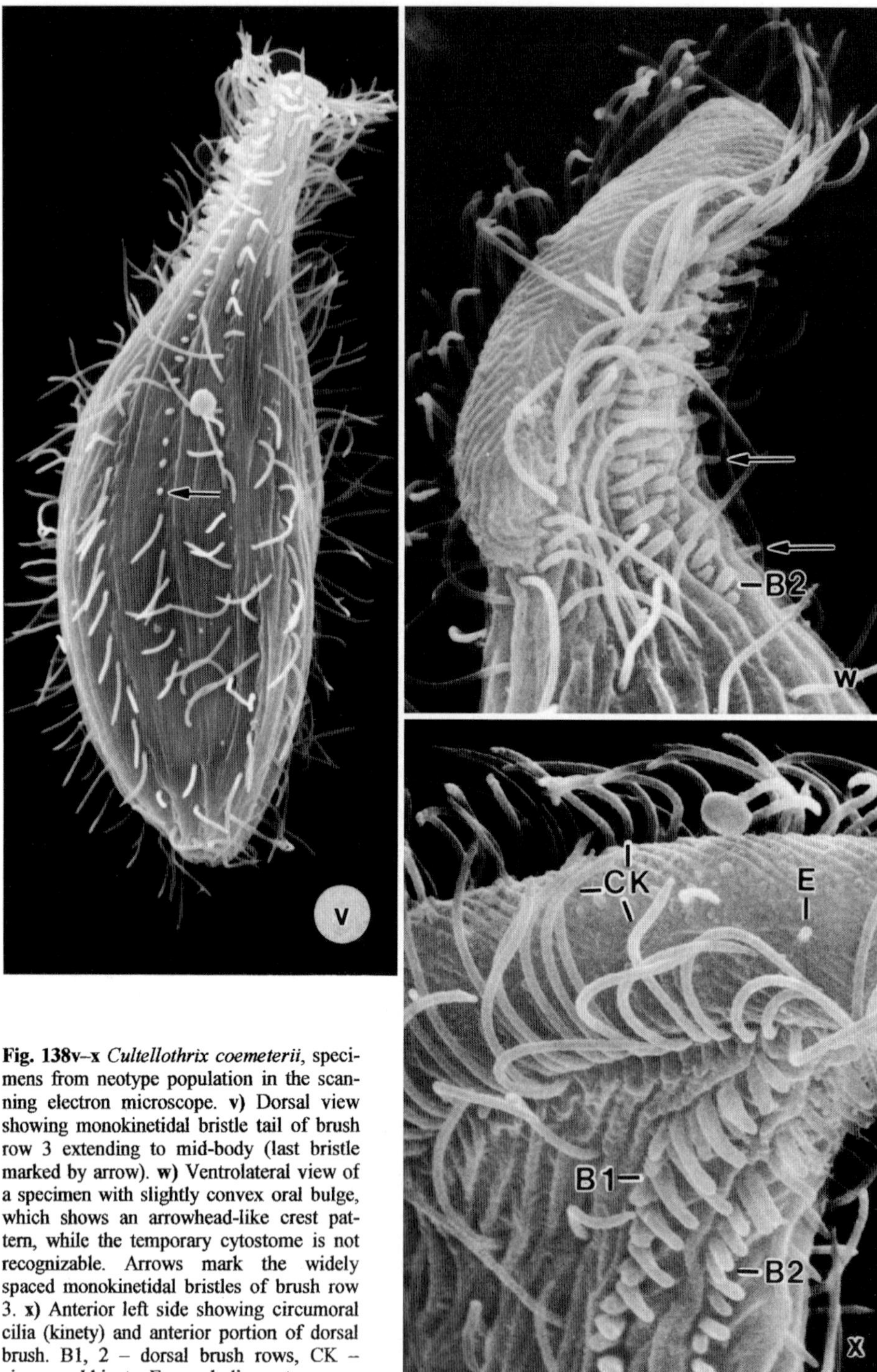

Fig. 138v–x *Cultellothrix coemeterii*, specimens from neotype population in the scanning electron microscope. **v)** Dorsal view showing monokinetidal bristle tail of brush row 3 extending to mid-body (last bristle marked by arrow). **w)** Ventrolateral view of a specimen with slightly convex oral bulge, which shows an arrowhead-like crest pattern, while the temporary cytostome is not recognizable. Arrows mark the widely spaced monokinetidal bristles of brush row 3. **x)** Anterior left side showing circumoral cilia (kinety) and anterior portion of dorsal brush. B1, 2 – dorsal brush rows, CK – circumoral kinety, E – exploding extrusome.

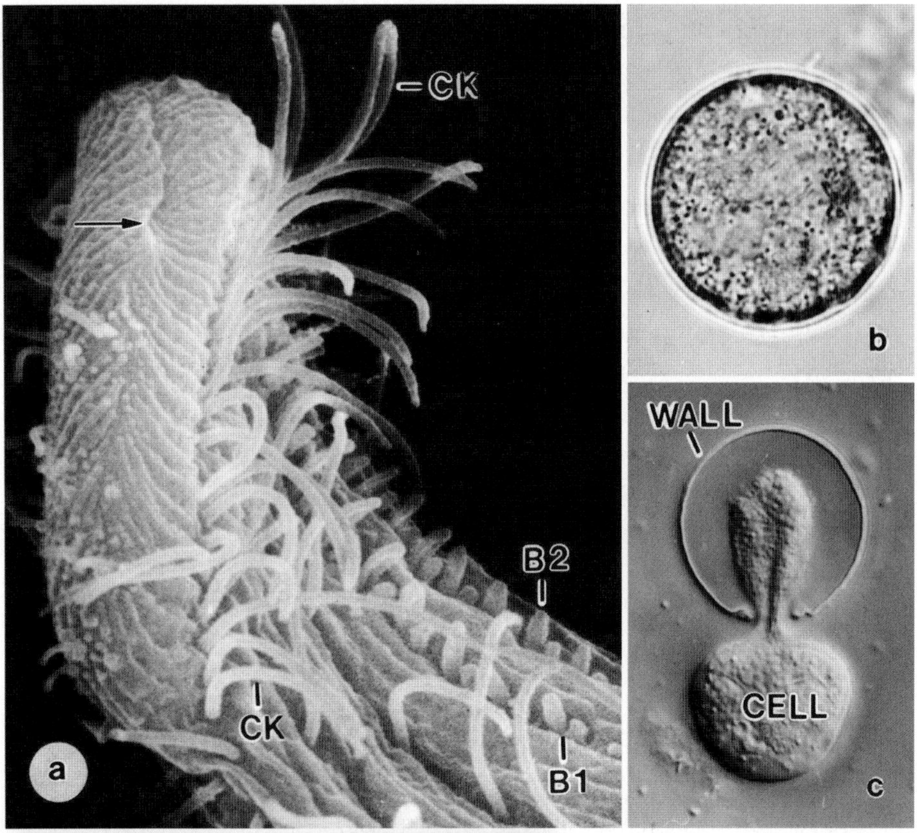

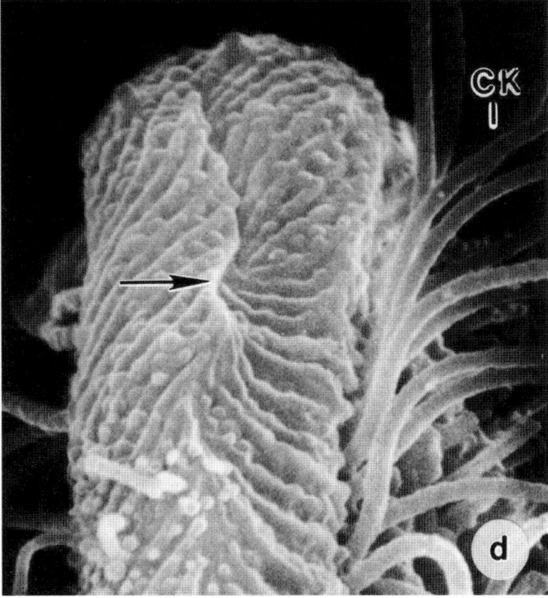

Fig. 139a–d *Cultellothrix coemeterii*, specimens from Austrian neotype population in the scanning electron microscope (a, d) and from life (b, c). **a, d)** Ventrolateral view of a specimen with distinct temporary cytostome (arrows) near the dorsal end of the slightly cuneate oral bulge. The bulge crests, which extend obliquely dorsally and form an arrowhead-like pattern because they abut in bulge midline, are caused by the transverse microtubular ribbons originating from the circumoral dikinetids. The crests form a whirl around the mouth entrance, where they extend a few micrometers into the bulge and support the wall of the cytostome. **b, c)** Resting cyst in bright field and interference contrast showing the thin, smooth cyst wall, when the content has been released due to slight coverslip pressure. B1, 2 – dorsal brush rows, CK – circumoral kinety (cilia).

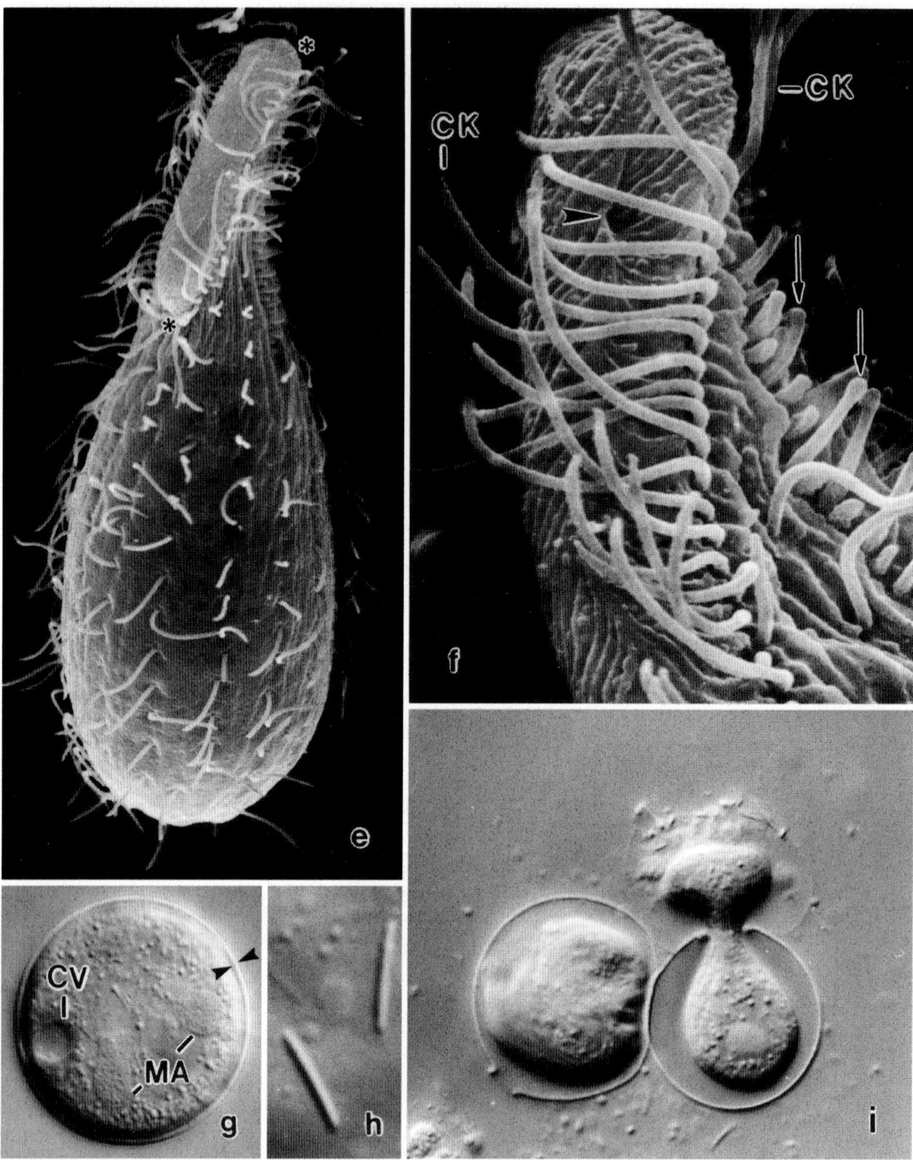

Fig. 139e–i *Cultellothrix coemeterii* in the scanning electron microscope (e, f) and from life (g–i). From FOISSNER at al. (2005a). **e)** Ventral view showing the oblong, flat oral bulge with ends marked by asterisks. **f)** Ventral view of a specimen with distinct temporary cytostome (arrowhead) near the dorsal bulge end. The bulge crests, which extend obliquely dorsally and form an arrowhead-like pattern because they abut in bulge midline, are caused by the transverse microtubular ribbons originating from the circumoral dikinetids. The crests form a whirl around the mouth entrance, where they extend into the bulge and support the wall of the cytostome. Arrows mark distally inflated dorsal brush bristles. **g, i)** Resting cysts are about 30 µm across and have a smooth, thin (≤ 1 µm) wall (opposed arrowheads), which becomes distinct in squashed cysts (t), where the wall breaks at any site. **h)** Resting cyst extrusomes, length 6 µm. CK – cilia of circumoral kinety, CV – contractile vacuole, MA – macronucleus.

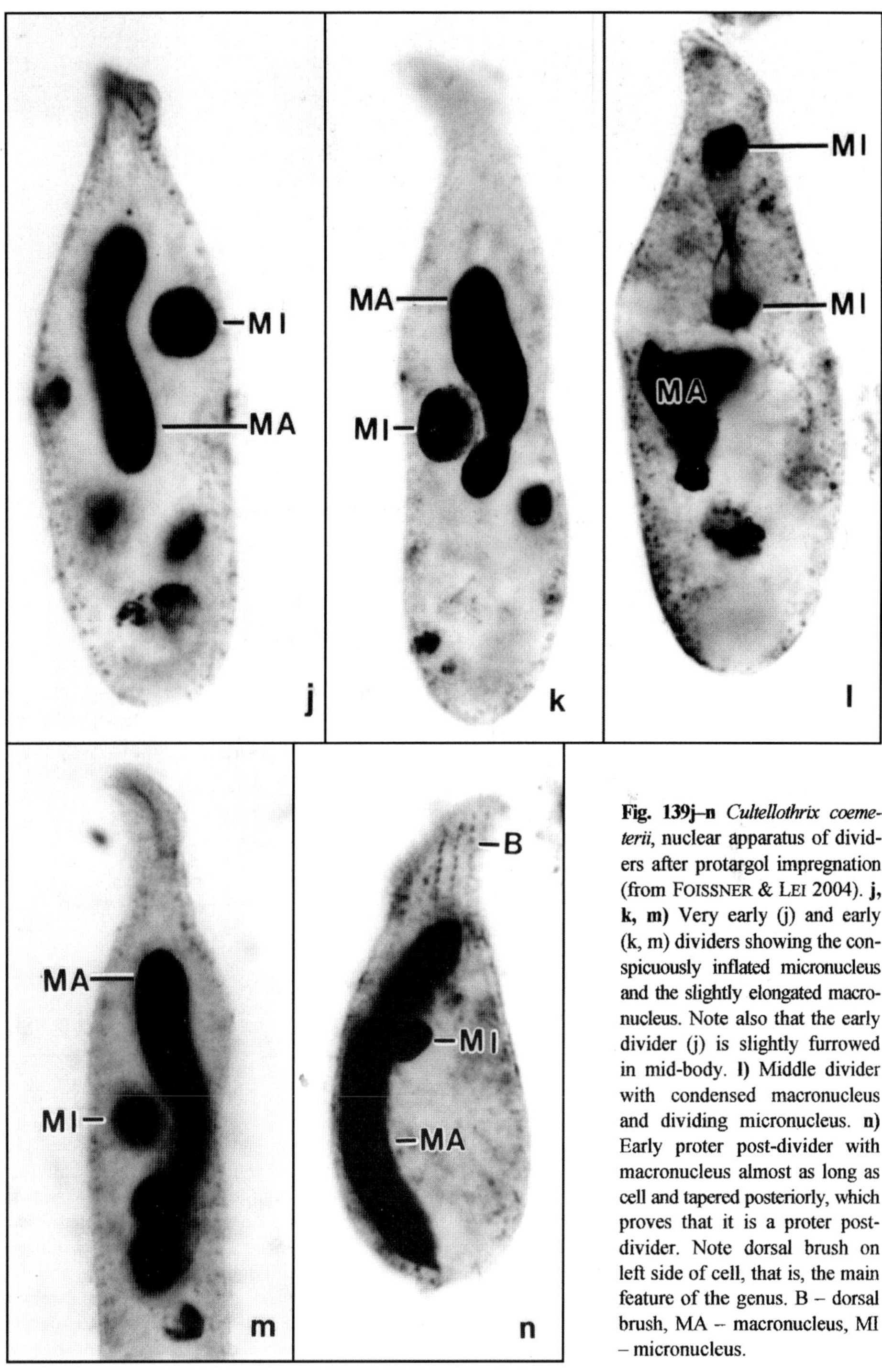

Fig. 139j–n *Cultellothrix coemeterii*, nuclear apparatus of dividers after protargol impregnation (from FOISSNER & LEI 2004). **j, k, m)** Very early (j) and early (k, m) dividers showing the conspicuously inflated micronucleus and the slightly elongated macronucleus. Note also that the early divider (j) is slightly furrowed in mid-body. **l)** Middle divider with condensed macronucleus and dividing micronucleus. **n)** Early proter post-divider with macronucleus almost as long as cell and tapered posteriorly, which proves that it is a proter post-divider. Note dorsal brush on left side of cell, that is, the main feature of the genus. B – dorsal brush, MA – macronucleus, MI – micronucleus.

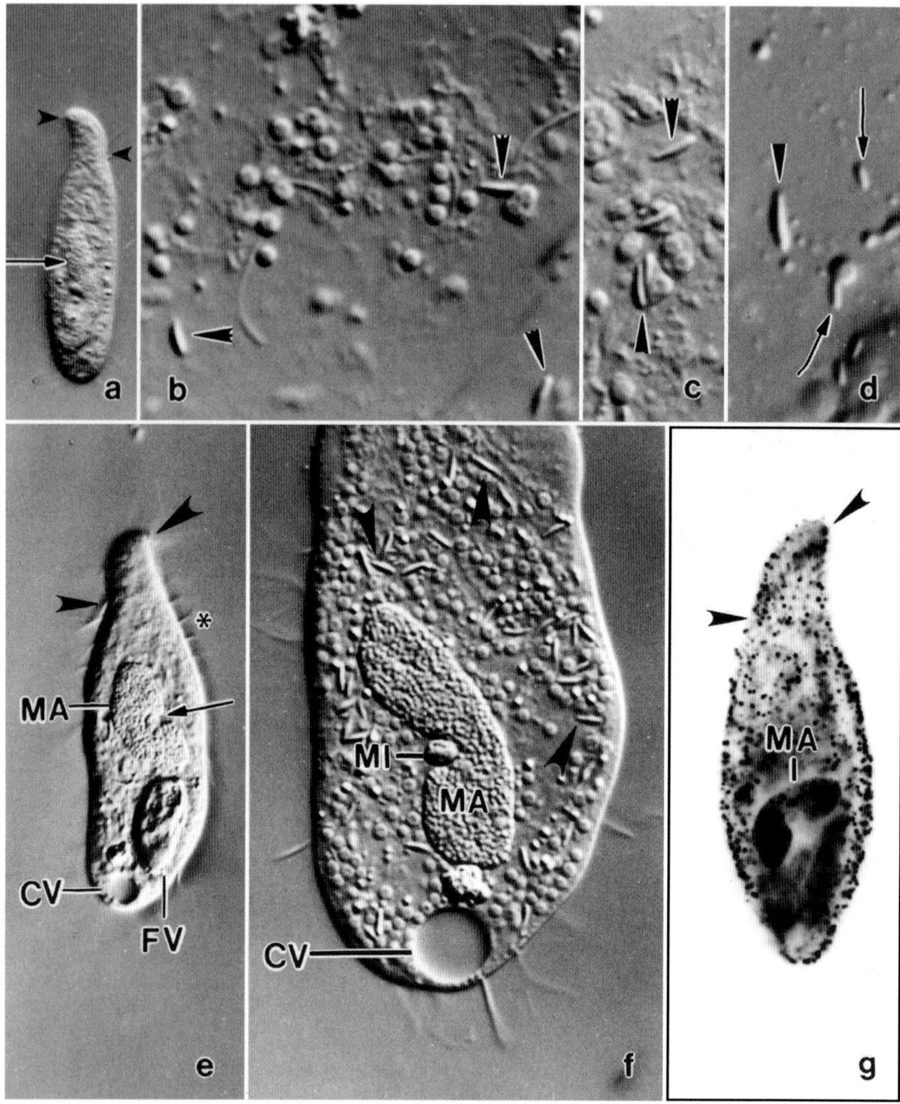

Fig. 140a–g *Cultellothrix paucistriata* nov. spec. from life (a–f) and after protargol impregnation (g). **a)** Right side view of a freely motile specimen showing body shape, oral bulge (ends marked by arrowheads), and macronucleus (arrow). **b–d)** *Cultellothrix paucistriata* has two types of oral bulge extrusomes: type I (arrowheads) is narrowly ovate and slightly curved and has a size of about 3 × 0.7 µm; type II (arrows) is oblong and about 1.5 µm long. **e)** A slightly squeezed specimen showing the nuclear apparatus (arrow marks micronucleus), the oral bulge (ends marked by arrowheads), the monokinetidal bristle tail of brush row 3 (asterisk), and a comparatively large food vacuole containing a small ciliate, viz., *Protocyclidium muscicola*. **f)** A heavily squeezed specimen showing the nuclear apparatus and many cytoplasmic extrusomes (arrowheads). **g)** Usually, mainly the cortical granules impregnate with protargol and hid the ciliary pattern. Arrowheads mark ends of oral bulge. CV – contractile vacuole, FV – food vacuole, MA – macronucleus, MI – micronucleus.

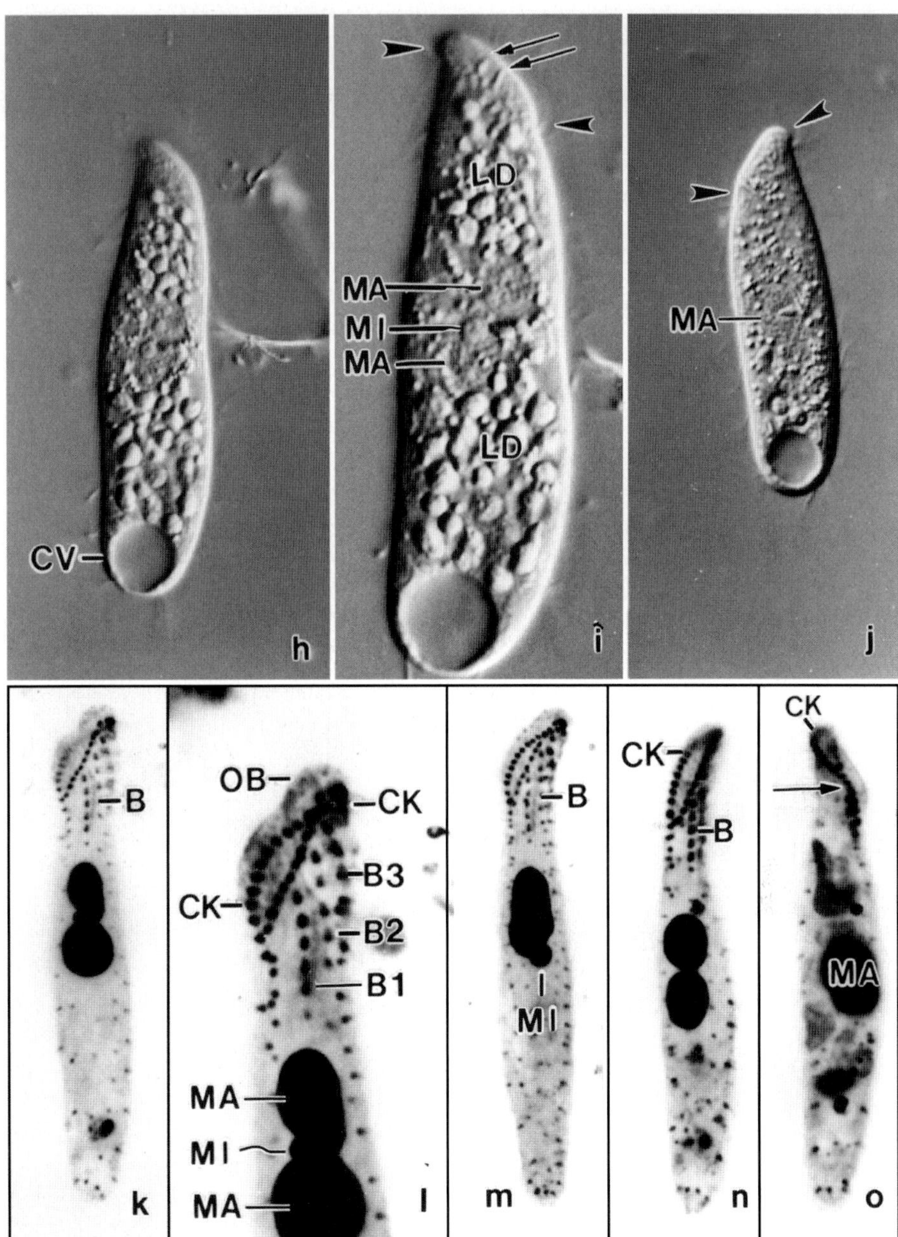

Fig. 140h–o *Cultellothrix tortisticha* nov. spec. from life (h–j) and after protargol impregnation (k–o). **h–j)** Slightly squeezed specimens showing oral bulge extrusomes (arrows), the nuclear apparatus and the oral bulge (ends marked by arrowheads). The cytoplasm of the left specimen is packed with lipid droplets. **k–n)** Left side views showing the typical (k, l, n), but highly variable (m, o) nuclear apparatus; the dorsal brush on the left side of the cell; and the propeller-like screwed circumoral kinety. **o)** Right side view. Arrow marks area where the circumoral kinety optically overlaps. B(1-3) – dorsal brush (rows), CK – circumoral kinety, CV – contractile vacuole, MA – macronucleus nodules, MI – micronucleus, OB – oral bulge.

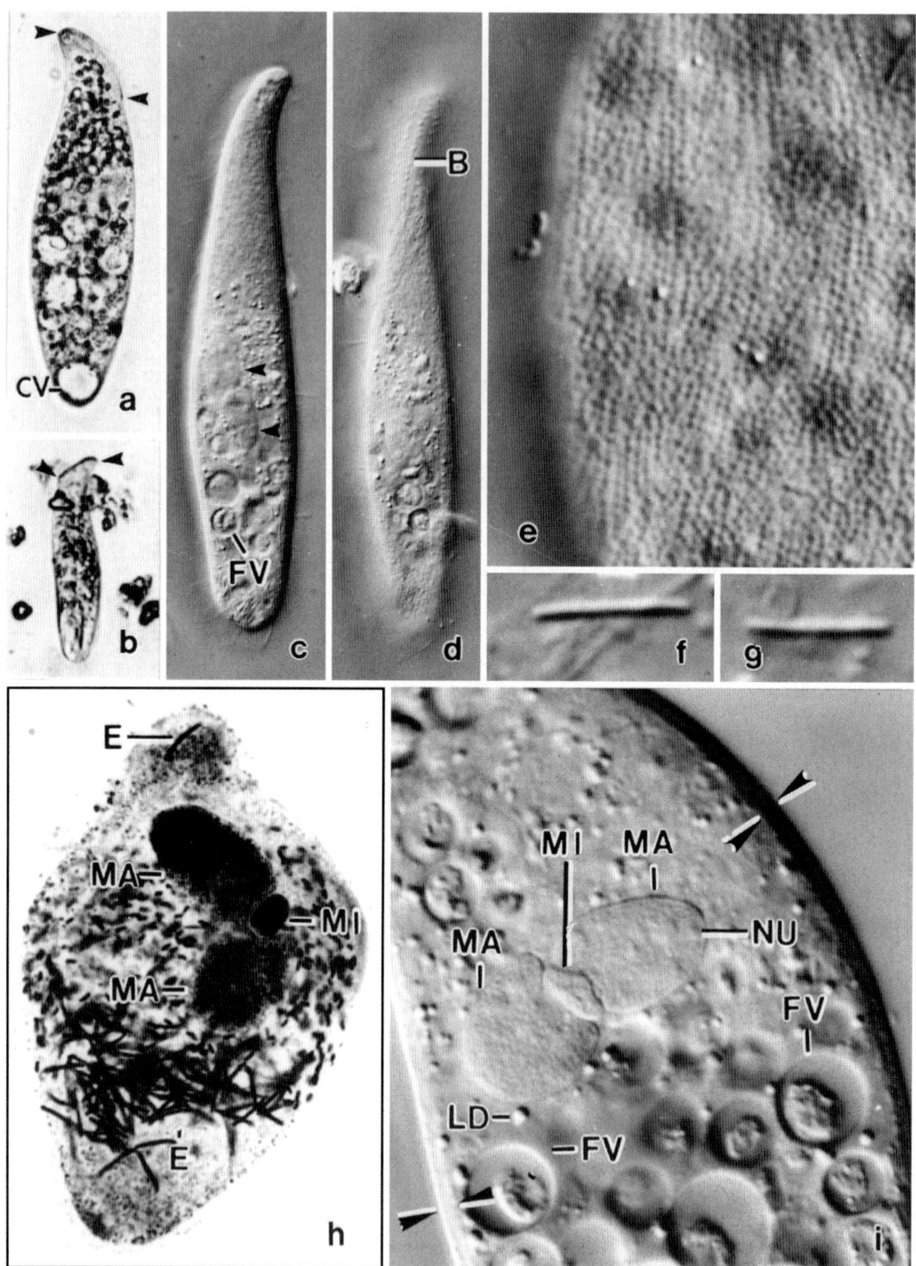

Fig. 141a–i *Cultellothrix atypica*, Austrian specimens from life (a–g, i) and after silver carbonate impregnation (h). **a, b)** Slightly squeezed (a) and freely motile (b) specimen showing body shape and oral bulge (ends marked by arrowheads). **c, d)** Dorsolateral and dorsal view of a freely motile specimen. Arrowheads mark macronucleus nodules. Note pleurostomatid body shape. **e)** Cortical alveolar pattern. **f, g)** Oral bulge extrusomes, length 5 μm. **h)** Nuclear apparatus. **i)** Cytoplasmic details. Opposed arrowheads mark the 1.5 μm thick cortex. B – dorsal brush, CV – contractile vacuole, E – extrusomes, FV – food vacuoles, LD – lipid droplet, MA – macronucleus nodules, MI – micronucleus, NU – nucleoli.

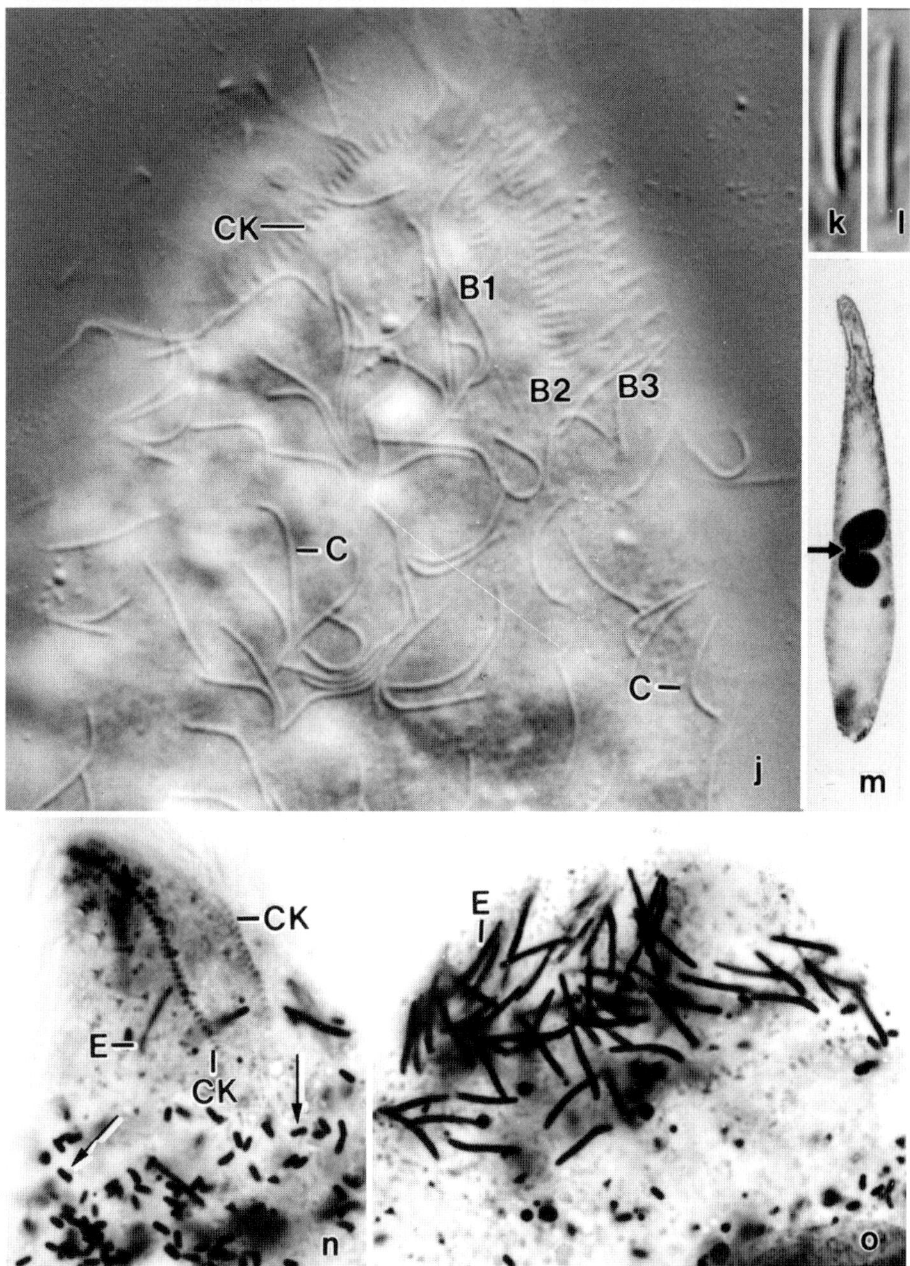

Fig. 141j–o *Cultellothrix atypica* (j–l, n, o) and *C. japonica* (m; from FOISSNER 1988) from life (j–l) and after protargol (m) and silver carbonate (n, o) impregnation. **j)** Anterior region of left side, showing the dorsal brush, longest bristles 4 µm. **k, l, o)** Oral bulge extrusomes of Austrian specimens, length 6 µm. **m)** Body shape and nuclear pattern. Arrow marks micronucleus. **n)** Ventrolateral view showing the cuneate circumoral kinety (CK) and many minute rods (arrows), possibly cortical mucocysts. B1-3 – dorsal brush rows, C – ordinary somatic cilia, CK – circumoral kinety, E – oral bulge extrusomes.

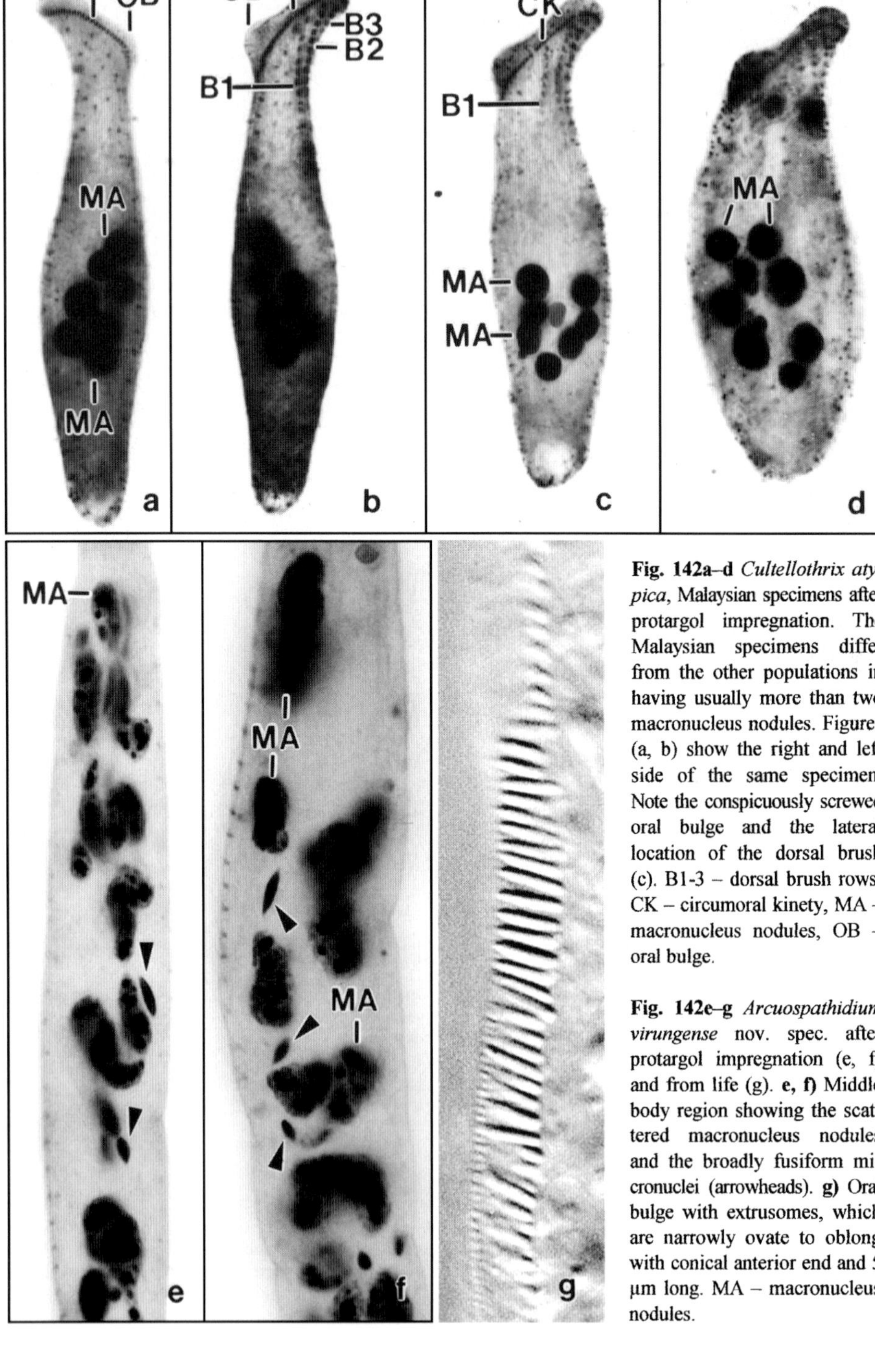

Fig. 142a–d *Cultellothrix atypica*, Malaysian specimens after protargol impregnation. The Malaysian specimens differ from the other populations in having usually more than two macronucleus nodules. Figures (a, b) show the right and left side of the same specimen. Note the conspicuously screwed oral bulge and the lateral location of the dorsal brush (c). B1-3 – dorsal brush rows, CK – circumoral kinety, MA – macronucleus nodules, OB – oral bulge.

Fig. 142e–g *Arcuospathidium virungense* nov. spec. after protargol impregnation (e, f) and from life (g). **e, f)** Middle body region showing the scattered macronucleus nodules and the broadly fusiform micronuclei (arrowheads). **g)** Oral bulge with extrusomes, which are narrowly ovate to oblong with conical anterior end and 5 μm long. MA – macronucleus nodules.

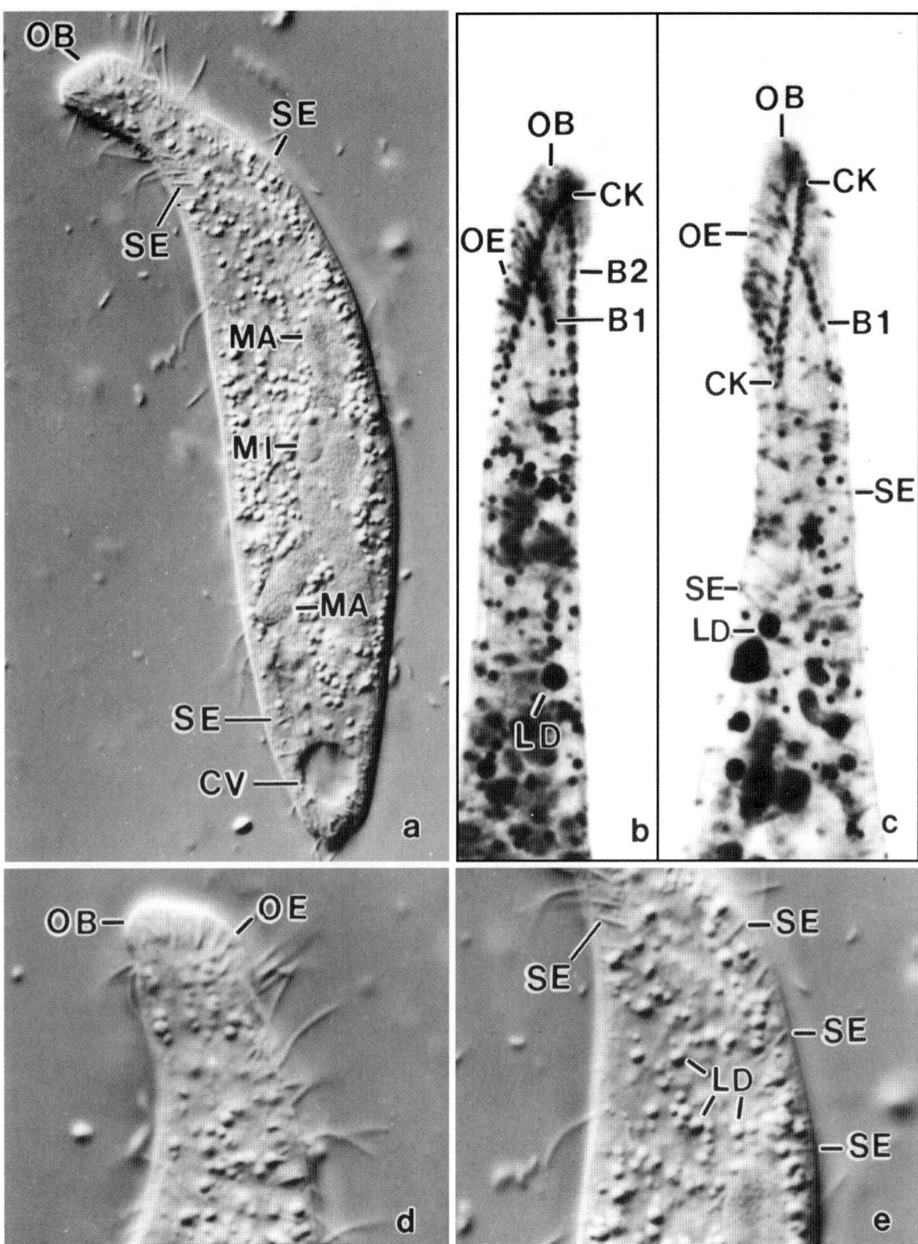

Fig. 143a–e *Armatospathula periarmata* nov. spec. from life (a, d, e) and after protargol impregnation (b, c). **a, d, e)** Overview and details of a slightly squashed and thus inflated specimen showing the main features of the species, especially the somatic extrusome rows (SE). **b, c)** Left side and ventrolateral view of anterior body region showing details of the ciliary pattern. B1, 2 – dorsal brush rows, CV – contractile vacuole, CK – circumoral kinety, LD – lipid droplets, MA – macronucleus, MI – micronucleus, OB – oral bulge, OE, SE – oral and somatic extrusomes.

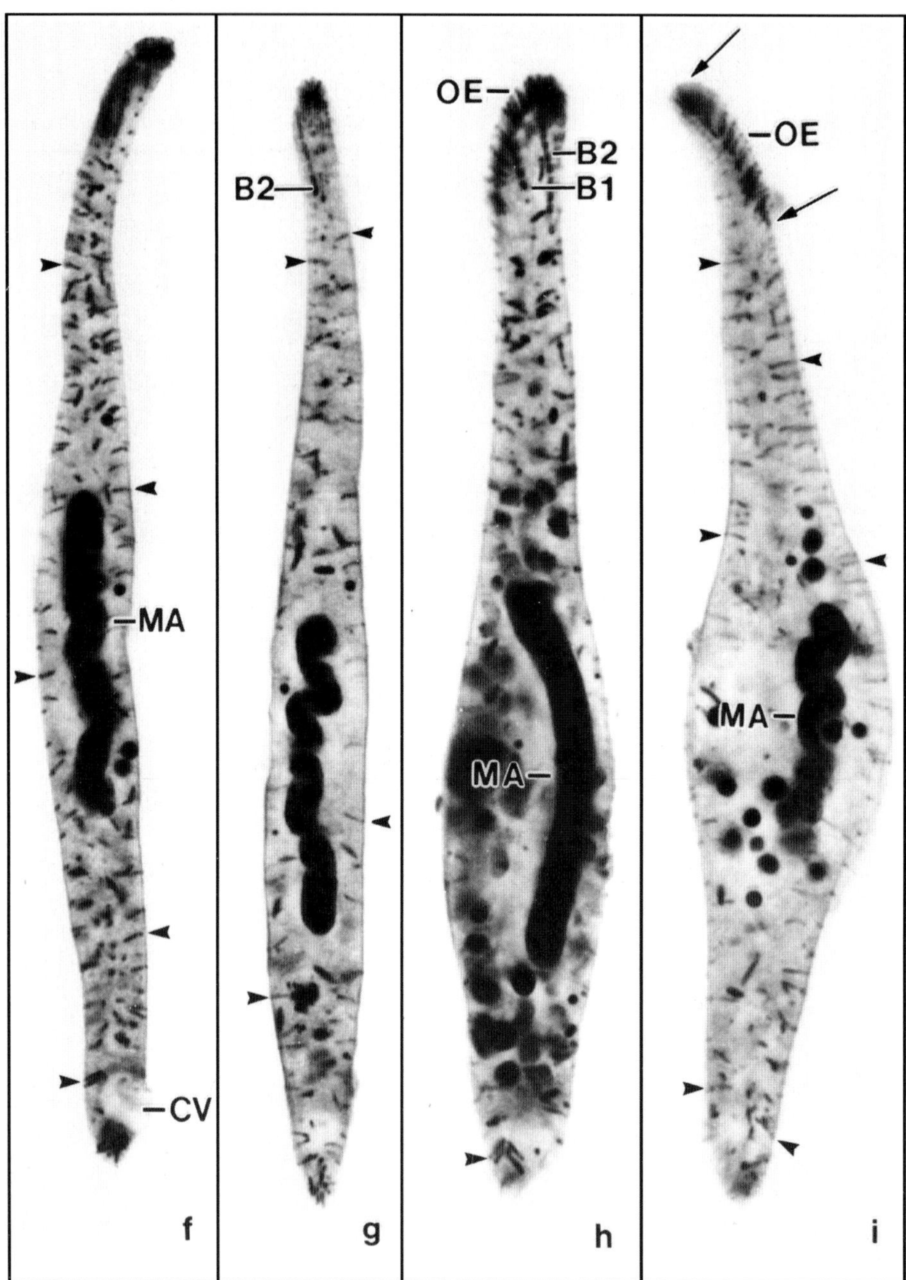

Fig. 143f–i *Armatospathula periarmata* nov. spec. after protargol impregnation. These overviews (for details, see figures 143b, c) show the main features of the species, viz., the slender shape; the somatic extrusome rows (arrowheads), and the rod-shaped macronucleus. Arrows mark oral bulge packed with extrusomes. B1, 2 – dorsal brush rows, CV – contractile vacuole, MA – macronucleus, OE – oral bulge extrusomes.

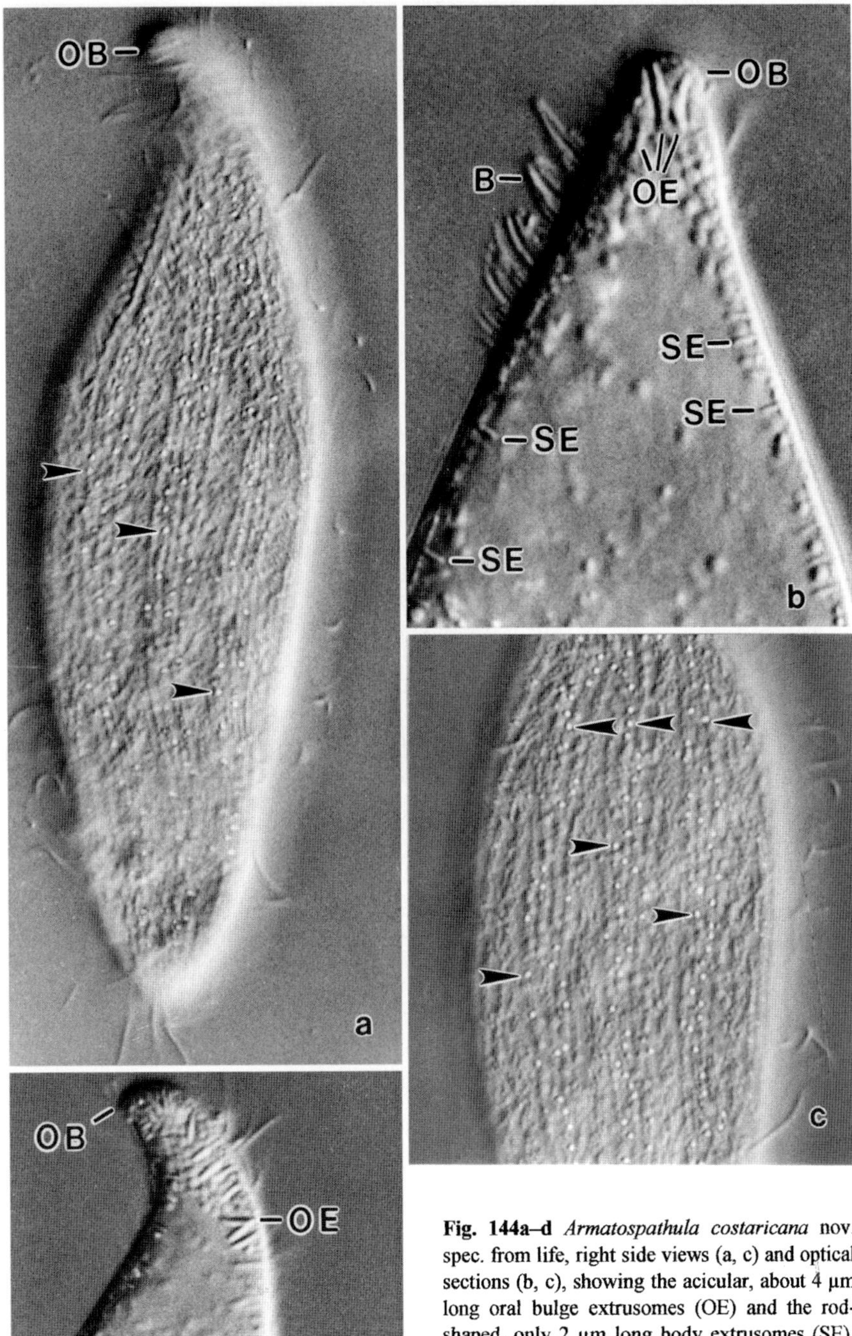

Fig. 144a–d *Armatospathula costaricana* nov. spec. from life, right side views (a, c) and optical sections (b, c), showing the acicular, about 4 µm long oral bulge extrusomes (OE) and the rod-shaped, only 2 µm long body extrusomes (SE), which are highly refractive and thus appear as conspicuous dot-stripes (arrowheads) in surface view (a, c). B – dorsal brush, OB – oral bulge, OE – oral bulge extrusomes, SE – somatic extrusomes.

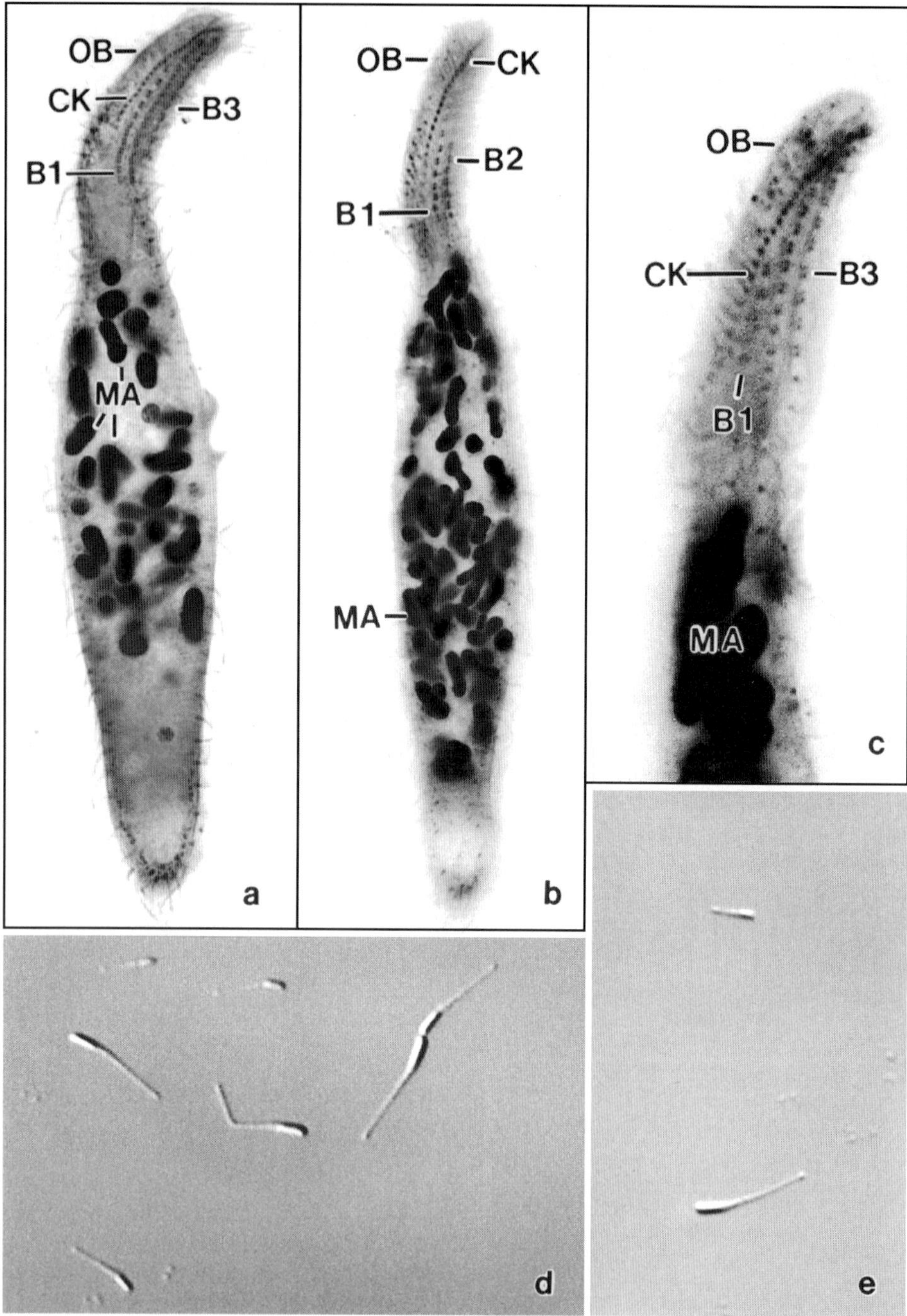

Fig. 145a–e *Armatospathula plurinucleata* nov. spec. after protargol impregnation (a–c) and from life (d, e). **a–c)** Left side (a) and ventrolateral (b, c) views showing the brush occupying the entire left side. **d, e)** Exploded toxicysts, 6 µm. B1-3 – dorsal brush rows, CK – circumoral kinety, MA – macronucleus nodules, OB – oral bulge.

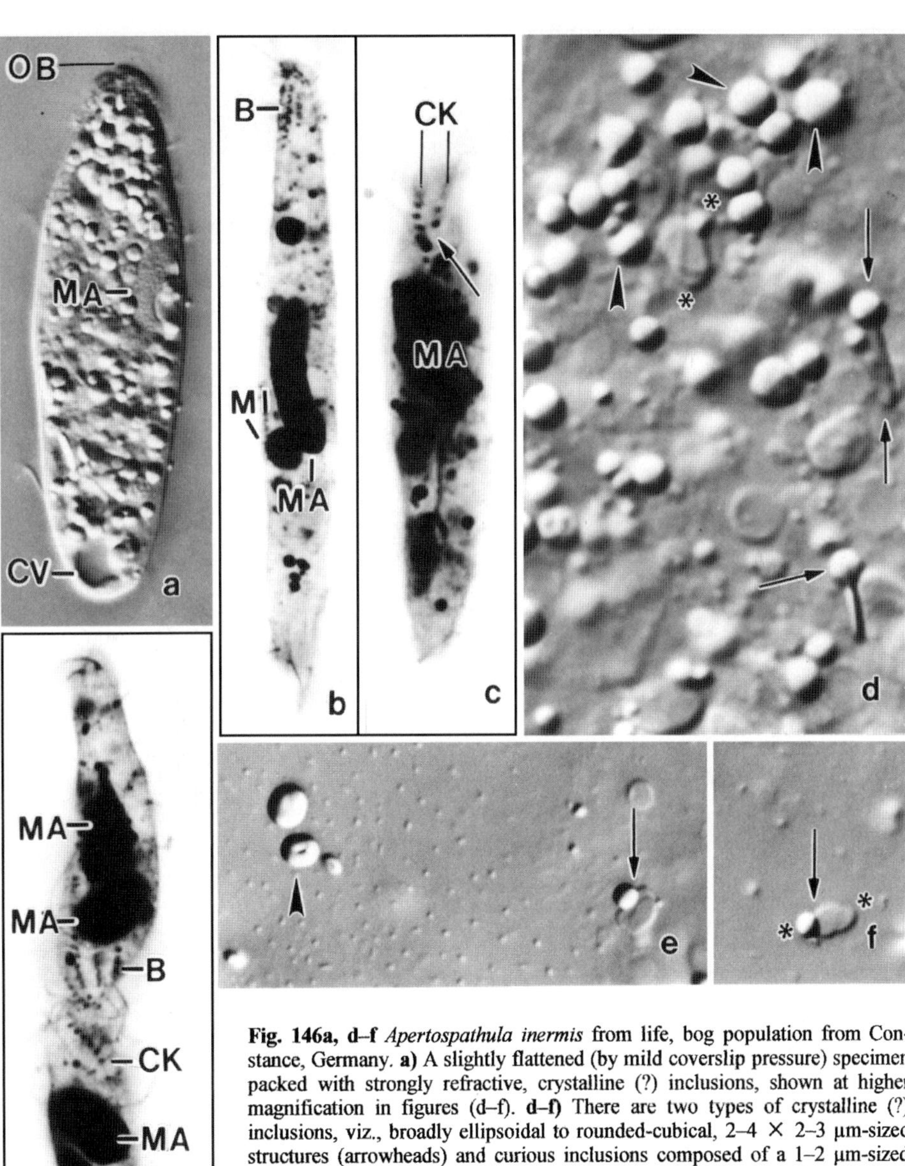

Fig. 146a, d–f *Apertospathula inermis* from life, bog population from Constance, Germany. **a)** A slightly flattened (by mild coverslip pressure) specimen packed with strongly refractive, crystalline (?) inclusions, shown at higher magnification in figures (d–f). **d–f)** There are two types of crystalline (?) inclusions, viz., broadly ellipsoidal to rounded-cubical, 2–4 × 2–3 µm-sized structures (arrowheads) and curious inclusions composed of a 1–2 µm-sized globule attached to a circular, ellipsoidal, or irregular thin plate of up to 10 µm in size (arrows); the second type appears drumstick-shaped when seen from the narrow side (d, arrows) and inflated when observed from the broad side (d–f, asterisks).

Fig. 146b, c, g *Apertospathula armata*, Greece specimens after protargol impregnation. **b, c)** Dorsolateral and ventral view of ciliary pattern and nuclear apparatus. Arrow marks ventrally opened circumoral kinety. **g)** Conjugating specimens attach with the oral bulge and have each two globular to ellipsoidal macronucleus nodules.

B – dorsal brush, CK – circumoral kinety, CV – contractile vacuole, MA – macronucleus, MI – micronucleus, OB – oral bulge.

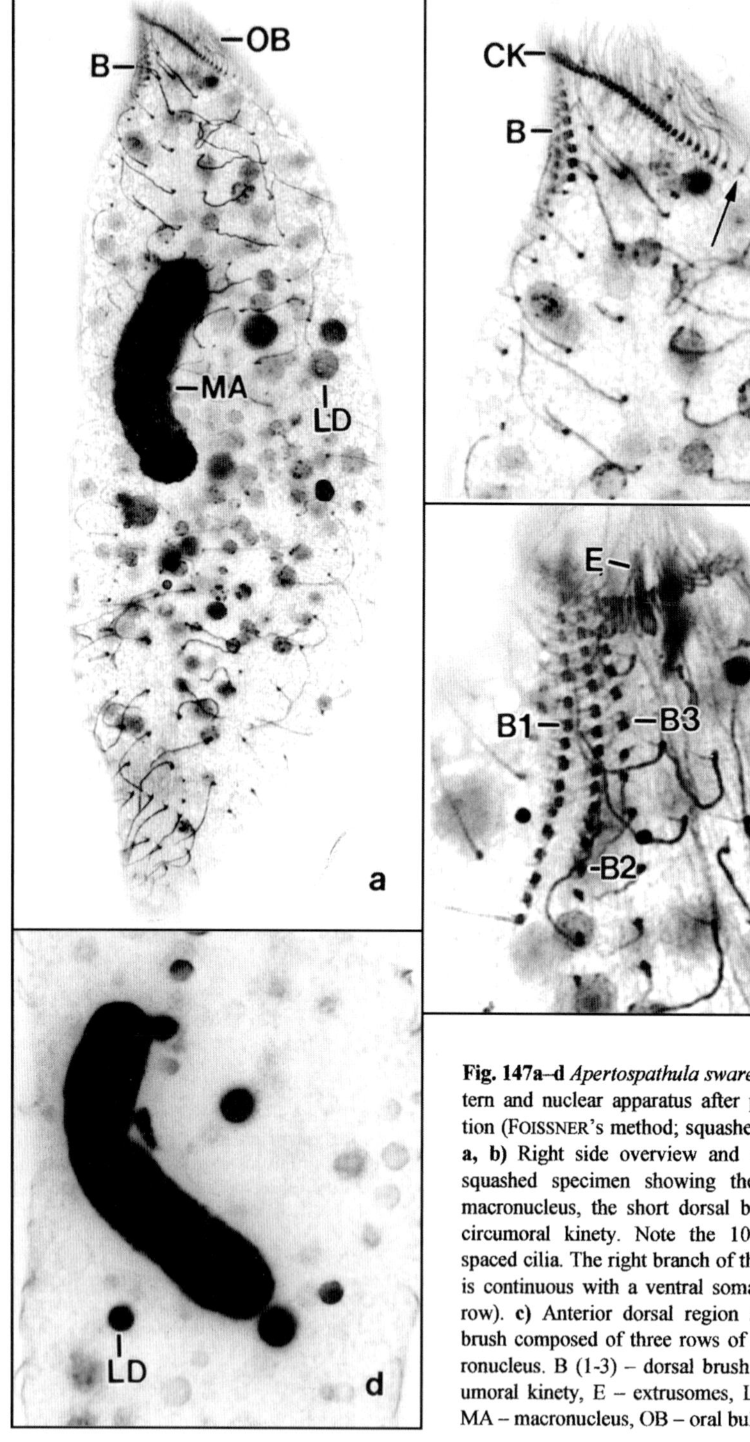

Fig. 147a–d *Apertospathula swarezewskyi*, ciliary pattern and nuclear apparatus after protargol impregnation (FOISSNER's method; squashed in wet condition). **a, b)** Right side overview and detail of a slightly squashed specimen showing the body shape, the macronucleus, the short dorsal brush, and the short circumoral kinety. Note the 10 μm long, widely spaced cilia. The right branch of the circumoral kinety is continuous with a ventral somatic ciliary row (arrow). **c)** Anterior dorsal region showing the dorsal brush composed of three rows of dikinetids. **d)** Macronucleus. B (1-3) – dorsal brush (rows), CK – circumoral kinety, E – extrusomes, LD – lipid droplets, MA – macronucleus, OB – oral bulge.

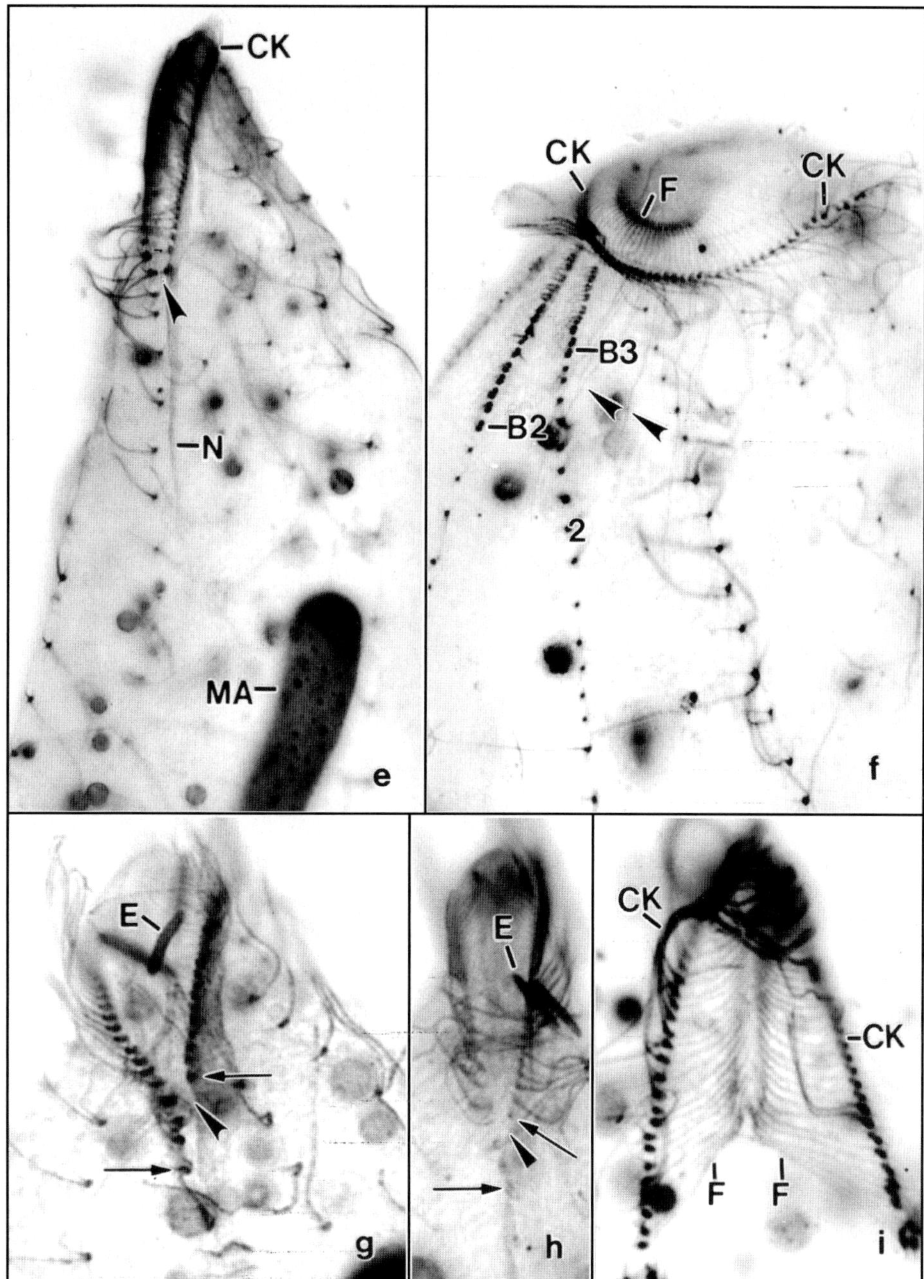

Fig. 147e–i *Apertospathula swarezewskyi* after protargol impregnation. **e, g, h)** Ventral views showing the open circumoral kinety (arrowheads) with the left branch longer than the right (arrows). **f, i)** Strongly flattened specimens (by coverslip pressure) showing the fine nematodesmata (arrowheads) and the distinct bulge fibres originating from the circumoral kinety, which consists of dikinetids clearly recognizable in figure (i). B2, 3 – dorsal brush rows, CK – circumoral kinety, E – extrusomes, F – fibres, MA – macronucleus, N – nematodesma bundle.

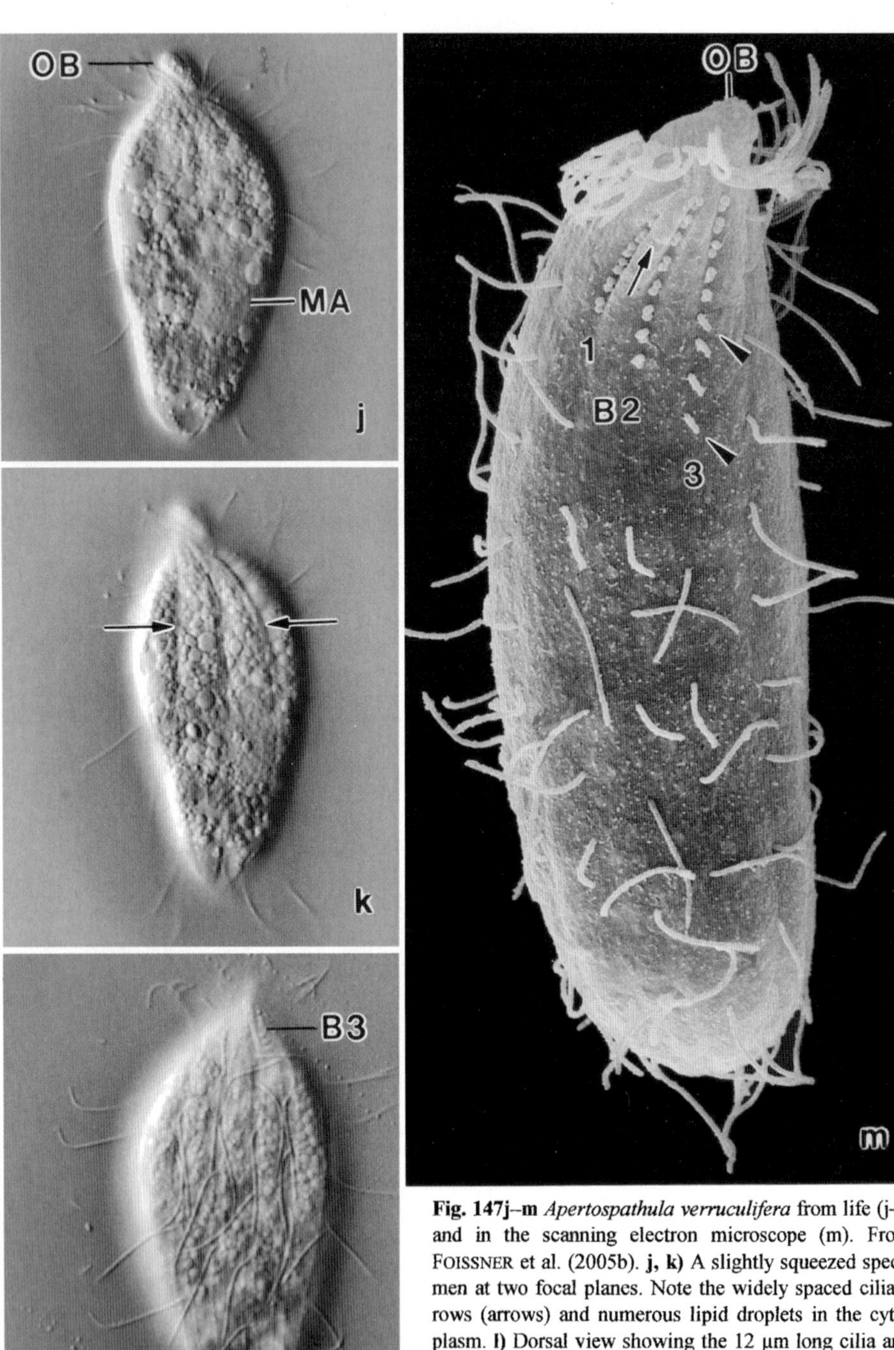

Fig. 147j–m *Apertospathula verruculifera* from life (j–l) and in the scanning electron microscope (m). From FOISSNER et al. (2005b). **j, k)** A slightly squeezed specimen at two focal planes. Note the widely spaced ciliary rows (arrows) and numerous lipid droplets in the cytoplasm. **l)** Dorsal view showing the 12 μm long cilia and the minute brush bristles. **m)** Dorsolateral view showing the dorsal brush and the short tail of brush row 3 (arrowheads). The palpus dorsalis (arrow) is collapsed. B(1-3) – dorsal brush (rows), MA – macronucleus, OB – oral bulge.

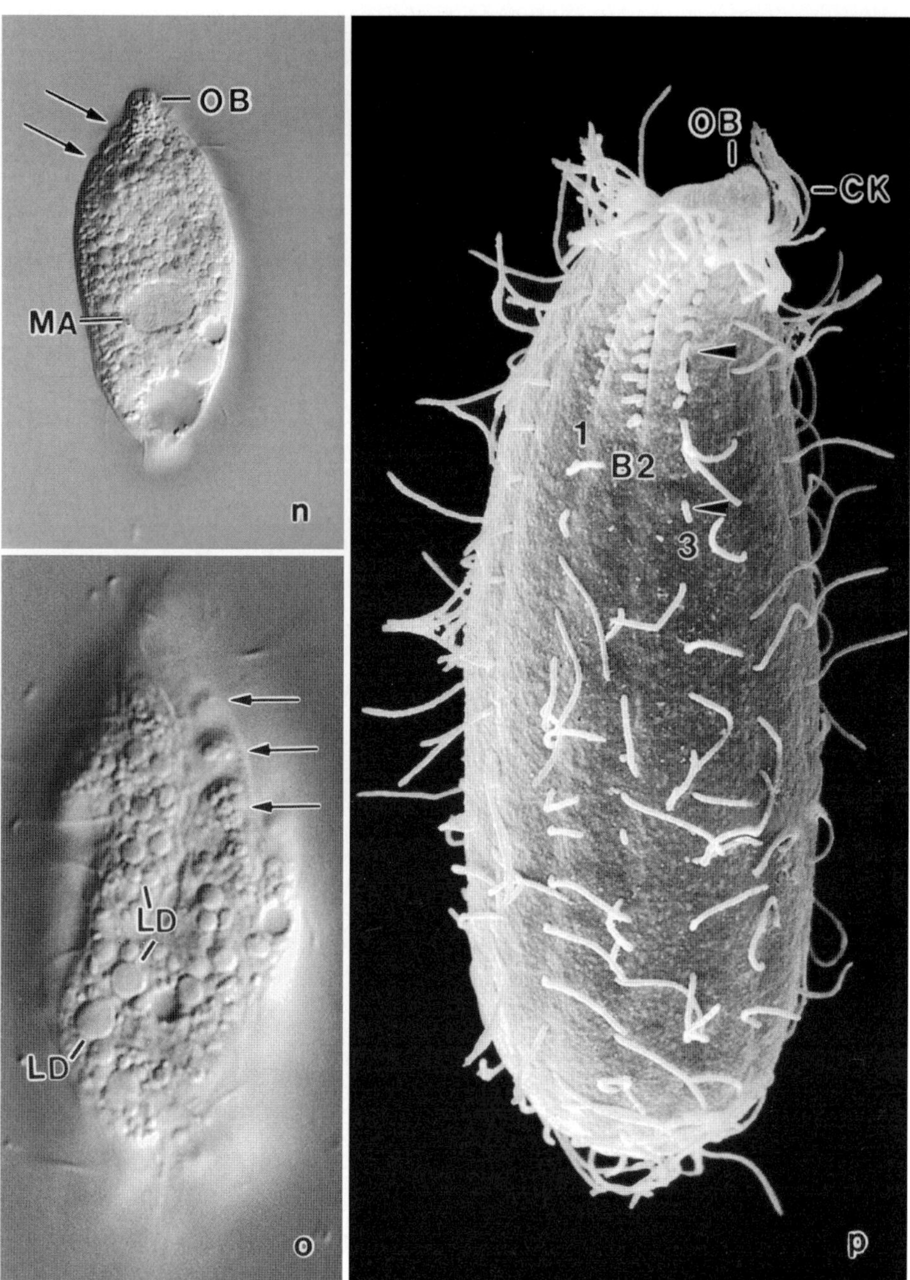

Fig. 147n–p *Apertospathula verruculifera* from life (n, o) and in the SEM (p). From FOISSNER et al. (2005b). The small cortex bolsters (arrows) in the anterior quarter and the palpus dorsalis are not recognizable in the scanning electron micrograph, showing that careful live observation is indispensable. Note that the anterior bristle of the brush dikinetids is longer than the posterior; the tail of row 3 is marked by arrowheads. CK – circumoral kinety, LD – lipid droplets, MA – macronucleus, OB – oral bulge.

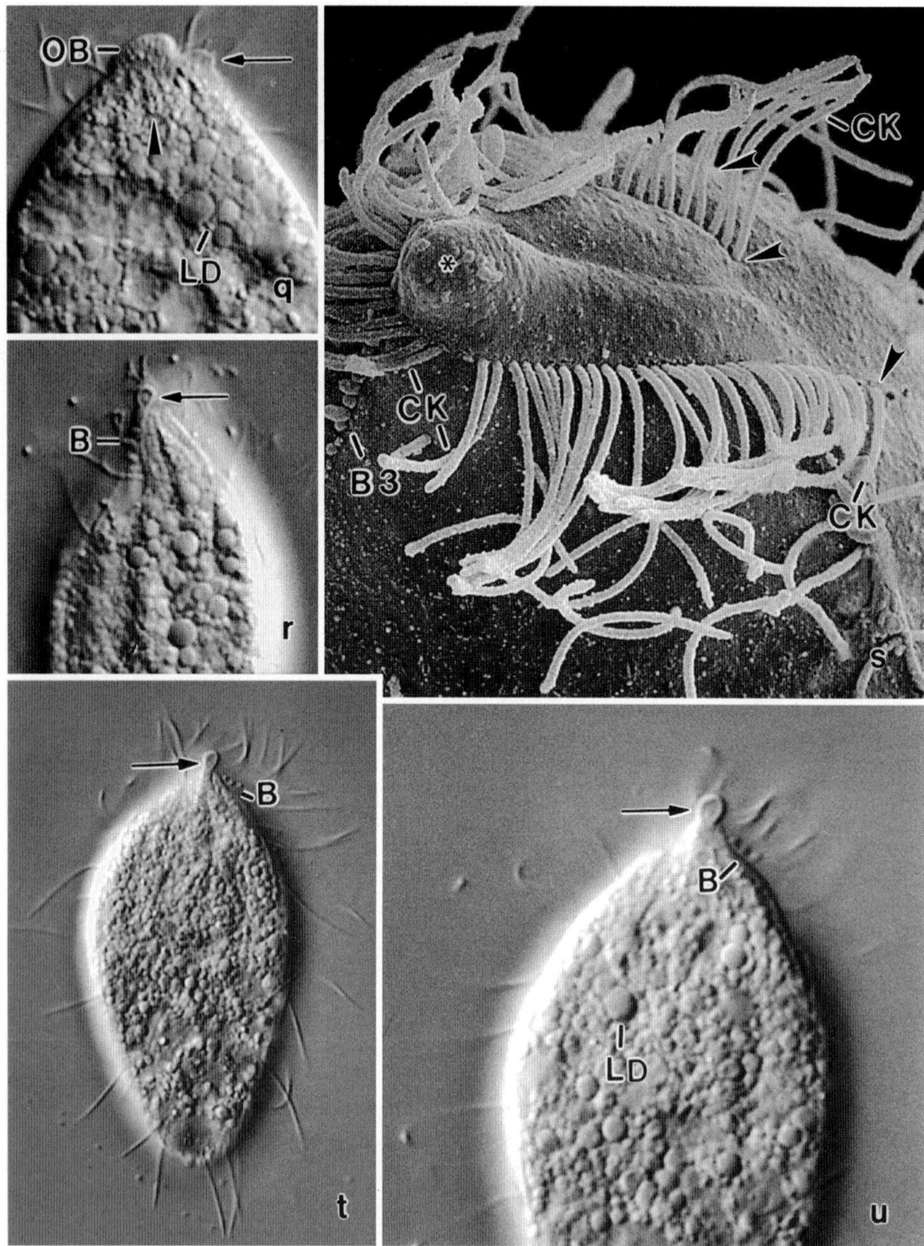

Fig. 147q–u *Apertospathula verruculifera* from life (q, r, t, u) and in the scanning electron microscope (s). From FOISSNER et al. (2005b). **q)** Dorsolateral view showing the palpus dorsalis (arrow) and the refractive granules (arrowhead) under the oral bulge. **r, t, u)** Dorsal views showing the palpus dorsalis (arrows) and the minute brush bristles. **s)** Frontolateral view showing the broad gap between the ends of the cicumoral kinety marked by arrowheads. Note the wart-like dorsal end of the oral bulge (asterisk). Likely, this is another species because it has much more circumoral dikinetids. B(3) – dorsal brush (row), CK – circumoral kinety, LD – lipid droplets, OB – oral bulge.

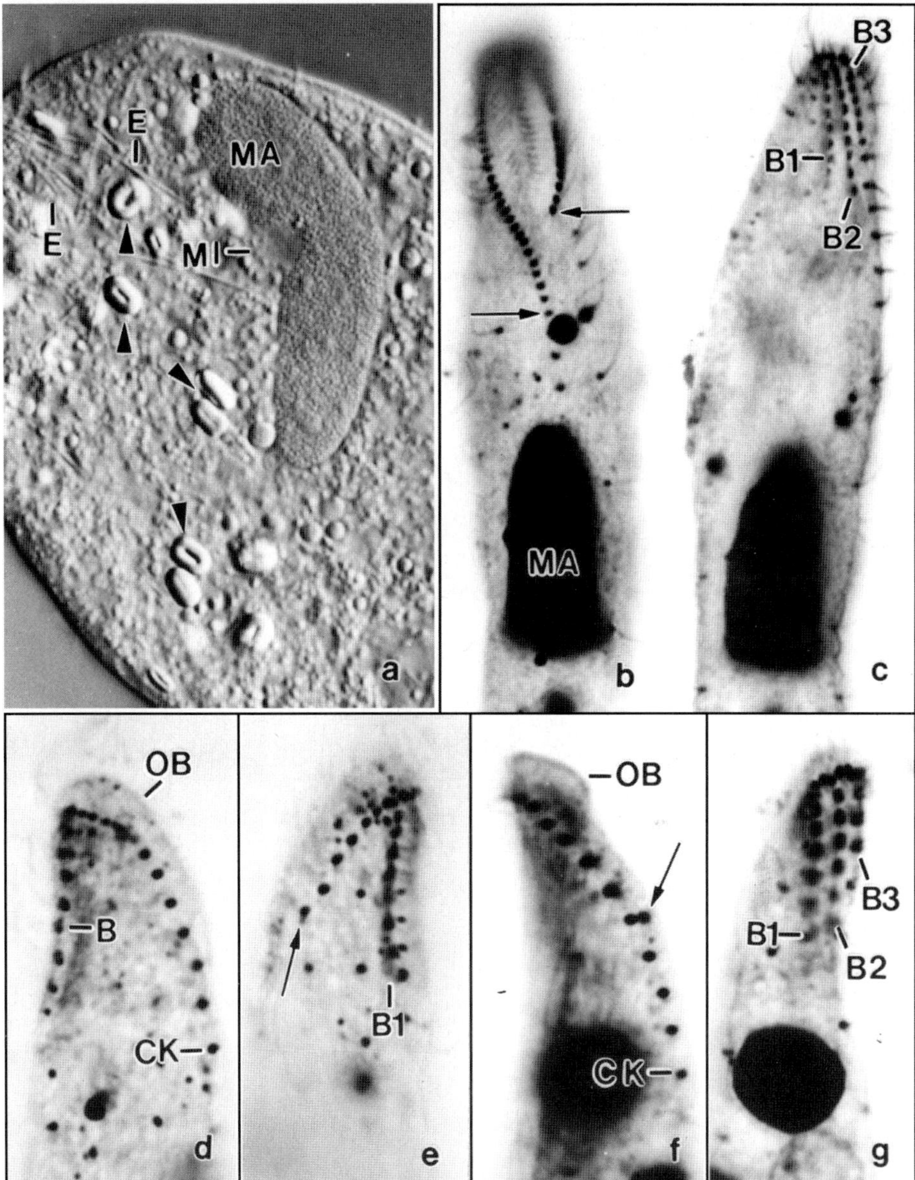

Fig. 148a–c *Apertospathula lajacola* nov. spec. from life (a) and after protargol impregnation (b, c). **a)** Squashed specimen showing some main organelles and curious inclusions (arrowheads). **b, c)** Ciliary pattern of ventral and dorsal side of same specimen. The right branch of the circumoral kinety is distinctly longer than the left (ends marked by arrows), and the kinety is widely open ventrally. **148d–g** *Longispatha elegans*, ciliary pattern in anterior body portion after protargol impregnation. **d, e)** Right and left side view showing the short left branch of the circumoral kinety (last kinetid marked by arrow). **f, g)** Ventral and dorsal view showing the long right branch of the circumoral kinety and the end of the left branch (arrow). From FOISSNER et al. (2005b).

B(1-3) – dorsal brush (rows), CK – circumoral kinety, E – extrusomes, MA – macronucleus, MI – micronucleus, OB – oral bulge.

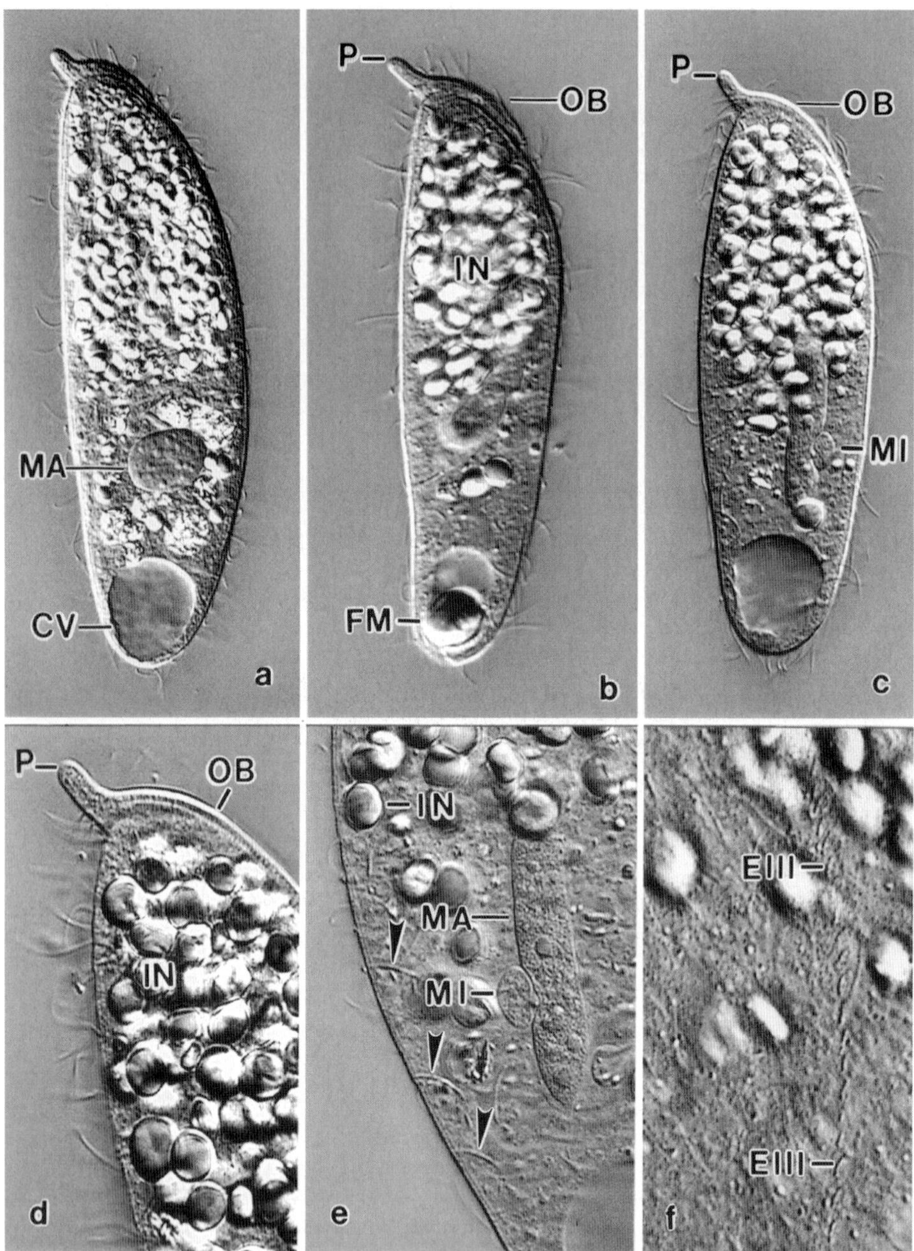

Fig. 149a–f *Rhinothrix porculus* from life (from FOISSNER et al. 2005b). **a–d)** Right side overviews of freely motile specimens showing shape variability, the conspicuous oral palp (P), the bright inclusions (IN) in anterior body half, and the indistinctness of the oral bulge (OB) and the extrusomes. **e)** Optical section showing the nuclear apparatus and the about 7 μm long body extrusomes (arrowheads). **f)** Surface view of ventral middle body region showing the short rows formed by the type III oral bulge extrusomes. CV – contractile vacuole, EIII – type III extrusomes, FM – fecal mass, IN – refractive inclusions, MA – macronucleus, MI – micronucleus, OB – oral bulge, P – palpus oralis.

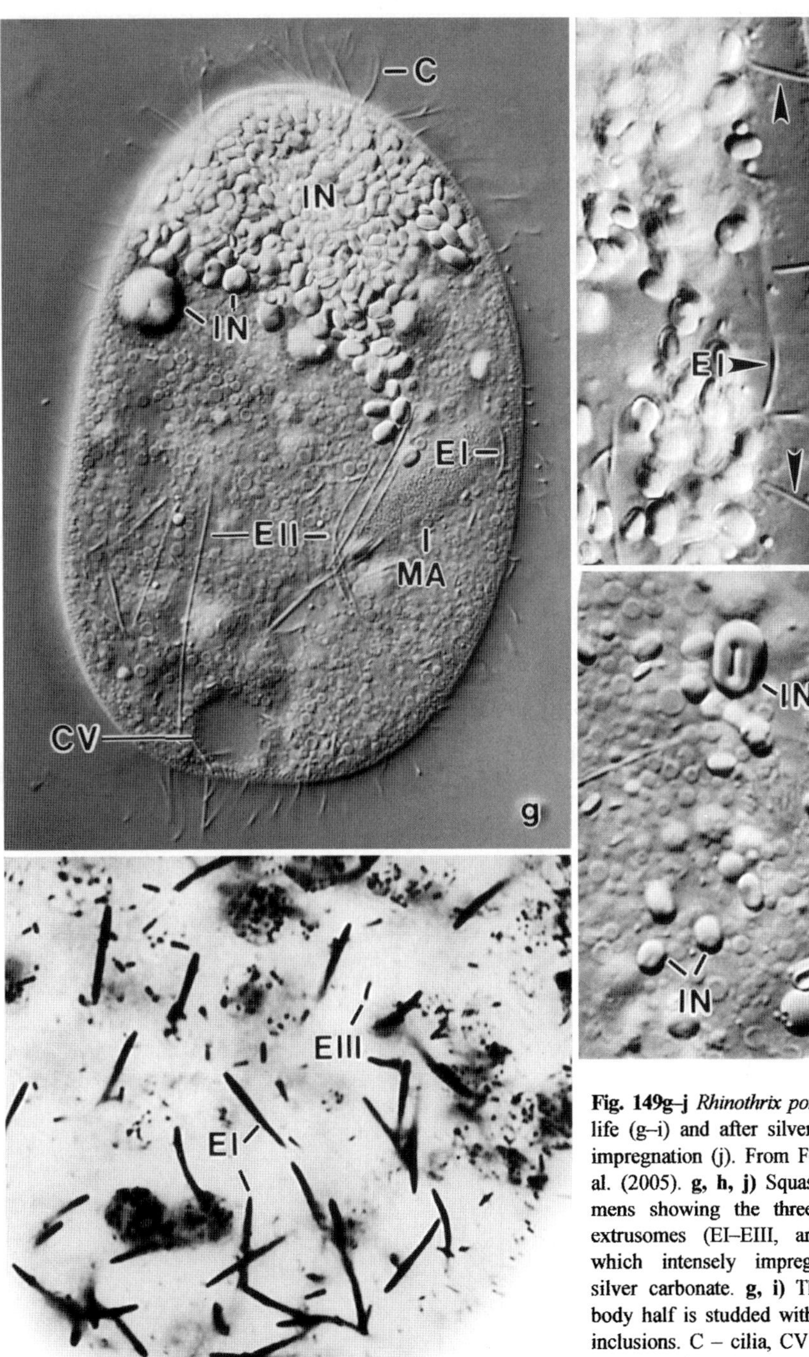

Fig. 149g–j *Rhinothrix porculus* from life (g–i) and after silver carbonate impregnation (j). From FOISSNER et al. (2005). **g, h, j)** Squashed specimens showing the three types of extrusomes (EI–EIII, arrowheads), which intensely impregnate with silver carbonate. **g, i)** The anterior body half is studded with refractive inclusions. C – cilia, CV – contractile vacuole, EI–III – extrusome types, IN – inclusions, MA – macronucleus.

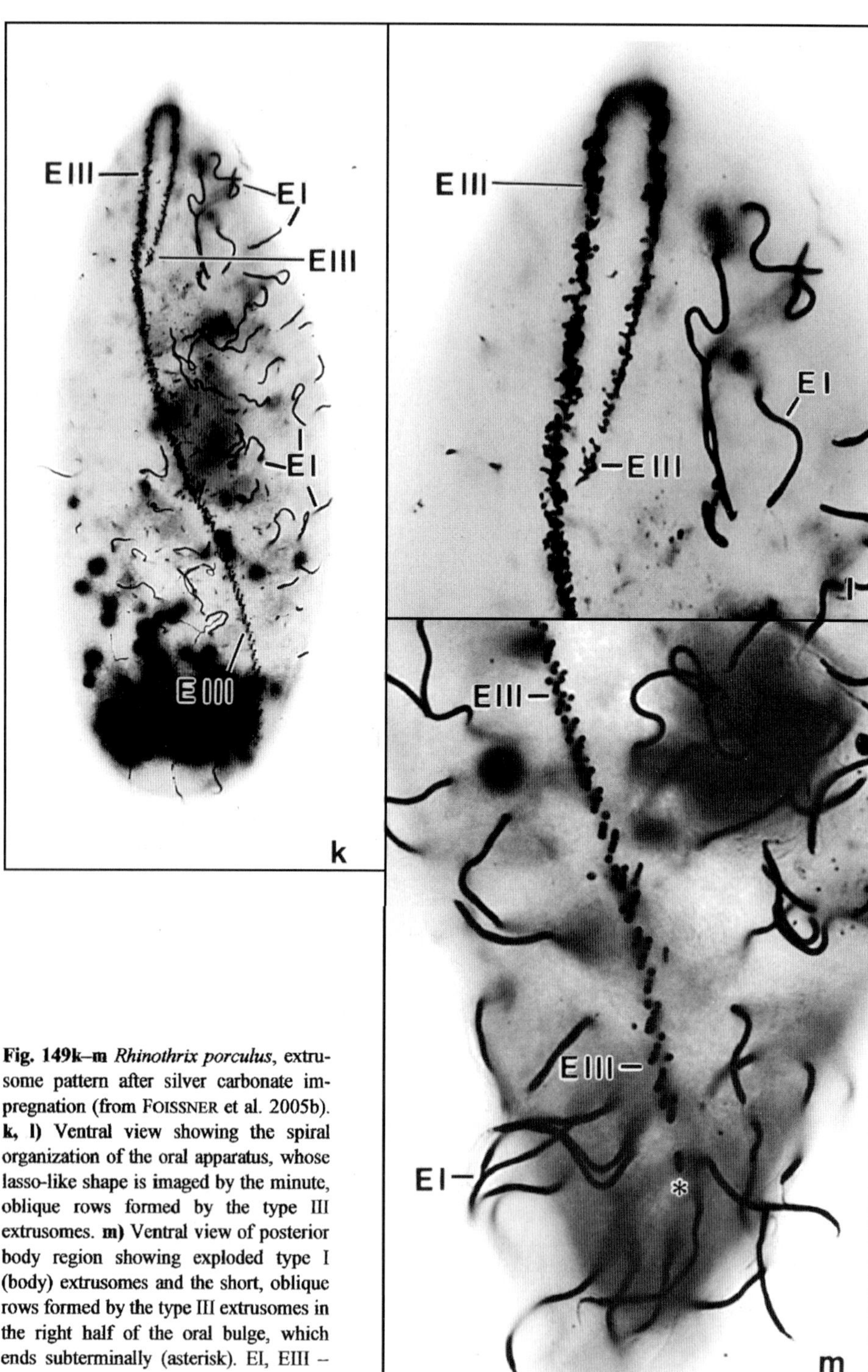

Fig. 149k–m *Rhinothrix porculus*, extrusome pattern after silver carbonate impregnation (from FOISSNER et al. 2005b). **k, l)** Ventral view showing the spiral organization of the oral apparatus, whose lasso-like shape is imaged by the minute, oblique rows formed by the type III extrusomes. **m)** Ventral view of posterior body region showing exploded type I (body) extrusomes and the short, oblique rows formed by the type III extrusomes in the right half of the oral bulge, which ends subterminally (asterisk). EI, EIII – extrusome types.

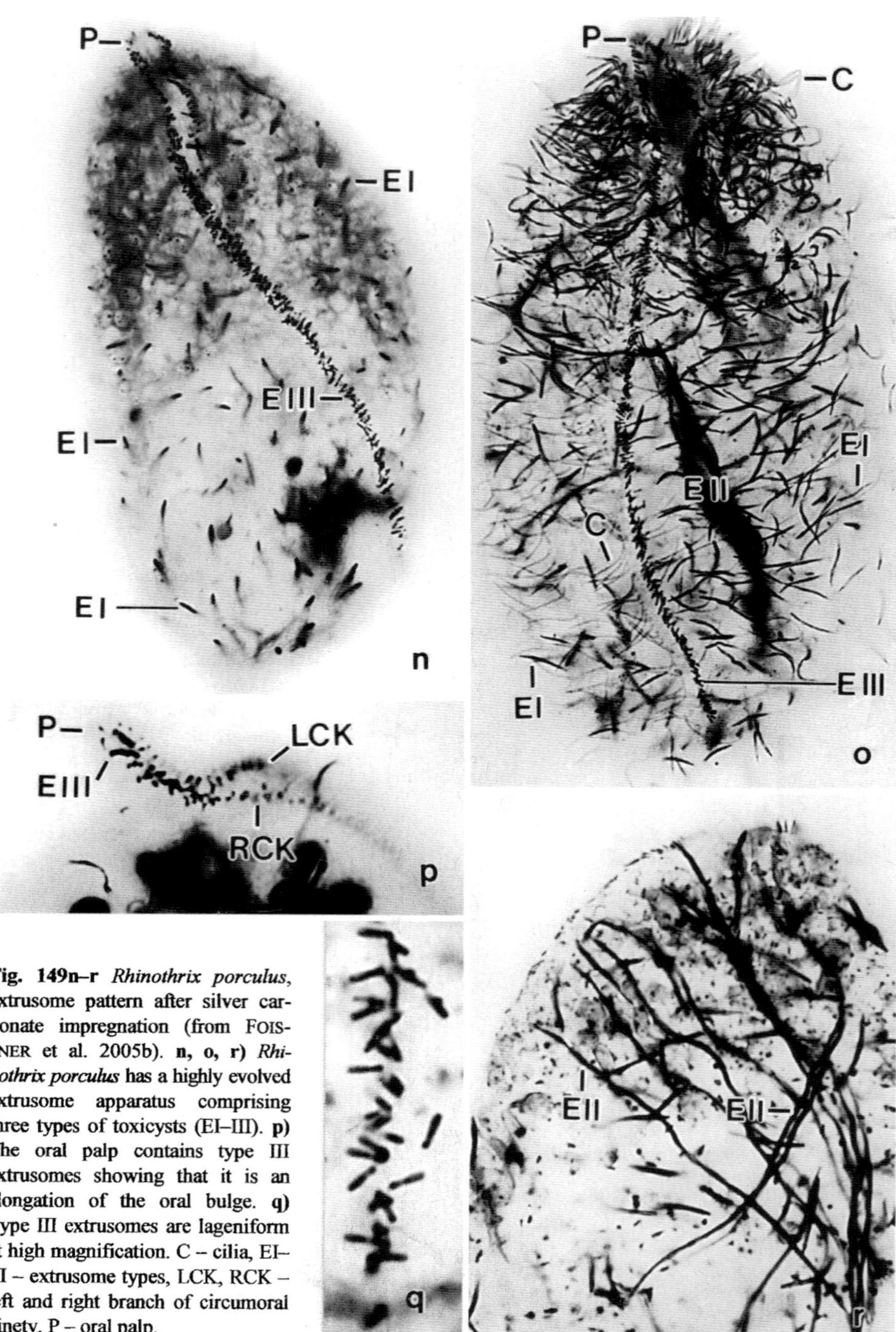

Fig. 149n–r *Rhinothrix porculus*, extrusome pattern after silver carbonate impregnation (from FOISSNER et al. 2005b). **n, o, r)** *Rhinothrix porculus* has a highly evolved extrusome apparatus comprising three types of toxicysts (EI–III). **p)** The oral palp contains type III extrusomes showing that it is an elongation of the oral bulge. **q)** Type III extrusomes are lageniform at high magnification. C – cilia, EI–III – extrusome types, LCK, RCK – left and right branch of circumoral kinety, P – oral palp.

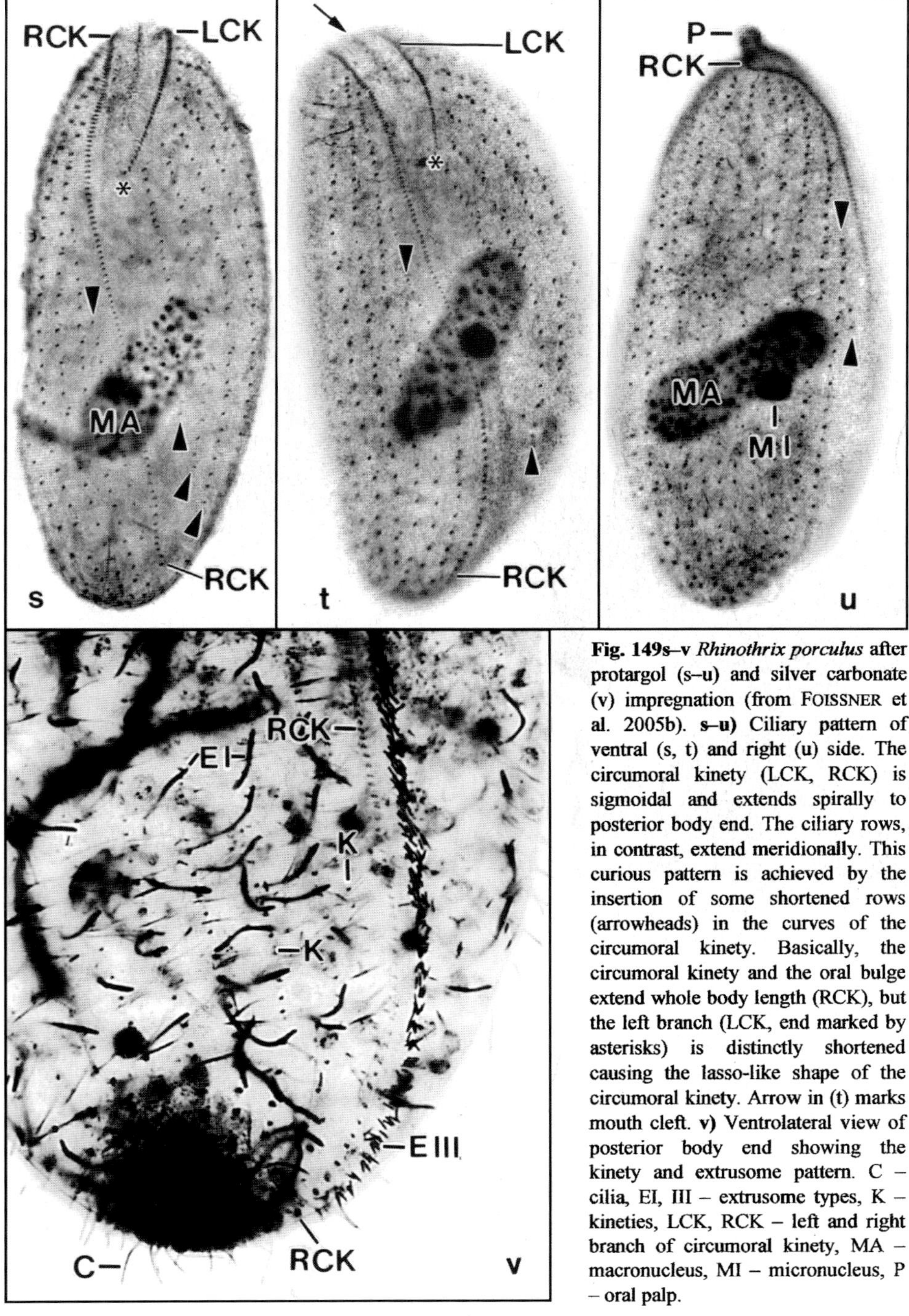

Fig. 149s–v *Rhinothrix porculus* after protargol (s–u) and silver carbonate (v) impregnation (from FOISSNER et al. 2005b). **s–u)** Ciliary pattern of ventral (s, t) and right (u) side. The circumoral kinety (LCK, RCK) is sigmoidal and extends spirally to posterior body end. The ciliary rows, in contrast, extend meridionally. This curious pattern is achieved by the insertion of some shortened rows (arrowheads) in the curves of the circumoral kinety. Basically, the circumoral kinety and the oral bulge extend whole body length (RCK), but the left branch (LCK, end marked by asterisks) is distinctly shortened causing the lasso-like shape of the circumoral kinety. Arrow in (t) marks mouth cleft. **v)** Ventrolateral view of posterior body end showing the kinety and extrusome pattern. C – cilia, EI, III – extrusome types, K – kineties, LCK, RCK – left and right branch of circumoral kinety, MA – macronucleus, MI – micronucleus, P – oral palp.

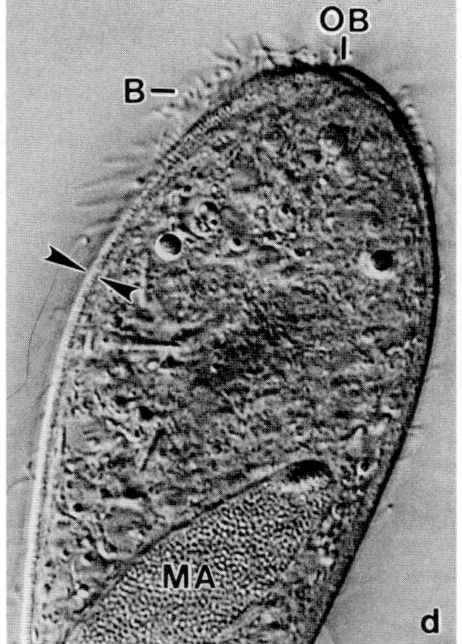

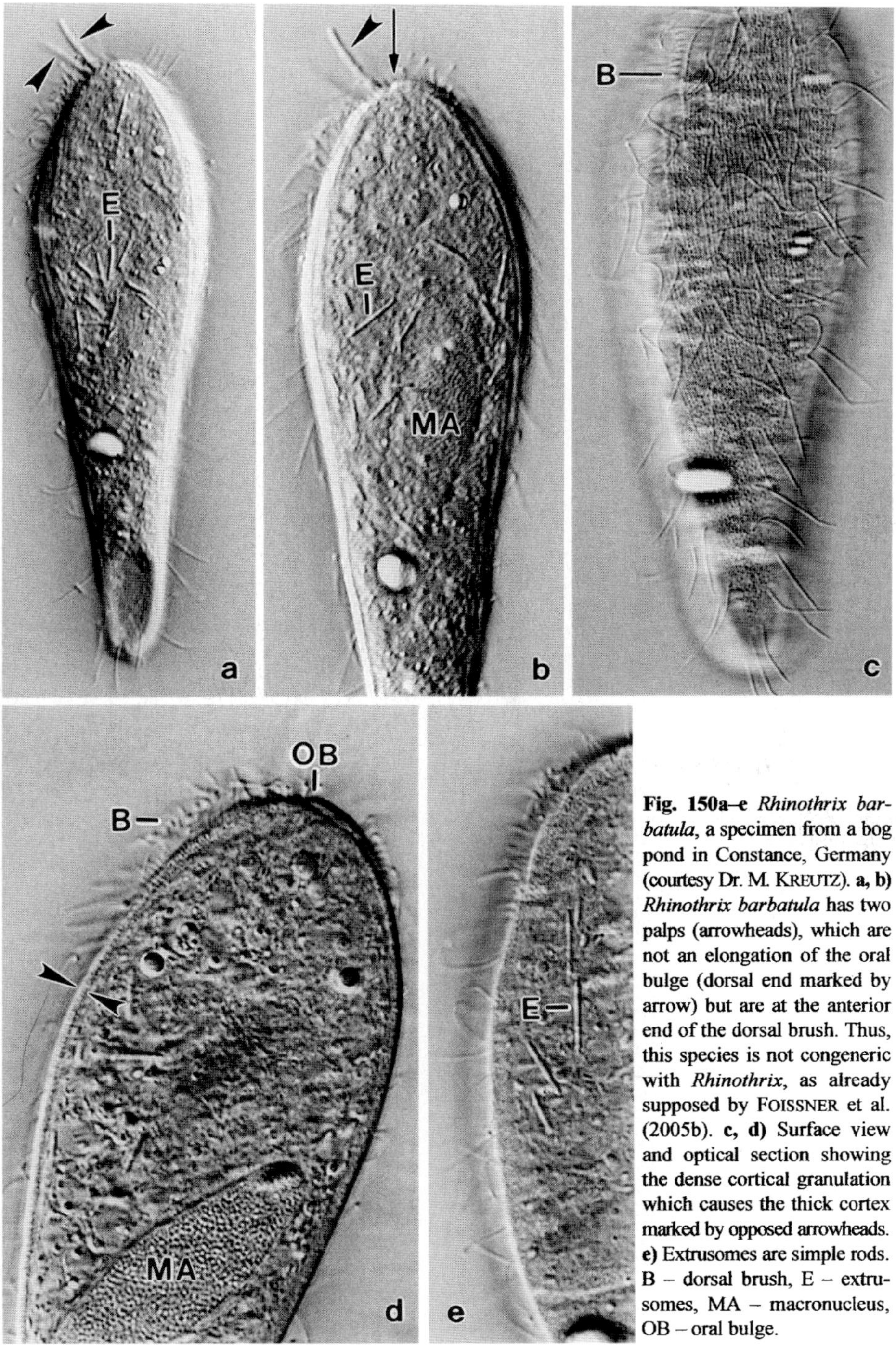

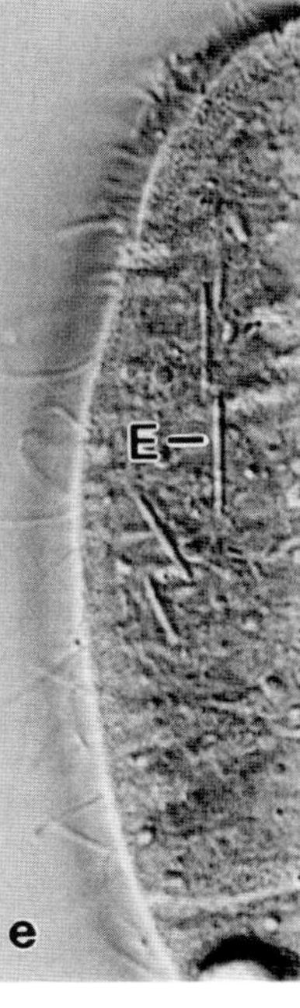

Fig. 150a–e *Rhinothrix barbatula*, a specimen from a bog pond in Constance, Germany (courtesy Dr. M. KREUTZ). **a, b)** *Rhinothrix barbatula* has two palps (arrowheads), which are not an elongation of the oral bulge (dorsal end marked by arrow) but are at the anterior end of the dorsal brush. Thus, this species is not congeneric with *Rhinothrix*, as already supposed by FOISSNER et al. (2005b). **c, d)** Surface view and optical section showing the dense cortical granulation which causes the thick cortex marked by opposed arrowheads. **e)** Extrusomes are simple rods. B – dorsal brush, E – extrusomes, MA – macronucleus, OB – oral bulge.

Monographiae Biologicae

Series Editor: **Prof. Henri J. Dumont**
State University of Ghent, Institute of Animal Ecology, Ghent, Belgium,
henri.dumont@rug.ac.be

Published titles within the series:

Volume 81 **Monograph of the Spathidiidae**
Foissner, W.; Xu, K.
2006, 486 p., Hardcover, ISBN: 1-4020-4210-8

Volume 80 **Nouragues**
Dynamics and Plant-Animal Interactions in a Neotropical Rainforest
Bongers, F.; Charles-Dominique, P.; Forget, P.-M.; Théry, M. (Eds.)
2002, 456 p., Hardcover, ISBN: 1-4020-0123-1

Volume 79 **Environmental Change and Response in East African Lakes**
Lehman, J.T. (Ed.),
1998, 250 p., Hardcover, ISBN: 0-7923-5118-5

Volume 78 **Monograph of the Oxytrichidae (Ciliophora, Hypotrichia)**
Berger, H.
1999, 1092 p., Hardcover, ISBN: 0-7923-5795-7

Volume 77 **The Pantanal of Poconé**
Biota and Ecology in the Northern Section of the World's Largest Pristine Wetland
Heckman, C.W.
1998, 592 p., Hardcover, ISBN: 0-7923-4863-X

Volume 76 **An Introduction to Saline Lakes on the Qinghai-Tibet Plateau**
Zheng Mianping
1997, 328 p., Hardcover, ISBN: 0-7923-4098-1

Volume 75 **Rocas Alijos**
Schmieder, Robert W.
1996, 520 p., Hardcover, ISBN: 0-7923-4056-6

Volume 74 **Tropical Rainforest Research - Current Issues**
Proceedings of the conference held in Bandar Seri Begawan, April 1993
Edwards, D.S.; Booth, W.E.; Choy, S.C.
1996, 570 p., Hardcover, ISBN: 0-7923-4038-8

For more information on our ongoing series, please visit www.springer.com